APPLIED FUNCTIONAL ANALYSIS AND VARIATIONAL METHODS IN ENGINEERING

J. N. Reddy

Clifton C. Garvin, Jr., Professor of Engineering
Virginia Polytechnic Institute and State University

McGraw-Hill Book Company

New York St. Louis San Francisco Auckland Bogotá Hamburg
Johannesburg London Madrid Mexico Montreal New Delhi
Panama Paris São Paulo Singapore Sydney Tokyo Toronto

This book was set in Times Roman.
The editor was Anne Murphy;
the cover was designed by Nadja Furlan-Lorbek;
the production supervisor was Diane Renda.
Project supervision was done by Santype International Limited.
Halliday Lithograph Corporation was printer and binder.

APPLIED FUNCTIONAL ANALYSIS AND VARIATIONAL METHODS IN ENGINEERING

1234567890 HALHAL 89876

ISBN 0-07-051348-1

Library of Congress Cataloging in Publication Data

Reddy, J. N. (Junuthula Narasimha), 1945–
Applied functional analysis and variational methods in engineering.

Bibliography: p.
Includes index.
1. Engineering mathematics. 2. Functional analysis. 3. Calculus of variations. I. Title.
TA330.R44 1986 620'.0042 85-11658
ISBN 0-07-051348-1

ABOUT THE AUTHOR

J. N. REDDY is the Clifton C. Garvin, Jr., Distinguished Professor of Engineering at Virginia Polytechnic Institute and State University. He obtained a Ph.D. in engineering mechanics in 1974. He authored two books, *An Introduction to the Finite Element Method* (McGraw-Hill, 1984) and *Energy and Variational Methods in Applied Mechanics.* He has coauthored three other books, including *An Introduction to the Mechanical Theory of Finite Elements* (with J. T. Oden). Dr. Reddy is author of over 75 archival publications in the fields of mechanics and applied mathematics. He serves on the editorial boards of seven professional journals, including *International Journal for Numerical Methods in Engineering*, *Computers and Structures*, and *International Journal of Non-Linear Mechanics.* Dr. Reddy received the 1984 Huber Research Prize of the American Society of Civil Engineers and the 1985 Alumni Award for Research Excellence at Virginia Polytechnic Institute.

To my motherland
India

CONTENTS

PREFACE

An increased interest is seen in recent years in the study of functional analysis among engineers and physicists who are theoretically inclined. This is because it is now widely accepted that functional analysis is a powerful tool in the solution of mathematical problems arising from physical situations. The main motivation which led me to the writing of this book came from the following observation: most engineers and physicists, whose interest is primarily in applications and who are without special training in mathematics, face a difficult task in bringing the tools of functional analysis to bear on the questions of the existence and uniqueness of solutions of mathematical problems, and the quality of their approximation by variational methods, including the finite-element method.

This book is intended to be a simple and easy introduction to functional analysis techniques that are useful in the study of differential equations arising in engineering analysis. Since most applications in engineering do not require extensive knowledge of functional analysis, only the concepts that are necessary for an engineer to equip him/herself for his/her study are presented here. In order to make the present book as accessible as possible, I have tried to avoid difficult topics while presenting concepts that are simple and useful. A greater amount of explanation and larger number of illustrative examples than is usually found in most books on the subject are also presented. In addition, it is shown how the functional analysis tools can be put to work in the formulation as well as the solution of engineering problems by the variational methods.

Readers of this book should be familiar with calculus and linear algebra, theory of ordinary and partial differential equations, vectors and matrices, and basic courses in fluid mechanics, heat transfer and mechanics of solids.

Following the introduction, the major equations of engineering are reviewed in Chapter 2. The equations developed in this chapter are studied in the later chapters from the existence and uniqueness of solutions point of view, and from their numerical solution point of view. Most graduate students of engineering are likely to have had a course in continuum mechanics or its equivalent, and therefore can skip the chapter in their first reading and refer to it whenever the need arises during the coverage of the other chapters.

In Chapter 3, an introduction to functional analysis is presented. Concepts from linear vector spaces, normed spaces and inner product spaces are systematically developed and illustrated via examples. Most of the concepts, for example, the notion of a vector, norm, inner product, orthogonality, projection, orthonormal bases, and generalized Fourier series, are rather intuitive. They are presented as natural generalizations of the corresponding concepts from the Euclidean space. This inductive rather than deductive approach should be welcomed by engineers and physicists. The abstract Banach and Hilbert spaces are introduced late in Chapter 3. This chapter constitutes a basic prerequisite for the rest of the book. Those, especially mathematics majors, who have had a course in advanced calculus and/or analysis can go straight to Chapter 4.

Chapter 4 is devoted to the discussion of linear functionals on Hilbert spaces, Sobolev spaces, generalized solutions of boundary-value problems, the minimum of a quadratic functional and concepts from calculus of variations (such as the first variation of a functional, natural and essential boundary conditions and the Euler equations). Most of these concepts are relatively familiar to engineers.

Chapter 5 deals with the questions of existence and uniqueness of linear algebraic equations, operator equations, and variational boundary-value and eigenvalue problems. The Lax–Milgram theorem and its generalizations are presented and existence and uniqueness results are included for field problems governed by the Poisson or Laplace equation, plane elasticity, plate bending, and Stokes flow problems. Much of the material presented in Chapter 5 is new to engineers and physicists.

In Chapter 6, several classical variational methods are described and used to determine the solution of various problems in engineering. These include the Ritz method, Bubnov–Galerkin method, least squares method, Kantorovich method, and Trefftz method. While these methods are familiar to most engineers, the general description and convergence results presented in the book should aid in a greater understanding of the applicability and limitations of the methods. Considerable attention is devoted to practical aspects, such as the selection of the basis of the approximation space and the convergence and stability of the numerical schemes.

Chapter 7 is devoted to the study of the finite-element method. The Ritz as well as weighted-residual finite-element models are introduced and their application to problems in one and two dimensions is demonstrated via several model problems. Applications to problems in heat transfer, fluid mechanics and solid mechanics are included. The questions of convergence and stability of various finite element models are also addressed.

Throughout the book, numerous example problems are presented, and exercise problems are included at appropriate intervals to test and extend the understanding of the concepts covered. The book can be used both as a text book in engineering and applied mathematics and as a reference for theoretically oriented engineers and physicists, and applied mathematicians.

If the book is used for a single course, Chapters 2 and 3 should be either required as prerequisites or reviewed quickly to allow sufficient time for the coverage of the remaining chapters.

The author's writings in the area of variational methods are profoundly influenced by the works of S. G. Mikhlin and K. Rectorys, among few others. The author expresses his sincere thanks to all those who have by their work, advice and support contributed to the writing of this book. Special thanks to K. Chandrashekhara, Paul Heylinger, and C. F. Liu for the proof reading during the preparation and production of the book. It is with great pleasure and appreciation the author acknowledges the patience in the skilful typing of the manuscript by Vanessa McCoy.

J. N. Reddy

CHAPTER

ONE

GENERAL CONCEPTS AND FORMULAE

1-1 INTRODUCTION

Almost all physical phenomena in nature can be described in terms of differential equations. In cases where both the given equation and the given domain are simple, the solution can be obtained in closed form (say, by means of the Fourier method, the Navier method, or Laplace transform method), often in the form of an infinite series. Efforts to develop new methods of exact solution to general partial differential equations are foiled by the irregular and geometrically complicated domains and the geometric and material nonlinearities of most practical problems. Consequently, various methods of finding suitable approximate solutions have been under continuous development since the last century. Among the methods of approximation, the finite-difference method and variational methods have dominated the applications to problems in engineering. In recent decades, the finite element method, an off-shoot of the classical variational method, has gained considerable popularity among practicing engineers as well as applied mathematicians.

The finite-difference method is endowed with the simplicity of principle (i.e., representation of derivatives of a function in terms of finite Taylor's series), and ease of formulation and implementation on a digital computer. The disadvantages of the method include: the difficulty in implementing the boundary conditions on irregular boundaries, and the difficulty in accurately representing irregular domains.

The classical variational methods (i.e., Ritz, Galerkin, and Kantorovich type methods that are applied to the whole domain at once) are based either on the minimization of a quadratic functional associated with the given problem or on the minimization, in some sense, of the error in the approximation. Despite the simplicity and accuracy, both in the solution and its derivatives, the classical variational methods were not regarded as competitive with the finite-difference method. This was mainly due to the difficulty in constructing approximation functions for irregular domains. The finite-element method overcomes the shortcoming of the classical variational methods by seeking the approximate solution on a collection of simple subdomains (called elements) on which the approximation functions can be generated systematically. The collection of elements, called a finite element mesh, replaces the original domain to a desired accuracy. The variational methods, including the finite-element method, are based on integral statements, which are obtained by multiplying the equation with a weight function and integrating over the domain. Therefore, they are not particularly suitable for problems with sharp gradients or discontinuities because the method tends to smoothen the discontinuity, which is not desirable when one is interested in predicting the magnitude of the gradient.

1-2 THE PRESENT STUDY

The phrase "variational formulation" in recent times is used in connection with generalized formulations of boundary- or initial-value problems. However, in the classical sense of the phrase, it also has to do with the minimization of a quadratic functional, which includes all of the intrinsic features of the problem, such as the governing equations, boundary and/or initial conditions, constraint conditions, and even jump conditions. Variational formulations, in either sense of the phrase, suggest new theories, provide a means for studying mathematical properties of solutions, and most importantly, provide natural means of approximation.

Variational formulations can be useful in three related ways. First, many problems of mechanics are posed in terms of finding the extremum (i.e., minima or maxima) and thus by their nature can be formulated in terms of variational statements. Second, there are problems that can be formulated by other means, such as vector mechanics (e.g., Newton's laws), but these can also be formulated by means of variational principles. Third, variational principles form a powerful basis for obtaining approximate solutions to practical problems, many of which are intractable otherwise.

As an example of problems posed in terms of finding the minimum, consider the so-called *brachistochrone* problem, posed by John Bernoulli in 1696. The problem can be stated as follows: determine the curve joining the two points A: $(0, 0)$ and B: (a, b) in a vertical plane such that a material particle of mass m, sliding without friction under its own weight, travels from point A to point B in the shortest (i.e., minimum) possible time (see Fig. 1-1). The variational formulation of the problem begins with the application of the principle of the conservation of energy. If $y = u(x)$ is the equation of the curve joining point A to point B, at some instant of time t, the velocity of the particle can be obtained by equating the energies at time $t = 0$ and time t:

$$mgb = mg(b - u) + \tfrac{1}{2}mv^2$$

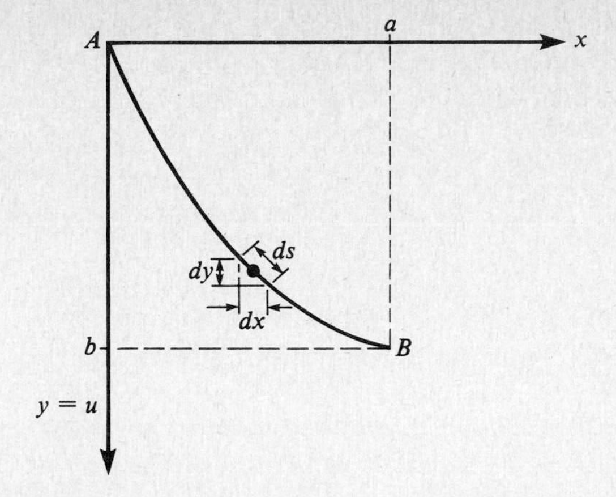

Figure 1-1 The brachistochrone problem.

which gives

$$mgu = \tfrac{1}{2}mv^2 \qquad \text{or} \qquad v = \sqrt{2gu} \tag{1-1}$$

where g is the acceleration due to gravity and v is the velocity of the particle. If s represents the distance along $y = u(x)$, measured from A, we have

$$\begin{aligned} v &= \frac{ds}{dt} \\ &= \frac{\sqrt{(dx)^2 + (du)^2}}{dt} \\ &= \sqrt{1 + (u')^2}\,\frac{dx}{dt} \end{aligned} \tag{1-2}$$

where u' denotes du/dx. The time taken by the particle in traveling a distance ds is

$$\begin{aligned} dt &= \frac{\sqrt{1 + (u')^2}}{v}\,dx \\ &= \sqrt{\frac{1 + (u')^2}{2gu}}\,dx \end{aligned} \tag{1-3}$$

The total time taken by a particle in going from point A to point B is given by

$$T = \int_0^a \sqrt{\frac{1 + (u')^2}{2gu}}\,dx \tag{1-4}$$

Thus, the brachistochrone problem is reduced to the following variational problem: find a function $u(x)$ within the class of functions with continuous derivatives [so that T in Eq. (1-4) can be evaluated] which satisfies the *end conditions*

$$u(0) = 0 \qquad u(a) = b \tag{1-5}$$

and which minimizes the integral expression in Eq. (1-4). The minimization of the integral expression (1-4) leads to the differential equation

$$2u\frac{d^2u}{dx^2} + \left(\frac{du}{dx}\right)^2 + 1 = 0 \tag{1-6}$$

The solution of Eqs. (1-6) and (1-5) is a cycloid, which is a curve generated by the motion of a point fixed to the circumference of a rolling wheel.

As an example of problems that can be formulated by vector mechanics, consider an elastic cable fixed at $x = 0$ and $x = L$, and subjected to transverse distributed load $f = f(x)$ (see Fig. 1-2). Consider an element of length Δx in the cable, and label the internal forces and externally applied force (f) of the element. Let T be the tension in the cable, and u be the transverse deflection in the cable. The tension is approximately uniform if the deflection u is small. Equilibrium of the forces on the element gives (i.e., using Newton's second law)

$$\sum F_x = 0: \quad -T\cos\theta + T\cos(\theta + \Delta\theta) = 0 \tag{1-7a}$$

$$\sum F_y = 0: \quad T\sin\theta - T\sin(\theta + \Delta\theta) - f\,\Delta x = 0 \tag{1-7b}$$

where θ is the slope of the deflection curve,

$$\theta = \frac{du}{dx}$$

For small values of θ, we have $\cos\theta = 1$ and $\sin\theta = \theta$. Then in the limit $\Delta x \to 0$ (equivalently limit $\Delta\theta \to 0$), Eq. (1-7a) is identically satisfied, and Eq. (1-7b) gives

$$\lim_{\Delta x \to 0} \frac{\Delta}{\Delta x}\left(-T\frac{du}{dx}\right) = f$$

or

$$-\frac{d}{dx}\left(T\frac{du}{dx}\right) = f \qquad 0 < x < L \tag{1-8}$$

The solution of Eq. (1-8) must also satisfy the end conditions

$$u(0) = u(L) = 0 \tag{1-9}$$

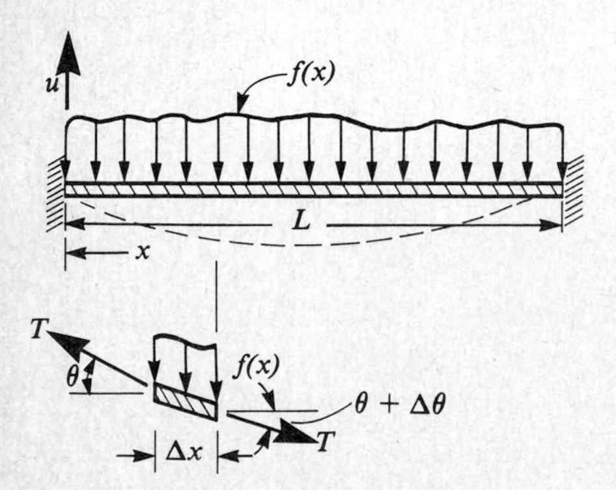

Figure 1-2 Small deflections of a cable.

The problem of the transverse deflection of a cable can also be formulated by means of a variational principle. The principle of minimum potential energy states that the function which satisfies the end conditions in Eq. (1-9) and minimizes the integral expression

$$I(u) = \int_0^L \left[\frac{T}{2} \left(\frac{du}{dx} \right)^2 + uf \right] dx \tag{1-10}$$

is the true configuration of the cable. In other words, the solution of Eqs. (1-8) and (1-9) is also the continuous function that satisfies Eq. (1-9) and minimizes the functional in Eq. (1-10). The expression in Eq. (1-10) is known as the *total potential energy* of the cable. The first term under the integral represents the strain energy, and the second term represents the work done by the distributed load per unit length of the cable.

The third use of variational formulations lies in the determination of solutions in an approximate numerical form. As an example, consider the cable problem discussed above. As a candidate for u, one can choose

$$u_1(x) = c_1 x(L - x) \tag{1-11}$$

where c_1 is a constant to be determined. Note that $u_1(x)$ is continuous in the interval $0 < x < L$ and satisfies the conditions in Eq. (1-9). The function u_1 represents an approximation to the true solution u if u_1 minimizes the functional in Eq. (1-10). Substituting u_1 for u into Eq. (1-10), one obtains

$$I = \int_0^L \left[\frac{T}{2} (L - 2x)^2 c_1^2 + c_1(Lx - x^2) f \right] dx \tag{1-12}$$

For a uniformly distributed load, $f = f_0$, the expression becomes

$$I(c_1) = \frac{T}{6} L^3 c_1^2 + \frac{L^3}{6} f_0 c_1 \tag{1-13}$$

Thus, Eq. (1-11) is an approximate solution to a cable with fixed ends and subjected to uniform loading if c_1 minimizes $I = I(c_1)$. From the calculus of ordinary functions, a necessary condition for a function to attain its minimum (or maximum) is that its first derivative with respect to its argument be zero. Therefore, one has

$$\frac{dI}{dc_1} = 0 = \tfrac{1}{3} T L^3 c_1 + \frac{f_0 L^3}{6} \tag{1-14}$$

or

$$c_1 = -\frac{f_0}{2T} \tag{1-15}$$

Therefore,

$$u_1(x) = -\frac{f_0}{2T} x(L - x) \tag{1-16}$$

In the present case, the approximate solution coincides with the exact solution.

The method outlined above for finding an approximate solution to the cable problem is due to W. Ritz (1913). In brief, the method involves seeking an approximate solution as a linear combination of appropriate functions ϕ_i (we do not enter into more details concerning the properties of these functions at this point) with undetermined parameters, c_i:

$$u_N = \sum_{i=1}^{N} c_i \phi_i \tag{1-17}$$

These parameters are determined such that the integral expression, $I(u)$, called the *variational form*, is minimized. The Ritz method converts the problem of minimizing the variational form $I(u)$ over an admissible set of functions to that of minimizing an algebraic function $I\ (c_1, c_2, \ldots, c_N)$ of parameters c_i $(i = 1, 2, \ldots, N)$:

$$\frac{\partial I}{\partial c_i} = 0 \qquad i = 1, 2, \ldots, N \tag{1-18}$$

The Ritz method, and other variational methods to be discussed in this study, provide very powerful means for determining approximate solutions to boundary-value problems of engineering. The main task in the application of these methods lies in the selection of the approximation functions ϕ_i, which are subject to certain admissibility conditions to insure the convergence of the approximate solution u_N to the exact solution u in the limit $N \to \infty$. A mathematically rigorous discussion of the properties of ϕ_i requires a knowledge of functional analysis. Thus, functional analysis and variational methods are closely related.

The purpose of the present book is to introduce the essential concepts from functional analysis, study variational formulations and underlying mathematical properties of boundary, initial and eigenvalue problems, and use variational methods to solve problems of engineering. A simple but mathematically rigorous treatment of the energy and variational formulations with applications to various fields of engineering, including elasticity, fluid mechanics and heat transfer, is undertaken in this work. A brief review of the equations of continuum mechanics is included in Chap. 2 for those who wish to familiarize themselves with the field equations of engineering.

1-3 SOME PRELIMINARY CONCEPTS AND FORMULAE

1-3-1 Domain (or Region) and Boundary

The object of most engineering analyses is to determine functions, such as u in Eq. (1-8), which satisfy a set of differential equations in a given domain and certain conditions on the boundary of the domain. The functions to be determined are called *dependent variables.* A *domain* is a collection of points in space with the property that if P is a point in the domain, then all points sufficiently close to P belong to the domain. The property implies that a domain consists only of internal points. If any two points of the domain can be joined by a line lying entirely within the domain, then it is said to be *convex* and *simply-connected.* The *boundary* of a domain is the set of points such that in any neighborhood of each point there are both points belonging to the domain and points not belonging to it (see Fig. 1-3*a*). Note from the definition of a domain that points on the boundary do not belong to the domain. In the present study we shall also consider domains that are multiply connected (see Fig. 1-3*b*). We shall use the symbol Ω to denote a domain, and Γ to denote its boundary (the combination $\Omega + \Gamma$ is denoted by $\bar{\Omega}$ and called the *closed domain*).

When the dependent variables are functions of one independent variable (say, x), the domain is a line segment, such as in Eq. (1-8): $\Omega = (0, L)$. The end points, $x = 0$ and $x = L$, of the domain are called *boundary points* of the domain Ω. When the dependent variables are functions of two independent variables (say, x and y), the domain is a (two-dimensional) surface, most often a plane, and the boundary is the closed line enclosing the domain. If the number of independent variables equals three (say x, y, and z), the domain is three-dimensional and the boundary is the surface enclosing the volume. Thus if we are solving the problem of the transverse deflection of a cable or heat transfer in a fin, the deflection or temperature must be defined in an interval, say $(0, L)$, where L is the length of the domain. If we are solving a plane elasticity problem, torsion of a cylindrical member, or flow through an axisymmetric channel, then the displacements, stress function, or velocities must be defined in a plane domain.

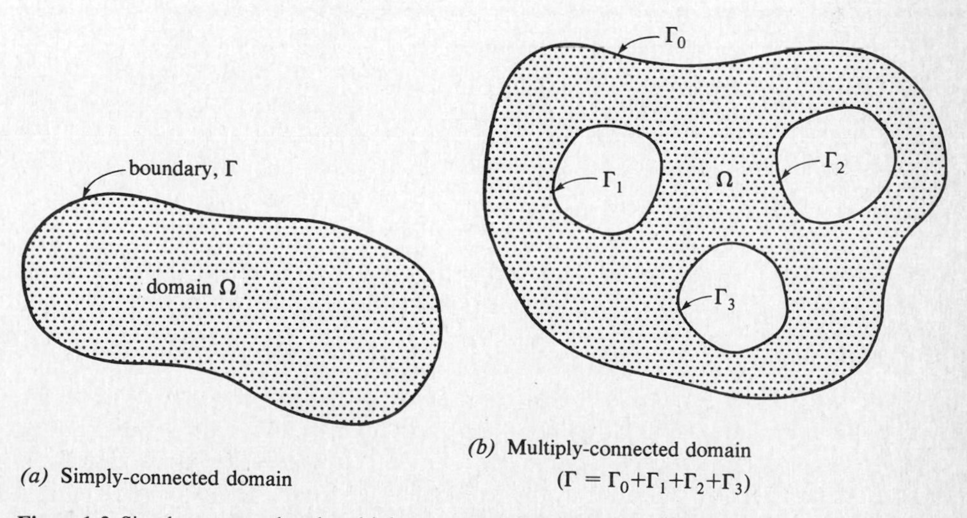

(a) Simply-connected domain

(b) Multiply-connected domain ($\Gamma = \Gamma_0+\Gamma_1+\Gamma_2+\Gamma_3$)

Figure 1-3 Simply-connected and multiply-connected domains in two dimensions.

1-3-2 Boundary- and Initial-Value Problems

A differential equation is said to describe a *boundary-value problem* if the dependent variable and possibly its derivatives are required to take specified values on the boundary. An *initial-value problem* is one in which the dependent variable and possibly its derivatives are specified initially. Initial-value problems are generally time-dependent problems in which the initial values (i.e., values at time $t = 0$) of the dependent variable and its time derivatives are specified. An example of a boundary-value problem is provided by Eqs. (1-8) and (1-9). An example of an initial-value problem is given by

$$\rho \frac{d^2u}{dt^2} + \alpha u = f \qquad 0 < t \leq t_0 \tag{1-19}$$

$$u(0) = u_0 \qquad \frac{du}{dt}(0) = v_0 \tag{1-20}$$

An example of a problem in which both initial and boundary conditions are specified is given by

$$-\frac{\partial}{\partial x}\left(a \frac{\partial u}{\partial x}\right) + \rho \frac{\partial u}{\partial t} = f(x, t) \qquad 0 < x < L \qquad 0 < t \leq t_0 \tag{1-21}$$

$$u(0, t) = g_1(t) \qquad \left(a \frac{\partial u}{\partial x}\right)\Bigg|_{x=L} = g_2(t) \qquad u(x, 0) = u_0(x) \tag{1-22}$$

When the specified boundary and initial values (that is, g_1, g_2, u_0, and v_0) are nonzero, the conditions are said to be *nonhomogeneous*; otherwise, they are called *homogeneous*. For example, $u(0) = g_1$ is a nonhomogeneous boundary condition, and the associated homogeneous boundary condition is $u(0) = 0$. Differential equations in which the right-hand side, f, is zero are called *homogeneous differential equations*.

The problem of determining the values of the constant λ such that

$$-\frac{d}{dx}\left(a \frac{du}{dx}\right) = \lambda u \qquad 0 < x < L \tag{1-23}$$

$$u(0) = 0 \qquad a\left(\frac{du}{dx}\right)\Bigg|_{x=L} = 0 \tag{1-24}$$

is called the *eigenvalue problem* associated with Eqs. (1-21) and (1-22). The values of λ are called *eigenvalues* and the associated nontrivial values of u are called *eigenvectors*.

By *classical* (or exact) *solution* of a differential equation we mean the function that identically satisfies the differential equation (i.e., the classical solution is sufficiently differentiable as required by the equation) and the specified boundary and/or initial conditions. By *variational solution* of a differential equation we mean the solution of an associated variational problem. In other words, the variational solution is not differentiable enough to satisfy the differential equation but is differentiable enough to satisfy a variational equation equivalent to the differential equation. For example, the classical solution of Eq. (1-8) is differentiable two times, whereas its variational solution [i.e., the solution of Eq. (1-10)] is required to be differentiable only once with respect to x.

1-3-3 Gradient and Divergence Theorems

In the forthcoming chapters we shall make use of the gradient and divergence theorems in two and three dimensions, and integration by parts in one dimension. Here we summarize the results for future reference.

Let ∇ and ∇^2 denote, respectively, the gradient and laplacian operators in a three-dimensional space:

$$\nabla = \hat{\mathbf{e}}_1 \frac{\partial}{\partial x_1} + \hat{\mathbf{e}}_2 \frac{\partial}{\partial x_2} + \hat{\mathbf{e}}_3 \frac{\partial}{\partial x_3} \tag{1-25}$$

$$\nabla^2 \equiv \nabla \cdot \nabla = \frac{\partial^2}{\partial x_1^2} + \frac{\partial^2}{\partial x_2^2} + \frac{\partial^2}{\partial x_3^2} \tag{1-26}$$

where $\hat{\mathbf{e}}_1$, $\hat{\mathbf{e}}_2$, and $\hat{\mathbf{e}}_3$ denote the unit basis vectors along the x_1, x_2, and x_3 (rectangular) coordinates, respectively. If $F(x_1, x_2, x_3)$ and $\mathbf{G}(x_1, x_2, x_3)$ are differentiable functions defined over a three-dimensional domain Ω, the following gradient and divergence theorems hold:

Gradient theorem:

$$\int_\Omega \text{grad } F \, dx_1 \, dx_2 \, dx_3 = \oint_\Gamma \hat{\mathbf{n}} F \, ds \quad \text{(vector form)} \tag{1-27}$$

$$\int_\Omega \frac{\partial F}{\partial x_i} \, dx_1 \, dx_2 \, dx_3 = \oint_\Gamma n_i F \, ds \quad \text{(component form)} \tag{1-28}$$

Divergence theorem:

$$\int_\Omega \text{div } (\mathbf{G}) \, dx_1 \, dx_2 \, dx_3 = \oint_\Gamma \hat{\mathbf{n}} \cdot \mathbf{G} \, ds \quad \text{(vector form)} \tag{1-29}$$

$$\int_\Omega \left(\frac{\partial G_1}{\partial x_1} + \frac{\partial G_2}{\partial x_2} + \frac{\partial G_3}{\partial x_3} \right) dx_1 \, dx_2 \, dx_3 = \oint_\Gamma (n_1 G_1 + n_2 G_2 + n_3 G_3) \, ds \quad \text{(component form)} \tag{1-30}$$

Here the dot denotes the scalar product of vectors, $\hat{\mathbf{n}}$ denotes the unit vector normal to the surface Γ of the domain Ω, n_i and G_i are the rectangular components of $\hat{\mathbf{n}}$ and $\mathbf{G}$, respectively, and the circle on the surface integral indicates that the integration is taken on the entire surface of the boundary. The direction cosines n_1, n_2, and n_3 of the unit normal $\hat{\mathbf{n}}$ can be written as (see Fig. 1-4)

$$n_1 = \cos (x_1, \hat{\mathbf{n}}) \qquad n_2 = \cos (x_2, \hat{\mathbf{n}}) \qquad n_3 = \cos (x_3, \hat{\mathbf{n}}) \tag{1-31}$$

where $\cos (x_i, \hat{\mathbf{n}})$ means the cosine of the angle between the x_i-coordinate and the unit vector $\hat{\mathbf{n}}$.

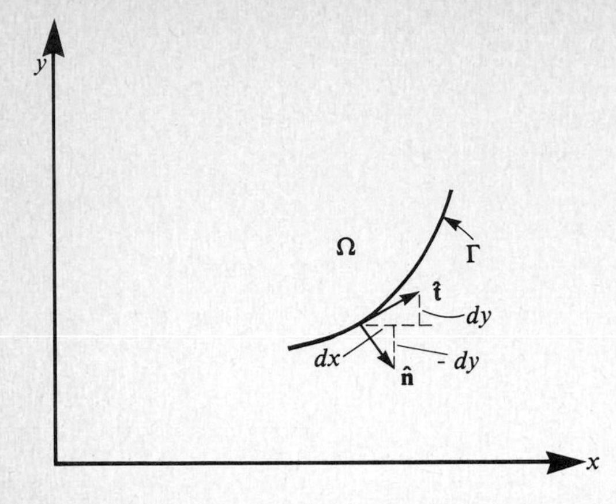

Figure 1-4 Unit normal and tangent vectors on the boundary of a two-dimensional domain.

The following identities, which can be derived using the gradient and divergence theorems, will be useful in the sequel: let F and G be scalar functions and $\mathbf{E}$ be a vector function defined in a three-dimensional domain Ω. Then we have

$$\int_\Omega \mathbf{E}\cdot\nabla F\, dx_1\, dx_2\, dx_3 = -\int_\Omega (\nabla\cdot\mathbf{E})F\, dx_1\, dx_2\, dx_3 + \oint_\Gamma (\hat{\mathbf{n}}\cdot\mathbf{E})F\, ds \qquad \text{(vector form)} \quad (1\text{-}32a)$$

$$\int_\Omega E_i \frac{\partial F}{\partial x_i}\, dx_1\, dx_2\, dx_3 = -\int_\Omega \frac{\partial E_i}{\partial x_i} F\, dx_1\, dx_2\, dx_3 + \oint_\Gamma n_i E_i F\, ds \qquad \text{(component form)} \quad (1\text{-}32b)$$

$$-\int_\Omega (\nabla^2 F)G\, dx_1\, dx_2\, dx_3 = \int_\Omega \nabla F\cdot\nabla G\, dx_1\, dx_2\, dx_3 - \oint_\Gamma \frac{\partial F}{\partial n} G\, ds \qquad (1\text{-}33)$$

$$\begin{aligned}\int_\Omega (\nabla^4 F)G\, dx_1\, dx_2\, dx_3 &= -\int_\Omega \nabla(\nabla^2 F)\cdot\nabla G\, dx_1\, dx_2\, dx_3 + \oint_\Gamma \frac{\partial}{\partial n}(\nabla^2 F)G\, ds \\ &= \int_\Omega (\nabla^2 F)(\nabla^2 G)\, dx_1\, dx_2\, dx_3 \\ &\quad + \oint_\Gamma \left[\frac{\partial}{\partial n}(\nabla^2 F)G - \nabla^2 F\frac{\partial G}{\partial n}\right] ds \end{aligned} \qquad (1\text{-}34)$$

where $\partial/\partial n$ denotes the *normal derivative* operator,

$$\frac{\partial}{\partial n} = \hat{\mathbf{n}}\cdot\nabla = n_1\frac{\partial}{\partial x_1} + n_2\frac{\partial}{\partial x_2} + n_3\frac{\partial}{\partial x_3} \qquad (1\text{-}35)$$

1-3-4 Summation Convention, and Kronecker Delta and Permutation Symbols

It is convenient to abbreviate a summation of terms by understanding that a repeated index means summation over all values of that index. Thus the summation

$$\mathbf{x} = x_1\hat{\mathbf{e}}_1 + x_2\hat{\mathbf{e}}_2 + x_3\hat{\mathbf{e}}_3 = \sum_{i=1}^{3} x_i\hat{\mathbf{e}}_i$$

can be abbreviated to

$$\mathbf{x} = x_i\hat{\mathbf{e}}_i \tag{1-36}$$

The repeated index (i.e., the index that appears twice) is a *dummy index* and therefore can be replaced by *any other symbol that has not already been used.* Thus we can also write the expression in Eq. (1-36) as

$$\mathbf{x} = x_i\hat{\mathbf{e}}_i = x_j\hat{\mathbf{e}}_j$$

In an orthonormal coordinate system (see Fig. 1-5), the basis vectors $\hat{\mathbf{e}}_i$ are orthogonal to each other and their lengths are unity:

$$\begin{array}{lll} \hat{\mathbf{e}}_1\cdot\hat{\mathbf{e}}_1 = 1 & \hat{\mathbf{e}}_1\cdot\hat{\mathbf{e}}_2 = 0 & \hat{\mathbf{e}}_1\cdot\hat{\mathbf{e}}_3 = 0 \\ \hat{\mathbf{e}}_2\cdot\hat{\mathbf{e}}_2 = 1 & \hat{\mathbf{e}}_2\cdot\hat{\mathbf{e}}_3 = 0 & \hat{\mathbf{e}}_3\cdot\hat{\mathbf{e}}_3 = 1 \end{array}$$

These six relations can be compactly expressed by the relation

$$\hat{\mathbf{e}}_i\cdot\hat{\mathbf{e}}_j = \delta_{ij} \tag{1-37}$$

where

$$\delta_{ij} = \begin{cases} 1 & i = j \\ 0 & i \neq j \end{cases} \tag{1-38}$$

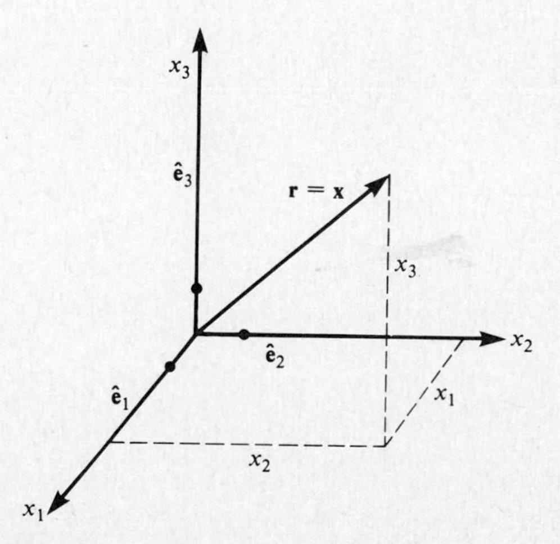

Figure 1-5 Cartesian base vectors in $\mathbb{R}^3$.

The symbol δ_{ij} is called the *Kronecker delta*. The dot product of two vectors **A** and **B** in a rectangular cartesian system can now be expressed in the form

$$\begin{aligned}\mathbf{A}\cdot\mathbf{B} &= (A_i\hat{\mathbf{e}}_i)\cdot(B_j\hat{\mathbf{e}}_j)\\ &= A_i B_j(\hat{\mathbf{e}}_i\cdot\hat{\mathbf{e}}_j)\\ &= A_i B_j \delta_{ij}\\ &= A_i B_i\end{aligned} \tag{1-39}$$

Similarly, the divergence of a vector function **A** is given by

$$\operatorname{div}\mathbf{A} = \left(\hat{\mathbf{e}}_i\frac{\partial}{\partial x_i}\right)\cdot(A_j\hat{\mathbf{e}}_j) = \frac{\partial A_j}{\partial x_i}\delta_{ij} = \frac{\partial A_i}{\partial x_i} \tag{1-40}$$

A use of the index notation is illustrated in the following example.

Example 1-1 Prove the identity

$$\operatorname{div}(r\mathbf{r}) = 4r$$

where $r = |\mathbf{r}| = \sqrt{x_i x_i}$.

We have

$$\begin{aligned}\left(\hat{\mathbf{e}}_i\frac{\partial}{\partial x_i}\right)\cdot(rx_j\hat{\mathbf{e}}_j) &= \frac{\partial}{\partial x_i}(rx_j)\,\delta_{ij}\\ &= \left(\frac{\partial r}{\partial x_i}x_j + r\frac{\partial x_j}{\partial x_i}\right)\delta_{ij}\\ &= \left(\frac{\partial r}{\partial x_i}x_j + r\,\delta_{ij}\right)\delta_{ij}\\ &= \left(\frac{\partial r}{\partial x_i}x_i + r\,\delta_{ii}\right)\end{aligned}$$

Now consider $\partial r/\partial x_i$:

$$\begin{aligned}\frac{\partial r}{\partial x_i} &= \frac{\partial}{\partial x_i}(x_j x_j)^{1/2}\\ &= \frac{1}{2}\frac{1}{(x_j x_j)^{1/2}}\frac{\partial}{\partial x_i}(x_j x_j)\\ &= \frac{1}{2r}\left(\frac{\partial x_j}{\partial x_i}x_j + x_j\frac{\partial x_j}{\partial x_i}\right)\\ &= \frac{1}{r}\delta_{ij}x_j\\ &= \frac{x_i}{r}\end{aligned}$$

Thus we have

$$\operatorname{div}(r\mathbf{r}) = \left(\frac{x_i x_i}{r} + 3r\right) = 4r$$

This completes the proof of the identity.

In an orthonormal system, the basis vectors $\hat{\mathbf{e}}_i$ also satisfy the following cross-product relations.

$$\hat{\mathbf{e}}_1 \times \hat{\mathbf{e}}_2 = \hat{\mathbf{e}}_3, \ \hat{\mathbf{e}}_2 \times \hat{\mathbf{e}}_1 = -\hat{\mathbf{e}}_3, \ \hat{\mathbf{e}}_2 \times \hat{\mathbf{e}}_3 = \hat{\mathbf{e}}_1$$

$$\hat{\mathbf{e}}_3 \times \hat{\mathbf{e}}_1 = \hat{\mathbf{e}}_2 \qquad \text{and} \qquad \hat{\mathbf{e}}_i \times \hat{\mathbf{e}}_j = 0 \qquad \text{if } i = j$$

or in short,

$$\hat{\mathbf{e}}_i \times \hat{\mathbf{e}}_j = \begin{cases} \hat{\mathbf{e}}_k & \text{if } i \neq j \neq k \quad \text{and} \quad i, j, k \text{ permute in cyclic order} \\ -\hat{\mathbf{e}}_k & \text{if } i \neq j \neq k \quad \text{and} \quad i, j, k \text{ permute not in a cyclic order} \\ 0 & \text{if any two of the indices are the same} \end{cases}$$

This can be conveniently expressed as (sum on repeated index is implied)

$$\hat{\mathbf{e}}_i \times \hat{\mathbf{e}}_j = \varepsilon_{ijk} \hat{\mathbf{e}}_k \tag{1-41}$$

where ε_{ijk} is the *permutation symbol* ($\varepsilon_{ijk} = \varepsilon_{kij} = \varepsilon_{jki}$)

$$\varepsilon_{ijk} = \begin{cases} 1 & \text{if } i, j, k \text{ are in cyclic order and } i \neq j \neq k \\ -1 & \text{if } i, j, k \text{ are not in cyclic order, and } i \neq j \neq k \\ 0 & \text{if any two of the indices are the same} \end{cases} \tag{1-42}$$

The permutation symbol can be used to write the cross-product of two vectors in an abbreviated form

$$\begin{aligned} \mathbf{A} \times \mathbf{B} &= (A_i \hat{\mathbf{e}}_i) \times (B_j \hat{\mathbf{e}}_j) \\ &= A_i B_j \varepsilon_{ijk} \hat{\mathbf{e}}_k \end{aligned} \tag{1-43}$$

$$\begin{aligned} \nabla \times \mathbf{A} &= \left(\hat{\mathbf{e}}_i \frac{\partial}{\partial x_i}\right) \times (A_j \hat{\mathbf{e}}_j) \\ &= \frac{\partial A_j}{\partial x_i} \varepsilon_{ijk} \hat{\mathbf{e}}_k \end{aligned} \tag{1-44}$$

The permutation symbol and the Kronecker delta are related by the *ε–δ identity*

$$\varepsilon_{ijk} \varepsilon_{mnk} = \delta_{im} \delta_{jn} - \delta_{in} \delta_{jm} \tag{1-45}$$

Note that at least one index should be common to the two permutation symbols in order to use the identity. The following two examples illustrate the use of the permutation symbol and the ε–δ identity.

Example 1-2 Prove the vector identity

$$\nabla \cdot (\mathbf{A} \times \mathbf{B}) = (\nabla \times \mathbf{A}) \cdot \mathbf{B} - (\nabla \times \mathbf{B}) \cdot \mathbf{A}$$

In any proof one must start with the expression on one side of the equality and arrive at the expression on the other side. Here we begin with the left-hand side expression

$$\begin{aligned}
\left(\hat{\mathbf{e}}_i \frac{\partial}{\partial x_i}\right) \cdot (A_j \hat{\mathbf{e}}_j \times B_k \hat{\mathbf{e}}_k) &= \hat{\mathbf{e}}_i \cdot \frac{\partial}{\partial x_i} (A_j B_k \varepsilon_{jkm} \hat{\mathbf{e}}_m) \\
&= \frac{\partial}{\partial x_i} (A_j B_k) \varepsilon_{jkm} \delta_{im} \\
&= \left(\frac{\partial A_j}{\partial x_i} B_k + A_j \frac{\partial B_k}{\partial x_i}\right) \varepsilon_{jki} \\
&= \frac{\partial A_j}{\partial x_i} \varepsilon_{ijk} B_k + \frac{\partial B_k}{\partial x_i} \varepsilon_{kij} A_j
\end{aligned}$$

In view of Eq. (1-44) we note that

$$\frac{\partial A_j}{\partial x_i} \varepsilon_{ijk} = k\text{th component of } \nabla \times \mathbf{A}$$

$$\frac{\partial B_k}{\partial x_i} \varepsilon_{kij} = -\frac{\partial B_k}{\partial x_i} \varepsilon_{ikj} = j\text{th component of } -(\nabla \times \mathbf{B})$$

Thus we have

$$\frac{\partial A_j}{\partial x_i} \varepsilon_{ijk} B_k + \frac{\partial B_k}{\partial x_i} \varepsilon_{kij} A_j = (\nabla \times \mathbf{A}) \cdot \mathbf{B} - (\nabla \times \mathbf{B}) \cdot \mathbf{A}$$

This completes the proof.

Example 1-3 Prove that

$$\begin{aligned}
\text{curl (curl } \mathbf{v}) &= \text{grad (div } \mathbf{v}) - \text{div (grad } \mathbf{v}) \\
&\equiv \nabla(\nabla \cdot \mathbf{v}) - \nabla^2 \mathbf{v}
\end{aligned}$$

where $\mathbf{v}$ is a vector function of position $\mathbf{r}(\equiv \mathbf{x})$.

We have

$$\left(\hat{\mathbf{e}}_i \frac{\partial}{\partial x_i}\right) \times \left(\varepsilon_{jkm} \frac{\partial v_k}{\partial x_j} \hat{\mathbf{e}}_m\right) = \varepsilon_{jkm} \frac{\partial^2 v_k}{\partial x_i\, \partial x_j} (\hat{\mathbf{e}}_i \times \hat{\mathbf{e}}_m)$$

$$= \varepsilon_{jkm}\, \varepsilon_{imn}\, \hat{\mathbf{e}}_n \frac{\partial^2 v_k}{\partial x_i\, \partial x_j}$$

$$= \varepsilon_{jkm}\, \varepsilon_{nim}\, \hat{\mathbf{e}}_n \frac{\partial^2 v_k}{\partial x_i\, \partial x_j}$$

$$= (\delta_{jn}\, \delta_{ki} - \delta_{ji}\, \delta_{kn})\hat{\mathbf{e}}_n \frac{\partial^2 v_k}{\partial x_i\, \partial x_j}$$

$$= \hat{\mathbf{e}}_j \frac{\partial^2 v_i}{\partial x_i\, \partial x_j} - \hat{\mathbf{e}}_k \frac{\partial^2 v_k}{\partial x_i\, \partial x_i}$$

$$= \hat{\mathbf{e}}_j \frac{\partial}{\partial x_j}\left(\frac{\partial v_i}{\partial x_i}\right) - \frac{\partial^2}{\partial x_i\, \partial x_i} (v_k\, \hat{\mathbf{e}}_k)$$

$$= \nabla(\nabla \cdot \mathbf{v}) - \nabla^2(\mathbf{v})$$

This completes the proof of the identity. The decomposition of $\nabla(\nabla \cdot \mathbf{v})$ into the sum of $\nabla \times (\nabla \times \mathbf{v})$ and $\nabla^2\mathbf{v}$ is useful in the study of vector potentials.

The summation notation will be used mostly in Chap. 2. In other chapters, the summation sign is used to indicate the range of indices.

PROBLEMS

Prove the following identities:

1-1 $\displaystyle\int_\Omega [F\nabla^2 G + \nabla F \cdot \nabla G]\, dx_1\, dx_2\, dx_3 = \oint_\Gamma F \frac{\partial G}{\partial n}\, ds$

1-2 $\displaystyle\int_\Omega [F\nabla^2 G - G\nabla^2 F]\, dx_1\, dx_2\, dx_3 = \oint_\Gamma \left[F \frac{\partial G}{\partial n} - G\frac{\partial F}{\partial n}\right] ds$

1-3 $\displaystyle\int_\Omega [F\nabla^4 G - \nabla^2 F \nabla^2 G]\, dx_1\, dx_2\, dx_3 = \oint_\Gamma \left[F \frac{\partial}{\partial n} (\nabla^2 G) - \nabla^2 G \frac{\partial F}{\partial n}\right] ds$

1-4 Verify the following identities:
(i) $\delta_{ij}\, \delta_{ik} = \delta_{jk}$
(ii) $\varepsilon_{ijk}\, \varepsilon_{ijk} = 6$
(iii) $A_{ij}\, \varepsilon_{ijk} = 0$ if A_{ij} is symmetric

1-5 Using the index notation, prove the following vector identities:
(i) curl $(\phi\mathbf{A}) = (\text{grad } \phi) \times \mathbf{A} + \phi$ curl $\mathbf{A}$
(ii) $(\mathbf{A} \times \mathbf{B}) \cdot (\mathbf{C} \times \mathbf{D}) = (\mathbf{A} \cdot \mathbf{C})(\mathbf{B} \cdot \mathbf{D}) - (\mathbf{A} \cdot \mathbf{D})(\mathbf{C} \cdot \mathbf{B})$
(iii) $(\mathbf{A} \times \mathbf{B}) \times (\mathbf{C} \times \mathbf{D}) = [\mathbf{A} \cdot (\mathbf{C} \times \mathbf{D})]\mathbf{B} - [\mathbf{B} \cdot (\mathbf{C} \times \mathbf{D})]\mathbf{A}$

1-6 Prove the following vector identities:

(i) div $(r^n\mathbf{r}) = (n + 3)r^n$
(ii) curl $(r^n\mathbf{r}) = 0$
(iii) grad $(r^n) = nr^{n-2}\mathbf{r}$

1-7 Show the relation between integration by parts and the gradient and divergence theorems by specializing the result in Prob. 1-1 for $F = 1$, and $G = G(x, y)$ in a rectangular domain. In other words, show that

$$\int_0^a \int_0^b \nabla^2 G \, dx \, dy = \int_0^b \left[\left(\frac{\partial G}{\partial x}\right)_{x=a} - \left(\frac{\partial G}{\partial x}\right)_{x=0} \right] dy + \int_0^a \left[\left(\frac{\partial G}{\partial y}\right)_{y=b} - \left(\frac{\partial G}{\partial y}\right)_{y=0} \right] dx$$

using both integration by parts and the gradient theorem.

CHAPTER

TWO

A REVIEW OF THE FIELD EQUATIONS OF ENGINEERING

2-1 INTRODUCTION

The major objective of this chapter is to present the main principles of continuum mechanics and to record the governing equations of a continuous medium for use in the subsequent chapters. The adjective *continuous* is used here to imply that all of the mathematical functions, such as density, and their derivatives up to the order entering into the mathematical description of the medium are continuous. In other words, the molecular structure of matter is disregarded, and the matter is assumed to contain no gaps or empty spaces. This assumption allows us to describe the macroscopic motion and deformation of a medium (whether solid, fluid, or gas), and to define stress at a point. The particles that constitute a medium are called *material points*. The simultaneous position of all the material points of a system is called the *configuration* of the system. In order to describe the configuration of a system, we require a coordinate system that is attached to some fixed system, called a *reference frame*.

The mechanics of continuous media deals with the set of all configurations that a system can assume under the action of external specified loads. This set is called the *configuration space* of the system, and each configuration is called a point or an element in the configuration space, which can be identified as a linear vector space with appropriate rules of vector addition and scalar multiplication (see Chap. 3). For example, each term in the Fourier series $w = \sum c_k \sin k\pi x/L$, which represents the deflection of a cable fixed at its ends, can be viewed as an element of the configuration space, because each term $c_k \sin k\pi x/L$ $(k = 1, 2, \ldots)$ represents a possible configuration of the cable under applied loads (see Fig. 2-1).

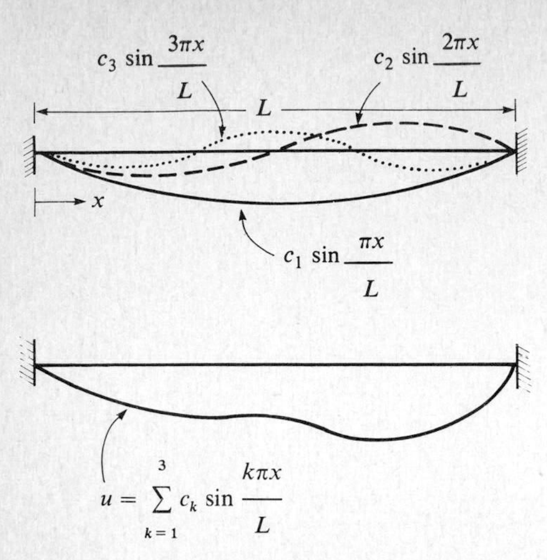

Figure 2-1 Possible configurations of a cable fixed at the ends.

If a system consists of a finite number of material points or rigid bodies, its configuration can be specified by a finite number of real variables, called *generalized coordinates.* For example, a free rigid body requires six coordinates: three cartesian coordinates and three rotational coordinates about the cartesian coordinates. A deformable cable does not possess a finite set of coordinates, but it might require infinitely many coordinates, $\sin k\pi x/L$ $(k = 1, 2, \ldots)$ to represent its configuration (i.e., deflection of the cable) $u = \sum c_k \sin k\pi x/L$. If the series is approximated by a finite number of terms, the cable configuration is effectively approximated by a finite number of generalized coordinates. Such approximations form the central topic of the discussion in Chaps. 6 and 7.

The equations governing a continuous medium can be classified into four basic categories: (1) kinematics, (2) kinetics and mechanical balance laws, (3) thermodynamic principles, and (4) constitutive laws. *Kinematics* is a study of the geometry of motion and deformation without consideration of the forces causing them. *Kinetics* is a study of the mechanical balance laws, such as conservation of mass and balance of momenta. The balance laws require the introduction of the stress and strain at a point. The thermodynamic principles of continua are concerned with the relations among heat, work, and properties of systems in equilibrium. Thermodynamics is considered when temperature and internal energy are involved in the description of kinetics. The constitutive laws involve relations between kinematic variables and kinetic variables, and define the constitutive (or material) behavior of the matter under consideration.

Since the objective here is to summarize the basic equations of continuum mechanics, only a brief discussion of the derivations and underlying concepts is given. For a more detailed account of the theory of a continuous medium, the reader is referred to a number of books suggested in the bibliography at the end of the present book.

2-2 KINEMATICS

2-2-1 Descriptions of Motion

We describe the motion of a continuous medium Ω as it deforms (i.e., changes in geometry occur) and moves through space. To describe the motion of a material body, we establish a fixed reference frame in three-dimensional space $\mathbb{R}^3$. A point in space, referred to the reference frame, is denoted by $\mathbf{x}$ or its cartesian coordinates x_i, $i = 1, 2, 3$.

There are two commonly used descriptions of motion in continuum mechanics. The *referential* (or lagrangian) *description* refers the motion to a reference configuration in which particle X occupies position $\mathbf{X}$. In solid mechanics the reference configuration is usually chosen as the unstressed state, the configuration to which the body will return when the forces causing the elastic deformation are removed. The coordinates $\mathbf{X} = (X_1, X_2, X_3)$ are called the *material coordinates* of the particle X. The *spatial* (or eulerian) *description* refers the motion to the present position $\mathbf{x}$ occupied by the particle X at time t. The spatial description focuses attention on a given region of space instead of on a given body of matter, and it is the description most used in fluid mechanics. Thus, in the lagrangian description the coordinate system $\mathbf{X}$ is fixed on a given body of matter and its motion at any time is referred to the material coordinates X_i (see Fig. 2-2):

$$\mathbf{x} = \mathbf{x}(X_1, X_2, X_3, t) \tag{2-1}$$

The motion described by Eq. (2-1) gives the position $\mathbf{x}$ occupied at time t by a particle whose position in the reference configuration was $\mathbf{X}$. In the eulerian description the coordinate system is fixed in space and the motion of the material particles passing through the fixed region of space is observed:

$$\mathbf{x} = \mathbf{x}(t) \qquad \mathbf{x}(0) = \mathbf{X} \tag{2-2}$$

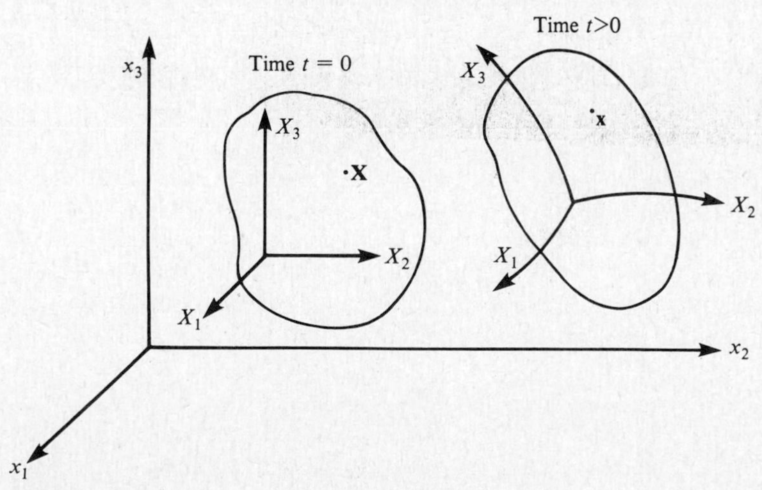

Figure 2-2 Deformation of a continuous medium.

We assume that the transformation in (2-1) is single-valued and differentiable with respect to its arguments, and that the determinant of the transformation (jacobian) is positive:

$$J \equiv \begin{vmatrix} \dfrac{\partial x_1}{\partial X_1} & \dfrac{\partial x_1}{\partial X_2} & \dfrac{\partial x_1}{\partial X_3} \\ \dfrac{\partial x_2}{\partial X_1} & \dfrac{\partial x_2}{\partial X_2} & \dfrac{\partial x_2}{\partial X_3} \\ \dfrac{\partial x_3}{\partial X_1} & \dfrac{\partial x_3}{\partial X_2} & \dfrac{\partial x_3}{\partial X_3} \end{vmatrix} > 0 \tag{2-3}$$

Therefore, the transformation (2-1) has a unique inverse (usually not known)

$$\mathbf{X} = \mathbf{X}(x_1, x_2, x_3, t) \tag{2-4}$$

It is informative to note that the material coordinates were introduced by Euler in 1761, although they are often referred to as lagrangian coordinates. The spatial coordinates, widely known as eulerian coordinates, were introduced by d'Alembert in 1752.

2-2-2 Material Derivative in Spatial Coordinates

The partial time derivative with the material coordinates $\mathbf{X}$ held constant should be distinguished from the partial time derivative with spatial coordinates $\mathbf{x}$ held constant. The derivative $\partial\phi/\partial t$ of a function ϕ in the spatial description $\phi = \phi(\mathbf{x}, t)$ signifies the *local rate of change* of ϕ. The time derivative of a function $\phi = \phi(\mathbf{x}, t)$ with $\mathbf{X}$ held constant signifies the *total rate of change* of ϕ.

Since the laws of dynamics deal with the time derivatives of properties of a particle and not with local rates of change of the properties, we need to calculate the time derivative $D\phi/Dt$, called the *material time derivative*, from a knowledge of the spatial description $\phi = \phi[\mathbf{x}(\mathbf{X}, t), t]$. The total time derivative of a property $\phi(\mathbf{x}, t)$ at a fixed place $\mathbf{x}$ with the material coordinates held constant is given by

$$\frac{D\phi}{Dt} = \left(\frac{\partial\phi}{\partial t}\right)_{\mathbf{x}=\text{const}} + \frac{\partial\phi}{\partial x_i}\left(\frac{\partial x_i}{\partial t}\right)_{\mathbf{X}=\text{const}} \tag{2-5}$$

where $(\partial\phi/\partial t)_{\mathbf{x}}$ denotes the local rate of change of ϕ and the summation on the repeated index is implied. It is the rate of change in ϕ observed at the fixed place $\mathbf{x}$, which is not the same as the rate of change of ϕ seen by an observer passing the place $\mathbf{x}$. By definition, $(\partial x_i/\partial t)_{\mathbf{X}=\text{const}} = Dx_i/Dt$ is the velocity $\mathbf{v}$ of the particle X. Equation (2-5) takes the form

$$\frac{D\phi}{Dt} = \frac{\partial\phi}{\partial t} + \frac{\partial\phi}{\partial x_i} v_i$$

$$= \frac{\partial\phi}{\partial t} + \mathbf{v}\cdot\text{grad }\phi \tag{2-6}$$

For example, the acceleration **a** is obtained by substituting **v** for ϕ:

$$\mathbf{a} = \frac{D\mathbf{v}}{Dt}$$

$$= \frac{\partial \mathbf{v}}{\partial t} + \mathbf{v} \cdot \text{grad } \mathbf{v} \tag{2-7a}$$

with the cartesian components

$$a_i = \frac{\partial v_i}{\partial t} + v_k \frac{\partial v_i}{\partial x_k} \tag{2-7b}$$

Thus the material derivative operator

$$\frac{D}{Dt} = \left(\frac{\partial}{\partial t} + \mathbf{v} \cdot \text{grad}\right) \tag{2-8}$$

can be applied to a scalar, a vector, or a tensor function of the spatial position **x** and time t.

2-2-3 Strain Tensors

Let us now consider the description of deformation. The word *deformation* refers to displacements and changes in the geometry experienced by points in a body. Let a body Ω occupy a space S. Referred to a rectangular cartesian frame of reference, every particle X in the body has a set of coordinates **X**. When the body is deformed under the action of external forces, particle X (which occupies place **X** in the undeformed body) moves to a new position **x** referred to the (x_1, x_2, x_3) system. Since (X_1, X_2, X_3) are fixed to the particle X, they also deform and become curvilinear in the deformed configuration (see Fig. 2-2). The displacement of a particle X is given by

$$\mathbf{u} = \mathbf{x} - \mathbf{X} \qquad \text{or} \qquad u_i = x_i - X_i \tag{2-9}$$

If the displacement of every particle in the body is known, we can construct the deformed configuration from the reference (or undeformed) configuration.

If the displacement is expressed in terms of the coordinates X_i in the reference configuration (lagrangian description), we have

$$u_i(X_1, X_2, X_3, t) = x_i(X_1, X_2, X_3, t) - X_i \tag{2-10}$$

If the displacement is expressed in terms of the current coordinates x_i (eulerian description), we have

$$u_i(x_1, x_2, x_3, t) = x_i - X_i(x_1, x_2, x_3, t) \tag{2-11}$$

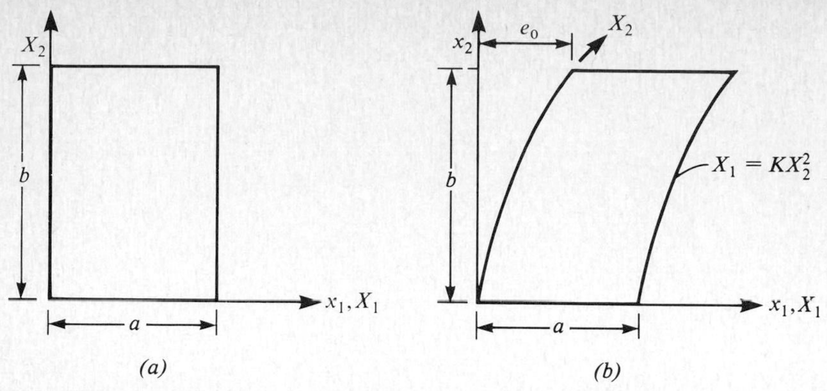

Figure 2-3 Deformation of a rectangular block.

Example 2-1 Consider a rectangular block of rubber with planeform dimensions a and b, as shown in Fig. 2-3a. Under the action of applied loads the rubber block deformed to the shape shown in Fig. 2-3b. The displacement of a particle X that occupied place $\mathbf{X}$ in the undeformed body can be calculated from the current configuration:

$$\left.\begin{aligned} x_1 &= X_1 + \frac{e_0}{b^2} X_2^2 \\ x_2 &= X_2 \\ x_3 &= X_3 \end{aligned}\right\} \quad \text{(lagrangian description)} \tag{2-12a}$$

$$\left.\begin{aligned} X_1 &= x_1 - \frac{e_0}{b^2} x_2^2 \\ X_2 &= x_2 \\ X_3 &= x_3 \end{aligned}\right\} \quad \text{(eulerian description)} \tag{2-12b}$$

The displacement field is given by

$$\left.\begin{aligned} u_1 &= \frac{e_0}{b^2} X_2^2 \\ u_2 &= 0 \\ u_3 &= 0 \end{aligned}\right\} \quad \text{(lagrangian description)} \tag{2-13a}$$

$$\left.\begin{aligned} u_1 &= \frac{e_0}{b^2} x_2^2 \\ u_2 &= 0 \\ u_3 &= 0 \end{aligned}\right\} \quad \text{(eulerian description)} \tag{2-13b}$$

This completes the example.

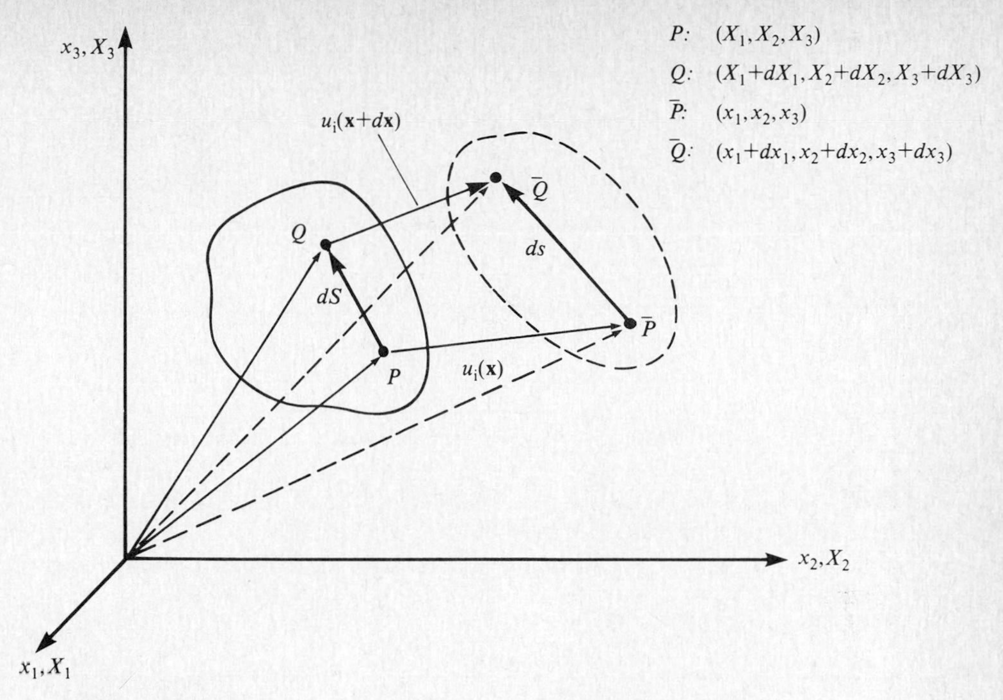

Figure 2-4 Relative displacements between two neighboring points in a body.

A rigid-body motion does not alter the shape but only the location of a body. The stretching and distortion (or, in short, change of distance between two particles) of a body can be determined by considering the change of length between any two arbitrary points of the body. Consider two neighboring points P: (X_1, X_2, X_3) and Q: $(X_1 + dX_1, X_2 + dX_2, X_3 + dX_3)$ in the undeformed configuration. The vector connecting P to Q is given by $d\mathbf{X}$. These points move to new places $\bar{P}$ and $\bar{Q}$, respectively, in the deformed body (see Fig. 2-4). Let the position vectors of $\bar{P}$ and $\bar{Q}$ be (x_1, x_2, x_3) and $(x_1 + dx_1, x_2 + dx_2, x_3 + dx_3)$, respectively. The vector connecting $\bar{P}$ and $\bar{Q}$ is given by $d\mathbf{x}$. The square of the length dS of $d\mathbf{X}$ is given by

$$(dS)^2 = d\mathbf{X} \cdot d\mathbf{X} = dX_i \, dX_i \tag{2-14}$$

and the square of the length ds of $d\mathbf{x}$ is given by

$$(ds)^2 = d\mathbf{x} \cdot d\mathbf{x} = dx_i \, dx_i \tag{2-15}$$

The difference in the squares of the distances between the material points $\mathbf{X}$ and $\mathbf{X} + d\mathbf{X}$ in the deformed and undeformed configurations is given by

$$(ds)^2 - (dS)^2 = dx_i \, dx_i - dX_i \, dX_i \tag{2-16}$$

The difference can be expressed in terms of either the material coordinates X_i or spatial coordinates x_i by means of the relations (2-1) and (2-4),

$$dx_i = \frac{\partial x_i}{\partial X_j}\, dX_j \qquad dX_i = \frac{\partial X_i}{\partial x_j}\, dx_j \tag{2-17}$$

$$(ds)^2 - (dS)^2 = \begin{cases} \left(\dfrac{\partial x_i}{\partial X_j}\dfrac{\partial x_i}{\partial X_k} - \delta_{jk}\right) dX_j\, dX_k & \text{(lagrangian description)} \\ \left(\delta_{jk} - \dfrac{\partial X_i}{\partial x_j}\dfrac{\partial X_i}{\partial x_k}\right) dx_j\, dx_k & \text{(eulerian description)} \end{cases} \tag{2-18}$$

where in writing the last equation several changes in the dummy subscripts were made.

Now we define the components of the lagrangian and eulerian strain tensors:

$$(ds)^2 - (dS)^2 = \begin{cases} 2L_{jk}\, dX_j\, dX_k & \text{(lagrangian)} \\ 2E_{jk}\, dx_j\, dx_k & \text{(eulerian)} \end{cases} \tag{2-19}$$

where the components of the lagrangian strain tensor **L** and eulerian strain tensor **E** are given by

$$\begin{aligned} L_{jk} &= \frac{1}{2}\left(\frac{\partial x_i}{\partial X_j}\frac{\partial x_i}{\partial X_k} - \delta_{jk}\right), \\ E_{jk} &= \frac{1}{2}\left(\delta_{jk} - \frac{\partial X_i}{\partial x_j}\frac{\partial X_i}{\partial x_k}\right) \end{aligned} \tag{2-20}$$

The strain tensor L_{ij} was introduced by Green and St. Venant and is called Green's strain tensor. The strain tensor E_{ij} was introduced by Cauchy for infinitesimal strains and by Almansi and Hamel for finite strains and is known as Almansi's strain tensor.

One can make several observations concerning the strain components. First, we note that L_{ij} and E_{ij} are symmetric, i.e., $L_{ji} = L_{ij}$ and $E_{ji} = E_{ij}$. Second, $(ds)^2 - (dS)^2 = 0$ if and only if $L_{ij} = E_{ij} = 0$. A deformation in which the length of every line element remains unchanged is a rigid-body motion. Therefore, a necessary and sufficient condition that a deformation of a body be a rigid-body motion is that all components of the strain tensor L_{ij} or E_{ij} be zero throughout the body.

We now return to Eq. (2-20), and express the strain components in terms of the gradients of displacement components using Eqs. (2-10) and (2-11). We have

$$\frac{\partial x_i}{\partial X_j} = \frac{\partial u_i}{\partial X_j} + \delta_{ij} \qquad \frac{\partial X_i}{\partial x_j} = \delta_{ij} - \frac{\partial u_i}{\partial x_j} \tag{2-21}$$

Substituting Eq. (2-21) into Eq. (2-20), we obtain

$$L_{jk} = \frac{1}{2}\left(\frac{\partial u_j}{\partial X_k} + \frac{\partial u_k}{\partial X_j} + \frac{\partial u_i}{\partial X_j}\frac{\partial u_i}{\partial X_k}\right) \tag{2-22a}$$

$$E_{jk} = \frac{1}{2}\left(\frac{\partial u_j}{\partial x_k} + \frac{\partial u_k}{\partial x_j} - \frac{\partial u_i}{\partial x_j}\frac{\partial u_i}{\partial x_k}\right) \tag{2-22b}$$

Typical strain components, in unabridged notation, are given by

$$L_{11} = \frac{\partial u_1}{\partial X_1} + \frac{1}{2}\left[\left(\frac{\partial u_1}{\partial X_1}\right)^2 + \left(\frac{\partial u_2}{\partial X_1}\right)^2 + \left(\frac{\partial u_3}{\partial X_1}\right)^2\right]$$

$$L_{12} = \frac{1}{2}\left(\frac{\partial u_1}{\partial X_2} + \frac{\partial u_2}{\partial X_1} + \frac{\partial u_1}{\partial X_1}\frac{\partial u_1}{\partial X_2} + \frac{\partial u_2}{\partial X_1}\frac{\partial u_2}{\partial X_2} + \frac{\partial u_3}{\partial X_1}\frac{\partial u_3}{\partial X_2}\right) \tag{2-23a}$$

$$E_{11} = \frac{\partial u_1}{\partial x_1} - \frac{1}{2}\left[\left(\frac{\partial u_1}{\partial x_1}\right)^2 + \left(\frac{\partial u_2}{\partial x_1}\right)^2 + \left(\frac{\partial u_3}{\partial x_1}\right)^2\right]$$

$$E_{12} = \frac{1}{2}\left(\frac{\partial u_1}{\partial x_2} + \frac{\partial u_2}{\partial x_1} - \frac{\partial u_1}{\partial x_1}\frac{\partial u_1}{\partial x_2} - \frac{\partial u_2}{\partial x_1}\frac{\partial u_2}{\partial x_2} - \frac{\partial u_3}{\partial x_1}\frac{\partial u_3}{\partial x_2}\right) \tag{2-23b}$$

If the displacement gradients are so small that their squares and products are negligible, then Almansi's strain tensor reduces to Cauchy's infinitesimal strain tensor

$$e_{ij} = \frac{1}{2}\left(\frac{\partial u_i}{\partial x_j} + \frac{\partial u_j}{\partial x_i}\right) \tag{2-24}$$

In addition, if the displacements themselves are very small, we have from Eq. (2-21)

$$\frac{\partial u_i}{\partial X_j} = \frac{\partial x_i}{\partial X_j} - \delta_{ij} \ll 1 \tag{2-25a}$$

and therefore

$$\frac{\partial(\ )}{\partial X_i} = \frac{\partial(\ )}{\partial x_j}\frac{\partial x_j}{\partial X_i} \approx \frac{\partial(\ )}{\partial x_i} \tag{2-25b}$$

Consequently, in the infinitesimal displacement case, the distinction between the lagrangian and eulerian strain tensor disappears.

Example 2-2 Consider the problem of Example 2-1. The components of the strain tensor are

$$L_{11} = 0 \qquad L_{22} = 2\left(\frac{e_0}{b^2}X_2\right)^2 \qquad L_{33} = 0$$

$$L_{12} = \frac{e_0}{b^2}X_2 \qquad L_{13} = 0 \qquad L_{23} = 0$$

$$E_{11} = 0 \qquad E_{22} = -2\left(\frac{e_0}{b^2}x_2\right)^2 \qquad E_{33} = 0$$

$$E_{12} = \frac{e_0}{b^2}x_2 \qquad E_{13} = 0 \qquad E_{23} = 0$$

The only nonzero infinitesimal strain components are

$$e_{12} = \frac{e_0}{b^2}x_2$$

2-2-4 Compatibility Equations

By definition, the components of the strain tensors can be computed from a differentiable displacement field. However, if the *six* components of strain are given and if we are required to find the *three* displacement components, the strains given should be such that the six differential equations relating the strains and displacements possess a solution. The existence of the solution is guaranteed if the strain components satisfy the following six compatibility conditions (given here only for infinitesimal strains case)

$$\frac{\partial^2 e_{ij}}{\partial x_m \, \partial x_n} + \frac{\partial^2 e_{mn}}{\partial x_i \, \partial x_j} - \frac{\partial^2 e_{im}}{\partial x_j \, \partial x_n} - \frac{\partial^2 e_{jn}}{\partial x_i \, \partial x_m} = 0 \tag{2-26}$$

For a two dimensional case, Eq. (2-26) reduces to the following single compatibility equation

$$\frac{\partial^2 e_{11}}{\partial x_2^2} + \frac{\partial^2 e_{22}}{\partial x_1^2} - 2\,\frac{\partial^2 e_{12}}{\partial x_1 \, \partial x_2} = 0 \tag{2-27}$$

2-3 KINETICS AND MECHANICAL BALANCE LAWS

2-3-1 Stress

Consider a material body Ω in its current configuration and let Γ be the surface of Ω. Forces acting on a material body can be classified as *internal* and *external.* The internal forces are those which resist the tendency of one part of the body to be separated from another part. The external forces can be classified into two kinds: body forces and surface forces. *Body forces* act on the elements of volume or mass inside the body. Examples of body forces are provided by gravitational and magnetic forces. *Surface forces* are contact forces acting on the bounding surface Γ. Examples of surface forces are provided by applied forces and contact forces (exerted by one body on another).

The surface force per unit area acting on an elemental area ds is called the *traction* or *stress vector* acting on the element. The concept also applies to a surface created by slicing the body with a plane (see Fig. 2-5). Consider the surface force $\Delta\mathbf{f}$ (which can be viewed as the resultant of forces) acting on a portion Δs of the surface Γ of the body Ω. The stress vector on Δs is the vector $\Delta\mathbf{f}/\Delta s$, which has the same direction as $\Delta\mathbf{f}$ but with magnitude equal to $|\Delta\mathbf{f}|/\Delta s$. The stress vector at a point P on Δs is given by

$$\mathbf{T} = \lim_{\Delta s \to 0} \frac{\Delta\mathbf{f}}{\Delta s} \tag{2-28}$$

Since the magnitude and direction of $\mathbf{T}$ depends on the orientation of the plane that created the surface, we shall denote it by $\mathbf{T}^{(\hat{\mathbf{n}})}$, where $\hat{\mathbf{n}}$ denotes the unit vector normal to the plane. In other words, $\mathbf{T}^{(\hat{\mathbf{n}})}$ is the stress vector at point P on a plane passing through the point and whose normal is $\hat{\mathbf{n}}$.

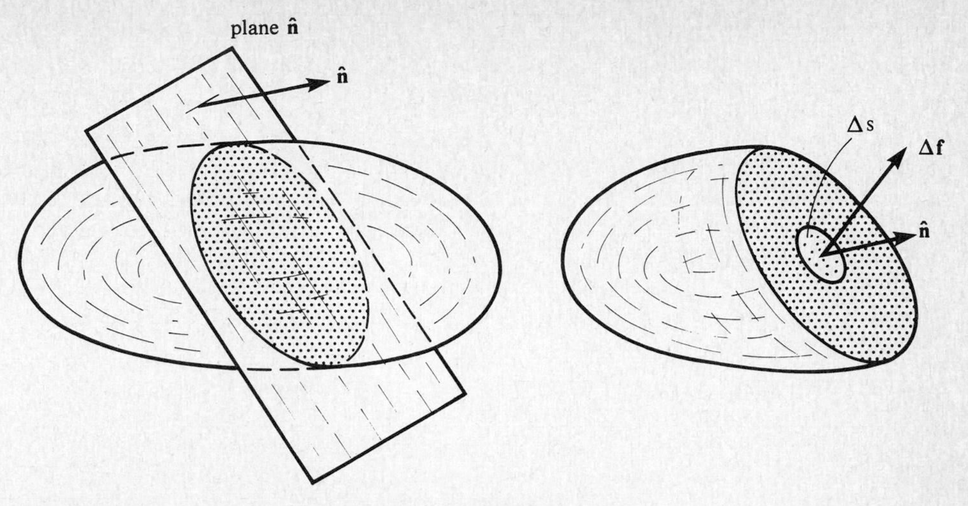

Figure 2-5 Representation of a stress vector at a point in the body.

The traction vectors on three mutually perpendicular planes, say planes perpendicular to the coordinate axes, at a point, are useful in determining the stress vector at that point on any other plane. Let $\mathbf{T}^{(i)}$ denote the stress vector at point P on a plane perpendicular to the x_i-axis (see Fig. 2-6). Each vector $\mathbf{T}^{(i)}$ can be resolved into components along the coordinate lines,

$$\mathbf{T}^{(i)} = \sigma_{ij}\hat{\mathbf{e}}_j \qquad (i = 1, 2, 3) \tag{2-29}$$

where σ_{ij} denotes the component of the stress vector $\mathbf{T}^{(i)}$ along the x_j-direction. The stress components σ_{ij} are shown on three perpendicular planes in Fig. 2-7. It should be noted that the cube that is used to indicate the stress components has

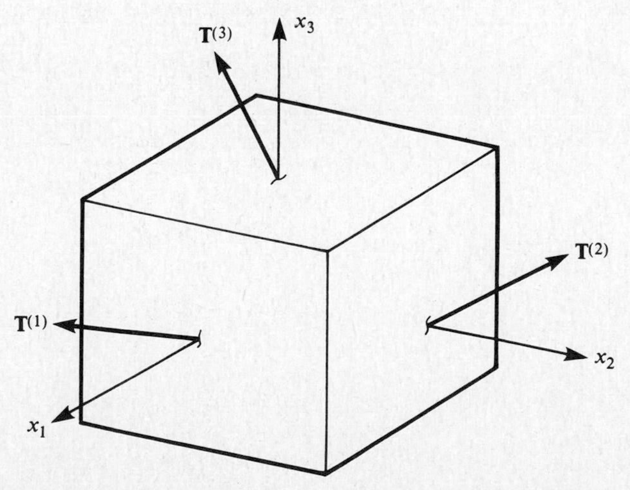

Figure 2-6 Stress vectors and their components on three mutually perpendicular planes at a point.

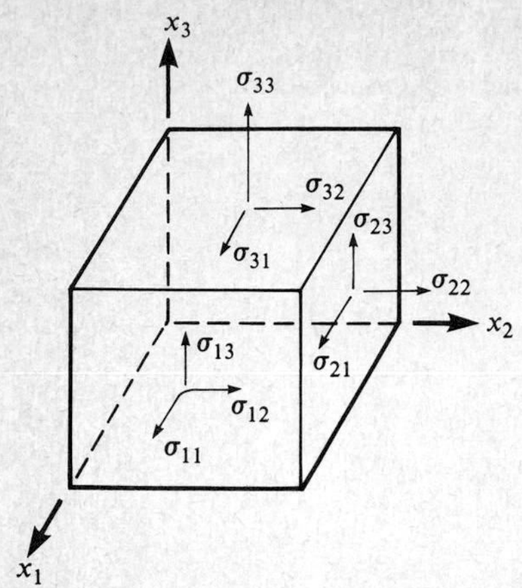

Figure 2-7 Components of a stress tensor on three mutually perpendicular planes at a point in the body.

no dimensions; it is a point cube so that all nine components are acting at the point P.

It can be shown that the traction vector $\mathbf{T}^{(\hat{\mathbf{n}})}$ on a plane with unit normal $\hat{\mathbf{n}}$ is related to the components σ_{ij} by the relation

$$\mathbf{T}^{(\hat{\mathbf{n}})} = \hat{\mathbf{n}} \cdot \boldsymbol{\sigma} = \boldsymbol{\sigma}^t \cdot \hat{\mathbf{n}} \quad \text{or} \quad \{\mathbf{T}^{(\hat{\mathbf{n}})}\} = [\sigma]^t\{\hat{\mathbf{n}}\} \tag{2-30}$$

where $\boldsymbol{\sigma}$ is the stress tensor (and $\boldsymbol{\sigma}^t$ is its transpose), called the *stress tensor* of *Cauchy*. The stress tensor associates with each unit normal $\hat{\mathbf{n}}$ the traction vector

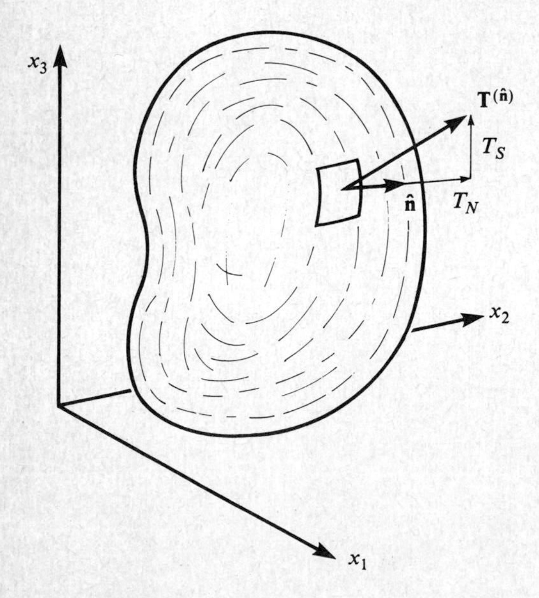

Figure 2-8 The normal and shearing components of a stress vector on the surface of a body.

$\mathbf{T}^{(\hat{\mathbf{n}})}$. The normal component T_N and shearing components T_S of the stress vector $\mathbf{T}^{(\hat{\mathbf{n}})}$ are given by (see Fig. 2-8)

$$\begin{aligned} T_N &= \mathbf{T}^{(\hat{\mathbf{n}})} \cdot \hat{\mathbf{n}} \\ &= (\boldsymbol{\sigma}^t \cdot \hat{\mathbf{n}}) \cdot \hat{\mathbf{n}} \qquad (2\text{-}31a) \\ &= \sigma_{ij} n_i n_j \end{aligned}$$

$$T_S^2 = |\mathbf{T}^{(\hat{\mathbf{n}})}|^2 - T_N^2 \qquad (2\text{-}31b)$$

2-3-2 CONSERVATION OF MASS: THE CONTINUITY EQUATION

Lagrangian form The principle of conservation of mass states that the total mass of a material body is invariant (i.e., unchanged) during the motion. This implies

$$\int_{\Omega_0} \rho(X_1, X_2, X_3, 0)\, dV_0 = \int_{\Omega} \rho(x_1, x_2, x_3, t)\, dV \qquad (2\text{-}32)$$

where Ω is the volume occupied at time t by the material that occupied Ω_0 at time $t = 0$ (note that attention is focused on a *fixed* mass of material), and dV_0 and dV denote the volume elements in Ω_0 and Ω, respectively. The second integral can be evaluated by transforming to the material coordinates by means of Eq. (2-1)

$$\begin{aligned} \rho(x_1, x_2, x_3, t) &= \rho[x_i(X_1, X_2, X_3, t), t] \\ dV &= J\, dV_0 \end{aligned} \qquad (2\text{-}33)$$

where J is the determinant of the jacobian matrix in Eq. (2-3). Then Eq. (2-32) becomes

$$\int_{\Omega_0} (\rho_0 - J\rho)\, dV_0 = 0 \qquad \text{for arbitrary volume } \Omega_0$$

where ρ_0 denotes the first integrand of Eq. (2-32). This gives the continuity equation in terms of the material coordinates

$$\rho J = \rho_0 \qquad (2\text{-}34)$$

Eulerian form Here we consider an arbitrary volume Ω fixed in space, bounded by surface Γ. The total mass M of the continuous medium that fills the volume at time t is given by

$$M = \int_{\Omega} \rho\, dV \qquad (2\text{-}35)$$

where ρ is the density of the medium

$$\rho = \rho(x_1, x_2, x_3, t)$$

The rate of increase of the total mass in the volume Ω is

$$\frac{\partial M}{\partial t} = \int_\Omega \frac{\partial \rho}{\partial t}\, dV \tag{2-36}$$

The flux or rate of mass *outflow* through an elemental surface ds of the body is

$$\begin{aligned} \text{outflow} &= \rho v_n\, ds \\ &= \rho \hat{\mathbf{n}} \cdot \mathbf{v}\, ds \end{aligned}$$

where $\hat{\mathbf{n}}$ is the unit outward normal on ds. The total mass outflow is given by

$$\begin{aligned} \text{outflow} &= \oint_\Gamma \rho \hat{\mathbf{n}} \cdot \mathbf{v}\, ds \\ &= \oint_\Gamma \hat{\mathbf{n}} \cdot (\rho \mathbf{v})\, ds \end{aligned} \tag{2-37}$$

Since no mass is created or destroyed inside Ω, the sum of the rate of increase of the total mass and the total mass outflow should be equal to zero,

$$\int_\Omega \frac{\partial \rho}{\partial t}\, dV + \oint_\Gamma \hat{\mathbf{n}} \cdot (\rho \mathbf{v})\, ds = 0 \tag{2-38}$$

Using the divergence theorem (1-29), the second integral in Eq. (2-38) can be transformed to an integral over the volume. We have

$$\int_\Omega \left[\frac{\partial \rho}{\partial t} + \nabla \cdot (\rho \mathbf{v}) \right] dV = 0 \tag{2-39}$$

Since the integral in Eq. (2-39) vanishes for an arbitrary choice of the volume Ω, it follows that the integrand must vanish,

$$\frac{\partial \rho}{\partial t} + \nabla \cdot (\rho \mathbf{v}) = 0 \tag{2-40}$$

The equation can be written in the following alternative forms

$$\frac{\partial \rho}{\partial t} + \mathbf{v} \cdot \nabla \rho + \rho \nabla \cdot \mathbf{v} = 0 \tag{2-41}$$

$$\frac{D\rho}{Dt} + \rho \nabla \cdot \mathbf{v} = 0 \tag{2-42}$$

For steady flow $\partial \rho / \partial t = 0$, and for a homogeneous material, $\nabla \rho = 0$. If the density is constant, the continuity equation takes the form

$$\nabla \cdot \mathbf{v} = 0 \tag{2-43}$$

2-3-3 Momentum Principles

Equations of motion and equilibrium Consider a given mass of the medium (i.e., material body), instantaneously occupying a volume Ω bounded by a surface Γ. Suppose that the body is acted upon by external forces $\mathbf{T}$ per unit surface area and $\mathbf{f}$ per unit volume. The momentum principle states that the rate of change of the total momentum of a given continuous medium equals the vector sum of all the external forces acting on the medium, provided Newton's third law of action and reaction governs the internal forces. The rate of change of total momentum of the given mass is

$$\frac{D}{Dt}\int_{\Omega} \rho \mathbf{v}\, dV$$

The momentum balance principle can then be expressed by

$$\frac{D}{Dt}\int_{\Omega} \rho \mathbf{v}\, dV = \oint_{\Gamma} \mathbf{T}\, ds + \int_{\Omega} \mathbf{f}\, dV \tag{2-44}$$

Using Eq. (2-30) for $\mathbf{T}$ and making use of the divergence theorem in Eq. (1-29), the surface integral in Eq. (2-44) can be transformed to a volume integral. Also, using the Reynolds transport theorem (see Prob. 2-23) the left-hand side of the equality in Eq. (2-44) can be rewritten in an alternative form. We have

$$\int_{\Omega}\left(\rho\frac{D\mathbf{v}}{Dt} - \nabla\cdot\boldsymbol{\sigma} - \mathbf{f}\right) dV = \mathbf{0} \tag{2-45}$$

for any arbitrary volume Ω. Thus, we have at each point of the volume

$$\rho\frac{D\mathbf{v}}{Dt} = \nabla\cdot\boldsymbol{\sigma} + \mathbf{f} \tag{2-46}$$

In rectangular coordinates, the equation takes the form

$$\rho\frac{Dv_i}{Dt} = \frac{\partial\sigma_{ji}}{\partial x_j} + f_i \tag{2-47}$$

When the medium is in static equilibrium, the acceleration $D\mathbf{v}/Dt$ in the lagrangian description is zero and Eq. (2-47) reduces to

$$\nabla\cdot\boldsymbol{\sigma} + \mathbf{f} = \mathbf{0} \qquad \text{or} \qquad \frac{\partial\sigma_{ji}}{\partial x_j} + f_i = 0 \tag{2-48}$$

Note that the equilibrium relations in Eq. (2-48) contain three equations relating six stress components ($\sigma_{ij} = \sigma_{ji}$), and therefore cannot be solved for all six components. Additional equations are required. These include the strain-displacement relations discussed in Sec. 2-2 and constitutive relations or stress-strain relations that define the nature of the particular material under consideration. Next we show that the principle of moment of momentum leads to the conclusion $\sigma_{ij} = \sigma_{ji}$.

The moment of momentum principle The principle states that, in the absence of distributed couples, the time rate of the total moment of momentum for a given material domain is equal to the vector sum of the moments of the external forces acting on the body. Expressed in an analytical form, the postulate takes the form

$$\oint_\Gamma (\mathbf{r} \times \mathbf{T})\, ds + \int_\Omega \mathbf{r} \times \mathbf{f}\, dV = \frac{D}{Dt} \int_\Omega \mathbf{r} \times (\rho \mathbf{v})\, dV \tag{2-49}$$

where $\mathbf{r}$ is the position vector. Using the Reynolds transport theorem and Eq. (2-46), the right-hand side of (2-49) can be written as

$$\begin{aligned}
\frac{D}{Dt} \int_\Omega \rho(\mathbf{r} \times \mathbf{v})\, dV &= \int_\Omega \rho \frac{D}{Dt} (\mathbf{r} \times \mathbf{v})\, dV \\
&= \int_\Omega \rho \left(\frac{D\mathbf{r}}{Dt} \times \mathbf{v} + \mathbf{r} \times \frac{D\mathbf{v}}{Dt} \right) dV \\
&= \int_\Omega \rho \left(\mathbf{r} \times \frac{D\mathbf{v}}{Dt} \right) dV \\
&= \int_\Omega \mathbf{r} \times [\nabla \cdot \boldsymbol{\sigma} + \mathbf{f}]\, dV
\end{aligned} \tag{2-50}$$

Upon substituting Eq. (2-50) into Eq. (2-49) the body force terms cancel on both sides. We have

$$\oint_\Gamma (\mathbf{r} \times \mathbf{T})\, ds = \int_\Omega \mathbf{r} \times (\nabla \cdot \boldsymbol{\sigma})\, dV \tag{2-51}$$

To further simplify the equation, we express it in rectangular component form:

$$\oint_\Gamma \varepsilon_{ijk}\, x_i\, T_j\, \hat{\mathbf{e}}_k\, ds = \int_\Omega \varepsilon_{ijk}\, x_i \frac{\partial \sigma_{mj}}{\partial x_m}\, \hat{\mathbf{e}}_k\, dV \tag{2-52}$$

Since $T_j = n_m \sigma_{mj}$, we have

$$\begin{aligned}
\oint_\Gamma \varepsilon_{ijk}\, x_i\, n_m\, \sigma_{mj}\, ds &= \int_\Omega \varepsilon_{ijk} \frac{\partial}{\partial x_m} (x_i \sigma_{mj})\, dV \\
&= \int_\Omega \left(\varepsilon_{ijk}\, \delta_{im}\, \sigma_{mj} + \varepsilon_{ijk}\, x_i \frac{\partial \sigma_{mj}}{\partial x_m} \right) dV
\end{aligned}$$

Substituting into Eq. (2-52), we obtain

$$\int_\Omega \varepsilon_{ijk}\, \sigma_{ij}\, dV = 0$$

or

$$\varepsilon_{ijk}\, \sigma_{ij} = 0 \qquad \text{for} \qquad k = 1, 2, 3 \tag{2-53}$$

This implies that the components of the stress tensor are symmetric:

$$\begin{aligned} k = 1:\quad & \sigma_{23} - \sigma_{32} = 0 \\ k = 2:\quad & \sigma_{31} - \sigma_{13} = 0 \\ k = 3:\quad & \sigma_{12} - \sigma_{21} = 0 \end{aligned} \tag{2-54}$$

2-4 THERMODYNAMIC PRINCIPLES

2-4-1 Introduction

The thermodynamic principles of a continuous medium are concerned with the balance of energy and limits on the direction of thermodynamic processes. Thermodynamics is a science that treats the relations among heat, work, and the properties of systems in equilibrium. Two alternative choices for a thermodynamic system exist: a *closed system* consisting of a given collection of continuous matter and a fixed *control surface* in space accounting for the flux of matter through it. The control surface approach is convenient for the study of heat engines or flow processes where the control surface can be identified with an actual container wall.

The first law of thermodynamics is commonly known as the balance of energy principle, and it can be regarded as a statement of the interconvertibility of heat and work. The law does not place any restrictions on the direction of the process. For instance, in the study of mechanics of particles and rigid bodies the kinetic energy and potential energy can be fully transformed from one to the other in the absence of friction and other dissipative mechanisms. However, when thermal phenomena are involved, the second law of thermodynamics imposes restrictions on the reversibility of energy from thermal to mechanical. For example, the kinetic energy of a flywheel can all be converted to internal energy by means of a friction brake; if the whole system is insulated, the internal energy causes the temperature of the system to rise. Although the first law does not restrict the reversal process, namely the conversion of heat to internal energy and internal energy to set the flywheel in motion, such a reversal cannot occur because the frictional dissipation is an *irreversible process*. The second law of thermodynamics places restrictions on such processes. To restrict the scope of the present coverage to reasonable limits, we will not consider irreversible processes in the present book. Therefore the second law will not be discussed.

2-4-2 The First Law of Thermodynamics: The Energy Equation

The first law of thermodynamics states that the time rate of change of the total energy (i.e., the sum of the kinetic energy and the internal energy) is equal to the sum of the rate of work done by the external forces and the change of heat content per unit time:

$$\frac{D}{Dt}(K + U) = W + H \tag{2-55}$$

Here K denotes the kinetic energy, U the internal energy, W the power input, and H the heat input of the system.

Kinetic and internal energies The kinetic energy of the system is given by

$$K = \tfrac{1}{2}\int_\Omega (\rho\mathbf{v}\cdot\mathbf{v})\, dV \tag{2-56}$$

If e is the energy per unit mass (or *specific internal energy*), the total internal energy of the system is given by

$$U = \int_\Omega \rho e\, dV \tag{2-57}$$

The kinetic energy K of a system is the energy associated with the macroscopically observable velocity of the continuum. The kinetic energy associated with the (microscopic) motions of molecules of the continuum is considered to be a part of the internal energy; the elastic strain energy and other forms of energy are also parts of internal energy.

Power input The power input, in the nonpolar case, consists of the rate at which the external surface tractions $\mathbf{T}$ per unit area and body forces $\mathbf{f}$ per unit volume are doing work on the mass system instantaneously occupying the volume Ω bounded by Γ:

$$\begin{aligned} W &= \oint_\Gamma \mathbf{T}\cdot\mathbf{v}\, ds + \int_\Omega \mathbf{f}\cdot\mathbf{v}\, dV \\ &= \oint_\Gamma (\hat{\mathbf{n}}\cdot\boldsymbol{\sigma})\cdot\mathbf{v}\, ds + \int_\Omega \mathbf{f}\cdot\mathbf{v}\, dV \\ &= \int_\Omega [\nabla\cdot(\boldsymbol{\sigma}\cdot\mathbf{v}) + \mathbf{f}\cdot\mathbf{v}]\, dV \\ &= \int_\Omega [(\nabla\cdot\boldsymbol{\sigma} + \mathbf{f})\cdot\mathbf{v} + \boldsymbol{\sigma} : \nabla\mathbf{v}]\, dV \end{aligned} \tag{2-58}$$

where : denotes the "double-dot" product $\boldsymbol{\sigma}^1 : \boldsymbol{\sigma}^2 = \sigma^1_{ij}\sigma^2_{ji}$. The expression in parentheses equals $\rho D\mathbf{v}/Dt$ by the equations of motion, Eq. (2-46). Also, by virtue of the symmetry of the stress tensor and the result of Exercise 2-4a, the last expression in the integrand of Eq. (2-58) can be replaced by $\boldsymbol{\sigma} : \mathbf{D}$. Hence,

$$W = \frac{D}{Dt}\int_\Omega \tfrac{1}{2}\rho\mathbf{v}\cdot\mathbf{v}\, dV + \int_\Omega \boldsymbol{\sigma} : \mathbf{D}\, dV. \tag{2-59}$$

Heat input The rate of heat input consists of conduction through the surface Γ and internal heat generation (possibly from a radiation field):

$$H = -\oint_{\Gamma} \mathbf{q} \cdot \hat{\mathbf{n}}\, ds + \int_{\Omega} \rho Q\, dV \tag{2-60}$$

where $\mathbf{q}$ is the heat flux vector normal to the surface, and Q is the internal heat source per unit mass.

Substituting Eqs. (2-56) to (2-60) into Eq. (2-55), we obtain

$$\frac{D}{Dt}\int_{\Omega} \rho(\tfrac{1}{2}\mathbf{v}\cdot\mathbf{v} + e)\, dV = \frac{D}{Dt}\int_{\Omega} \tfrac{1}{2}\rho\mathbf{v}\cdot\mathbf{v}\, dV + \int_{\Omega} \boldsymbol{\sigma} : \mathbf{D}\, dV - \oint_{\Gamma} \mathbf{q}\cdot\hat{\mathbf{n}}\, ds + \int_{\Omega} \rho Q\, dV \tag{2-61}$$

or

$$\int_{\Omega}\left(\rho \frac{De}{Dt} - \boldsymbol{\sigma} : \mathbf{D} + \nabla\cdot\mathbf{q} - \rho Q\right) dV = 0 \tag{2-62}$$

for an arbitrary volume Ω. Thus the conservation of energy implied by the first law of thermodynamics is given by

$$\begin{aligned} \rho \frac{De}{Dt} &= \boldsymbol{\sigma} : \mathbf{D} + \rho Q - \nabla\cdot\mathbf{q} \\ \rho \frac{De}{Dt} &= \sigma_{ij} D_{ij} + \rho Q - \frac{\partial q_i}{\partial x_i} \end{aligned} \tag{2-63}$$

2-5 CONSTITUTIVE LAWS

2-5-1 General Comments

The kinematic relations and the mechanical and thermodynamic principles reviewed in the previous sections are applicable to any continuum irrespective of its physical constitution. It is a common experience that two pieces of continuous material of the same geometry but different constitution respond differently under identical external forces. This implies that the response of a given continuous body depends on its internal constitution. Here we consider equations characterizing the individual material and its reaction to applied loads. These equations are called *constitutive equations.* The formulation of the constitutive equations for a given material is guided by certain rules. We will not discuss them here but we will review the linear constitutive relations for solids and fluids.

2-5-2 Linear Elastic Solids: Generalized Hooke's Law

A material body is said to be *ideally elastic* when the body recovers (under isothermal conditions) its original form completely upon removal of the forces causing deformation, and there is a one-to-one relationship between the state of stress and state of strain. The generalized Hooke's law relates the nine components of stress to the nine components of strain,

$$\sigma_{ij} = C_{ijkl} e_{kl} \tag{2-64}$$

where e_{kl} are the infinitesimal strain components, σ_{ij} are the Cauchy stress components, and C_{ijkl} are the material constants. The nine equations in Eq. (2-64) contain 81 constants. However, due to the symmetry of both σ_{ij} and e_{kl}, it follows that

$$C_{ijkl} = C_{jikl} \qquad C_{ijkl} = C_{ijlk} \tag{2-65}$$

and there are only 36 constants or moduli. In matrix form Eq. (2-64) can be expressed in the following convenient form

$$\begin{Bmatrix} \sigma_{11} \\ \sigma_{22} \\ \sigma_{33} \\ \sigma_{23} \\ \sigma_{31} \\ \sigma_{12} \end{Bmatrix} = \begin{bmatrix} c_{11} & c_{12} & c_{13} & c_{14} & c_{15} & c_{16} \\ c_{21} & c_{22} & c_{23} & c_{24} & c_{25} & c_{26} \\ c_{31} & c_{32} & c_{33} & c_{34} & c_{35} & c_{36} \\ c_{41} & c_{42} & c_{43} & c_{44} & c_{45} & c_{46} \\ c_{51} & c_{52} & c_{53} & c_{54} & c_{55} & c_{56} \\ c_{61} & c_{62} & c_{63} & c_{64} & c_{65} & c_{66} \end{bmatrix} \begin{Bmatrix} e_{11} \\ e_{22} \\ e_{33} \\ 2e_{23} \\ 2e_{31} \\ 2e_{12} \end{Bmatrix} \tag{2-66}$$

or

$$\sigma_i = c_{ij} e_i \tag{2-67}$$

where $\sigma_1 = \sigma_{11}$, $\sigma_2 = \sigma_{22}$, $\sigma_3 = \sigma_{33}$, $\sigma_4 = \sigma_{23}$, $\sigma_5 = \sigma_{13}$, $\sigma_6 = \sigma_{12}$, $e_1 = e_{11}$, $e_2 = e_{22}$, $e_3 = e_{33}$, $e_4 = 2e_{23}$, $e_5 = 2e_{13}$, $e_6 = 2e_{12}$.

The number of independent constants is further reduced from 36 to 21 when a strain-energy function exists (and we have $c_{ij} = c_{ji}$). When one plane of elastic symmetry exists, the number of elastic constants is reduced to 13, and when three mutually orthogonal planes of elastic symmetry exist, the number is reduced to 9. Finally, when there exist no preferred directions in the material (i.e., the material is isotropic), the number of independent constants reduces to 2. In this special case, Eq. (2-64) can be expressed in the form

$$\sigma_{ij} = [\lambda \delta_{ij} \delta_{kl} + \mu(\delta_{ik} \delta_{jl} + \delta_{il} \delta_{jk})] e_{kl} \tag{2-68}$$

where λ and μ are the Lamé elastic constants and δ_{ij} is the Kronecker delta. The Lamé constants are related to the shear modulus G, Young's modulus E, and Poisson's ratio ν as follows

$$\mu = G = \frac{E}{2(1+\nu)} \qquad \lambda = \frac{\nu E}{(1+\nu)(1-2\nu)} \tag{2-69}$$

The coefficients c_{ij} for isotropic materials have the following meaning:

$$c_{11} = c_{22} = c_{33} = \lambda + 2\mu \qquad c_{12} = c_{13} = c_{23} = \lambda \qquad c_{44} = c_{55} = c_{66} = \mu \tag{2-70}$$

and all other c_{ij} are zero.

2-5-3 Newtonian Fluids

The fluids whose constitution is described by linear constitutive relations are called newtonian fluids. For a newtonian fluid the constitutive equations are given by

$$\sigma_{ij} = -P\delta_{ij} + C_{ijkl} D_{kl} \tag{2-71a}$$

where P is the thermodynamic pressure, and D_{kl} are the components of the rate of deformation tensor,

$$D_{kl} = \tfrac{1}{2}(v_{k,\,l} + v_{l,\,k}) \tag{2-71b}$$

The pressure P is related to the density ρ and temperature θ through the *kinetic equation of state*

$$F(P, \rho, \theta) = 0 \tag{2-72}$$

An example of the equation of state is provided by the perfect-gas law

$$P = \rho R\theta \tag{2-73}$$

where R is the gas constant for the particular gas.

For isotropic fluids, Eq. (2-71a) takes the form

$$\sigma_{ij} = -P\delta_{ij} + [\lambda\delta_{ij}\,\delta_{kl} + \mu(\delta_{ik}\,\delta_{jl} + \delta_{il}\,\delta_{jk})]D_{kl} \tag{2-74}$$

In terms of the *deviatoric components* σ'_{ij} and D'_{ij}

$$\sigma'_{ij} \equiv \sigma_{ij} + \bar{P}\,\delta_{ij} \qquad D'_{ij} \equiv D_{ij} - \tfrac{1}{3}D_{kk}\,\delta_{ij} \tag{2-75}$$

Eq. (2-74) becomes

$$\sigma'_{ij} = (\bar{P} - P)\delta_{ij} + (\lambda + \tfrac{2}{3}\mu)D_{kk}\,\delta_{ij} + 2\mu D'_{ij} \tag{2-76}$$

Here $\bar{P}$ denotes the hydrostatic pressure, $\bar{P} = -\tfrac{1}{3}\sigma_{kk}$. Since $D'_{ii} = \sigma'_{ii} = 0$, we have from Eq. (2-76)

$$(\bar{P} - P) + (\lambda + \tfrac{2}{3}\mu)D_{kk} = 0 \tag{2-77}$$

and Eq. (2-76) takes the simple form

$$\sigma'_{ij} = 2\mu D'_{ij} \tag{2-78}$$

It should be noted that μ in the case of fluids represents the viscosity of the fluid.

2-5-4 Thermoelasticity

In writing the constitutive relations for solids and fluids, we implicitly assumed that the variations in the elastic constants with temperature are negligible. While we neglect the variations in the elastic constants, we are required to consider the thermal expansion of the material. When the thermal expansions (i.e., geometric changes due to heating) are prevented by boundary conditions or other constraints, the body develops thermal stresses in addition to stresses caused by other mechanical loads. In such cases, the linear thermoelastic constitutive equations for solids are given by

$$\sigma_{ij} = -\alpha_{ij}(\theta - \theta_0) + C_{ijkl}e_{kl} \tag{2-79}$$

where θ_0 is the reference temperature (hence, $\theta - \theta_0$ is the change in temperature) and α_{ij} are the cartesian components of the thermal conductivity tensor $\boldsymbol{\alpha}$. For an isotropic body Eq. (2-79) takes the simple form

$$\sigma_{ij} = -\alpha\delta_{ij}(\theta - \theta_0) + \lambda\delta_{ij}e_{kk} + 2\mu e_{ij} \tag{2-80}$$

In the case of newtonian fluids, temperature changes are accounted for via buoyancy forces. For both solids and fluids, consideration of temperature effects requires the solution of the energy equation.

A linear constitutive relation between heat flux $\mathbf{q}$ and temperature gradient $\nabla\theta$ is given by the Fourier heat conduction law:

$$\mathbf{q} = -\boldsymbol{\alpha}\cdot\nabla\theta \qquad \text{or} \qquad q_i = -\alpha_{ij}\frac{\partial\theta}{\partial x_j} \tag{2-81}$$

where α_{ij} are the components of the thermal conductivity tensor. For an isotropic solid or fluid, the heat flux is necessarily in the same direction as the temperature gradient, giving $\alpha_{ij} = \alpha\delta_{ij}$ and

$$\mathbf{q} = -\alpha\nabla\theta \qquad \text{or} \qquad q_i = -\alpha\frac{\partial\theta}{\partial x_i} \tag{2-82}$$

With $\alpha > 0$, Eq. (2-82) requires that heat flows from regions of high temperature toward regions of lower temperature.

Finally, the constitutive equation relating the internal energy e to the density ρ and temperature θ is expressed in functional form by the *caloric equation of state*

$$I(e, \rho, \theta) = 0 \tag{2-83}$$

This completes the description of the constitutive equations.

2-6 SPECIALIZATION OF THE EQUATIONS TO LINEAR FIELD PROBLEMS

2-6-1 Introductory Comments

The kinematic relations, mechanical and thermodynamic principles, and linear constitutive equations presented in the previous sections are applicable to any continuous body whose constitutive behavior is linear. Here we summarize these equations for linear field problems.

Kinematic relations:

$$e_{ij} = \tfrac{1}{2}(u_{i,j} + u_{j,i})$$

$$D_{ij} = \tfrac{1}{2}(v_{i,j} + v_{j,i}) \equiv \frac{De_{ij}}{Dt} \tag{2-84}$$

Balance equations:

$$\frac{D\rho}{Dt} = -\rho \frac{\partial v_i}{\partial x_i} \qquad \rho_0 = \rho J$$

$$\rho \frac{Dv_j}{Dt} = \frac{\partial \sigma_{ij}}{\partial x_i} + f_j \tag{2-85}$$

$$\sigma_{ij} = \sigma_{ji}$$

$$\rho \frac{De}{Dt} = \sigma_{ij} D_{ij} + \rho Q - \frac{\partial q_i}{\partial x_i}$$

Constitutive equations:

$$\sigma_{ij} = -\alpha_{ij}(\theta - \theta_0) + C_{ijkl} e_{kl} \qquad \text{or} \qquad \sigma_{ij} = -P\,\delta_{ij} + C_{ijkl} D_{kl}$$

$$q_i = -\alpha_{ij} \frac{\partial \theta}{\partial x_j} \tag{2-86}$$

$$F(\rho, \theta, P) = 0$$

$$I(\rho, \theta, e) = 0$$

Thus, there are 22 unknowns and 22 equations to be solved in conjunction with appropriate initial and boundary conditions. The dependent variables and specified data (i.e., specified functions) are as follows:

Variables	Number	Data
u_i or v_i $(i = 1, 2, 3)$	3	—
e_{ij} or D_{ij} $(i, j = 1, 2, 3)$	6	—
ρ	1	
σ_{ij} $(i, j = 1, 2, 3)$	6	f_i
e, q_i $(i = 1, 2, 3)$	4	Q
P	1	C_{ijkl}
θ	1	α_{ij}

In most problems of interest, the number of dependent unknowns (hence, the number of equations) can be reduced. In the following sections we consider two special cases of the field equations, Eqs. (2-84) to (2-86).

2-6-2 Linear Theory of Elasticity

The formulation uses the lagrangian description in terms of material coordinates X_i of a particle in the natural (i.e., undeformed) state. Under the assumptions that the displacement gradients are small compared to unity, that the generalized Hooke's law is valid, and that the equations of motion or equilibrium are satisfied in the undeformed reference configuration, the equations governing a linear elastic solid under isothermal conditions can be expressed as

$$e_{ij} = \frac{1}{2}\left(\frac{\partial u_i}{\partial X_j} + \frac{\partial u_j}{\partial X_i}\right)$$

$$\sigma_{ij} = C_{ijkl}\, e_{kl} \tag{2-87}$$

$$\frac{\partial \sigma_{ij}}{\partial X_j} + f_i = \rho \frac{\partial^2 u_i}{\partial t^2}$$

where ρ, f_i, and σ_{ij} are understood to be evaluated as functions of $\mathbf{X}$. There are 15 equations for 6 stresses, 6 strains, and 3 displacements. These equations can be solved, with the aid of appropriate boundary and initial conditions, for the 15 variables, and one need not consider the energy equation and the equations of state.

When thermal expansions are involved, the constitutive equations must be modified as follows:

$$\sigma_{ij} = -\alpha_{ij}(\theta - \theta_0) + C_{ijkl}\, e_{kl} \tag{2-88}$$

where θ is the temperature and θ_0 is the reference temperature. However, this introduces an additional variable, θ, into the formulation. This necessitates the use of the energy equation. In ideal elasticity, one assumes that all of the input work is converted into internal energy in the form of recoverable stored elastic strain energy $e = e(e_{ij})$. However, when the temperature is included, we have $e = e(e_{ij}, \theta)$, and

$$\frac{\partial e}{\partial t} = \frac{\partial e}{\partial e_{ij}} \frac{\partial e_{ij}}{\partial t} + \frac{\partial e}{\partial \theta} \frac{\partial \theta}{\partial t}$$

$$\equiv \left(\frac{1}{\rho} \sigma_{ij} - \theta \frac{\partial \sigma_{ij}}{\partial \theta}\right) \frac{\partial e_{ij}}{\partial t} + c \frac{\partial \theta}{\partial t} \tag{2-89}$$

where c denotes the specific heat. Consequently, the energy equation simplifies to

$$\rho \theta \alpha_{ij} \frac{\partial u_{i,j}}{\partial t} + \rho c \frac{\partial \theta}{\partial t} + \frac{\partial q_i}{\partial X_i} = \rho Q \tag{2-90}$$

The first term in (2-90) is usually small and is often neglected. Using the Fourier heat conduction law, the energy equation can be expressed in the form

$$\rho c \frac{\partial \theta}{\partial t} - \frac{\partial}{\partial X_i}\left(\alpha_{ij} \frac{\partial \theta}{\partial X_j}\right) = \rho Q \tag{2-91}$$

Equation (2-91) provides the additional equation for the determination of the temperature θ.

Solution of 15 differential equations (2-87) for 15 unknowns is a formidable task. Alternatively, one can formulate the problem in terms of fewer unknowns and fewer equations. One such alternative is provided by the displacement formulation, which consists of rewriting Eq. (2-87) in terms of the three displacements. For an isotropic body these are given by Navier's equation of motion,

$$(\lambda + \mu) \frac{\partial^2 u_j}{\partial X_i \, \partial X_j} + \mu \frac{\partial^2 u_i}{\partial X_j \, \partial X_j} + f_i = \rho \frac{\partial^2 u_i}{\partial t^2} \tag{2-92}$$

For a two-dimensional orthotropic elastic body the displacement formulation is described by the equations

$$\begin{aligned} &\frac{\partial}{\partial X_1}\left(c_{11} \frac{\partial u_1}{\partial X_1} + c_{12} \frac{\partial u_2}{\partial X_2}\right) + c_{33} \frac{\partial}{\partial X_2}\left(\frac{\partial u_1}{\partial X_2} + \frac{\partial u_2}{\partial X_1}\right) + f_1 = \rho \frac{\partial^2 u_1}{\partial t^2} \\ &c_{33} \frac{\partial}{\partial X_1}\left(\frac{\partial u_1}{\partial X_2} + \frac{\partial u_2}{\partial X_1}\right) + \frac{\partial}{\partial X_2}\left(c_{12} \frac{\partial u_1}{\partial X_1} + c_{22} \frac{\partial u_2}{\partial X_2}\right) + f_2 = \rho \frac{\partial^2 u_2}{\partial t^2} \end{aligned} \tag{2-93}$$

We shall discuss the form of the boundary conditions next. The boundary conditions are of two types:

$$\begin{aligned} &\textit{Geometric} \text{ (or essential):} \qquad && u_i = \hat{u}_i \\ &\textit{Static} \text{ (or natural):} \qquad && \sigma_{ij} n_j = \hat{t}_i \end{aligned} \tag{2-94}$$

Only one of the two boundary conditions can be specified at a point on the boundary.

2-6-3 Newtonian Fluid Mechanics

The field equations of an isotropic newtonian fluid obeying the Fourier heat conduction equation and the caloric equation of state are given by

$$\begin{aligned} \frac{D\rho}{Dt} + \rho \nabla \cdot \mathbf{v} &= 0 \\ \nabla : \boldsymbol{\sigma} + \mathbf{f} &= \rho \frac{D\mathbf{v}}{Dt} \\ \boldsymbol{\sigma} : \mathbf{D} + \rho Q &= -\nabla \cdot (\alpha \nabla \theta) + \rho \frac{De}{Dt} \\ \sigma_{ij} &= -P \, \delta_{ij} + \lambda \, \delta_{ij} D_{kk} + 2\mu D_{ij} \\ F(P, \rho, \theta) &= 0 \\ I(e, \rho, \theta) &= 0 \end{aligned} \tag{2-95}$$

where D_{ij} are understood to be known in terms of v_i. These 13 equations in terms of 13 unknowns can be reduced to seven equations in seven unknowns by eliminating the components σ_{ij} through use of the constitutive equations. The equations of motion and energy equation take the form

$$\rho \frac{D\mathbf{v}}{Dt} = -\nabla P + (\lambda + \mu)\nabla(\nabla \cdot \mathbf{v}) + \mu\nabla^2\mathbf{v} + \mathbf{f} \tag{2-96}$$

$$\rho \frac{De}{Dt} = \nabla \cdot (\alpha\nabla\theta) - P(\nabla \cdot \mathbf{v}) + \rho Q + \underline{(\lambda + \tfrac{2}{3}\mu)D_{ii}D_{jj} + \mu(2D'_{ij}D'_{ij})} \tag{2-97}$$

where D'_{ij} are the components of the deviatoric strain-rate tensor

$$D'_{ij} = D_{ij} - \tfrac{1}{3}D_{kk}\delta_{ij}. \tag{2-98}$$

The underlined term represents the dissipation function, Φ. Equations (2-96) are the generalized Navier–Stokes equations. These equations are nonlinear due to the fact that the material derivative of v_i contains nonlinear terms,

$$\rho \frac{Dv_i}{Dt} = \rho\left[\frac{\partial v_i}{\partial t} + v_k \frac{\partial v_i}{\partial x_k}\right].$$

The usual Navier-Stokes equations of motion for an incompressible fluid ($\nabla \cdot \mathbf{v} = D_{ii} = 0$) are given by

$$-\nabla P + \mu\nabla^2\mathbf{v} + \mathbf{f} = \rho\left(\frac{\partial \mathbf{v}}{\partial t} + \mathbf{v} \cdot \nabla\mathbf{v}\right) \tag{2-99}$$

For the steady and low-speed (so that the inertia and convective terms can be neglected) flow of an incompressible fluid, the governing equations take the form (ρ = constant)

$$\begin{gathered} \frac{\partial v_1}{\partial x_1} + \frac{\partial v_2}{\partial x_2} = 0 \\ -\frac{\partial P}{\partial x_i} + \mu \frac{\partial^2 v_i}{\partial x_k\, \partial x_k} + f_i = 0 \end{gathered} \tag{2-100}$$

which can be solved, with appropriate boundary conditions of the problem, for the three velocity components and the pressure.

The boundary conditions for fluid flow are of the type

$$\begin{aligned} &\textit{essential:} && v_i = \hat{v}_i \\ &\textit{natural:} && n_i(\sigma_{ij} - P\,\delta_{ij}) = \hat{t}_j \end{aligned} \tag{2-101}$$

Only one element of the pairs $(\hat{v}_i, \hat{t}_j)$ can be specified at a point. The classification of boundary conditions into essential and natural type are discussed in Chap. 4.

2-6-4 Heat Transfer in Solids

In the study of conduction and convection heat transfer in solids, one is required to solve Eq. (2-91) for temperature and then use it in Eq. (2-88) to determine the thermal stresses. The subject of heat transfer is primarily concerned with the study of conduction and radiation, i.e., the solution of Eq. (2-91).

In the absence of radiation effects, Eq. (2-91) takes the form

$$\rho c \frac{\partial \theta}{\partial t} - \frac{\partial}{\partial X_i}\left(\alpha_{ij} \frac{\partial \theta}{\partial X_j}\right) = \rho Q \tag{2-102}$$

Here Q is assumed to contain no radiation related terms. The boundary conditions are of the type

$$\text{essential:} \qquad \theta = \hat{\theta}$$

$$\text{natural:} \qquad n_i \alpha_{ij} \frac{\partial \theta}{\partial X_j} + h(\theta - \theta_\infty) = \hat{q}_n \tag{2-103}$$

where h is the film heat-transfer coefficient and θ_∞ is the ambient temperature. The term $h(\theta - \theta_\infty)$ represents convective heat transfer from the body to the surrounding medium, which is at temperature θ_∞. When no convective boundary is involved in the problem, the term $h(\theta - \theta_\infty)$ is set to zero.

If the surface of a heat conducting body is exposed to a high-temperature source, it receives heat by radiation. The rate of heat transfer by radiation between two surfaces at absolute temperatures θ_1 and θ_2 and separated by a vacuum is expressed by the Stefan–Boltzmann law

$$q = C_2 \theta_2^4 - C_1 \theta_1^4 \tag{2-104}$$

where C_1 and C_2 are constants that depend on the relative orientation of the two surfaces, the distance separating them, and the absorption and reflection properties of the surfaces. If θ_∞ and θ are the absolute temperatures of the source and surface of the body, respectively, the natural boundary condition in Eq. (2-103) is modified to account for the radiative heat transfer

$$n_i \alpha_{ij} \frac{\partial \theta}{\partial X_j} + C(\theta_\infty^4 - \theta^4) = 0 \tag{2-105}$$

We shall return to the equations developed in this chapter for specific applications and solutions of the equations.

PROBLEMS

2-1 Find the material time derivative of the following velocity fields:

(*a*) $\mathbf{v} = 2x_1 t\hat{\mathbf{e}}_1 + 3x_2 t\hat{\mathbf{e}}_2$

(*b*) $\mathbf{v} = C \dfrac{x_1}{t} \hat{\mathbf{e}}_1 + Dx_1^2 t\hat{\mathbf{e}}_2$ $\quad$ C and D are constants

2-2 Consider a rectangular block of dimensions a and b (and unit thickness) in its undeformed configuration (see Fig. P2*a*). The deformed configuration of the body is shown in Fig. P2*b*. Describe the deformed configuration by writing the position of an arbitrary material point in the body in terms of

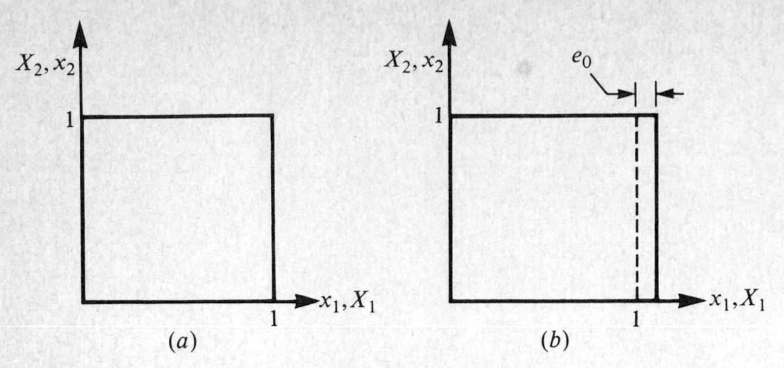

Figure P2a **Figure P2b**

its original position, i.e., $x_i = x_i(X_1, X_2, X_3)$, and find the displacements and the components of the lagrangian strain tensor.

2-3 Repeat Prob. 2-2 above for the problem in Fig. P3.

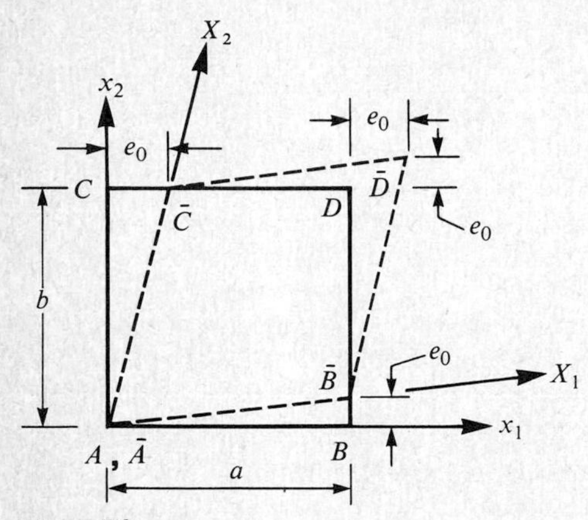

Figure P3

2-4 It is useful to split the velocity-gradient tensor **F** into symmetric and antisymmetric parts:

$$\dot{\mathbf{F}} = \mathbf{D} + \boldsymbol{\omega} \qquad \text{or} \qquad \frac{\partial v_i}{\partial x_j} = D_{ij} + \omega_{ij}$$

where

$$D_{ij} = \frac{1}{2}\left(\frac{\partial v_i}{\partial x_j} + \frac{\partial v_j}{\partial x_i}\right) \qquad \omega_{ij} = \frac{1}{2}\left(\frac{\partial v_i}{\partial x_j} - \frac{\partial v_j}{\partial x_i}\right)$$

Show that:

(a) $\sigma_{ij} v_{i,j} = \sigma_{ij} D_{ij}$ for $\sigma_{ij} = \sigma_{ji}$

(b) $\omega_{ij} = \frac{1}{2}\varepsilon_{ijk}\,\varepsilon_{mnk}\dfrac{\partial v_m}{\partial x_n}$

(c) $\Omega_k = -\frac{1}{2}\varepsilon_{ijk}\,\omega_{ij}$ where $\boldsymbol{\Omega} = \frac{1}{2}(\text{curl } \mathbf{v})$.

2-5 (*Torsion problem*) Given the displacement field, $u_1 = -\alpha x_2 x_3$, $u_2 = \alpha x_1 x_3$, $u_3 = \alpha\phi(x_1, x_2)$ determine the components of the infinitesimal strain tensor. Assume that α is a constant.

2-6 (*Plate bending*) Given the displacement field

$$u_1 = u_0(x_1, x_2) + x_3\,\psi(x_1, x_2)$$
$$u_2 = v_0(x_1, x_2) + x_3\,\phi(x_1, x_2)$$
$$u_3 = w_0(x_1, x_2)$$

determine the components of the linear strain tensor in terms of the functions u_0, v_0, w_0, ψ, and ϕ.

2-7 Show that in order to have legitimate displacements corresponding to a given infinitesimal strain tensor **e**, the strain tensor **e** must satisfy the compatibility relation (equivalent to Eq. (2-26))

$$\nabla \times (\nabla \times \mathbf{e})^T = 0 \qquad \text{or} \qquad \varepsilon_{imp}\,\varepsilon_{jnq}\,e_{ij,\,mn} = 0$$

Hint: Begin with the requirement $u_{i,\,jk} = u_{i,\,kj}$ and use the results of Prob. 2-4 to obtain $\varepsilon_{imp}\,\varepsilon_{jnq}\,e_{ij,\,mn} = 0$.

2-8 Is it necessary to check the compatibility conditions when the displacements are given? Does the displacement field given in Prob. 2-5 satisfy the compatibility condition?

2-9 Check whether the following strains satisfy the compatibility conditions

(*a*) $e_{11} = \alpha(x_1^2 + x_2^2 + x_3^2) \qquad e_{12} = 2\alpha x_1 x_2 x_3$

$e_{22} = x_2^2 + x_3^2 \qquad e_{33} = x_3^2 \qquad e_{13} = e_{23} = 0$

(*b*) $e_{11} = \alpha x_3(x_1^2 + x_2^2) \qquad e_{12} = 2\alpha x_1 x_2 x_3$

$e_{22} = x_2^2 x_3 \qquad e_{13} = e_{23} = e_{33} = 0$

2-10 Given the strain tensor $\mathbf{e} = E_r\,\hat{\mathbf{e}}_r\hat{\mathbf{e}}_r + E_\theta\,\hat{\mathbf{e}}_\theta\hat{\mathbf{e}}_\theta$ in the cylindrical coordinate system $(\hat{\mathbf{e}}_r, \hat{\mathbf{e}}_\theta, \hat{\mathbf{e}}_z)$, with E_r and E_θ being functions of r only, determine the compatibility conditions between E_r and E_θ by requiring curl (curl $\mathbf{e})^t = 0$. Here $\mathbf{e}^t$ denotes the transpose of **e**.

$$\textit{Hint: } \vec{\nabla} = \hat{\mathbf{e}}_r \frac{\partial}{\partial r} + \frac{\hat{\mathbf{e}}_\theta}{r}\frac{\partial}{\partial \theta} + \hat{\mathbf{e}}_z \frac{\partial}{\partial z} \qquad \frac{\partial \hat{\mathbf{e}}_r}{\partial \theta} = \hat{\mathbf{e}}_\theta \qquad \frac{\partial \hat{\mathbf{e}}_\theta}{\partial \theta} = -\hat{\mathbf{e}}_r$$

2-11 Derive the Cauchy stress formula $\mathbf{T}^{(\hat{\mathbf{n}})} = \hat{\mathbf{n}} \cdot \boldsymbol{\sigma}$ for a continuous body by considering a tetrahedral element of the continuum.

2-12 Find the stress vector and its normal and shearing components on the plane indicated for the following states of stress:

(*a*) $\begin{bmatrix} 4 & -4 & 0 \\ -4 & 0 & 0 \\ 0 & 0 & 3 \end{bmatrix}$ on plane $2x_1 + 3x_2 + \sqrt{3}\,x_3 = 0$

(*b*) $\begin{bmatrix} 2 & -1 & 1 \\ -1 & 0 & 1 \\ 1 & 1 & 2 \end{bmatrix}$ on plane $x_1 - 2x_2 + x_3 = 2$

(*c*) $\begin{bmatrix} 1 & -3 & \sqrt{2} \\ -3 & 1 & -\sqrt{2} \\ \sqrt{2} & -\sqrt{2} & 4 \end{bmatrix}$ on plane $2x_1 - 2x_2 + x_3 = 3$

2-13 Show that the eigenvalues of a real symmetric matrix are real.

2-14 Show that the eigenvectors associated with the distinct eigenvalues of a real symmetric matrix are orthogonal.

2-15 Find the principal stresses and principal planes associated with the stress tensor in Prob. 2-12(*a*).

2-16 Find the principal stresses and principal planes for the states of stress given in Prob. 2-12(*c*).

2-17 Verify that the eigenvectors associated with the stress components in Prob. 2-12(*b*) are mutually orthogonal.

2-18 (*Deviatoric components of stress*) Define the alternative stress components, called the deviatoric stress components, by $\sigma'_{ij} = \sigma_{ij} - \frac{1}{3}\sigma_{kk}\,\delta_{ij}$. Show that the characteristic polynomial associated with the deviatoric stress components is given by $-(\lambda')^3 - I'_2\lambda' + I'_3 = 0$, where I'_2 and I'_3 are the invariants $I'_2 = -\frac{1}{2}\sigma'_{ij}\sigma'_{ij}$, $I'_3 = \det \sigma'_{ij}$. Further, show that the eigenvalues λ'_i are related to the eigenvalues λ_i of σ_{ij} by $\lambda_i = \lambda'_i + \frac{1}{3}\sigma_{kk}$, $i = 1, 2, 3$.

2-19 (Continuation of Prob. 2-18) Use the transformation

$$\lambda' = 2(-\tfrac{1}{3}I'_2)^{1/2} \cos \alpha \tag{a}$$

to rewrite the characteristic polynomial of the deviatoric stress tensor in terms of α (more precisely, in terms of $\cos 3\alpha$), and show that the roots (i.e., eigenvalues) of the polynomial are given by

$$\lambda'_i = 2(-\tfrac{1}{3}I'_2)^{1/2} \cos \alpha_i, \tag{b}$$

where α_i $(i = 1, 2, 3)$ are given by

$$\alpha_1 = \tfrac{1}{3}\left\{\cos^{-1}\left[\frac{I'_3}{2}\left(\frac{-3}{I'_2}\right)^{3/2}\right]\right\} \qquad \alpha_2 = \alpha_1 + \frac{2\pi}{3} \qquad \alpha_3 = \alpha_1 - \frac{2\pi}{3} \tag{c}$$

2-20 The principal stresses at a point in a body are given by $\lambda_1 = 4$, $\lambda_2 = 2$, $\lambda_3 = 1$, and the principal planes of the first two principal stresses are given by $\hat{\mathbf{n}}^{(1)} = \dfrac{1}{\sqrt{2}}(0, 1, -1)$, $\hat{\mathbf{n}}^{(2)} = \dfrac{1}{\sqrt{2}}(0, 1, 1)$.

Determine the state of stress at the point.

Hint: Note that the principal stresses are distinct, and that $(\hat{\mathbf{n}}^{(1)}, \hat{\mathbf{n}}^{(2)}, \hat{\mathbf{n}}^{(3)})$ can be viewed as the basis of a new coordinate system in which the state of stress is known.

2-21 Repeat Prob. 2-20 for the following two sets of data:

(*a*) $\lambda_1 = 5$, $\lambda_2 = 4$, $\lambda_3 = 1$, $\mathbf{n}^{(1)} = \frac{1}{2}(1, -\sqrt{3}, 0)$, $n_1^{(2)} = 0$, $n_2^{(2)} = 0$

(*b*) $\lambda_1 = 3$, $\lambda_2 = 2$, $\lambda_3 = -1$, $\mathbf{n}^{(1)} = \dfrac{1}{\sqrt{2}}(1, 0, 1)$, $\mathbf{n}^{(2)} = \dfrac{1}{\sqrt{3}}(-1, 1, n_3^{(2)})$.

2-22 Given the displacement field $u_1 = 0$, $u_2 = c_1x_3$, $u_3 = c_2x_2$ (c_1, c_2 are constants), determine:

(*a*) the strain components e_{mn} of the infinitesimal strain tensor;

(*b*) the maximum normal strain and its direction; and

(*c*) the axial (normal) strain in a line element originally at 45° counterclockwise from the x_3 axis.

Hint: Use the strain transformation equations, $\bar{e}_{ij} = a_{im}\,a_{jn}\,e_{mn}$.

2-23 (*Reynolds transport theorem*) Show that the material time derivative of the volume integral of $\rho\phi$, where ρ is the density and ϕ is any property of the mass (measured per unit mass) occupying the spatial volume V (and bounded by surface S) is

$$\frac{D}{Dt}\int_V \rho\phi \, dV = \int_V \frac{\partial}{\partial t}(\rho\phi)\, dV + \oint_S \phi\rho\mathbf{v}\cdot\hat{\mathbf{n}}\, ds$$

Further, show (using the conservation of mass equation) that

$$\frac{D}{Dt}\int_\Omega \rho\phi\, dV = \int_\Omega \rho \frac{D\phi}{Dt}\, dV$$

2-24 Express the velocity gradient as a sum of the *rate-of-deformation tensor* (or rate-of-strain tensor) and the *spin tensor.*

$$\frac{\partial v_i}{\partial x_j} = D_{ij} + \Omega_{ij}$$

where
$$D_{ij} = \frac{1}{2}\left(\frac{\partial v_i}{\partial x_j} + \frac{\partial v_j}{\partial x_i}\right)$$

$$\Omega_{ij} = \frac{1}{2}\left(\frac{\partial v_i}{\partial x_j} - \frac{\partial v_j}{\partial x_i}\right)$$

and show that
$$\sigma_{ij}\frac{\partial v_i}{\partial x_j} = \sigma_{ij} D_{ij}$$

2-25 Express the energy equation in the following alternate form

$$\rho \frac{D}{Dt}\left(e + \frac{P}{\rho} + \frac{v^2}{2}\right) = \frac{\partial P}{\partial t} + (\sigma'_{ij} v_j)_{,i} + f_i v_i - q_{i,i} + \rho Q$$

where σ'_{ij} are the deviatoric stress components: $\sigma'_{ij} = \sigma_{ij} + P\,\delta_{ij}$.

Hint: Use the continuity equation in the form $v_{i,i} = \rho \dfrac{D}{Dt}(1/\rho)$, and write

$$(\sigma_{ij} v_j)_{,i} = -\rho P \frac{D}{Dt}(1/\rho) - v_i P_{,i} + (\sigma'_{ij} v_j)_{,i}$$

2-26 A vector field $\mathbf{v}$ satisfying the condition $\nabla \cdot \mathbf{v} = 0$ is called *solenoidal.* Show that any vector field of the type $\mathbf{v} = \nabla \times \boldsymbol{\psi}$ is solenoidal.

2-27 For plane elasticity problems in which $\sigma_{13} = \sigma_{23} = \sigma_{33} = 0$ (called *plane stress* problems) and there are no body forces, it is assumed that there exists a twice-differentiable function ϕ such that

$$\sigma_{11} = \frac{\partial^2 \phi}{\partial x_2^2} \qquad \sigma_{22} = \frac{\partial^2 \phi}{\partial x_1^2} \qquad \sigma_{12} = -\frac{\partial^2 \phi}{\partial x_1\, \partial x_2}$$

Show that the function ϕ, called the *Airy stress function*, satisfies the equilibrium equations.

2-28 For an ideal nonviscous fluid the stress components are of the form $\sigma_{ij} = -P\,\delta_{ij}$, where P is the hydrostatic pressure. Show that the equations of motion take the form

$$-\frac{1}{\rho}\nabla P + \bar{\mathbf{f}} = \frac{\partial \mathbf{v}}{\partial t} + \mathbf{v}\cdot\nabla\mathbf{v},\ \bar{\mathbf{f}} = \mathbf{f}/\rho.$$

This equation is known as Euler's equation of motion.

2-29 Multiply the continuity equation, Eq. (2-42) by the velocity $\mathbf{v}$ and add the result to the left-hand side of the equation of motion, Eq. (2-46). After the use of vector identities, obtain the result $\dfrac{\partial}{\partial t}(\rho\mathbf{v}) + \text{div}\,(\rho\mathbf{v}\mathbf{v} - \boldsymbol{\sigma}) = \mathbf{f}$. This equation is called the conservation form of the momentum equation (2-46).

2-30 Using the momentum equation, Eq. (2-46), establish the identity

$$\frac{\rho}{2}\frac{D}{Dt}(\mathbf{v}\cdot\mathbf{v}) = \mathbf{v}\cdot\text{div}\,\boldsymbol{\sigma} + \mathbf{v}\cdot\mathbf{f}$$

2-31 Show that in the absence of heat sources and temperature changes, the energy equation is identically satisfied for all linear elastic solids for which the internal energy function e exists such that $\rho(\partial e/\partial e_{ij}) = \sigma_{ij}$.

2-32 Show that for linear thermoelasticity, the internal energy function $e = e(e_{ij}, \theta)$ is related to the stress components and temperature by

$$de = \left(\frac{1}{\rho}\sigma_{ij} - \theta\frac{\partial\sigma_{ij}}{\partial\theta}\right)de_{ij} + \frac{\partial e}{\partial\theta}\,d\theta$$

$$= \frac{1}{\rho}\sigma_{ij}\,de_{ij} - \theta\frac{\partial\sigma_{ij}}{\partial\theta}\,du_{i,j} + \frac{\partial e}{\partial\theta}\,d\theta$$

2-33 For plane stress problems show that the compatibility conditions can be expressed in the form (in the absence of the body forces) $\nabla^2(\sigma_{11} + \sigma_{22}) = 0$. Then show that the Airy stress function ϕ satisfies the biharmonic equation $\nabla^4\phi = 0$.

2-34 Consider a thin cantilevered beam (see Fig. P34) subjected to uniform shearing stress $\sigma_{21} = \tau_0$ along its upper surface ($x_2 = h$). The surfaces $x_2 = -h$ and $x_1 = L$ are stress-free. Show that the Airy stress function

$$\phi = \frac{\tau_0}{4}\left(x_2x_2 - \frac{x_1x_2^2}{h} - \frac{x_1x_2^3}{h^2} + \frac{Lx_2^2}{h} + \frac{Lx_2^3}{h^2}\right)$$

satisfies the governing equation of the problem. Determine whether it also satisfies the boundary conditions.

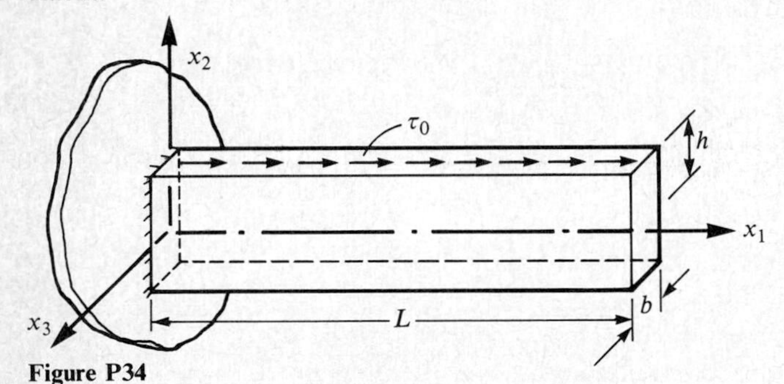

Figure P34

2-35 Consider an isotropic, narrow cantilevered beam of rectangular cross-section, loaded at its free end by a concentrated force P. Beginning with the following state of stress

$$\sigma_{11} = -\frac{Px_1x_2}{I} \qquad \sigma_{22} = 0 \qquad \sigma_{12} = -\frac{P}{2I}(h^2 - x_2^2)$$

where I is the moment of inertia and $2h$ is the beam depth, determine the displacement field. Assume that the plane state of stress exists and that $u_3 = 0$. Determine the constants of integration in the displacement field by using the boundary conditions $u_1 = \partial u_1/\partial x_2 = u_2 = 0$ at $x_1 = L$.

2-36 Modify the Navier's equations of elasticity for thermoelasticity problems, i.e., derive $(\lambda + \mu)\nabla(\nabla\cdot\mathbf{u}) + \mu\nabla^2\mathbf{u} - \alpha\nabla\theta + \mathbf{f} = \mathbf{0}$ in Ω, $u_i = \hat{u}_i$ on Γ_1, $n_j[\mu(u_{i,j} + u_{j,i}) + \lambda u_{k,k}\,\delta_{ij} - \alpha\theta\,\delta_{ij}] = \hat{t}_i$ on Γ_2, where $\Gamma_1 + \Gamma_2 = \Gamma$ is the total boundary of Ω.

2-37 For plane stress thermoelasticity problems, derive the governing equations for the Airy stress function ϕ and temperature θ.

2-38 For an incompressible fluid show that the pressure P equals the mean normal pressure $\bar{P}$.

2-39 For the laminar flow of an incompressible fluid between parallel plates (in the x_1x_3 plane), assume $v_3 = 0$ everywhere and derive the following equations of equilibrium (for steady flow)

$$-\frac{1}{\rho}\frac{\partial P}{\partial x_1} + \nu\frac{\partial^2 v_1}{\partial x_3^2} = 0 \qquad \nu = \mu/\rho$$

$$-\frac{1}{\rho}\frac{\partial P}{\partial x_3} - g = 0 \qquad \gamma = \rho g \text{ is the specific weight}$$

2-40 (Continuation of Prob. 2-39). If the upper plate is moving to the right at speed V_0, determine the velocity component v_1.

2-41 For plane flow parallel to the x_1x_2 plane, the Navier–Stokes equations for an incompressible fluid under gravitational body force $\mathbf{f} = -\text{grad } H$ can be expressed in terms of the stream function ψ and vorticity ω

$$v_1 = \frac{\partial \psi}{\partial x_2} \qquad v_2 = -\frac{\partial \psi}{\partial x_1} \qquad \omega = \left(\frac{\partial v_2}{\partial x_1} - \frac{\partial v_1}{\partial x_2}\right)$$

Show that the continuity equation is satisfied identically by ψ and that ω and ψ satisfy the equations $\mu \nabla^2 \omega = \rho(D\omega/Dt)$, $\omega = -\nabla^2 \psi$.

Using the method of separation of variables, obtain the analytical form of the solution to the heat transfer problems in Probs. 2-42 to 2-46.

2-42 $\alpha \nabla^2 \theta = 0$ in a unit square
$\theta = 0$ at $x_1 = 0$ for any x_2
$\theta = 0$ at $x_2 = 0$ for any x_1
$\theta = 0$ at $x_1 = 1$ for any x_2
$\theta = 1$ at $x_2 = 1$ for any x_1

2-43 Same as in Prob. 2-42, except that the last boundary condition is modified to read $\theta = \sin \pi x_1$ at $x_2 = 1$ for any x_1.

2-44 $\nabla^2 \theta = 1$ in a unit square
$\theta = 0$ on the boundary

2-45 Same as in Prob. 42, except that the sides $x_1 = 0$ and $x_2 = 0$ are insulated, and the other two sides are maintained at zero temperature.

2-46 $\nabla^2 \theta = \dfrac{\partial \theta}{\partial t}$ in a unit square

$$\theta(0, x_2, t) = \theta(x_1, 0, t) = \theta(x_1, 1, t) = 0$$

$$\frac{\partial \theta}{\partial x_1}(1, x_2, t) = -h\theta(1, x_2, t), \; \theta(x_1, x_2, 0) = 1$$

CHAPTER

THREE

CONCEPTS FROM FUNCTIONAL ANALYSIS

3-1 INTRODUCTION

3-1-1 General Comments

The subject of functional analysis is a generalization of concepts and methods from elementary analysis, algebra, and geometry. For example, the definition of a functional, of the extremum of a functional, and the conditions for the existence of an extremum in the calculus of variations are entirely analogous to the definitions of a function, of the extremum of a function, and the conditions for the existence of an extremum in the differential calculus. Another example of the analogy between functional analysis and geometry is provided by the representation of functions with respect to an orthogonal basis, which resemble orthogonal bases in the euclidean space. Also, the decomposition of a function into a Fourier series corresponds to the decomposition of a geometric vector (i.e., vector in three-dimensional euclidean space) into its components. Functional analysis has its origins in the calculus of variations, the theory of differential equations, and the theory of approximations.

Before we begin the study of the elements of functional analysis, let us pause for a moment to review the reasons for the present study. In the study of solutions to boundary and initial-value problems of mathematical physics and engineering that are unrelated, one finds that these problems share a common mathematical structure. Then it is useful to study a particular problem in the context of the general. The mathematical treatment of a general problem leads to a formal treatment that is more abstract than the treatment of a specific problem.

For example, in the approximate solution of boundary-value problems, we are led ultimately to the solution of a set of algebraic equations. If we know the conditions for the existence of solution of a general set of algebraic equations, it is possible to determine whether the given set of equations at hand possess a solution. In the present study, we shall be concerned with the solution of boundary-value problems of engineering and science, and therefore in the study of the existence and uniqueness of solutions of boundary-value problems and their variational approximations. To this end, we study the function spaces which aid in finding answers to questions like: Does the problem possess a solution? If it does, is it unique? In what function space does the solution exist? What methods are available to solve the problem approximately? How do we select the approximations in the solution space, and what is the error in the approximation? The concepts necessary for answering these questions precisely constitute a part of functional analysis.

To keep the scope of the present study within the limits of the subject of the book, only the essential concepts from functional analysis that have direct use in the study of boundary and initial-value problems and their solution by variational methods are included here. More specifically, we deal with linear vector spaces, normed and inner product spaces, and operators and functionals in inner product spaces. The reader who is familiar with these concepts can skip this chapter and go to Chap. 4.

3-1-2 Notation

In any subject of sciences and engineering use of a formal language (i.e., terminology), logic and notation results in abstraction. For example, to restate the trivial fact, "all bears are animals," in terms of the set theory, we write

$$x \in B \subset A \text{ (means: } x \text{ an element in } B \text{ which is a subset of } A) \tag{3-1}$$

where A is the set of all animals and B is the set of all bears. Similarly, we write

$$f \in L_2(\Omega) \tag{3-2}$$

to indicate that "f is a member of the set (actually a vector space) $L_2(\Omega)$ of all square-integrable functions defined over the domain Ω":

$$\int_\Omega |f(x)|^2 \, dx < \infty \tag{3-3}$$

Obviously, we do not want to repeat the above statement in quotes every time we encounter such a function; we simply say that it belongs to $L_2(\Omega)$. Such notation is convenient to express complicated thoughts, and perhaps long statements, more succinctly. Of course, there is a tendency to shy away from books and papers that use such an abstract language. However, the same abstraction makes it simple to express complicated thoughts.

We shall use the following notation, standard in set theory, in our study:

$$
\begin{aligned}
&\in && \text{means "a member (or element) of"} \\
&\notin && \text{means " not a member of"} \\
&B \subset A && \text{means "} B \text{ is a subset of } A\text{"} \\
&A \supset B && \text{means "} B \text{ is contained in } A\text{"} \\
&A \cup B && \text{means "set of elements that belong to } A \textit{ or } B\text{"} \\
&A \cap B && \text{means "set of elements that belong to } A \textit{ and } B\text{"} \\
&A - B && \text{means "set of elements that belong to } A \textit{ but not} \text{ to } B\text{"} \\
&\varnothing && \text{means "empty set or set containing } \textit{no} \text{ elements"}
\end{aligned}
\tag{3-4}
$$

The complement of a set A that is a subset of a larger set S is the set of elements in S that do not belong to A. That is, the *complement of A with respect to S*, denoted A/S, is the difference of S and A. Obviously, $S/S = \varnothing$, $\varnothing/S = S$.

Figure 3-1 shows geometrical representations, called Venn diagrams, of various set operations defined in Eq. (3-4).

We shall frequently use the following sets of numbers:

$$
\begin{aligned}
\mathbb{R} &= \{x\colon\ x \text{ is a real number}, -\infty < x < \infty\} \\
\mathbb{C} &= \{z\colon\ z \text{ is a complex number}, z = x + iy, i = \sqrt{-1}, -\infty < x, y < \infty\}
\end{aligned}
\tag{3-5}
$$

In addition to the above sets, we shall use the following notation for intervals of real numbers, which are themselves sets of real numbers:

$$
\begin{aligned}
[a, b] &= \{x\colon\ x \text{ is real}, a \le x \le b\} \\
[a, b) &= \{x\colon\ x \text{ is real}, a \le x < b\} \\
(a, b] &= \{x\colon\ x \text{ is real}, a < x \le b\} \\
(a, b) &= \{x\colon\ x \text{ is real}, a < x < b\}
\end{aligned}
\tag{3-6}
$$

The set $[a, b]$ is called a *closed interval* and (a, b) is called an *open interval.*

3-1-3 Supremum and Infimum of Sets

A set $A \subset \mathbb{R}$ is said to be *bounded from above* if there exists a real number such that $a \le \mu_0$ for all $a \in A$. The real number μ_0 is said to be an *upper bound* of the set A. Similarly, a set A is said to be *bounded from below* if there exists a real number γ_0 such that $a \ge \gamma_0$ for all $a \in A$. The real number γ_0 is said to be a *lower bound* of the set A. If a set A is bounded from above and from below, we say that A is bounded. An upper (lower) bound, $M(m)$, for A is said to be the *maximum* (*minimum*) of $A \subset \mathbb{R}$ if $M \in A$ ($m \in A$). It should be noted that even a bounded set need not have a maximum or a minimum.

Every nonempty set of real numbers bounded from above has a least upper bound, and every nonempty set of real numbers bounded from below has a greatest lower bound. It should be pointed out that the above statements do not hold

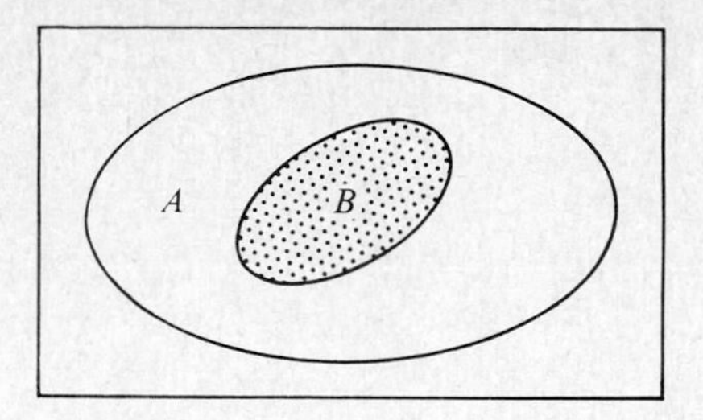

(*a*) *Subset* B of A
$(B \subset A)$

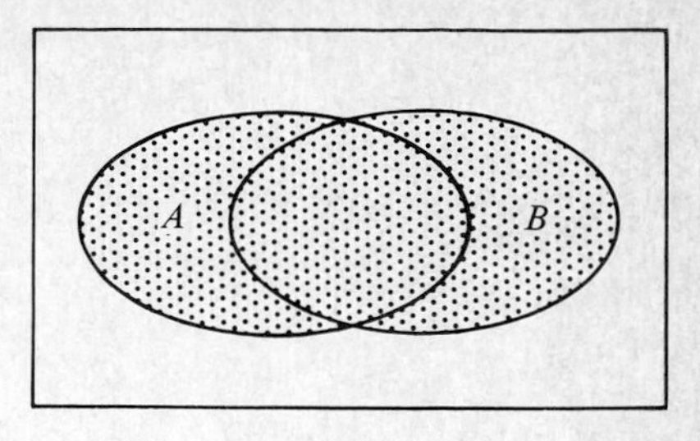

(*b*) *Union* of A and B
$(A \cup B)$

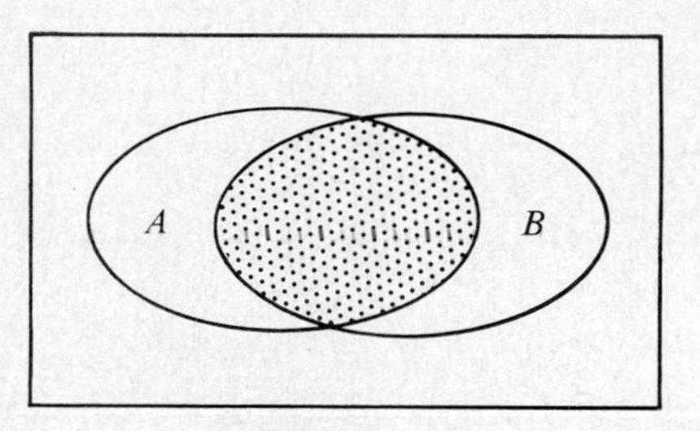

(*c*) Nonempty *intersection*,
$A \cap B \neq \varnothing$

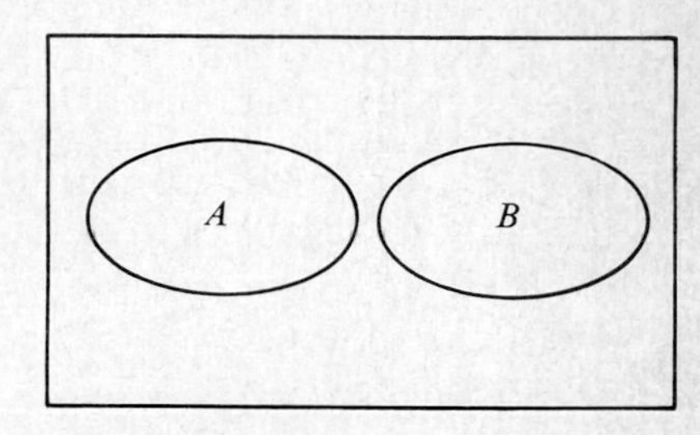

(*d*) Empty intersection,
$A \cap B = \varnothing$

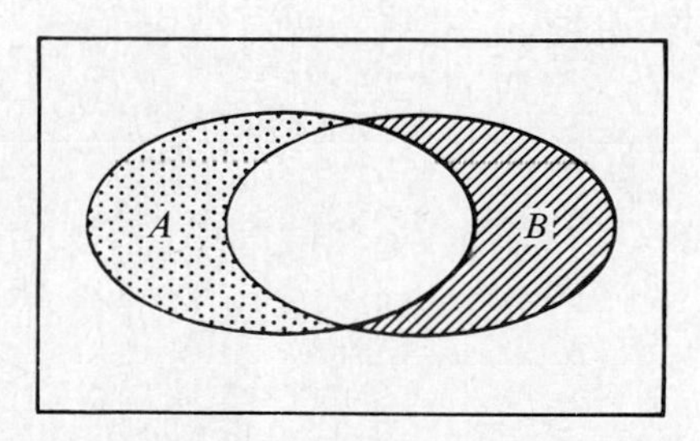

(*e*) *Difference* of A and B
$A-B$, $B-A$

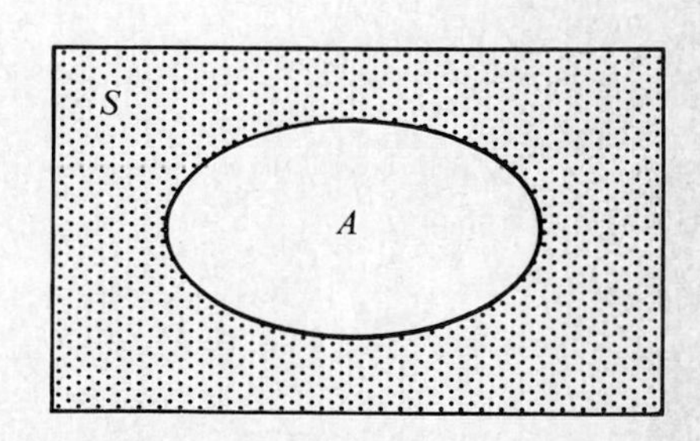

(*f*) *Complement* of A with
respect to S (A/S)

Figure 3-1 Geometrical representation of various set operations.

for the set of rational numbers Q. Proofs of these statements can be found in Kolmogorov and Fomin (1957).

Let A be a set that is bounded from above, and let X be the set of all upper bounds of A. The "least upper bound" of A is the minimum of X. This minimum of X is called the *supremum* of A and is denoted by *sup* A. Similarly, for a non-empty set A that is bounded below, let Y denote the set of all lower bounds of A. The "greatest lower bound" of A is the maximum of Y. The maximum of Y is called the *infimum* of A and is denoted by *inf* A. We shall consider several examples to illustrate these ideas.

Example 3-1 Let A and B be the subsets of the real-number field:

$$A = \{a\colon\ a \in \mathbb{R},\ 0 \leq a \leq 1\} \qquad B = \{b\colon\ b \in \mathbb{R},\ 0 < b < 1\}$$

The set X_A of all upper bounds and the set Y_A of all lower bounds of A are given by

$$X_A = \{x\colon\ 1 \leq x < \infty\} \qquad Y_A = \{y\colon\ -\infty < y \leq 0\}$$

Note that $X_A = X_B$ and $Y_A = Y_B$, where X_B and Y_B are sets of upper bounds and lower bounds for B. The maximum of A is 1 and the minimum of A is 0. However, B has no maximum or minimum. The least upper bound of A is 1 and the greatest lower bound of A is 0, and $\sup B = \sup A$ and $\inf B = \inf A$.

Example 3-2 Let Q be the set of rational numbers,

$$Q = \{q\colon\ q = m/n, \text{ where } m \text{ and } n \text{ are integers, } n \neq 0\}$$

Let $A \subset Q$ be the set,

$$A = \{a \in Q,\ 0 < a < \sqrt{2}\}$$

The set A has neither a maximum nor a supremum. The infimum of A is 0.

Example 3-3 Let X be the set of functions $f_n(t) = \sin n\pi t$, $n = 0, 1, 2, \ldots$, and $-\infty < t < \infty$. Since $-1 \leq f_n(t) \leq 1$ for all values of n and t, the set $\{f_n(t)\}$ has a maximum and a minimum.

3-1-4 Functions

Every student of engineering is familiar with the word *function*. When we think of $f(x)$, we understand that f is a function that transforms x from a set X to $y \equiv f(x)$ in another set Y. For example, consider a rubber strip, X, of unit length and stretch it "uniformly" to twice its original length and call the resulting strip Y. We can write the correspondence between points y in Y and points x in X as: $x \in X$ corresponds to the number $y \in Y$ such that $y = 2x$. The correspondence can also be described, more concisely, as the set $\mathbb{R}^2 \subset X \times Y$ defined by

$$\mathbb{R}^2 = \{(x, y)\colon\ y = 2x\} = \{(x, 2x)\colon\ x \in X\} \tag{3-7}$$

Yet another way of looking at the correspondence is to think of it being a relation between the points of X and Y.

A function is a relation (or correspondence) that is "single-valued." That is, if $y_1 = 2x$ and $y_2 = 2x$ for x in X, then $y_1 = y_2$. In other words, a function transforms each element from one set into only one element in the other. However, a function can map two distinct elements in X into the same element in Y. A formal definition of a function follows. Let X and Y be two sets, and suppose that f is a *rule* that assigns to each element in X *exactly one* element of Y. Then f is called a *function* defined on X with values in Y. The set X is called the *domain* of f, denoted by $\mathscr{D}(f)$. If $y = f(x)$, we say that y is the *image* of x. The *range* of a function $f\colon X \to Y$ (means f maps set X into set Y), denoted $\mathscr{R}(f)$, is the set of all images of f. If $\mathscr{R}(f) \subset Y$, f is said to map X *into* Y. If $\mathscr{R}(f) = Y$, then f is said to map X *onto* Y.

If the domain of a function f does not contain two elements with the same image, then f is called a *one-to-one* function or mapping. In other words, f is a one-to-one mapping from X into Y if and only if every point in $\mathscr{R}(f)$ has only one pre-image in X. When a function is one-to-one and onto its range, a mapping of the range into the domain, called the *inverse*, can be defined. When the range of a function contains only one element, the function is called a *constant function* (because every element in the domain is mapped into the same element).

The *composite of two functions*, $f\colon X \to Y$ and $g\colon Y \to Z$, is defined by

$$(g \cdot f)(a) = g(f(a)) \tag{3-8}$$

In general we have $g \cdot f \neq f \cdot g$ (in fact, $g \cdot f$ may exist even if $f \cdot g$ does not). The composition of functions is associative:

$$(h \cdot g) \cdot f = h \cdot (g \cdot f) \tag{3-9}$$

If $S \subset X$ then its *direct image* under the mapping f is given by

$$f(S) = \{f(a)\colon\ a \in S \subset X\} \tag{3-10}$$

Here it is understood that $f(S)$ is a set rather than an element. Clearly, $f(S)$ is a subset of $\mathscr{R}(f)$. Figure 3-2 shows various classifications of functions.

3-2 LINEAR VECTOR SPACES

3-2-1 Introduction

A generalization of the familiar concept of the ordinary physical vector (to study more general objects) forms the subject of linear vector spaces, or simply linear spaces. Recall that in previous discussions $f\colon X \to Y$ denoted a transformation from set X into set Y. The sets X and Y in practical applications are sets of points in euclidean space, or sets of functions of position. Therefore an algebraic structure (i.e., addition and subtraction of elements of a set) is needed to work

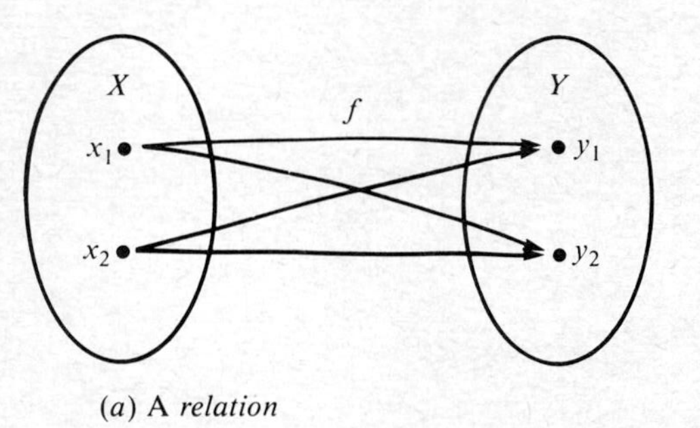

(a) A *relation*

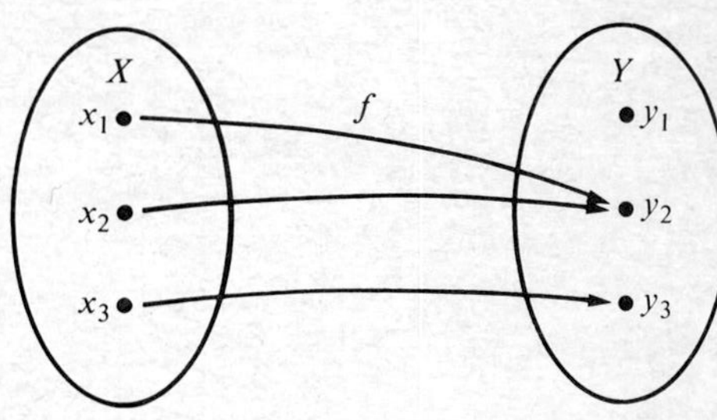
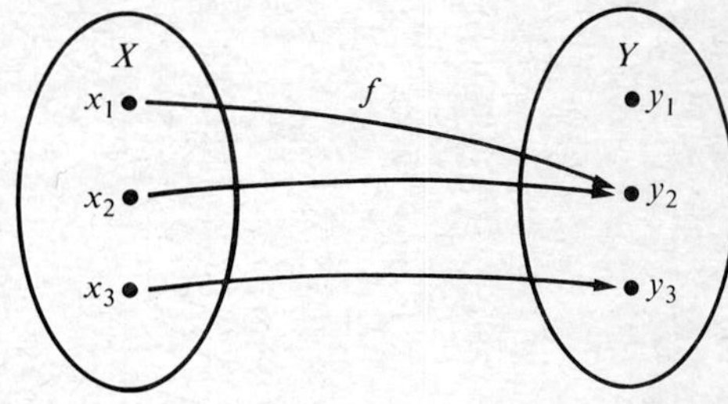

(b) A *function* (*into*); not onto Y

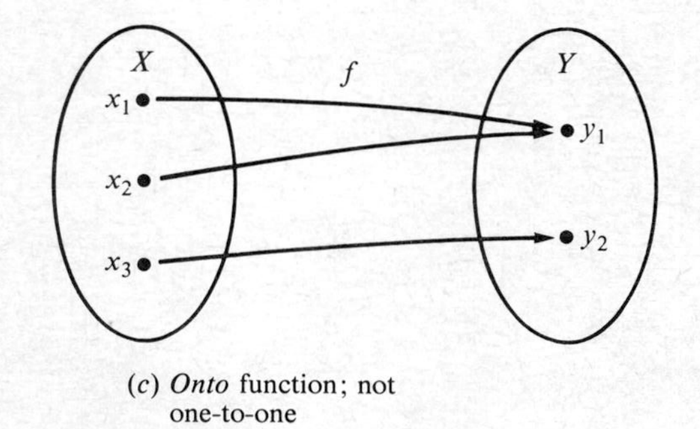

(c) *Onto* function; not one-to-one

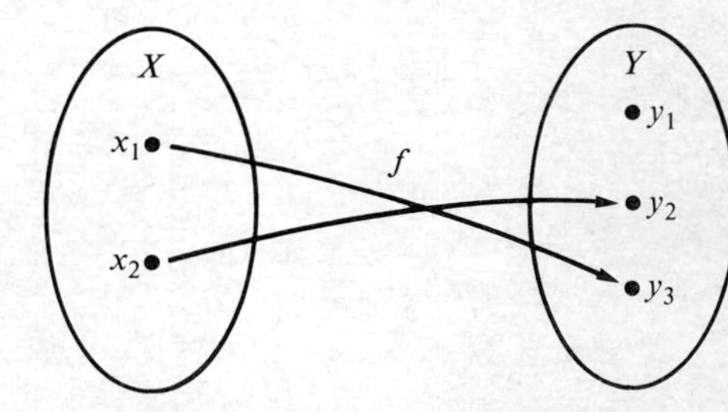

(d) *One-to-one* function; not onto Y

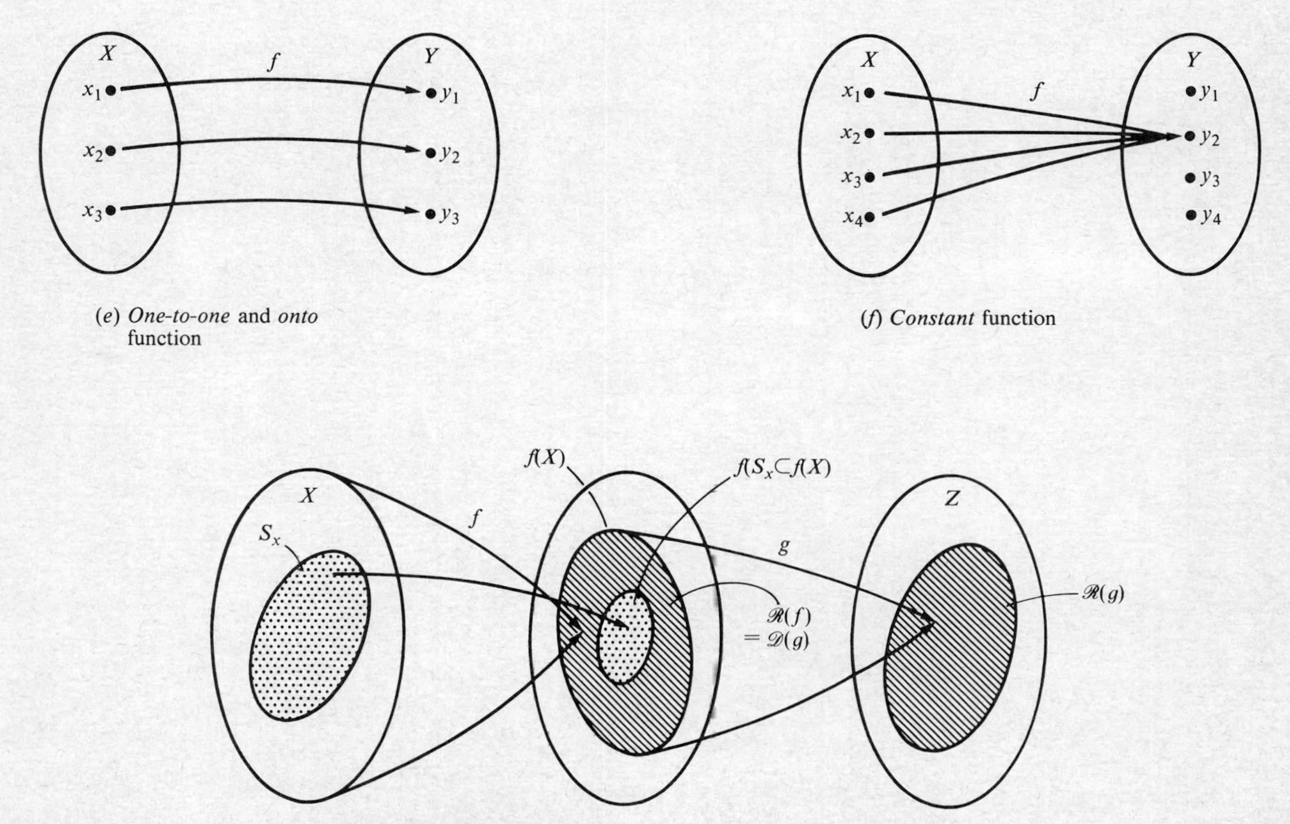

(*e*) *One-to-one* and *onto* function

(*f*) *Constant* function

Figure 3-2 Classification of functions and composition of functions.

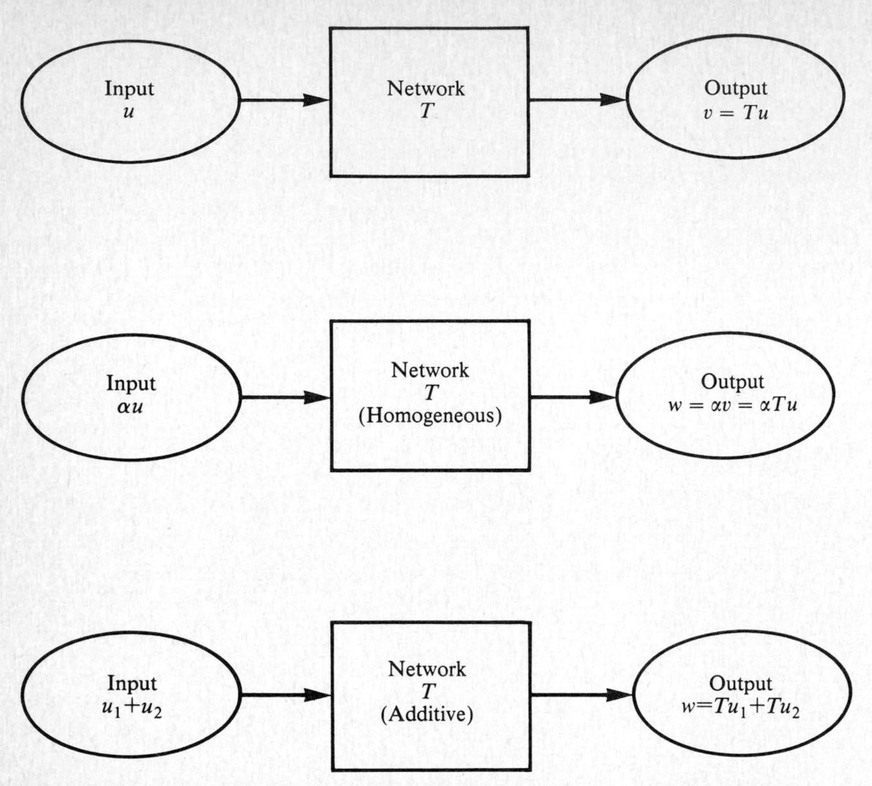

Figure 3-3 A network that transforms an input linearly.

with the elements of sets. For example, consider an electrical network (see Fig. 3-3). A network receives an input and produces an output. The network can be viewed as a transformation which transforms a given input (single-valued) into an output. One can combine two inputs by *adding* them and putting their sum through the system. This will produce an output. If the output is the sum of the outputs of the individual inputs the system is said to be *additive*. One can also change the input by multiplying its magnitude by a *constant* (scalar) factor. If the resulting output is also a multiple of the same factor, the system is said to be *homogeneous*. If the system is both additive and homogeneous, it is said to be *linear*. Further, one can also measure the difference between two inputs. Thus, it is useful to equip the sets with a certain algebraic structure so that the sum of two elements and the scalar multiplication of an element from a set are defined. When sets are equipped with an algebraic structure, they are called vector spaces, and their elements are called vectors—a word carried from geometric vectors (directed line segments).

The rules of vector addition and scalar multiplication, to be given shortly, are analogous to those of geometric vectors in euclidean space. The " direction " of a geometric vector takes on a new interpretation in the case of an abstract vector, which can be a point in the euclidean space or a function. The norm and inner products are generalizations of magnitude and scalar products for geometric vectors.

3-2-2 Linear Spaces and Subspaces

A collection of *vectors*, u, v, w, ... is called a *real linear vector space* (or simply vector space or linear space) V over the real number field $\mathbb{R}$ if the following rules are satisfied by the elements of the vector space:

Vector addition To every pair of vectors u and v there corresponds a unique vector $u + v$ in V, called the *sum* of u and v, with the properties:

(1*a*) $u + v = v + u$ (commutative);

(1*b*) $(u + v) + w = u + (v + w)$ (associative);

(1*c*) there exists a unique vector, Θ, independent of u such that $u + \Theta = u$ for $u \in V$ (existence of an identity element); (3-11)

(1*d*) to every u there exists a unique vector (that depends on u), denoted I_u (or $-u$) such that $u + I_u = \Theta$ (existence of the additive inverse element).

Scalar multiplication To every vector u and every real number $\alpha \in \mathbb{R}$, there corresponds a unique vector αu in V, called the *product* of u and α, such that

(2*a*) $\alpha(\beta u) = (\alpha\beta)u$ (associative);

(2*b*) $(\alpha + \beta)u = \alpha u + \beta u$ (distributive with respect to scalar addition);

(2*c*) $\alpha(u + v) = \alpha u + \alpha v$ (distributive with respect to vector addition); (3-12)

(2*d*) $1 \cdot u = u \cdot 1$.

The vector, Θ, is called the *identity element*, and the vector I_u associated with each element $u \in V$ is called the *inverse* element of u. In order to check whether a given set V qualifies as a linear vector space, one must first define the rules of vector addition and scalar multiplication of a vector over the set. Then the *closure property* must be verified: If $u, v \in V$ then $\alpha u + \beta v \in V$ for all scalars α, $\beta \in \mathbb{R}$.

Note that the above axioms are identical to those which are satisfied by the physical vectors. However, there is one exception: magnitude is not a basic property of an abstract vector. Thus, the set of all physical vectors studied in any undergraduate course forms a special linear vector space provided the abstract vector allows us to define its magnitude (see Sec. 3.4).

The set of all *ordered n-tuples* $(a_1, a_2, \ldots, a_n)$ of real numbers $a_1, a_2, \ldots, a_n$ is called the *Cartesian Space*, and is denoted by $\mathbb{R}^n$. The cartesian space is a linear vector space with respect to the usual rules of addition and scalar multiplication. An example of $\mathbb{R}^3$ follows.

Example 3-4 1. (Ordinary physical vector) Consider the set V of geometric vectors in $\mathbb{R}^3$. An element $\mathbf{a} \in V$ represents an ordered triple (a_1, a_2, a_3). The set of geometric vectors is a linear vector space with respect to the vector addition and scalar multiplication defined by

$$\begin{aligned} &\textit{addition} \;\; \mathbf{a} + \mathbf{b} \equiv (a_1 + b_1, a_2 + b_2, a_3 + b_3) \\ &\textit{scalar multiplication} \;\; \alpha\mathbf{a} \equiv (\alpha a_1, \alpha a_2, \alpha a_3) \end{aligned} \tag{3-13}$$

where $\mathbf{a}, \mathbf{b} \in V$ and $\alpha \in \mathbb{R}$. Clearly, both operations are *closed* in V (i.e., the vector sum and scalar multiple of a vector are also in V). It is a simple matter to verify the axioms of a linear vector space. Note that the vector addition defined above is, geometrically, equivalent to the *parallelogram law of vector addition.* The identity element Θ is the *zero* vector $\mathbf{0} = (0, 0, 0)$, and the inverse of an element is its negative.

2. (Second order tensors) Consider the set V of all dyadics in $\mathbb{R}^3$. An element $\mathbf{a}$ in V represents an ordered 3×3 array of real numbers

$$\mathbf{a} = \begin{bmatrix} a_{11} & a_{12} & a_{13} \\ a_{21} & a_{22} & a_{23} \\ a_{31} & a_{32} & a_{33} \end{bmatrix} \tag{3-14}$$

We define the vector addition and scalar multiplication of a vector by the following rules:

$$\mathbf{a} + \mathbf{b} \equiv \begin{bmatrix} a_{11} + b_{11} & a_{12} + b_{12} & a_{13} + b_{13} \\ a_{21} + b_{21} & a_{22} + b_{22} & a_{23} + b_{23} \\ a_{31} + b_{31} & a_{32} + b_{32} & a_{33} + b_{33} \end{bmatrix} \tag{3-15a}$$

$$\alpha\mathbf{a} \equiv \begin{bmatrix} \alpha a_{11} & \alpha a_{12} & \alpha a_{13} \\ \alpha a_{21} & \alpha a_{22} & \alpha a_{23} \\ \alpha a_{31} & \alpha a_{32} & \alpha a_{33} \end{bmatrix} \tag{3-15b}$$

The vector addition and scalar multiplication are closed in V by virtue of the closure property of real numbers. The identity element is the zero dyadic represented by the zero matrix, and the inverse of an element is represented by the negative matrix. Thus, V is a linear vector space. As will be seen later, a dyadic represents a linear transformation from $\mathbb{R}^n$ into itself (i.e., the set of all linear transformations is a linear vector space).

Several examples of linear vector spaces are given below.

Example 3-5 1. Let P be the set of polynomials in an independent variable x over the real number field. Define the vector addition to be the usual addition of polynomials, and the multiplication as the usual multiplication of a polynomial by an element of $\mathbb{R}$. Then P is a linear vector space. A typical element in P is of the form, $p(x) = a_0 + a_1 x + a_2 x^2 + \cdots$

2. Let V be the set of all m by n matrices defined over the real number field. Then V is a linear vector space with respect to the addition and scalar

multiplication of matrices defined in Eq. (3-15). The identity element is the $m \times n$ matrix of zeros, and the inverse element is the negative of a matrix.

3. The set $C[a, b]$ of all continuous real-valued functions on the interval $[a, b]$, with the vector addition and scalar multiplication defined by the rules,

$$\begin{aligned}(f+g)(x) &\equiv f(x)+g(x)\\ (\alpha f)(x) &\equiv \alpha f(x)\end{aligned} \tag{3-16}$$

is a linear vector space.

4. Let P be the set of polynomials with real coefficients and *positive* constant term, and let the vector addition and scalar multiplication be as defined in Eq. (3-16). Then, P is *not* a linear space because if $p(x)$ is in P then $-p(x)$ is not in P.

5. Consider the set V of elements $\{\mathbf{a}, \mathbf{b}, \mathbf{c}, \mathbf{d}, \ldots\}$. Each element in V is an ordered pair of real numbers, $\mathbf{a} = (a_1, a_2)$. Suppose that the vector addition and scalar multiplication are defined by

$$\begin{aligned}\mathbf{a}+\mathbf{b} = (a_1, a_2) + (b_1, b_2) &\equiv (a_1 b_1, a_2 b_2)\\ \alpha\mathbf{a} &\equiv (\alpha a_1^2, \alpha a_2^2)\end{aligned} \tag{3-17}$$

Then the addition is commutative and associative. The identity element can be determined from

$$\mathbf{a} + \boldsymbol{\Theta} \equiv (a_1\theta_1, a_2\theta_2) = (a_1, a_2)$$

which implies that $\boldsymbol{\Theta} = (1, 1)$. Similarly, the inverse element can be determined from

$$\mathbf{a} + \mathbf{I}_a \equiv (a_1 I_{a1}, a_2 I_{a2}) = \boldsymbol{\Theta} = (1, 1)$$

which gives $\mathbf{I}_a = (1/a_1, 1/a_2)$.

Now check the associativity of scalar multiplication of a vector.

$$\begin{aligned}\alpha(\beta\mathbf{a}) &= \alpha(\beta a_1^2, \beta a_2^2) = (\alpha(\beta a_1^2)^2, \alpha(\beta a_2^2)^2)\\ &= (\alpha\beta^2 a_1^4, \alpha\beta^2 a_2^4) \neq (\alpha\beta a_1^2, \alpha\beta a_2^2) = (\alpha\beta)\mathbf{a}\end{aligned}$$

Thus, V is *not* a linear vector space with respect to the rules of vector addition and scalar multiplication defined above.

6. Let V be the set of all functions with continuous second derivatives defined over the interval $[0, L]$ and satisfying the differential equation

$$\frac{d^2u}{dx^2} - 2\frac{du}{dx} - 3u = 0$$

The set V is a linear vector space with respect to the rules of addition and scalar multiplication given in Eq. (3-16).

A *linear subspace* S of a given linear vector space V over $\mathbb{R}$ is a nonempty subset of V which is itself a linear vector space with respect to the operations of addition and scalar multiplication defined over V. Clearly, the vector space containing the zero element is a subspace of V, along with V itself.

A *proper subspace* of a linear space V is a subspace other than V. A *nontrivial subspace* is a subspace containing non-zero elements. The *trivial subspace* is the vector space containing the zero vector only, $\{0\}$.

If S_1 and S_2 are subspaces of a vector space V, their intersection $S_1 \cap S_2$ is a subspace of V. Also, the set $S_1 + S_2$ of all $u_1 + u_2$, with $u_1 \in S_1$ and $u_2 \in S_2$, is a subspace of V. If

$$S_1 + S_2 = V \text{ and } S_1 \cap S_2 = \{0\} \tag{3-18}$$

then each of S_1 and S_2 is said to be the *complement* of the other with respect to V. Whenever Eq. (3-18) holds, we will call V the *direct sum* of S_1 and S_2, and write

$$V = S_1 \oplus S_2 \tag{3-19}$$

To determine whether a given subset of a linear vector space is a subspace, one must check the existence of the identity element, and verify the closure property of vectors in the subspace. In other words, a subset S of a linear space V is a subspace if and only if the zero element $\theta \in S$ and $\alpha u + \beta v \in S$ for all $u, v \in S$ and scalars $\alpha, \beta \in \mathbb{R}$. We now consider several examples of subspaces.

Example 3-6 1. Let $V = \mathbb{R}^3$, and consider the set $S_2 = \{\mathbf{x} \in \mathbb{R}^3\colon x_1^2 + x_2^2 + x_3^2 = r^2\}$, where r is a real number. The zero element is not in S_2 unless $r = 0$. Also, for *any* nonzero value of r the sum $\alpha\mathbf{x} + \beta\mathbf{y}$ is not in S_2. In other words, the set of the (euclidean) metrics of all elements in $\mathbb{R}^3$ is not a subspace.

2. Let $V = C^2[0, L]$ be the set of all real-valued, twice differentiable functions $u(x)$ defined on $0 \leq x \leq L$. The set $C^2[0, L]$ is a linear vector space with the usual definitions of addition and scalar multiplication [see Eq. (3-16)]. Let S_f be the subset

$$S_f = \left\{u \in C^2[0, L]\colon \frac{d^2u}{dx^2} - 4u = f,\ f \text{ is a real number}\right\}$$

Clearly, the zero element is not in S_f unless f is equal to zero. Further, if u_1 and u_2 are in S_f, then $\alpha u_1 + \beta u_2 \notin S_f$. Of course, S_0 (i.e., for $f = 0$) is a subspace of $C^2[0, L]$. This means that the set S_0 of all solutions to the homogeneous differential equation is a linear vector space but S_f is not. Also note that S_0 consists of infinitely many elements of the type $u = k_1 e^{-2x} + k_2 e^{2x}$, where k_1 and k_2 are real numbers.

3. Let $V = C^2[0, L]$ and define the set

$$S = \{p(x) \in C^2[0, L]\colon p(0) = \alpha,\ p(L) = \beta \text{ for any } \alpha, \beta \in \mathbb{R}\}$$

The set is not a linear vector space for any nonzero α or β, and it is a subspace for $\alpha = \beta = 0$. The subspace contains infinitely many elements, which are divisible by $x(x - L)$.

4. Let $V = \mathbb{R}^n$ and consider any homogeneous system of linear equations in n unknowns with real coefficients

$$\sum_{j=1}^{n} a_{ij} x_j = 0 \qquad (i = 1, 2, \ldots, n)$$

for $a_{ij} \in \mathbb{R}$, and $\mathbf{x} = (x_1, x_2, \ldots, x_n) \in \mathbb{R}^n$. The set S of all solutions of the homogeneous system is a subspace of $\mathbb{R}^n$, and is called the *solution space*.

Example 3-7 1. The intersection of set S_0 defined in part 2 and S defined in part 3 (with $\alpha \equiv \beta = 0$) of Example 3-6 is $\{0\}$. This means that the solution to the differential equation $d^2u/dx^2 - 4u = 0$ that satisfies the conditions $u(0) = u(L) = 0$ is the trivial solution.

2. Let V be the set of all polynomials with real coefficients. It was already shown that V is a vector space. Let the subspaces S_1 and S_2 be defined by

$$S_1 = \{p(x) \in V\colon\ p(1) = 0\} \qquad S_2 = \{p(x) \in V\colon\ p(2) = 0\}$$

The intersection $S_1 \cap S_2$ contains all polynomials divisible by $(x - 1)$ $(x - 2)$.

Product spaces Let U and V be two linear vector spaces. The *product space* $W \equiv U \times V$ is defined to be the set of ordered pairs of vectors (u, v), $u \in U$ and $v \in V$:

$$W = \{w = (u, v)\colon\ u \in U, v \in V\} \tag{3-20}$$

Example 3-8 Consider the transverse motion of a cable of length L, fixed at its ends. Let $C[0, L]$ denote the set of all real-valued, continuous functions $y(x, t)$ defined on the closed interval $0 \le x \le L$ for any time t, $0 \le t \le t_0$.

The transverse deflection $y(\cdot, t)$ (i.e., configuration) of the cable at any given time t can be viewed as a point in the vector space $C[0, L]$. Hence, the motion of the cable can be viewed as a curve $y \in C[0, L]$ parameterized by t. Note that not every function in $C[0, L]$ can qualify as the curve y because all possible motions y should vanish at $x = 0$, and $x = L$ (see Fig. 3-4). Thus, all

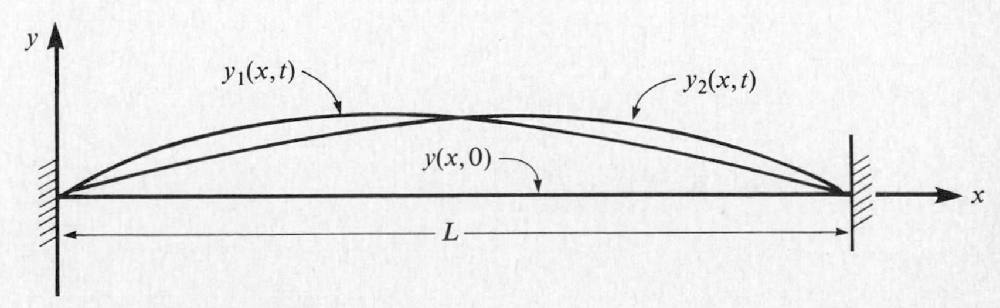

Figure 3-4 A cable fixed at both ends.

possible deflection curves of the cable form only a subspace of $C[0, L]$. Therefore, the subspace

$$S(t) = \{y(\cdot, t): y \in C[0, L], y(0, t) = y(L, t) = 0\}$$

contains all possible deflections of the cable. Of these deflections, we are interested in only the one that satisfies the governing differential equation of the cable [a dynamic version of Eq. (1-8)].

Suppose that $v \equiv dy/dt$ is the rate of change of deflection with time. Then $v(\cdot, t)$ (assumed to exist) can be viewed as a point in the vector space $C[0, L]$ at any time t. The ordered pair $(y, dy/dt) = (y, v)$ can be thought of as a point in the product space $C[0, L] \times C[0, L]$.

3-2-3 Linear Dependence and Independence of Vectors

Let V be a linear space. If an element u in V can be expressed as $u = \sum \alpha_i u_i$, $\alpha_i \in \mathbb{R}$, and $u_i \in V$, we say that u is a *linear combination* of u_i's. An expression of the form $\sum \alpha_i u_i = 0$ is called a *linear relation* among the u_i's. A relation with all $\alpha_i = 0$ is called a *trivial relation*, and a relation with at least one coefficient nonzero is called a *nontrivial linear relation.*

A set of vectors is said to be *linearly dependent* if there exists a nontrivial linear relation among them. Otherwise, the set is said to be *linearly independent.* The linear dependence of abstract vectors is equivalent to the condition that geometric vectors are coplanar.

The following observations are simple consequences of the definitions of linear dependence and independence of vectors:

1. If $(u_1, u_2, \ldots, u_n)$ are dependent, at least one vector is a linear combination of the others;
2. If the zero vector belongs to a set of vectors, the set is linearly dependent (since for $\alpha \in \mathbb{R}$, and $0 \in V$ we have $\alpha 0 = 0$);
3. A set consisting of exactly one nonzero vector is linearly independent (since for $\alpha \in \mathbb{R}$, and $u \in V$, $\alpha u = 0$ implies $\alpha = 0$);
4. If $(u_1, u_2, \ldots, u_n)$ is a linearly dependent set, so is any set which includes $(u_1, u_2, \ldots, u_n)$;
5. If $(u_1, u_2, \ldots, u_n)$ is a linearly independent set, any subset of it is also linearly independent.

Example 3-9 1. Let P be the (vector) space of polynomials in x, and let $p_1(x) = 1 + x + x^2$, $p_2(x) = x^2 - x - 2$, $p_3(x) = x^2 + x - 1$, and $p_4(x) = x - 1$. Then the linear combination of the vector p_i, $i = 1, 2, 3, 4$ is given by

$$\begin{aligned}\sum_{i=1}^{4} p_i \alpha_i &= \alpha_1(1 + x + x^2) + \alpha_2(x^2 - x - 2) + \alpha_3(x^2 + x - 1) + \alpha_4(x - 1) \\ &= (\alpha_1 - 2\alpha_2 - \alpha_3 - \alpha_4) + (\alpha_1 - \alpha_2 + \alpha_3 + \alpha_4)x + (\alpha_1 + \alpha_2 + \alpha_3)x^2\end{aligned}$$

Now suppose that $\sum_{i=1}^{4} p_i \alpha_i = 0$. This implies that

$$\alpha_1 - 2\alpha_2 - \alpha_3 - \alpha_4 = 0$$
$$\alpha_1 - \alpha_2 + \alpha_3 + \alpha_4 = 0$$
$$\alpha_1 + \alpha_2 + \alpha_3 = 0$$

which has infinitely many solutions. For example, $\alpha_1 = 3$, $\alpha_2 = 2$, $\alpha_3 = -5$, and $\alpha_4 = 4$ is a solution. Thus, the linear combination of p_i is non-trivial and therefore is a dependent set. Indeed, we can express $p_3(x)$ in terms of p_1, p_2, and p_4:

$$p_3(x) = (3p_1(x) + 2p_2(x) + 4p_4(x))/5$$

2. The set $p_1(x) = 1$, $p_2(x) = x - 1$, $p_3 = 1 + x + x^2$ is linearly independent, because

$$\alpha p_1(x) + \beta p_2(x) + \gamma p_3(x) = 0$$

implies that $\alpha = \beta = \gamma = 0$.

3-2-4 Span, Basis and Dimension

Span One can show that the set of all linear combinations of k vectors $\{u_1, u_2, \ldots, u_k\}$ of a vector space V makes up a subspace of V. This subspace is known as the *linear manifold generated* or *spanned* by $\{u_1, u_2, \ldots, u_k\}$, and it will be denoted by $\{u_i\}$. For example, consider the vector space P of polynomials with real coefficients (see Part 1 of Example 3-5). For any positive n, the set of functions $\{1, x, x^2, \ldots, x^n\}$ is linearly independent. The subspace spanned by $\{1, x, x^2, \ldots, x^n\}$ is the vector space that contains all polynomials of degree less than or equal to n.

A vector space can have more than one span. For example, the vector $u = (\alpha, \beta, \gamma) \in \mathbb{R}^3$ can be expressed as a linear combination of the vectors $\{u_1, u_2, u_3\}$, $\{u_4, u_5, u_3\}$, or $\{u_1, u_2, u_3, u_6\}$, where $u_1 = (1, 2, 3)$, $u_2 = (0, 1, 2)$, $u_3 = (0, 0, 1)$, $u_4 = (1, 0, 0)$, $u_5 = (0, 1, 0)$, and $u_6 = (1, 1, 2)$. We have

$$\begin{aligned}(\alpha, \beta, \gamma) &= \alpha u_1 + (\beta - 2\alpha)u_2 + (\gamma - 2\beta + \alpha)u_3 \\ &= \alpha u_4 + \beta u_5 + \gamma u_3 \\ &= 2\alpha u_1 + (\beta - 3\alpha)u_2 + (\gamma + 2\alpha - 2\beta)u_3 - \alpha u_6\end{aligned}$$

Note that the sets $\{u_1, u_2, u_3\}$ and $\{u_4, u_5, u_3\}$ are linearly independent whereas the set $\{u_1, u_2, u_3, u_6\}$ is linearly dependent.

Example 3-10 Let P_2 be the set of polynomials of degree less than or equal to 2. We wish to know whether the set $\{p_1, p_2, p_3\}$, where

$$p_1 = -2 + x - x^2, \; p_2 = 3 - x + 2x^2, \; p_3 = 1 - 2x - x^2$$

spans P_2. We equate a linear combination of p_1, p_2, and p_3 to an arbitrary element from P_2:

$$\alpha p_1 + \beta p_2 + \alpha p_3 = a_0 + a_1 x + a_2 x^2$$

The element $p = a_0 + a_1 x + a_2 x^2$ is an arbitrary element in P_2 if and only if there is no relationship between the coefficients a_0, a_1, and a_2. By equating like powers of x, we get

$$-2\alpha + 3\beta + \gamma = a_0, \; \alpha - \beta - 2\gamma = a_1, \; -\alpha + 2\beta - \gamma = a_2$$

The determinant of the coefficient matrix of these algebraic equations is zero, therefore no solution exists. The set does not span P_2.

Basis A linearly independent set spanning a linear space V is called a *basis* or *base* of V.

A basis is essentially a coordinate system (not necessarily rectangular). Since a vector space can be spanned by more than one linearly independent set, there can be more than one basis for a vector space. For example, in three-dimensional euclidean space $\mathbb{R}^3$ the cartesian base vectors, cylindrical base vectors, and spherical base vectors each form a basis (that is, an arbitrary vector can be represented by any one of these bases). In fact, any three non-coplanar vectors form a basis in three-dimensional euclidean space. For example, the set $\{(1, 0, 0), (0, 1, 0), (0, 0, 1)\}$ is a basis for $\mathbb{R}^3$. First, the set is linearly independent since

$$\alpha(1, 0, 0) + \beta(0, 1, 0) + \gamma(0, 0, 1) = (\alpha, \beta, \gamma)$$

and the linear combination is zero only if $\alpha = \beta = \gamma = 0$. Second, if (α, β, γ) is any vector in $\mathbb{R}^3$ it can be expressed in terms of the basis as shown above. Another basis of $\mathbb{R}^3$ is given by the set $\{(1, 0, 0), (1, 1, 0), (1, 1, 1)\}$. An example of bases of polynomial vector spaces is given below.

Example 3-11 Consider the vector space P_2 of polynomials of degree less than or equal to 2. The space is spanned by the set $\{1, x, x^2\}$. Since the set is linearly independent, it is a basis for P_2. Any arbitrary element, $p(x) = a_0 + a_1 x + a_2 x^2$, in P_2 is represented by the basis. Also, the set $\{1, 1 + x, 1 + x + x^2\}$ forms a basis for P_2. To verify this we check whether it spans P_2 and is linearly independent. We have

$$\begin{aligned} a_0 + a_1 x + a_2 x^2 &= \alpha \cdot 1 + \beta(1 + x) + \gamma(1 + x + x^2) \\ &= (\alpha + \beta + \gamma) + (\beta + \gamma)x + \gamma x^2 \end{aligned}$$

Equating the coefficients of 1, x, and x^2 we get

$$\begin{aligned} \alpha + \beta + \gamma &= a_0 \\ \beta + \gamma &= a_1 \\ \gamma &= a_2 \end{aligned}$$

whose solution is $\gamma = a_2$, $\beta = a_1 - a_2$, $\alpha = a_0 - a_1$. Hence, $\{1,\ 1 + x,\ 1 + x + x^2\}$ spans P_2. To check the linear independence, we set

$$\alpha \cdot 1 + \beta(1 + x) + \gamma(1 + x + x^2) = 0$$

and find that $\alpha = \beta = \gamma = 0$. Therefore the set is linearly independent.

One can show that the set $\{p_1, p_2, p_3\}$, where

$$p_1 = 2 - x + x^2 \qquad p_2 = x - x^2 \qquad p_3 = -1 + x$$

spans P_2 and is linearly independent, therefore it is also a basis for P_2.

Dimension A linear space V is called *n-dimensional* if it possesses a set of n linearly independent vectors, but every set of $n + 1$ vectors is a dependent set. A linear space with a finite basis is called a *finite-dimensional* vector space, and the number of elements in a basis is called the *dimension* of the space. If there is no upper bound to the number of linearly independent vectors in a vector space, then the vector space is said to be an *infinite-dimensional* vector space. For example the space P of all polynomials is infinite-dimensional.

The following statements can be proven and are left as exercises to the reader:

1. If a vector space has one basis with a finite number of elements, then all other bases are finite and have the same number of elements.
2. In a finite-dimensional vector space, every spanning set contains a basis.
3. In a finite-dimensional vector space any linearly dependent set of vectors can be extended to a basis.

Thus, any vector $\mathbf{u}$ in a n-dimensional vector space V can be expressed as

$$\mathbf{u} = \alpha_1 \mathbf{u}_1 + \alpha_2 \mathbf{u}_2 + \cdots + \alpha_n \mathbf{u}_n \tag{3-21}$$

where $\{\mathbf{u}_1, \mathbf{u}_2, \ldots, \mathbf{u}_n\}$ is a set of linearly independent vectors that spans V. We refer to the vectors $\mathbf{u}_1, \mathbf{u}_2, \ldots, \mathbf{u}_n$ as *base vectors*, to the scalars $\alpha_1, \alpha_2, \ldots, \alpha_n$ as the *coordinates* of $\mathbf{u}$ relative to the basis $\{\mathbf{u}_1, \mathbf{u}_2, \ldots, \mathbf{u}_n\}$, and to the vectors $\alpha_i \mathbf{u}_i$ as the *ith component* of $\mathbf{u}$.

The set of all vectors in a n-dimensional vector space V_n which can be expressed as linear combinations of a given set of k vectors forms a subspace V_r of the vector space V_n. The integer r, the dimension of V_r, is called the *rank* of the given set of k vectors. Obviously, r is less than or equal to both k and n. If the given set of vectors happens to be linearly independent, then its rank equals its dimension ($r = k$). The following theorem on dimensions of finite-dimensional spaces is useful.

Theorem 3-1 If S_1 and S_2 are subspaces of a finite-dimensional vector space, then $\dim(S_1 + S_2) + \dim(S_1 \cap S_2) = \dim S_1 + \dim S_2$.

Example 3-12 1. Let S be the subset in $\mathbb{R}^3$ spanned by the vectors (1, 0, 1) and (0, 1, 1). Any vector in S has the form (the reader should verify this)

$$S = \{\mathbf{x} = (x_1, x_2, x_3): x_3 = x_1 + x_2\}$$

The set $\{(1, 0, 1), (0, 1, 1)\}$ is linearly independent and therefore constitutes a basis of S. The dimension of S is 2.

2. Let S be the subspace of $\mathbb{R}^4$ that consists of all vectors $\mathbf{x} = (x_1, x_2, x_3, x_4)$ such that $x_1 + x_2 - x_3 = 0$. We wish to find a basis for S. From inspection, we note that (1, 0, 1, 1) (0, 1, 1, 1) and (1, 1, 2, 0) are members of S. These vectors are linearly independent and therefore form a basis of S. The dimension of S is 3.

3. Let S_1 be the subspace of $\mathbb{R}^4$ spanned by $\{(1, 1, 0, 0), (1, 0, 1, 1)\}$. Let S_2 be the space of $\mathbb{R}^4$ spanned by $\{(2, -1, 3, 3), (0, 1, -1, -1)\}$. We first characterize the subspaces S_1 and S_2. Let $(a_1, a_2, a_3, a_4) \in S_1$. Then

$$(a_1, a_2, a_3, a_4) = \alpha_1(1, 1, 0, 0) + \alpha_2(1, 0, 1, 1)$$

This gives

$$\alpha_1 + \alpha_2 = a_1, \alpha_1 = a_2, \alpha_2 = a_3, \alpha_2 = a_4$$

Thus, S_1 is defined by

$$S_1 = \{\mathbf{x} \in \mathbb{R}^4: x_1 = x_2 + x_3, x_4 = x_3\}$$

A typical element of S_1 is given by $x = (x_1, x_2, x_1 - x_2, x_1 - x_2)$. Next let $(b_1, b_2, b_3, b_4) \in S_2$. Then

$$(b_1, b_2, b_3, b_4) = \beta_1(2, -1, 3, 3) + \beta_2(0, 1, -1, -1)$$

which gives

$$2\beta_1 = b_1 \qquad \beta_2 - \beta_1 = b_2 \qquad 3\beta_1 - \beta_2 = b_3 \qquad 3\beta_1 - \beta_2 = b_4$$

solving we get $b_3 = b_4$, $b_1 = b_2 + b_3$. Thus, S_2 is the same subspace as S_1. In other words, both sets span the same subspace of $\mathbb{R}^4$. It can be shown that the set $\{(1, 1, 0, 0), (1, 0, 1, 1)\}$ is linearly independent, therefore $S_1 = S_2$ is a two-dimensional subspace of $\mathbb{R}^4$ and the set $\{(1, 1, 0, 0), (1, 0, 1, 1)\}$ is a basis for S_1.

4. Let S_1 be the vector space spanned by the set $\{(1, 0, 2), (1, 2, 2)\}$, and S_2 be the vector space spanned by $\{(1, 1, 0), (0, 1, 1)\}$. It can be shown that these two sets are linearly independent. Hence S_1 and S_2 are two-dimensional vector spaces (subspaces of $\mathbb{R}^3$). Both are subspaces of dimension 2. Indeed, S_1 and S_2 represent planes in a three-dimensional space. Their intersection (which cannot be empty since the zero vector lies in both planes) is either a line or, if they coincide, a plane. Let us find a basis for $S_1 \cap S_2$. Any $u = (\alpha, \beta, \gamma) \in S_1 \cap S_2$ can be expressed as

$$u = (\alpha, \beta) = \alpha_1(1, 0, 2) + \alpha_2(1, 2, 2) \text{ since } u \in S_1$$

$$u = (\alpha, \beta) = \alpha_3(1, 1, 0) + \alpha_4(0, 1, 1) \text{ since } u \in S_2$$

The components α_1 and α_2 in S_1 are related to the components α_3 and α_4 in S_2 by

$$\alpha_1(1, 0, 2) + \alpha_2(1, 2, 2) = \alpha_3(1, 1, 0) + \alpha_4(0, 1, 1)$$

This leads to

$$\alpha_1 + \alpha_2 = \alpha_3 \qquad 2\alpha_2 = \alpha_3 + \alpha_4 \qquad 2(\alpha_1 + \alpha_2) = \alpha_4$$

whose solution is,

$$\alpha_2 = -3\alpha_1 \qquad \alpha_3 = -2\alpha_1 \qquad \alpha_4 = -4\alpha_1$$

Thus,

$$\begin{aligned}(\alpha, \beta, \gamma) &= \alpha_1[(1, 0, 2) - 3(1, 2, 2)] = \alpha_1(-2, -6, -4) \text{ in } S_1 \\ &= -2\alpha_1(1, 1, 0) - 4\alpha_1(0, 1, 1) = \alpha_1(-2, -6, -4) \text{ in } S_2\end{aligned}$$

Hence, $(-2, -6, -4)$ or $(1, 3, 2)$ is a basis for $S_1 \cap S_2$. Note also that $\{(1, 3, 2), (1, 0, 2)\}$ is a basis of S_1, and $\{(1, 1, 0), (1, 3, 2)\}$ is a basis of S_2.

In the next example we show the procedure to determine the dimension of the *solution space* of a differential equation.

Example 3-13 1. Let $V = C^2[0, 1]$ be the linear vector space of all real-valued twice differentiable functions $u(x)$ defined on $[0, 1]$ such that

$$\frac{d^2u}{dx^2} - k^2u = 0 \qquad k > 0 \tag{3-22}$$

We know that e^{-kx} and e^{kx} are in V because $u = e^{-kx}$ and $u = e^{kx}$ satisfy the equation. It can be easily verified that the pair is linearly independent. Consider the linear combination

$$c_1 e^{-kx} + c_2 e^{kx} = 0$$

Since we need another equation to determine c_1 and c_2, we obtain it by differentiating the above linear combination,

$$k(-c_1 e^{-kx} + c_2 e^{kx}) = 0$$

The solution of these two equations is given by $c_1 = c_2 = 0$. Hence $\{e^{-kx}, e^{kx}\}$ is a linearly independent set.

Next consider a set $\{u_1, u_2, u_3\}$ of three elements from V. Again we set

$$\begin{aligned} c_1 u_1 + c_2 u_2 + c_3 u_3 &= 0 \\ c_1 \frac{du_1}{dx} + c_2 \frac{du_2}{dx} + c_3 \frac{du_3}{dx} &= 0 \\ c_1 \frac{d^2u_1}{dx^2} + c_2 \frac{d^2u_2}{dx^2} + c_3 \frac{d^2u_3}{dx^2} &= 0 \end{aligned} \tag{3-23}$$

In matrix form, we have ($u_1' \equiv du_1/dx$, etc.)

$$\begin{bmatrix} u_1 & u_2 & u_3 \\ u_1' & u_2' & u_3' \\ u_1'' & u_2'' & u_3'' \end{bmatrix} \begin{Bmatrix} c_1 \\ c_2 \\ c_3 \end{Bmatrix} = \begin{Bmatrix} 0 \\ 0 \\ 0 \end{Bmatrix} \tag{3-24}$$

These equations have a zero solution if and only if the determinant

$$D \equiv \begin{vmatrix} u_1 & u_2 & u_3 \\ u_1' & u_2' & u_3' \\ u_1'' & u_2'' & u_3'' \end{vmatrix} = u_1(u_2' u_3'' - u_2'' u_3') + u_2(u_1'' u_3' - u_3'' u_1') + u_3(u_2'' u_1' - u_2' u_1'') \tag{3-25}$$

is nonzero. Since each u_i satisfies the differential equation $u_i'' - k^2 u_i = 0$, we have

$$D = k^2 u_1(u_2' u_3 - u_2 u_3') + k^2 u_2(u_1 u_3' - u_3 u_1') + k^2 u_3(u_2 u_1' - u_2' u_1) = 0 \tag{3-26}$$

Thus there are many solutions to the set, and therefore *any* three elements in V are linearly dependent. In other words, the differential equation has only two linearly independent solutions. Therefore, the dimension of V is 2, $\{e^{kx}, e^{-kx}\}$ is the basis of V, and any other element in V is a linear combination of elements e^{kx} and e^{-kx}.

PROBLEMS 3-1

3-1-1 Let V be the set of elements $\mathbf{a}$ of the form $\mathbf{a} = (a_1, a_2)$. Define the vector addition and the scalar multiplication by

$$\mathbf{a} + \mathbf{b} = (a_1 b_1, a_2 b_2)$$

$$\alpha \mathbf{a} = (\alpha a_1, \alpha a_2) \qquad \alpha \in \mathbb{R}$$

Is V a vector space on $\mathbb{R}$? Determine the identity and inverse elements if they exist.

3-1-2 Consider the set V of elements $\{\mathbf{a}, \mathbf{b}, \mathbf{c}, \ldots\}$, with each element of the form $\mathbf{a} = (a_1, a_2, a_3)$. Define the vector addition and scalar multiplication by

$$\mathbf{a} + \mathbf{b} = (a_2 b_3 - b_2 a_3, b_1 a_3 - a_1 b_3, a_1 b_2 - b_1 a_2)$$

$$\alpha \mathbf{a} = (\alpha a_1, \alpha a_2, \alpha a_3) \qquad \alpha \in \mathbb{R}$$

What axioms of the vector space are violated by the vector addition and scalar multiplication?

3-1-3 Determine whether the following sets qualify as linear vector spaces:

(*a*) P_1 = set of polynomials of degree less than or equal to n,

(*b*) P_2 = set of polynomials of degree equal to n.

3-1-4 Consider the equation $\nabla^2 u + \lambda u = 0$ where $\nabla^2 = \partial^2/\partial x^2 + \partial^2/\partial y^2$ and λ is a constant.

(*a*) Show that the solutions u form a vector space V.

(*b*) Show that the solutions u depending only on x form a vector space of V. Give the dimension and basis of the subspace.

3-1-5 Prove that the intersection of any two subspaces of a vector space is also a subspace.

3-1-6 Determine which of the following subsets of $\mathbb{R}^n$ are subspaces:

(*a*) $S_1 = \{\mathbf{x} \in \mathbb{R}^n\colon x_n \geq 0\}$
(*b*) $S_2 = \{\mathbf{x} \in \mathbb{R}^n\colon x_1 - x_2 = 0\}$
(*c*) $S_3 = \{\mathbf{x} \in \mathbb{R}^n\colon x_1 + x_2 = 1\}$
(*d*) $S_4 = \{\mathbf{x} \in \mathbb{R}^4\colon 4x_1 + 3x_2 - 2x_3 - x_4 = 0, -x_1 + 4x_2 + 3x_3 - 2x_4 = 0\}$

3-1-7 Let P be the space of all polynomials with real coefficients. Determine which of the following subsets of P are subspaces.

(*a*) $S_1 = \{p(x)\colon p(1) = 0\}$
(*b*) $S_2 = \{p(x)\colon \text{constant term is unity}\}$

3-1-8 Show that the sum of two subspaces S_1 and S_2 of a vector space is a subspace.

3-1-9 Let W be a linear vector space and let U and V be subspaces of W. Prove that the following assertions are equivalent:

(*a*) If $W = U \oplus V$, then the subspaces U and V have only the zero (null) element of the space in common.

(*b*) If every element $w \in W$ can be represented in the form $w = u + v$, $u \in U$ and $v \in V$, and if $U \cap V = \{0\}$, then $W = U \oplus V$.

3-1-10 If $W = U \oplus V$, show that $w \in W$ can be *uniquely* represented in the form $w = u + v$, $u \in U$ and $v \in V$.

3-1-11 Determine the sum and intersection of the following subspaces of $\mathbb{R}^3$. Also state whether the sum is a direct sum.

(*a*) $U = \{\mathbf{x} \in \mathbb{R}^3\colon x_1 = x_2 = x_3\}$, $V = \{\mathbf{x} \in \mathbb{R}^3\colon x_1 = 0\}$
(*b*) $U = \{\mathbf{x} \in \mathbb{R}^3\colon x_3 = 0\}$, $V = \{\mathbf{x} \in \mathbb{R}^3\colon x_1 = 0\}$
(*c*) $U = \{\mathbf{x} \in \mathbb{R}^3\colon x_1 + x_2 + x_3 = 0\}$, $V = \{\mathbf{x} \in \mathbb{R}^3\colon x_1 = x_3\}$
(*d*) $U = \{\mathbf{x} \in \mathbb{R}^3\colon x_1 = x_2 = 0\}$, $V = \{\mathbf{x} \in \mathbb{R}^3\colon x_1 = x_3\}$

3-1-12 Determine which of the following sets in $\mathbb{R}^3$ are linearly dependent, and if so, express one vector as a linear combination of the others:

(*a*) $\{(-1, 1, 0), (-1, 1, 1), (-2, -1, 1), (1, 1, 1)\}$
(*b*) $\{(1, 0, 0), (1, 1, 0), (1, 1, 1)\}$
(*c*) $\{(1, 0, 0), (0, 1, 0), (0, 0, 1)\}$
(*d*) $\{(2, 1, 0), (1, 1, -2), (0, 0, 0)\}$

3-1-13 Let $\{u_i\}_{i=1}^n$ be a set of vectors of the form, $u_i = (g_{1i}, g_{2i}, \ldots, g_{ni})$. Show that the vectors are linearly dependent only if the determinant of the matrix of coefficients g_{ij} vanishes:

$$[G] = \begin{bmatrix} g_{11} & g_{12} \cdots & g_{1n} \\ g_{21} & g_{22} \cdots & g_{2n} \\ \cdots & \cdots & \cdots \\ g_{n1} & g_{n2} \cdots & g_{nn} \end{bmatrix}$$

The converse can also be shown to be true.

3-1-14 Determine which of the following sets span $\mathbb{R}^3$:

(*a*) $\{(1, 3, -1)(-4, 3, -5), (2, 1, 1)\}$
(*b*) $\{(1, 1, 1), (-2, -1, 2), (-1, 1, 1), (-1, 1, 0)\}$
(*c*) $\{(-1, 1, 0), (-1, 1, 1), (-2, -1, 1), (1, 1, 1)\}$
(*d*) $\{(1, 1, 0), (1, 1, -2), (1, 0, -1)\}$

3-1-15 Let S be the subspace spanned by the set $\{(1, 0, 1, 0), (0, 1, 0, 0), (0, 0, 0, 1)\}$. Determine (i.e., characterize) the subspace S.

3-1-16 Let S be the subspace spanned by $\{(1, -1, 1, 1), (2, -1, 4, 5)\}$. Determine the subspace S.

3-1-17 Show that the vector space P of *all* polynomials over the field of real numbers cannot be generated by a finite number of vectors.

3-1-18 Determine a basis for the linear vector space containing solutions of the differential equation

$$\frac{d^4u}{dx^4} - k^4u = 0 \qquad k > 0.$$

3-1-19 Let $S_1 = \{(1, 2, 3, 6), (4, -1, 3, 6), (5, 1, 6, 12)\}$, and $S_2 = \{(1, -1, 1, 1), (2, -1, 4, 5)\}$ be the subspaces of $\mathbb{R}^4$. Find the basis for $S_1 + S_2$ and $S_1 \cap S_2$.

3-1-20 Find a basis for the solution space of the following pair of linear equations

$$x_1 + 2x_2 + x_3 = 0$$
$$3x_1 + x_2 = 0$$

3-1-21 Find a basis for the solution space of the following linear equations

$$x_1 + 3x_2 + x_3 - x_4 = 0$$
$$-2x_1 + 2x_2 - x_3 + x_4 = 0$$

3-1-22 Find the basis and dimension of the solution space of the differential equation

$$\frac{d^2u}{dx^2} + 2\frac{du}{dx} - u = 0$$

3-1-23 Find a basis for the subspace of $\mathbb{R}^4$ consisting of all elements (x_1, x_2, x_3, x_4) such that $x_1 = x_2 = -x_3$.

3-3 LINEAR TRANSFORMATIONS AND FUNCTIONALS

3-3-1 Introduction

Most problems of engineering are described mathematically by differential or integral equations. These equations relate physical quantities of interest that are to be determined to the sources or data of the problem. The relationships between the dependent variables of the problem and the source terms can be viewed as a functional relationship. More specifically, consider the differential equation [see Eq. (1-8)]

$$-\frac{d}{dx}\left(a\frac{du}{dx}\right) = f \qquad 0 < x < L \tag{3-27}$$

where $a = a(x)$ and $f = f(x)$ are known functions of position x and u is the dependent variable to be determined such that it satisfies the differential equation and certain appropriate boundary conditions. The variables a and f are part of the data. Equation (3-27) represents a transformation between the space of functions u that are twice differentiable and the space of functions f that are continuous. The transformation (or operator) T in the present case is given by

$$T \equiv -\frac{d}{dx}\left(a\frac{d}{dx}\right) \tag{3-28}$$

Since we wish to study (i.e., solve) a particular equation in the context of the general, we focus our attention on operator equations of the form

$$Tu = f \tag{3-29}$$

The operator equation (3-29) has solutions under certain conditions on the spaces involved as well as the operator. Toward studying the properties (e.g., existence of solution, etc.) of operator equations, we first study various concepts akin to operators.

3-3-2 Linear Transformations

The notion of a function from one set into another can be extended to vector spaces. A transformation T from a linear vector space U into another linear vector space V is a correspondence which assigns to each element u in U a unique element $v = Tu$ in V. We use the terms "transformation," "mapping," and "operator" interchangeably. Strictly speaking, a linear transformation must specify its domain space, its range space, and the definition of the transformation.

A transformation T of a vector space U into a vector space V, where U and V have the same scalar field, is said to be *linear* if

$$\begin{aligned} &\text{(i)}\ \ T(\alpha u) = \alpha T(u) \text{ for all } u \in U, \text{ scalars } \alpha \text{ (homogeneous)} \\ &\text{(ii)}\ \ T(u_1 + u_2) = T(u_1) + T(u_2), \text{ for all } u_1, u_2 \in U \text{ (additive)} \end{aligned} \tag{3-30}$$

Otherwise it is said to be a *nonlinear transformation.* Conditions (i) and (ii) can be combined into one: a transformation T is linear if

$$T(\alpha u_1 + \beta u_2) = \alpha T(u_1) + \beta T(u_2) \tag{3-31}$$

for all $u_1, u_2 \in U$ and scalars α and β. Roughly speaking, properties (i) and (ii) are equivalent to saying that the image of the sum is the sum of the images and the image of the product (of a vector with a scalar) is the product of the image of the vector with the scalar.

Since linear transformations are essentially functions defined on elements of a vector space, the concept of composition of functions, and one-to-one and onto definitions can be naturally extended to linear transformations. If $T_1 : U \to V$ and $T_2 : V \to W$ are two linear transformations, then their composition $T_2 T_1$ is also linear:

$$\begin{aligned} (T_2 T_1)(\alpha u_1 + \beta u_2) &= T_2(\alpha T_1(u_1) + \beta T_1(u_2)) \\ &= \alpha T_2(T_1(u_1)) + \beta T_2(T_1(u_2)) \\ &= \alpha(T_2 T_1)(u_1) + \beta(T_2 T_1)(u_2) \end{aligned}$$

A one-to-one transformation T [i.e., for $u \neq v$ it necessarily follows that $T(u) \neq T(v)$] is called a *monomorphism.* If U is a vector space and S is a subspace of U, and $T: S \to V$, the set $T(S)$ denotes the collection of images of elements from S; $T(S)$ is called the *image* of T, denoted Im (T). Obviously, Im $(T) \subset V$. If Im $(T) = V$, the transformation is *onto* and is called an *epimorphism.* A transformation that is both a monomorphism and an epimorphism is called an *isomorphism.*

The *null space*, also known as the *kernel* of the transformation $T: U \to V$ and denoted $\mathcal{N}(T)$, is defined by

$$\mathcal{N}(T) = \{u:\ u \in U,\ Tu = 0\} \tag{3-32}$$

The null space is the subset of elements of U which have the zero image. It is easy to verify that this set is a linear vector space. The following result is useful in the study of operator equations.

Theorem 3-2 A linear transformation, $T: U \to V$, is one-to-one if and only if its null space is trivial, $\mathcal{N}(T) = \{0\}$.

PROOF If T is a monomorphism (i.e., one-to-one), then $u \in \mathcal{N}(T)$ implies that $u = 0$ (since $T0 = 0$ and $Tu = 0$). Hence, $\mathcal{N}(T)$ is the null set, $\{0\}$. Conversely, if $\mathcal{N}(T) = \{0\}$, then $Tu_1 = Tu_2$ implies that (since T is linear)

$$T(u_1 - u_2) = 0 \qquad \text{or} \qquad u_1 - u_2 \in \mathcal{N}(T) \to u_1 - u_2 = 0 \qquad \text{or} \qquad u_1 = u_2$$

Thus, a linear transformation is one-to-one if and only if its null space is trivial, $\mathcal{N}(T) = \{0\}$.

The dimension of the subspace, $\mathcal{R}(T) = \text{Im}\,(T) \subset V$ is called the *rank* of the transformation, $T: U \to V$. The rank of T is smaller than the dimension of U as well as V. The dimension of the null space T is called the *nullity* of T. The sum of the rank and nullity of the transformation is equal to the dimension of U.

Example 3-14 1. Let $U = C^2[a, b]$ be the space of twice-differential functions over $[a, b]$, and $V = C[a, b]$ be the set of continuous functions on $[a, b]$. Let T be the transformation

$$Tu = c_1 u + c_2 \frac{d^2 u}{dx^2} + c_3 \int_a^b K(x, y) u(y)\, dy$$

where c_1, c_2, and c_3 are functions of x. We will show that T is linear. For any $u_1, u_2 \in U$ and real numbers α and β, we have

$$\begin{aligned}
T(\alpha u_1 + \beta u_2) &= c_1(\alpha u_1 + \beta u_2) + c_2 \frac{d^2}{dx^2}(\alpha u_1 + \beta u_2) \\
&\quad + c_3 \int_a^b K(x, y)[\alpha u_1(y) + \beta u_2(y)]\, dy \\
&= c_1 \alpha u_1 + c_2 \alpha \frac{d^2 u_1}{dx^2} + c_3 \alpha \int_a^b K(x, y) u_1(y)\, dy \\
&\quad + c_1 \beta u_2 + c_2 \beta \frac{d^2 u_2}{dx^2} + c_3 \beta \int_a^b K(x, y) u_2(y)\, dy \\
&= \alpha T(u_1) + \beta T(u_2)
\end{aligned}$$

Note that the operator T is the sum of three linear operators, $T(u) = T_1(u) + T_2(u) + T_3(u)$:

$$T_1(u) = c_1 u \qquad T_2(u) = c_2 \frac{d^2u}{dx^2} \qquad T_3(u) = c_3 \int_a^b K(x, y)u(y)\,dy$$

2. Let $T: C[0, 1] \to \mathbb{R}$ be

$$Tu = c \qquad c = \text{constant}$$

The operator is *not* linear because it does not satisfy either of the two properties in Eq. (3-30).

3. Consider a rectangular m by n matrix $[A] = [a_{ij}]$. The expression

$$y_i = \sum_{j=1}^{n} a_{ij} x_j \qquad i = 1, 2, \ldots, m$$

defines an operator A that transforms an element $\mathbf{x} = (x_1, x_2, \ldots, x_n)$ of the n-dimensional euclidean space $\mathbb{R}^n$ into an element $\mathbf{y} = (y_1, y_2, \ldots, y_m)$ of the m-dimensional euclidean space $\mathbb{R}^m$. Clearly, A is a linear (matrix) operator.

Example 3-15 Consider the spring-dashpot-mass system (an idealization, for example, of the motion of a piston and piston-rod in an automobile engine) shown in Fig. 3-5. We assume that there is viscous friction between the piston (mass) and the cylinder wall it slides on, and that the piston rod behaves elastically. In the mathematical modeling of the system we assume that the elastic and viscous responses are linear (i.e., the restoring force is linearly proportional to the displacement and the time rate of the displacement, respectively). We wish to model the motion of the system for times $t > 0$. At time $t = 0$, we assume that $x(0) = 0$ and $(dx/dt)(0) = 0$. The differential equation relating the displacement $x(t)$ and the applied force $f(t)$ is given by

$$Tx \equiv m \frac{d^2x}{dt^2} + \eta \frac{dx}{dt} + kx = f \tag{3-33}$$

For $f \in C[0, \infty)$, the equation describes a mapping T from $C^2[0, \infty)$ into $C[0, \infty)$. Clearly T is linear. It is one-to-one and onto its range.

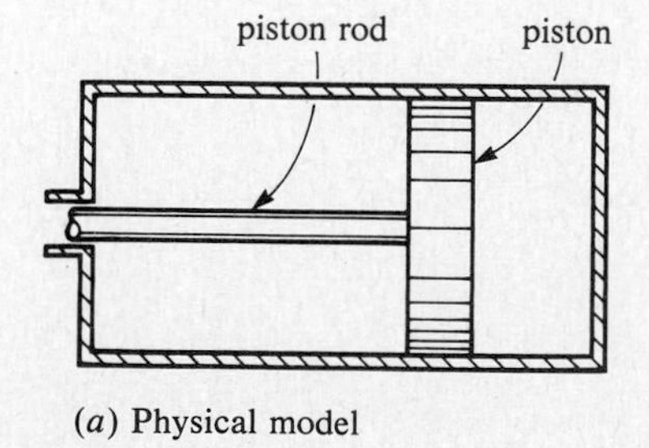

(*a*) Physical model

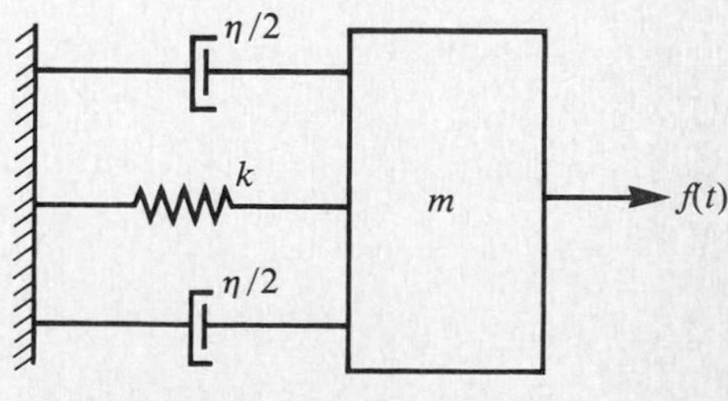

(*b*) Mathematical model

Figure 3-5 A mathematical model for a spring-dashpot-mass system.

Next consider the mapping $S: C[0, \infty) \to C[0, \infty)$ defined by

$$Sf \equiv \frac{1}{m(\lambda_1 - \lambda_2)} \int_0^t [e^{\lambda_1(t-\tau)} - e^{\lambda_2(t-\tau)}] f(\tau)\, d\tau \tag{3-34}$$

where λ_1 and λ_2 are the roots of the equation

$$m\lambda^2 + \eta\lambda + k = 0 \tag{3-35}$$

We assume that m, η, and k are chosen such that λ_1 and λ_2 are real and distinct. The composition of T and S is given by

$$\begin{aligned} TSf &= m \frac{d^2}{dt^2}(Sf) + \eta \frac{d}{dt}(Sf) + kSf \\ &= f + \frac{1}{m(\lambda_1 - \lambda_2)} \sum_{i=1}^{2} (-1)^{i+1}(m\lambda_i^2 + \eta\lambda_i + k) \int_0^t e^{\lambda_i(t-\tau)} f(\tau)\, d\tau \\ &= f \end{aligned}$$

where Leibniz' theorem for differentiation of an integral is used to arrive at the last step:

$$\frac{d}{dt} \int_{A(t)}^{B(t)} f(t, \tau)\, d\tau = \int_{A(t)}^{B(t)} \frac{\partial}{\partial t} f(t, \tau)\, d\tau + f(t, B(t)) \frac{dB}{dt} - f(t, A(t)) \frac{dA}{dt} \tag{3-36}$$

Equation $TSf = f$ implies that $TS = I$, where I is the *identity operator*, $If = f$ for every $f \in C[0, \infty)$. Therefore, T is an inverse of S and vice versa.

3-3-3 Linear Transformations on Finite-Dimensional Spaces

Linear transformations on finite-dimensional spaces are encountered in the approximate solution of boundary-value problems of continuous systems. Most approximate methods seek solutions of continuous systems in finite-dimensional subspaces of infinite-dimensional vector spaces in which the true solution lies.

Linear transformations on finite-dimensional vector spaces can be represented by matrices. Let U and V be finite-dimensional vector spaces, and let $T: U \to V$ be a linear operator. Let $\{\phi_1, \phi_2, \ldots, \phi_n\}$ and $\{\psi_1, \psi_2, \ldots, \psi_m\}$ be bases for U and V, respectively. Then $u \in U$ and $v \in V$ can be expressed uniquely as

$$u = \sum_{i=1}^{n} \alpha_i \phi_i \qquad v = \sum_{j=1}^{m} \beta_j \psi_j \tag{3-37}$$

Thus for any $u \in U$, $T(u) = v$ can be written as

$$T(u) \equiv \sum_{i=1}^{n} \alpha_i T(\phi_i) = \sum_{j=1}^{m} \beta_j \psi_j \tag{3-38}$$

Since $T(\phi_i) \in V$, we can write

$$T(\phi_i) = \sum_{j=1}^{m} t_{ji} \psi_j \tag{3-39}$$

We have from (3-38) and (3-39)

$$\sum_{j=1}^{m} \beta_j \psi_j - \sum_{i=1}^{n} \alpha_i \left(\sum_{j=1}^{m} t_{ji} \psi_j \right) = 0$$

or

$$\sum_{j=1}^{m} \left(\beta_j - \sum_{i=1}^{n} t_{ji} \alpha_i \right) = 0 \tag{3-40}$$

In matrix form, we have

$$\begin{Bmatrix} \beta_1 \\ \beta_2 \\ \vdots \\ \beta_m \end{Bmatrix} = \begin{bmatrix} t_{11} & t_{12} & \cdots & t_{1n} \\ t_{21} & t_{22} & \cdots & t_{2n} \\ \\ t_{m1} & \cdots\cdots & & t_{mn} \end{bmatrix} \begin{Bmatrix} \alpha_1 \\ \alpha_2 \\ \vdots \\ \alpha_n \end{Bmatrix} \tag{3-41}$$

The matrix $[t_{ij}]$ is said to represent the linear transformation relative to the bases $\{\phi_i\}^n$ and $\{\psi_i\}^m$.

Example 3-16 1. Let $U = P_3$ be the space of polynomials of degree ≤ 3, and let $V = P_1$ be the space of polynomials of degree ≤ 1. Let $D = d^2/dx^2$ be the linear operator. The matrix representing the transformation D can be represented with respect to any pair of bases in P_3 and P_1.

Let the bases in P_3 and P_1 be

$$\{\phi_i\} = \{1, x, x^2, x^3\}, \{\psi_i\} = \{1, x\}$$

Then we have, from Eq. (3-39)

$$D(\phi_1) = 0 = 0 \cdot \psi_1 + 0 \cdot \psi_2 = \sum_{j=1}^{2} d_{j1} \psi_j$$

$$D(\phi_2) = 0 = 0 \cdot \psi_1 + 0 \cdot \psi_2 = \sum_{j=1}^{2} d_{j2} \psi_j$$

$$D(\phi_3) = 2 = 2 \cdot \psi_1 + 0 \cdot \psi_2 = \sum_{j=1}^{2} d_{j3} \psi_j$$

$$D(\phi_4) = 6x = 0 \cdot \psi_1 + 6 \cdot \psi_2 = \sum_{j=1}^{2} d_{j4} \psi_j$$

or

$$\begin{Bmatrix} \beta_1 \\ \beta_2 \end{Bmatrix} = \begin{bmatrix} 0 & 0 & 2 & 0 \\ 0 & 0 & 0 & 6 \end{bmatrix} \begin{Bmatrix} \alpha_1 \\ \alpha_2 \\ \alpha_3 \\ \alpha_4 \end{Bmatrix} \qquad [d] = \begin{bmatrix} 0 & 0 & 2 & 0 \\ 0 & 0 & 0 & 6 \end{bmatrix}$$

where $[d]$ represents the transformation relative to the bases $\{1, x, x^2, x^3\}$ and $\{1, x\}$.

2. Consider the rotation transformation T of the xy-plane through an angle θ about the z-axis. The point (x, y, z) is mapped into the point (x', y', z). The transformation $T\colon \mathbb{R}^3 \to \mathbb{R}^3$ can be represented by

$$T(x, y, z) = (x \cos \theta - y \sin \theta, x \sin \theta + y \cos \theta, z)$$

Note that

$$T(\hat{e}_1) \equiv T(1, 0, 0) = (\cos \theta, \sin \theta, 0) = \cos \theta\, \hat{e}_1 + \sin \theta\, \hat{e}_2 + 0 \cdot \hat{e}_3$$

$$T(\hat{e}_2) \equiv T(0, 1, 0) = (-\sin \theta, \cos \theta, 0) = -\sin \theta\, \hat{e}_1 + \cos \theta\, \hat{e}_2 + 0 \cdot \hat{e}_3$$

$$T(\hat{e}_3) \equiv T(0, 0, 1) = (0, 0, 1) = 0 \cdot \hat{e}_1 + 0 \cdot \hat{e}_2 + \hat{e}_3$$

Hence the matrix T relative to the basis $(\hat{e}_1, \hat{e}_2, \hat{e}_3)$, is given by

$$[t_{ij}] = \begin{bmatrix} \cos \theta & \sin \theta & 0 \\ -\sin \theta & \cos \theta & 0 \\ 0 & 0 & 1 \end{bmatrix}$$

It turns out that the inverse of the matrix $[T]$ is equal to its transpose.

3-3-4 Linear, Bilinear, and Quadratic Forms

Linear transformations that map a given linear vector space or a product of two linear vector spaces into the real numbers $\mathbb{R}$ are of considerable interest in the study of the variational formulation of operator equations. For instance, the total potential energy expression in solid mechanics problems is an example of a quadratic functional that maps the (product) space of displacements into a scalar, which is known as the energy of the system. The first variation of a quadratic functional is an example of a bilinear form. Numerous other examples of linear, bilinear, and quadratic forms will be encountered during our study of variational methods. Therefore, it is essential for us to study the properties of linear, bilinear, and quadratic forms.

Linear functional Let U be a linear vector space over the real number field $\mathbb{R}$. A linear transformation l of U into $\mathbb{R}$ is called a *linear form* or *linear functional.*

Since linear functionals are a special case of linear transformations, the concepts and results given in preceding pages remain valid for linear functionals. The set of all linear functionals on a linear vector space is itself a vector space, called the *dual* or *conjugate space* of U. The dual space is denoted by U'. When U is a finite-dimensional vector space of dimension n, then the dual space U' is a finite-dimensional vector space of dimension n.

The basis of the dual space U' of U has a special relation to the basis of U. Let $\{\phi_i\}$ be the basis of U. Define the linear functional l_i by

$$u = \sum_{i=1}^{n} \alpha_i \phi_i \qquad l_i(u) \equiv \alpha_i \in \mathbb{R} \tag{3-42}$$

We shall call l_i the ith coordinate function. We now show that l_i is a linear functional. Indeed, for $u = \sum_i \alpha_i \phi_i$, $v = \sum_i \beta_i \phi_i$, and for any scalars μ and λ, we have

$$l_j(\mu u + \lambda v) = l_j\left(\sum_i (\mu\alpha_i + \lambda\beta_i)\phi_i\right) = \mu\alpha_j + \lambda\beta_j = \mu l_j(u) + \lambda l_j(v)$$

Note that the relationship between $\{l_i\}$ and $\{\phi_i\}$ is characterized by

$$l_i(\phi_j) = \delta_{ij} \qquad \text{for all } i, j \tag{3-43}$$

Since ϕ_j are linearly independent, l_i are also linearly independent. Therefore the set $\{l_i\}$ form a basis for the dual space U', and we shall call $\{l_i\}$ the *dual* (or conjugate) *basis.*

Example 3-17 The linear functionals in $\mathbb{R}^n$ are of the form,

$$f(\mathbf{x}) = \sum_{i=1}^{n} f_i x_i$$

where x_i are components of the vector $\mathbf{x}$ in $\mathbb{R}^n$, and f_i are any numbers which can be thought of as components of the vector f. Clearly, the scalar product of ordinary vectors is a special case of the above statement. Indeed, every linear functional on $\mathbb{R}^n$ can be written as a scalar product,

$$f(\mathbf{x}) = (f, \mathbf{x})$$

Example 3-18 1. Let $V = L_2[0, 1]$ and define a linear functional on V by

$$l(v) = \int_0^1 f(x)v(x)\, dx \tag{3-44}$$

where f is an arbitrary function. The functional l is linear because it is homogeneous,

$$l(\alpha v) = \int_0^1 f(x)\alpha v(x)\, dx = \alpha \int_0^1 f(x)v(x)\, dx = \alpha l(v)$$

and additive,

$$\begin{aligned} l(v_1 + v_2) &= \int_0^1 f(x)(v_1 + v_2)\, dx = \int_0^1 f(x)v_1(x)\, dx + \int_0^1 f(x)v_2(x)\, dx \\ &= l(v_1) + l(v_2). \end{aligned}$$

2. Let S be the n-dimensional subspace of $L_2[0, 1]$, spanned by the set $\{\phi_i\}_{i=1}^n$, and l be the linear functional defined by Eq. (3-44). Then for $v \in S$, we have

$$\begin{aligned} l(v) &= \int_0^1 f(x) \sum_{i=1}^n \alpha_i \phi_i(x)\, dx \\ &= \sum_{i=1}^n \alpha_i \int_0^1 f(x)\phi_i(x)\, dx \\ &= \sum_{i=1}^n \alpha_i b_i \end{aligned} \tag{3-45}$$

where
$$b_i = \int_0^1 f(x)\phi_i(x)\, dx$$

Equation (3-45) is a representation of the linear functional $l(\cdot)$ on S.

Bilinear forms Let U and V be two vector spaces with the same field of scalars. The operator $B: U \times V \to \mathbb{R}$ that maps pairs (u, v), $u \in U$, $v \in V$, into the field of scalars is called a bilinear functional, denoted $B(u, v)$, if $B(u, v)$ satisfies the condition [which is analogous to Eq. (3-31)]

$$B(\alpha u_1 + \beta u_2, \mu v_1 + \lambda v_2) = \alpha\mu B(u_1, v_1) + \alpha\lambda B(u_1, v_2) + \beta\mu B(u_2, v_1) + \beta\lambda B(u_2, v_2) \tag{3-46}$$

for all $u_1, u_2 \in U$, $v_1, v_2 \in V$, and scalars α, β, μ, and λ.

Analogous to the representation of linear operators on finite-dimensional spaces, one can represent bilinear forms on finite-dimensional spaces by matrices. Consider a bilinear form $B(\cdot, \cdot)$ on finite-dimensional vector spaces U and V. Let U and V be m- and n-dimensional vector spaces, respectively. Let $\{\phi_i\}$ be a basis for U and let $\{\psi_i\}$ be a basis for V. For any u in U and any v in V, we have

$$u = \sum_{i=1}^m \alpha_i \phi_i \qquad v = \sum_{i=1}^n \beta_i \psi_i \tag{3-47}$$

for $\alpha_i, \beta_i \in \mathbb{R}$. Then

$$\begin{aligned} B(u, v) &= B\left(\sum_{i=1}^m \alpha_i \phi_i, \sum_{j=1}^n \beta_j \psi_j \right) = \sum_{i=1}^m \sum_{j=1}^n \alpha_i \beta_j B(\phi_i, \psi_j) \\ &= \sum_{i=1}^m \sum_{j=1}^n \alpha_i \beta_j b_{ij} = \{\alpha\}[B]\{\beta\}^T, \; b_{ij} = B(\phi_i, \psi_j) \end{aligned} \tag{3-48}$$

The matrix $[B] = [b_{ij}]$ represents the bilinear form $B(u, v)$ with respect to the bases $\{\phi_i\}$ and $\{\psi_i\}$.

A bilinear form $B(\cdot, \cdot)$: $U \times U \to \mathbb{R}$ is said to be *symmetric* if $B(u, v) = B(v, u)$, for all $u, v \in U$. If $B(u, u) = 0$ for all $u \in U$, we say that the bilinear form is *skew-symmetric*. Note that if $B(\cdot, \cdot)$ is symmetric, we have

$$b_{ij} = B(\phi_i, \phi_j) = B(\phi_j, \phi_i) = b_{ji} \tag{3-49}$$

Similarly, if $B(\cdot, \cdot)$ is skew-symmetric, we have

$$0 = B(u + v, u + v) = B(u, u) + B(u, v) + B(v, u) + B(v, v)$$

or

$$b_{ij} = -b_{ji}$$

Every bilinear form can be represented uniquely as a sum of a symmetric bilinear form and a skew-symmetric bilinear form

$$\begin{aligned} B(u, v) &= \tfrac{1}{2}[B(u, v) + B(v, u)] + \tfrac{1}{2}[B(u, v) - B(v, u)] \\ &= B_s(u, v) + B_{ss}(u, v) \end{aligned} \tag{3-50}$$

Example 3-19 Consider the operator equation

$$Au \equiv -\nabla^2 u = f \qquad \text{in } \Omega \in \mathbb{R}^3 \tag{3-51}$$

where $A: C^2(\Omega) \rightarrow L_2(\Omega)$. Define $B(u, v)$, $u, v \in C^2(\Omega)$, by

$$\begin{aligned} B(u, v) &= \int_\Omega Auv\, dx = -\int_\Omega \nabla^2 uv\, dx \\ &= \int_\Omega \operatorname{grad} u \cdot \operatorname{grad} v\, dx - \oint_\Gamma \frac{\partial u}{\partial n} v\, ds \end{aligned}$$

It is clear that $B(u, v)$ is linear both in u and v, but it is not symmetric in general. However, if $A: S \subset C^2(\Omega) \rightarrow L_2(\Omega)$, where S contains functions that vanish on the boundary Γ of Ω, then

$$B(u, v) = \int_\Omega \operatorname{grad} u \cdot \operatorname{grad} v\, dx \tag{3-52}$$

which is symmetric. If S is a finite-dimensional subspace spanned by the set $\{\phi_i\}_{i=1}^N$, we have $u = \sum_{i=1}^N \alpha_i \phi_i$, $v = \sum_{j=1}^N \beta_j \phi_j$, and

$$B(u, v) = \sum_{i=1}^N \sum_{j=1}^N b_{ij} \alpha_i \beta_j \tag{3-53a}$$

where

$$b_{ij} = B(\phi_i, \phi_j) \tag{3-53b}$$

Note that all ϕ_i should vanish on Γ so that $B(u, v)$ in Eq. (3-52) is valid.

If we take f to the left side of Eq. (3-51), multiply with $\phi_i \in S$, and integrate over Ω, we obtain

$$\int_\Omega \operatorname{grad} \phi_i \cdot \operatorname{grad} u\, dx - \int_\Omega \phi_i f\, dx = 0$$

or

$$\sum_{j=1}^N b_{ij} \alpha_j - f_i = 0 \tag{3-54}$$

where b_{ij} is given by Eq. (3-53b), and f_i is given by

$$f_i = \int_\Omega f \phi_i\, dx$$

Thus the solution of Eq. (3-51), subject to $u = 0$ on Γ, in the finite-dimensional subspace S is given by

$$u = \sum_{j=1}^{N} \alpha_j \phi_j \tag{3-55}$$

where α_j are obtained by solving the algebraic equations in (3-54). Variational methods of approximation to be discussed in Chap. 6 are based on the ideas presented in this example.

Quadratic form Let U be a vector space, and let $B(\cdot\,,\,\cdot)$ be a bilinear form on U. A *quadratic form* is a functional $Q(u)$ on U, such that

$$Q(\alpha u) = \alpha^2 Q(u) \tag{3-56}$$

An example of $Q(u)$ is given by setting $v = u$ in Eq. (3-52): $Q(u) = B(u, u)$.

Note that if $B(\cdot, \cdot)$ is represented as the sum of a symmetric $[B_s(\cdot, \cdot)]$ and a skew-symmetric $[B_{ss}(\cdot, \cdot)]$ bilinear form we can write

$$Q(u) \equiv B(u, u) = B_s(u, u) + B_{ss}(u, u) = B_s(u, u) \tag{3-57}$$

That is, $Q(u)$ is completely determined by the symmetric part of the bilinear form. Also, two different bilinear forms with the same symmetric part must generate the same quadratic form.

PROBLEMS 3-2

3-2-1 Determine whether the following transformation is linear (check for homogeneity and additivity).

$$T : U = C[0, T] \to V = C[0, T]$$

$$Tu = \begin{cases} u_0 & u(t) \geq u_0 \\ u(t) & -u_0 < u(t) < u_0 \\ -u_0 & u(t) \leq -u_0 \end{cases}$$

where $u_0 > 0$ is a real number.

3-2-2 Describe the range of the transformation in Prob. 3-2-1. Is T onto and/or one-to-one?

Identify the transformations in the differential equations given in Probs. 3-2-3 to 3-2-8. Also, identify the domain and range spaces and determine whether they are linear.

3-2-3 *Bending of elastic beams on elastic foundation*

$$kw + \frac{d^2}{dx^2}\left(b\,\frac{d^2 w}{dx^2}\right) = f \qquad \text{in } 0 < x < L$$

where w is the transverse deflection, b is the flexural rigidity, k is the foundation modulus, and f is the distributed transverse load.

3-2-4 *Heat transfer in an orthotropic medium:*

$$-\frac{\partial}{\partial x}\left(K_1 \frac{\partial u}{\partial x}\right) - \frac{\partial}{\partial y}\left(K_2 \frac{\partial u}{\partial y}\right) = f \qquad \text{in } \Omega \subset \mathbb{R}^2$$

where u is the temperature, K_1 and K_2 are conductivities in the x and y directions, and f is the heat generation (see Sec. 2-6-4). The equation also arises in other fields of engineering; the meaning of the variables is different in different fields (see Table 4-2).

3-2-5 *Laminar flow of a viscous fluid (the Stokes equations):*

$$\left.\begin{aligned} -2\mu \frac{\partial^2 u}{\partial x^2} - \mu \frac{\partial}{\partial y}\left(\frac{\partial u}{\partial y} + \frac{\partial v}{\partial x}\right) + \frac{\partial P}{\partial x} &= f_1 \\ -\mu \frac{\partial}{\partial x}\left(\frac{\partial u}{\partial y} + \frac{\partial v}{\partial x}\right) - 2\mu \frac{\partial^2 v}{\partial y^2} + \frac{\partial P}{\partial y} &= f_2 \\ \frac{\partial u}{\partial x} + \frac{\partial v}{\partial y} &= 0 \end{aligned}\right\} \quad \text{in } \Omega \subset \mathbb{R}^2$$

where μ is the viscosity of the fluid, u and v are the velocity components along the x and y axes, P is the pressure, and f_1 and f_2 are the body forces along the x and y directions, respectively (see Sec. 2-6-3). Write the equations in the matrix operator form.

3-2-6 *Expansion of a gas behind a piston:*

$$\frac{\partial u}{\partial t} + u \frac{\partial u}{\partial x} + \frac{c^2}{\rho} \frac{\partial \rho}{\partial x} = 0$$

$$\frac{\partial \rho}{\partial t} + u \frac{\partial \rho}{\partial x} + \rho \frac{\partial u}{\partial x} = 0$$

where $c = \rho^{(1-\nu)/2}$, t denotes time, x the spatial coordinate, u the velocity, ρ the density, and ν the ratio of specific heats.

3-2-7 *Poisson–Boltzmann equation* (arises in the theory of colloids and plasmas):

$$\frac{d^2\phi}{dx^2} = e^{\phi} - e^{-\phi}, \qquad a \le x \le b$$

where ϕ is the nondimensional electric potential.

3-2-8 *Hilbert integral operator:*

$$Tu = \frac{1}{x} \int_0^x u(s)\, ds$$

3-2-9 Consider the temperature $T(x, t)$ distribution in an infinite bar as a function of time t and position x when the bar is being heated along its length by a distributed heat source $q(x, t)$. Assume that $q(x, t)$ is an element in the linear space V made up of all bounded continuous, real-valued functions defined on the tensor product $(-\infty, \infty) \times [0, \infty)$. Then the temperature is given by

$$T(x, t) = \int_0^{\infty} \int_{-\infty}^{\infty} K(x - y, t - s) q(y, s)\, dy\, ds \tag{a}$$

where

$$K(x, t) = \begin{cases} c \dfrac{\exp(-x^2/4t)}{\sqrt{2}} & t > 0 \\ 0 & t \le 0 \end{cases} \tag{b}$$

and c is a constant. Show that Eq. (*a*) represents a linear transformation of V into itself.

3-2-10 Let T_1 and T_2 be linear operators defined on $\mathbb{R}^3$ by

$$T_1\mathbf{x} = (x_1 - x_3, x_2 + x_3, 0)$$

$$T_2\mathbf{x} = (2x_1 - x_2, x_1, x_2 - x_3)$$

Determine (*a*) $T_1 + T_2$, (*b*) $T_1 T_2$, (*c*) $T_2 T_1$.

3-2-11 Let $D = d/dt$ be the differential operator on the space P of polynomials in a real variable t. Suppose that the transformation T is defined on P by $Tp(t) = tp(t)$ for any $p(t) \in P$. Show that (*a*) $DT = TD + I$, (*b*) $(TD)^2 = T^2D^2 + TD$. Here I is the identity operator.

3-2-12 Prove that the range and kernel of a linear transformation are linear vector spaces.

3-2-13 Find the null space of the linear transformation $T\colon \mathbb{R}^3 \to \mathbb{R}^3$ described by the algebraic equations:

$$2x_1 - x_2 - x_3 = 1$$

$$x_1 + 2x_2 + x_3 = 1$$

$$4x_1 - 7x_2 - 5x_3 = 1$$

3-2-14 Find the null spaces of the linear transformations in Prob. 10.

3-2-15 Let $V = P_3$ be the vector space of all third degree polynomials in one variable x. Represent the following transformations by a matrix with respect to the basis $\{1, x, x^2, x^3\}$:

(*a*) the linear transformation T_1 which associates the polynomial $p(x + 1)$ with each polynomial $p(x)$;

(*b*) the linear transformation T_2 which associates the polynomial dp/dx with each polynomial $p(x)$;

(*c*) the sum $T_1 + T_2$ and composition T_1T_2 of the transformations defined in (*a*) and (*b*).

Rewrite the functionals in Probs. 3-2-16 to 3-2-18 as sums of bilinear and linear forms.

3-2-16

$$0 = \int_0^L \left(a \frac{dv}{dx}\frac{du}{dx} - vf \right) dx - v(0)P_1 - v(L)P_2$$

where a, f, P_1 and P_2 are known quantities.

3-2-17

$$0 = \int_\Omega \left(K_1 \frac{\partial v}{\partial x}\frac{\partial u}{\partial x} + K_2 \frac{\partial v}{\partial y}\frac{\partial u}{\partial y} + K_3 vu - vf \right) dx\,dy - \oint_\Gamma vq\,ds$$

where K_1, K_2, K_3, f, and q are known functions of position.

3-2-18

$$0 = D \int_0^a \int_0^b \left\{ \nabla^2 v \nabla^2 w + (1 - \nu)\left[2 \frac{\partial^2 v}{\partial x\,\partial y}\frac{\partial^2 w}{\partial x\,\partial y} - \frac{\partial^2 v}{\partial x^2}\frac{\partial^2 w}{\partial y^2} - \frac{\partial^2 v}{\partial y^2}\frac{\partial^2 w}{\partial x^2} \right] \right\} dx\,dy - \int_0^a \int_0^b vq\,dx\,dy$$

Check whether the functionals given in Probs. 3-2-19 to 3-2-21 qualify as linear, bilinear, or quadratic forms.

3-2-19

$$F(u) = \int_\Omega \left[K_1\left(\frac{\partial u}{\partial x}\right)^2 + K_2\left(\frac{\partial u}{\partial y}\right)^2 + K_3 u^2 \right] dx\,dy \quad (K_1, K_2, \text{ and } K_3 \text{ are constants})$$

3-2-20

$$F(u, v, P) = \int_\Omega \mu\left[\left(\frac{\partial u}{\partial x}\right)^2 + \left(\frac{\partial v}{\partial y}\right)^2 + \frac{1}{2}\left(\frac{\partial u}{\partial x} + \frac{\partial v}{\partial y}\right)^2 - P\left(\frac{\partial u}{\partial x} + \frac{\partial v}{\partial y}\right)\right] dx\,dy$$

where μ is a constant.

3-2-21

$$F(w, M) = \int_0^L \left(\frac{dw}{dx}\frac{dM}{dx} + \frac{1}{EI} M^2 + wf \right) dx$$

3-2-22

$$F(u, v) = \int_0^t u(t)v(t - s)\,ds \qquad u = u(t) \qquad v = v(t) \qquad 0 < t < T$$

Is $F(u, v)$ symmetric?

3-2-23 Consider the bilinear and linear forms

$$B(u, v) = \int_0^L \alpha \frac{d^2u}{dx^2}\frac{d^2v}{dx^2}\,dx \qquad l(v) = \int_0^L \beta v(x)\,dx$$

Determine the associated matrices with respect to the basis

$$\phi_1 = 1 - 3\xi^2 + 2\xi^3 \qquad \phi_2 = -3\xi L(1-\xi)^2 \qquad \phi_3 = 3\xi^2 - 2\xi^3$$

$$\phi_4 = -\xi L(\xi^2 - \xi) \qquad \xi = x/L$$

3-2-24 (*Projection operators*) A linear transformation P of a linear vector space V into itself is called a projection operator if $P^2 = P$. Let $V = \mathbb{R}^2$ and S_1 and S_2 be the subspaces, $S_1 = \{\mathbf{x} \in \mathbb{R}^2 : x_1 = 0\}$, $S_2 = \{\mathbf{x} \in \mathbb{R}^2 : x_2 = 0\}$. Show that the operator P defined by $P(\mathbf{x}^1 + \mathbf{x}^2) = \mathbf{x}^1$ for every $\mathbf{x}^1 = (0, x_2) \in S_1$ and $\mathbf{x}^2 = (x_1, 0) \in S_2$ is a projection.

3-2-25 Let $V = L_2[-\pi, \pi]$. Show that

$$P(x) = \int_{-\pi}^{\pi} K(t, x)x(s)\,dx \qquad K(t, x) = \frac{1}{2\pi}\sum_{n=-10}^{n=10} e^{in(t-s)}$$

represents a projection on V. Here $i = \sqrt{-1}$.

3-2-26 Let P be a projection operator on a linear space V. (*a*) Show that $I - P$, where I is the identity operator, is also a projection. (*b*) What is the kernel of $I - P$? (*c*) Show that $P(I - P) = 0$.

3-2-27 Let $V = \mathbb{R}^3$, and consider the transformation $T(\mathbf{x}) = (x_1, x_2, 0)$. Show that (*a*) T is a projection (of $\mathbb{R}^3$ to $\mathbb{R}^2$), (*b*) $\mathscr{R}(T) = \mathbb{R}^2$, (*c*) $\|\mathbf{x}_0 - T\mathbf{x}_0\| < \|\mathbf{x}_0 - \mathbf{x}\|$ for all $\mathbf{x} \neq T\mathbf{x}_0$ in $\mathbb{R}^2$, where $\|\cdot\|$ denotes the euclidean norm. Interpret the result geometrically.

3-2-28 Let V be a linear vector space and S_1 and S_2 be two disjoint linear subspaces of V such that $V = S_1 + S_2$. Show that there exists a projection P such that $\mathscr{R}(P) = S_1$ and $\mathscr{N}(P) = S_2$. Equivalently, $V = \mathscr{R}(P) + \mathscr{N}(P)$; $\mathscr{R}(P) \cap \mathscr{N}(P) = \{0\}$.

3-4 Theory of Normed Spaces

3-4-1 Norm and Normed Spaces

Many of the vector spaces we deal with in this study are function spaces. The use of vector spaces in engineering analysis lies in the study of the existence and uniqueness of solutions to boundary-value problems and their approximations. This study requires us to equip vector spaces with a measure of the "length" of a vector, the "distance" between two vectors, and the "inner product" of two vectors. These concepts are completely analogous to the concepts of length, distance, and scalar products of geometric vectors. Unlike geometric vectors, vectors from function spaces allow many alternatives for the measure of length, distance, and inner products. For example, the length of an element from vector space $C[0, 1]$ can be defined by the *net* area under the function or by the maximum of all possible values the function could take in the interval $[0, 1]$. However, all measures of length should obey certain basic rules, called *norm axioms*. We will study these axioms and other properties of normed spaces. Inner product spaces will be studied in Sec. 3-5.

Norm of a vector Let V be a linear vector space over the real-number field $\mathbb{R}$. A *norm* on vector space V is a function that transforms every element $u \in V$ into a real number, $\|u\|$, such that $\|u\|$ satisfies the following conditions:

$$\begin{aligned} &\text{(i)}\ (a)\ \|u\| \geq 0, \text{ and } (b)\ \|u\| = 0 \text{ if and only if } u = 0 \text{ (positive);} \\ &\text{(ii)}\ \|\alpha u\| = |\alpha|\,\|u\|,\ \alpha \in \mathbb{R} \text{ (homogeneous);} \\ &\text{(iii)}\ \|u + v\| \leq \|u\| + \|v\|,\ u, v \in V \text{ (triangle inequality).} \end{aligned} \tag{3-58}$$

If $\|u\|$ satisfies (i*a*), (ii), and (iii) only, we call it a *seminorm* of the vector u, and denote it by $|u|$. Associated with each norm, we can define a metric (or distance)

$$d(u, v) \equiv \|u - v\| \tag{3-59}$$

The metric $d(u, v)$ defined above is called *natural metric induced by the norm.* Note that norm $\|\cdot\|$ in a vector space V is a special functional, namely a mapping from the product space $V \times V$ into the space of real numbers:

$$\|\cdot\| : V \times V \rightarrow \mathbb{R}$$

Clearly, the axioms of a norm are satisfied by the length (or magnitude) of a geometric vector, $\mathbf{x}$ in $\mathbb{R}^n$:

$$\|\mathbf{x}\| \equiv |\mathbf{x}| = \sqrt{\mathbf{x}\cdot\mathbf{x}} = \sqrt{\sum_{i=1}^{n} x_i^2} \tag{3-60}$$

In this case the length of a vector, $\mathbf{x}$, is given by $d(\mathbf{x}, 0) = \|\mathbf{x}\| = |\mathbf{x}|$.

In order to check whether the real number $\|u\|$ associated with $u \in V$ qualifies as a norm (or a pseudonorm), one should verify the conditions in (3-58). In these conditions, it is often difficult to establish that the triangle inequality holds. In most cases use is made of certain standard inequalities associated with sums, and integrals of real-valued functions. These inequalities are known as the Hölder and Minkowski inequalities. We list them here for our reference [see Reddy and Rasmussen (1982) and Naylor and Sell (1971)].

Hölder inequalities $(1/p + 1/q = 1,\ 1 < p < \infty)$

(*a*) *Finite sums:*

$$\sum_{i=1}^{n} |x_i y_i| \leq \left(\sum_{i=1}^{n} |x_i|^p\right)^{1/p} \left(\sum_{i=1}^{n} |y_i|^q\right)^{1/q} \tag{3-61}$$

(*b*) *Infinite sums:* For $\sum_{i=1}^{\infty} |x_i|^p < \infty$ and $\sum_{i=1}^{\infty} |y_i|^q < \infty$, we have

$$\sum_{i=1}^{\infty} |x_i y_i| \leq \left(\sum_{i=1}^{\infty} |x_i|^p\right)^{1/p} \left(\sum_{i=1}^{\infty} |y_i|^q\right)^{1/q} \tag{3-62}$$

(*c*) *Integrals:* For $\int_\Omega |x|^p\, dt < \infty$ and $\int_\Omega |y|^q\, dt < \infty$, we have

$$\int_\Omega |xy|\, dt \leq \left(\int_\Omega |x|^p\, dt\right)^{1/p} \left(\int_\Omega |y|^q\, dt\right)^{1/q} \tag{3-63}$$

For $p = q = 2$, the above inequalities are known commonly as the *Schwarz inequalities.*

Minkowski inequalities $(1 \le p \le \infty)$
(*a*) *Finite sums:*

$$\left(\sum_{i=1}^{n} |x_i \pm y_i|^p\right)^{1/p} \le \left(\sum_{i=1}^{n} |x_i|^p\right)^{1/p} + \left(\sum_{i=1}^{n} |y_i|^p\right)^{1/p} \tag{3-64}$$

(*b*) *Infinite sums:* For $\sum_{i=1}^{\infty} |x_i|^p < \infty$ and $\sum_{i=1}^{\infty} |y_i|^p < \infty$, we have

$$\left(\sum_{i=1}^{\infty} |x_i \pm y_i|^p\right)^{1/p} \le \left(\sum_{i=1}^{\infty} |x_i|^p\right)^{1/p} + \left(\sum_{i=1}^{\infty} |y_i|^p\right)^{1/p} \tag{3-65}$$

(*c*) *Integrals:* For $\int_\Omega |x|^p\,dt < \infty$ and $\int_\Omega |y|^p\,dt < \infty$, we have

$$\left(\int_\Omega |x \pm y|^p\,dt\right)^{1/p} \le \left(\int_\Omega |x|^p\,dt\right)^{1/p} + \left(\int_\Omega |y|^p\,dt\right)^{1/p} \tag{3-66}$$

The Minkowski inequalities follow from the Hölder inequalities.

Example 3-20 Let $V = C[0, t_0]$ be the set of all real-valued, continuous functions of t in the closed interval $[0, t_0]$. Define the norm, called the *sup-norm,*

$$\|x\|_\infty = \sup\,\{|x(t)| : \quad 0 \le t \le t_0\} \tag{3-67}$$

Conditions (i) and (ii) of a norm are obviously satisfied by $\|x\|_\infty$. To prove the triangle inequality, consider

$$\begin{aligned}
\|x - y\|_\infty &= \sup_{0 \le t \le t_0} |x(t) - z(t) + z(t) - y(t)| \\
&\le \sup_{0 \le t \le t_0} \{|x(t) - z(t)| + |z(t) - y(t)|\} \\
&\le \sup_{0 \le t \le t_0} |x(t) - z(t)| + \sup_{0 \le t \le t_0} |z(t) - y(t)| \\
&= \|x - z\|_\infty + \|z - y\|_\infty
\end{aligned}$$

Thus $\|\cdot\|_\infty$ satisfies the triangle inequality, and hence $\|\cdot\|_\infty$ is a norm on $C[0, t_0]$. The associated metric is given by

$$d_\infty(x, y) = \sup\,\{|x(t) - y(t)| : \quad 0 \le t \le t_0\} \tag{3-68}$$

Another norm can be defined on $C[0, t_0]$, called the L_2-*norm,*

$$\|x\|_0 = \left[\int_0^{t_0} |x(t)|^2\,dt\right]^{1/2} \tag{3-69}$$

By means of the Minkowski inequality for integrals (3-66), it can be readily verified that the triangle inequality is satisfied by $\|\cdot\|_0$.

Since $C[0, t_0]$ is a space of continuous functions $x(t)$, $t \in [0, t_0]$, the sup-norm $\|\cdot\|$ gives the largest point-wise value of the function over the interval, while the L_2-norm gives the net area under the function. The sup-norm and the L_2-norm are shown in Fig. 3-6.

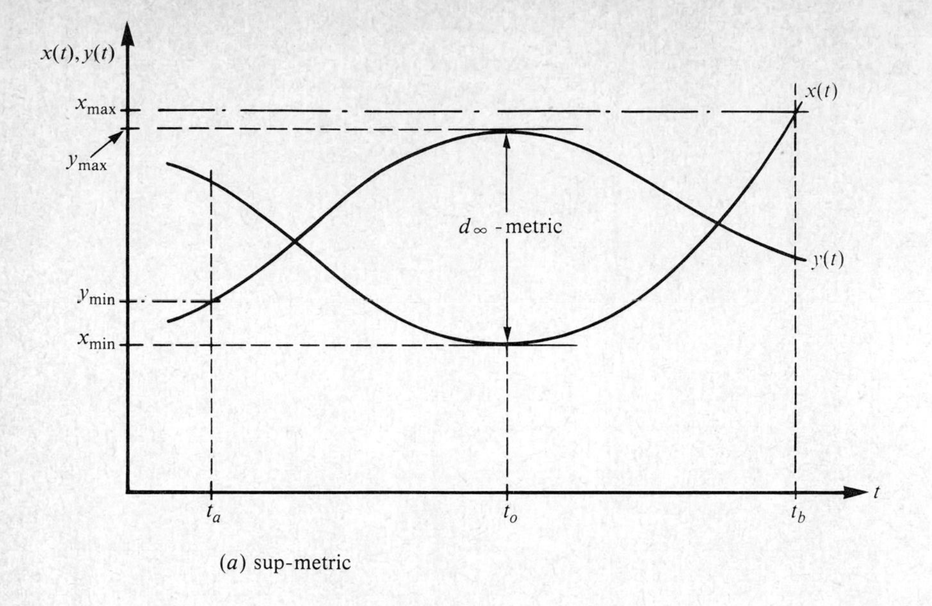

(a) sup-metric

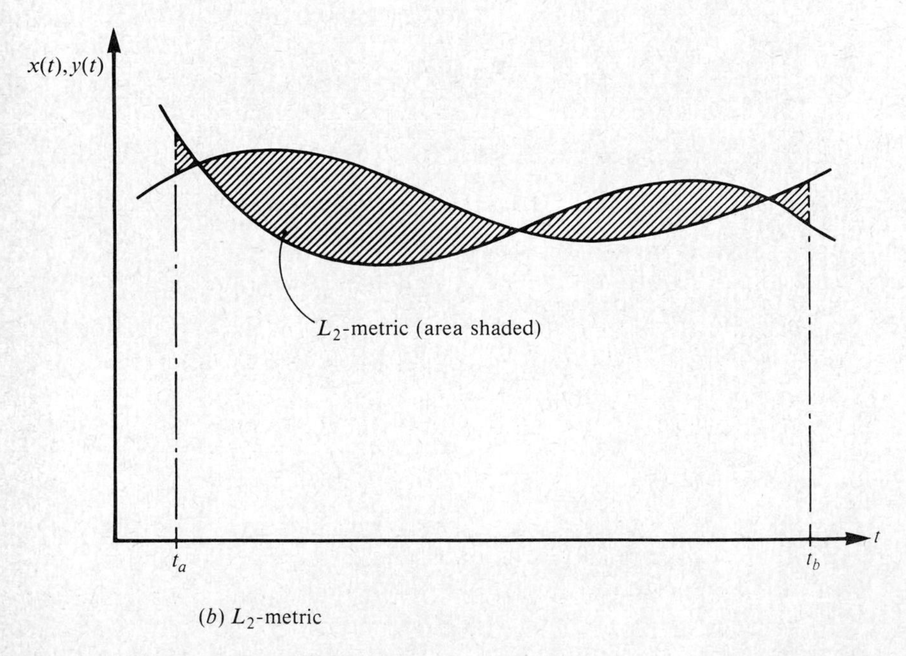

(b) L_2-metric

Figure 3-6 Various measures of the difference of two functions.

Example 3-21 The L_2-norm of a function $u(x, y) = \sin \pi x \sin \pi y$ from $C(\Omega)$, where Ω is a unit square

$$\Omega = \{\mathbf{x} = (x, y), \quad 0 < x < 1, \ 0 < y < 1\}$$

is given by

$$\begin{aligned}\|u\|_0 &= \left[\int_0^1 \int_0^1 \sin^2 \pi x \sin^2 \pi y \, dx \, dy\right]^{1/2} \\ &= \left[\int_0^1 \sin^2 \pi x \, dx \int_0^1 \sin^2 \pi y \, dy\right]^{1/2} \\ &= \tfrac{1}{2}\end{aligned}$$

On the other hand, the sup-norm of the function $u(x, y) = \sin \pi x \sin \pi y$ is given by

$$\begin{aligned}\|u\|_\infty &= \sup_{\substack{0 \le x \le 1 \\ 0 \le y \le 1}} \{|\sin \pi x \sin \pi y|\} = u(1/2, 1/2) \\ &= 1.\end{aligned}$$

The distance between two functions $u = \sin \pi x \sin \pi y$ and $v = 1/2$ is given by

$$\begin{aligned}\|u - v\|_0 &= \left[\int_0^1 \int_0^1 (\sin \pi x \sin \pi y - \tfrac{1}{2})^2 \, dx \, dy\right]^{1/2} \\ &= \left[\frac{1}{4} + \frac{1}{4} - \frac{4}{\pi^2}\right]^{1/2} = (0.0947)^{1/2} \\ &= 0.3077\end{aligned}$$

$$\|u - v\|_\infty = \sup_{\substack{0 \le x \le 1 \\ 0 \le y \le 1}} \{|\sin \pi x \sin \pi y - \tfrac{1}{2}|\} = \tfrac{1}{2}$$

The value $\|u - v\|_\infty = 1/2$ occurs at the points $(x, y) = (0, y), (x, 0), (1/2, 1/2)$.

Normed linear spaces A linear vector space on which a norm can be defined is called a *normed linear space*. A linear subspace S of a normed linear space V is a linear subspace equipped with the norm of V. Examples of normed spaces are given below.

Example 3-22 (*Lebesgue norm*, $\|\cdot\|_p$) Consider the space $L_p[0, T]$, $1 \le p < \infty$, consisting of all scalar-valued measurable functions [for an account of the Lebesgue measure theory, see Kolmogorov and Fomin (1957)] $u(t)$ defined over $0 \le t \le T$ such that

$$\|u\|_p = \left[\int_0^T |u(t)|^p \, dt\right]^{1/p} < \infty$$

Using the Minkowski inequality for integrals one can show that $\|u\|_p$ satisfies the triangle inequality. Other axioms of the norm can be easily verified. Therefore, $\|u\|_p$ is a norm on $L_p[0, T]$ space.

Example 3-23 (*Sobolev norm*) Let $C^m(\Omega)$, where Ω is an open bounded set in $\mathbb{R}^n$, denote the set of all real-valued functions with continuous derivatives up to and including order m defined on Ω, and let $C^\infty(\Omega)$ denote the set of infinitely differentiable continuous functions. We define on $C^m(\Omega)$ the norm

$$\|u\|_{m,p} = \left[\int_\Omega \sum_{|\alpha| \le m} |D^\alpha u|^p \, dx\right]^{1/p} \tag{3-70}$$

for $1 \le p < \infty$ and for all $u \in C^m(\Omega)$ for which $\|u\|_{m,p}$ is finite. Here $\boldsymbol{\alpha}$ denotes a n-tuple of nonnegative integers:

$$\begin{aligned} \boldsymbol{\alpha} &= (\alpha_1, \alpha_2, \ldots, \alpha_n) \qquad |\boldsymbol{\alpha}| = \sum_{i=1}^{n} \alpha_i \qquad \alpha_i \ge 0 \\ D^\alpha &= (\partial^{|\alpha|}/\partial x_1^{\alpha_1}\, \partial x_2^{\alpha_2} \cdots \partial x_n^{\alpha_n}) \end{aligned} \tag{3-71}$$

The triangle inequality can be verified by means of the Minkowski inequality for finite sums. The norm $\|u\|_{m,p}$ is called the *Sobolev norm.* Note also that for $m = 0$, we have $\|u\|_p$, the Lebesgue norm, and for $m = 0$ and $p = 2$ we get the L_2-norm. For $m = 1$, $n = 2$, and any $1 \le p < \infty$, we have $[\boldsymbol{\alpha} = (\alpha_1, \alpha_2), \alpha_1, \alpha_2 = 0, 1]$

$$\|u\|_{1,p} = \left\{\int_{\Omega \subset \mathbb{R}^2} \left[|u|^p + \left|\frac{\partial u}{\partial x}\right|^p + \left|\frac{\partial u}{\partial y}\right|^p\right] dx\, dy\right\}^{1/p} \tag{3-72}$$

The $\|u\|_{1,2}$-norm of $u = \sin \pi x \sin \pi y$ on Ω (see Example 3-21) is given by

$$\begin{aligned} \|u\|_{1,2} &\equiv \left\{\int_0^1 \int_0^1 \left[u^2 + \left(\frac{\partial u}{\partial x}\right)^2 + \left(\frac{\partial u}{\partial y}\right)^2\right] dx\, dy\right\}^{1/2} \\ &= \left\{\int_0^1 \int_0^1 [\sin^2 \pi x \sin^2 \pi y + \pi^2(\cos^2 \pi x \sin^2 \pi y \right. \\ &\qquad \left. + \sin^2 \pi x \cos^2 \pi y)]\, dx\, dy\right\}^{1/2} \\ &= [\tfrac{1}{2}\cdot\tfrac{1}{2} + \pi^2(\tfrac{1}{2}\cdot\tfrac{1}{2} + \tfrac{1}{2}\cdot\tfrac{1}{2})]^{1/2} \\ &= \tfrac{1}{2}(1 + 2\pi^2)^{1/2} = 2.277 \end{aligned}$$

Two norms $\|\cdot\|_1$ and $\|\cdot\|_2$ on a normed linear space V are said to be *equivalent* if there exist positive numbers c_1 and c_2, independent of $u \in V$, such that the following double inequality holds:

$$c_1\|u\|_1 \le \|u\|_2 \le c_2\|u\|_1 \tag{3-73}$$

Norms on product spaces can be defined in terms of the norms of the spaces involved in the product. Let $(U, \|\cdot\|_U)$ and $(V, \|\cdot\|_V)$ be two normed spaces. The product space $W \equiv U \times V$ can be endowed with any one of the following norms:

$$\text{(i)} \qquad \|(u, v)\|_W = \|u\|_U + \|v\|_V$$

$$\text{(ii)} \qquad \|(u, v)\|_W = [\|u\|_U^2 + \|v\|_V^2]^{1/2} \tag{3-74}$$

$$\text{(iii)} \qquad \|(u, v)\|_W = \max\{\|u\|_U, \|v\|_V\}$$

The addition of vectors and scalar multiplication of vectors in W are defined by

$$(u_1, v_1) + (u_2, v_2) = (u_1 + u_2, v_1 + v_2)$$
$$\alpha(u, v) = (\alpha u, \alpha v) \tag{3-75}$$

for all $u_1, u_2, u \in U$ and $v_1, v_2, v \in V$.

3-4-2 Continuous Linear Transformations

Our objective in studying the normed vector spaces is to deal with the questions of existence, accuracy and convergence of approximate solutions to various boundary-value problems. In the approximate solution of boundary-value problems, one has a procedure (e.g., the Ritz–Galerkin or finite-element methods) for generating a sequence of approximations, which we hope come closer and closer to the true solution of the problem. To show that the original equation and its approximate analogues have solutions, and the sequence of approximations is convergent, we should define continuity and nature of convergence in the normed vector space in which the solutions lie. The notion of continuity and convergence in normed spaces is essentially identical to that in ordinary analysis. Therefore, we first recall the definitions of continuity and convergence of real numbers, and then extend them to normed linear spaces.

Continuity The notion of continuity in normed spaces is essentially the same as that used in real analysis. Recall that a function $f(x)$ that maps set X into set Y is said to be continuous at a point $x_0 \in X$ if for every $\varepsilon > 0$ there exists a $\delta > 0$ such that

$$|f(x_0) - f(x)| < \varepsilon \qquad \text{whenever } |x - x_0| < \delta \tag{3-76}$$

The function f is said to be continuous if it is continuous at each point in the domain of definition.

Now if T is a mapping of elements from a normed vector space U into another normed space V, the mapping T is said to be *continuous* if there exists a real number $M > 0$ such that

$$\|Tu_1 - Tu_2\|_V \leq M\|u_1 - u_2\|_U \qquad \text{for all } u_1, u_2 \in U \tag{3-77}$$

where $\|\cdot\|_V$ and $\|\cdot\|_U$ denote the norms in spaces V and U, respectively. An operator $T\colon U \to V$ is said to be *bounded* if there exists a real number $M > 0$ such that

$$\|Tu\|_V \leq M\|u\|_U \qquad \text{for all } u \in U \tag{3-78}$$

Note that continuity of an operator depends on the norm in both of the spaces involved.

For linear transformations, the continuity and boundedness mean the same. If T is linear and bounded, we have

$$\|Tu_1 - Tu_2\|_V = \|T(u_1 - u_2)\|_V \leq M\|u_1 - u_2\|_U \qquad \text{for all } u_1, u_2 \in U$$

Therefore, T is continuous in U.

The *norm*, $\|T\|$, of a bounded linear transformation, $T: U \to V$, where U and V are normed spaces, is defined by

$$\|T\| = \inf \{M: \|Tu\|_V \leq M\|u\|_U \qquad \text{for all } u \in U\} \tag{3-79}$$

Other equivalent norms are

$$\|T\| = \sup \{\|Tu\|_V : \|u\|_U \leq 1\} \tag{3-80}$$

$$\|T\| = \sup \left\{\frac{\|Tu\|_V}{\|u\|_U} : \|u\|_U \neq 0\right\} \tag{3-81}$$

For example, in the case of an electrical input and output system, the number $\|Tu\|/\|u\|$ can be viewed as the "amplification" of T for the input u, and $\|T\|$ as the maximum amplification of the system.

It can be shown (see Prob. 3-3-17) that the set of linear bounded operators from a vector space into another forms a vector space. Further, if the domain and range spaces are normed spaces, the space of bounded linear operators is also a normed space. Recall that in the case when the range space is the set of real numbers the space of linear bounded operators is called the conjugate (or dual) space.

Since functionals are a special case of transformations, all of the results presented here for transformations are also valid for functionals. We now consider two examples of linear transformations.

Example 3-24 Consider the integration operator defined by

$$y(t) = \int_0^t x(\tau)\, d\tau \tag{3-82}$$

from $C[0, 1]$ into itself. The operator is defined on the whole space. Continuity follows from the Schwarz inequality.

Example 3-25 Now consider the differential operator $D = d/dt$,

$$y(t) = \frac{dx(t)}{dt} \tag{3-83}$$

The operator is not defined on the whole of $C[0, 1]$ and when $Dx(t)$ exists, then $y(t)$ does not necessarily belong to $C[0, 1]$. However, if we take the domain of D to be the linear manifold $\mathscr{D}$ of functions with continuous first derivatives, then the range is also contained in $C[0, 1]$. The operator D is not continuous from $\mathscr{D}$ into $C[0, 1]$. For example, let $x_n(t) = \sin n\pi t$; then $\|x_n\|^2 = \frac{1}{2}$, where $\|\cdot\|$ is the norm induced by the L_2-norm in (3-69). We have $\|Dx_n\|^2 = (n\pi)^2/2$, and $\|Dx_n\|/\|x_n\|$ is unbounded as $n \to \infty$.

Most engineering analyses ultimately lead to the solution of equations on finite-dimensional spaces. Therefore, it is of interest to study continuity of linear operators on finite-dimensional spaces. Toward this we give a basic result that states that each individual component of a vector is less than or equal to a constant (finite in value) multiple of the norm of the vector. The proof can be found in Naylor and Sell (1971).

Theorem 3-3 Let V be a finite-dimensional normed linear space and let $\{\phi_k\}_{k=1}^n$ be a basis for V. Then each coefficient α_i, $i = 1, 2, \ldots, n$, in the expansion

$$u = \alpha_1 \phi_1 + \alpha_2 \phi_2 + \cdots + \alpha_n \phi_n = \sum_{i=1}^{n} \alpha_i \phi_i \tag{3-84}$$

is bounded in the sense that there exists a constant $M > 0$ such that

$$|\alpha_i| \leq M\|u\| \tag{3-85}$$

for $i = 1, 2, \ldots, n$ and all $u \in V$.

The result in Eq. (3-85) can be easily understood in the case of $V = \mathbb{R}^n$, where $\phi_i = \hat{\mathbf{e}}_i$. In this case Eq. (3-85) holds with $M = 1$ because the length of a geometric vector is always greater than or equal to the magnitude of any of its components.

One of the major simplifications in going from infinite-dimensional spaces to finite-dimensional spaces is that *all* linear transformations are continuous. This result is very useful in the study of the existence and uniqueness of solutions to equations in finite-dimensional spaces.

Theorem 3-4 Let $T: U \to V$ be a linear transformation where U and V are normed linear spaces. If U is finite dimensional, then T is continuous.

PROOF Let $\{\phi_1, \phi_2, \ldots, \phi_n\}$ be a basis for U. For any $u \in U$, we have

$$\begin{aligned}
u &= \alpha_1 \phi_1 + \alpha_2 \phi_2 + \cdots + \alpha_n \phi_n \\
Tu &= \alpha_1 T(\phi_1) + \alpha_1 T(\phi_2) + \cdots + \alpha_n T(\phi_n) \\
\|Tu\| &\leq |\alpha_1| \|T(\phi_1)\| + |\alpha_2| \|T(\phi_2)\| + \cdots + |\alpha_n| \|T(\phi_n)\| \\
&\leq C_0(|\alpha_1| + |\alpha_2| + \cdots + |\alpha_n|)
\end{aligned}$$

where $C_0 = \max \{\|T\phi_i\|, i = 1, 2, \ldots, n\}$. By Theorem 3-3, there exists a constant $\tilde{M}$ such that

$$|\alpha_1| + |\alpha_2| + \cdots + |\alpha_n| \leq \tilde{M}\|u\|$$

which gives

$$\|Tu\| \leq C_0 \tilde{M}\|u\| = \hat{M}\|u\|$$

Hence T is bounded.

Note that in Theorem 3-4, the range space is not required to be finite-dimensional. Note also that the finite-dimensionality of U is used in the proof via Theorem 3-3. The following example verifies the assertion of Theorem 3-4.

Example 3-26 1. Consider the linear transformation $T\colon \mathbb{R}^3 \to \mathbb{R}^3$ defined by $T(\mathbf{x}) = (x_1 + x_2, x_2 - x_3, x_3)$. Using the euclidean norm in Eq. (3-60), we write

$$\begin{aligned}\|T(\mathbf{x})\|^2 &= (x_1 + x_2)^2 + (x_2 - x_3)^2 + x_3^2 \\ &= x_1^2 + 2x_2^2 + 2x_3^2 + 2x_1x_2 - 2x_2x_3 \\ &\leq 2x_1^2 + 4x_2^2 + 3x_3^2 \\ &\leq 4(x_1^2 + x_2^2 + x_3^2) = 4\|\mathbf{x}\|^2\end{aligned}$$

Thus T is bounded (with $M = 2$).

2. Consider the linear transformation $T\colon \mathbb{R}^2 \to \mathbb{R}^2$ represented by the matrix,

$$[T] = \begin{bmatrix} t_{11} & t_{12} \\ t_{12} & t_{22} \end{bmatrix} \qquad t_{11} > 0,\ t_{22} > 0 \tag{3-86}$$

we have

$$[T]\{x\} = \begin{Bmatrix} t_{11}x_1 + t_{12}x_2 \\ t_{12}x_1 + t_{22}x_2 \end{Bmatrix}$$

$$\|T\mathbf{x}\| = [(t_{11}^2 + t_{12}^2)x_1^2 + (t_{12}^2 + t_{22}^2)x_2^2 + 2t_{12}(t_{11} + t_{22})x_1x_2]^{1/2}$$

$$\begin{aligned}\|T\| &= \sup \frac{\|T\mathbf{x}\|}{\|\mathbf{x}\|} \\ &= \sup \frac{[(t_{11}^2 + t_{12}^2)x_1^2 + (t_{12}^2 + t_{22}^2)x_2^2 + 2t_{12}(t_{11} + t_{22})x_1x_2]^{1/2}}{[x_1^2 + x_2^2]^{1/2}}\end{aligned}$$

To simplify this expression, we use the transformation $x_1 = r\cos\theta$ and $x_2 = r\sin\theta$.

$$\begin{aligned}\|T\| &= \sup\, [\tfrac{1}{2}(t_{11}^2 + 2t_{12}^2 + t_{22}^2) + \tfrac{1}{2}(t_{11}^2 - t_{22}^2)\cos 2\theta \\ &\qquad + t_{12}(t_{11} + t_{22})\sin 2\theta]^{1/2} \\ &\leq [\tfrac{1}{2}(t_{11}^2 + 2t_{12}^2 + t_{22}^2) + \tfrac{1}{2}(t_{11} + t_{22})\sqrt{(t_{11} - t_{22})^2 + 4t_{12}^2}]^{1/2} \\ &= \tfrac{1}{2}[t_{11} + t_{22} + \sqrt{(t_{11} - t_{22})^2 + 4t_{12}^2}]\end{aligned} \tag{3-87}$$

It can be verified that $\|T\|$ is precisely the largest eigenvalue of the symmetric matrix $[T]$. In general, every linear operator T given by matrix (t_{ij}) on $\mathbb{R}^n$ is bounded. Its norm depends on the norm used in $\mathbb{R}^n$:

(i)
$$\|\mathbf{x}\|_\infty = \max_{1 \le i \le n} |x_i|$$

$$\|T\| = \max_{1 \le i \le n} \sum_{k=1}^{n} |t_{ik}| \tag{3-88a}$$

(ii)
$$\|\mathbf{x}\| = \sum_{i=1}^{n} |x_i|$$

$$\|T\| = \max_{1 \le k \le n} \sum_{i=1}^{n} |t_{ik}| \tag{3-88b}$$

(iii)
$$\|\mathbf{x}\|_0 = \left[\sum_{i=1}^{n} |x_i|^2\right]^{1/2}$$

$$\|T\| = \sqrt{\lambda_1} \tag{3-88c}$$

where λ_1 is the largest eigenvalue of the matrix $[T][T]^T$ ($[T]^T$ denotes the transpose of $[T]$). If $[T]$ is symmetric, $\|T\| = \mu_1$, where μ_1 is the largest eigenvalue of the matrix $[T]$.

The concept of inverse transformation is the same as that of a function. A transformation T (linear or not) has an inverse defined on its range if and only if T is one-to-one. From the previous discussion we know that if T is linear, then it is one-to-one if and only if the null space of the transformation is trivial.

A linear operator $T: U \to V$ is said to be *bounded below*, if there exists a constant $C > 0$ such that

$$\|Tu\|_V \ge C\|u\|_U \qquad u \in U \tag{3-89}$$

The following theorem establishes that a bounded below operator has a continuous inverse defined on its range. Note that T is not required to be continuous.

Theorem 3-5: (Bounded inverse theorem) Let $T: U \to V$ be a linear transformation, where U and V are normed linear spaces. If T is bounded below, then T has a continuous inverse T^{-1} defined on its range $\mathscr{R}(T)$. In other words, the inverse operator T^{-1} is bounded:

$$\|T^{-1}v\|_U \le \frac{1}{C}\|v\|_V$$

PROOF To show that T^{-1} exists, we must show that T is one-to-one, or equivalently, that $\mathscr{N}(T) = \{0\}$. This follows immediately from $\|Tu\|_V \ge C\|u\|_U$, which implies that if $u \ne 0$, then $Tu \ne 0$. To show that T^{-1}:

$\mathscr{R}(T) \to U$ is continuous (or bounded), let $v \in \mathscr{R}(T)$; then there is an $u \in U$ such that $Tu = v$ and $u = T^{-1}v$. We have

$$\|T^{-1}v\|_U = \|u\|_U \leq \frac{1}{C}\|Tu\|_V = \frac{1}{C}\|v\|_V$$

for all v in $\mathscr{R}(T)$. Hence T^{-1} is bounded.

The next example shows that the boundedness of an operator depends on the norm used in thc domain space.

Example 3-27 1. The differential operator, $D = d/dx$ was shown to be unbounded from $C[0, 1]$ into itself. It is also unbounded from $C_0[0, 1]$ (the space of functions from $C[0, 1]$ which vanish at $x = 0$ and $x = 1$) into itself. However, the operator is bounded below on $C_0[0, 1]$ with the sup-norm. For $u \in C_0[0, 1]$, we have (since $u(y) = 0$ at $y = 0$)

$$u(x) = \int_0^x \frac{du}{dy}\, dy \leq \sup_{x \in (0,1)} \left|\frac{du}{dx}\right| = \|Du\|_\infty$$

$$\sup_{x \in (0,1)} |u(x)| \leq \|Du\|_\infty \qquad \text{or} \qquad \|u\|_\infty \leq \|Du\|_\infty$$

The differential operator $D = d/dx$ is also bounded below with respect to the L_2-norm. We have

$$u(x) = \int_0^x \frac{du}{dy}\, dy \leq \left[\int_0^x \left|\frac{du}{dy}\right|^2 dy\right]^{1/2}$$

$$\int_0^1 |u(x)|^2\, dx \leq \int_0^1 \left|\frac{du}{dx}\right|^2 dx$$

or

$$\|u\|_0 \leq \|Du\|_0$$

where $\|\cdot\|_0$ denotes the L_2-norm.

2. Let

$$Tu = \frac{d}{dx}\left[a(x)\frac{du}{dx}\right] + b(x)u \qquad 0 < x < 1$$

be an unbounded (Sturm–Liouville) operator defined on the linear set of twice continuously differentiable functions, such that $u(0) = u(1) = 0$. The inverse of T on $C[0, 1]$ is given by

$$T^{-1}(f) = \int_0^1 G(x, y) f(y)\, dy$$

where $G(x, y)$ is Green's function. It can be shown that T^{-1} is a bounded linear operator defined on the entire space $C[0, 1]$.

3-4-3 Complete Normed Spaces: Banach Spaces

Completeness Another property of normed vector spaces that plays an important role in variational methods is completeness. All normed linear spaces can be divided into two kinds: those in which all convergent sequences have limit points in the space, and those in which all convergent sequences do not have limit points in the space. The first kind is called "complete," and the second kind is called "incomplete." To give a formal definition of a complete space, we need to introduce the concept of convergence and the Cauchy sequence.

Convergence Recall from a course on real analysis that a sequence of functions $\{f_n\}$ is said to *converge uniformly* to f_0 on a domain (of points) Ω if for each $\varepsilon > 0$ there is a number $N > 0$, depending on ε but not on x, such that

$$|f_n(x) - f_0(x)| < \varepsilon \qquad \text{for all } n > N \text{ and all } x \in \Omega \tag{3-90}$$

We write

$$\lim_{n\to\infty} f_n(x) = f_0 \qquad x \in \Omega \tag{3-91}$$

to indicate that the sequence $\{f_n\}$ converges uniformly to the limit f_0 in Ω.

The definition can be extended to functions in normed vector spaces. A sequence of vectors $\{u_n\}$ in a normed vector space $(V, \|\cdot\|)$ is said to *converge* (*strongly*) to u_0 if for each $\varepsilon > 0$ there is a number $N > 0$, independent of u, such that

$$\|u_n - u_0\| < \varepsilon \qquad \text{for all } n > N \tag{3-92}$$

Obviously, the limit point value depends on the space as well as the norm used. Further, note that the definition of convergence requires us, in order to verify whether a given sequence is convergent, to insert the limit element u_0 into Eq. (3-92). It is convenient to have a criterion that does not require the limit point u_0 to check whether the sequence converges. Such a criterion will be discussed in connection with completeness of spaces.

Example 3-28 1. Consider the sequence $u_n(x) = x^n$ in $C[0, 1]$. The sequence converges to u_0, where

$$u_0(x) = \begin{cases} 0 & 0 \le x < 1 \\ 1 & x = 1 \end{cases}$$

We thus have a sequence of continuous functions converging to a discontinuous limit:

$$\sup_{0 \le x \le 1} |u_n(x) - u_0(x)| = 1$$

Hence, the sequence does not converge uniformly on [0, 1].

2. Consider the sequence $u_n(x) = n^2 x e^{-nx}$ in $C[0, 1]$. The sequence converges to 0 for $0 \le x \le 1$:

$$\sup_{0 \le x \le 1} |u_n(x) - 0| = 0$$

Thus the sequence converges uniformly to zero in the sup-norm. However,

$$\lim_{n \to \infty} \int_0^1 u_n(x)\, dx = 1$$

and therefore its limit in the L_1-norm is unity.

3. Let $V = C[0, 1]$ with the L_2-norm. Consider the sequence $\{u_n\}$ in $(V, \|\cdot\|_0)$,

$$u_n(x) = \begin{cases} 1 - nx & 0 \le x \le 1/n \\ 0 & 1/n < x \le 1 \end{cases}$$

We claim that the sequence converges and the limit is $u_0(x) = 0$. We have

$$\|u_n - u_0\| = \left[\int_0^1 |u_n(x) - u_0(x)|^2\, dx \right]^{1/2}$$

$$= \left[\int_0^{1/n} (1 - nx)^2\, dx \right]^{1/2}$$

$$= \left(\frac{1}{3n} \right)^{1/2}$$

Therefore

$$\lim_{n \to \infty} \|u_n - u_0\| = 0$$

and the sequence $\{u_n\}$ converges uniformly to the limit $u_0 = 0$.

A *Cauchy sequence* of elements of a normed space V is a sequence $\{u_n\}$ such that for any $\varepsilon > 0$ there is a number $N > 0$ such that

$$\|u_n - u_m\| < \varepsilon \qquad \text{for all } n, m > N \tag{3-93}$$

Note again that the definition of a Cauchy sequence involves consideration of elements from the sequence only, whereas the definition of convergence involves consideration of elements outside the sequence, namely, the limit of the sequence. Therefore, it is often easier to prove that a given sequence is a Cauchy sequence than to show that it is convergent. It is important to note that not all Cauchy sequences are convergent. However, all convergent sequences in a normed space are Cauchy.

Theorem 3-6 Every convergent sequence in a normed space is a Cauchy sequence.

PROOF If $\{u_n\}$ is a convergent sequence, with $\{u_n\} \to u_0$, then for any $\varepsilon > 0$ there is a number M such that

$$\|u_n - u_0\| < \frac{\varepsilon}{2}$$

for all $n > M$. Then by the triangle inequality

$$\|u_n - u_m\| \leq \|u_n - u_0\| + \|u_0 - u_m\| < \varepsilon$$

for all $n, m > M$.

A normed space is called *complete* if every Cauchy sequence is convergent. A complete normed space is called a *Banach space.* A linear subspace of a Banach space is itself a Banach space if and only if the subspace is complete.

The importance of complete spaces in the theory of approximation should be obvious: to solve a given equation, one can construct a sequence of approximations to the solution, and then show that it is a Cauchy sequence. Then, since the space is complete, we deduce that the sequence converges to a member of the space.

Every incomplete normed space can be completed by adding the limit points of all Cauchy sequences in the space. If V is an incomplete normed space, one constructs a new space $\bar{V}$ that is complete and contains V as a subspace. The space $\bar{V}$ is called the *completion* of V. Typically, $\bar{V}$ contains the set K of all Cauchy sequences in V and the limit points of all Cauchy sequences. We note that the set of real numbers $\mathbb{R}$ with the usual norm (i.e., absolute value) and the euclidean space $\mathbb{R}^n$ with the euclidean norm form complete normed spaces. However, the space of rational numbers with the absolute value norm is not a complete normed space.

Example 3-29 1. Let $V = C[0, 1]$ with the *sup-metric.* We shall show that $(C[0, 1], \|\cdot\|_\infty)$ is complete, hence a Banach space. Let $\{x_n(t)\}$ be an arbitrary Cauchy sequence in $(C[0, 1], \|\cdot\|_\infty)$. That is, given any $\varepsilon > 0$, there is an $N(\varepsilon)$ such that

$$|x_n(t) - x_m(t)| \leq \|x_n - x_m\|_\infty \leq \varepsilon \qquad \text{for all } t, \text{ and } m, n > N$$

Since t is arbitrary, the sequence $\{x_n(\cdot)\}$ converges pointwise to a function $x(\cdot)$. But $N = N(\varepsilon)$, being independent of t, implies that $\{x_n(\cdot)\}$ converges uniformly to $x(\cdot)$. Since $x(t)$ is the limit of a uniformly convergent sequence of continuous functions, it is also continuous on $[0, 1]$. Therefore $x(t) \in C[0, 1]$ and $\lim_{n\to\infty} \{x_n - x\} \to 0$.

2. Consider the space $C[-1, 1]$ again. Now we use a different norm than the one used above, namely the L_2-norm:

$$\|x - y\|_0 = \left[\int_{-1}^{1} [x(t) - y(t)]^2 \, dt\right]^{1/2}$$

We can show that $(C[-1, 1], \|\cdot\|_0)$ is a normed space that is *not* complete. We show this by a counter example (note that in any case, proving something holds requires a general proof whereas proving that something does not hold requires an example of contradiction). Consider the sequence of continuous functions from $C[-1, 1]$,

$$x_k(t) = \frac{1}{2} + \frac{1}{\pi} \arctan kt \qquad -1 \leq t \leq 1$$

This sequence is a Cauchy sequence, $\lim_{m,n\to\infty} \int_{-1}^{1} \{x_m(t) - x_n(t)\}^2 \, dt = 0$. However, the limit of this sequence is the *discontinuous* function

$$x(t) = \begin{cases} 1, \; 0 < t \leq 1 \\ 1/2, \qquad t = 0 \\ 0, \; -1 \leq t < 0 \end{cases}$$

3. (Space $L_2[0, 1]$) Let $V = C[0, 1]$ with the L_2-norm. Then the completion of V is $L_2[0, 1]$, which is the space of all square-integrable (in the Lebesgue sense) functions $u(x)$ defined on $[0, 1]$ with the property

$$\int_0^1 |u(x)|^2 \, dx < \infty$$

The space $C[0, 1]$ of continuous functions is not extensive enough to accommodate the L_2-metric. The space is completed (not described here) by adding the limits of all non-convergent Cauchy sequences to form $L_2[0, 1]$. The $L_2(\Omega)$ space contains discontinuous functions that the square-integrable over the domain Ω. For example, the function $u(x) = x^{-1/3}$ belongs to $L_2[0, 1]$, but $u(x) = x^{-2/3}$ does not (why?).

There are many nonzero functions $u(x)$ such that

$$\int_\Omega |u(x)|^2 \, dx = 0$$

An example of such function is provided by a function that vanishes at a finite number of points in the domain Ω. Then how do we distinguish between two such elements? If we cannot, then $L_2(\Omega)$ is not a normed space (because many nonzero elements have zero norm). To circumvent this problem, we understand that two functions u and v which satisfy the equation

$$\int_\Omega |u - v|^2 \, dx = 0$$

are the same (said to be equal *almost everywhere*; *a.e.*). By doing so we define $L_2(\Omega)$ space to be the space of square integrable functions, with the understanding that two functions are identical if they are *equal almost everywhere.*

4. [*Sobolev space*, $W^{m,p}(\Omega)$] This is a continuation of Example 3-23. The space $C^m(\Omega)$ is not complete with respect to the Sobolev norm $\|\cdot\|_{m,p}$. The

completion of $C^m(\Omega)$ with respect to the norm $\|\cdot\|_{m,p}$ is called the *Sobolev space of order* (m, p), denoted by $W^{m,p}(\Omega)$. Hence the Sobolev space is a Banach space. Of course, the Lebesgue space $L_p(\Omega)$ is a special case of $W^{m,p}$ for $m = 0$ and $L_2(\Omega)$ for $p = 2$. The comments made in part 3 of this example for $L_2(\Omega)$ also apply to $L_p(\Omega)$, the space of pth integrable functions.

It is interesting to know that all finite-dimensional normed spaces are Banach spaces. To stress this point, we state it as a theorem.

Theorem 3-7 Any finite-dimensional normed linear space is complete (hence a Banach space).

PROOF Let $\{\phi_i\}_{i=1}^n$ be any basis for V and let $\{u_k\}_{k=1}^m$ be any Cauchy sequence in V. We must show that the sequence $\{u_k\}$ is convergent in V. Since u_k is in V, we can express u_k as

$$u_k = \alpha_{k1}\phi_1 + \alpha_{k2}\phi_2 + \cdots + \alpha_{kn}\phi_n$$

for $k = 1, 2, \ldots, m$. From Theorem 3-3 we know that there exists an $M > 0$ such that

$$|\alpha_{ki} - \alpha_{li}| \leq M\|u_k - u_l\|$$

for $i = 1, 2, \ldots, n$. This shows that each sequence of real numbers $\{\alpha_{ki}\}$ is a Cauchy sequence. Since the space of real numbers is complete, the sequence $\{\alpha_{ki}\}$ is convergent. Let α_{0i} be the limit of $\{\alpha_{kj}\}$ (i.e., $\alpha_{0j} = \lim_{k\to\infty} \alpha_{kj}$), $j = 1, 2, \ldots, n$, and let $u_0 = \alpha_{01}\phi_1 + \alpha_{02}\phi_2 + \cdots + \alpha_{0n}\phi_n$. Then,

$$\begin{aligned}\|u_k - u_0\| &= \|(\alpha_{k1} - \alpha_{01})\phi_1 + (\alpha_{k2} - \alpha_{02})\phi_2 + \cdots + (\alpha_{kn} - \alpha_{0n})\phi_n\| \\ &\leq |\alpha_{k1} - \alpha_{01}|\,\|\phi_1\| + |\alpha_{k2} - \alpha_{02}|\,\|\phi_2\| + \cdots + |\alpha_{kn} - \alpha_{0n}|\,\|\phi_n\|,\end{aligned}$$

from which it follows that $u_k \to u_0$ as $k \to \infty$, that is, $\{u_k\}$ converges to u_0. This completes the proof of the theorem.

Recall that the space of linear bounded operators, $T: U \to V$, where U and V are normed spaces, is also a normed space. If the range space V is complete, then the space of bounded linear operators is also complete, and hence a Banach space. In particular, the space of bounded linear transformations on a normed space is complete (because the set of real numbers is complete). Note that the completeness of the domain space is not required for the space of bounded linear operators to be a Banach space.

3-4-4 Closure, Denseness, and Separability

Closure Recall that a subset S of a normed space is said to be *closed* if it contains all its limit points. Closed subsets are useful in the solution of equations, where we seek approximate solutions by constructing sequences of approximations, all of which belong to a set S of functions with certain properties. If S is a closed set and if the sequence is proved to converge, the limit also belongs to S, giving a convergent sequence of approximations in the solution set S. Every open set can be closed in the same sense as every incomplete space can be completed. For any set S in a normed space, the *closure* of S, denoted $\bar{S}$, is the union of S with the set of all limit points of S. Obviously $S \subset \bar{S}$, and $S = \bar{S}$ if S is closed.

The linear space $C[-1, 1]$ of all functions continuous in the interval $[-1, 1]$ is not a closed space in $L_2[-1, 1]$.

Denseness When the limit points are added to an open set S (to form $\bar{S}$), the new points, being the limit points, are all close to points of S. Given an arbitrary vector $v \in \bar{S}$ we can find an element in S that is as close as we wish to v. Then S is said to be dense in $\bar{S}$. A formal definition of a dense set is given below.

Let S_1 and S_2 be two subsets of a normed space V such that $S_1 \subset S_2$. Then S_1 is dense in S_2 if for each $v \in S_2$ and each $\varepsilon > 0$ there is a $u \in S_1$ with

$$\|v - u\|_V < \varepsilon \tag{3-94}$$

An example of a dense set is provided by the set Q of rational numbers which is dense in $\mathbb{R}$. The space $C(\Omega)$ is a dense subspace of $L_2(\Omega)$. Also, the set of all polynomials is dense in $L_2(\Omega)$.

Separability A set S is called *countable* if it is possible to arrange its elements into a sequence, i.e., if it is possible to establish a one-to-one correspondence between the elements of S and positive integers. For example, the set of rational numbers is countable while the set of all real numbers is not. All sets with a finite number of elements are countable.

A normed space V is called *separable* if there exists a countable set which is dense in V. The set of real numbers with the usual norm forms a separable normed space since the set of rational numbers is dense and countable. Note that a separable normed space can contain more than one set which is dense and countable. The space $\mathbb{R}^n$ is separable. The space $L_2(\Omega)$ is separable because the set of all polynomials with rational coefficients is dense and countable in $L_2(\Omega)$.

Two normed spaces $(V_1, \|\cdot\|_1)$ and $(V_2, \|\cdot\|_2)$ are said to be *isomorphic* to one another if there exists a one-to-one mapping T of V_1 onto V_2 such that

$$\|u_1 - u_2\|_1 = \|Tu_1 - Tu_2\|_2 \tag{3-95}$$

The mapping T is called an *isomorphism.*

If two normed spaces are isometric, they can be viewed as being essentially the same normed space in two different guises. The names given to points in the two spaces are different, but the names are unimportant as far as distance, continuity, and convergence are concerned.

Example 3-30 Consider once again the space $C[0, 1]$ with the L_2-norm. The space $(C[0, 1], \|\cdot\|_0)$ is not complete, as was shown in Example 3-29. Let S be the two-dimensional subspace of all functions $u(x) \in C[0, 1]$ of the form

$u(x) = c_1 \sin 2\pi x + c_2 \cos 2\pi x$, where c_1 and c_2 are real numbers. We show that the subspace S is closed. This is done by showing that S is isomorphically equivalent to the space $\mathbb{C}$ of complex numbers, which is a complete normed space with respect to the norm

$$\|z\| = \|x + iy\| \equiv \frac{1}{\sqrt{2}}(x^2 + y^2)^{1/2}$$

We have for $u \in S$,

$$\|u\|_2^2 = \int_0^1 (c_1^2 \sin^2 2\pi x + c_2^2 \cos^2 2\pi x + 2c_1 c_2 \sin 2\pi x \cdot \cos 2\pi x)\, dx$$
$$= \tfrac{1}{2}(c_1^2 + c_2^2)$$

By identifying the isometry $T: S \to \mathbb{C}$ to be $T(u(x)) = c_1 + ic_2$, we conclude that S and $\mathbb{C}$ are isometric and that S is complete since $\mathbb{C}$ is complete.

PROBLEMS 3-3

3-3-1 Consider the space $L_p[0, 1]$, $0 < p < 1$, of all functions $u(x)$ with

$$\|u\| = \int_0^1 |u(x)|^p\, dx < \infty$$

Does $\|u\|$ qualify as a norm on $L_p[0, 1]$?

3-3-2 Show that

$$\|u\| = \left[\int_a^b |u|^2 p(x)\, dx\right]^{1/2}$$

where $p(x)$ is a fixed continuous positive function on $[a, b]$, is a norm on $L_2[a, b]$. Show that this norm is equivalent to the usual $L_2[a, b]$-norm. Assume that $0 < c_1 \leq p(x) \leq c_2$ for $x \in [a, b]$.

3-3-3 A set S in $\mathbb{R}^n$ is *convex* if, given two points x and y in S, the line segment $z(\alpha) = (1 - \alpha)x + \alpha y (0 \leq \alpha \leq 1)$ is in S. A real-valued function f on a convex set S is said to be *convex* on S if for every pair of points x and y in S the inequality $f[(1 - \alpha)x + \alpha y] \leq (1 - \alpha)f(x) + \alpha f(y)$ holds. Show that the following functions are convex:

(*a*) $f(x) = x^n - nb \log nx$, $b \geq 0$, n is a positive integer, $0 < x < \infty$.

(*b*) $f(x) = [1 - |x|^2]^{-1/2}$, $|x| < 1$.

Compute the norms of the following functions in the interval indicated in Probs. 3-3-4 to 3-3-11:

3-3-4 $u(x) = \sin \pi x - x$, $L_2[0, 1]$-norm.

3-3-5 $u(x) = x^{1/3}$, $L_2[0, 1]$-norm.

3-3-6 $u(x) = \cos \pi x + 2x - 1$, $W_{2,1}[0, 1]$-norm.

3-3-7 $u(x) = \sqrt{1 + x^2} - [1 + \sqrt{2}(1 - x)]$, semi $W_{2,1}[0, 1]$-norm.

3-3-8 $u(x, y) = xy(1 - x)(1 - y)$, $L_2(\Omega)$-norm, where Ω is the unit square, $0 \leq (x, y) \leq 1$.

3-3-9 $u(x, y) = \cos \pi x/2 \cos \pi y/2$, $L_2(\Omega)$-norm, where Ω is the unit square, $0 \leq (x, y) \leq 1$.

3-3-10 Seminorm of the function $u(x) = \sin \pi x - x$ in the $W_{2,1}[0, 1]$ space.

3-3-11 Seminorm of the function $u(x, y) = \cos \pi x/2 \cos \pi y/2$ in the $W_{2,1}[0, 1]$ space.

3-3-12 Let $\Omega = (a, b)$ be an open bounded interval in $\mathbb{R}$ and let $C^1(\Omega)$ denote the collection of all real-valued functions with continuous first derivatives which are pth integrable on Ω. Define for $1 \le p < \infty$ and $u \in C^1(\Omega)$

$$|u|_{1,p} = \left\{ \int_\Omega \left| \frac{du}{dt} \right|^p dt \right\}^{1/p}$$

Show that $|\cdot|_{1,p}$ is a norm on the subspace $C_0^1(\Omega)$

$$C_0^1(\Omega) = \{u \in C^1(\Omega): \ u = 0 \text{ on the boundary of } \Omega\}$$

3-3-13 Show that every linear operator T given by a matrix (t_{ij}) on $\mathbb{R}^n$ is bounded, and that the norm of $\|T\|$, when the sup-norm is used, is given by

$$\|T\| = \max_{1 \le i \le n} \sum_{j=1}^{n} |t_{ij}|$$

3-3-14 *Integral operators* If a linear integral operator T with a continuous kernel $K(x, y)$ is considered as an operator from $C[0, 1]$ into $C[0, 1]$, show that it is bounded, and that its norm is given by

$$\|T\| = \max_{0 \le x, y \le 1} \int_0^1 \int_0^1 |K(x, y)| \, dx \, dy$$

3-3-15 *Hilbert integral operator* Show that the operator

$$Tu = \int_0^\infty \frac{u(y)}{x + y} \, dy$$

is bounded from $L_p[0, \infty)$ into $L_p[0, \infty)$ for $1 < p < \infty$, and its norm is equal to

$$\|T\| = \frac{\pi}{\sin(\pi/p)}$$

3-3-16 *Hardy integral operator* Show that the operator

$$Tu = \frac{1}{x} \int_0^x u(y) \, dy$$

is bounded from $L_p[0, \infty)$ into $L_p[0, \infty)$ for $p > 1$, and its norm is equal to

$$\|T\| = \frac{p}{p - 1}$$

3-3-17 Show that the set $\mathscr{L}(U, V)$ of all continuous linear transformations $T: U \to V$, where U and V are normed spaces, is a normed space with respect to the operations

$$(T_1 + T_2)u = T_1 u + T_2 u$$

$$(\alpha T)u = \alpha(Tu)$$

3-3-18 Show that the following linear and bilinear forms are bounded on $W_{1,2}[a, b]$:

$$l(u) = \int_a^b f(x)u \, dx$$

$$B(u, v) = \int_a^b \left[r(x) \frac{du}{dx} \frac{dv}{dx} + s(x)uv \right] dx$$

where $r(x)$ and $s(x)$ are positive continuous functions in $[a, b]$.

3-3-19 If $T: U \rightarrow V$ is a continuous linear transformation, then prove that the null space of the transformation is a closed subspace.

3-3-20 (Closed operators) Let U and V be two Banach spaces, and let S be a linear subspace of U. A linear operator $T: U \rightarrow V$ is said to be *closed* if whenever $u_n \rightarrow u$ in U and $v_n = Tu_n$ in V one has $u \in S$ and $v = Tu$. Show that every bounded linear operator is closed.

3-5 THEORY OF INNER PRODUCT SPACES

3-5-1 Inner Product and Inner Product Spaces

The concept of a norm provides us with a way of measuring the length of a vector, or the difference between two vectors of a linear vector space, as discussed in Sec. 3-4. In order to have a means of measuring the "angle" between or "orientation" of two vectors, we need to introduce the concept of an inner product, which is analogous to the scalar product of geometric vectors.

Inner product An *inner product* on a linear vector space V is a bilinear form on $V \times V$ that associates with each pair of vectors $u, v \in V$, a scalar, denoted (u, v), that satisfies the following axioms:

$$\begin{array}{ll} \text{(i)} & (u, v) = (v, u) \text{ (symmetry)} \\ \text{(ii)} & \left.\begin{array}{l} (a)\ (\alpha u, v) = \alpha(u, v) \text{ (homogeneous)} \\ (b)\ (u_1 + u_2, v) = (u_1, v) + (u_2, v) \text{ (additive)} \end{array}\right\} \text{(bilinear)} \\ \text{(iii)} & (u, u) > 0 \text{ and } (u, u) = 0 \text{ if and only if } u = 0 \text{ (positive definite)} \end{array} \tag{3-96}$$

for every $u, u_1, u_2, v \in V$, and $\alpha \in \mathbb{R}$. Note that (u, v) merely denotes the inner product of two vectors from the same vector space; one should define, consistent with the axioms in Eq. (3-96), the meaning of the symbol (u, v) in actual use.

One can associate a norm with every inner product by defining

$$\|u\| = \sqrt{(u, u)} \tag{3-97}$$

The norm thus obtained is called the *natural metric* generated by the inner product.

Cauchy–Schwarz inequality. A special case of the Hölder inequality for $p = q = 2$ is known as the Cauchy–Schwarz inequality. The inequality is stated here for inner product spaces. Let u and v be any elements of an inner product space V. Then the Cauchy–Schwarz inequality is given by

$$|(u, v)| \leq \sqrt{(u, u)(v, v)} = \|u\| \, \|v\| \tag{3-98}$$

The equality holds only if u and v are linearly dependent.

The following lemma states the continuity of the inner product.

Lemma 3-1 If $u_n \rightarrow u$, then $(u_n, v) \rightarrow (u, v)$ for any v. If

$$u = \sum_{i=1}^{\infty} u_n, \text{ then } \sum_{i=1}^{\infty} (u_n, v) = (u, v) \text{ for any } v$$

PROOF It is sufficient to prove the first statement because the second statement follows from the first. We have

$$|(u_n, v) - (u, v)| = |(u_n - u, v)|$$
$$\leq \|u_n - u\| \, \|v\|$$

Since $u_n \to u$, we have $(u_n, v) \to (u, v)$.

We now consider some examples of (u, v).

Example 3-31 1. Let $V = \mathbb{R}^n$, and define for $\mathbf{x}, \mathbf{y} \in \mathbb{R}^n$, the inner product

$$(\mathbf{x}, \mathbf{y}) \equiv \mathbf{x} \cdot \mathbf{y} = \sum_{i=1}^{n} x_i y_i \tag{3-99}$$

which satisfies the axioms in Eq. (3-96). It is important to note that $\mathbf{x}$ and $\mathbf{y}$ denote directed line segments in the n-dimensional euclidean space, and therefore the product is a real number.

2. (*The* $L_2(\Omega)$*-inner product*) Let $V = L_2(\Omega)$, $\Omega \subset \mathbb{R}^3$, and define

$$(u, v)_0 = \int_\Omega uv \, d\mathbf{x} \tag{3-100}$$

for all $u, v \in L_2(\Omega)$. Properties (i) and (ii) are clearly satisfied by $(u, v)_0$. To check property (iii) set $v = u$ in Eq. (3-100),

$$(u, u)_0 = \int_\Omega u^2 \, d\mathbf{x} \geq 0$$

The equality holds only if $u = 0$. Thus $(u, u)_0$ is an inner product, called the L_2-inner product. The natural norm generated by the inner product is the L_2-norm, $\|u\|_0$.

For example, the $L_2(\Omega)$ inner product of $u = \sin \pi x \sin \pi y$ and $v = x$ on a unit square is given by

$$(u, v)_0 = \int_0^1 \int_0^1 \sin \pi x \sin \pi y \cdot x \, dx \, dy$$
$$= \int_0^1 x \sin \pi x \, dx \cdot \int_0^1 \sin \pi y \, dy$$
$$= \frac{1}{\pi} \cdot \frac{2}{\pi}$$
$$= 0.2026$$

3. Let $V = W_{1,2}(\Omega)$ be the Sobolev space of order (1, 2). The space $W_{1,2}$ contains functions along with their first derivatives that are square integrable over $\Omega \subset \mathbb{R}^3$:

$$\int_\Omega |u|^2 \, d\mathbf{x} < \infty \qquad \int_\Omega \left| \frac{\partial u}{\partial x} \right|^2 d\mathbf{x} < \infty \qquad \int_\Omega \left| \frac{\partial u}{\partial y} \right|^2 d\mathbf{x} < \infty \tag{3-101}$$

If we define (u, v) on $W_{1,2}(\Omega)$ by

$$(u, v) = \int_\Omega \left(\frac{\partial u}{\partial x} \frac{\partial v}{\partial x} + \frac{\partial u}{\partial y} \frac{\partial v}{\partial y} \right) d\mathbf{x} \tag{3-102}$$

(u, v) does not satisfy axiom (iii). This can be seen by taking $u = c_1$, where c_1 is a constant (which belongs to $W_{1,2}(\Omega)$ because a constant is differentiable many times). We get

$$(u, u) = \int_\Omega (0^2 + 0^2)\, d\mathbf{x} = 0$$

but u is not zero. Thus (u, u) defined by (3-102) is not an inner product on space $W_{1,2}(\Omega)$. But it is an inner product on the space $W^0_{1,2}(\Omega)$ of functions from $W_{1,2}(\Omega)$ which vanish on the boundary of Ω. The space $W^0_{1,2}(\Omega)$ consists of non-constant continuous functions that are square integrable, because a constant (function) cannot vanish on the boundary of Ω unless the constant is zero.

4. [Inner product in $W_{m,2}(\Omega)$] Let $V = W_{m,2}(\Omega)$, $\Omega \subset \mathbb{R}^3$, and define

$$(u, v)_m = \sum_{|\boldsymbol{\alpha}| \le m} \int_\Omega D^{\boldsymbol{\alpha}} u D^{\boldsymbol{\alpha}} v \, d\mathbf{x} \tag{3-103}$$

where [see Eq. (3-71)]

$$\boldsymbol{\alpha} = (\alpha_1, \alpha_2, \alpha_3) \qquad |\boldsymbol{\alpha}| = \sum_{i=1}^{3} \alpha_i \ge 0$$

for all $i = 1, 2, 3$ and

$$D^{\boldsymbol{\alpha}} = \partial^{|\boldsymbol{\alpha}|} / \partial x_1^{\alpha_1} \, \partial x_2^{\alpha_2} \, \partial x_3^{\alpha_3}$$

The inner product satisfies the axioms in Eq. (3-96).

Consider a special case of Eq. (3-103) for $m = 1$:

$$(u, v)_1 = \int_\Omega \left(uv + \frac{\partial u}{\partial x_1} \frac{\partial v}{\partial x_1} + \frac{\partial u}{\partial x_2} \frac{\partial v}{\partial x_2} + \frac{\partial u}{\partial x_3} \frac{\partial v}{\partial x_3} \right) dx_1 \, dx_2 \, dx_3$$

Clearly, $(u, v)_1$ is linear in u and v, and is also symmetric, $(u, v)_1 = (v, u)_1$ and $(u, v)_1 \ge 0$. Furthermore, if $(u, u)_1 = 0$, then

$$\int_\Omega |u|^2 \, dx_1 \, dx_2 \, dx_3 = 0 \qquad \int_\Omega \left| \frac{\partial u_1}{\partial x_1} \right|^2 dx_1 \, dx_2 \, dx_3 = 0, \text{ etc.}$$

Since u is continuous, $\int_\Omega |u|^2 \, dx_1 \, dx_2 \, dx_3 = 0$ implies that $u = 0$. Therefore $(u, v)_m$ defined in Eq. (3-103) is an inner product on $W_{m,2}(\Omega)$.

The *natural* norm is given by Eq. (3-97).

Inner product spaces A linear vector space on which an inner product can be defined is called an *inner product space*. A linear subspace S of an inner product space V is a subspace with the inner product of V. Several examples of inner product spaces can be found in Example 3-31.

Since we can associate a norm with each inner product, every inner product space is also a normed linear space. It should be obvious to the reader that the converse does not hold in general. In other words, the set of inner product vector spaces is a proper subset of the set of normed spaces. Consequently, the concepts of continuity, convergence, and completeness presented in Sec. 3-4 are also valid in inner product spaces. In addition to these concepts, inner product spaces are equipped with additional properties. We shall study them here.

3-5-2 Orthogonal Vectors, Complements, and Projections

Orthogonality The concept of orthogonality (or perpendicularity) from the space of geometric vectors can be generalized to the elements of normed vector spaces with the aid of an inner product. Two vectors u and v in an inner product space V are said to be *orthogonal* if

$$(u, v) = 0 \tag{3-104}$$

A set of nonzero vectors, $\{u_1, u_2, \ldots\}$, is called an orthogonal set if each pair of the set is orthogonal:

$$(u_i, u_j) = 0 \qquad \text{for } i \neq j \tag{3-105}$$

If two vectors u and v of an inner product space V are orthogonal, then Pythagoras' theorem holds even in function spaces:

$$\|(u + v)\|^2 = (u + v, u + v) = (u, u) + 2(u, v) + (v, v) = \|u\|^2 + \|v\|^2 \tag{3-106}$$

From the definition of orthogonality, it is obvious that if $(u, v) = 0$ for *every* $v \in V$ then it follows that $u = 0$. This result is known as the *fundamental lemma of variational calculus*:

Lemma 3-2 Let V be an inner product space. If $(u, v) = 0$ for all $v \in V$, then $u = 0$.

A set of vectors $\{u_n\}$ in an inner product space are said to be *orthonormal* if

$$(u_i, u_j) = \delta_{ij} \tag{3-107}$$

where δ_{ij} is the Kronecker delta.

Theorem 3-8 An orthonormal set of nonzero vectors is linearly independent.

Proof Let $\{u_i\}$ be an orthonormal set. Consider the linear combination

$$\alpha_1 u_1 + \alpha_2 u_2 + \cdots + \alpha_n u_n = 0$$

We wish to show that the only solution of this equation is $\alpha_1 = \alpha_2 = \cdots = \alpha_n = 0$. Indeed we have

$$0 = (0, u_1) = (\alpha_1 u_1 + \alpha_2 u_2 + \cdots + \alpha_n u_n, u_1) = \alpha_1(u_1, u_1) + \alpha_2(u_2, u_1) + \cdots$$

Since $(u_i, u_1) = \delta_{i1}$, we have

$$0 = \alpha_1(u_1, u_1) \qquad \text{or} \qquad \alpha_1 = 0 \qquad (\text{since } u_1 \neq 0)$$

Similarly, we get $\alpha_2 = \alpha_3 = \cdots = \alpha_n = 0$.

It should be noted that any nonzero set of orthogonal vectors can be changed into an orthonormal set by replacing each vector u_i with $u_i/\|u_i\|$. For example the set $\{\cos nx\}_{n=0}^{\infty}$ in $L_2[-\pi, \pi]$ is orthogonal. After normalization we obtain the orthonormal system (see Prob. 3-4-11):

$$\frac{1}{\sqrt{2\pi}}, \frac{1}{\sqrt{\pi}} \cos x, \frac{1}{\sqrt{\pi}} \cos 2x, \ldots$$

Example 3.32 1. Consider the $L_2[-1, 1]$-inner product of the functions $u = 1$ and $v = a + bx$, where a and b are constants:

$$(u, v)_0 = \int_{-1}^{1} 1 \cdot (a + bx)\, dx$$
$$= 2a$$

Thus $u = 1$ and $v = a + bx$ are orthogonal in $L_2[-1, 1]$ if $a = 0$. The functions $\bar{u} = u/\|u\|_0 = 1/\sqrt{2}$ and $\bar{v} = v/\|v\|_0 = \sqrt{3/2}\,x$ are orthonormal. Similarly, one can show that

$$\phi_1 = \frac{1}{\sqrt{2}} \qquad \phi_2 = \sqrt{\frac{3}{2}}\,x \qquad \phi_3 = \sqrt{\frac{5}{8}}\,(3x^2 - 1) \tag{3-108}$$

are orthonormal.

2. Consider the $W_{1,2}[-1, 1]$-inner product of the functions $u = x$ and $v = ax + bx^2$:

$$(u, v)_1 = \int_{-1}^{1} [x \cdot (ax + bx^2) + 1 \cdot (a + 2bx)]\, dx$$
$$= \frac{2a}{3} + 2a$$

Thus the functions are orthogonal in $W_{1,2}[-1, 1]$ only if $a = 0$. The functions $\bar{u} = \sqrt{\frac{3}{8}}\,x$ and $\bar{v} = \sqrt{\frac{15}{46}}\,x^2$ are orthonormal with respect to the $W_{1,2}[-1, 1]$-inner product.

Some geometrical aspects of inner product spaces are presented in the following paragraphs.

Orthogonal complements Let V be an inner product space and let M be any subset of V. The *orthogonal complement* of M, denoted $M^{\perp}$, is defined by

$$M^{\perp} = \{u \in V: (u, v) = 0 \qquad \text{for all } v \in M\} \tag{3-109}$$

That is, $M^{\perp}$ is made up of all vectors that are orthogonal to every vector in M. Obviously, if $M = \varnothing$ (empty), then $M^{\perp} = V$.

Example 3-33 1. Let $V = \mathbb{R}^3$ and $M = \{\mathbf{A}\}$, where $\mathbf{A}$ is a nonzero vector. The orthogonal complement of M is the plane through the origin perpendicular to the vector **A** (see Fig. 3-7).

2. Let $V = \mathbb{R}^3$ and $M = \{\mathbf{A}, \mathbf{B}\}$, where **A** and **B** are nonzero nonparallel vectors. Then $M^{\perp}$ is the set of vectors that are perpendicular to both **A** and

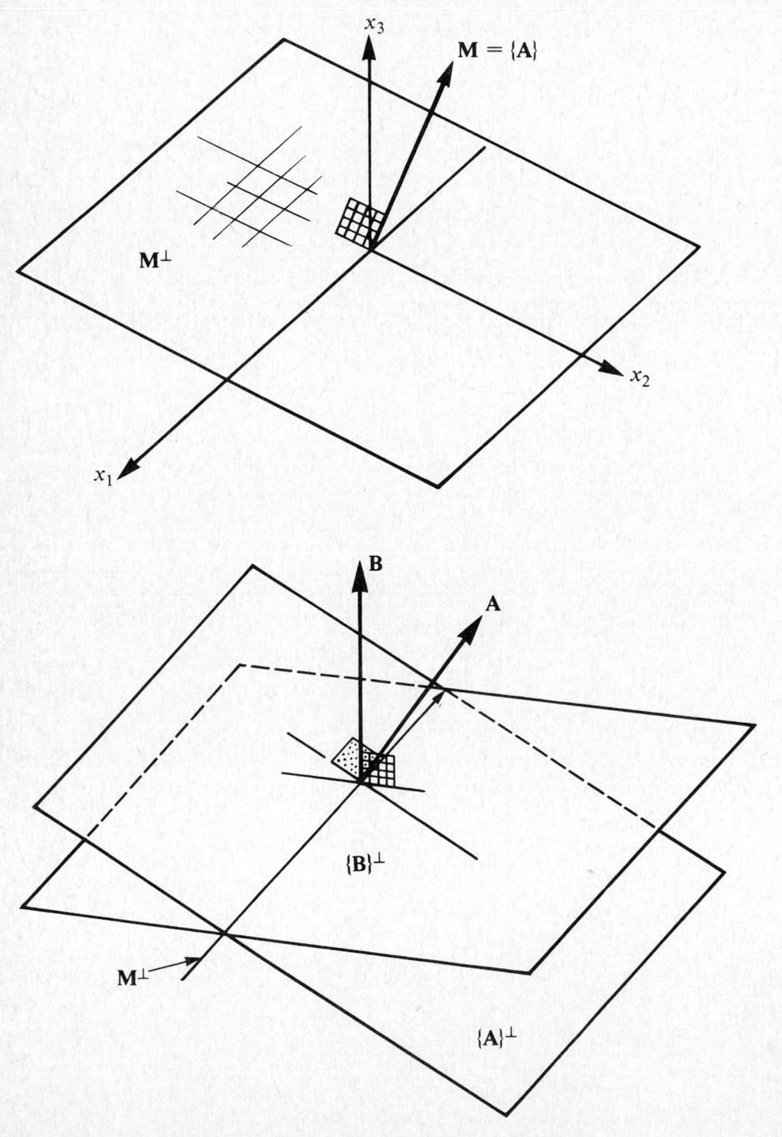

Figure 3-7 Orthogonal complements in Euclidean space $\mathbb{R}^3$.

B, i.e., the intersection of the orthogonal complements of vectors **A** and **B**. The intersection is a line through the origin (see Fig. 3-7).

3. Let S be the subspace of $\mathbb{R}^3$ spanned by the set $\{(1, 0, 1), (0, 2, 3)\}$. A typical element of S can be expressed as

$$(x_1, x_2, x_3) = \alpha(1, 0, 1) + \beta(0, 2, 3)$$

which gives

$$\alpha = x_1 \qquad \beta = \frac{x_2}{2} \qquad \alpha + 3\beta = x_3$$

Therefore, a typical element of S is of the form $(x_1, x_2, x_1 + \frac{3}{2}x_2)$. The orthogonal complement of S can be constructed as follows. Let $(x_1, x_2, x_3) \in S^\perp$. Then for $(y_1, y_2, y_3) \in S$ we have

$$\begin{aligned} 0 &= x_1 y_1 + x_2 y_2 + x_3 y_3 = x_1 y_1 + x_2 y_2 + x_3(y_1 + \tfrac{3}{2}y_2) \\ &= (x_1 + x_3)y_1 + (x_2 + \tfrac{3}{2}x_3)y_2 \end{aligned}$$

Since y_1 and y_2 are linearly independent (i.e., y_1 and y_2 are not related to each other), we have $x_1 + x_3 = 0$ and $x_2 + \frac{3}{2}x_3 = 0$. Therefore $S^\perp$ is given by

$$\begin{aligned} S^\perp &= \{\mathbf{x} = (x_1, x_2, x_3)\colon x_1 = -x_3, x_2 = -\tfrac{3}{2}x_3\} \\ &= \{\mathbf{x} \in \mathbb{R}^3, \mathbf{x} = (-x_3, -\tfrac{3}{2}x_3, x_3)\} \end{aligned}$$

and $(2, 3, -2)$ forms a basis for $S^\perp$.

Orthogonal projections A projection P (see Prob. 3-2-24) on an inner product space V is said to be *orthogonal* if its range and null space are orthogonal, $\mathcal{R}(P) \perp \mathcal{N}(P)$:

$$(u, v) = 0 \qquad \text{for all } u \in \mathcal{R}(P) \text{ and } v \in \mathcal{N}(P) \tag{3-110}$$

From the definition it follows that if P is an orthogonal projection, then so is $I - P$. Also, orthogonal projections are continuous, as will be shown in Theorem 3-9.

The following theorem establishes some of the properties of orthogonal projections on inner product spaces.

Theorem 3-9 Let V be an inner product space and let P be an orthogonal projection on V. Then

(i) Each element $w \in V$ can be written uniquely as $w = u + v$, where $u \in \mathcal{R}(P)$ and $v \in \mathcal{N}(P)$.
(ii) $\|w\|^2 = \|u\|^2 + \|v\|^2$.
(iii) $\mathcal{N}(P)$ and $\mathcal{R}(P)$ are closed linear subspaces of V.
(iv) $\mathcal{N}(P) = \mathcal{R}(P)^\perp$ and $\mathcal{R}(P) = \mathcal{N}(P)^\perp$.

PROOF (i) From Prob. 3-2-28 (which does not require P to be orthogonal) we know that V is a direct sum of the linear spaces $\mathscr{R}(P)$ and $\mathscr{N}(P)$:

$$V = \mathscr{R}(P) + \mathscr{N}(P), \mathscr{R}(P) \cap \mathscr{N}(P) = \{0\} \tag{3-111}$$

Therefore, every element w in V is of the form $w = u + v$, $u \in \mathscr{R}(P)$ and $v \in \mathscr{N}(P)$. To show that the representation is unique, let $w = u_1 + v_1 = u_2 + v_2$, where $u_1, u_2 \in \mathscr{R}(P)$ and $v_1, v_2 \in \mathscr{N}(P)$. Then $u_1 - u_2 = v_2 - v_1$. Since $u_1 - u_2 \in \mathscr{R}(P)$, $v_2 - v_1 \in \mathscr{N}(P)$ and $\mathscr{R}(P) \cap \mathscr{N}(P) = \{0\}$, we have $u_1 - u_2 = v_1 - v_2 = 0$. Hence, the representation is unique.

(ii) Since $w = u + v$, $u \in \mathscr{R}(P)$ and $v \in \mathscr{N}(P)$, we have

$$\|w\|^2 = (u + v, u + v) = \|u\|^2 + \|v\|^2 + 2(u, v)$$

Since $\mathscr{R}(P)$ is orthogonal to $\mathscr{N}(P)$, we have $(u, v) = 0$, and the result follows.

(iii) By Prob. 3-3-19, the null space of a continuous linear transformation is closed. Therefore, it is sufficient to show that P is continuous. Indeed, from parts (i) and (ii) we have $\|Pw\|^2 = \|P(u + v)\|^2 = \|Pu\|^2 = \|u\|^2 \leq \|w\|^2$. Thus P is continuous. In fact if $P \neq 0$, $\|P\| = 1$. Similarly, the operator $I - P$ is continuous. The proof of (iii) follows from the fact that $\mathscr{N}(P)$ and $\mathscr{R}(P)$ are the null spaces of the continuous operators P and $I - P$, respectively.

(iv) Since $\mathscr{N}(P) \perp \mathscr{R}(P)$, it is clear that $\mathscr{N}(P) \subset \mathscr{R}(P)^{\perp}$. Now we wish to show that $\mathscr{N}(P) \supset \mathscr{R}(P)^{\perp}$ so that we have $\mathscr{N}(P) = \mathscr{R}(P)^{\perp}$. Consider an element w in $\mathscr{R}(P)^{\perp}$. Then there exists a unique $u_0 \in \mathscr{R}(P)$ and $v_0 \in \mathscr{N}(P)$ such that $w = u_0 + v_0$. Since $w \in \mathscr{R}(P)^{\perp}$, we have $(w, u) = 0$ for all $u \in \mathscr{R}(P)$. Then $0 = (w, u) = (u_0, u)$ for all $u \in \mathscr{R}(P)$. This implies that $u_0 = 0$ and $w = v_0 \in \mathscr{N}(P)$, which shows that $\mathscr{R}(P)^{\perp} \subset \mathscr{N}(P)$. A similar argument proves that $\mathscr{R}(P) = \mathscr{N}(P)^{\perp}$.

3-5-3 Complete Inner Product Spaces: Hilbert Spaces

From previous discussions and examples it is clear that the geometry of normed and inner product vector spaces is much like the familiar two- and three-dimensional euclidean geometry. The geometry of Hilbert spaces is even closer to the euclidean geometry.

A complete (in its natural metric) inner product space is called a *Hilbert space.*

Example 3-34 1 (Euclidean space $\mathbb{R}^n$) The n-dimensional euclidean space is a Hilbert space. The inner product in $\mathbb{R}^n$ is defined by (3-99). Using the fact that $\mathbb{R}$ is complete we prove that $\mathbb{R}^n$ is complete. Let $\mathbf{x}^k = (x_1^k, x_2^k, \ldots, x_n^k)$ be a Cauchy sequence in $\mathbb{R}^n$. Then for any $\varepsilon > 0$, there exists an N such that

$$d(\mathbf{x}^m, \mathbf{x}^p) \equiv (\mathbf{x}^m - \mathbf{x}^p, \mathbf{x}^m - \mathbf{x}^p) = \left[\sum_{i=1}^{n} (x_i^m - x_i^p)^2 \right]^{1/2} \leq \varepsilon$$

whenever $m, p > N$. This implies that

$$|x_1^m - x_1^p| \leq \varepsilon, \ |x_2^m - x_2^p| \leq \varepsilon, \ldots, |x_n^m - x_n^p| \leq \varepsilon \qquad \text{for } m, p > N$$

That is each of the sequences x_n^k must converge, since $\mathbb{R}$ is complete, as $k \to \infty$. Let $\lim_{k\to\infty} x_i^k = x_i$; then $\lim_{k\to\infty} \mathbf{x}^k = (x_1, x_2, \ldots, x_n)$, and the space $\mathbb{R}^n$ is complete. Hence it is a Hilbert space. The Schwarz inequality becomes

$$\left[\sum_{k=1}^{n} x_k y_k\right]^2 \leq \left[\sum_{k=1}^{n} x_k^2\right]\left[\sum_{k=1}^{n} y_k^2\right]$$

The completeness of $\mathbb{R}^n$ implies the completeness of all finite-dimensional linear vector spaces because every finite-dimensional linear space is isomorphic to $\mathbb{R}^n$.

2 [Hilbert spaces, $H^m(\Omega)$, $m = 0, 1, 2, \ldots$] Since the Sobolev space $W^{m,p}(\Omega)$ is a Banach space, $W^{m,2}(\Omega) = H^m(\Omega)$ is a Hilbert space with the inner product defined in part 4 of Example 3-31. It should be pointed out once again that the Hilbert space $H^m(\Omega)$ is a special case of the Sobolev space $W^{m,p}(\Omega)$. Of course, $W^{m,p}(\Omega)$ is not a Hilbert space for $p \neq 2$. Also $H^0(\Omega) = L_2(\Omega)$.

The Hilbert space $H_0^m(\Omega)$ is a linear subspace of functions from $H^m(\Omega)$ that vanish along with their derivatives up to order $m - 1$ on the boundary of Ω.

Separable Hilbert Spaces In the preceding sections we have seen that the Hilbert spaces are much like the euclidean spaces. The so-called separable Hilbert spaces are even closer to the euclidean spaces. The separability ensures that the infinite-dimensional Hilbert space is not too "large." In other words, separable Hilbert spaces are Hilbert spaces with countable orthonormal bases. We give the precise definition of separable Hilbert spaces below.

Let H be a Hilbert space, and let S be a set of vectors in H. The set $\mathscr{L}(S)$ of all finite linear combinations of vectors from the set S is itself a linear subspace, which is in general not closed. If $\mathscr{L}(S)$ contains only a finite number of linearly independent vectors, then $\mathscr{L}(S)$ is a finite-dimensional linear subspace and hence is closed. If $\mathscr{L}(S)$ is infinite-dimensional, $\mathscr{L}(S)$ and its closure $\overline{\mathscr{L}(S)}$ differ. For example, the set $S = \{1, x, x^2, \ldots\}$ generates the space P of all polynomials, and the closure of P is the whole of $L_2[a, b]$.

A Hilbert space H is said to be *separable* if there exists a countable set of elements $\{\phi_n\}$ in H whose finite linear combinations are dense in H. In other words, given an element u in H and an $\varepsilon > 0$, there exists an index N (which usually depends on ε) and scalars $\alpha_1, \alpha_2, \ldots, \alpha_N$ such that

$$\left\| u - \sum_{i=1}^{N} \alpha_i \phi_i \right\| < \varepsilon \tag{3-112}$$

The set $\{\phi_n\}$ is called the *spanning set*. A separable Hilbert space contains a linearly independent spanning set of elements, which can be converted to an orthonormal set. For this reason, some authors define a separable Hilbert space to be a Hilbert space with a countable orthonormal basis. One must note that

Eq. (3-112) does not guarantee that u can be expanded in an infinite series unless it is possible to choose the scalars α_i in (3.112) independently of ε.

A finite-dimensional normed linear space is separable since any basis serves as a countable (finite) spanning set. The Hilbert spaces $H^m(\Omega)$, $m = 0, 1, 2, \ldots$, are separable since the countable set $\{1, x, x^2, \ldots\}$ is a spanning set.

In the following theorem we establish some properties of orthogonal complements in Hilbert spaces.

Theorem 3-10 Let H be a Hilbert space and S be a linear subspace of H. Then one has

(i) $S^\perp = \{0\}$ if and only if S is dense in H.
(ii) If S is closed and $S^\perp = \{0\}$, then $S = H$.
(iii) $S^{\perp\perp} \equiv (S^\perp)^\perp = \bar{S}$, where $\bar{S}$ is the closure of S.
(iv) If S is closed, then $S^{\perp\perp} = S$.

PROOF (i) Let S be dense in H. For all $u \in S$ and $v \in S^\perp$ we have by the Pythagorean theorem

$$\|u - v\|^2 = \|u\|^2 + \|v\|^2 \geq \|v\|^2$$

Since S is dense in H, we can choose u such that $\|u - v\|$ can be made arbitrarily small (for every $v \in S^\perp \subset H$). Hence $v = 0$ or $S^\perp = \{0\}$. The necessity can be proved by considering $S^\perp = \{0\}$. We have $S^{\perp\perp} = \{0\}^\perp = H$. By part (iii) (to be proved) $H = \bar{S}$, and hence S is closed. Parts (ii) and (iv) follow directly from (i).

(iii) Since $M \subset \bar{M}$, $M \subset M^{\perp\perp}$, and $M^{\perp\perp}$ is closed, we have $\bar{M} \subset M^{\perp\perp}$. Now let us show that $M^{\perp\perp} \subset \bar{M}$. Let $w \in M^{\perp\perp}$. Then $w \perp (M^{\perp\perp})^\perp = M^\perp$. Since $\bar{M}^\perp \subset M^\perp$, we have $w \in \bar{M}^\perp$ and hence $w \in \bar{M}$. Since w is an arbitrary element in $M^{\perp\perp}$ it follows that $M^{\perp\perp} \subset \bar{M}$. Thus we have $\bar{M} = M^{\perp\perp}$

In three-dimensional space, any vector can be projected onto a given plane in the space. More specifically, given $\mathbf{x}$ in $\mathbb{R}^3$, there is a vector $\mathbf{y}$ in the plane and vector $\mathbf{z}$ normal to the plane such that $\mathbf{x} = \mathbf{y} + \mathbf{z}$ (see Fig. 3-8). The vector $\mathbf{y}$ is called the projection of $\mathbf{x}$ onto the plane. This representation in euclidean space $\mathbb{R}^3$ is also valid in function spaces, especially in Hilbert spaces. The following theorem establishes this result (see also Theorem 3-9).

Theorem 3-11: The projection theorem Let S be any closed linear subspace of a Hilbert space H. Then the following statements are valid.

(i) $H = S + S^\perp$.
(ii) Each $u \in H$ can be expressed uniquely as $u = v + w$, where $v \in S$ and $w \in S^\perp$.
(iii) There is one and only one orthogonal projection P with $\mathscr{R}(P) = S$.

PROOF (i) First we prove that $W = S + S^\perp$ is a closed linear subspace of H. Let $\{w_n\}$ be a convergent sequence in W with the limit w. If we show that w is

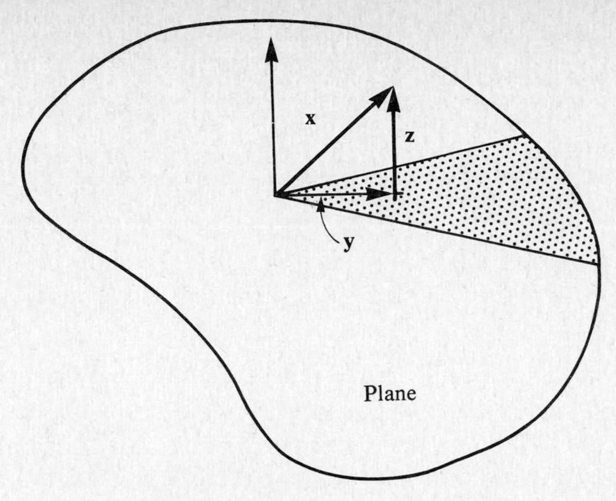

Figure 3-8 Projection of a vector in $\mathbb{R}^3$ onto a plane.

in W, then we have shown that W is closed (hence complete). By definition, w_n is of the form $w_n = u_n + v_n$ where $u_n \in S$ and $v_n \in S^\perp$. Since $S \subset S^\perp$ we also know that $u_n \perp v_n$, and

$$\|w_n - w_m\|^2 = \|u_n - u_m\|^2 + \|v_n - v_m\|^2 \geq \|u_n - u_m\|^2, \text{ etc.}$$

by the Pythagorean theorem. Hence $\{u_n\}$ and $\{v_n\}$ are Cauchy sequences in S and $S^\perp$, respectively. Since S and $S^\perp$ are complete, the limits u and v of the sequences $\{u_n\}$ and $\{v_n\}$, respectively, exist and $u \in S$ and $v \in S^\perp$. Then it follows from the continuity of addition that $w = u + v$ and that $W = S + S^\perp$ is closed.

Since $S \subset W$ and $S^\perp \subset W$, we have (see Prob. 3-4-13) $W^\perp \subset S^\perp$ and $W^\perp \subset S^{\perp\perp}$. This implies that $W^\perp \subset S^\perp \cap S^{\perp\perp}$. Again by Prob. 3-4-13 it follows that $W^\perp = \{0\}$, proving that S and $S^\perp$ are disjoint. Part (i) follows immediately from part (ii) of Theorem 3-10.

(ii) Since S and $S^\perp$ are disjoint, each $w \in W = S + S^\perp$ can be expressed as $w = u + v$, where $u \in S$ and $v \in S^\perp$. To prove that this representation is unique, assume that

$$w = u_1 + v_1 = u_2 + v_2 \qquad \text{for } u_1, u_2 \in S \qquad \text{and} \qquad v_1, v_2 \in S^\perp$$

Then $u_1 - u_2 = v_2 - v_1$, and since $u_1 - u_2 \in S$, $v_2 - v_1 \in S^\perp$, and $S \cap S^\perp = \{0\}$, it follows that $u_1 - u_2 = 0$ and $v_1 - v_2 = 0$. Hence the representation is unique. Also, by the Pythagorean theorem one has

$$\|w\|^2 = \|u\|^2 + \|v\|^2$$

(iii) Define the operator $P: H \to H$ $(H = S + S^\perp)$ by

$$P(u + v) = u$$

for all $u \in S$ and $v \in S^{\perp}$. It can be easily verified that P is an orthogonal projection with $\mathscr{R}(P) = S$. We wish to show that P is unique. If Q is another orthogonal projection with $\mathscr{R}(Q) = S$, then one has $\mathscr{N}(Q) = S^{\perp}$ and

$$P(u) = u = Q(u) \qquad u \in S$$

$$P(v) = 0 = Q(v) \qquad v \in S^{\perp}$$

Therefore, $P(u + v) = Q(u + v)$ for all $w = u + v \in H$. This implies that $P = Q$. This completes the proof of the theorem.

It should be noted that the completeness of H is essential in the above theorem. The completeness of H was used in proving that $S + S^{\perp}$ is a closed linear subspace of H. Note also that the name projection theorem is derived from the fact that the projection operator P partitions H into S and $S^{\perp}$. When H is not complete, there may not exist an orthogonal projection operator with the property $\mathscr{R}(P) = S$, where S is a closed subspace of H. On the other hand if S is not closed then $S + S^{\perp}$ is not closed and we cannot establish that $H = S + S^{\perp}$ and $S \cap S^{\perp} = \{0\}$. Recall that every finite-dimensional linear subspace of a normed space is closed.

Next we present a result that plays an important role in approximation theory.

Theorem 3-12: Best approximation Let H be a Hilbert space, and let S be a closed linear subspace of H. Further, let P be the orthogonal projection with $\mathscr{R}(P) = S$. Then for every $u_0 \in H$ the following inequality holds:

$$\|u_0 - Pu_0\| \leq \|u_0 - u\|$$

or equivalently,

$$\|u_0 - Pu_0\| = \inf \{\|u_0 - u\| : u \in S\}$$

In other words, for every element $u_0 \in H$ the nearest element in S is given by the projection of u_0 in S, Pu_0.

PROOF Since $u_0 = Pu_0 + v$, where $Pu_0 \in S$ and $v \in S^{\perp}$, we have $v = u_0 - Pu_0 \in S^{\perp}$. For all $u \in S$, we have $Pu_0 - u \in S$ and therefore

$$(u_0 - Pu_0, Pu_0 - u)_H = 0$$

Hence,

$$\begin{aligned} \|u_0 - u\|^2 &= \|u_0 - Pu_0 + Pu_0 - u\|^2 \\ &= \|u_0 - Pu_0\|^2 + \|Pu_0 - u\|^2 + 2(u_0 - Pu_0, Pu_0 - u) \\ &\geq \|Pu_0 - u_0\|^2 \end{aligned}$$

The inequality is a strict inequality except for $u = u_0 = Pu_0$. Note also that the projection operator P is continuous, since we also have

$$\|u_0 - u\| \geq \|Pu_0 - u\| = \|P(u_0 - u)\|$$

where we have used the fact that $P(u) = u$ for all $u \in S$.

Example 3-35 Let $H = L_2[0, \pi]$ and S be the set of all constant functions in the interval $[0, \pi]$. It can be shown that S is a linear space. Let us use the $L_2[0, \pi]$ metric for the space S. We can show that S is complete. It is sufficient to prove that every Cauchy sequence has its limit in S. Let

$$u_1 = c_1 \qquad u_2 = c_2, \qquad \ldots, \qquad u_n = c_n$$

be a Cauchy sequence of elements from S, where c_i $(i = 1, 2, \ldots, n)$ are real constants. We have

$$\|u_m - u_n\|_0 = \left[\int_0^\pi |c_m - c_n|^2 \, dx\right]^{1/2}$$

$$= \sqrt{\pi}\,|c_m - c_n|$$

Therefore $\{u_1, u_2, \ldots, u_n\}$ is a Cauchy sequence in S if and only if $\{c_1, c_2, \ldots, c_n\}$ is a Cauchy sequence in $\mathbb{R}$. By the Bolzano–Cauchy criterion from real analysis, the Cauchy sequence $\{c_n\}$ has a limit in $\mathbb{R}$. Denote this limit by c_0. Then it immediately follows from the above equality that the sequence $\{u_n\}$ has a limit in S, namely $u_0 = c_0$, because

$$\lim_{n\to\infty} \|u_0 - u_n\| = \lim_{n\to\infty} \sqrt{\pi}\,|c_0 - c_n| = 0$$

Consequently, every Cauchy sequence in S has a limit in S. Therefore, the space S is complete and it is a linear subspace of $L_2[0, \pi]$.

From Theorem 3-11 it follows that every function w in $L_2[0, \pi]$ can be decomposed into the sum

$$w = u + v \qquad u \in S \quad \text{and} \quad v \in S^\perp$$

For example, $w = \sin x$ can be decomposed as

$$\sin x = a + (\sin x - a)$$

The constant $a \in S$ is determined from the orthogonality of $v = \sin x - a$ to every element c in S:

$$(c, v) = \int_0^\pi c(\sin x - a)\, dx = 0$$

$$= c(2 - a\pi)$$

Since c is arbitrary, we have $a = 2/\pi$. Thus, the projection of $w = \sin x$ into S is the function $u(x) = 2/\pi$.

Define the projection operator P on $H = L_2[0, \pi]$ by

$$Pw = P(u + v) = u \qquad u \in S$$

$$Pv = 0 \qquad v \in S^{\perp}$$

and compute $\|w - Pw\|_0$ and $\|w - u\|_0$:

$$\left\| \sin x - \frac{2}{\pi} \right\|_0 = \left[\int_0^{\pi} \left(\sin x - \frac{2}{\pi} \right)^2 dx \right]^{1/2}$$

$$= \left[\int_0^{\pi} \left(\sin^2 x + \frac{4}{\pi^2} - \frac{4}{\pi} \sin x \right) dx \right]^{1/2}$$

$$= \left(\frac{\pi}{2} + \frac{4}{\pi} - \frac{8}{\pi} \right)^{1/2} \approx 0.545$$

$$\|\sin x - c\|_0 = \left[\int_0^{\pi} (\sin^2 x + c^2 - 2c \sin x)\, dx \right]^{1/2}$$

$$= \left[\frac{\pi}{2} + c^2\pi - 4c \right]^{1/2}$$

3-5-4 Orthonormal Bases and Generalized Fourier Series

From a course on analysis we know that every continuous function $u(x)$ on $(0, \pi)$ can be represented in the form

$$u(x) = \sum_{n=0}^{\infty} (a_n \sin nx + b_n \cos nx)$$

In the language of inner product spaces one can say that every element of $C[0, \pi]$ can be represented as a (infinite) linear combination of the set $\{1, \cos x, \sin x, \cos 2x, \sin 2x, \ldots\}$. Thus, we can say the set forms a basis for the space.

An *orthogonal basis* for an inner product space V is an orthogonal set $\{\phi_i\}$ such that for any $u \in V$ there exist scalars α_i such that

$$u = \sum_{i=1}^{\infty} \alpha_i \phi_i \tag{3-113}$$

If the elements ϕ_i are orthonormal, it is called an *orthonormal basis.*

An orthonormal basis $\{\phi_i\}$ in a Hilbert space H is called *maximal* or *complete* if and only if there is not a unit vector ϕ_0 in H such that $\{\phi_0, \phi_1, \phi_2, \ldots\}$ is an orthonormal set. In other words, the sequence $\{\phi_i\}$ in H is complete if and only if the only vector orthogonal to each of the ϕ_i's is the null vector.

If $\{\phi_i\}$ is an orthonormal basis in a Hilbert space H, then the numbers $\alpha_i = (u, \phi_i)_H$ are called the *Fourier coefficients* of the element u with respect to the system $\{\phi_i\}$ and $\sum \alpha_i \phi_i$ is called the Fourier series of the element u. The linear combination in Eq. (3-113) gives the best approximation of u in comparison with other combinations $\sum c_i \phi_i$.

The Bessel Inequality Let $\{\phi_n\}$ be an orthonormal set in an inner product space V. For any finite subset $\{\phi_1, \phi_2, \ldots, \phi_N\}$ from $\{\phi_n\}$, we have

$$0 \leq \left\| u - \sum_{i=1}^{N} (u, \phi_i)\phi_i \right\|^2 = \left(u - \sum_{i=1}^{N} (u, \phi_i)\phi_i, u - \sum_{j=1}^{N} (u, \phi_j)\phi_j \right)$$

$$= \|u\|^2 - 2\sum_{i=1}^{N} (u, \phi_i)(u, \phi_i) + \sum_{i,j=1}^{N} (u, \phi_i)(u, \phi_j)(\phi_i, \phi_j)$$

$$= \|u\|^2 - \sum_{i=1}^{N} |(u, \phi_i)|^2$$

In arriving at the last step we have employed the orthonormal property of $\{\phi_n\}$. Thus we have

$$\|u\|^2 \geq \sum_{i=1}^{N} |(u, \phi_i)|^2 \tag{3-114}$$

and since the left side does not depend on N, the inequality, known as the *Bessel inequality*, holds for countable sums

$$\|u\|^2 \geq \sum_{i} |(u, \phi_i)|^2 \tag{3-115}$$

Next, we study certain properties of the linear combinations of the form $\sum_i \alpha_i \phi_i$, where $\{\phi_i\}$ is an orthonormal set. We establish a necessary and sufficient condition for an infinite series of orthogonal vectors to converge in a Hilbert space.

Lemma 3-3 Let $\{\phi_n\}$ be a countably infinite orthonormal set in a Hilbert space H. Then the following statements hold:

(i) The infinite series $\sum_{n=1}^{\infty} \alpha_n \phi_n$, where α_n are scalars, converges if and only if the series of real numbers $\sum_{n=1}^{\infty} |\alpha_n|^2$ converges.

(ii) If $\sum_{n=1}^{\infty} \alpha_n \phi_n$ converges and

$$u = \sum_{n=1}^{\infty} \alpha_n \phi_n = \sum_{n=1}^{\infty} \beta_n \phi_n$$

then $\alpha_n = \beta_n$ for all n and $\|u\|^2 = \sum_{n=1}^{\infty} |\alpha_n|^2$.

PROOF (i) First consider the "if" part. Suppose that the series $\sum_{n=1}^{\infty} \alpha_n \phi_n$ is convergent and let $u = \sum_{n=1}^{\infty} \alpha_n \phi_n$. Equivalently,

$$\lim_{N \to \infty} \left\| u - \sum_{n=1}^{N} \alpha_n \phi_n \right\|^2 = 0$$

We have, since the inner product is continuous,

$$(u, \phi_m) = \left(\sum_{n=1}^{\infty} \alpha_n \phi_n, \phi_m \right) = \sum_{n=1}^{\infty} \alpha_n(\phi_n, \phi_m) = \alpha_m \qquad m = 1, 2, \ldots$$

In view of the Bessel inequality we obtain

$$\sum_{m=1}^{\infty} |(u, \phi_m)|^2 = \sum_{m=1}^{\infty} |\alpha_m|^2 \leq \|u\|^2$$

which shows that $\sum_{m=1}^{\infty} |\alpha_m|^2$ converges. It should be noted that the completeness of H is not used in this part of the proof.

Next consider the "only if" part. Suppose that $\sum_{n=1}^{\infty} |\alpha_n|^2$ converges. Consider the finite sum $S_n = \sum_{i=1}^{n} \alpha_i \phi_i$. We have

$$\|S_n - S_m\|^2 = \left(\sum_{i=m+1}^{n} \alpha_i \phi_i, \sum_{j=m+1}^{n} \alpha_j \phi_j \right) = \sum_{i=m+1}^{n} |\alpha_i|^2$$

which shows that $\{S_n\}$ is a Cauchy sequence. Since H is complete, the sequence of partial sums $\{S_n\}$ is convergent in H and therefore the series $\sum_n \alpha_n \phi_n$ converges.

(ii) We first prove that $\|u\|^2 = \sum_{n=1}^{\infty} |\alpha_n|^2$. Consider

$$\|u\|^2 - \sum_{n=1}^{N} |\alpha_n|^2 = (u, u) - \sum_{n=1}^{N} \sum_{m=1}^{N} (\alpha_n \phi_n, \alpha_m \phi_m)$$

$$= \left(u, u - \sum_{n=1}^{N} \alpha_n \phi_n \right) + \left(\sum_{n=1}^{N} \alpha_n \phi_n, u - \sum_{n=1}^{N} \alpha_n \phi_n \right)$$

$$\leq \left\| u - \sum_{n=1}^{N} \alpha_n \phi_n \right\| \left\{ \|u\| + \left\| \sum_{n=1}^{N} \alpha_n \phi_n \right\| \right\}$$

Since $\sum_{n=1}^{\infty} \alpha_n \phi_n$ converges to u, the right-hand side converges to zero, giving the desired result.

If $u = \sum_{n=1}^{\infty} \alpha_n \phi_n = \sum_{n=1}^{\infty} \beta_n \phi_n$, then

$$0 = \lim_{N \to \infty} \left[\sum_{i=1}^{N} (\alpha_n - \beta_n)\phi_n \right] \to 0^2 = \sum_{n=1}^{\infty} \|\alpha_n - \beta_n\|^2$$

implying that $\alpha_n = \beta_n$ for all n. This completes the proof of the lemma.

In the following lemma we characterize the orthogonal projections on a Hilbert space in terms of an orthonormal basis.

Lemma 3-4 Let V be an inner product space, and let S be the finite-dimensional linear subspace spanned by the finite orthonormal set $\{\phi_i\}_{i=1}^{n}$. Then the orthogonal projection of V into S is given by

$$Pu = \sum_{i=1}^{n} (u, \phi_i)\phi_i \tag{3-116}$$

PROOF First we note that P is linear. Next we prove that P is a projection:

$$P(Pu) = \sum_{i=1}^{N} (u, \phi_i)P\phi_i = \sum_{i=1}^{n} (u, \phi_i) \sum_{j=1}^{n} (\phi_i, \phi_j)\phi_j$$

$$= \sum_{i=1}^{n} (u, \phi_i)\phi_i = Pu$$

Further, by definition, $\mathscr{R}(P) \subset S$. On the other hand, for $u = \sum_{i=1}^{n} \alpha_i \phi_i \in S$, we have $Pu = u$ (and $\alpha_i = (u, \phi_i)$), so $\mathscr{R}(P) = S$.

Now it remains to be shown that P is orthogonal. Let $v \in \mathscr{N}(P)$, and $u \in \mathscr{R}(P)$ since $u \in \mathscr{R}(P) = S$ we have $Pu = u$ and

$$(v, u) = (v, Pu) = \left(v, \sum_{i=1}^{n} (u, \phi_i)\phi_i\right)$$

$$= \sum_{i=1}^{n} (v, \phi_i)(u, \phi_i)$$

$$= \left(\sum_{i=1}^{n} (v, \phi_i)\phi_i, u\right) = (Pv, u)$$

Since $v \in \mathscr{N}(P)$, we have $Pv = 0$ and hence $\mathscr{N}(P) \perp \mathscr{R}(P)$. This proves that P is orthogonal.

The result of Lemma 3-3 can be stated in alternative terms as follows: Given an element u and an orthonormal sequence $\{\phi_i\}$ in a Hilbert space H, the series $\sum_{i=1}^{\infty} (u, \phi_i)\phi_i$ converges to an element $u_0 \equiv Pu$ in the closed subspace S spanned by the ϕ_i's. The difference vector $u - u_0$ is orthogonal to S.

We remark that the above results can be carried on to closed linear subspaces (not necessarily finite-dimensional) spanned by a countable orthonormal set. We state this fact in the following lemma.

Lemma 3-5 Let S be the closed linear subspace generated by a countable orthonormal set $\{\phi_i\}$ in a Hilbert space H. Then every vector $u \in S$ can be written uniquely as

$$u = \sum_{i} (u, \phi_i)\phi_i \tag{3-117}$$

PROOF The uniqueness of (3-117) follows from Lemma 3-3. We wish to show that every vector $u \in S$ can be represented by (3-117). For $u \in S$ we have (since S is closed)

$$u = \lim_{N \to \infty} \sum_{i=1}^{M} \alpha_i \phi_i \qquad M \geq N$$

From Theorem 3-11 and Lemma 3-3 it follows that

$$\left\| u - \sum_{i=1}^{M} (u, \phi_i)\phi_i \right\| \le \left\| u - \sum_{i=1}^{M} \alpha_i \phi_i \right\|$$

and, as $N \to \infty$, we get the desired result.

We come now to the important topic of this section, namely the generalized Fourier representation of functions. Much of the preceding discussion of this section might have already prompted the reader to compare the representations in (3-113) and (3-117) with the usual Fourier series representations. The following theorem gives the fundamental properties of orthonormal bases.

Theorem 3-13: Fourier series theorem For any orthonormal set $\{\phi_n\}$ in a Hilbert space H, the following statements are *equivalent:*

(*a*) (Fourier series expansion). Every u in H can be represented by

$$u = \sum_i (u, \phi_i)\phi_i$$

(*b*) (Parseval equality). For any pair of vectors $u, v \in H$ one has

$$(u, v) = \sum_{i=1} (u, \phi_i)(v, \phi_i) \tag{3-118}$$

(*c*) For any $u \in H$ one has

$$\|u\|^2 = \sum_{i=1} |(u, \phi_i)|^2 \tag{3-119}$$

(*d*) Any linear subspace S of H that contains $\{\phi_i\}$ is dense in H.

PROOF $(a) \to (b)$. This follows immediately from (3-117) and the fact that $\{\phi_i\}$ is orthonormal.

$(b) \to (c)$. Follows from (3-118) by setting $v = u$.

$(a) \to (d)$. The statement (*d*) is equivalent to saying the orthogonal projection onto $\bar{S}$, the closure of S, is the identity. In view of Lemma 3-5, statement (*d*) is equivalent to statement (*a*).

The Fourier series representation (3-117) can be interpreted as the sum of projections of u along each vector ϕ_i in the orthonormal set $\{\phi_i\}$. The scalars (u, ϕ_n) are the *coordinates*, also known as the *Fourier coefficients*, of u with respect to the orthonormal basis $\{\phi_i\}$.

Example 3-36 (Fourier trigonometric series) Let $H = L_2[-\pi, \pi]$. The set

$$\{\phi_i\}_{i=1}^{\infty} = \{\cos x, \sin x, \cos 2x, \sin 2x, \ldots, \cos nx, \sin nx, \ldots\}$$

is not maximal because there exists $\phi_0 = 1$ which is orthogonal to $\{\phi_i\}_{i=1}^{\infty}$. The set

$$\{\phi_i\} = \left\{\frac{2}{\sqrt{\pi}}, \frac{1}{\sqrt{\pi}} \cos x, \frac{1}{\sqrt{\pi}} \sin x, \ldots, \frac{1}{\sqrt{\pi}} \cos nx, \frac{1}{\sqrt{\pi}} \sin nx, \ldots\right\}$$

is maximal in $L_2[-\pi, \pi]$. By the Fourier series theorem any function $u(x)$ in $L_2[-\pi, \pi]$ can be expressed in the form

$$u(x) = \frac{2a_0}{\sqrt{\pi}} + \frac{1}{\sqrt{\pi}} \sum_{i=1}^{\infty} (a_i \cos ix + b_i \sin ix) \tag{3-120}$$

where

$$a_0 = \left(u, \frac{2}{\sqrt{\pi}}\right)_0 \qquad a_i = (u, \cos ix)_0 \qquad b_i = (u, \sin ix) \qquad (i = 1, 2, \ldots)$$

The $L_2[-\pi, \pi]$-norm of u is given in terms of the coefficients of the Fourier series by

$$\|u\|_0^2 = a_0^2 + \sum_{i=1}^{\infty} (a_i^2 + b_i^2)$$

PROBLEMS 3-4

3-4-1 Prove the following relations in an inner product space:

(*a*) The parallelogram law:

$$\|u + v\|^2 + \|u - v\|^2 = 2(\|u\|^2 + \|v\|^2)$$

(*b*)
$$(u, v) = \tfrac{1}{4}[\|u + v\|^2 - \|u - v\|^2]$$

(*c*)
$$|\,\|u\| - \|v\|\,| \le \|u - v\|$$

3-4-2 Let V be the linear vector space of continuous complex-valued functions defined on an interval $[0, 1]$. Define for $u, v \in V$,

$$(u, v) \equiv \int_0^1 u(x)\bar{v}(x)\, dx$$

where $\bar{v}(x)$ is the complex conjugate of $v(x)$ (i.e., if $v = a + ib$, then $\bar{v} = a - ib$). Show that $(\cdot, \cdot)$ is an inner product on V. If S is the linear subspace of V spanned by $\{\sin \pi x\}$, find a function in S that is orthogonal to $f(x) = (a + ib) \sin \pi x$.

3-4-3 Consider the linear vector space V of continuous real-valued functions $u(x, t)$ on $\Omega[0, 1] \times [0, T]$. Define for any pair of elements $u, v \in V$,

$$(u, v) = \int_0^1 \int_0^T u(x, T - t)v(x, t)\, dt\, dx$$

Show that $(\cdot, \cdot)$ is an inner product on V.

Compute the inner product of the following pairs of functions in the spaces indicated in Probs. 3-4-4 to 3-4-8.

3-4-4 $u = x - x^2$, $v = \sin \pi x$ in $L_2[0, 1]$

3-4-5 $u = (1 + x)$, $v = 3x^2 - 1$, in $L_2[-1, 1]$

3-4-6 $u = x$, $v = \cos x$, in $H^1[0, \pi]$

3-4-7 $u = \sin \pi x$, $v = 1 + x$ in $H^1[0, 1]$

3-4-8 $u = x^2 + y^2$, $v = (2 - x^2 - y^2)$, in $L_2(\Omega)$, Ω is a unit square.

3-4-9 Determine the subspace of vectors in $\mathbb{R}^3$ that are orthogonal to:
(*a*) (0, 1, 1) (*b*) (2, −1, 1) (*c*) (1, 2, 3)

3-4-10 Determine a vector in the space P of polynomials of degree 2 such that the vector is orthogonal to the polynomials $p_1 = x - x^2$ and $p_2 = 1 + 4x + x^2$ in the $L_2[0, 1]$ space.

3-4-11 *Gram–Schmidt orthonormalization.* Let $\{\phi_1, \phi_2, \ldots, \phi_n\}$ be any linearly independent set of vectors in an inner product space V. Construct an orthonormal set $\{\psi_1, \psi_2, \ldots, \psi_n\}$ using the following procedure: Let $\psi_1 \equiv \phi_1/\|\phi_1\|$. Next choose the second vector ϕ_2 and require that the vector $\psi_2' = \phi_2 - \alpha\psi_1$ is orthogonal to ψ_1:

$$0 = (\psi_1, \phi_2 - \alpha\psi_1) = (\psi_1, \phi_2) - \alpha\|\psi_1\|^2$$

or $\alpha = (\psi_1, \phi_2)$. Then the second element of the orthonormal set is given by $\psi_2 = \psi_2'/\|\psi_2'\|$. Continue the procedure to obtain the $(r+1)$th element

$$\psi_{r+1}' = \phi_{r+1} - (\psi_1, \phi_{r+1})\psi_1 - (\psi_2, \phi_{r+1})\psi_2 - \cdots - (\psi_r, \phi_{r+1})\psi_r$$

$$\psi_{r+1} = \psi_{r+1}'/\|\psi_{r+1}'\|$$

3-4-12 Use the Gram–Schmidt orthonormalization to construct the orthonormal set associated with the linearly independent functions $\{1, x, x^2\}$ in $L_2[-1, 1]$.

3-4-13 Let V be an inner product space and let S be any subset of V. Prove the following properties of orthogonal complements. Here S_1 and S_2 denote nonempty subsets of V.
(i) If $S_1 \subset V$, then $V^\perp \subset S^\perp$.
(ii) If $S_1 \subset S_2$, then $S_2^\perp \subset S_1^\perp$.
(iii) If $u \in S_1 \cap S_1^\perp$, then $u = 0$.
(iv) If S_1 is a dense subset of V, then $S_1^\perp = \{0\}$.

3-4-14 Let $V = C[-1, 1]$ be the space of all continuous functions on the interval $[-1, 1]$ with the L_2-inner product. Let S be the subspace of "odd" functions $f(x)$ such that $f(-x) = -f(x)$. Determine the orthogonal complement of S.

3-4-15 Let $\{P_1, P_2, \ldots, P_n\}$ be a collection of orthogonal projections with $P_iP_j = 0$ for $i \neq j$. Show that $T = \sum_{i=1}^m P_i$ is an orthogonal projection.

3-4-16 Let S be the linear subspace of $L_2[0, 1]$ whose elements are all constants in the interval $[0, 1]$. Decompose the function $u(x) \in L_2[0, 1]$ into the sum $u = \alpha + v$ where $\alpha \in S$ and $v(x) \in S^\perp$. Take $u(x)$ to be
(*a*) $u(x) = x^2$ (*b*) $u(x) = 1 - x$ (*c*) $u(x) = x(1 - x)$

3-4-17 Let $P\colon U \to S$, where U is a Hilbert space and S is a subspace of U, and define the operator P, for all $u \in U$ and all $v \in S$, by $(u - Pu, v) = 0$. Prove that (*a*) P is an orthogonal projection, (*b*) $I - P$ is an orthogonal projection, and (*c*) the norm of the remainder, $\|u - Pu\|$ is the smallest distance from u to the subspace.

3-4-18 Consider the set S of all constant functions in $L_2[0, 1]$.
(*a*) Show that S is a complete linear subspace of $L_2[0, 1]$.
(*b*) Decompose the function $u = x^2 \in L_2[0, 1]$ into the sum $u = v + w$, where $v \in S$ and $w \in S^\perp$. Characterize the space S.

3-4-19 Calculate the quadratic polynomial, $p(x) = a + bx + cx^2$ which gives the best fit to $f(x) = e^x$ over the interval $[0, 1]$ in the following sense: $\int_0^1 [f(x) - p(x)]^2\,dx$ is a minimum.

3-4-20 Expand the function $f(x) = \frac{1}{2}(2\pi - x)\sin x$ in $L_2[0, 2\pi]$ in terms of the Fourier coefficients.

3-4-21 Consider the functional $B(u, v) = \frac{1}{4}(\|u + v\|^2 - \|u - v\|^2)$ on an inner-product space. Show that
(*a*) $B(\cdot, \cdot)$ is bilinear in u and v,
(*b*) for fixed u, show that $f_u(v) = B(u, v)$ is a bounded linear functional, and
(*c*) $B(u, v)$ satisfies the axioms of an inner product.

3-4-22 Determine whether the operator $T \equiv -\nabla^2 : H \to H$, $H = H_0^1(\Omega) \cap H^2(\Omega)$ and $\Omega \subset \mathbb{R}^2$, is bounded. Is the operator bounded below?

CHAPTER

FOUR

VARIATIONAL FORMULATIONS OF BOUNDARY-VALUE PROBLEMS

4-1 LINEAR FUNCTIONALS AND OPERATORS ON HILBERT SPACES

4-1-1 Introduction

In this chapter we will be concerned with variational formulations of operator equations of the form

$$Au = f \text{ in } \Omega \tag{4-1}$$

where A is a linear or nonlinear operator from an inner product space U into another inner product space V. Equation (4-1) is an abstract form of the equations of science and engineering. An example of the equation is given below.

Example 4-1 Consider the partial differential equation

$$-\nabla^2 u \equiv -\left(\frac{\partial^2 u}{\partial x^2} + \frac{\partial^2 u}{\partial y^2}\right) = f \text{ in } \Omega \tag{4-2a}$$

$$u = 0 \text{ on } \Gamma \tag{4-2b}$$

where $\Omega \subset \mathbb{R}^2$ is a plane and Γ is the boundary of Ω. Equations (4-2a, b) are known as the *Dirichlet problem* for the Poisson equation. For the moment we assume that $f \in C(\bar{\Omega})$, $\bar{\Omega} = \Omega + \Gamma$.

The *classical solution* of the problem (4-2) means the function $u(x, y)$, which is continuous in the closed domain $\bar{\Omega}$, satisfies Eq. (4-2*a*) in the open domain Ω and is equal to zero on the boundary Γ. By assumption $f \in C(\bar{\Omega})$, the solution u belongs to $C^2(\bar{\Omega})$, the space of continuous functions with continuous partial derivatives up to second-order inclusive, and equals zero on Γ. The set $\mathscr{D}_A$ of these admissible functions,

$$\mathscr{D}_A = \{u(\mathbf{x}) \in C^2(\bar{\Omega}), \mathbf{x} \in \Omega \subset \mathbb{R}^2, u = 0 \text{ on } \Gamma\}$$

forms a linear space, because if u_1 and u_2 are arbitrary functions in $\mathscr{D}_A$, then the linear combination $\alpha u_1 + \beta u_2$, for any scalars α and β, also belongs to $\mathscr{D}_A$. Note that if the boundary condition in (4-2*b*) is nonhomogeneous (for example, $u = g$ on Γ) then the set $\mathscr{D}_A$ is not a linear space.

Thus the given problem can be stated as follows: Find $u \in \mathscr{D}_A$ such that Eq. (4-2*a*) is satisfied. Note that the boundary condition (4-2*b*) is included in the specification of the space $\mathscr{D}_A$. The space $\mathscr{D}_A$ is called the *domain of the definition of operator A*. The operator $A = -\nabla^2$ assigns to every function $u \in \mathscr{D}_A$ a function $v = -\nabla^2 u$ continuous in Ω. The set of all functions $v = -\nabla^2 u$ is also a linear space, called the range of $A = -\nabla^2$, and is denoted $\mathscr{R}(A)$.

Besides differential operators, matrix operators will also be encountered in our study. For example, the numerical solution of differential equations by a variational method leads to the solution of algebraic equations in matrix form. Therefore, we will be interested in determining the conditions under which the operator equation (4-1) has a solution. We study such conditions in Chap. 5.

There are several approaches to determine the solution of Eq. (4-1). When the given equation and given domain are very simple, one can obtain solutions in closed form using the Fourier method or separation of variables technique. When the given equation and/or the given domain are complicated so that classical methods of solution cannot be used, approximate methods are to be used. In the present study, we deal with the variational methods, which are based on variational statements of Eq. (4-1). In this chapter we study the variational formulations of various operator equations, some of which admit functional formulation and some only a weak formulation.

In our theoretical considerations we assume that the domain under consideration Ω is bounded and its boundary Γ is sufficiently smooth, or more specifically, *Lipschitzian*. Examples of Lipschitz boundaries are circles, annuli, triangles, rectangles, spheres, cubes, etc. Two-dimensional domains with cuspidal points and three-dimensional domains with corresponding singularities are not lipschitzian. A mathematical definition of the lipschitzian is considerably involved. Roughly speaking, the Lipschitz boundary Γ of an N-dimensional domain Ω allows us to define functions $f(x_1, x_2, \ldots, x_{N-1})$ which are continuous in the $(N - 1)$ dimensional cubes defined on Γ. The Lipschitz boundary Γ has the (outward) normal $\mathbf{n}$ almost everywhere (to avoid possible confusion, N is used to denote the dimensions of the domain in this part of the book). The components $n_1(s), n_2(s), \ldots$ of the unit vector $\mathbf{n}$ are bounded measurable functions on Γ. Some non-lipschitzian domains in two dimensions are shown in Fig. 4-1.

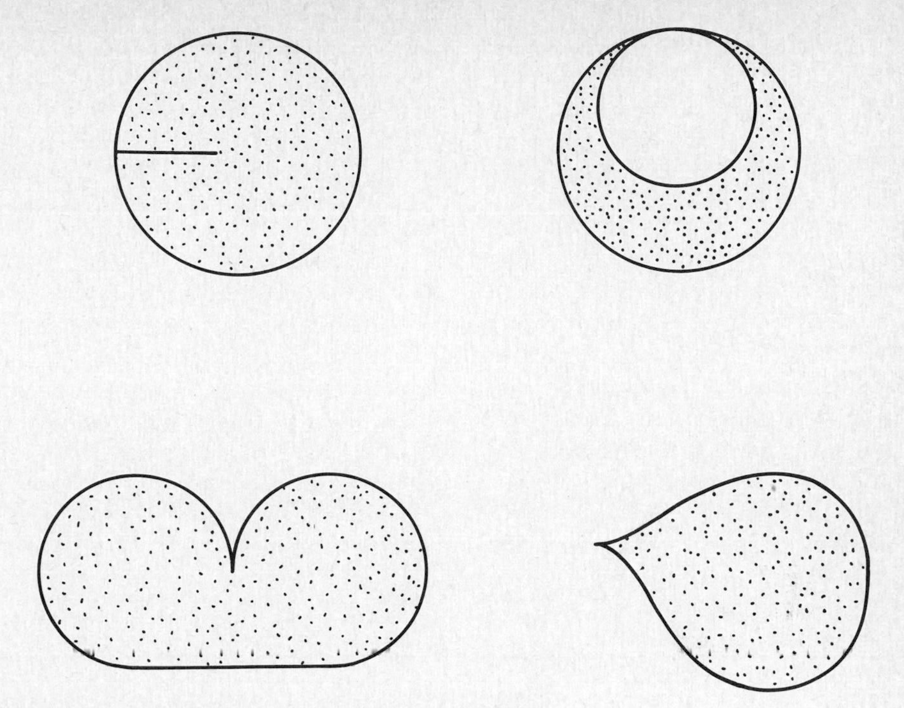

Figure 4-1 Some non-lipschitzian domains in two dimensions.

4-1-2 Representation of Linear Functionals

Since a linear functional is a special kind of linear operator, the concept of continuity for linear functionals is the same as that for linear operators (see Sec. 3-3). In a normed linear vector space, the concepts of continuity and boundedness are equivalent for linear functionals.

Bounded linear functionals on Hilbert spaces have simple representations. Consider an inner product space V and let v_0 be a fixed element in V. Consider the operator l defined by

$$l(u) = (v_0, u)$$

for every u in V. Clearly, l is a functional. It is linear because

$$l(\alpha u + \beta v) = (v_0, \alpha u + \beta v) = \alpha(v_0, u) + \beta(v_0, v)$$
$$= \alpha l(u) + \beta l(v)$$

Also, by the Schwarz inequality we have

$$|l(u)| = |(v_0, u)| \leq \|v_0\| \|u\| = M\|u\|$$

where $M = \|v_0\|$. Hence l is bounded. Further, note that $\|l\| \leq \|v_0\|$. However, we also have $|l(v_0)| = \|v_0\|^2$ or $\|l\| \geq \|v_0\|$. Thus the operator norm of l is given by

$\|l\| = \|v_0\|$. From this discussion it is clear that we can associate a bounded linear functional with each element of an inner product space. However, the converse is not true in general. The Riesz representation theorem states that the converse holds in Hilbert spaces. A geometrical interpretation of the theorem for $\mathbb{R}^3$, which is a Hilbert space, is given below.

Consider the space $\mathbb{R}^3$, and let f be a continuous linear functional $f\colon \mathbb{R}^3 \to \mathbb{R}$. In general, f takes positive values in a region $\Omega^+ \subset \mathbb{R}^3$ and negative values in a region $\Omega^- \subset \mathbb{R}^3$. Since f is linear, the set Ω of points where f vanishes (that is, Ω is the null space of f) is a subspace of $\mathbb{R}^3$. Any curve joining a point in Ω^+ to a point Ω^- must pass through Ω, because a continuous function cannot pass from a positive to a negative value without passing through zero. Thus Ω is a plane that separates Ω^+ from Ω^- in $\mathbb{R}^3$. Then any vector $\mathbf{x} \in \mathbb{R}^3$ can be resolved into a vector $\mathbf{y}$ in the plane Ω and a vector $\mathbf{z}$ perpendicular to the plane such that $\mathbf{x} = \mathbf{y} + \mathbf{z}$ (see Fig. 3-8). Therefore,

$$f(\mathbf{x}) = f(\mathbf{y} + \mathbf{z}) = f(\mathbf{z}) = \|\mathbf{z}\| f(\hat{\mathbf{e}})$$

where $\hat{\mathbf{e}}$ is a unit vector perpendicular to the plane (that is, $\mathbf{z} = \|\mathbf{z}\|\hat{\mathbf{e}}$). Then

$$\begin{aligned}(\mathbf{z}, \hat{\mathbf{e}}) &= \|\mathbf{z}\|(\hat{\mathbf{e}}, \hat{\mathbf{e}}) \\ &= \|\mathbf{z}\|\end{aligned}$$

and

$$\begin{aligned}f(\mathbf{x}) &= (\mathbf{z}, \hat{\mathbf{e}}) f(\hat{\mathbf{e}}) \\ &= (\mathbf{z}, \mathbf{u})\end{aligned}$$

where $\mathbf{u} = \hat{\mathbf{e}} f(\hat{\mathbf{e}})$. Thus, every linear functional $f(\mathbf{x})$ in $\mathbb{R}^3$ can be represented as an inner product. This result holds for Hilbert spaces, as stated in the following theorem.

Theorem 4-1: The Riesz representation theorem Let l be a bounded linear functional on a Hilbert space H. Then there is a unique vector v_0 in H such that

$$l(w) = (v_0, w)$$

for all w in H.

The vector $v_0 \in H$ is called the *representation* of l. It should be noted that l is a linear functional on H while v_0 is an element in H.

PROOF Let $S = \{u \in H\colon l(u) = 0\}$. If $S = H$, then l is identically zero, and $v = 0$ gives $l(u) = (u, v)$ for all u. If $S \neq H$, we must show that the orthogonal complement $S^\perp$ is one-dimensional so that for every $w \in H$ we have $(H = S + S^\perp)$

$$l(w) = l(u + v) = l(v) \qquad u \in S,\, v \in S^\perp$$

and the discussion of the preceding paragraph applies, that is, $l(w) = (v, v_0) = (w, v_0)$, where $v_0 = l(\hat{\mathbf{e}})\hat{\mathbf{e}}$ and $\hat{\mathbf{e}}$ is a unit basis vector of $S^\perp$.

To show that $S^\perp$ is one-dimensional, it is sufficient to show that every pair of vectors in $S^\perp$ is linearly dependent. Let $u, v \in S^\perp$. Then $\alpha u + \beta v \in S^\perp$ for any scalars α and β. In particular, the element $w \equiv l(v)u - l(u)v$ is in $S^\perp$. However, the element w is also in S because $l(w) = 0$. Therefore, being orthogonal to itself, $w = 0$. This implies that a linear combination of any two vectors from $S^\perp$ is zero with the constants nonzero. Therefore, $S^\perp$ is one-dimensional.

To prove the uniqueness, let $l(w) = (w, u_1) = (w, u_2)$ for all $w \in H$. Then

$$0 = (w, u_1) - (w, u_2)$$

$$0 = (w, u_1 - u_2)$$

For all $w = u_1 - u_2$ we get $\|u_1 - u_2\| = 0$ which gives $u_1 = u_2$.

As discussed earlier, the collection of all linear functionals l defined on a normed linear space V forms a Banach space, V^*, which is called the *conjugate* or the *dual* of the space V. For example, the space $L_p(\Omega)$ of all pth integrable functions defined on a bounded domain Ω is the dual of $L_q(\Omega)$ for $[(1/p) + (1/q)] = 1$. A linear functional defined on a Hilbert space is spanned by elements of the same space. Therefore, a Hilbert space is self-dual. In particular, the n-dimensional euclidean space $\mathbb{R}^n$ is self-dual.

Since the conjugate V^* of a normed space is also a normed space, it is possible to construct $V^{**} \equiv (V^*)^*$, the conjugate of the conjugate. The space V^{**} is the space of linear functionals l^* defined on V^*, whose elements are themselves linear functionals l defined on V. When $V = V^{**}$, the space V is called *reflexive*. We remark that the spaces $\mathbb{R}^n$ and $L_p(\Omega)$, $p > 1$, are reflexive, whereas the space $C(\Omega)$ is not reflexive.

4-1-3 Adjoint Operators

Consider a bounded linear operator T which maps a normed linear space U into a normed linear space V. Let $\phi(v)$ be a linear functional defined on V. Then, $\phi(v)$ is defined for $v = Tu$, $u \in U$. We have

$$\phi(v) = \phi(Tu) \equiv l(u)$$

where $l(u)$ is a functional defined on U. Clearly, $l(u)$ is linear. Hence, the functional $l \in U^*$ corresponds to every $\phi \in V^*$. The collection of all correspondences so constructed forms a certain operator T^* with domain V^* and range contained in U^*. The operator T^* is called the *adjoint* of T. The equality $\phi(v) = l(u)$ is expressed as $l = T^*\phi$.

The expression $l(u) = \langle f, u \rangle$ is a bilinear functional of the two variables $u \in U$ and $f \in U^*$. When U is a Hilbert space, we have $U = U^*$ and $\langle \cdot, \cdot \rangle$ becomes the inner product, and also in the general case when $U \neq U^*$ and $f \in U^*$ the expression $\langle f, u \rangle$ is called the duality pairing. In the following paragraphs, we consider adjoint operators on Hilbert spaces.

Consider a bounded operator T on a Hilbert space H (that is, $T: H \to H$). For a fixed element v in H, the inner product (Tu, v) in H can be regarded as a number which varies with u. Therefore, $(Tu, v) = l(u)$ is a linear functional on H. Since T is bounded, using the Schwarz inequality, we can show that $l(u)$ is bounded. Therefore, by the Riesz representation theorem there exists a unique element v_0 in H such that

$$(Tu, v) = (u, v_0)$$

for all u in H. This implies that given a $v \in H$ there is a unique v_0 associated with element v. In other words, there exists a mapping T^* of H into itself such that $v_0 = T^*v$. We call T^* the *adjoint* (not to be confused with the adjoint of a matrix) of T. We have

$$(Tu, v) = (u, T^*v) \qquad \text{for all } u, v \in H \tag{4-3}$$

It can be easily verified that the adjoint of a linear bounded operator is also linear and bounded with $\|T^*\| = \|T\|$. Furthermore, the adjoint is uniquely defined. The following properties of adjoint operators should be noted:

1. $(T_1 + T_2)^* = T_1^* + T_2^*$,
2. $(\alpha T_1)^* = \alpha T_1^*$, α is a real number,
3. $(T_1 T_2)^* = T_2^* T_1^*$,
4. $(T^*)^* = T$,
5. $(T^{-1})^* = (T^*)^{-1}$, if T^{-1} exists.

The uniqueness of T^* is easy to prove. If T_1^* and T_2^* are both adjoints of T, then from Eq. (4-3) we have

$$(Tu, v) = (u, T_1^* v) = (u, T_2^* v)$$

or

$$(u, (T_1^* - T_2^*)v) = 0 \qquad \text{for all } u, v \in H$$

This implies, since u is arbitrary, $(T_1^* - T_2^*)v = 0$ giving $T_1^* = T_2^*$. To show that T^* is bounded, we write

$$\begin{aligned} \|T^*v\|^2 &= (T^*v, T^*v) \\ &= (TT^*v, v) \\ &\le \|TT^*v\| \, \|v\| \\ &\le \|T\| \, \|T^*v\| \, \|v\| \end{aligned}$$

or

$$\|T^*v\| \le \|T\| \, \|v\|$$

Hence T^* is bounded. We have

$$\|T^*\| \le \|T\|$$

On the other hand, we have

$$\|T^{**}\| \leq \|T^*\|$$

$$\|T\| \leq \|T^*\|$$

Therefore, it follows that $\|T\| = \|T^*\|$.

A bounded linear operator T is said to be *self-adjoint* if $T = T^*$:

$$(Tu, v) = (u, Tv) \tag{4-4}$$

Note that any bounded operator $A: H \to H$ of the form

$$A = T^*T \tag{4-5}$$

is self-adjoint, because

$$(Au, v) = (T^*Tu, v) = (Tu, Tv) = (u, Av)$$

We now turn to unbounded operators on Hilbert spaces. Let T be a linear operator on a dense subset $\mathscr{D}_T$ of a real Hilbert space H. The adjoint operator T^* is defined for all those elements v for which the functional (Tu, v) is bounded, by Eq. (4-3). The domain of T^* is defined to be the space

$$\mathscr{D}_T^* = \{v\colon\ v \in H,\ (Tu, v) = (u, v_0) \text{ for some } v_0 \text{ in } H \text{ and } u \in \mathscr{D}_T\}$$

so that the mapping $v \to v_0$ is uniquely defined. Note that the domains $\mathscr{D}_T$ and $\mathscr{D}_T^*$ are, in general, different for unbounded operators.

A linear operator T defined on dense subspace $\mathscr{D}_T$ of a Hilbert space H is said to be *closed* if for every sequence $\{u_n\}$ in $\mathscr{D}_T$, with the property that both limits $u = \lim u_n$ and $v = \lim Tu_n$ exist, one has $u \in \mathscr{D}_T$ and $v = Tu$. We remark that the domain $\mathscr{D}_T^*$ of the adjoint T^* of a closed operator T is dense in H.

A linear operator $T\colon \mathscr{D}_T \to H$ is said to be *self-adjoint* if $\mathscr{D}_T = \mathscr{D}_T^*$ and $Tu = T^*u$ for all $u \in \mathscr{D}_T$. If T is bounded, $\mathscr{D}_T = \mathscr{D}_T^* = H$. An unbounded self-adjoint operator is always closed.

A linear operator T is said to be *normal* if $T^*T = TT^*$. Clearly, every self-adjoint operator is normal.

Example 4-2 1. Let $\mathscr{D}_T$ be the subspace of $L_2[0, 1]$ consisting of continuously differentiable functions $u(x)$ on $0 \leq x \leq 1$ with $u(0) = 0$, and T be the first-order differential operator, $Tu = du/dx$. We wish to calculate T^*. We write

$$\int_0^1 Tuv\,dx = \int_0^1 \frac{du}{dx} v\,dx = \int_0^1 u\left(-\frac{dv}{dx}\right) dx + (uv)\Big|_0^1$$

Since $u(0) = 0$, one of the boundary terms vanishes. The second boundary term vanishes if we define $\mathscr{D}_T^*$ to be the subset of $L_2[0, 1]$ consisting of continuously differentiable functions $v(x)$ on $0 \leq x \leq 1$ with $v(1) = 0$. Then we have $T^* = -d/dx$. Since $T^* \neq T$ and $\mathscr{D}_T^* \neq \mathscr{D}_T$, T is *not* self-adjoint.

2. Let $\mathscr{D}_T$ be the subspace of $H^1[0, 1]$ of functions $u(x)$ with continuous first derivatives on $0 \le x \le 1$ with $u(0) = u(1) = 0$, and let $T = d^2/dx^2$. The adjoint of T is given by

$$(Tu, v) = \int_0^1 \left(\frac{d^2u}{dx^2}\right) v\, dx = \int_0^1 \frac{du}{dx}\left(-\frac{dv}{dx}\right) dx + \left(\frac{du}{dx} v\right)\Big|_0^1$$

$$= \int_0^1 u\left(\frac{d^2v}{dx^2}\right) dx + \left[\frac{du}{dx} v - u \frac{dv}{dx}\right]_0^1$$

If we define $\mathscr{D}_T^*$ to be the subspace of $H^1[0, 1]$ of functions $v(x)$ with continuous first derivatives on $0 \le x \le 1$ with $v(0) = v(1) = 0$, we have

$$(Tu, v) = \int_0^1 u \frac{d^2v}{dx^2} dx = (u, T^*v)$$

or $T^* = d^2/dx^2 = T$. Hence, T is self-adjoint.

The notion of an adjoint operator can be extended to linear operators defined from one Hilbert space into another. Let $T\colon \mathscr{D}_T \subset H_1 \to H_2$ be a linear operator, and let v be a fixed element in H_2. As u varies over H_1, $(Tu, v)_2$ takes on various numerical values, so that $(Tu, v)_2$ is a linear functional (not necessarily bounded) in u. The operator $T^*\colon H_2 \to H_1$ is the adjoint of T if the following relation holds:

$$(Tu, v)_2 = (u, T^*v)_1 \tag{4-6}$$

Here $(\cdot, \cdot)_1$ and $(\cdot, \cdot)_2$ denote the inner products in H_1 and H_2, respectively. We have to be more precise about the domain of T^*. We define $\mathscr{D}_T^*$ as follows:

$$\mathscr{D}_T^* = \{v\colon v \in H_2, (Tu, v)_2 = (u, v_0)_1 \text{ for some } v_0 \text{ in } H_1 \text{ and all } u \text{ in } \mathscr{D}_T\}$$

Example 4-3 Let $\mathscr{D}_T$ be the subspace of $L_2(\Omega)$ consisting of continuously differentiable functions u on a bounded domain Ω in three-dimensional euclidean space, $\mathbb{R}^3$. Let T be the gradient operator mapping u in $\mathscr{D}_T$ into a continuous vector-valued function $\mathbf{v} = (v_1, v_2, v_3)$ in the product space $H_2 = L_2(\Omega) \times L_2(\Omega) \times L_2(\Omega)$. We choose the inner product in H_2 to be the integral of the vector dot product:

$$(\mathbf{v}, \mathbf{w})_2 = \int_\Omega \mathbf{v} \cdot \mathbf{w}\, d\mathbf{x}$$

We wish to calculate the adjoint of $T = \text{grad}$. We have

$$(Tu, \mathbf{v})_2 = \int_\Omega (\text{grad } u) \cdot \mathbf{v}\, d\mathbf{x}$$

$$= \int_\Omega [\text{div } (u\mathbf{v}) - u \text{ div } \mathbf{v}]\, d\mathbf{x}$$

$$= \int_\Omega u(-\text{div } \mathbf{v})\, dx + \oint_\Gamma \hat{\mathbf{n}} \cdot (u\mathbf{v})\, ds$$

Here $d\mathbf{x} = dx_1\, dx_2\, dx_3$ and $\hat{\mathbf{n}}$ is the unit outward normal to the surface Γ. It follows that $\mathscr{D}_T^*$ is the subspace of H_2 with $\hat{\mathbf{n}}\cdot\mathbf{v} = 0$ on Γ, and $T^* = -\text{div}$. The inner product $(\cdot, \cdot)_1$ in $\mathscr{D}_T$ is taken to be the L_2-inner product.

4-1-4 Symmetric, Positive, and Positive-Definite Operators

An operator A, linear in its domain $\mathscr{D}_A$, is called *symmetric* in $\mathscr{D}_A \subset H$ if for every pair of elements u, v from $\mathscr{D}_A$ we have

$$(Au, v) = (u, Av) \tag{4-7}$$

Clearly, every self-adjoint operator (that is, $A = A^*$) is always a symmetric operator.

A symmetric operator is said to be *positive in its domain* $\mathscr{D}_A$ if for all u in $\mathscr{D}_A$ the following relation holds:

$$(Au, u) \geq 0 \qquad \text{and} \qquad (Au, u) = 0 \qquad \text{implies } u = 0 \text{ in } \mathscr{D}_A \tag{4-8}$$

Further, if we can find a constant $\gamma > 0$ such that for all u in $\mathscr{D}_A$ the relation

$$(Au, u) \geq \gamma \|u\|^2 \tag{4-9}$$

holds, then the operator A is called *positive-definite in* $\mathscr{D}_A$.

Example 4-4 Let $H = L_2(\Omega)$, where $\Omega \subset \mathbb{R}^2$ is a plane domain. Let $\mathscr{D}_A$ be the set of functions from $C^2(\Omega)$ which vanish on the boundary Γ. Since $C^2(\Omega)$ is dense in $L_2(\Omega)$, $\mathscr{D}_A$ is dense in $L_2(\Omega)$. Let A be the differential operator, $A \equiv -(\partial^2/\partial x^2 + \partial^2/\partial y^2) \equiv -\nabla^2$. First we show that A is symmetric on $\mathscr{D}_A$. Consider

$$\begin{aligned}(Au, v)_0 &= \int_\Omega -\left(\frac{\partial^2 u}{\partial x^2} + \frac{\partial^2 u}{\partial y^2}\right) v\, dx\, dy = \int_\Omega (-\nabla^2 u) v\, dx\, dy \\ &= \int_\Omega -[\nabla\cdot(\nabla u v) - \nabla u\cdot\nabla v]\, dx\, dy\end{aligned}$$

Using the divergence theorem

$$\int_\Omega \nabla\cdot\mathbf{f}\, dx\, dy = \oint_\Gamma \hat{\mathbf{n}}\cdot\mathbf{f}\, ds$$

where $\hat{\mathbf{n}}$ is the unit outward normal, we obtain

$$\begin{aligned}(Au, v)_0 &= \int_\Omega \nabla u\cdot\nabla v\, dx\, dy - \oint_\Gamma (\hat{\mathbf{n}}\cdot\nabla u) v\, ds \\ &= \int_\Omega \left(\frac{\partial u}{\partial x}\frac{\partial v}{\partial x} + \frac{\partial u}{\partial y}\frac{\partial v}{\partial y}\right) dx\, dy - \oint_\Gamma \left(\frac{\partial u}{\partial x} n_x + \frac{\partial u}{\partial y} n_y\right) v\, ds\end{aligned}$$

Since $v = 0$ on Γ, the boundary term vanishes. Due to the symmetry of the right side in u and v, we immediately have $(Au, v)_0 = (Av, u)_0 = (u, Av)_0$.

Next we prove that A is positive. For u in $\mathscr{D}_A$ we have

$$(Au, u)_0 = \int_\Omega \left[\left(\frac{\partial u}{\partial x}\right)^2 + \left(\frac{\partial u}{\partial y}\right)^2\right] dx\, dy \geq 0$$

If $(Au, u)_0 = 0$, then it follows that (since $\partial u/\partial x = \partial u/\partial y = 0$) $u =$ constant. Since $u = 0$ on Γ it follows that this constant is zero, and $u = 0$. Thus $(Au, u)_0 = 0$ implies $u = 0$. This proves that A is positive.

To prove that A is positive-definite on $\mathscr{D}_A$, we invoke the *Friedrichs inequality* [see Necas (1967)]: For u in $C^1(\Omega)$, the following inequality holds:

$$\int_\Omega u^2(\mathbf{x})\, d\mathbf{x} \leq c_1 \sum_{k=1}^{2} \int_\Omega \left(\frac{\partial u}{\partial x_k}\right)^2 d\mathbf{x} + c_2 \oint_\Gamma u^2(s)\, ds \tag{4-10}$$

where c_1 and c_2 are nonnegative constants dependent on Ω but independent of u. For u in $\mathscr{D}_A$ we have $u = 0$ on Γ; hence the second term on the right side of (4-10) is zero. Consequently,

$$(Au, u)_0 \geq \frac{1}{c_1} \int_\Omega u^2(x)\, dx = \frac{1}{c_1} \|u\|_0$$

Thus A is positive-definite on $\mathscr{D}_A$.

PROBLEMS 4-1

4-1-1 Let A and B be linear operators, and let A^* and B^* denote the adjoints of A and B, respectively. Prove the following identities: (*a*) $(A + B)^* = A^* + B^*$; (*b*) $(AB)^* = B^*A^*$; (*c*) $\|AA^{**}\| = \|A\|^2$, A is a bounded operator.

4-1-2 Let A be a linear operator mapping $\mathbb{R}^n$ into itself: $\mathbf{y} = A\mathbf{x}$, where $\mathbf{x} = (x_1, x_2, \ldots, x_n)$ and $\mathbf{y} = (y_1, y_2, \ldots, y_n)$ in $\mathbb{R}^n$. Show that the matrix representing A^* is the transpose of the matrix representing A.

4-1-3 Find the adjoint of the first-order differential operator with respect to the scalar product

$$(u, v) \equiv \int_0^T u(t)v(T - t)\, dt$$

4-1-4 Let $A\colon H \to H$ be a bounded linear operator on a Hilbert space H. Show that there exists a unique operator $A^*\colon H \to H$ such that

$$(u, A^*v) = (Au, v) \qquad \text{for all } u, v \in H$$

Show that A^* is linear and bounded, $\|A^*\| = \|A\|$, and $(A^*)^* = A$.

Hint: Use the Riesz representation theorem.

4-1-5 Let A be a bounded linear operator on a Hilbert space H. Define $\mathscr{R}(A) = \{Au : u \in H\}$ and $\mathscr{N}(A) = \{u \in H : Au = 0\}$. Prove that $\mathscr{N}(A)$ is a subspace and $\mathscr{R}(A)$ a vector subspace of H, and that $[\mathscr{R}(A)]^\perp = \mathscr{N}(A^*)$.

4-1-6 (*Elastic bending of a clamped beam*) Consider the fourth-order differential equation,

$$\frac{d^2}{dx^2}\left(b(x)\frac{d^2 w}{dx^2}\right) + kw = f(x) \qquad 0 < x < L$$

which governs the transverse deflection w of a beam of flexural rigidity $b(x)$ on an elastic foundation (foundation modulus $k > 0$) and subjected to the distributed loading $f(x)$. Assuming the boundary conditions of a clamped beam, $w = dw/dx = 0$ at $x = 0$ and $x = L$, show that the operator is self-adjoint on its domain.

4-1-7 Prove that a continuous projection P on a Hilbert space H is orthogonal if and only if it is self-adjoint.

4-1-8 Let H be a Hilbert space, and let S be a closed linear subspace of H. If P_1 and P_2 are the orthogonal projections of H onto S and $S^{\perp}$, respectively, show that the operator $T = \lambda_1 P_1 + \lambda_2 P_2$ is self adjoint if and only if λ_1 and λ_2 are real numbers.

4-1-9 Show that the operator T defined in the preceding problem is normal.

4-1-10 Show that the differential operator $D = d/dx : C^1[a, b] \to C[a, b]$, where $C^1[a, b]$ is the space of continuously differentiable functions, is bounded with respect to the $L_2[a, b]$-norm.

Show that the operators in Probs. 4-1-11 to 4-1-14 are positive on the domain of the operator.

4-1-11
$$-\frac{d^2u}{dx^2} + (1 + \cos^2 x)u = \sin x \qquad 0 < x < \pi$$
$$u(0) = u(\pi) = 0$$

The domain of the operator $A = -(d^2/dx^2) + 1 + \cos^2 x$ is defined by $\mathscr{D}_A = \{u : u \in C^2[0, \pi], u(0) = u(\pi) = 0\}$.

4-1-12
$$\frac{d^2}{dx^2}\left(EI\,\frac{d^2u}{dx^2}\right) = f \qquad (E > 0, I > 0) \qquad 0 < x < L$$
$$u(0) = u(L) = u''(0) = u''(L) = 0 \qquad (u'' = d^2u/dx^2)$$

The domain of the operator is the linear set of functions which are continuous with their derivatives up to the fourth order inclusive in the interval $[0, L]$ and satisfy the boundary conditions.

4-1-13
$$-\left(\frac{\partial^2 u}{\partial x^2} + \frac{\partial^2 u}{\partial y^2}\right) = f \qquad 0 < x < 1 \qquad 0 < y < 1$$
$$u(x, 0) = u(x, 1) = u(0, y) = u(1, y) = 0$$

The domain of the operator is the linear set of functions which are continuous with their derivatives up to the second order inclusive in the domain $\Omega \subset \mathbb{R}^2$ and satisfy the boundary conditions.

4-1-14
$$\nabla^2\nabla^2 w = f \text{ in } \Omega \subset \mathbb{R}^2$$
$$w = \partial w/\partial n = 0 \text{ on } \Gamma$$

The domain of the operator $A = \nabla^2\nabla^2 = \partial^4/\partial x^4 + 2(\partial^4/\partial x^2\,\partial y^2) + (\partial^4/\partial y^4)$ is defined by
$$\mathscr{D}_A = \{w : w \in C^4(\Omega) \qquad w = \partial w/\partial n = 0 \text{ on } \Gamma\}$$

4-2 SOBOLEV SPACES AND THE CONCEPT OF GENERALIZED SOLUTION

4-2-1 Sobolev Spaces, $H^m(\Omega)$

Many of the functional-analytical techniques developed in this book are for the solution of differential equations. A basic difficulty in working with differential operators is that they are unbounded, and are not defined on the whole of

Banach spaces $C(\Omega)$ or $L_2(\Omega)$. As pointed out earlier, differential operators can be made bounded and continuous by using a different norm. For example, differentiation is a continuous operator on the space $C[a, b]$ of continuously differentiable functions with the norm

$$\|u\|_1 = \left[\int_a^b (|u|^2 + |u'|^2)\, dx\right]^{1/2}$$

However, with respect to that norm the space $C[a, b]$ is not complete.

We shall take a distributional approach to the development of special complete spaces (or Hilbert spaces), known as the Sobolev spaces. While the spaces of differentiable ordinary functions are incomplete, the spaces of differentiable generalized functions are complete. These spaces of differentiable generalized functions are the Sobolev spaces. For example, $H^1(\Omega)$, the Sobolev space of order 1, is a space of distribution on $\Omega \subset \mathbb{R}^2$ with the norm

$$\|u\|_1 = \left[\int_\Omega \left(|u|^2 + \left|\frac{\partial u}{\partial x_1}\right|^2 + \left|\frac{\partial u}{\partial x_2}\right|^2\right) d\mathbf{x}\right]^{1/2}$$

This norm is meaningful only for those distributions in $L_2(\Omega)$ whose derivatives are square-integrable. To describe the simplest of Sobolev spaces, we first introduce the concepts of generalized derivative and distribution.

Generalized derivative A function $\phi: \mathbb{R}^N \rightarrow \mathbb{R}$ is said to be *smooth* or infinitely differentiable if its derivatives of all orders exist and are continuous. The set of all smooth functions with *compact support* in $\Omega \subset \mathbb{R}^N$ is denoted by $C_0^\infty(\Omega)$. The subscript "0" on $C_0^\infty(\Omega)$ indicates that functions $\phi \in C_0^\infty(\Omega)$ vanish near the boundary of Ω. The *support* of a function $f(\mathbf{x})$, $\mathbf{x} \in \Omega \subset \mathbb{R}^N$, is defined to be the closure of the set of points in $\mathbb{R}^N$ at which f is nonzero.

A *test function* ϕ is a smooth function with compact support, $\phi \in C_0^\infty(\Omega)$. We hardly need to consider specific test functions. They are needed for theoretical purposes only.

A function $u \in C^m(\Omega)$ is said to have the $\boldsymbol{\alpha}$th *generalized derivative* $D^{\boldsymbol{\alpha}}u$, $1 \leq |\boldsymbol{\alpha}| \leq m$, if the following relation (generalized Green's formula) holds:

$$\int_\Omega D^{\boldsymbol{\alpha}} u \phi \, dx = (-1)^{|\boldsymbol{\alpha}|} \int_\Omega u D^{\boldsymbol{\alpha}} \phi \, dx \qquad \text{for every } \phi \in C_0^\infty(\Omega) \tag{4-11}$$

For $u \in C_0^\infty(\Omega)$, the generalized derivatives are derivatives in the ordinary (i.e., classical) sense.

Distribution A set of test functions $\{\phi_n\}$ is said to converge to a test function ϕ_0 in $C_0^\infty(\Omega)$ if there is a bounded set $\Omega_0 \subset \Omega$ containing the supports of $\phi_0, \phi_1, \phi_2, \ldots$, and if ϕ_n and all its generalized derivatives converge to ϕ_0 and its derivatives, respectively. A functional f on $C_0^\infty(\Omega)$ is *continuous* if it maps every convergent sequence in $C_0^\infty(\Omega)$ into a convergent sequence in $\mathbb{R}$, i.e., if $f(\phi_n) \rightarrow f(\phi_0)$ whenever $\phi_n \rightarrow \phi_0$ in $C_0^\infty(\Omega)$.

A continuous linear functional on $C_0^\infty(\Omega)$ is called a *distribution* or *generalized function.*

An example of the distribution is provided by the delta distribution δ, defined by

$$\langle \delta, \phi \rangle = \phi(0) \qquad \text{for all } \phi \in C_0^\infty(\Omega) \tag{4-12}$$

The notation $\langle f, \phi \rangle$ is used to denote the "action" of the distribution f on the test function ϕ.

The operations of addition and scalar multiplication of distributions are defined as follows: if f and g are distributions and α and β are scalars, we define the distribution $\alpha f + \beta g$ to be the functional $\alpha\langle f, \phi \rangle + \beta\langle g, \phi \rangle$ for all $\phi \in C_0^\infty(\Omega)$. We are now ready to introduce the Sobolev spaces.

Recall that the space $C^\infty(\Omega)$ is an inner product space with respect to the $L_2(\Omega)$-inner product. But it is incomplete with respect to the norm generated by the inner product

$$(u, v)_m = \int_\Omega \sum_{|\boldsymbol{\alpha}| \leq m} D^{\alpha}u D^{\alpha}v \, d\mathbf{x} \tag{4-13}$$

where u and v, along with their derivatives up to m, are square integrable in the Lebesgue sense

$$\int_\Omega |D^\alpha u|^2 \, d\mathbf{x} < \infty \qquad \text{for } |\boldsymbol{\alpha}| \leq m$$

The Lebesgue definition of the integral, together with its basic properties, can be found in Kolmogorov and Fomin (1957). However, to understand and use the results here it is not necessary for the reader to become acquainted with the details of the Lebesgue integration. The space $C^\infty(\Omega)$ can be completed by adding the limit points of all Cauchy sequences in $C^\infty(\Omega)$. It turns out that the distributions are these limit points.

Two Cauchy sequences $\{\phi_n\}$ and $\{\psi_n\}$ are said to be *equivalent* if $\|\phi_n - \psi_n\|_0 \to 0$ as $n \to \infty$. The set S of all distributions generated by Cauchy sequences in $C^\infty(\Omega)$ is itself a vector space: if $f, g \in S$ are generated by $\{\phi_n\}$ and $\{\psi_n\}$, respectively, then the distribution $\alpha f + \beta g$ is generated by the Cauchy sequence $(\alpha\phi_n + \beta\psi_n)$. When the space S is endowed with the inner product

$$(f, g)_m = \lim_{n \to \infty} (\phi_n, \psi_n)_m \tag{4-14}$$

where $(\cdot, \cdot)_m$ is defined by Eq. (4-13), the resulting inner product space is the Sobolev space, $W_2^m(\Omega) = H^m(\Omega)$. In other words, $W_2^m(\Omega)$ is the space of all distributions generated by Cauchy sequences in $C^\infty(\Omega)$, and it also contains all functions along with their generalized derivatives up to and including the order required in the inner product $(\cdot, \cdot)_m$, are square-integrable in Ω. Clearly, the Sobolev space $W_2^m(\Omega)$ is a Hilbert space. We shall use the notation

$$W_2^m(\Omega) = H^m(\Omega)$$

$$W_2^0(\Omega) = H^0(\Omega) = L_2(\Omega)$$

interchangeably. Also, we have $H^{m_1}(\Omega) \subset H^{m_2}(\Omega)$ for $m_1 \geq m_2 \geq 0$.

The space $H_0^m(\Omega)$ is defined as the closure of $C_0^\infty(\bar{\Omega})$ in the norm

$$\|u\|_m^2 = \sum_{|\alpha| \leq m} \|D^\alpha u\|_0^2 \tag{4-15}$$

where $\| \cdot \|_0$ denotes the $L_2(\Omega)$-inner product.

It is possible to define the Sobolev spaces $H^m(\Omega)$ for both fractional and negative values of m. For $0 < s < [s] + \theta$, $0 < \theta < 1$, where $[s]$ denotes the integral part of s (e.g., if $s = 2.65$ then $[s] = 2$ and $\theta = 0.65$), we define $H^s(\Omega)$ to be the closure of $C^\infty(\Omega)$ with respect to the norm

$$\|u\|_s^2 = \|u\|_{[s]}^2 + \sum_{|\alpha| = [s]} \|D^\alpha u\|_\theta^2 \tag{4-16a}$$

where
$$\|u\|_\theta^2 = \int_\Omega \int_\Omega \frac{|u(t) - u(\tau)|^2}{\|t - \tau\|^{N+2\theta}} \, dt \, d\tau \tag{4-16b}$$

and N is the dimension of the domain Ω. Similarly, $H_0^s(\Omega)$ denotes the closure of $C_0^\infty(\Omega)$ with respect to the norm in Eq. (4-16a).

The spaces $H^s(\Omega)$ and $H_0^s(\Omega)$, for $s < 0$, are defined to be the closure of $C^\infty(\Omega)$ and $C_0^\infty(\Omega)$, respectively, with respect to the norm

$$\|u\|_s = \sup_{u \in H^{-s}(\Omega)} \frac{\int_\Omega uv \, d\mathbf{x}}{\|u\|_{-s}} \tag{4-17}$$

4-2-2 Traces of Functions from $H^m(\Omega)$

In the previous section we introduced the Sobolev space $W_2^m(\Omega) = H^m(\Omega)$ as a complete space of functions which with their generalized derivatives up to and including order m belong to $L_2(\Omega)$. For every continuous function $u(\mathbf{x}) \in H^m(\Omega)$, $\mathbf{x} \in \bar{\Omega}$, the function $u(s)$, $s \in \Gamma$, is called the *trace of the function* $u(\mathbf{x})$ on Γ. Since the trace of a function $u(\mathbf{x}) \in C^\infty(\bar{\Omega})$ is continuous on Γ, it is square integrable on Γ. Therefore, $u(s) \in L_2(\Gamma)$. If $u \in L_2(\Omega)$ is continuous on $\bar{\Omega}$, then the trace of u is defined by the value of u on Γ. However, when $u \in L_2(\Omega)$ is not continuous in Ω, then we must define what we mean by the trace of u. We approximate the function u in Ω by a sequence of functions $\{u_k(\mathbf{x})\}$ that are sufficiently smooth in $\bar{\Omega}$, and define the trace $u(s)$ of $u(\mathbf{x})$ as the limit of the corresponding sequence of traces $\{u_k(s)\}$. The convergence of the traces in $L_2(\Gamma)$ is not always guaranteed. The following theorem guarantees the convergence of the sequence $\{u_k(s)\}$ when $u(\mathbf{x}) \in H^1(\Omega)$.

Theorem 4-2: Extension theorem Let Ω be a domain with a Lipschitz boundary. Then there exists a unique bounded operator T which maps the space $H^1(\Omega)$ into the space $L_2(\Gamma)$. For $u \in C^\infty(\Omega)$ we have $Tu(\mathbf{x}) = u(s)$ and

$$\|Tu\|_m \leq c(m)\|u\|_m \qquad \text{for any } m \tag{4-18}$$

In the discussion of boundary-value problems, we will have boundary conditions of the form $u = g(s)$ on Γ. If $u \in H^1(\Omega)$ then it should be understood that $g(s)$ is the trace of $u(\mathbf{x})$ on the boundary and $g \in L_2(\Gamma)$.

If $u \in H^m(\Omega)$, then $D^\alpha u \in H^1(\Omega)$ for $|\boldsymbol{\alpha}| \leq m - 1$. Then by Theorem 4-2, to each of $D^\alpha u$ there corresponds a function $D^\alpha u(s) \in L_2(\Gamma)$. The function $D^\alpha u(s)$ is called the trace of the function $D^\alpha u(\mathbf{x})$. This allows us to introduce, for $m \geq 2$, the derivative of $u \in H^m(\Omega)$ with respect to the normal to the boundary Γ. For $u \in H^m(\Omega)$ we define

$$\gamma_1 u \equiv \frac{\partial u}{\partial n} = \sum_{j=1}^{N} \frac{\partial u}{\partial x_j}(s) n_j(s) \tag{4-19}$$

where n_j ($j = 1, 2, \ldots, N$) denotes the direction cosines of the unit normal and N is the dimension of the domain $\Omega \subset \mathbb{R}^N$. Similarly, the higher-order normal derivative $\gamma_i u \equiv \partial^i u / \partial n_i$ can be defined for any $u \in H^m(\Omega)$, $m \geq i + 1$. The following result is very useful in the sequel.

Theorem 4-3: The Trace theorem The trace operators γ_j can be extended to continuous linear operators mapping $H^m(\Omega)$ onto $H^{m-j-1/2}(\Gamma)$,

$$\|\gamma_j u\|_{m-j-1/2} \leq c_j \|u\|_m \qquad 0 \leq j \leq m - 1 \tag{4-20}$$

where $c_j > 0$ are constants independent of u. The kernel (null space) of the trace operator γ_j is $H_0^m(\Omega)$,

$$\gamma_j(H_0^m(\Omega)) = 0 \qquad j = 0, 1, \ldots, m - 1 \tag{4-21}$$

4-2-3 The Friedrichs and Poincaré Inequalities

We state (without proof) the Friedrichs [see Eq. (4-10)], Poincaré and other inequalities, which are very useful in establishing the positive-definite property of an operator (or a bilinear form). See Necas (1967) for additional details.

The Friedrich inequality Let Ω be a domain with a Lipschitz boundary Γ, and let Γ_1 be its (open) part with a positive Lebesgue measure. Then there exists a constant $c_2 > 0$, depending only on the given domain and on Γ_1, such that for every $u \in H^1(\Omega)$ we have

$$\|u\|_1^2 \leq c_2 \left\{ \sum_{j=1}^{N} \int_\Omega \left(\frac{\partial u}{\partial x_j} \right)^2 d\mathbf{x} + \int_{\Gamma_1} |u(s)|^2 \, ds \right\} \tag{4-22a}$$

Also, for $u \in H^2(\Omega)$, we have

$$\|u\|_2^2 \leq c_2(\Omega) \left\{ \sum_{|\alpha|=2} \int_\Omega |D^\alpha u|^2 \, d\mathbf{x} + \int_\Gamma |u(s)|^2 \, ds \right\} \tag{4-22b}$$

Note that for $u \in H_0^m(\Omega)$, inequalities (4-22a) and (4-22b) hold without the boundary terms.

The Poincaré inequality Let Ω be a domain with a Lipschitz boundary Γ. Then there exists a constant $c_3 > 0$ such that, for $u \in H^m(\Omega)$, we have

$$\|u\|_m^2 \leq c_3(\Omega)\left\{\sum_{|\alpha|=m} \int_\Omega |D^\alpha u|^2 \, d\mathbf{x} + \sum_{|\alpha|<m} \left(\int_\Omega D^\alpha u \, d\mathbf{x}\right)^2\right\} \tag{4-23a}$$

In particular, for $m = 1$, we have

$$\|u\|_1^2 \leq c_3(\Omega)\left\{\sum_{j=1}^{N} \int_\Omega \left|\frac{\partial u}{\partial x_j}\right|^2 d\mathbf{x} + \left(\int_\Omega u \, dx\right)^2\right\} \tag{4-23b}$$

We now prove Friedrich's inequality (4-10) for $u \in H_0^1(\Omega)$.

Let $\Omega \subset \mathbb{R}^2$ be an open bounded region. Let $u(x_1, x_2)$ be a continuous function which is continuously differentiable in $\bar{\Omega} = \Omega + \Gamma$ and vanishes on Γ. The region $\bar{\Omega}$ can be enclosed in a square region K (see Fig. 4-2):

$$K = \{(x_1, x_2)\colon\ 0 \leq x_1 \leq a,\ 0 \leq x_2 \leq a\}$$

The function $u(\mathbf{x})$ can be extended onto the set K by setting $u = 0$ on $K - \Omega$. Hence, u is continuously differentiable in K and vanishes on the boundary of K. Now consider the identity

$$u(\mathbf{x}) = \int_0^{x_1} \frac{\partial u}{\partial \xi}(\xi, x_2) \, d\xi$$

Using the Schwarz inequality, we write

$$|u(\mathbf{x})|^2 = \left|\int_0^{x_1} \frac{\partial u}{\partial \xi}(\xi, x_2)\, d\xi\right|^2 \leq \left\{\left[\int_0^{x_1} 1^2 \, d\xi\right]^{1/2}\left[\int_0^{x_1} \left|\frac{\partial u}{\partial \xi}\right|^2 d\xi\right]^{1/2}\right\}^2$$

$$\leq x_1\left[\int_0^{x_1} \left|\frac{\partial u}{\partial \xi}\right|^2 d\xi\right] \leq a \int_0^{a} \left|\frac{\partial u}{\partial \xi}\right|^2 d\xi$$

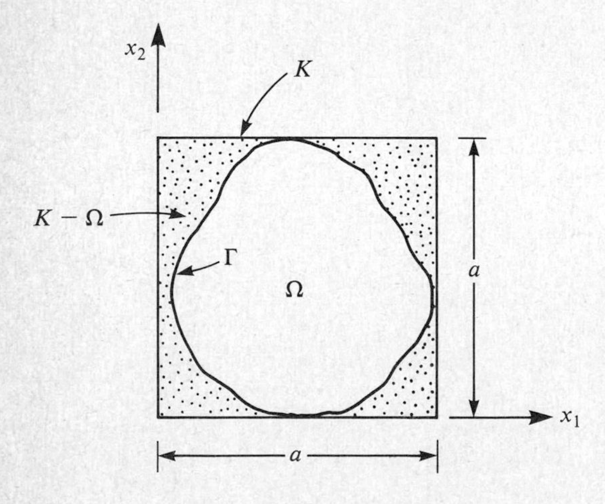

Figure 4-2 Extension of an arbitrary domain in two dimensions to a square domain.

Integrating with respect to x_1 and x_2 over K, we obtain

$$\begin{aligned}\int_0^a \int_0^a |u(\mathbf{x})|^2\,dx_1\,dx_2 &\le a \int_0^a \int_0^a \left(\int_0^a \left| \frac{\partial u}{\partial \xi} \right|^2 d\xi \right) dx_1\,dx_2 \\ &= a^2 \int_0^a \int_0^a \left| \frac{\partial u}{\partial \xi} \right|^2 d\xi\,dx_2 \\ &= a^2 \int_K \left| \frac{\partial u}{\partial x_1} \right|^2 dx_1\,dx_2\end{aligned}$$

Similarly, we obtain

$$\int_0^a \int_0^a |u(\mathbf{x})|^2\,dx_1\,dx_2 \le a^2 \int_K \left| \frac{\partial u}{\partial x_2} \right|^2 dx_1\,dx_2$$

From the above two inequalities and from the fact that $u = 0$ on $K - \Omega$, we have

$$\int_\Omega |u|^2\,dx_1\,dx_2 \le \frac{a^2}{2} \int_\Omega \left(\left| \frac{\partial u}{\partial x_1} \right|^2 + \left| \frac{\partial u}{\partial x_2} \right|^2 \right) dx_1\,dx_2$$

or

$$\|u\|_0 \le c(\Omega)\,|u|_1 \tag{4-24}$$

where c is a constant which depends on Ω (which dictates the value of a) only. We can write

$$c_2^2 \int_\Omega \left[|u|^2 + \left| \frac{\partial u}{\partial x_1} \right|^2 + \left| \frac{\partial u}{\partial x_2} \right|^2 \right] dx_1\,dx_2 \le \int_\Omega \left(\left| \frac{\partial u}{\partial x_1} \right|^2 + \left| \frac{\partial u}{\partial x_2} \right|^2 \right) dx_1\,dx_2$$

or

$$c_2 \|u\|_1 \le |u|_1 \tag{4-25}$$

where $c_2^2 = \min\,(1, 2/a^2)$. Equation (4-25) is a special case of Eq. (4-10).

For $u \in H_0^1(\Omega)$, the seminorm $|u|_1$ is equivalent to the norm $\|u\|_1$. This can be easily seen from Eq. (4-25) and the inequality

$$|u|_m \le c_1 \|u\|_m \qquad u \in H^m(\Omega) \tag{4-26}$$

Thus, we have

$$c_2 \|u\|_1 \le |u|_1 \le c_1 \|u\|_1 \qquad \text{for all } u \in H_0^1(\Omega) \tag{4-27}$$

4-2-4 Weak (or Generalized) Solutions

Consider the Dirichlet problem for the Poisson equation

$$-\nabla^2 u = f \text{ in } \Omega \subset \mathbb{R}^3 \qquad u = 0 \text{ on } \Gamma \tag{4-28}$$

For $f \in C(\Omega)$, the classical solution u of Eq. (4-28) belongs to $C^2(\Omega)$ and vanishes on Γ. Multiplying Eq. (4-28) by an arbitrary function $\phi(\mathbf{x})$ with compact support (so that $\phi = 0$ on Γ) in Ω, and integrating the result, we obtain

$$-\int_\Omega \phi \nabla^2 u\,d\mathbf{x} = \int_\Omega \phi f\,d\mathbf{x} \tag{4-29}$$

Using the gradient theorem [see Eq. (1-33)], the left side of Eq. (4-29) can be integrated by parts to obtain

$$\sum_{i=1}^{3} \int_{\Omega} \frac{\partial \phi}{\partial x_i} \frac{\partial u}{\partial x_i} \, d\mathbf{x} = \int_{\Omega} \phi f \, d\mathbf{x} \qquad \text{for every } \phi \in C_0^{\infty}(\Omega) \tag{4-30}$$

If $f \in C(\Omega)$, then Eq. (4-28) does not have a classical solution [that is, $u \notin C^2(\Omega)$]. For such a case it is necessary to generalize the concept of the solution in an appropriate way. Note that for $f \in L_2(\Omega)$, Eq. (4-30) makes sense if $\partial u/\partial x_i$ belongs to $L_2(\Omega)$. If $u \in H_0^1(\Omega)$ and if the derivatives $\partial u/\partial x_i$ are considered in the generalized sense, then it follows from the definition of the Sobolev spaces that $\partial u/\partial x_i$ belongs to $L_2(\Omega)$. Then $u \in H_0^1(\Omega)$ is said to be the *weak* or *generalized solution* of Eq. (4-28) if, for $f \in L_2(\Omega)$, u satisfies Eq. (4-30). In other words, a generalized solution of Eq. (4-28) is a distribution $u \in H_0^1(\Omega)$ such that Eq. (4-29) or equivalently, Eq. (4-30), is satisfied for all $\phi \in C_0^{\infty}(\Omega)$ and a given distribution f in $L_2(\Omega)$.

We note that $C_0^{\infty}(\Omega)$ is a dense subspace of $H_0^1(\Omega)$ [because $H_0^1(\Omega)$ is the closure of $C_0^{\infty}(\Omega)$]. Therefore, Eq. (4-30) is equivalent to finding $u \in H_0^1(\Omega)$ such that

$$(\nabla u, \nabla \phi)_0 = (f, \phi)_0 \qquad \text{for all } \phi \in C_0^{\infty}(\Omega) \tag{4-31a}$$

where $(\cdot, \cdot)_0$ is the L_2-inner product. Equation (4-31a) makes sense when ϕ is any member of $H_0^1(\Omega)$. Since $C_0^{\infty}(\Omega)$ is a dense subspace of $H_0^1(\Omega)$, it follows that Eq. (4-31a) is equivalent to

$$(\nabla u, \nabla \phi)_0 = (f, \phi)_0 \qquad \text{for all } \phi \in H_0^1(\Omega) \tag{4-31b}$$

Equation (4-31) is called a *variational* or *weak form* of Eq. (4-28).

In the next section we will show that the variational problem (4-31) is equivalent, in some cases, to the minimization of an associated quadratic functional. The equivalence holds for operator equations in which the operator is positive and symmetric.

4-3 THE MINIMUM OF A QUADRATIC FUNCTIONAL

4-3-1 The Minimum Functional Theorem

When the operator A in Eq. (4-1) is positive (hence, symmetric) on its domain $\mathscr{D}_A$ and $f \in H$, where H is a Hilbert space, the solution u of Eq. (4-1) can be shown to be the function that minimizes a quadratic functional, $Q(u)$. Conversely, if we can determine the function u that minimizes $Q(u)$ on H, then in effect we have the solution to Eq. (4-1). This is a significant result because it is often easier to find, as will be shown in Chap. 6, the minimizing function than to solve Eq. (4-1) exactly. In the following theorem we establish the equivalence between the operator equation (4-1) and an associated quadratic functional.

Theorem 4-4 Let $A\colon \mathscr{D}_A \subset H \to H$, where H is a Hilbert space, be a linear positive operator on $\mathscr{D}_A$ and let $f \in H$. Then the quadratic functional

$$Q(u) \equiv (Au, u) - 2(f, u) \tag{4-32}$$

where $(\cdot, \cdot)$ is an inner product in H, assumes its minimum value at $u_0 \subset \mathscr{D}_A$ if and only if u_0 is also a solution of Eq. (4-1).

Proof "If" part. Let u_0 be the solution of Eq. (4-1). Then $f = Au_0$ and we have

$$\begin{aligned} Q(u) &= (Au, u) - 2(Au_0, u) \\ &= (A(u - u_0), u) - (Au_0, u) \end{aligned}$$

Since the inner product is symmetric and the operator is symmetric, we can write

$$\begin{aligned} Q(u) &= (A(u - u_0), u) - (Au, u_0) + (Au_0, u_0) - (Au_0, u_0) \\ &= (A(u - u_0), u - u_0) - (Au_0, u_0) \\ &= Q(u_0) + (A(u - u_0), u - u_0) \end{aligned}$$

Since A is positive, we have

$$\begin{aligned} (A(u - u_0), u - u_0) &> 0 \qquad \text{for every } u \in \mathscr{D}_A \\ &= 0 \qquad \text{if and only if } u - u_0 = 0 \text{ in } \mathscr{D}_A \end{aligned}$$

from which we can conclude that

$$Q(u) \geq Q(u_0) \tag{4-33}$$

where the equality holds if and only if $u = u_0$. Inequality (4-33) implies that the quadratic functional $Q(u)$ assumes its minimal value at the solution $u_0 \in \mathscr{D}_A$ of the equation $Au_0 = f$; any other function $u \in \mathscr{D}_A$ makes $Q(u)$ larger than $Q(u_0)$.

Next we prove the "only if" part. Let the functional $Q(u)$ assume its minimal value at $u_0 \in \mathscr{D}_A$. This means for any arbitrary element $u \in \mathscr{D}_A$, $Q(u) \geq Q(u_0)$. In particular, for $u = u_0 + \alpha v$, $v \in \mathscr{D}_A$ and α a real number, we have

$$Q(u_0 + \alpha v) \geq Q(u_0)$$

By definition of Q, we have

$$\begin{aligned} Q(u_0 + \alpha v) &= (A(u_0 + \alpha v), u_0 + \alpha v) - 2(f, u_0 + \alpha v) \\ &= (Au_0, u_0) + 2\alpha(Au_0, v) + \alpha^2(Av, v) - 2\alpha(f, v) - 2(f, u_0) \end{aligned}$$

where the symmetry of A is used in arriving at the second step. Since $u_0 \in \mathscr{D}_A$ and $f \in H$ are fixed elements, it is clear from the last equation that, for fixed $v \in \mathscr{D}_A$, $Q(u_0 + \alpha v)$ is a quadratic function of α. Then $Q(u_0 + \alpha v)$ has a minimum at $\alpha = 0$ (that is, Q has a minimum at $u = u_0$) if the first derivative of $Q(u_0 + \alpha v)$ with respect to α is zero at $\alpha = 0$:

$$\begin{aligned} 0 &= \left[\frac{d}{d\alpha} Q(u_0 + \alpha v)\right]_{\alpha=0} \\ &= 2(Au_0, v) - 2(f, v) \\ &= 2(Au_0 - f, v) \end{aligned}$$

Then by Lemma 3-2, it follows that (since v is an arbitrary element of $\mathscr{D}_A$)

$$Au_0 - f = 0 \qquad \text{or} \qquad Au_0 = f$$

Thus, we proved that the element $u_0 \in \mathscr{D}_A$ that minimizes the quadratic functional (4-32) is a solution of Eq. (4-1).

Theorem 4-4 is very significant in that it establishes an equivalence between the operator equation $Au = f$ in $\mathscr{D}_A$ with the minimum of the quadratic functional $Q(u) = (Au, u) - 2(f, u)$. This variational formulation (i.e., expressing the operator equation as the problem of minimizing a quadratic functional) facilitates the mathematical analysis, such as establishing existence and uniqueness results, and naturally leads to the determination of approximate solutions. In the later sections we will present the procedures to construct the variational statements from operator equations. The existence and uniqueness studies are undertaken in Chap. 5, and the variational methods are considered in Chap. 6.

We consider a specific example that satisfies the hypothesis and results of Theorem 4-4.

Example 4-5 Consider the differential equation

$$-\frac{d}{dx}\left(a\frac{du}{dx}\right) = f \qquad 0 < x < L \tag{4-34a}$$

with the boundary conditions

$$u(0) = u(L) = 0 \tag{4-34b}$$

We assume that $a(x)$ and its first derivative are continuous in the interval $[0, L]$ and that

$$a(x) > 0 \text{ in } [0, L]$$

Equation (4-34a) arises, for example, in connection with the transverse deflection of a cable fixed at the ends at a distance of L apart. Here u denotes the transverse deflection, a is the tension in the cable, and f is the continuous transverse loading (see Table 4-1).

Here we choose $H = L_2[0, L]$ and $A = -(d/dx)(a(d/dx))$. Then $\mathscr{D}_A$ is the linear space of functions that are continuous with their second derivative inclusive in the interval $[0, L]$ and satisfy the conditions in Eq. (4-34b). Since

Table 4-1 Some examples of second-order equation in one dimension†

$$-\frac{d}{dx}\left(a\frac{du}{dx}\right)=f \qquad \text{for } 0<x<L$$

Essential boundary condition: $u|_{x=0}=g_0$; natural boundary condition: $\left(a\frac{du}{dx}\right)\Big|_{x=L}=h_0$

Field	Primary variable u	a	Source term f	Secondary variable h_0
1. Transverse deflection of a cable	Transverse deflection	Tension in cable	Distributed transverse load	Axial force (generally unknown)
2. Axial deformation of a bar	Longitudinal displacement	EA, E = modulus, A = area of cross section	Friction or contact force on surface of bar	Axial force
3. Heat transfer along a fin in heat exchanger	Temperature	Thermal conductivity	Heat generation	Heat flux
4. Flow through pipes	Hydrostatic pressure	$\pi D^4/128\mu$, D = diameter μ = viscosity	Flow source (generally zero)	Flow rate
5. Laminar incompressible flow through a channel under constant pressure gradient	Velocity	Viscosity	Pressure gradient	Axial stress
6. Flow through porous media	Fluid head	Coefficient of permeability	Fluid flux	Flow (seepage)
7. Electrostatics	Electrostatic potential	Dielectric constant	Charge density	Electric flux

† From Reddy (1984a)

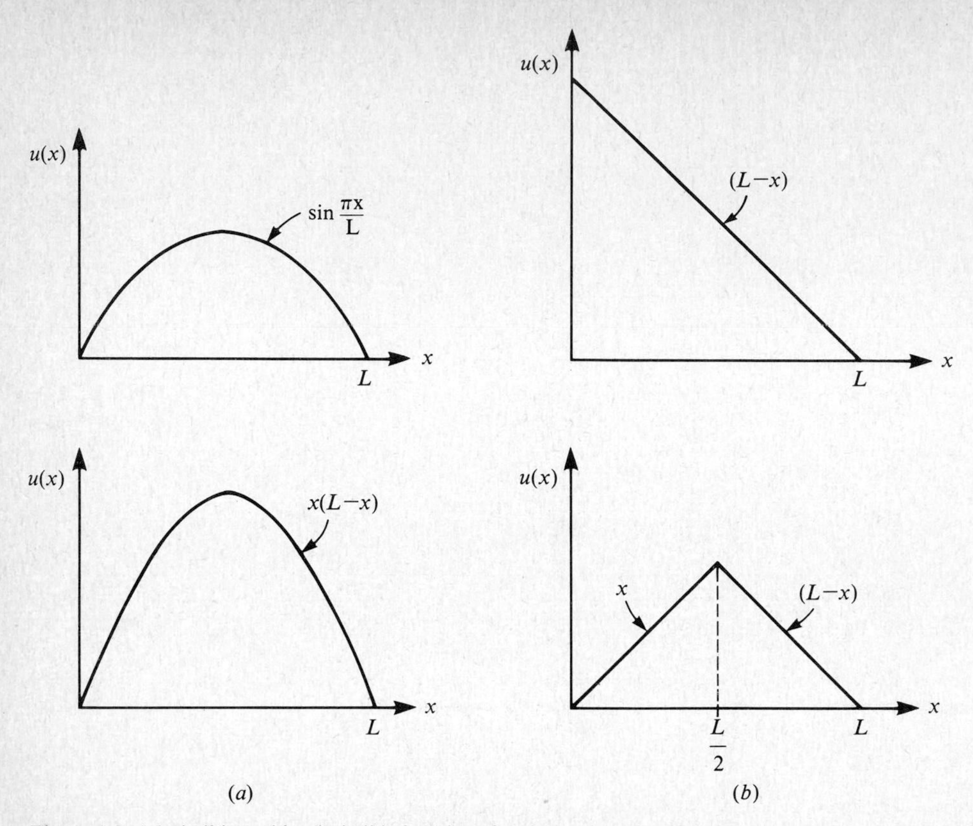

Figure 4-3 Admissible and inadmissible functions for the problem in Eqs. (4-34).

the set of all continuous functions is dense in $L_2[0, L]$, it follows that $\mathscr{D}_A$ is dense in $L_2[0, L]$.

Examples of functions that are members of $\mathscr{D}_A$ and that are not members of $\mathscr{D}_A$ are shown in Fig. 4-3*a* and Fig. 4-3*b*, respectively. The function $u = L - x$, although it belongs to $C^2[0, L]$, is not admissible because it does not vanish at $x = 0$. The second function in Fig. 4-3*b* satisfies the conditions $u(0) = u(L) = 0$, but it does not belong to $C^2[0, L]$ (in fact, it does not even belong to $C^1[0, L]$ because du/dx is discontinuous at the point $x = L/2$). Hence, it is not an admissible function.

We now prove that A is symmetric and positive on $\mathscr{D}_A$. For every $u, v \in \mathscr{D}_A$ we have

$$(Au, v)_0 = \int_0^L -\frac{d}{dx}\left(a\frac{du}{dx}\right)v\,dx = \int_0^L a\frac{du}{dx}\frac{dv}{dx}\,dx - \left[a\frac{du}{dx}v\right]_0^L$$
$$= \int_0^L a\frac{du}{dx}\frac{dv}{dx}\,dx \tag{4-35a}$$

The boundary terms $[a(du/dx)v]_0^L$ are equal to zero because $v \in \mathscr{D}_A$ satisfies the conditions in (4-34*b*). In a similar way, we get

$$(u, Av)_0 = \int_0^L u\left[-\frac{d}{dx}\left(a\frac{dv}{dx}\right)\right] dx = \int_0^L a\frac{du}{dx}\frac{dv}{dx}\, dx \tag{4-35b}$$

Thus, $(Au, v)_0 = (u, Av)_0$, that is, A is symmetric. By setting $v = u$ in Eq. (4-35*a*) we obtain

$$(Au, u)_0 = \int_0^L a\left(\frac{du}{dx}\right)^2 dx \geq 0 \qquad \text{for } u \in \mathscr{D}_A \tag{4-36}$$

Moreover, if $(Au, u)_0 = 0$, it follows from Eq. (4-36) that (because $a > 0$),

$$\frac{du}{dx} = 0 \text{ in } [0, L]$$

or $u = c$, constant in $[0, L]$. Since $u(0) = u(L) = 0$ it follows that the constant is equal to zero. In other words, $(Au, u)_0 = 0$ if and only if $u = 0$. Therefore A is positive.

Since A is positive, then the quadratic functional $Q(u)$ is given by

$$\begin{aligned} Q(u) &= (Au, u)_0 - 2(f, u)_0 \\ &= \int_0^L \left[a\left(\frac{du}{dx}\right)^2 - 2fu\right] dx \end{aligned} \tag{4-37}$$

The functional $Q(u)$ represents twice the total *potential energy* of the cable: the first part of the integral represents double the elastic strain energy, and the second part represents double the potential energy of the external forces.

Let u_0 be the solution of Eq. (4-34*a*) in $\mathscr{D}_A$. Substituting $f = -(d/dx)(a(du_0/dx))$ into Eq. (4-37), we get

$$\begin{aligned} Q(u) &= \int_0^L \left[a\left(\frac{du}{dx}\right)^2 + 2\frac{d}{dx}\left(a\frac{du_0}{dx}\right)u\right] dx \\ &= \int_0^L \left[a\left(\frac{du}{dx}\right)^2 - 2a\frac{du_0}{dx}\frac{du}{dx}\right] dx \\ &= \int_0^L \left[a\left(\frac{du}{dx} - \frac{du_0}{dx}\right)^2 - a\left(\frac{du_0}{dx}\right)^2\right] dx \\ &\leq \int_0^L \left[a\left(\frac{du}{dx} - \frac{du_0}{dx}\right)^2\right] dx \end{aligned}$$

Thus, $Q(u)$ is minimal in $\mathscr{D}_A$ if and only if $(du/dx) - (du_0/dx) = 0$ in $\mathscr{D}_A$. Since both u and u_0 satisfy conditions in Eq. (4-34*b*) it follows that $u - u_0$ in $\mathscr{D}_A$. This verifies the "if" part of Theorem 4-4.

Conversely, let $u_0 \in \mathscr{D}_A$. Let v be an arbitrary element in $\mathscr{D}_A$ and let α be an arbitrary real number. Then the condition of minimum becomes

$$\begin{aligned}
0 &= \frac{d}{d\alpha} Q(u_0 + \alpha v)\bigg|_{\alpha=0} \\
&= \left\{\frac{d}{d\alpha}\int_0^L \left[a\left(\frac{du_0}{dx} + \alpha\frac{dv}{dx}\right)^2 - 2f(u_0 + \alpha v)\right] dx\right\}_{\alpha=0} \\
&= \int_0^L \left(2a\frac{du_0}{dx}\frac{dv}{dx} - 2fv\right) dx \\
&= \int_0^L 2\left[-\frac{d}{dx}\left(a\frac{du_0}{dx}\right) - f\right] v\, dx
\end{aligned}$$

for every $v \in \mathscr{D}_A$. Then by Lemma 3-2, it follows that $Au_0 = f$ in $[0, L]$, proving the "only if" part of Theorem 4-4.

Theorem 4-4 is a general statement of the principle of the minimum total potential energy in elasticity. In mechanics, the quadratic functional is often known as the *total potential energy* (the sum of the elastic strain energy and the work done by external forces), and one derives the associated operator equations, i.e., the *Euler equations*, using Theorem 4-4.

4-3-2 The Energy Space

A close examination of Theorem 4-4 shows that there is no guarantee that Eq. (4-1) has a solution or that $Q(u)$ has a minimum in $\mathscr{D}_A$. The theorem gives the relationship between the equation and the associated quadratic functional under the assumption that the equation has a solution in $\mathscr{D}_A$ or that the quadratic functional attains its minimum in $\mathscr{D}_A$. In other words, Theorem 4-4 is valid under the conditions that the operator equation (4-1) has a solution $u_0 \in \mathscr{D}_A$ or the quadratic functional in Eq. (4-32) assumes its minimum on $\mathscr{D}_A$. Neither of these conditions are always met. For example, if f in Example 4-5 is not continuous in $[0, L]$, say equal to a constant in $(0, a)$ and zero in (a, L), then the solution of Eq. (4-34) does not belong to $\mathscr{D}_A$ defined in Example 4-5 because for every $u \in \mathscr{D}_A$, Au is a continuous function in $[0, L]$. Therefore, the functional $Q(u)$ does not assume its minimum in $\mathscr{D}_A$. This difficulty can be overcome by identifying a suitable domain $\mathscr{D}_A$ that is large enough to accommodate discontinuous functions from $L_2[0, L]$. The selection of the new domain is by no means simple, because $\mathscr{D}_A$ should be such that $Q(u)$ attains, in some sense, its minimum on the new $\mathscr{D}_A$. In this section we study the construction of the so-called energy space, H_A, as the domain of the operator A. This space is sufficiently large to accommodate the solutions to problems with "sufficiently" discontinuous data.

Consider the operator equation (4-1) on a linear space $\mathscr{D}_A$ dense in a Hilbert space H. Let A be positive-definite on $\mathscr{D}_A$: there exists a constant $\mu_0 > 0$ such that

$$(Au, u) \geq \mu_0 \|u\|^2 \qquad \text{for every } u \in \mathscr{D}_A \tag{4-38}$$

where $(\cdot, \cdot)$ and $\|\cdot\|$ denote the inner product and norm, respectively, in H. Since A is symmetric and positive definite, (Au, u) qualifies as an inner product on $\mathscr{D}_A$:

$$(u, v)_A \equiv (Au, v) \qquad \text{for all } u, v \in \mathscr{D}_A \tag{4-39}$$

It is easy to verify that $(\cdot, \cdot)_A$ satisfies the axioms of an inner product.

As an example, consider the operator A and domain $\mathscr{D}_A$ defined in Example 4-5. It was shown that the operator is positive-definite on $\mathscr{D}_A$, and therefore

$$(u, v)_A = (Au, v) = \int_0^L a \frac{du}{dx}\frac{dv}{dx}\, dx, \qquad a > 0$$

is an inner product on $\mathscr{D}_A$.

The natural norm generated by the inner product $(\cdot, \cdot)_A$ is given by

$$\|u\|_A = (u, u)_A \tag{4-40}$$

Therefore $\mathscr{D}_A$ is an inner product space. In general, $\mathscr{D}_A$ is not complete. As mentioned in Sec. 3-4, this space can be completed by the addition of the limit points (which are in H) of all Cauchy sequences in $\mathscr{D}_A$ to $\mathscr{D}_A$. Let us denote the resulting space by H_A. The details of the construction of H_A are quite involved, and the interested reader is referred to the excellent monograph by Rektorys (1980). The space H_A is a Hilbert space and is called the *energy space*. The name is appropriate especially in applied mechanics because the norm (4-40) in H_A signifies the elastic strain energy of the problem described by the operator equation $Au = f$.

The energy space H_A is much "broader" than the set $\mathscr{D}_A$. This is because H_A contains elements that have one-to-one correspondence with all sequences that are fundamental, with respect to the energy norm, in $\mathscr{D}_A$ and hence in H: for u_0, $v_0 \in H_A$, we have

$$(u_0, v_0)_A = \lim_{n \to \infty} (u_n, v_n)_A \tag{4-41}$$

where $\{u_n\}$ and $\{v_n\}$ ($u_n, v_n \in \mathscr{D}_A$) are sequences from classes which correspond in the above sense to u_0 and v_0, respectively. The elements of H_A along with their derivatives of order m, where $2m$ is the order of the positive-definite operator A, are continuous and square-integrable, whereas the elements of $\mathscr{D}_A$ are continuous, along with their derivatives up to and including order $2m$. For example, consider the operator A in Example 4-5. In this case $H = L_2[0, L]$ and $\mathscr{D}_A$ consists of functions $u(x)$ which are continuous with their first and second derivatives in the closed interval $[0, L]$ and satisfy the boundary conditions

$$u(0) = u(L) = 0$$

The inner product $(u, v)_A$ is given by the relation

$$(u, v)_A = \int_0^L \frac{du}{dx}\frac{dv}{dx}\, dx$$

The energy space H_A consists of functions from $H = L_2[0, L]$ which do not belong to $\mathscr{D}_A$ but for which the integral in the definition of the inner product $(u, v)_A$ is meaningful. More specifically, H_A contains functions u whose first (generalized) derivatives are square-integrable in $(0, L)$, and satisfy the boundary conditions $u(0) = u(L) = 0$; that is, $H_A = H_0^1(0, L)$.

In general, for a $2m$th order positive-definite differential operator defined over a bounded domain Ω, we take $H = L_2(\Omega)$, $\mathscr{D}_A = \{u\colon Au$ exists, and u and its derivatives satisfy the boundary conditions involving the traces of $D^\alpha u$, $|\alpha| \leq m - 1\}$, and H_A contains functions that have generalized derivatives up to and including order m that are square-integrable and satisfy the homogeneous form of the boundary conditions satisfied by the functions in $\mathscr{D}_A$.

Returning to the minimum of a quadratic functional in the energy space, we rewrite the functional $Q(u)$ in terms of the inner product in H_A,

$$Q(u) = (u, u)_A - 2(f, u) \qquad u \in \mathscr{D}_A \tag{4-42a}$$

Since $(\cdot, \cdot)_A$ and $(\cdot, \cdot)$ are defined for all $u \in H_A \subset H$, we can *extend* the definition of $Q(u)$ to the entire space H_A,

$$Q(u) = (u, u)_A - 2(f, u) \qquad u \in H_A \tag{4-42b}$$

Now we will show that the functional extended in this way attains its minimum on H_A, and that the element u_0 at which the minimum is attained in H_A is uniquely determined by the element $f \in H$. First we show that $l(u) \equiv (f, u)$ is bounded in H_A. For every fixed $f \in H$, by the Schwarz inequality we have

$$|(f, u)| \leq \|f\|\, \|u\|$$

Also, for $u \in H_A$ we have from Eq. (4-38)

$$\|u\| \leq \frac{1}{\sqrt{\mu_0}} \|u\|_A$$

Therefore

$$|(f, u)| \leq \frac{\|f\|}{\sqrt{\mu_0}} \|u\|_A = M\|u\|_A \tag{4-43}$$

for every $f \in H$ and $u \in H_A$.

Since (f, u) is bounded in H_A, by the Riesz representation theorem (Theorem 4-1), there exists an element $u_0 \in H_A$, uniquely determined by the element $f \in H$, such that

$$(u_0, u)_A = (f, u) \qquad \text{for all } u \in H_A \tag{4-44}$$

Equation (4-44) is a restatement of the operator equation (4-1) in the space H_A. It is called a *weak* or *variational statement* of Eq. (4-1), and the solution u_0 of Eq. (4-44) is called a *weak, variational* or *generalized solution* of Eq. (4-1).

To show that $Q(u)$ attains its minimum on H_A, we use Eq. (4-44) for (f, u) in Eq. (4-42*b*),

$$\begin{aligned} Q(u) &= (u, u)_A - 2(u_0, u)_A \\ &= (u - u_0, u - u_0)_A - (u_0, u_0)_A \\ &= \|u - u_0\|_A^2 - \|u_0\|_A^2 \end{aligned}$$

Since $\|\cdot\|_A$ is a norm on H_A, $\|u - u_0\|_A > 0$ for $u \neq u_0$ and $\|u - u_0\|_A = 0$ if and only if $u = u_0$ in H_A. Then it follows that $Q(u) \leq \|u - u_0\|_A^2$, and $Q(u)$ assumes its minimum in H_A if and only if $u = u_0$, the element determined by Eq. (4-44). The minimum is given by $Q(u_0) = -\|u_0\|_A^2$. We state this fact in the following theorem.

Theorem 4-5 Let $A: \mathscr{D}_A \to H$ be a positive-definite operator on the linear space $\mathscr{D}_A$ which is dense in the Hilbert space H, and let H_A be the energy space defined earlier. Then the functional Q assumes its minimum at u_0, which is uniquely determined by the equation (4-44). Further, the solution u_0 *continuously depends on the data* f in the following sense:

$$\|u_0\|_A \leq \frac{\|f\|}{\sqrt{\mu_0}} \tag{4-45}$$

This inequality follows from Eqs. (4-43) and (4-44).

The inequality (4-45) implies that if f is "small" in the norm of H, then the generalized solution u_0 is also "small" in the norm of H_A. For example, if the data differs (e.g., due to the approximation of the domain Ω in a numerical method) slightly in the norm of H, then the corresponding generalized solution differs slightly in the norm of H_A. This error in the generalized solution due to the error in the data can be estimated by means of Eq. (4-45):

$$\|u_0 - u_h\| \leq \frac{1}{\sqrt{\mu_0}} \|f - f_h\| \tag{4-46}$$

where u_h is the generalized solution of the equation $Au_h = f_h$. We shall return to this error estimate in Chap. 6.

In general, for a 2*m*th order differential operator [see Eq. (3-71) for the notation]

$$Au = \sum_{|\boldsymbol{\beta}|,\, |\boldsymbol{\alpha}| \leq m} (-1)^{|\boldsymbol{\alpha}|} D^{\boldsymbol{\alpha}}(a_{\boldsymbol{\alpha\beta}} D^{\boldsymbol{\beta}} u)$$

the energy space H_A is defined by

$$H_A = \{u \in H^m(\Omega): D^{\boldsymbol{\alpha}} u = 0 \text{ on } \Gamma \qquad 0 \leq |\boldsymbol{\alpha}| \leq m\} = H_0^m(\Omega)$$

and the quadratic functional associated with the operator equation

$$Au = f \text{ in } \Omega$$
$$D^{\alpha}u = 0 \text{ on } \Gamma \qquad |\alpha| \leq m - 1$$

is given by

$$I(u) = \frac{1}{2} \sum_{|\alpha|,\, |\beta| \leq m} \int_{\Omega} a_{\alpha\beta} D^{\alpha}u D^{\beta}u \, d\mathbf{x} - \int_{\Omega} fu \, d\mathbf{x}$$

4-3-3 Concepts from Variational Calculus

The variational operator In the above discussions we assumed that the boundary-value problem under consideration had homogeneous boundary conditions. When the boundary conditions are nonhomogeneous, the set $\mathscr{D}_A$ is not a vector space. However, the space of all variations v is always a linear vector space, $\mathscr{S}$, called *the space of admissible variations.*

In the calculus of variations it is a common practice to use δu to denote a variation of u. Thus δ can be viewed as an operator that changes a function u into δu, and du/dx into $\delta(du/dx)$, etc. with the meaning,

$$\delta u = \alpha v \qquad \delta\left(\frac{du}{dx}\right) = \alpha \frac{dv}{dx} \tag{4-47}$$

where α is a small arbitrary real number. The operator δ is called the *variational operator*, and δu is called the *first variation* of u. Note that δu differs from $v \in \mathscr{S}$ by only a multiplicative constant. The variational operator proves to be very convenient in the calculation of the variation of a functional.

In analogy with the total differential operator d of ordinary functions (i.e., functions of independent coordinates), the *variational operator δ acts like the total differential operator with respect to the independent variables of the system.* To see this consider the total differential of $F = F(x, u, u')$

$$dF = \frac{\partial F}{\partial x} dx + \frac{\partial F}{\partial u} du + \frac{\partial F}{\partial u'} du' \tag{4-48}$$

The first variation of F is defined by

$$\begin{aligned} \delta F &\equiv \alpha \left[\frac{d}{d\alpha} F(x, u + \alpha v, u' + \alpha v') \right]_{\alpha=0} \\ &= \alpha\left(\frac{\partial F}{\partial u} v + \frac{\partial F}{\partial u'} v' \right) = \frac{\partial F}{\partial u} \delta u + \frac{\partial F}{\partial u'} \delta u' \end{aligned} \tag{4-49}$$

Since x is fixed during the variation of u to $u + \delta u$ (that is, δu denotes the variation of u at an arbitrarily fixed x), $dx = 0$ in Eq. (4-48), and the analogy between

Eq. (4-48) and (4-49) becomes apparent. The following relations of variational calculus, again analogous to those of differential calculus, can be shown to hold for δ. Let $F_1 = F_1(u)$, $F_2 = F_2(u)$, and $G = G(u, v)$ where u and v are dependent variables of the problem. We have

$$\begin{aligned}
&(a) \qquad \delta(F_1 \pm F_2) = \delta F_1 \pm \delta F_2 \\
&(b) \qquad \delta(F_1 F_2) = \delta F_1 F_2 + F_1\, \delta F_2 \\
&(c) \qquad \delta\left(\frac{F_1}{F_2}\right) = (\delta F_1 F_2 - F_1\, \delta F_2)/F_2^2 \\
&(d) \qquad \delta(F_1)^n = n(F_1)^{n-1}\, \delta F_1 \\
&(e) \qquad \delta G = \frac{\partial G}{\partial u}\, \delta u + \frac{\partial G}{\partial v}\, \delta v
\end{aligned} \tag{4-50}$$

The variational operator can be interchanged with differential and integral operators:

$$\begin{aligned}
&(a) \qquad \frac{d}{dx}(\delta u) = \frac{d}{dx}(\alpha v) = \alpha \frac{dv}{dx} = \delta\left(\frac{du}{dx}\right) \\
&(b) \qquad \delta\left(\int_\Omega u\, dx\right) = \alpha \int_\Omega v\, dx = \int_\Omega \alpha v\, dx = \int_\Omega \delta u\, dx
\end{aligned} \tag{4-51}$$

Next we study the consequence of the vanishing of the first variation of a functional.

The Euler equations Consider the problem of determining the minimum of the functional

$$I(u) = \int_a^b F(x, u, u')\, dx \qquad u' \equiv du/dx \tag{4-52}$$

on $\mathscr{D}$, which consists of functions with their second derivatives continuous in $[a, b]$, and satisfy the end conditions

$$u(a) = 0 \qquad u(b) = 0 \tag{4-53}$$

Obviously, set $\mathscr{D}$ is a linear vector space. For each $u \in \mathscr{D}$, $F(x, u, u')$ exists and is continuously differentiable with respect to its arguments.

Suppose that $I(u)$ assumes its minimum at $u_0 \in \mathscr{D}$. Then for any $u \in \mathscr{D}$ we have

$$I(u) \geq I(u_0) \tag{4-54}$$

In particular Eq. (4-54) holds for $u = u_0 + \alpha v$, $v \in \mathscr{S}$, the space of admissible variations (see Fig. 4-4)

$$\mathscr{S} = \{v \in H^2(a, b)\colon\ v(a) = v(b) = 0\} = \mathscr{D} \tag{4-55}$$

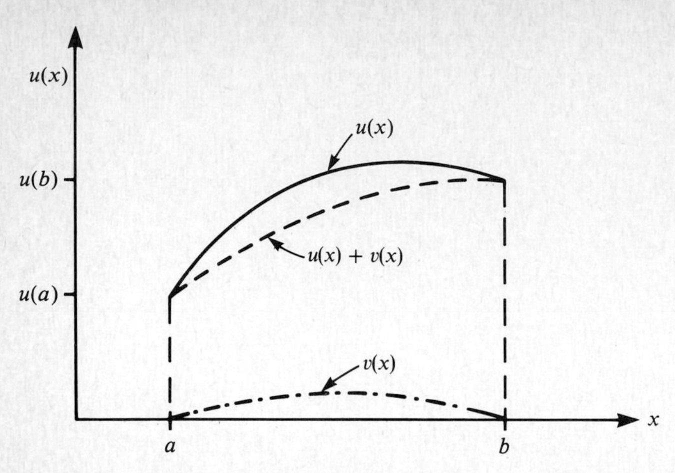

Figure 4-4 A function $u(x)$ with specified nonzero end conditions and a function $v(x)$ from the space of admissible variations.

For fixed $u_0 \in \mathscr{D}$, $I(u_0 + \alpha v)$ is a function of α. Then $I(u_0 + \alpha v)$ has a minimum at $\alpha = 0$ (i.e., equivalently at u_0) if the first derivative of $I(u_0 + \alpha v)$ with respect to α is zero at $\alpha = 0$:

$$\frac{d}{d\alpha} I(u_0 + \alpha v)\bigg|_{\alpha=0} = 0 \tag{4-56}$$

This is precisely the definition of the first variation (and the Gateaux derivative) of I,

$$\delta I(u_0 ; v) \equiv \alpha\left[\frac{d}{d\alpha} I(u_0 + \alpha v)\right]_{\alpha=0} \tag{4-57}$$

Thus, a necessary condition for a functional $I(u)$ to have a minimum at $u_0 \in \mathscr{D}$ is that Eq. (4-56) holds, or equivalently, the first variation of I at $u = u_0$ is zero:

$$\delta I(u_0 ; v) = 0 \tag{4-58}$$

For the problem at hand, condition (4-58) gives

$$\begin{aligned} 0 = \delta I &= \int_a^b \delta F\, dx \\ &= \int_a^b \left(\frac{\partial F}{\partial u_0}\,\delta u + \frac{\partial F}{\partial u_0'}\,\delta u'\right) dx \\ &= \int_a^b \left[\frac{\partial F}{\partial u_0} - \frac{d}{dx}\left(\frac{\partial F}{\partial u_0'}\right)\right] \delta u\, dx + \left[\delta u\,\frac{\partial F}{\partial u_0'}\right]_a^b \end{aligned} \tag{4-59}$$

The boundary term vanishes because $\delta u \in \mathscr{S} = \mathscr{D}$. We get

$$0 = (Au_0, \delta u)_H \tag{4-60}$$

where $(\cdot, \cdot)_H$ denotes the inner product in $H = L_2[a, b]$, and $A: H \to H$ is the operator

$$Au_0 = \frac{\partial F}{\partial u_0} - \frac{d}{dx}\left(\frac{\partial F}{\partial u_0'}\right) \tag{4-61}$$

Whether A is linear or not depends on the form of F. In any case, since δu is arbitrary in (a, b), by virtue of Lemma 3-2, Eq. (4-60) gives

$$Au_0 = 0 \tag{4-62a}$$

or

$$\frac{\partial F}{\partial u_0} - \frac{d}{dx}\left(\frac{\partial F}{\partial u_0'}\right) = 0 \tag{4-62b}$$

Thus, the true minimizer u_0 of the functional (4-52) in $\mathscr{D}$ is the solution of Eq. (4-62).

Equation (4-62) is called the *Euler equation* of the functional in (4-52). The name *Euler–Lagrange equation* is also often used in reference to Eq. (4-62).

Natural and essential boundary conditions We reconsider the problem of minimizing the functional in Eq. (4-52) with the following alternative boundary condition

$$u(a) = 0 \tag{4-63}$$

The domain $\mathscr{D}$ of the functional I in this case consists of functions u with their second derivatives continuous in $[a, b]$, and satisfy the condition (4-63). Once again the space of admissible variations $\mathscr{S}$ coincides with the domain space.

The necessary condition (4-58) for the minimum gives the expression in Eq. (4-59). Since $\delta u \in \mathscr{S} = \mathscr{D}$, we have $\delta u(a) = 0$ and

$$0 = \int_a^b \left[\frac{\partial F}{\partial u_0} - \frac{d}{dx}\left(\frac{\partial F}{\partial u_0'}\right)\right] \delta u \, dx + \delta u(b) \frac{\partial F}{\partial u_0'}(b) \tag{4-64}$$

Since δu is arbitrary both in (a, b) and at $x = b$, and δu in (a, b) is independent of $\delta u(b)$ (this is to say that δu satisfies different conditions inside the domain and on the boundary), it follows from Eq. (4-64) that the coefficients of δu in (a, b) and at $x = b$ be zero separately:

$$\frac{\partial F}{\partial u_0} - \frac{d}{dx}\left(\frac{\partial F}{\partial u_0'}\right) = 0 \qquad \text{in } a < x < b \tag{4-65}$$

$$\frac{\partial F}{\partial u_0'}(b) = 0 \tag{4-66}$$

The boundary condition in Eq. (4-66) is considered as part of the Euler equations.

The boundary conditions that are obtained as an Euler equation (i.e., the boundary conditions that are obtained from a functional) are called the *natural boundary conditions.* The boundary conditions that are included in the specification of the domain space are called the *essential boundary conditions.* Essential boundary conditions are also known as *Dirichlet* boundary conditions, and natural boundary conditions are known as *Neumann* boundary conditions. In mechanics, the natural boundary conditions are also known as *kinetic* (*or force*) *boundary conditions*, and the essential boundary conditions are known as the *kinematic* (or *geometric*) *boundary conditions.* In the present case the essential and natural boundary conditions are given, respectively, by Eqs. (4-63) and (4-66). Thus, by definition, the natural boundary conditions are always included in the quadratic functional (or the weak formulation) of the problem.

Example 4-6 Consider the problem of minimizing the total potential energy functional of a cantilever beam (see Fig. 4-5). The total potential energy functional Π of a solid body is equal to the difference of the strain energy functional and the total work done on the body by applied external forces:

$$\Pi(w) = \int_0^L \frac{EI}{2}\left(\frac{d^2w}{dx^2}\right)^2 dx - \left(\int_0^L fw\,dx + w(L)F + \frac{dw}{dx}(L)M\right) \tag{4-67}$$

The *principle of the minimum total potential energy* [see Reddy (1984*b*)] states that of all admissible configurations the beam can assume only the one that makes the total potential energy a minimum corresponds to the equilibrium configuration. The energy space H_A in this case is given by

$$H_A = \left\{w\colon w, \frac{dw}{dx} \text{ and } \frac{d^2w}{dx^2} \text{ are continuous in } [0, L] \text{ and belong to } L_2(0, L), \text{ and } w(0) = \frac{dw}{dx}(0) = 0\right\} = \mathscr{S} \tag{4-68}$$

Setting $\delta\Pi$ to zero, we obtain

$$0 = \int_0^L \left[EI\frac{d^2w}{dx^2}\frac{d^2\delta w}{dx^2} - \delta wf\right] dx - \delta w(L)F - \delta\left(\frac{dw}{dx}(L)\right)M \tag{4-69}$$

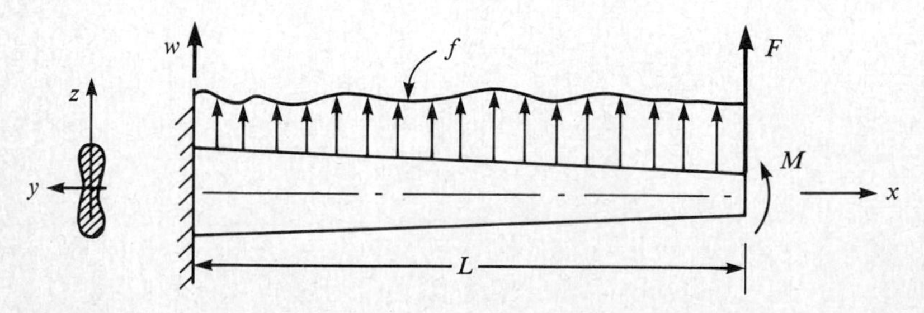

Figure 4.5 A clamped beam subjected to bending loads f, F, and M.

Integrating the first term twice by parts to relieve δw of all derivatives we obtain

$$0 = \int_0^L \left[\frac{d^2}{dx^2}\left(EI \frac{d^2 w}{dx^2} \right) - f \right] \delta w \, dx + \left[EI \frac{d^2 w}{dx^2} \delta\left(\frac{dw}{dx}\right) - \frac{d}{dx}\left(EI \frac{d^2 w}{dx^2} \right) \delta w \right]_0^L - \delta w(L) F - \delta\left(\frac{dw}{dx}(L) \right) M \tag{4-70}$$

The Euler equation in (0, L) is given by setting the coefficient of δw in (0, L) to zero:

$$\frac{d^2}{dx^2}\left(EI \frac{d^2 w}{dx^2} \right) = f \qquad 0 < x < L \tag{4-71}$$

which is the equation of equilibrium governing the (Euler) bending of beams. The operator A is given by

$$Aw = \frac{d^2}{dx^2}\left(EI \frac{d^2 w}{dx^2} \right) \tag{4-72}$$

which can be shown to be positive (for $EI > 0$) in H_A. The inner product in H_A for this problem is defined by

$$(u, v)_A = \int_0^L EI \frac{d^2 u}{dx^2} \frac{d^2 v}{dx^2} \, dx \tag{4-73}$$

Now consider the boundary terms in Eq. (4-70). Collecting the coefficients of $\delta w(0)$, $\delta w(L)$, $\delta(dw/dx(0))$, and $\delta(dw/dx(L))$ separately, we have

$$0 = \delta w(0) \left[\frac{d}{dx}\left(EI \frac{d^2 w}{dx^2} \right) \right]_{x=0} + \delta w(L) \left[-F - \frac{d}{dx}\left(EI \frac{d^2 w}{dx^2} \right) \right]_{x=L} + \delta\left(\frac{dw}{dx}(0) \right) \left[-EI \frac{d^2 w}{dx^2} \right]_{x=0} + \delta\left(\frac{dw}{dx}(L) \right) \left[-M + EI \frac{d^2 w}{dx^2} \right]_{x=L} \tag{4-74}$$

Since $\delta w(0) = \delta(dw/dx(0)) = 0$, by setting the coefficients of $\delta w(L)$ and $\delta(dw/dx(L))$ to zero separately, we obtain the additional Euler equations:

$$\begin{aligned} \delta w(L)&: \quad -F - \left[\frac{d}{dx}\left(EI \frac{d^2 w}{dx^2} \right) \right]_{x=L} = 0 \\ \delta\left(\frac{dw}{dx}(L) \right)&: \quad -M + \left(EI \frac{d^2 w}{dx^2} \right)\Big|_{x=L} = 0 \end{aligned} \tag{4-75}$$

Thus, the essential boundary conditions are given by

$$w(0) = \frac{dw}{dx}(0) = 0 \tag{4-76}$$

The natural boundary conditions are given by Eq. (4-75).

Note that the essential boundary conditions are the boundary conditions resulting from the geometric constraints on the beam (i.e., clamped at $x = 0$), and the natural boundary conditions resulted from the application of the generalized forces F and M at $x = L$. Equation (4-75) states that F should be equal to the shear force $-d/dx(EI(d^2w/dx^2))$ at $x = L$, and that M should be equal to the bending moment $EI(d^2w/dx^2)$ at $x = L$.

In any problem one has only one of the following three possible combinations of boundary conditions:

(*a*) all are of essential type;
(*b*) all are of natural type;
(*c*) some of them are of natural type and the remaining are of essential type (4-77)

Problems in which all of the boundary conditions are of essential type are called *Dirichlet problems* (*or* boundary-value problems of the *first kind*), and those in which all of the boundary conditions are of natural type are called *Neumann problems* (*or* boundary-value problems of the *second kind*). *Mixed problems* (*or* boundary-value problems of the *third kind*) are those in which both essential and natural boundary conditions are specified. Examples of the three kinds of boundary-value problems are provided, respectively, by a beam clamped at both ends, a beam that is (geometrically) free at both ends, and a cantilever beam.

The minimum character of the total potential energy of an elastic structure can be easily established. Here we consider the small displacement theory of an elastic structure. The strain energy of a linear elastic body is given by

$$U = \int_{\Omega} U_0 \, d\mathbf{x} \tag{4-78}$$

where U_0 denotes the strain energy per unit volume (i.e., the *strain energy density*) of the body, and is given by

$$U_0 = \tfrac{1}{2}\sigma_{ij} e_{ij} = \tfrac{1}{2}C_{ijkl} e_{ij} e_{kl} \qquad C_{ijkl} > 0 \tag{4-79}$$

where C_{ijkl} are the material coefficients. Let $\mathbf{u}^0 \in \mathscr{D}$ be the solution of Eq. (2-92) for the steady case. Then for any $\mathbf{u} = \mathbf{u}^0 + \alpha\mathbf{v}$, $\mathbf{v} \in \mathscr{S}$, we have

$$\Pi(\mathbf{u}) = \int_{\Omega} \tfrac{1}{8}C_{ijkl}[u^0_{i,j} + u^0_{j,i} + \alpha(v_{i,j} + v_{j,i})][u^0_{k,l} + u^0_{l,k} + \alpha(v_{k,l} + v_{l,k})] \, d\mathbf{x}$$
$$- \int_{\Omega} f_i(u^0_i + \alpha v_i) \, d\mathbf{x} - \int_{\Gamma_2} \hat{t}_i(u^0_i + \alpha v_i) \, ds \tag{4-80}$$

$$= \Pi(\mathbf{u}^0) + \alpha \int_{\Omega} [\tfrac{1}{2}C_{ijkl}(g_{ij}\varepsilon^0_{kl} + \varepsilon^0_{ij} g_{kl} + \alpha g_{ij} g_{kl}) - f_i v_i] \, d\mathbf{x}$$
$$- \alpha \int_{\Gamma_2} \hat{t}_i v_i \, ds \tag{4-81}$$

where $$\varepsilon^0_{ij} = \tfrac{1}{2}(u^0_{i,j} + u^0_{j,i}) \qquad g_{ij} = \tfrac{1}{2}(v_{i,j} + v_{j,i})$$

Using $f_i = -\sigma^0_{ij,j} = -C_{ijkl}\varepsilon^0_{kl,j}$ and $\hat{t}_i = C_{ijkl}\varepsilon^0_{kl}n_j$ in Eq. (4-81), and integrating by parts we get

$$\Pi(\mathbf{u}) = \Pi(\mathbf{u}^0) + \frac{\alpha^2}{2}\int_\Omega C_{ijkl}g_{ij}g_{kl}\,d\mathbf{x} \tag{4-82}$$

Since the second term on the right side is positive and equal to zero only if $\mathbf{u} = \mathbf{u}^0$ it follows that

$$\Pi(\mathbf{u}) \geq \Pi(\mathbf{u}^0) \tag{4-83}$$

which we set out to prove.

4-3-4 Nonhomogeneous Boundary Conditions

In the preceding discussions we often assumed that the boundary conditions of a boundary-value problem are homogeneous. This allowed us to identify the domain of the operator to be a linear space (see Examples 4-1 and 4-5). For example, consider the Dirichlet problem

$$-\nabla^2 u = f \text{ in } \Omega$$

$$u = 0 \text{ on } \Gamma$$

Several examples of problems in which the Laplace operator occurs are listed in Table 4-2. The solution space S consists of functions u that have first and second partial derivatives in Ω and satisfy the condition $u = 0$ on Γ. The set S is a linear vector space because if $u, v \in S$ then $\alpha u + \beta v \in S$. However, if the homogeneous boundary condition $u = 0$ on Γ is replaced by $u = g$ on Γ, then S is not a linear set because $u + v = 2g$ on Γ so that the condition $u + v = g$ on Γ is not satisfied.

In order to utilize the vector space theory described in this book for the study of boundary-value problems with nonhomogeneous boundary conditions, we must transform problems with nonhomogeneous boundary conditions to problems with homogeneous boundary conditions. The procedure is described here.

Consider the linear operator equation

$$Au = f \text{ in } \Omega \tag{4-84a}$$

with linear nonhomogeneous boundary conditions

$$Bu = g \text{ on } \Gamma \tag{4-84b}$$

where $f \in L_2(\Omega)$ and $g \in L_2(\Gamma)$. Suppose that we can find a function w which satisfies the specified boundary condition $Bw = g$ on Γ and is sufficiently smooth so that $Aw \in L_2(\Omega)$. Then $z = u - w$ satisfies the equations

$$\begin{aligned} Az &= f - Aw \text{ in } \Omega \\ Bz &= 0 \text{ on } \Gamma \end{aligned} \tag{4-85}$$

Table 4-2 Some examples of the Poisson equation $-\nabla\cdot(k\nabla u) = f$†

Natural boundary condition: $k\dfrac{\partial u}{\partial n} + h(u - u_\infty) = q$; essential boundary condition: $u = \hat{u}$

Field of application	Primary variable u	Material constant k	Source variable f	Secondary variables $q, \dfrac{\partial u}{\partial x}, \dfrac{\partial u}{\partial y}$
1. Heat transfer	Temperature T	Conductivity k	Heat source Q	Heat flow q [comes from conduction $k(\partial T/\partial n)$ and convection $h(T - T_\infty)$]
2. Irrotational flow of an ideal fluid	Stream function ψ	Density ρ	Mass production σ (normally zero)	velocities: $\dfrac{\partial\psi}{\partial x} = -v, \dfrac{\partial\psi}{\partial y} = u$
	Velocity potential ϕ	Density ρ	Mass production σ (normally zero)	$\dfrac{\partial\phi}{\partial x} = u, \dfrac{\partial\phi}{\partial y} = v$
3. Ground-water flow	Piezometric head ϕ	Permeability K	Recharge Q (or pumping, $-Q$)	Seepage q $q = K\dfrac{\partial\phi}{\partial n}$ velocities: $u = -K\dfrac{\partial\phi}{\partial x}$ $v = -K\dfrac{\partial\phi}{\partial y}$
4. Torsion of constant cross-section members	Stress function ϕ	$k = \dfrac{1}{G}$, G = shear modulus	$f = 2\theta$, θ = angle of twist per unit length	$\dfrac{\partial\phi}{\partial x} = -\tau_{zy}$ $\dfrac{\partial\phi}{\partial y} = \tau_{zx}$ τ_{zx}, τ_{zy} are shear stresses
5. Electrostatics	Scalar potential ϕ	Dielectric constant ε	Charge density ρ	Displacement flux density D_n
6. Magnetostatics	Magnetic potential ϕ	Permeability μ	Charge density ρ	Magnetic flux density B_n
7. Transverse deflection of elastic membranes	Transverse deflection u	$k = T$, T = tension in membrane	Transversely distributed load	Normal force q

† From Reddy (1984a)

Thus, we are able to transform the given operator equation with nonhomogeneous boundary conditions to one with homogeneous boundary conditions. Once z is determined from Eq. (4-85), the solution to Eqs. (4-84a) and (4-84b) is given by $u = z + w$.

In the weak formulation of the boundary-value problems, the natural boundary conditions are included in the variational formulation and the essential boundary conditions are included in the definition of the solution space. Thus it is only necessary to transform boundary-value problems with nonhomogeneous boundary conditions to those with homogeneous essential boundary conditions. We consider two examples here.

Example 4-7 Consider the Dirichlet problem for the Laplace equation on the rectangle, $\Omega = \{(x_1, x_2): 0 < x_1 < a, 0 < x_2 < b\}$

$$-\nabla^2 u = 0 \text{ in } \Omega$$

with the boundary condition on Γ (see Fig. 4-6)

$$u = \sin \frac{\pi x}{a} \text{ for } y = b \text{ and } 0 < x < a$$

$$u = 0 \text{ on the remaining boundary}$$

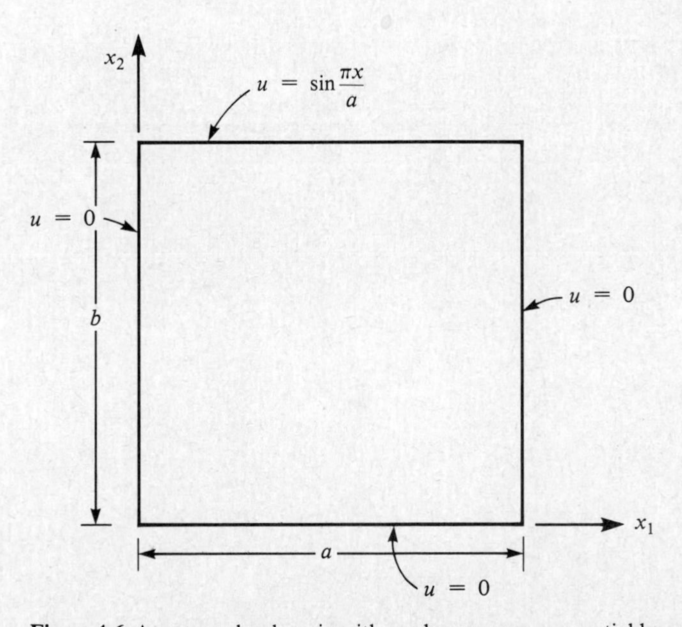

Figure 4-6 A rectangular domain with nonhomogeneous essential boundary conditions.

In this case w is given by

$$w = y \sin \frac{\pi x}{a}$$

which is twice-differentiable with respect to x and y, and satisfies the boundary conditions. The transformed problem becomes

$$-\nabla^2 z = \nabla^2 w = -\left(\frac{\pi}{a}\right)^2 y \sin \frac{\pi x}{a} \text{ in } \Omega$$

$$z = 0 \text{ on } \Gamma$$

Example 4-8 Consider the nonhomogeneous Stokes problem [see Eq. (2-100) and Prob. 3-2-5]

$$\begin{aligned} -\nu\nabla^2\mathbf{u} + \operatorname{grad} P &= \mathbf{f} \text{ in } \Omega \\ \operatorname{div} \mathbf{u} &= 0 \text{ in } \Omega \\ \mathbf{u} &= \mathbf{g} \text{ on } \Gamma \end{aligned} \tag{4-86a}$$

where $\mathbf{u}$ is the velocity vector, P is the pressure (divided by density ρ), $\mathbf{f}$ is the body force vector per unit mass, and ν is the kinematic viscosity. Note that P does not have any boundary condition because it is included in the specification of boundary stresses, and since $\mathbf{u}$ is specified, we cannot specify boundary stresses.

Let $\mathbf{w} \in \mathbf{H}^2(\Omega) \equiv [H^1(\Omega)]^N$, where N is the dimension of the domain [$\mathbf{w} = (w_1, w_2, \ldots, w_N)$], be such that $\mathbf{w} = \mathbf{g}$ on Γ. Also let $\mathbf{w}_1 \in \mathbf{H}_0^1(\Omega) \cap \mathbf{H}^2(\Omega)$ such that $\operatorname{div} \mathbf{w}_1 = -\operatorname{div} \mathbf{w}$. Set $\mathbf{z} = \mathbf{u} - \mathbf{w} - \mathbf{w}_1$ in Eq. (4-86a) to obtain the following homogeneous Stokes problem:

$$\begin{aligned} -\nu\nabla^2\mathbf{z} + \operatorname{grad} P &= \mathbf{f} - \nu(\nabla^2\mathbf{w} + \nabla^2\mathbf{w}_1) \\ \operatorname{div} \mathbf{z} &= 0 \text{ in } \Omega \\ \mathbf{z} &= \mathbf{0} \text{ on } \Gamma \end{aligned} \tag{4-86b}$$

The weak problem involves finding $\mathbf{u} - \mathbf{z} \in \mathbf{H}$ such that

$$B(\mathbf{v}, \mathbf{u}) = (\mathbf{v}, \hat{\mathbf{f}})_0 \qquad \text{for every } \mathbf{v} \in \mathbf{H} \tag{4-86c}$$

holds, where

$$B(\mathbf{v}, \mathbf{u}) = \sum_{i=1}^{N} \int_\Omega \operatorname{grad} v_i \cdot \operatorname{grad} u_i \, d\mathbf{x}$$

$$\hat{\mathbf{f}} = \mathbf{f} - \nu(\nabla^2\mathbf{w} + \nabla^2\mathbf{w}_1)$$

$$\mathbf{H} = \{\mathbf{v} \in \mathbf{H}_0^1(\Omega) \colon \operatorname{div} \mathbf{v} = 0 \text{ in } \Omega\} \tag{4-86d}$$

PROBLEMS 4-2

4-2-1 (*The Friedrichs inequality* in one dimension) Let S be a linear set of functions continuous with their first derivatives in the closed interval $[a, b]$. Then show, for $u \in S$, that

$$\int_a^b |u(x)|^2\, dx \le c_1 \int_a^b \left|\frac{du}{dx}(x)\right|^2 dx + c_2 |u(b)|^2$$

$$\int_a^b |u(x)|^2\, dx \le c_3 \int_a^b \left|\frac{du}{dx}(x)\right|^2 dx + c_4 |u(a)|^2$$

Hint: See Rektorys (1980, Chap. 18).

4-2-2 (*The Poincaré inequality* in one dimension) Let S be the linear set of functions with continuous first derivatives in $[a, b]$. Show that there exist constants c_1 and c_2 such that the inequality

$$\int_a^b |u(x)|^2\, dx \le c_1 \int_a^b \left|\frac{du}{dx}(x)\right|^2 dx + c_2 \left[\int_a^b u(x)\, dx\right]^2$$

holds for all $u \in S$.

Use Eq. (4-32) to construct the quadratic functional of Probs. 4-2-3 and 4-2-4.

4-2-3 $(d^2/dx^2)(EI(d^2u/dx^2)) = f\,(E > 0,\, I > 0)$, $\quad 0 < x < L$; $\quad u(0) = u(L) = u''(0) = u''(L) = 0 \quad (u'' = d^2u/dx^2)$. The domain of the operator is the linear set of functions which are continuous with their derivatives up to the fourth order inclusive in the interval $[0, L]$ and satisfy the boundary conditions.

4-2-4 $-(\partial^2 u/\partial x^2 + \partial^2 u/\partial y^2) = f$, $0 < x < 1$, $0 < y < 1$; $u(x, 0) = u(x, 1) = u(0, y) = u(1, y) = 0$. The domain of the operator is the linear set of functions which are continuous with their derivatives up to the second order inclusive in the domain $\Omega \subset \mathbb{R}^2$ and satisfy the boundary conditions.

For the functionals given in Probs. 4-2-5 to 4-2-8, determine the Euler equations and classify the boundary conditions into natural and essential type:

4-2-5
$$I(u) = \frac{1}{2}\int_0^1\int_0^1 \left[\left(\frac{\partial u}{\partial x}\right)^2 + \left(\frac{\partial u}{\partial y}\right)^2 - 2fu\right] dx\, dy + \int_0^1 \hat{t}_1(y)u(0, y)\, dy + \int_0^1 \hat{t}_2(x)u(x, 0)\, dx$$

4-2-6
$$I(u, v, w) = \int_\Omega \left\{-\alpha\left[\left(\frac{\partial u}{\partial x}\right)^2 + \left(\frac{\partial v}{\partial y}\right)^2 + \frac{1}{2}\left(\frac{\partial u}{\partial y} + \frac{\partial v}{\partial x}\right)^2\right]\right.$$
$$\left. + w\left(\frac{\partial u}{\partial x} + \frac{\partial v}{\partial y}\right) - f_1 u - f_2 v\right\} dx\, dy + \int_{\Gamma_2} \hat{t}_1 u\, ds + \int_{\Gamma_2} \hat{t}_2 v\, ds$$

4-2-7
$$I(u) = \int_0^1 \left[\frac{1}{48}\left(\frac{du}{dx}\right)^4 + \left(\frac{du}{dx}\right)^2 + u^6 - 6u\right] dx$$

4-2-8 *Shear deformation theory of beams*

$$I(u, w, \psi) = \int_0^t \int_0^L \int_A \left[\frac{\rho}{2}\left(\frac{\partial u}{\partial t} + z\frac{\partial \psi}{\partial t}\right)^2 + \frac{\rho}{2}\left(\frac{\partial w}{\partial t}\right)^2 - \frac{G}{2}\left(\psi + \frac{\partial w}{\partial x}\right)^2\right.$$
$$\left. - \frac{E}{2}\left(\frac{\partial u}{\partial x} + z\frac{\partial \psi}{\partial x}\right)^2\right] dA\, dx\, dt + \int_0^t Pw(L)\, dt$$

where ρ, A, E, and P are constants.

4-2-9 The equations governing the slow and steady laminar flow of an incompressible fluid in two dimensions can be variationally formulated as one of minimizing the functional

$$I_0(u, v) = \int_\Omega \left\{v\left[\left(\frac{\partial u}{\partial x}\right)^2 + \left(\frac{\partial v}{\partial y}\right)^2 + \frac{1}{2}\left(\frac{\partial u}{\partial y} + \frac{\partial v}{\partial x}\right)^2\right] + f_1 u + f_2 v\right\} dx\, dy - \int_{\Gamma_2} (\hat{t}_1 u + \hat{t}_2 v)\, ds$$

on the space

$$H^* = \left\{(u, v): u, v \in H^2(\Omega), \frac{\partial u}{\partial x} + \frac{\partial v}{\partial y} = 0 \text{ in } \Omega, u = v = 0 \text{ on } \Gamma_1 = \Gamma - \Gamma_2\right\}$$

Here (u, v) denote the components of the velocity vector, ν is the kinematic viscosity, (f_1, f_2) are the body force components, and $(\hat{t}_1, \hat{t}_2)$ are the components of the specified boundary stress vector. Derive the Euler equations.

Hint: Note that

$$\int_\Omega \mathbf{F} \cdot \mathbf{u}\, d\mathbf{x} = 0 \text{ for any } \mathbf{u} = (u, v) \in H^*$$

implies that there exists a P such that $\mathbf{F} = \text{grad } P$.

4-2-10 The total potential energy of the linear bending of a clamped orthotropic circular plate is given by

$$\Pi(w) = \frac{1}{2}\int_0^a \left[D_{11}\left(\frac{d^2w}{dr^2}\right)^2 + \frac{2D_{12}}{r}\frac{dw}{dr}\frac{d^2w}{dr^2} + D_{22}\left(\frac{1}{r}\frac{dw}{dr}\right)^2\right] r\, dr - \int_0^a wf\, dr$$

where r is the radial coordinate, a is the radius of the plate, f is the distributed transverse load, and D_{ij} are the plate stiffnesses. Use the principle of the total potential energy to determine the governing differential equation of the plate.

4-2-11 (*Reissner's variational principle for circular plates*) Determine the Euler equations of the functional

$$R(u, w, \psi) = \frac{\pi}{2}\int_0^a \left[A_{11}\left(\frac{du}{dr}\right)^2 + \frac{2A_{12}}{r} u \frac{du}{dr} + A_{22}\left(\frac{u}{r}\right)^2 + A_{55}\left(\psi + \frac{dw}{dr}\right)^2\right.$$

$$+ 2M_r \frac{d\psi}{dr} + 2M_\theta \frac{\psi}{r} + \bar{D}_{22} M_r^2 - 2\bar{D}_{12} M_r M_\theta$$

$$\left. + \bar{D}_{11} M_\theta^2 - 2fw \right] r\, dr$$

where A_{ij} and $\bar{D}_{ij}$ are the plate stiffnesses, (u, w, ψ) are the displacements, (M_r, M_θ) are the moments, and f is the transverse distributed load.

Give the variational formulation of the equations in Probs. 4-2-12 to 4-2-16. Identify the energy space associated with each problem:

4-2-12 *Bending of a clamped beam*

$$-\frac{d}{dx}\left(a\frac{dw}{dx}\right) + \frac{d^2}{dx^2}\left(EI\frac{d^2w}{dx^2}\right) = f \qquad 0 < x < L$$

$$w(0) = w'(0) = w(L) = w'(L) = 0$$

4-2-13 *Two-dimensional convective heat transfer*

$$-\nabla \cdot (k\nabla u) + f = 0 \text{ in } \Omega \subset \mathbb{R}^2$$

$$k\frac{\partial u}{\partial n} + h(u - u_\infty) + q_n = 0 \text{ on } \Gamma_2$$

$$u = \hat{u} \text{ on } \Gamma_1$$

4-2-14 *Elasticity*

$$(\lambda + \mu) \text{ grad div } \mathbf{u} + \mu\nabla^2\mathbf{u} = \mathbf{f} \text{ in } \Omega$$

$$\mathbf{u} = \mathbf{g} \text{ on } \Gamma_1 \qquad \text{and} \qquad \mathbf{t} = \hat{\mathbf{t}} \text{ on } \Gamma_2$$

4-2-15 *Elastic-plastic torsion of a cylindrical bar*

$$-\nabla^2\Phi = 2\theta \text{ in } \Omega$$

$$|\operatorname{grad} \Phi| \leq 1 \text{ in } \Omega$$

$$\Phi = 0 \text{ on } \Gamma$$

where Φ denotes the Prandtl stress function and θ is the twist per unit length.

4-2-16 *Stokes flow in two dimensions*

$$-2\mu \frac{\partial^2 u}{\partial x^2} - \mu \frac{\partial}{\partial y}\left(\frac{\partial u}{\partial y} + \frac{\partial v}{\partial x}\right) - \gamma \frac{\partial}{\partial x}\left(\frac{\partial u}{\partial x} + \frac{\partial v}{\partial y}\right) = f_x$$

$$-\mu \frac{\partial}{\partial x}\left(\frac{\partial u}{\partial y} + \frac{\partial v}{\partial x}\right) - 2\mu \frac{\partial^2 v}{\partial y^2} - \gamma \frac{\partial}{\partial y}\left(\frac{\partial u}{\partial x} + \frac{\partial v}{\partial y}\right) = f_y$$

where μ and γ are constants.

4-4 PROBLEMS WITH EQUALITY CONSTRAINTS

4-4-1 Introduction

In Sec. 4-3 we studied variational formulations of differential equations with appropriate boundary conditions. The number of differential equations was always equal to the number of dependent unknowns. It is not uncommon in engineering to find problems in which the number of equations exceed the number of dependent unknowns. The extra equations are called *subsidiary conditions* or *constraint equations.* Sometimes, the problem can be posed as one of extremizing a given functional subjected to a constraint. The constraint can be in the form of an algebraic equation or an integral expression among (all or some of) the dependent unknowns. The subject of optimization is rich in such examples.

As an example, consider the *isoperimetric problem* from calculus of variations: among all continuous curves $y(x)$ between two fixed points (a, y_a) and (b, y_b) and having a given length, find that which maximizes the area under the curve. Stated in analytical terms, find $y \in H^2[0, L]$ such that $y(a) = y_a$ and $y(b) = y_b$, and

$$J(y) = \int_a^b y(x)\, dx \text{ is a maximum} \tag{4-87a}$$

$$G(y) \equiv \int_a^b \sqrt{1 + \left(\frac{dy}{dx}\right)^2}\, dx - L = 0 \tag{4-87b}$$

Equation (4-87*b*) is called a *constraint.* A candidate that possibly gives the maximum of J should necessarily satisfy the end conditions, which can also be viewed as constraints,

$$y(a) = y_a \qquad y(b) = y_b \tag{4-87c}$$

and the constraint (4-87*b*).

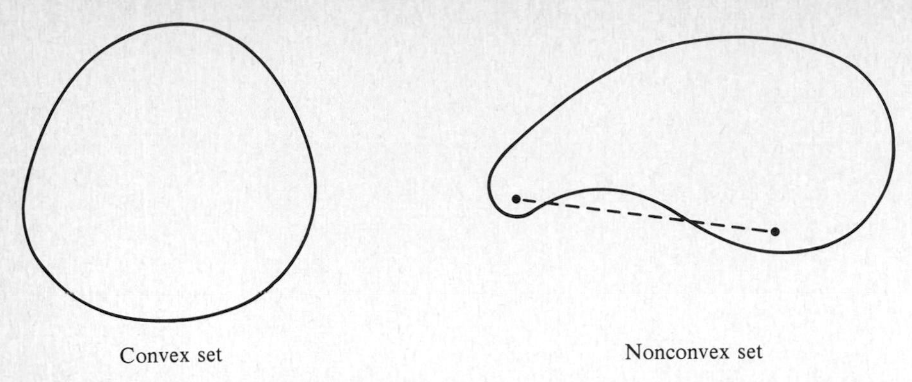

Figure 4-7 Convex and nonconvex sets.

Another example of a constrained variational problem is provided by the second equation in Eq. (4-86*a*) of Example 4-8. The constraint is the divergence-free condition on the velocity field: div $\mathbf{u} = 0$. In the numerical solution of the weak problem (4-86*c*) we seek approximate solutions in a finite-dimensional subspace $\mathbf{H}$ of admissible vectors, which is defined in Eq. (4-86*d*). Construction of such a space is rather difficult. The difficulty can be overcome by posing the problem as an unconstrained variational problem. In this section we study the Lagrange multiplier method and the penalty function method. Both methods can be used to include the constraints of a given problem into its weak formulation, and thus remove the constraint on the solution space.

Recall that a set X in a linear space is said to be *convex* if, given $x_1, x_2 \in X$, all points of the form $\alpha x_1 + (1 - \alpha)x_2$ with $0 \leq \alpha \leq 1$ are in X. Geometrically this means that given two points in a convex set, the line segment between them is also in the set (see Fig. 4-7). Note, in particular, that all subspaces and linear manifolds are convex. The empty set is considered convex.

The basic problem considered in this section is

$$\begin{aligned} &\text{minimize } \; J(u) \\ &\text{subject to } u \in \mathscr{D},\; G(u) = 0 \end{aligned} \tag{4-88}$$

where $\mathscr{D}$ is a convex subset of a vector space U, J is a real-valued convex functional on $\mathscr{D}$, and G is a convex mapping from $\mathscr{D}$ into a normed space V. In the coming sections we will study alternative formulations of this problem.

4-4-2 The Lagrange Multiplier Method

To gain some geometric insight into the problem in Eq. (4-88), let us consider the case in which J and G are real-valued functions defined on $\mathbb{R}^n$. In particular, let us consider the problem

$$\begin{aligned} &\text{minimize: } J(x_1, x_2) \\ &\text{subject to the constraint } G(x_1, x_2) = 0 \end{aligned}$$

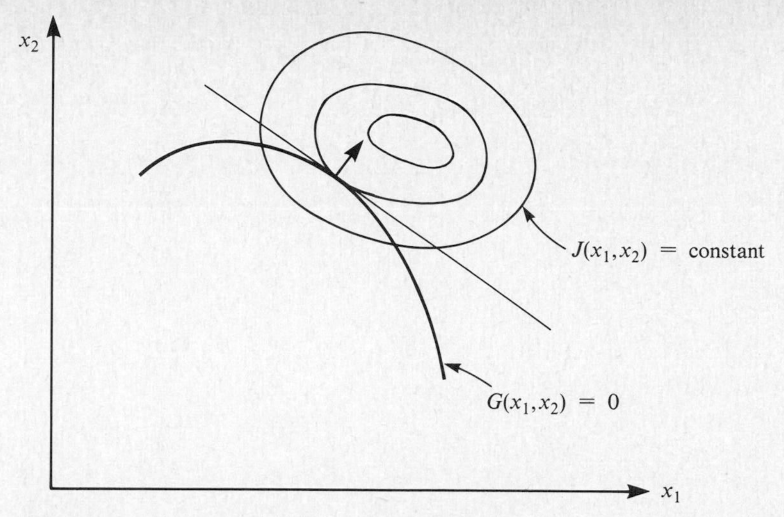

Figure 4-8 Constrained extremum in two dimensions.

We can plot the curve described by the constraint $G(x_1, x_2) = 0$ in the x_1x_2-plane, and for every constant we can plot the curve defined by $J(x_1, x_2) =$ constant (see Fig. 4-8). From Fig. 4-8 it is clear that the solution is the point at which the curve intersects the curve of constant J with the least value. If both curves are continuously differentiable at the point, then they will be tangent to each other. Thus, they will possess a common tangent and a common normal at the solution point. Since the normal to the curve $\phi(x_1, x_2) =$ constant is proportional to its gradient, $\mathbf{n} = \nabla\phi$, it follows that ∇J and ∇G are related,

$$\begin{aligned} \nabla J &= \bar{\lambda}\nabla G \qquad \bar{\lambda} = \text{constant} \\ \text{or grad } (J + \lambda G) &= 0 \qquad (\lambda = -\bar{\lambda}) \end{aligned} \tag{4-89}$$

which is a necessary condition for the problem to have a solution.

The idea just described for an elementary problem forms the basis of the Euler–Lagrange multiplier theorem to be presented shortly. The constant λ is called the *Lagrange multiplier*. Leonard Euler (1707–1783) had discovered how to solve constrained extremum problems using what amounts to Eq. (4-89) by the year 1741. Joseph Lagrange (1736–1813) studied Euler's results and later formulated the multiplier theorem for the special case in which J and G are functions of n real variables. With the development of the calculus of functionals on normed spaces in the twentieth century, the multiplier theorem has taken on a more general and mathematically rigorous form.

It is possible in some problems to have more than one constraint. For example, consider the problem of minimizing $J(x_1, x_2, x_3)$ subject to the constraints $G_1(x_1, x_2, x_3) = 0$ and $G_2(x_1, x_2, x_3) = 0$. At the solution $\mathbf{x}^0 = (x_1^0, x_2^0,$

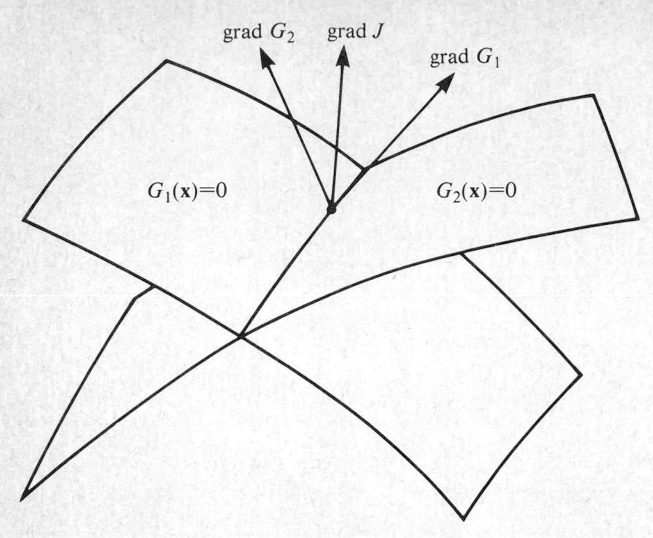

Figure 4-9 Extremum with two constraints.

x_3^0), if it exists, the gradient of J must lie in the plane generated by the gradients of G_1 and G_2; hence

$$\text{grad } (J + \lambda_1 G_1 + \lambda_2 G_2) = 0$$

Figure 4-9 contains a geometric interpretation of the problem.

Now we are ready to state, without proof, the Euler–Lagrange multiplier theorem. Interested readers can consult the books by Hestenes (1966) and Luenberger (1969).

Theorem 4-6: The Euler–Lagrange multiplier theorem Consider the constrained minimization problem in Eq. (4-88). Suppose that J and G are continuously Gateaux differentiable and J has a local minimum under the constraint $G(u) = 0$ at the regular point u_0 (that is, $u_0 \in \mathscr{D}$ is such that $\delta J(u_0)$ maps U onto $\mathbb{R}$). Then there exists an element $\lambda \in V'$, where V' is the dual space of V, such that the *Lagrangian functional*

$$L(u, \lambda) = J(u) + \langle \lambda, G(u) \rangle \tag{4-90a}$$

is stationary at u_0, i.e.,

$$\delta J(u_0) + \lambda \delta G(u_0) = 0 \tag{4-90b}$$

where $\langle \cdot, \cdot \rangle$ denotes the duality pairing on $V' \times V$.

Note that the constraint condition $G(u) = 0$ as well as Eq. (4-90b) can be recovered by setting the partial variations $\delta_\lambda L$ and $\delta_u L$ to zero, respectively:

$$\begin{aligned} \delta_\lambda L((u, \lambda); \delta\lambda) &= \langle \delta\lambda, G(u) \rangle = 0 \to G(u) = 0 \\ \delta_u L((u, \lambda); \delta u) &= \delta J(u; \delta u) + \langle \lambda, \delta G(u; \delta u) \rangle \\ &= 0 \to \delta J(u) + \lambda \delta G(u) = 0 \end{aligned} \tag{4-91}$$

Thus, Theorem 4-6 provides a method for the inclusion of the constraint into the variational problem. Instead of seeking the minimum of J on a subset of all vectors in $\mathscr{D}$ which satisfy the constraint $G(u) = 0$, the Lagrange multiplier method seeks the stationary point (u, λ) [i.e., the point (u, λ) that satisfies Eqs. (4-91)] of the lagrangian functional L. Equations (4-91) are the well known Euler–Lagrange equations. Of course, the theorem can be applied to problems with more than one constraint.

We now consider several examples to illustrate the use of the multiplier theorem.

Example 4-9 Consider the real-valued functions on $\mathscr{D} = \mathbb{R}$,

$$J(x) = x^2 \qquad G(x) = x^2 + 2x + \tfrac{3}{4}$$

We wish to find vectors (numbers) in $\mathbb{R}$ that satisfy the condition $G(x) = 0$ and minimize J. The situation is geometrically represented in Fig. 4-10.

A direct calculation (not possible for continuum problems) shows that the constraint is satisfied for $x = -\frac{1}{2}$ and $x = -\frac{3}{2}$. However, J does not have its local minimum at either point. If the domain $\mathscr{D}$ is restricted to the interval $[-1/2, -1/4]$, then J has its minimum at $x_0 = -\frac{1}{2}$. The minimum value of J is $\frac{1}{4}$.

The same result can be obtained using the Lagrange multiplier method. The lagrangian function is given by

$$L(x, \lambda) = x^2 + \lambda(x^2 + 2x + \tfrac{3}{4})$$

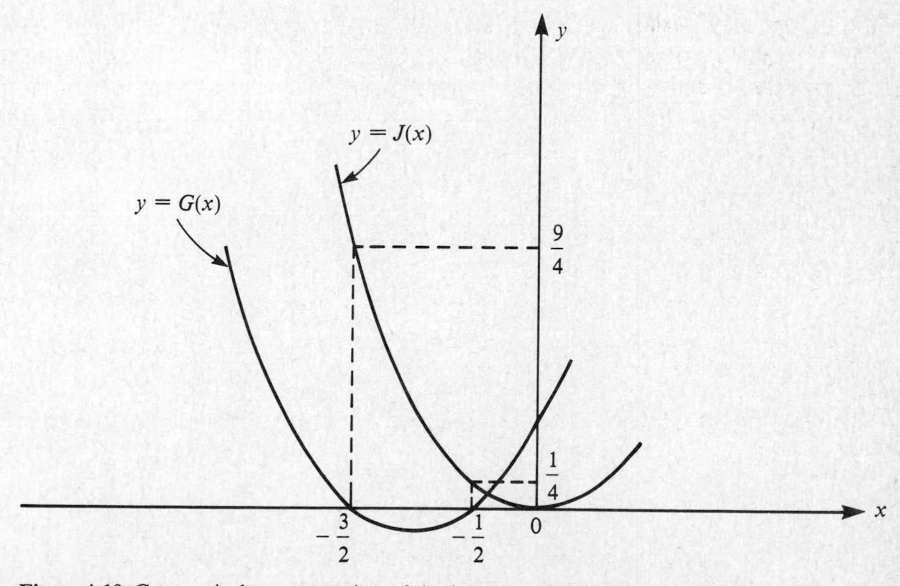

Figure 4-10 Geometrical representation of the functions $J(x)$ and $G(x)$ in Example 4-9.

Equation (4-90*b*) in the present case becomes

$$2x_0 + \lambda(2x_0 + 2) = 0 \qquad \text{or} \qquad x_0 = -\frac{\lambda}{1+\lambda}$$

Substituting this expression into the constraint, we obtain

$$\left(-\frac{\lambda}{1+\lambda}\right)^2 + 2\left(-\frac{\lambda}{1+\lambda}\right) + \frac{3}{4} = 0$$

or

$$\lambda^2 + 2\lambda - 3 = 0$$

This gives two solutions, $\lambda = 1$ and $\lambda = -3$. For $\lambda = 1$ we get $x_0 = -\frac{1}{2}$ and for $\lambda = -3$, we obtain $x_0 = -\frac{3}{2}$.

This example illustrates two points: first, the variation of the functional need not vanish at a local extremum vector that satisfies the constraint; second, the Lagrange multiplier theorem gives only a necessary condition which must be satisfied by any possible minimum and/or maximum vector. To determine whether the extremum vector minimizes or maximizes the given functional J, one must calculate higher-order variations of J.

Example 4-10 Consider the problem of minimizing the functional J: $L_2[1, 2] \to \mathbb{R}$,

$$J(u) \equiv \int_1^2 x\,|u(x)|^2\,dx$$

on the vector space $C[1, 2]$ subject to the constraint (G: $C[1, 2] \to \mathbb{R}$),

$$G(u) \equiv \int_1^2 u(x)\,dx - \ln 2 = 0$$

for any u in $C[1, 2]$. The lagrangian functional is given by

$$L(u, \lambda) = J(u) + \lambda G(u)$$

and Eq. (4-90*b*) gives

$$\int_1^2 [2xu_0(x) + \lambda]\,\delta u(x)\,dx = 0$$

Since the variation δu is arbitrary [one can take $\delta u(x) = 2xu_0(x) + \lambda$], we obtain

$$2xu_0(x) + \lambda = 0$$

or

$$u_0(x) = -\frac{\lambda}{2x}$$

Substituting this expression for $u(x)$ into the constraint, we obtain

$$-\int_1^2 \frac{\lambda}{2x}\,dx - \ln 2 = 0 \qquad \text{or} \qquad \lambda = -2$$

Hence the extremum vector $u_0(x)$ is given by

$$u_0(x) = \frac{1}{x} \qquad 1 \le x \le 2$$

Let us see whether $u_0(x) = 1/x$ minimizes or maximizes J. To show that u_0 minimizes J, we must establish that

$$J(u_0) \le J(u)$$

for any admissible vector u [i.e., satisfy $G(u) = 0$] in $C[1, 2]$. Let us take $u = u_0 + \eta$. In order that u is an admissible vector, we must have

$$G(u_0 + \eta) = 0$$

which implies that

$$\int_1^2 \eta(x)\, dx = 0$$

Now consider the value of J at $u_0 + \eta$,

$$\begin{aligned} J(u_0 + \eta) &= \int_1^2 x(u_0 + \eta)^2\, dx \\ &= \int_1^2 x u_0^2\, dx + \int_1^2 x(\eta^2 + 2\eta u_0)\, dx \\ &= J(u_0) + \int_1^2 x\eta^2\, dx \end{aligned}$$

where we used the fact that $xu_0 = 1$ and $\int_1^2 \eta(x)\, dx = 0$. Since the second term on the right side of the equality is positive for $x \in [1, 2]$, the inequality (it is an equality only if $\eta = 0$),

$$J(u_0 + \eta) \ge J(u_0)$$

holds for any admissible vector $u = u_0 + \eta$ in $C[1, 2]$. This inequality shows that any function other than u_0 would increase the value of J.

The next example is concerned with the Lagrange multiplier formulation of the isoperimetric problem described by Eqs. (4-87a) and (4-87b).

Example 4.11 The lagrangian functional for the isoperimetric problem is given by

$$\begin{aligned} L(y, \lambda) &= J(y) + \lambda G(y) \\ &= \int_a^b y(x)\, dx + \lambda \left[\int_a^b \sqrt{1 + (y')^2}\, dx - L \right] \end{aligned} \tag{4-92}$$

The variation δy of y satisfies the conditions

$$\delta y(a) = \delta y(b) = 0 \tag{4-93}$$

The Euler–Lagrange equations are given by setting $\delta_y L$ and $\delta_\lambda L$ to zero:

$$\delta_y L(y, \lambda)(\delta y) = \int_a^b \delta y\, dx + \int_a^b \frac{\lambda y' \delta y'}{\sqrt{1 + (y')^2}}\, dx$$

$$= \int_a^b \left\{1 - \frac{d}{dx}\left[\frac{\lambda y'}{\sqrt{1 + (y')^2}}\right]\right\} \delta y\, dx + \left[\frac{\lambda y'}{\sqrt{1 + (y')^2}}\, \delta y\right]_a^b$$

The second term on the right side of the equality vanishes by virtue of Eq. (4-93). Since δy is arbitrary in (a, b), we obtain the Euler–Lagrange equation

$$1 - \frac{d}{dx}\left[\frac{\lambda y'}{\sqrt{1 + (y')^2}}\right] = 0 \qquad y' \equiv \frac{dy}{dx} \tag{4-94}$$

Equations (4-94) and $G(y) = 0$ together provide the solution (y, λ) to the problem. Integration of Eq. (4-94) leads to the equation of a circle of radius λ,

$$(x - x_0)^2 + (y - y_0)^2 = \lambda^2$$

where x_0 and y_0 are the constants of integration, which can be determined using the end conditions in Eq. (4-87*c*). Thus, the solution is an arc of a circle of radius λ, and the length is equal to L.

From the above discussion it is clear that the Lagrange multiplier method gives the differential equations for the unknown variables. These equations, in most practical cases, do not permit their exact solution, and we are forced to seek an approximate solution to the problem. The variational methods to be discussed in Chap. 6 utilize the multiplier theorem directly, that is, $\delta L = 0$, without ever using the Euler–Lagrange equations.

The next illustrative example of the multiplier theorem deals with another continuous system, namely the Stokes problem. The main difference between this example and the previous ones is that the Lagrange multiplier is a function.

Example 4-12 Consider the Stokes problem in its general form

$$\left.\begin{aligned} -\mu\left[2\frac{\partial^2 u_1}{\partial x_1^2} + \frac{\partial}{\partial x_2}\left(\frac{\partial u_1}{\partial x_2} + \frac{\partial u_2}{\partial x_1}\right)\right] + \frac{\partial P}{\partial x_1} &= f_1 \\ -\mu\left[\frac{\partial}{\partial x_1}\left(\frac{\partial u_1}{\partial x_2} + \frac{\partial u_2}{\partial x_1}\right) + 2\frac{\partial^2 u_2}{\partial x_2^2}\right] + \frac{\partial P}{\partial x_2} &= f_2 \\ \frac{\partial u_1}{\partial x_1} + \frac{\partial u_2}{\partial x_2} &= 0 \end{aligned}\right\} \quad \text{in } \Omega \subset \mathbb{R}^2 \tag{4-95}$$

$$u_1 = 0 \qquad u_2 = 0 \text{ on } \Gamma \tag{4-96}$$

where (u_1, u_2) are the components of the velocity vector, P is the pressure, (f_1, f_2) are the components of the body force vector, and μ is the viscosity. Equations (4-95) differ from those in Eq. (4-86*a*) in that the divergence-free condition is *not* used to simplify the equations. This is done with foresight to preserve the physical form of the natural boundary conditions (which involve the specification of tractions).

The variational problem associated with Eqs. (4-95) and (4-96) can be stated as follows: find $(u_1, u_2) \in \mathbf{H}(\Omega)$ such that

$$B(\mathbf{v}, \mathbf{u}) = l(\mathbf{v}) \qquad \text{for all } \mathbf{v} = (v_1, v_2) \in \mathbf{H}(\Omega) \tag{4-97}$$

holds, where

$$B(\mathbf{v}, \mathbf{u}) = \mu \int_\Omega \left[2\frac{\partial v_1}{\partial x_1}\frac{\partial u_1}{\partial x_1} + 2\frac{\partial v_2}{\partial x_2}\frac{\partial u_2}{\partial x_2} + \left(\frac{\partial v_1}{\partial x_2} + \frac{\partial v_2}{\partial x_1}\right)\left(\frac{\partial u_1}{\partial x_2} + \frac{\partial u_2}{\partial x_1}\right)\right] d\mathbf{x}$$

$$l(\mathbf{v}) = \int_\Omega (f_1 v_1 + f_2 v_2)\, d\mathbf{x} \tag{4-98}$$

$$\mathbf{H}(\Omega) = \{\mathbf{v} \in [H_0^1(\Omega) \times H_0^1(\Omega)] \colon \operatorname{div} \mathbf{v} = 0 \text{ in } \Omega\} \tag{4-99}$$

Since $B(\mathbf{v}, \mathbf{u})$ is symmetric, we can use Eq. (4.32) to obtain the associated functional on $\mathbf{H}(\Omega)$. We have

$$\begin{aligned} J(\mathbf{u}) &= \tfrac{1}{2}B(\mathbf{u}, \mathbf{u}) - l(\mathbf{u}) \\ &= \frac{\mu}{2}\int_\Omega \left[2\left(\frac{\partial u_1}{\partial x_1}\right)^2 + 2\left(\frac{\partial u_2}{\partial x_2}\right)^2 + \left(\frac{\partial u_1}{\partial x_2} + \frac{\partial u_2}{\partial x_1}\right)^2\right] d\mathbf{x} \\ &\quad - \int_\Omega (f_1 u_1 + f_2 u_2)\, d\mathbf{x} \end{aligned} \tag{4-100}$$

The variational problem can now be restated as one of minimizing $J(\mathbf{u})$ on $\mathbf{H}(\Omega)$, or equivalently

$$\begin{aligned} &\text{minimize:} \qquad J(\mathbf{u}) \text{ on } [H_0^1(\Omega) \times H_0^1(\Omega)] \\ &\text{subject to the constraint } G(\mathbf{u}) \equiv \operatorname{div} \mathbf{u} = 0 \end{aligned} \tag{4-101}$$

This constrained minimum problem can be reformulated as an unconstrained stationary variational problem using the Lagrange multiplier method. We note that

$$J(\mathbf{u})\colon H_0^1(\Omega) \times H_0^1(\Omega) \to \mathbb{R} \qquad \text{and} \qquad G(\mathbf{u})\colon H_0^1(\Omega) \times H_0^1(\Omega) \to L_2(\Omega) \tag{4-102a}$$

Thus, by the multiplier theorem there exists a $\lambda \in L_2(\Omega)$ [because $L_2(\Omega)$ is self-dual] such that the lagrangian functional

$$\begin{aligned} L(\mathbf{u}, \lambda) &= J(\mathbf{u}) + \langle \lambda, G(\mathbf{u}) \rangle \\ &= J(\mathbf{u}) + \int_\Omega \lambda \operatorname{div} \mathbf{u}\, d\mathbf{x} \end{aligned} \tag{4-102b}$$

is stationary at the local minimum $\mathbf{u} \in H_0^1(\Omega) \times H_0^1(\Omega)$. Note that in the present case λ, being in $L_2(\Omega)$, is a function (in a distributional sense). The weak problem associated with Eq. (4-102) is given by

$$\begin{aligned} \delta J(\mathbf{u}; \mathbf{v}) + \langle \lambda, G(\mathbf{v}) \rangle = 0 \qquad & \text{for every } \mathbf{v} \in H_0^1(\Omega) \times H_0^1(\Omega) \\ \langle \mu, G(\mathbf{u}) \rangle = 0 \qquad & \text{for every } \mu \in V' = L_2(\Omega) \end{aligned} \tag{4-103}$$

where $\delta J(\mathbf{u}; \mathbf{v})$ denotes the first variation of J evaluated at $\mathbf{u}$, and $\delta J(\mathbf{u}; \mathbf{v}) = B(\mathbf{v}, \mathbf{u})$.

The Euler equations are given by (with this we made a complete circle to come back to the governing differential equations!)

$$\left.\begin{aligned} &\delta u_1\colon -\mu\left[2\frac{\partial^2 u_1}{\partial x_1^2} + \frac{\partial}{\partial x_2}\left(\frac{\partial u_1}{\partial x_2} + \frac{\partial u_2}{\partial x_1}\right)\right] - \frac{\partial \lambda}{\partial x_1} = f_1 \\ &\delta u_2\colon -\mu\left[\frac{\partial}{\partial x_1}\left(\frac{\partial u_1}{\partial x_2} + \frac{\partial u_2}{\partial x_1}\right) + 2\frac{\partial^2 u_2}{\partial x_2^2}\right] - \frac{\partial \lambda}{\partial x_2} = f_2 \\ &\delta \lambda\colon \frac{\partial u_1}{\partial x_1} + \frac{\partial u_2}{\partial x_2} = 0 \end{aligned}\right\} \text{ in } \Omega \tag{4-104}$$

A comparison of Eqs. (4-104) with Eqs. (4-95) shows that λ is equal to the negative of the pressure

$$\lambda = -P \tag{4-105}$$

This completes the example.

In the variational formulations we studied thus far, we included essential boundary conditions in the definition of the domain of the bilinear form, and natural boundary conditions (if nonhomogeneous) were included in the linear form. Now it is possible to include the essential boundary conditions, whether homogeneous or not, in the weak formulation of the problem. The next example illustrates this idea.

Example 4-13 Consider the mixed boundary-value problem for the Poisson equation,

$$\begin{gathered} -\nabla^2 u = f \text{ in } \Omega \\ u = g \text{ on } \Gamma_1 \qquad \text{and} \qquad \frac{\partial u}{\partial n} = h \text{ on } \Gamma_2 \end{gathered} \tag{4-106}$$

where $\Gamma_1 + \Gamma_2 = \Gamma$ is the total boundary of $\Omega \subset \mathbb{R}^2$. The usual weak formulation involves finding $u \in H^1(\Omega)$ such that $u - w \in V$ and

$$\int_\Omega \left(\frac{\partial v}{\partial x_1}\frac{\partial u}{\partial x_1} + \frac{\partial v}{\partial x_2}\frac{\partial u}{\partial x_2}\right) dx_1\, dx_2 = \int_\Omega vf\, dx_1\, dx_2 + \int_{\Gamma_2} vh\, ds \tag{4-107}$$

for all $v \in V$, where

$$V = \{v \in H^1(\Omega): v = 0 \text{ on } \Gamma_1\} \tag{4-108}$$

and w is a fixed element in $H^1(\Omega)$ such that $w = g$ on Γ_1.

We now use the Lagrange multiplier method to reformulate the weak problem of Eq. (4-106). First note that the functional corresponding to Eq. (4-107) is given by

$$J(u) = \frac{1}{2}\int_\Omega \left[\left(\frac{\partial u}{\partial x_1}\right)^2 + \left(\frac{\partial u}{\partial x_2}\right)^2\right] dx_1\, dx_2 - \int_\Omega uf\, dx_1\, dx_2 - \int_{\Gamma_2} uh\, ds \tag{4-109}$$

for all u in $H^1(\Omega)$ and admissible variations δu in V. Next we include the constraint $G(u) \equiv u - g = 0$ on Γ_1 using the Lagrange multiplier method. Here G is an operator mapping $u \in H^1(\Omega)$ into its trace in $L_2(\Gamma_1)$. The lagrangian functional is given by

$$\begin{aligned} L(u, \lambda) &= J(u) + \langle \lambda, G(u) \rangle \\ &= J(u) + \int_{\Gamma_1} \lambda(u - g)\, ds \end{aligned} \tag{4-110}$$

The lagrangian functional is defined on the product space $H^1(\Omega) \times L_2(\Gamma)$. Thus the new variational problem involves seeking the stationary points $u \in H^1(\Omega)$ and $\lambda \in L_2(\Gamma)$ such that

$$\delta L(u, \lambda; \delta u, \delta\lambda) = 0 \tag{4-111}$$

for all $\delta u \in V$ and $\delta\lambda \in L_2(\Gamma)$. In explicit form, Eq. (4-111) amounts to finding $(u, \lambda) \in H \equiv H^1(\Omega) \times L_2(\Gamma)$ such that

$$\begin{gathered} \int_\Omega \left(\frac{\partial v}{\partial x_1}\frac{\partial u}{\partial x_1} + \frac{\partial v}{\partial x_2}\frac{\partial u}{\partial x_2}\right) dx_1\, dx_2 + \int_{\Gamma_1} v\lambda\, ds = \int_\Omega vf\, dx_1\, dx_2 + \int_{\Gamma_2} vh\, ds \\ \int_{\Gamma_1} \mu(u - g)\, ds = 0 \end{gathered} \tag{4-112}$$

hold for all $(v, \mu) \in H$.

To see the physical meaning of the Lagrange multiplier λ, we compute the Euler equations from Eq. (4-112):

$$\begin{aligned} &-\nabla^2 u = f \text{ in } \Omega \text{ (from the coefficient of } v \text{ in } \Omega) \\ &\lambda + \frac{\partial u}{\partial n} = 0 \text{ on } \Gamma_1 \text{ (from the coefficient of } v \text{ on } \Gamma_1) \\ &\frac{\partial u}{\partial n} - h = 0 \text{ on } \Gamma_2 \text{ (from the coefficient of } v \text{ on } \Gamma_2) \\ &u - g = 0 \text{ on } \Gamma_1 \text{ (from the coefficient of } \mu \text{ on } \Gamma_1) \end{aligned} \tag{4-113}$$

Thus, the Lagrange multiplier is the negative of the normal derivative of u [or trace of $u \in H^1(\Omega)$] on Γ_1. This completes the example.

The last example of this section is concerned with the variational formulation of the equations governing the *classical theory of plates.* In the classical theory of plates it is assumed that straight lines normal to the undeformed midplane remain straight and perpendicular to the midsurface after deformation (*Kirchhoff assumption*). To formulate the problem, first we develop the theory without using the Kirchhoff assumption.

Example 4-14 The displacement field (u_1, u_2, u_3) of a plate not undergoing inplane displacements can be represented, under certain simplifying assumptions, by

$$\begin{aligned} u_1 &= x_3 \psi_1(x_1, x_2) \\ u_2 &= x_3 \psi_2(x_1, x_2) \\ u_3 &= w(x_1, x_2) \end{aligned} \tag{4-114}$$

where ψ_1 and ψ_2 represent rotations of normals to the midplane about the x_2- and x_1-axis, respectively, and w is the transverse deflection. The Kirchhoff assumption amounts to the conditions

$$\psi_1 = -\frac{\partial w}{\partial x_1} \qquad \psi_2 = -\frac{\partial w}{\partial x_2} \tag{4-115}$$

The principle of minimum total potential energy can be used directly to obtain the variational formulation for the displacement field in Eq. (4-114). We have

$$\begin{aligned} \Pi_s(w, \psi_1, \psi_2) = \int_{-h/2}^{h/2} \int_\Omega \frac{1}{2} (\sigma_{11} e_{11} + \sigma_{22} e_{22} + \sigma_{33} e_{33} + 2\sigma_{12} e_{12} \\ + 2\sigma_{13} e_{13} + 2\sigma_{23} e_{23}) \, d\mathbf{x} \\ - \int_\Omega wq \, dx_1 \, dx_2 + \int_{\Gamma_2} w\hat{Q}_n \, ds - \int_{\Gamma_4} \psi_n \hat{M}_n \, ds - \int_{\Gamma_6} \psi_s \hat{M}_s \, ds \end{aligned} \tag{4-116}$$

where the expression under the first integral denotes the strain energy density owing to bending, and the remaining integrals correspond to the work done by applied external forces. The total boundary Γ of the midplane Ω is divided into three pairs of disjoint portions (see Fig. 4-11):

$$\Gamma = \Gamma_1 + \Gamma_2 = \Gamma_3 + \Gamma_4 = \Gamma_5 + \Gamma_6 \tag{4-117}$$

where Γ_2, Γ_4, and Γ_6 denote the portions on which the transverse shear force $\hat{Q}_n$, normal bending moment $\hat{M}_n$ and twisting bending moment $\hat{M}_s$ are specified. On the complementary portions (that is, $\Gamma - \Gamma_2 = \Gamma_1$, $\Gamma - \Gamma_4 = \Gamma_3$, and $\Gamma - \Gamma_6 = \Gamma_5$) the displacement w, normal rotation ψ_n and tangential rotation ψ_s are specified (assumed to be zero for simplicity of formulation). The normal and tangential rotations are related to ψ_1 and ψ_2 by

$$\psi_n = \psi_1 n_1 + \psi_2 n_2 \qquad \psi_s = -\psi_1 n_2 + \psi_2 n_1 \tag{4-118}$$

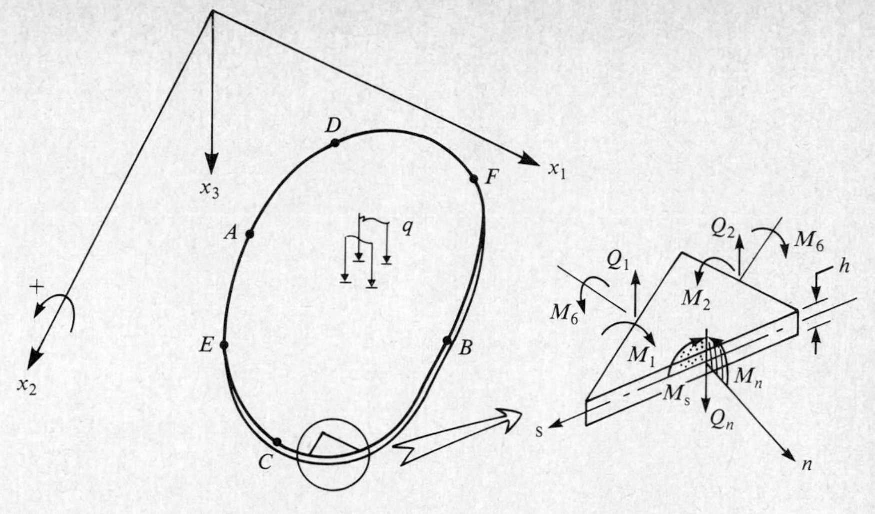

Figure 4-11 A thin plate in bending.

Using the strain-displacement equations (2-24) and the stress-strain equations (of an isotropic plane stress case), we obtain

$$
\begin{aligned}
\Pi_s = \frac{1}{2}\int_\Omega \bigg\{\int_{-h/2}^{h/2} & [c(e_{11} + ve_{22})e_{11} + c(e_{22} + ve_{11})e_{22} \\
& + 4Ge_{12}^2 + 4Ge_{23}^2 + 4Ge_{13}^2]\, dx_3\bigg\}\, dx_1\, dx_2 \\
& - \int_\Omega qw\, dx_1\, dx_2 + \int_{\Gamma_2} w\hat{Q}_n\, ds - \int_{\Gamma_4} \psi_n \hat{M}_n\, ds - \int_{\Gamma_6} \psi_s \hat{M}_s\, ds \\
= \frac{1}{2}\int_\Omega \bigg\{ & D\left[\left(\frac{\partial \psi_1}{\partial x_1}\right)^2 + \left(\frac{\partial \psi_2}{\partial x_2}\right)^2 + 2v\frac{\partial \psi_1}{\partial x_1}\frac{\partial \psi_2}{\partial x_2}\right] \\
& + G\left[\frac{h^3}{12}\left(\frac{\partial \psi_1}{\partial x_2} + \frac{\partial \psi_2}{\partial x_1}\right)^2 + Kh\left(\psi_1 + \frac{\partial w}{\partial x_1}\right)^2 + Kh\left(\psi_2 + \frac{\partial w}{\partial x_2}\right)^2\right]\bigg\}\, dx_1\, dx_2 \\
& - \int_\Omega qw\, dx_1\, dx_2 + \int_{\Gamma_2} w\hat{Q}_n\, ds - \int_{\Gamma_4} \psi_n \hat{M}_n\, ds - \int_{\Gamma_6} \psi_s \hat{M}_s\, ds
\end{aligned} \tag{4-119}
$$

where
$$D = \int_{-h/2}^{h/2} cx_3^2\, dx_3 = \frac{ch^3}{12} \qquad c = \frac{E}{1 - v^2} \tag{4-120}$$

and K is the shear correction coefficient. Since $2e_{13} = \psi_1 + \partial w/\partial x_1$ and $2e_{23} = \psi_2 + \partial w/\partial x_2$, the Kirchhoff assumption (4-115) results in the neglect of the transverse shear strains. The functional in Eq. (4-119) accounts for the energy due to transverse shear strains, and hence is called the total potential energy functional for a *shear deformation theory of plates* [see Reddy (1984b)].

To obtain the total potential energy functional for the classical plate theory, we impose the conditions in Eq. (4-115) on Eq. (4-119). We obtain

$$\Pi_c(w) = \frac{1}{2}\int_\Omega \left\{D\left(\frac{\partial^2 w}{\partial x_1^2} + \frac{\partial^2 w}{\partial x_2^2}\right)^2 - 2(1-\nu)\left[\frac{\partial^2 w}{\partial x_1^2}\frac{\partial^2 w}{\partial x_2^2} - \left(\frac{\partial^2 w}{\partial x_1\,\partial x_2}\right)^2\right]\right\} dx_1\,dx_2$$
$$- \int_\Omega qw\,dx_1\,dx_2 + \int_{\Gamma_2} w\hat{Q}_n\,ds - \int_{\Gamma_4}\frac{\partial w}{\partial n}\hat{M}_n\,ds - \int_{\Gamma_6}\frac{\partial w}{\partial s}\hat{M}_s\,ds \quad (4\text{-}121)$$

where $$\frac{\partial w}{\partial n} = n_1\frac{\partial w}{\partial x_1} + n_2\frac{\partial w}{\partial x_2} \qquad \frac{\partial w}{\partial s} = n_1\frac{\partial w}{\partial x_2} - n_2\frac{\partial w}{\partial x_1} \quad (4\text{-}122)$$

The usual variational problem for the classical plate theory involves seeking $w \in H = \{w \subset H^2(\Omega)\colon w = 0 \text{ on } \Gamma_1,\ \partial w/\partial n = \partial w/\partial s = 0 \text{ on } \Gamma_3\}$ such that $\Pi_c(w)$ is a minimum on H. Note that there are no constraints involved in the problem (other than the satisfaction of homogeneous essential boundary conditions).

The classical plate problem can be formulated alternatively as a constrained minimization problem. The advantage of such a formulation, as will be seen, is that we can seek solutions in much larger space [in $H^1(\Omega)$ instead of in $H^2(\Omega)$]. This is an important consideration from the finite-element modeling viewpoint. The disadvantage in using the Lagrange multiplier method to handle the constraints is that the minimum problem is replaced by a stationary problem. The constrained minimum problem can be stated as follows: Find $(w, \psi_1, \psi_2) \in \mathbf{H} = \{(w, \psi_1, \psi_2) \in H^1(\Omega) \times H^1(\Omega) \times H^1(\Omega)\colon w = 0 \text{ on } \Gamma_1,\ \psi_n = 0 \text{ on } \Gamma_3 \text{ and } \psi_s = 0 \text{ on } \Gamma_5\}$ such that

$$\Pi_0(w, \psi_1, \psi_2) = \frac{1}{2}\int_\Omega \left\{D\left[\left(\frac{\partial \psi_1}{\partial x_1}\right)^2 + \left(\frac{\partial \psi_2}{\partial x_2}\right)^2 + 2\nu\frac{\partial \psi_1}{\partial x_1}\frac{\partial \psi_2}{\partial x_2}\right]\right.$$
$$\left. + \frac{Gh^3}{12}\left(\frac{\partial \psi_1}{\partial x_2} + \frac{\partial \psi_2}{\partial x_1}\right)^2 - 2wq\right\} dx_1\,dx_2$$
$$+ \int_{\Gamma_2} w\hat{Q}_n\,ds - \int_{\Gamma_4} \psi_n \hat{M}_n\,ds - \int_{\Gamma_6} \psi_s \hat{M}_s\,ds \quad (4\text{-}123)$$

is a minimum on **H** subject to the constraints in (4-115)

$$G_1(w, \psi_1) \equiv \frac{\partial w}{\partial x_1} + \psi_1 = 0 \qquad G_2(w, \psi_2) \equiv \frac{\partial w}{\partial x_2} + \psi_2 = 0 \quad (4\text{-}124)$$

The Lagrange multiplier formulation of this problem requires us to construct the lagrangian functional

$$\Pi_L(w, \psi_1, \psi_2, \lambda_1, \lambda_2) = \Pi_0(w, \psi_1, \psi_2)$$
$$+ \int_\Omega \left[\lambda_1\left(\frac{\partial w}{\partial x_1} + \psi_1\right) + \lambda_2\left(\frac{\partial w}{\partial x_2} + \psi_2\right)\right] dx_1\,dx_2 \quad (4\text{-}125)$$

which attains its stationary value at the solution $(w, \psi_1, \psi_2, \lambda_1, \lambda_2)$. The Euler–Lagrange equations, in addition to the constraint equations (4-124), are

$$\left.\begin{aligned} -D\left[\frac{\partial}{\partial x_1}\left(\frac{\partial \psi_1}{\partial x_1} + \nu \frac{\partial \psi_2}{\partial x_2}\right) + \frac{1-\nu}{2}\frac{\partial}{\partial x_2}\left(\frac{\partial \psi_1}{\partial x_2} + \frac{\partial \psi_2}{\partial x_1}\right)\right] + \lambda_1 = 0 \\ -D\left[\frac{1-\nu}{2}\frac{\partial}{\partial x_1}\left(\frac{\partial \psi_1}{\partial x_2} + \frac{\partial \psi_2}{\partial x_1}\right) + \frac{\partial}{\partial x_2}\left(\frac{\partial \psi_2}{\partial x_2} + \nu \frac{\partial \psi_1}{\partial x_1}\right)\right] + \lambda_2 = 0 \\ \frac{\partial \lambda_1}{\partial x_1} + \frac{\partial \lambda_2}{\partial x_2} + q = 0 \end{aligned}\right\} \quad \text{in } \Omega \tag{4-126}$$

$$\begin{aligned} -Q_n \equiv \lambda_1 n_1 + \lambda_2 n_2 &= -\hat{Q}_n && \text{on } \Gamma_2 \\ M_n \equiv M_1 n_1^2 + 2M_6 n_1 n_2 + M_2 n_2^2 &= \hat{M}_n && \text{on } \Gamma_4 \\ M_s \equiv (M_2 - M_1)n_1 n_2 + M_6(n_1^2 - n_2^2) &= \hat{M}_s && \text{on } \Gamma_6 \end{aligned} \tag{4-127}$$

where M_1, M_2, and M_6 are the bending moments

$$\begin{aligned} M_1 &= -D\left(\frac{\partial \psi_1}{\partial x_1} + \nu \frac{\partial \psi_2}{\partial x_2}\right) = D\left(\frac{\partial^2 w}{\partial x_1^2} + \nu \frac{\partial^2 w}{\partial x_2^2}\right) \\ M_2 &= -D\left(\frac{\partial \psi_2}{\partial x_2} + \nu \frac{\partial \psi_1}{\partial x_1}\right) = D\left(\frac{\partial^2 w}{\partial x_2^2} + \nu \frac{\partial^2 w}{\partial x_1^2}\right) \\ M_6 &= \frac{-D(1-\nu)}{2}\left(\frac{\partial \psi_1}{\partial x_2} + \frac{\partial \psi_2}{\partial x_1}\right) = D(1-\nu)\frac{\partial^2 w}{\partial x_1\,\partial x_2} \end{aligned} \tag{4-128}$$

In arriving at the boundary conditions, the inverse of relations (4-118) and (4-122) is used.

A comparison of Eqs. (4-124) and (4-126) with the Euler equations of Π_c in Eq. (4-121) shows that the Lagrange multipliers λ_1 and λ_2 are the shear stress resultants,

$$\begin{aligned} \lambda_1 &= -Q_1 \equiv KGh\left(\frac{\partial w}{\partial x_1} + \psi_1\right) \\ \lambda_2 &= -Q_2 \equiv KGh\left(\frac{\partial w}{\partial x_2} + \psi_2\right) \end{aligned} \tag{4-129}$$

Note that, in view of the constraints, the Lagrange multipliers take the value of zero, and we have $\Pi_L = \Pi_c(w, \psi_1, \psi_2)$ with $\psi_1 = -\partial w/\partial x_1$ and $\psi_2 = -\partial w/\partial x_2$ (note that in this case Π_0 is the same as Π_c). This is consistent with the fact that the transverse shear stresses are zero in the classical plate theory. However, one should not set λ_1 and λ_2 to zero directly in Eq. (4-126). Differentiating the first equation with respect to x_1, the second equation with respect to x_2, adding the resulting equations and substituting for $\partial\lambda_1/\partial x_1 + \partial\lambda_2/\partial x_2 = -q$ from the third equation in Eq. (4-126), we obtain

$$\frac{\partial^2 M_1}{\partial x_1^2} + 2\frac{\partial^2 M_6}{\partial x_1\,\partial x_2} + \frac{\partial^2 M_2}{\partial x_2^2} = q \tag{4-130}$$

In view of Eq. (4-128), Eq. (4-130) becomes

$$D\left[\frac{\partial^2}{\partial x_1^2}\left(\frac{\partial^2 w}{\partial x_1^2} + \nu\frac{\partial^2 w}{\partial x_2^2}\right) + 2(1-\nu)\frac{\partial^4 w}{\partial x_1^2\,\partial x_2^2} + \frac{\partial^2}{\partial x_2^2}\left(\frac{\partial^2 w}{\partial x_2^2} + \nu\frac{\partial^2 w}{\partial x_1^2}\right)\right] = q \tag{4-131}$$

which, when simplified, gives the well-known biharmonic equation.

A comment on the boundary conditions for Eq. (4-131) is in order. For the shear deformation theory of plates, as described by the functional Π_s, the essential boundary conditions involve specifying

$$w = \hat{w} \text{ on } \Gamma_1 \qquad \psi_n = \hat{\psi}_n \text{ on } \Gamma_3 \qquad \psi_s = \hat{\psi}_s \text{ on } \Gamma_5 \tag{4-132}$$

and the natural boundary conditions are given by Eq. (4-127). For the classical plate theory, Eq. (4-132) amounts to specifying at each point of the boundary one quantity from each of the pairs (w, Q_n), $(\partial w/\partial n, M_n)$ and $(\partial w/\partial s, M_s)$. But these are too many for the fourth-order equation. To alleviate this problem, we assume (motivated by physical considerations) that $\Gamma_1 = \Gamma_5$ and $\Gamma_2 = \Gamma_6$, and combine the second and fourth boundary integrals in Eq. (4-121) to obtain

$$\int_{\Gamma_2} w\hat{Q}_n\,ds - \int_{\Gamma_2}\frac{\partial w}{\partial s}\hat{M}_s\,ds = \int_{\Gamma_2} w\left(\hat{Q}_n + \frac{\partial\hat{M}_s}{\partial s}\right)ds - [w\hat{M}_s] \tag{4-133}$$

where $[w\hat{M}_s]$ is the value of the enclosed quantity at the end points of the boundary Γ_2. For $\Gamma_2 = \Gamma$ this quantity is obviously equal to zero. In most cases this term is omitted even when $\Gamma_2 \neq \Gamma$, assuming that its contribution to two isolated points does not greatly influence the solution. Thus, the boundary conditions for the classical plate theory are given by

$$\begin{aligned} &\textit{Essential:} \quad \text{specify } w \text{ and } \partial w/\partial n \\ &\textit{Natural:} \quad \text{specify } V_n \text{ and } M_n \end{aligned} \tag{4-134}$$

where $V_n \equiv Q_n + \partial M_s/\partial s$. The boundary condition $V_n = \hat{V}_n$ is known as the *Kirchhoff free-edge condition.* This completes the example.

In closing this section, the reader should be reminded that we considered constraints of equality type only. For additional details on the Lagrange multiplier method, the reader is asked to consult any of many books on optimization [see, e.g., Hestenes (1966, 1975) and Luenberger (1969) among others].

4-4-3 The Penalty Function Method

An alternative approach to the use of Lagrange multipliers for constrained extremization which is approximate but frequently useful is the *method of penalty functions*. The basic idea of the penalty function method can be described for the constrained minimization problem:

$$\begin{aligned} &\text{minimize } J(u) \\ &\text{subject to the constraints } G_i(u) = 0 \qquad i = 1, 2, \ldots, n \end{aligned} \tag{4-135}$$

Solution of the problem by the Lagrange multiplier method requires introduction of a Lagrange multiplier with each constraint, which might entail considerable computational effort. We recognize that in practice it might be allowable to obtain a small but nonzero value of G_i in return for reduced computational effort. Thus we are led to consider the minimization of the unconstrained approximate functional

$$P(u) = J(u) + K \sum_{i=1}^{n} (G_i(u), G_i(u))_V \tag{4-136}$$

for some large positive constant K. Naturally, by solving an approximate problem, we can only expect to obtain an approximate solution of the original problem. For sufficiently large K it can be reasoned that the solution to Eq. (4-136) will be close to the actual solution. If K is a very large positive constant, then a minimum is obtainable only if the product $K \sum_{i=1}^{n} (G_i, G_i)_V$ is small, and hence minimizing P will be equivalent to minimizing a function not substantially different from J while ensuring that the constraint is nearly satisfied.

In the practical implementation of the penalty function method, we are driven on one hand to select K to be as large as possible to enhance the accuracy, and on the other hand to keep K relatively small so that when calculating the gradients the penalty terms do not completely outweigh the original functional. A common technique is to progressively solve Eq. (4-136) for a sequence of K's which gradually tend toward a large number. The ultimate value of K is a function of the word length in the computer being used to solve the problem. The sequence of approximate solutions can then be expected to converge to the solution of the original constrained problem.

At first sight the penalty function method might appear to be a crude scheme for overcoming the difficulties associated with the solution of constrained problems. However, a closer examination shows that the method has an intimate relation with other methods, e.g., the Lagrange multiplier method. Indeed, as will be shown later, the penalty function method provides an approximation for Lagrange multipliers also. Finally, we note that the particular form $\frac{1}{2}K(G, G)_V$ used for the *penalty functional* is only for convenience of introducing the concept and that any nonnegative functional which vanishes only when the constraint is satisfied would suffice.

Specifically, we introduce a *penalization operator* G which has the following properties:

$$\begin{aligned}
&G \text{ maps } V \text{ into } V', \text{ and is Lipschitz continuous}\\
&\text{Kernel of } G = \mathscr{K}\\
&G \text{ is monotone}
\end{aligned}\tag{4-137}$$

In the center of extremization of functionals, G is the gradient of a functional Ψ, called a penalty functional.

Before we discuss the method further, it would be useful to consider a specific example.

Example 4-15 Consider the problem of finding the dimensions of the rectangle having the *smallest perimeter* among all rectangles with given fixed area A. Let x_1 be the length and x_2 be the width of any such rectangle. Then the function to be minimized is the perimeter $2(x_1 + x_2)$ and the constraint is $x_1 x_2 = A$. Let

$$J(x_1, x_2) = 2(x_1 + x_2) \qquad G(x_1, x_2) = x_1 x_2 - A \tag{4-138}$$

for any vector $\mathbf{x} = (x_1, x_2)$ in $\mathbb{R}^2$. We must find a minimum point for J in the set

$$D = \{\mathbf{x} = (x_1, x_2) \in \mathbb{R}^2 \colon x_1 > 0 \text{ and } x_2 > 0\} \tag{4-139}$$

because the dimensions x_1 and x_2 cannot be negative, subject to the constraint $G(x_1, x_2) = 0$.

The augmented functional for the problem is given by

$$\begin{aligned}
P(x_1, x_2) &= J(x_1, x_2) + \frac{K}{2}\,[G(x_1, x_2)]^2\\
&= 2(x_1 + x_2) + \frac{K}{2}\,(x_1 x_2 - A)^2
\end{aligned}\tag{4-140}$$

The critical point of P is given by solving the equations

$$\begin{aligned}
\frac{\partial P}{\partial x_1} &= 2 + K(x_1 x_2 - A)x_2 = 0\\
\frac{\partial P}{\partial x_2} &= 2 + K(x_1 x_2 - A)x_1 = 0
\end{aligned}\tag{4-141}$$

For each $K > 0$ the point $[x_1(K), x_2(K)]$ converges to the minimum point $(\sqrt{A}, \sqrt{A})$ of J subject to $G = 0$. Indeed, from Eq. (4-141), we obtain

$$x_1 = x_2 \qquad \text{and} \qquad x_1^3 - A x_1 + \frac{2}{K} = 0$$

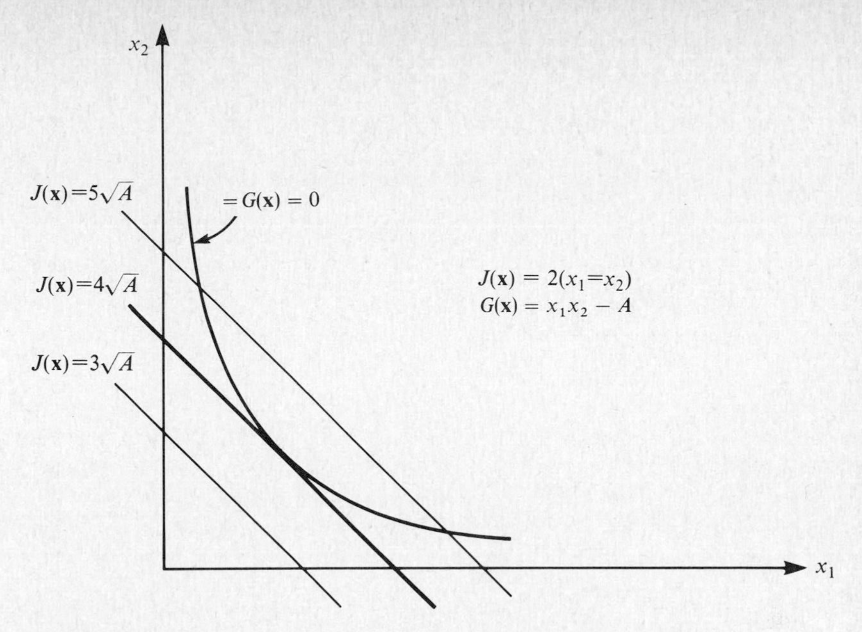

Figure 4-12 Geometric representation of the extremum problem of Example 4-15.

In the limit $K \to \infty$, we obtain $x_1^2 - A = 0$ or $x_1 = \sqrt{A}$ (note that the roots $x_1 = 0$ and $x_1 = -\sqrt{A}$ are not in the admissible set D). The situation is depicted in Fig. 4-12.

The following theorem for the finite-dimensional case establishes the conditions under which the sequence of penalty problems of the form

$$\begin{aligned} &\text{minimize } P(\mathbf{x}(K_n)) \\ &\text{subject to } G(\mathbf{x}(K_n)) = 0 \end{aligned} \qquad 0 < K_n < K_{n+1} \tag{4-142}$$

converges to the original problem [see Hestenes (1975)].

Theorem 4-7 Consider the problem of minimizing a function $J(\mathbf{x})$ on a compact set $\mathscr{D}$ subject to a constraint of the form $G(\mathbf{x}) = 0$. Let J and G^2 be continuous on $\mathscr{D}$ with $G^2 \geq 0$, and assume that the set S of points $\mathbf{x}$ in $\mathscr{D}$ having $G(\mathbf{x}) = 0$ is not empty so that there exists a point $\mathbf{x}_0$ in $\mathscr{D}$ which minimizes J on S. We assume that $\mathbf{x}_0$ is unique. If $\mathbf{x}(K)$ is a minimum point of the augmented function

$$P(\mathbf{x}, K) = J(\mathbf{x}) + \frac{K}{2}[G(\mathbf{x})]^2$$

on $\mathscr{D}$, then

$$\lim_{K \to \infty} \mathbf{x}(K) = \mathbf{x}_0 \qquad \lim_{K \to \infty} K[G(\mathbf{x})]^2 = 0 \tag{4-143}$$

Further, if $0 \le K \le K_0$, we have

$$P[\mathbf{x}(K), K] \le P[\mathbf{x}(K_0), K_0] \le J(\mathbf{x}_0)$$
$$J[\mathbf{x}(K)] \le J[\mathbf{x}(K_0)] \le J(\mathbf{x}_0) \tag{4-144}$$
$$0 \le G^2(\mathbf{x}(K_0)) \le G^2(\mathbf{x}(K))$$

PROOF Since $\mathbf{x}(K)$ minimizes P on $\mathscr{D}$ and $G^2(\mathbf{x}_0) = 0$, we have

$$P(\mathbf{x}(K), K) = J(\mathbf{x}(K)) + \frac{K}{2} G^2(\mathbf{x}(K)) \le P(\mathbf{x}_0, K) = J(\mathbf{x}_0) \tag{4-145}$$

which in turn implies that

$$J(\mathbf{x}(K)) \le J(\mathbf{x}_0)$$

Now suppose that $K \le K_0$. Then

$$0 \le P[\mathbf{x}(K_0), K] - P[\mathbf{x}(K), K] = \frac{K}{2} [G^2(\mathbf{x}(K_0)) - G^2(\mathbf{x}(K))] + J(\mathbf{x}(K_0)) - J(\mathbf{x}(K)) \tag{4-146}$$

$$0 \le P[\mathbf{x}(K), K_0] - P[\mathbf{x}(K_0), K_0] = \frac{K_0}{2} [G^2(\mathbf{x}(K)) - G^2(\mathbf{x}(K_0))] + J(\mathbf{x}(K)) - J(\mathbf{x}(K_0)) \tag{4-147}$$

Adding these two inequalities, we find that

$$0 \le (K_0 - K)[G^2(\mathbf{x}(K)) - G^2(\mathbf{x}(K_0))]$$

which implies that $G^2(\mathbf{x}(K)) \ge G^2(\mathbf{x}(K_0))$. Using this in Eq. (4-146), we obtain (4-144) for $K \le K_0$.

Next we prove the statement in Eq. (4-143). Consider a sequence $\{K_n\}$ tending to $+\infty$, and set $\mathbf{x}_n = \mathbf{x}(K_n)$. Since $G^2(\mathbf{x}) \ge 0$, Eq. (4-145) implies that

$$\limsup_{n\to\infty} \left[J(\mathbf{x}_n) + \frac{K_n}{2} G^2(\mathbf{x}_n) \right] \le J(\mathbf{x}_0)$$
$$\limsup_{n\to\infty} J(\mathbf{x}_n) \le J(\mathbf{x}_0) \qquad \lim_{n\to\infty} G^2(\mathbf{x}_n) = 0 \tag{4-148}$$

Therefore, any limit $\bar{\mathbf{x}} = \mathbf{x}(K_0)$ of a convergent subsequence of $\{\mathbf{x}_n\}$ must have $J(\bar{\mathbf{x}}) \le J(\mathbf{x}_0)$, $G^2(\bar{\mathbf{x}}) = 0$. Since $\mathscr{D}$ is compact, $\bar{\mathbf{x}}$ is in $\mathscr{D}$, and since $\mathbf{x}_0$ is the unique minimum of $J(\mathbf{x}_0)$ on $\mathscr{D}$ subject to $G^2 = 0$, it follows that $\bar{\mathbf{x}} = \mathbf{x}_0$. Since $\{\mathbf{x}_n\}$ is bounded and every convergent subsequence of $\{\mathbf{x}_n\}$ converges to $\mathbf{x}_0$, we must have

$$\lim_{n\to\infty} \mathbf{x}_n = \lim_{n\to\infty} \mathbf{x}(K_n) = \mathbf{x}_0$$

From this it follows that relations (4-143) hold. This completes the proof of the theorem.

The results of the above theorem can be extended to the case of multi-equality constraints. Let J, $G_1, G_2, \ldots, G_m$ be continuous functions on a compact set $\mathscr{D}$. Suppose that a point $\mathbf{x}_0$ in $\mathscr{D}$ affords a strict minimum to J on $\mathscr{D}$ subject to the constraints

$$G_i(\mathbf{x}) = 0 \qquad (i = 1, 2, \ldots, m)$$

Then $\lim_{K \to \infty} \mathbf{x}(K) = \mathbf{x}_0$, where $\mathbf{x}(K)$ is the minimum point of the augmented function

$$P(\mathbf{x}, K) = J(\mathbf{x}) + \frac{K}{2}\left[G_1^2(\mathbf{x}) + G_2^2(\mathbf{x}) + \cdots + G_m^2(\mathbf{x})\right] \tag{4-149}$$

In addition, the inequalities in Eq. (4-144) hold.

Next we consider the convergence of the penalty function method for continuous problems, i.e., problems posed on function spaces other than the euclidean space. We consider the problem of the minimum of the functional $J(u)$ in the Hilbert space H_1 subject to the constraint

$$G(u) = 0$$

where $G(u)$, in general, is a nonlinear operator from H_1 into some Hilbert space H_2. Instead of this constrained minimization problem, we consider the minimum of the augmented functional

$$P_n(u, K_n) = J(u) + \frac{K_n}{2}\|G(u)\|_2^2 \tag{4-150}$$

over the whole of H_1 for some $K_n > 0$. Here $\|\cdot\|_2$ denotes the norm in H_2 (*not* the $H^2(\Omega)$-norm). The next theorem provides the conditions for the existence of the unique point of the local minimum of P_n and the convergence of the penalty solution to the true solution [see Polyak (1971)].

Theorem 4-8 Let the following assumptions be satisfied.

1. There exists a local point of minimum u_0 of the original problem,

$$J(u_0) \le J(u) \qquad \text{for all } u \in H_1$$
$$G(u_0) = 0$$

2. The first and second Gateaux derivatives of J and G exist in a neighborhood of u_0, and $\delta^2 J$ and $\delta^2 G$ satisfy the Lipschitz conditions

$$\|\delta^2 J(u) - \delta^2 J(v)\|_2 \le c_1 \|u - v\|_1$$
$$\|\delta^2 G(u) - \delta^2 G(v)\|_2 \le c_2 \|u - v\|_1$$

where c_1 and c_2 are constants independent of u and v.

3. The adjoint of the linear operator $T \equiv \delta G(u_0)\colon H_1 \to H_2$ is bounded below

$$\|T^* v\| \ge M\|v\| \qquad M > 0 \text{ for all } v \in H_2'$$

where * denotes the adjoint.

Under these assumptions, the Lagrange multiplier Theorem 4-6 is applicable. Hence, there exists a $\lambda_0 \in H'_2$ such that

$$\delta_u L(u_0, \lambda_0) \equiv \delta J(u_0) + T^* \lambda_0 = 0$$

4. The linear self-adjoint operator $A \equiv \delta^2 L(u_0, \lambda_0)$ is positive-definite,

$$(Au, u)_1 \equiv (\delta^2 J(u_0)u, u)_1 + \langle \lambda_0, \delta^2 G(u_0)(u, u) \rangle$$
$$\geq \alpha \|u\|_1^2 \qquad \alpha > 0$$

Under the above four assumptions and sufficiently large $K_n > 0$, there exists a u_n which is the unique point of local minimum of $P_n(u, K_n)$ in a neighborhood of u_0, and the following estimates hold (which show the convergence of the solution u_n and Lagrange multiplier λ_n to the actual solution u_0 and λ_0, respectively):

$$\begin{aligned} \|u_n - u_0\|_1 &\leq \frac{c_3}{2K_n} \|\lambda\|_2 \\ \|K_n G(u_n) - \lambda_0\|_2 &\leq \frac{c_4}{2K_n} \|\lambda\|_2 \end{aligned} \tag{4-151}$$

PROOF The proof of the theorem is considerably involved, and interested readers can consult the paper by Polyak (1971).

The estimate in Eq. (4-151) shows that an approximation to the Lagrange multiplier is given by

$$\lambda_n = K_n G(u_n) \tag{4-152}$$

We now consider several examples of continuous problems. We begin with the isoperimetric problem of Example 4-11.

Example 4-16 The augmented functional for the isoperimetric problem is given by

$$\begin{aligned} P_n(y, K_n) &= J(y) + \frac{K_n}{2} [G(y)]^2 \\ &= \int_a^b y(x)\, dx + \frac{K_n}{2} \left[\int_a^b \sqrt{1 + (y')^2}\, dx - L \right]^2 \end{aligned} \tag{4-153}$$

The penalty weak form of the isoperimetric problem is given by

$$\begin{aligned} 0 &= \delta P_n(y, K_n) \\ &= \int_a^b \delta y(x)\, dx + K_n \left(\int_a^b \sqrt{1 + (y')^2}\, dx - L \right) \cdot \int_a^b \frac{y'\, \delta y'}{\sqrt{1 + (y')^2}}\, dx \end{aligned}$$

The Euler equations are given by

$$0 = 1 - \frac{d}{dx}\left[\frac{y'}{\sqrt{1+(y')^2}}\right]K_n\left[\int_a^b \sqrt{1+(y')^2}\,dx - L\right] \tag{4-154}$$

A comparison of Eq. (4-154) with Eq. (4-94) shows that the approximate Lagrange multiplier can be obtained using Eq. (4-152).

The next example is concerned with the penalty function formulation of the Stokes problem of Example 4-12.

Example 4-17 The augmented functional for the penalty function formulation is given by

$$P_n(\mathbf{u}, K_n) = J(\mathbf{u}) + \frac{K_n}{2}\int_\Omega (\text{div } \mathbf{u})^2\, d\mathbf{x} \tag{4-155}$$

where $J(\mathbf{u})$ is defined in Eq. (4-100). The weak problem corresponding to Eq. (4-155) involves finding $\mathbf{u} \in \mathbf{H}_k^1(\Omega)$ such that

$$B_k(\mathbf{v}, \mathbf{u}) = (\mathbf{f}, \mathbf{v})_0 \qquad \text{for all } \mathbf{v} \in \mathbf{H}_k^1(\Omega) \tag{4-156}$$

where

$$\mathbf{H}_k^1(\Omega) = \{\mathbf{u} \in \mathbf{H}_0^1(\Omega)\colon \|\mathbf{u}\|_k^2 = |\mathbf{u}|_1^2 + K_n\|G(\mathbf{u})\|_0^2\} \tag{4-157}$$

and $B_k(\mathbf{v}, \mathbf{u})$ is given by

$$B_k(\mathbf{v}, \mathbf{u}) = B(\mathbf{v}, \mathbf{u}) + K_n(G(\mathbf{v}), G(\mathbf{u}))_0 \tag{4-158}$$

$B(\mathbf{v}, \mathbf{u})$ being the bilinear form defined in Eq. (4-98).

The Lagrange multiplier can be computed from

$$\lambda_n = K_n \text{ div } \mathbf{u}_n \tag{4-159}$$

where $\mathbf{u}_n$ is the solution of the penalty problem (4-158). The same result can be obtained by comparing the Euler equations of the functional P_n in Eq. (4-155) with Eq. (4-104). The Euler equations of the penalty functional are

$$\begin{aligned} \delta u\colon \mu\left[2\frac{\partial^2 u_1}{\partial x_1^2} + \frac{\partial}{\partial x_2}\left(\frac{\partial u_1}{\partial x_2} + \frac{\partial u_2}{\partial x_1}\right)\right] + K_n\frac{\partial}{\partial x_1}\left(\frac{\partial u_1}{\partial x_1} + \frac{\partial u_2}{\partial x_2}\right) &= f_1 \text{ in } \Omega \\ \delta v\colon \mu\left[\frac{\partial}{\partial x_1}\left(\frac{\partial u_1}{\partial x_2} + \frac{\partial u_2}{\partial x_1}\right) + 2\frac{\partial^2 u_2}{\partial x_2^2}\right] + K_n\frac{\partial}{\partial x_2}\left(\frac{\partial u_1}{\partial x_1} + \frac{\partial u_2}{\partial x_2}\right) &= f_2 \text{ in } \Omega \end{aligned} \tag{4-160}$$

This completes the example.

The last example of this section deals with the mixed boundary-value problem for the Poisson equation (see Example 4-13).

Example 4-18 Consider the mixed boundary-value problem in Eq. (4-106), and its variational formulation in Eqs. (4-107) and (4-108). The penalty function formulation of the problem involves seeking the minimum of the augmented functional [see Babuska (1973b)],

$$P_n(u, K_n) = J(u) + \frac{K_n}{2} \int_{\Gamma_1} (u - g)^2 \, ds \tag{4-161}$$

where $J(u)$ is as defined in Example 4-13. We have from $\delta P_n = 0$ the weak form

$$\int_\Omega \left(\frac{\partial v}{\partial x_1} \frac{\partial u}{\partial x_1} + \frac{\partial v}{\partial x_2} \frac{\partial u}{\partial x_2} \right) dx_1 \, dx_2 + K_n \int_{\Gamma_1} vu \, ds$$
$$= \int_\Omega fv \, dx_1 \, dx_2 + \int_{\Gamma_2} vh \, ds + K_n \int_{\Gamma_1} vg \, ds \tag{4-162}$$

for all $v \in H^1(\Omega)$.

The Euler equations resulting from the penalty formulation are

$$\begin{aligned} -\nabla^2 u &= f \text{ in } \Omega \\ \frac{\partial u}{\partial n} - h &= 0 \text{ on } \Gamma_2 \\ \frac{\partial u}{\partial n} + K_n(u - g) &= 0 \text{ on } \Gamma_1 \end{aligned} \tag{4-163}$$

This completes the example.

The inclusion of essential boundary conditions of a given problem in its variational form (as in Examples 4-13 and 4-18) might not seem very interesting to the practitioners of variational methods, especially the finite-element method. Indeed, for second-order equations in two dimensions, it is rather academic to use such variational forms to construct finite-element models. However, in the finite-element analysis of fourth-order two-dimensional equations, the satisfaction of the interelement continuity conditions on the variables in the essential boundary conditions is not easy. In such cases, the idea presented in Examples 4-13 and 4-18 can be used for computational advantage. We shall return to this idea in Chap. 5.

PROBLEMS 4-3

4-3-1 Consider the constrained minimization problem: minimize $f(\mathbf{x})$, subject to $\mathbf{x} \in \Omega$, $G(\mathbf{x}) \leq 0$ where Ω is a convex subset of vector space X, f is a real-valued convex functional on Ω, and G is a convex mapping from Ω into a normed space Z having positive cone P. Define the set $\Gamma \subset Z$ by

$$\Gamma = \{z \in Z\colon \text{ there exists an } \mathbf{x} \in \Omega \text{ with } G(\mathbf{x}) \leq z\}$$

On the set Γ, define the *primal functional* ω,

$$\omega(z) = \inf \{f(\mathbf{x})\colon \mathbf{x} \in \Omega, G(\mathbf{x}) \leq z\}$$

Then the original problem becomes one of determining the value $\omega(0)$. Prove that (*a*) the set Γ is convex, and (*b*) the functional $\omega(z)$ is convex and decreasing; i.e., if $z_1 \le z_2$, then $\omega(z_1) \le \omega(z_2)$.

4-3-2 (continuation of Prob. 4-3-1) Let $\mu_0 = \inf f(\mathbf{x})$ subject to $\mathbf{x} \in \Omega$, $G(\mathbf{x}) \le 0$ and μ_0 be finite. Show that there is an element $z'_0 \ge 0$ in Z' (the dual of Z) such that $\mu_0 = \inf_{x \in \Omega} \{f(\mathbf{x}) + \langle G(\mathbf{x}), z'_0 \rangle\}$ and if $\mathbf{x}_0$ is the element in Ω that gives the infimum of $f(\mathbf{x})$ and satisfies the constraint $G(\mathbf{x}) \le 0$, that $\langle G(\mathbf{x}_0), z'_0 \rangle = 0$.

4-3-3 (continuation of Prob. 4-3-2) Prove that the lagrangian $L(\mathbf{x}, z') = f(\mathbf{x}) + \langle G(\mathbf{x}), z' \rangle$ has a *saddle point* at $(\mathbf{x}_0, z'_0)$; that is, $L(\mathbf{x}_0, z') \le L(\mathbf{x}_0, z_0) \le L(\mathbf{x}, z'_0)$ for all $\mathbf{x} \in \Omega$ and $z' \in Z'$ $(z' \ge 0)$.

Solve the constrained minimization problems given in Probs. 4-3-4 to 4-3-9 using the Lagrange multiplier method:

4-3-4 $J(x) = x^2 \qquad G(x) = 1 + x$

4-3-5 $J(\mathbf{x}) = 2(x_1 + x_2) \qquad G(\mathbf{x}) = x_1 x_2 - A \qquad x_1, x_2 > 0$

4-3-6 $J(\mathbf{x}) = x_1(1 + x_2) \qquad G(\mathbf{x}) = x_1 + x_2^2$

4-3-7 $J(\mathbf{x}) = x_1^2 + x_2^2 \qquad G(\mathbf{x}) = x_1^2 - (x_2 - 1)^3$

4-3-8 $J(\mathbf{x}) = x_1^2 + x_2^2 + 2x_3^2 - x_1 - x_2 x_3 \qquad G(\mathbf{x}) = x_1 + x_2 + x_3 - 35$

4-3-9 Find the minimum value of the functional $J(u) = \int_0^1 u(x)\, dx$ on $C[0, 1]$ subject to the constraint $G(u) \equiv \int_0^1 (u^2 + xu)\, dx - 47/12 = 0$. Show that $J(u_0)$ is a maximum, where u_0 is the solution to the problem.

4-3-10 Find the maximum value of the functional $J(u) = \int_0^1 xu(x)\, dx$ on $C[0, 1]$ subject to the constraint $G(u) \equiv \int_0^1 u^2\, dx - \frac{1}{12} = 0$. Verify that $J(u_0)$ is a minimum.

4-3-11 Find the Euler–Lagrange equations of the functionals obtained in Probs. 4-3-9 and 4-3-10.

4-3-12 *Queen Dido's problem* Let $A = (a, 0)$ and $B = (b, 0)$ be two fixed points on the x_1-axis with $a < b$, and let L be any given fixed length satisfying $b - a < L < (b - a)\pi/2$. Let $x_2 = y(x_1)$ be any curve of length L connecting points A and B with $y(x_1) \ge 0$, $a \le x_1 \le b$. Show that a circular arc encloses the greatest area among all such curves of length L.

Hint: Maximize the integral $J(y) = \int_a^b y(x_1)\, dx_1$ subject to the *three* constraints,

$$G_1(y) \equiv \int_a^b \sqrt{1 + \left(\frac{dy}{dx_1}\right)^2}\, dx_1 - L = 0$$

$$G_2(y) = y(a) = 0$$

$$G_3(y) = y(b) = 0$$

and obtain a differential equation of the form

$$\left(\frac{dy}{dx_1}\right)^2 = (x_1 - a)^2/[c^2 - (x_1 - a)^2] \qquad c = \text{constant}$$

4-3-13 By minimizing the functional

$$J(u) = \left[\int_0^1 \left(\frac{du}{dx}\right)^2 dx \Big/ \int_0^1 u^2\, dx\right]$$

subject to the constraint conditions $u(0) = u(1) = 1$, prove the inequality

$$\int_0^1 u^2\, dx \le \left(\frac{4}{\pi^2}\right) \int_0^1 \left(\frac{du}{dx}\right)^2 dx$$

for all continuously differentiable functions $u(x)$ which satisfy the conditions $u(0) = u(1) = 1$.

4-3-14 (*The Hu-Washizu variational principle*) In deriving the total potential energy principle it was assumed that the displacements and strains are related by a kinematic relationship and that the displacement field satisfies the specified boundary conditions on Γ_1. The principle gives the equilibrium equations and traction boundary conditions as the Euler equations. Include the strain-displacement equations (2-24) and the displacement boundary conditions $u_i = \hat{u}_i$ on Γ_1 into the variational statement by treating the equations as constraints. Let λ_{ij} $(i, j = 1, 2, 3)$ and μ_i $(i = 1, 2, 3)$ be the Lagrange multipliers associated with the strain-displacement equations and the displacement boundary conditions, respectively, and obtain the functional

$$H(u_i, \varepsilon_{ij}, \lambda_{ij}, \mu_i) = \int_\Omega \{[\tfrac{1}{2}(u_{i,j} + u_{j,i} + u_{m,i}u_{m,j}) - \varepsilon_{ij}]\lambda_{ij} + U_0(\varepsilon_{ij}) - f_i u_i\}\, d\mathbf{x}$$

$$- \int_{\Gamma_1} \mu_i(u_i - \hat{u}_i)\, ds - \int_{\Gamma_2} \hat{t}_i u_i\, ds$$

Show that the Lagrange multipliers λ_{ij} and μ_i are equal to stress components and boundary tractions, respectively [see Reddy (1984b)].

4-3-15 Rewrite Eqs. (4-128) in the form

$$\frac{\partial^2 w}{\partial x_1^2} = S(M_1 - \nu M_2) \tag{1}$$

$$\frac{\partial^2 w}{\partial x_2^2} = S(M_2 - \nu M_1) \tag{2}$$

$$\frac{\partial^2 w}{\partial x_1\, \partial x_2} = S(1 + \nu)M_6 \tag{3}$$

where $S = 1/D(1 - \nu)$. Use these three equations and Eq. (4-130) as the fourth equation to construct a lagrangian functional,

$$\delta L(w, M_1, M_2, M_6; \lambda_1, \lambda_2, \lambda_3, \lambda_4) = \sum_{i=1}^{4} \int_\Omega \lambda_i \text{ Eq. } (i)\, d\mathbf{x}$$

Identify the Lagrange multipliers by comparing the Euler–Lagrange equations with the *varied* form [e.g., the varied form of Eq. (1) is $\partial^2 \delta w/\partial x_1^2 = S(\delta M_1 - \nu \delta M_2)$] of the four equations we started with. Then derive the functional $L(w, M_1, M_2, M_6)$ from δL.

4-3-16–4-3-19 Use the penalty method to determine the solution to Probs. 4-3-4 and 4-3-6 to 4-3-8. Check whether in the limit $K \to \infty$ the penalty solution converges to the correct solution.

4-3-20 Assume that $u_1 = u_1(x_1, x_2)$, $u_2 = 0$ and $\mathbb{R} = \{\mathbf{x} = (x_1, x_2)\colon 0 < x_1 < 1, -b < x_2 < b\}$ in Eq. (4-155) and obtain the penalty functional and corresponding Euler equations. Solve the resulting partial differential equation for u_1. The boundary conditions are $u_1(x_1, -b) = u_2(x_1, b) = 0$ for all x_1: $\partial u_1/\partial x_1 = 1$ at $x = 0$ and $\partial u_1/\partial x_1 = 0$ at $x = 1$.

4-3-21 Consider the problem of the elastic-perfectly plastic torsion of a cylindrical bar. The variational problem consists of minimizing the functional $J(\Phi) = \int_\Omega \operatorname{grad} \Phi \cdot \operatorname{grad} \Phi\, d\mathbf{x} - 2 \int_\Omega \theta\Phi\, d\mathbf{x}$ subject to the constraint $|\operatorname{grad} \Phi| \le 1$.

Consider the penalty functional $P_k\colon W_0^{1,4}(\Omega) \to \mathbb{R}$

$$P_K(\Phi) = J(\Phi) + \frac{K}{4} \int_\Omega \{(1 - |\operatorname{grad} \Phi|^2)\}^2\, d\mathbf{x}$$

where $W_0^{1,4}(\Omega)$ is the Sobolev space of order $m = 1$, $p = 4$th integrable functions which vanish on Γ. Show that the Euler equations of the functional $P_K(\Phi)$ are given by

$$-\nabla^2\Phi - K \sum_{i=1}^{2} \frac{\partial}{\partial x_i}\left[(1 - |\operatorname{grad} \Phi|^2) \frac{\partial \Phi}{\partial x_i}\right] = f \text{ in } \Omega$$

$$u = 0 \text{ on } \Gamma$$

In this case the penalization operator G is the nonlinear operator $G\colon W_0^{1,4}(\Omega) \to W^{-1,4/3}(\Omega)$ defined by $(G(\Phi), v) \equiv \int_\Omega (1 - |\operatorname{grad} \Phi|^2) \operatorname{grad} \Phi \cdot \operatorname{grad} v\, d\mathbf{x}$ for $v, \Phi \in W_0^{1,4}(\Omega)$.

CHAPTER
FIVE

EXISTENCE AND UNIQUENESS OF SOLUTIONS

5-1 INTRODUCTION

In Chaps. 3 and 4, we have seen that problems of engineering can be cast in terms of operator equations, $Au = f$, with appropriate boundary conditions, and in Chap. 4 we discussed the variational formulation of such equations. Solution of these equations is the primary objective of any engineering analysis. Not all problems of the type $Au = f$ have solutions. The existence and uniqueness of solution of the operator equation or its variational problem can be established under certain conditions on the operator and the smoothness of the domain and its boundary and regularity of the data. The present chapter deals with the questions of existence and uniqueness of operator equations and their variational problems.

In the present study of the existence and uniqueness of solutions, we consider three types of problems associated with the operator equation, $Au = f$:

(i) A is a linear algebraic operator. An example of A is provided by the set of linear algebraic equations:

$$
\begin{aligned}
a_{11}\alpha_1 + a_{12}\alpha_2 + \cdots + a_{1n}\alpha_n &= f_1 \\
a_{21}\alpha_1 + a_{22}\alpha_2 + \cdots + a_{2n}\alpha_n &= f_2 \\
\cdots\cdots\cdots\cdots\cdots\cdots\cdots\cdots\cdots\cdots\cdots \\
a_{n1}\alpha_1 + a_{n2}\alpha_2 + \cdots + a_{nn}\alpha_n &= f_n
\end{aligned}
\tag{5-1}
$$

where $[A] = [a_{ij}]$ is called the coefficient matrix,

$$
A = \begin{bmatrix} a_{11} & a_{12} & \cdots & a_{1n} \\ a_{21} & a_{22} & \cdots & a_{2n} \\ \vdots & & & \\ a_{n1} & a_{n2} & \cdots & a_{nn} \end{bmatrix} \qquad f = \begin{Bmatrix} f_1 \\ f_2 \\ \vdots \\ f_n \end{Bmatrix}
\tag{5-2}
$$

Such equations arise, for example, in the solution of the operator equations $Au = f$ by variational methods.

(ii) A is a differential operator. Many examples of this case were seen in earlier discussions.

(iii) The variational problem of the operator equation:

$$B(v, u) = l(v) \tag{5-3}$$

where $B(\cdot, \cdot)$ and $l(\cdot)$ are bilinear and linear forms associated with the equation $Au = f$ (see Chap. 4).

As will be seen later, the variational solution of the operator equation $Au = f$ in finite dimensional spaces leads to algebraic equations listed in (i).

5-2 LINEAR ALGEBRAIC EQUATIONS

5-2-1 Vector Interpretation of Algebraic Equations

Before we study the solvability conditions (i.e., conditions under which a solution exists and is unique) for systems of linear equations, we give the vector interpretation of systems of linear equations (5-1). Let $l_1(u) \equiv f_1, l_2(u) \equiv f_2, \ldots, l_m(u) \equiv f_m$ be m given linear functionals defined over an n-dimensional vector space U_n. Let $\{\phi_i\}_{i=1}^n$ be the basis of U_n. Then we have

$$l_k(u) = \sum_{i=1}^{n} \alpha_i l_k(\phi_i) \equiv \sum_{i=1}^{n} a_{ki} \alpha_i = f_k \qquad (k = 1, 2, \ldots, m) \tag{5-4}$$

Equation (5-4) contains m linear algebraic equations in n unknows, α_i. Solving this system is equivalent to finding the components α_i $(i = 1, 2, \ldots, n)$ such that the m linear functionals l_k assume the m given values, f_k.

The collection of elements $l_k(u) = f_k$, for $u \in U_n$, is an m-dimensional vector space, V_m. Let $\{\psi_j\}_{j=1}^m$ be a basis of V_m and let

$$\mathbf{A}_i = \sum_{k=1}^{m} a_{ki} \psi_k \qquad (i = 1, 2, \ldots, n); \qquad \mathbf{f} = \sum_{k=1}^{m} f_k \psi_k \tag{5-5}$$

It is informative to notice from Eq. (5-5) that

$$\mathbf{A}_i = \{a_{1i}, a_{2i}, \ldots, a_{mi}\} \qquad (i = 1, 2, \ldots, n) \tag{5-6}$$

is the m-dimensional vector whose elements comprise the ith column of the coefficient matrix

$$[A] = [a_{ij}] = [\mathbf{A}_1 \mathbf{A}_2 \cdots \mathbf{A}_n] \tag{5-7}$$

Then the system of equations in (5-4) reduces to the vector equation

$$\mathbf{f} = \sum_{i=1}^{n} \alpha_i \mathbf{A}_i \qquad \text{or} \qquad \{\mathbf{f}\} = [A]\{\boldsymbol{\alpha}\} \tag{5-8}$$

With this interpretation, we see that Eqs. (5-1) are compatible if and only if the vector $\mathbf{f}$ is representable as a linear combination of the vectors $(\mathbf{A}_1, \mathbf{A}_2, \ldots, \mathbf{A}_m)$ in the array a_{ki}, which represents the linear operator with respect to the bases $\{\phi_i\}$ and $\{\psi_k\}$.

5-2-2 Solvability Conditions

Any system of linear equations in which all f_k are zero is called *homogeneous*. Consider the following homogeneous equations associated with operator A:

Homogeneous equation: $$A\boldsymbol{\alpha} = \mathbf{0} \tag{5-9}$$

Adjoint homogeneous equation: $$A^*\boldsymbol{\beta} = \mathbf{0} \tag{5-10}$$

where A^* is the adjoint of A. For the linear algebraic equations, $A^* = A^T$, the transpose of A.

The homogeneous adjoint equations can also be written in the form

$$(\mathbf{A}_i, \boldsymbol{\beta}) = 0 \qquad (i = 1, 2, \ldots, n) \tag{5-11}$$

where $(\cdot, \cdot)$ denotes the inner product in euclidean space. From Eqs. (5-8) and (5-11), we deduce the following result, known as the *solvability condition*: The nonhomogeneous equation $A\boldsymbol{\alpha} = \mathbf{f}$ possesses a solution $\boldsymbol{\alpha}$ if and only if the vector $\mathbf{f}$ is orthogonal to all vectors $\boldsymbol{\beta}$ that are the solutions of the homogeneous adjoint equation, $A^*\boldsymbol{\beta} = \mathbf{0}$. In analytical form this statement can be expressed as

$$(\mathbf{f}, \boldsymbol{\beta}) = 0 \tag{5-12}$$

We now consider two cases of linear equations, and discuss the existence and uniqueness of solutions of linear equations.

First consider the case $V_m = U_m$ $(m = n)$. The set of equations in (5-8) has a unique solution only if the vectors $(\mathbf{A}_1, \mathbf{A}_2, \ldots, \mathbf{A}_m)$ are linearly independent (i.e., the rows or columns of the matrix a_{ki} are linearly independent). This follows from the fact that every vector $\mathbf{f}$ in V_n can be expressed as a unique linear combination of the vectors $\mathbf{A}_i$, which, being linearly independent, form the basis of $U_m = V_m$.

Next consider the case when $U_n \subset V_m$ $(m > n)$ with vectors $\mathbf{A}_i$ and the vector $\mathbf{f}$ belonging to the vector space V_m. The elements of V_m which can be expressed as a linear combination of n arbitrary vectors $\mathbf{A}_i$ of V_m form a subspace of U_n. Let r be the rank of the set of vectors $\mathbf{A}_i$, and let V_r be the corresponding subspace. Since r of the vectors $\mathbf{A}_i$ are linearly independent, the other $(n - r)$ vectors of the set can be expressed as a linear combination of the r vectors:

$$\mathbf{A}_{r+k} = \sum_{i=1}^{r} c_i^{r+k} \mathbf{A}_i \qquad (k = 1, 2, \ldots, n - r) \tag{5-13}$$

where c_i^{r+k} are constants independent of $\mathbf{A}_i$ $(i = 1, 2, \ldots, n)$. For a given vector $\mathbf{f}$, there exist two alternatives:

(i) The vector $\mathbf{f}$ does not belong to V_r; the system then has no solution.

(ii) The vector **f** belongs to V_r and there exists a unique set of numbers α_i^0 such that

$$\mathbf{f} = \sum_{i=1}^{r} \alpha_i^0 \mathbf{A}_i \tag{5-14}$$

In the second case, from Eqs. (5-13) and (5-14) we have

$$\begin{aligned}\mathbf{f} &= \sum_{i=1}^{r} \alpha_i \mathbf{A}_i + \sum_{k=1}^{n-r} \alpha_{r+k} \mathbf{A}_{r+k} \\ &= \sum_{i=1}^{r} \left(\alpha_i + \sum_{k=1}^{n-r} c_i^{r+k} \alpha_{r+k} \right) \mathbf{A}_i \end{aligned} \tag{5-15}$$

Comparing Eq. (5-15) with (5-14), we see that a necessary and sufficient condition for the numbers α_i to be a solution of (5-4) is that they satisfy the equations

$$\alpha_i + \sum_{k=1}^{n-r} c_i^{r+k} \alpha_{r+k} = \alpha_i^0 \qquad i = 1, 2, \ldots, r \tag{5-16}$$

The remaining $(n - r)$ unknowns, α_{r+k}, can be chosen arbitrarily. We summarize these results in the following theorem.

Theorem 5-1 If V_r is the subspace of V_m generated by the vectors $\mathbf{A}_i$ associated with the given linear system (5-4), then one of the following two alternatives holds:

(i) either vector **f** does not belong to V_r and the linear system has no solution; or

(ii) vector **f** belongs to V_r and the linear system has solutions (not unique): the values of $(n - r)$ of the unknowns can be chosen arbitrarily and the values of the remaining r unknowns are determined by (5-16).

The results of Theorem 5-1 can be restated in alternative terms as follows:

1. If (5-9) has only the trivial (i.e., zero) solution, it follows that $\det A \neq 0$ (otherwise, the trivial solution cannot be determined) and hence $\det A^* \neq 0$. Therefore, the adjoint homogeneous equation (5-10) also has only the trivial solution. Moreover, the solvability conditions are automatically satisfied for any **f** (since the only solution of (5-10) is $\boldsymbol{\beta} = \mathbf{0}$), and the nonhomogeneous equation (5-1) has one and only one solution, $\boldsymbol{\alpha} = A^{-1}\mathbf{f}$, where A^{-1} is the inverse of the matrix A.
2. If (5-9) has nontrivial solutions, then $\det A = 0$. This in turn implies that the rows (or columns) of A are linearly dependent. If these linear dependencies are also reflected in the column vector **f** (i.e., e.g., if the third row of A is the sum of the first and second rows, we must have $f_3 = f_1 + f_2$ in order to have any solutions), then there is a hope of having a solution to the system. If there are

$r(\geq n)$ number of independent solutions to (5-1), A is said to have a r-*dimensional null space* (i.e., nullity of A is r). It can be shown that A^* also has a r-dimensional null space, which is in general different from that of A. A necessary and sufficient condition for (5-1) to have solutions is provided by the *solvability condition*, $(\mathbf{f}, \boldsymbol{\beta}) \equiv \sum_{i=1}^{n} f_i \beta_i = 0$, where $\boldsymbol{\beta}$ is the solution of Eq. (5-10), or equivalently, $\boldsymbol{\beta} \in \mathcal{N}(A^*)$.

We now consider two examples of the solvability of linear equations.

Example 5-1 This example has three cases:

1. Consider the following pair of equations in two unknowns α_1 and α_2:

$$\begin{aligned} 3\alpha_1 - 2\alpha_2 &= 4 \\ 2\alpha_1 + \alpha_2 &= 5 \end{aligned} \quad \text{or} \quad \begin{bmatrix} 3 & -2 \\ 2 & 1 \end{bmatrix} \begin{Bmatrix} \alpha_1 \\ \alpha_2 \end{Bmatrix} = \begin{Bmatrix} 4 \\ 5 \end{Bmatrix} \qquad (A\boldsymbol{\alpha} = \mathbf{f})$$

We note that $\det A = 3 + 4 = 7 \neq 0$. The solution is then given by

$$\begin{Bmatrix} \alpha_1 \\ \alpha_2 \end{Bmatrix} = \begin{bmatrix} \frac{1}{7} & \frac{2}{7} \\ -\frac{2}{7} & \frac{3}{7} \end{bmatrix} \begin{Bmatrix} 4 \\ 5 \end{Bmatrix} = \begin{Bmatrix} 2 \\ 1 \end{Bmatrix}$$

The solution of the adjoint equations is trivial, $\boldsymbol{\beta} = \mathbf{0}$, and therefore, the solvability condition is identically satisfied.

2. Next consider the pair of equations

$$\begin{aligned} 6\alpha_1 + 4\alpha_2 &= 4 \\ 3\alpha_1 + 2\alpha_2 &= 2 \end{aligned} \quad \text{or} \quad \begin{bmatrix} 6 & 4 \\ 3 & 2 \end{bmatrix} \begin{Bmatrix} \alpha_1 \\ \alpha_2 \end{Bmatrix} = \begin{Bmatrix} 4 \\ 2 \end{Bmatrix} \qquad (A\boldsymbol{\alpha} = \mathbf{f})$$

We have $\det A = 0$, because row 1 (R1) is equal to two times row 2 (R2). However, we also have $2f_2 = f_1$. Consequently, we have one linearly independent solution, say $\boldsymbol{\alpha}^{(1)}$, and the other depends on $\boldsymbol{\alpha}^{(1)}$:

$$\boldsymbol{\alpha}^{(1)} = (2, -2)$$

Note that there are many dependent solutions to the pair. For example, (2, −2), (4, −5), (−2, 4), etc., are solutions of $A\boldsymbol{\alpha} = \mathbf{f}$. The solution to the adjoint homogeneous equation

$$\begin{bmatrix} 6 & 3 \\ 4 & 2 \end{bmatrix} \begin{Bmatrix} \beta_1 \\ \beta_2 \end{Bmatrix} = \begin{Bmatrix} 0 \\ 0 \end{Bmatrix}$$

is given by $\beta_2 = -2\beta_1$. Note that $(\mathbf{f}, \boldsymbol{\beta}) \equiv f_1\beta_1 + f_2\beta_2 = 4(-\frac{1}{2}\beta_2) + 2\beta_2 = 0$, hence the solvability condition is satisfied.

3. Finally, consider the pair of equations

$$\begin{aligned} 6\alpha_1 + 4\alpha_2 &= 3 \\ 3\alpha_1 + 2\alpha_2 &= 2 \end{aligned} \quad \text{or} \quad \begin{bmatrix} 6 & 4 \\ 3 & 2 \end{bmatrix} \begin{Bmatrix} \alpha_1 \\ \alpha_2 \end{Bmatrix} = \begin{Bmatrix} 3 \\ 2 \end{Bmatrix}$$

We note that $\det A = 0$, because $2R_2 = R_1$. However, $2f_2 \neq f_1$. Hence the pair of equations is inconsistent, and therefore no solutions exist.

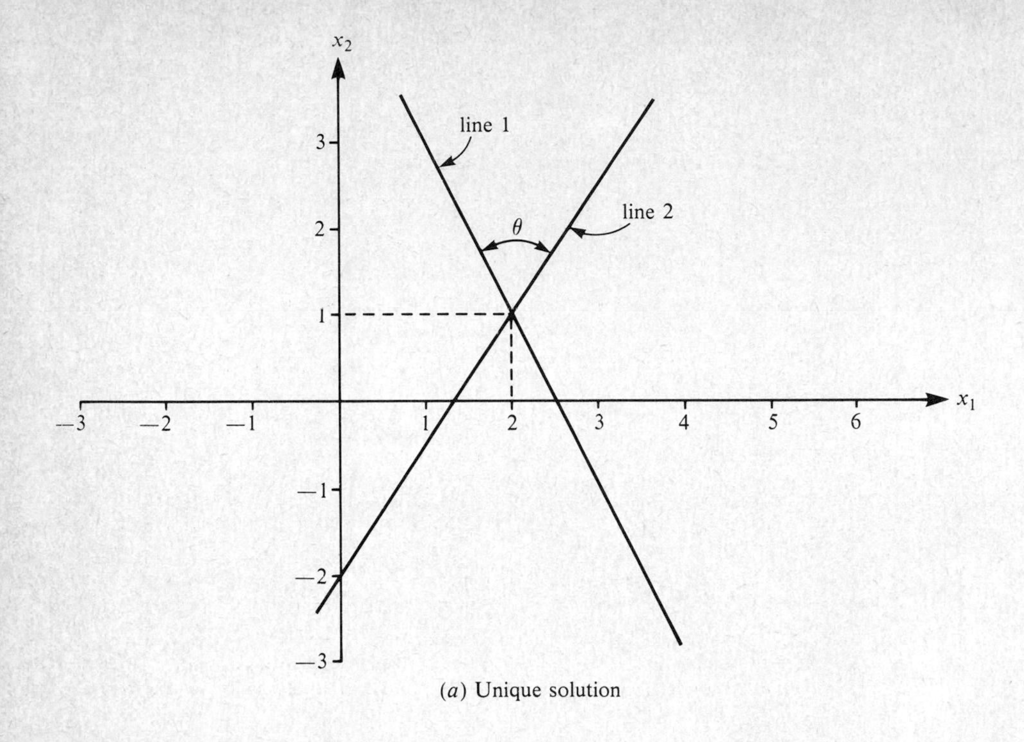

(*a*) Unique solution

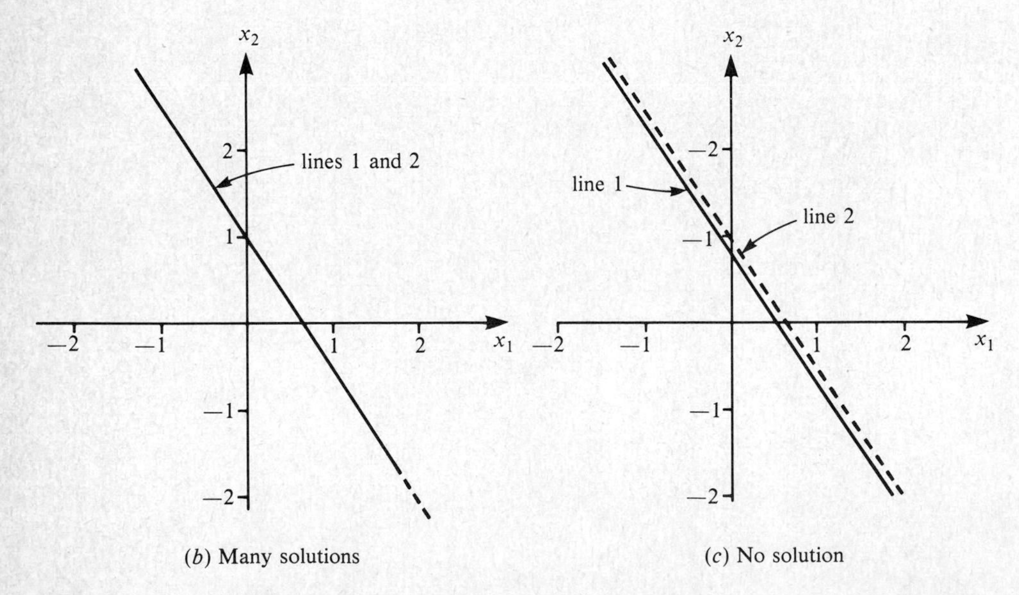

(*b*) Many solutions

(*c*) No solution

Figure 5-1 Geometric interpretation of the solution of two simultaneous algebraic equations in a plane.

Geometrically, we can interpret these three pairs of equations as pairs of straight lines in $\mathbb{R}^2$ with $\alpha_i = x_i$, $i = 1, 2$; see Fig. 5-1. In part 1, the lines represented by the two equations intersect at the point $(x_1, x_2) = (2, 1)$. In part 2, the lines coincide, or intersect at an infinite number of points, and hence many solutions exist. In part 3, the lines do not intersect at all showing that no solutions exist. From this geometric interpretation, one can see that when the lines are nearly parallel (i.e., the angle θ is nearly zero), the determinant of A is nearly zero (because $\tan\theta = (a_{11}a_{22} - a_{12}a_{21})/(a_{11}a_{21} + a_{12}a_{22})$) and therefore it is difficult to obtain an accurate numerical solution. In such cases the system of equations is said to be *ill-conditioned*. While these observations can be generalized to a system of n equations, the geometric interpretation becomes complicated.

Example 5-2 This example has three cases:

1. Consider the following set of three equations in three unknowns:

$$\begin{aligned} \alpha_1 + \alpha_2 + \alpha_3 &= 2 \\ \alpha_1 - \alpha_2 - 3\alpha_3 &= 3 \qquad \text{or} \qquad A\boldsymbol{\alpha} = \mathbf{f} \\ 3\alpha_1 + \alpha_2 - \alpha_3 &= 1 \end{aligned}$$

The adjoint homogeneous equations become

$$\begin{aligned} \beta_1 + \beta_2 + 3\beta_3 &= 0 \\ \beta_1 - \beta_2 + \beta_3 &= 0 \qquad \text{or} \qquad A^*\boldsymbol{\beta} = \mathbf{0} \\ \beta_1 - 3\beta_2 - \beta_3 &= 0 \end{aligned}$$

Solving for $\boldsymbol{\beta}$, we obtain $\beta_1 = 2\beta_2 = -2\beta_3$. Hence, the null space of A^* is defined by

$$\mathcal{N}(A^*) = \{(2a, a, -a), \ a \text{ is a real number}\}$$

Clearly $\mathcal{N}(A^*)$ is one-dimensional. The null space of A is given by

$$\mathcal{N}(A) = \{(a, -2a, a), \ a \text{ is a real number}\}$$

Note that $\mathcal{N}(A^*) \neq \mathcal{N}(A)$, but their dimension is the same. Clearly $(2, 1, -1)$ is a solution of $A^*\boldsymbol{\beta} = \mathbf{0}$ while $(1, -2, 1)$ is a solution of $A\boldsymbol{\alpha} = \mathbf{0}$. The solvability condition gives

$$(\mathbf{f}, \boldsymbol{\beta}) = 2 \times 2 + 3 \times 1 + 1 \times (-1) \neq 0$$

and therefore $A\boldsymbol{\alpha} = \mathbf{f}$ has *no* solution.

2. Reconsider the linear equations above with $\mathbf{f} = \{-1, 3, 1\}^T$. Then the solvability condition is clearly satisfied. Hence there is one linearly independent solution to $A\boldsymbol{\alpha} = \mathbf{f}$ (note that $-2\text{R}_1 + \text{R}_3 = \text{R}_2$ and $-2f_1 + f_3 = f_2$):

$$\boldsymbol{\alpha} \equiv (\alpha_1, \alpha_2, \alpha_3) = (1, -2, 0)$$

Only one of the three α's is arbitrary (not determined) and the remaining two α's are given in terms of the arbitrary α. For example, if α_1 is arbitrary, we have

$$\alpha_2 = -2\alpha_1 \qquad \text{and} \qquad \alpha_3 = \alpha_1 - 1$$

3. Consider the following set of equations

$$\begin{aligned} \alpha_1 + \alpha_2 + \alpha_3 &= 1 \\ \alpha_1 + \alpha_2 - 3\alpha_3 &= 2 \qquad \text{or} \qquad A\boldsymbol{\alpha} = \mathbf{f} \\ 3\alpha_1 + \alpha_2 - \alpha_3 &= 3 \end{aligned}$$

We have $\det A \neq 0$. It can be easily verified that $\mathcal{N}(A) = \mathcal{N}(A^*) = \{(0, 0, 0)\}$. The unique solution to $A\boldsymbol{\alpha} = \mathbf{f}$ is given by

$$\alpha_1 = \tfrac{3}{4} \qquad \alpha_2 = \tfrac{1}{2} \qquad \alpha_3 = -\tfrac{1}{4}.$$

5-3 LINEAR OPERATOR EQUATIONS

5-3-1 Introduction

In the preceding section we have discussed the solvability conditions (i.e., conditions for the existence and uniqueness of solutions) to linear algebraic equations (equivalently, operator equations on finite-dimensional linear spaces). Since finite-dimensional linear vector spaces are closed (hence complete), one may suspect that the results presented do not carry over to infinite-dimensional inner product spaces without some restrictions on the operator as well as the associated vector spaces. This section deals with the discussion of the solvability conditions for abstract operator equations.

To motivate the discussion of the solvability of operator equations, consider the following Neumann problem of determining u such that

$$-\frac{d^2u}{dx^2} = f \qquad 0 < x < 1; \qquad u'(0) = u'(1) = 0$$

The operator $A = -d^2/dx^2$ is symmetric on the set $\mathcal{D}_A$ of admissible functions

$$\mathcal{D}_A = \{u : u \in C^2[0, 1],\ u'(0) = u'(1) = 0\}$$

However, A is not positive on $\mathcal{D}_A$ because

$$(Au, u)_0 = 0 \text{ does not imply } u = 0$$

This loss of positivity of operator A has interesting consequences on the solvability of the problem posed above. Integrating the differential equation for $f = 1$, $-u'' = 1$, we obtain

$$u' = -x + c$$

Boundary condition $u'(0) = 0$ gives $c = 0$. On the other hand, the condition $u'(1) = 0$ gives $c = 1$. Clearly, these are contradicting requirements. Hence, the problem has *no* solution.

Now consider the case $f = -1 + 2x$. The solution of the equation is

$$u = \frac{x^2}{2} - \frac{x^3}{3} + c_1 x + c_2$$

Condition $u'(0) = 0$ or $u'(1) = 0$ both give $c_1 = 0$. But the constant c_2 is not determined by the boundary conditions. Thus, in this case the problem has *many* solutions, i.e., solution exists but is not unique.

Next consider the same operator equation but with the Dirichlet boundary conditions $u(0) = u(1) = 0$. The problem has a unique solution for $f = 1$ and $f = -1 + 2x$, respectively:

$$u = -\frac{x^2}{2} + x \qquad \text{and} \qquad u = -\frac{x^3}{3} + \frac{x^2}{2} - \frac{x}{6}$$

The difference between the Neumann and Dirichlet problems is that the operator is not positive in the former case while it is in the latter case.

The above discussion illustrates that the existence and uniqueness of solutions to a differential operator equation depend on the data as well as the positivity of the operator. In this section we study the solvability conditions for linear operator equations.

5-3-2 Solvability Conditions

Let A be a linear operator on a domain $\mathscr{D}_A$ dense in a Hilbert space H. We wish to know for what *data* f can we solve the nonhomogeneous equation

$$Au = f \qquad u \in \mathscr{D}_A \tag{5-17}$$

The existence of solutions to Eq. (5-17) largely depends on the solution of the homogeneous adjoint equation,

$$A^*v = 0 \tag{5-18}$$

Let u be the solution of Eq. (5-17) and v be the solution of Eq. (5-18). Taking the inner product of Eq. (5-17) with v, we get

$$(Au, v) = (f, v)$$

By the definition of the adjoint and Eq. (5-18) it follows that a *necessary* condition for the operator equation (5-17) to have solutions is that the data f should be orthogonal to the null space of A^*:

$$(f, v) = 0 \qquad \text{for all } v \in \mathscr{N}(A^*) \tag{5-19}$$

Clearly, the condition is identically satisfied for $v = 0 \in \mathscr{N}(A^*)$. We wish to know whether there are any nonzero elements in $\mathscr{N}(A^*)$. In particular we ask ourselves the following questions. Is Eq. (5-19) sufficient for solvability of Eq. (5-17)? If Eq. (5-17) has solutions, how many are there?

The following theorem establishes the sufficiency of (5-19) for closed, bounded below linear operators.

Theorem 5-2 Let A be a closed, bounded below linear operator on a Hilbert space H. Then Eq. (5-17) has solutions if and only if

$$\mathscr{R}(A) = \mathscr{N}(A^*)^{\perp} \tag{5-20}$$

PROOF We first prove that the range of a closed, bounded below operator is closed (that is, $\bar{\mathscr{R}}(A) = \mathscr{R}(A)$). Let $f_n \in \mathscr{R}(A)$ with $f_n \to f$ and $Av_n = f_n$. We wish to show that $f \in \mathscr{R}(A)$. Define $u_n = v_n - Pv_n$, where P is the projection operator from $\mathscr{D}_A$ onto $\mathscr{N}(A)$. Then $u_n \in \mathscr{D}_A \cap \mathscr{N}(A)^{\perp}$ and $Au_n = f_n$. We have, since A is bounded below,

$$\|u_n - u_m\| \leq \frac{1}{c} \|A(u_n - u_m)\| = \frac{1}{c} \|f_n - f_m\|$$

which shows that $\{u_n\}$ is a Cauchy sequence in $\mathscr{D}_A$ and therefore $u_n \to u$ in H. Since A is closed, $Au_n \to f$ and $u_n \to u$ implies that $u \in \mathscr{D}_A$ and $Au = f$. Hence, $\mathscr{R}(A)$ is closed.

Next we show that condition (5-20) is equivalent to (5-19). Since $\mathscr{R}(A)$ is closed $\mathscr{R}(A)^{\perp\perp} = \mathscr{R}(A) = \mathscr{N}(A^*)^{\perp}$ or $\mathscr{R}(A)^{\perp} = \mathscr{N}(A^*)$. This implies that $(Au, v) = (f, v) = 0$ for all $u \in \mathscr{D}_A$ and $v \in \mathscr{N}(A^*)$.

"*if*" *part*. Let $f \in \mathscr{R}(A)$. Then there exists an element $u \in \mathscr{D}_A$ such that $Au = f$. Since A is bounded below, A has a continuous inverse A^{-1} defined on its range, and, therefore, $u = A^{-1}f$. Moreover, we have

$$\|u\| = \|A^{-1}f\| \leq c\|f\|$$

"*only if*" *part*. Let $u = A^{-1}f$. Hence $Au = f$ and $(Au, v) = (f, v)$ for any $v \in \mathscr{R}(A)$. Clearly $(Au, v) = (u, A^*v) = 0$ for any $v \in \mathscr{N}(A^*)$. In other words $u \in \mathscr{N}(A^*)^{\perp}$. This completes the proof of the theorem.

We now consider some examples.

Example 5-3 (*Transverse deflection of a cable*) This example has three parts.

1. Consider the second-order differential equation,

$$Au \equiv -\frac{d}{dx}\left(a\frac{du}{dx}\right) = -f \qquad 0 < x < L \tag{5-21a}$$

subject to the Dirichlet boundary conditions,

$$u(0) = 0 \qquad u(L) = 0 \tag{5-21b}$$

Here $a = a(x)$ and $f = f(x)$ are given functions of x. We have $\mathscr{D}_A \subset L_2(0, L)$. More precisely, $\mathscr{D}_A = H_0^1(0, L) \cap H^2(0, L)$. Since A is self-adjoint (with respect to the L_2-inner product), it is also bounded below. The homogeneous adjoint equation becomes

$$-\frac{d}{dx}\left(a\frac{dv}{dx}\right) = 0 \qquad v \in \mathscr{D}_{A^*} \tag{5-22}$$

where $\mathscr{D}_{A^*} = \mathscr{D}_A$. For constant values of a, the solution v to the homogeneous adjoint equation is of the form $v = C_1 + C_2 x$. However, since $v \in H_0^1(\Omega)$ we have $v = 0$, so that $\mathscr{N}(A^*) = \{0\}$. Hence, the differential equation has a unique solution.

2. (*Axial deformation of a bar*) Let us consider the same differential equation as in the previous example, with the following mixed boundary conditions:

$$u(0) = 0 \qquad \left(a \frac{du}{dx}\right)\bigg|_{x=L} = 0$$

The domains of A and A^* are given by $\mathscr{D}_A = H_*^1(0, L) \cap H^2(0, L)$, where

$$H_*^1(0, L) = \{u : u \in H^1(0, L),\ u(0) = 0,\ a(du/dx)(L) = 0\}$$

and

$$\mathscr{D}_{A^*} = H_+^1(0, L) \cap H^2(0, L)$$

where $H_+^1(0, L) = H_*^1(0, L)$ (A is self-adjoint). Again the null space of A^* is trivial and hence the solution is unique.

3. We now look at the same differential equation as in part 1 of this example, but with the Neumann boundary conditions

$$\left(a \frac{du}{dx}\right)\bigg|_{x=0} = -P_1 \qquad \left(a \frac{du}{dx}\right)\bigg|_{x=L} = P_2$$

The domains $\mathscr{D}_A$ and $\mathscr{D}_{A^*}$ can be easily identified as before, with

$$H_*^1(0, L) = \{u : u \in H^1(0, L),\ a(du/dx)(0) = -P_1,\ a(du/dx)(L) = P_2\}$$

and

$$H_+^1(0, L) = \{v : v \in H^1(0, L),\ a(dv/dx)(0) = a(dv/dx)(L) = 0\}$$

The null space $\mathscr{N}(A^*)$ contains the constant element, hence the solvability condition becomes

$$\int_0^L f\,dx + P_1 + P_2 = 0 \tag{5-23}$$

Note that this equation represents the equilibrium of the forces. When condition (5-23) is satisfied, solutions exist but are not unique. We shall return to the Neumann problems in Sec. 5-4-5.

It should be noted that the domain $\mathscr{D}_A$ defined in part 3 of the above example is not a linear subspace. It would be a linear space if $P_1 = P_2 = 0$. It is possible to reformulate the problem so that the boundary conditions of a linear operator equation become homogeneous (see Sec. 4-3-4).

5-3-3 The Contraction Mapping Theorem

Next we consider a special type of nonlinear operator equation. Consider the equation

$$u = Tu \tag{5-24}$$

where $T : V \to V$, V is a normed space, and u is an unknown element of V. Solutions of Eq. (5-24) are called *fixed points* of the mapping T, meaning that they are the elements of the space which are unchanged by the operation of T.

A close examination of Eq. (5-24) suggests a method of constructing a sequence of approximations to the exact solution u: let u_0 be a first approximation to the solution u. If u_0 were the exact solution, then Tu_0 equals u_0. If u_0 is not equal to the exact solution, then Tu_0 will not equal either u_0 or u. However, we might expect Tu_0 closer to u than u_0 is. Therefore, we set the second approximation u_1 equal to Tu_0; then u_1 is a better approximation than u_0. Repeating this operation, we get at the beginning of the $(n + 1)$st iteration,

$$u_{n+1} = Tu_n \qquad (n = 0, 1, \ldots) \tag{5-25}$$

The question we ask is whether the sequence $\{u_n\}$ converges to the exact solution. We must find conditions under which the iterative procedure in Eq. (5-25) is convergent. Two such conditions are apparent from the iterative procedure: the space from which the sequence $\{u_n\}$ is taken should be complete, and the operator T should be such that when applied to two elements of the space it transforms the elements into new elements which are closer together than the original pair, that is, $\|Tu_1 - Tu_2\| < \alpha\|u_1 - u_2\|$, $0 < \alpha < 1$. Such an operator is called a "contraction." A formal definition of a contraction operator is given below.

A mapping $T : S \to V$, where S is a subset of normed space V, is called a *contraction mapping*, or *contraction*, if there exists a positive number α, $0 < \alpha < 1$ such that

$$\|Tu - Tv\| \leq \alpha\|u - v\| \qquad \text{for all } u, v \in S \tag{5-26}$$

Note that the constant α should be strictly positive and less than 1.

Thus, a contractive mapping "contracts distances": the distance between the images Tu and Tv is smaller, by scale factor of α, than the distance between the elements u and v. When α takes any positive value (i.e., not necessarily between zero and one), the mapping is called *Lipschitz continuous.* It is clear that a Lipschitz continuous (and hence, a contractive) mapping is continuous.

Example 5-4 This example has three parts.

1. Let $T : \mathbb{R} \to \mathbb{R}$ and $Tx = (1 + x)^{1/3}$. Then finding the solution of $x = Tx$ amounts to finding the roots of the equation $x^3 - x - 1 = 0$. The operator T is a contraction on the subset $S = [1, 2]$ because $[\alpha = (3)^{1/3} - 1]$,

$$|Tx - Ty| = |(1 + x)^{1/3} - (1 + y)^{1/3}| \leq \alpha|x - y|$$

for all $x, y \in S$ holds.

2. Let $f: \mathbb{R}^+ \to \mathbb{R}^+$ and $f(x) = x + e^{-x}$, where $\mathbb{R}^+$ is the set of positive numbers. For any $x, y \in \mathbb{R}^+$, we have by the Mean Value Theorem,

$$\|f(x) - f(y)\| = \|x - y\| \, |f'(z)| \qquad x < z < y$$

Note that $\alpha = |f'(z)| < 1$ for all $z > 0$. However, α depends on z (hence, on x and y) and therefore f is not a contraction. We can prove this also by contradiction. Suppose that there exists an $\alpha < 1$ satisfying the definition Eq. (5-26). Then

$$\|f(x) - f(y)\| = \|x - y\| \cdot (1 - e^{-z}) \geq a\|x - y\|$$

if both x and y are greater than $-\ln(1 - a)$. This contradicts the definition.

3. Let T be the operator mapping a unit sphere into a unit sphere

$$Tu = \int_0^1 u^2(x)\, dx$$

We have

$$\|Tu - Tv\| = \int_0^1 |u^2(x) - v^2(x)|\, dx$$

$$\leq 2\left[\int_0^1 |u(x) - v(x)|^2\, dx\right]^{1/2}$$

Hence T is not a contraction on a unit sphere. However, T is a contraction in the sphere $\|u\| \leq \frac{1}{4}$.

If T in Eq. (5-25) is a contractive mapping on a subset S of a normed space V, then the iteration sequence $\{u_n\}$ is a Cauchy sequence. If V is complete, the Cauchy sequence converges to the limit point in V. However, the limit in general need not lie in the domain S of the mapping T, and therefore is not a candidate for a solution of the equation $Tu = u$. If S is a closed set, that is, S contains all its limit points, then the limit of the sequence $\{u_n\}$ must lie in S, and we can expect the iterative procedure to converge to the true solution. These ideas are expressed in the following *Banach Fixed Point Theorem*, which gives the sufficient conditions for the existence and uniqueness of solutions.

Theorem 5-3 (Contraction mapping theorem) Let $T : S \to S$ be a contraction mapping of a closed subset S of a Banach space V. Then

(i) the equation $u = Tu$ has one and only one solution $u \in S \subset V$ and

(ii) the unique solution u of the equation $u = Tu$ can be obtained as the limit of the sequence of elements $u_n \in S$, $u = \lim_{n\to\infty} u_n$, where $u_n = Tu_{n-1}$, $n = 1, 2, \ldots$ and $u_0 \in S$ can be chosen arbitrarily.

PROOF For any $u_0 \in S$ and $u_n \in S$ we have

$$\begin{aligned}\|u_{n+1} - u_n\| &= \|Tu_n - Tu_{n-1}\| \\ &\leq \alpha\|u_n - u_{n-1}\| = \alpha\|Tu_{n-1} - Tu_{n-2}\| \\ &\leq \alpha^2\|u_{n-1} - u_{n-2}\| \\ &\cdots\cdots\cdots\cdots \\ &\leq \alpha^n\|u_1 - u_0\|\end{aligned}$$

Therefore, for any $m > n$, the triangle inequality gives

$$\begin{aligned}\|u_m - u_n\| &\leq \|u_m - u_{m-1}\| + \|u_{m-1} - u_{m-2}\| + \cdots + \|u_{n+1} - u_n\| \\ &\leq (\alpha^{m-1} + \alpha^{m-2} + \cdots + \alpha^n)\|u_1 - u_0\| \\ &\leq \left(\frac{\alpha^n}{1-\alpha}\right)\|u_1 - u_0\|\end{aligned}$$

Hence

$$\lim_{m,\, n \to \infty} \|u_m - u_n\| = 0$$

Thus $\{u_n\}$ is a Cauchy sequence. Since the space is complete, $\{u_n\}$ is convergent. Let $\bar{u}$ be the limit of the sequence. Since S is closed, $\bar{u} \in S$.

Next, we must show that $T\bar{u} = \bar{u}$. For any n, we have

$$\begin{aligned}\|T\bar{u} - \bar{u}\| &\leq \|T\bar{u} - Tu_n\| + \|Tu_n - \bar{u}\| \\ &\leq \alpha\|\bar{u} - u_n\| + \|u_{n+1} - \bar{u}\|\end{aligned}$$

The right-hand side tends to zero as $n \to \infty$, and therefore $T\bar{u} - \bar{u} = 0$, as required.

The uniqueness of the solution can be proved as follows: Let $T\bar{u} = \bar{u}$ and $T\bar{v} = \bar{v}$. Then

$$\|\bar{u} - \bar{v}\| = \|T\bar{u} - T\bar{v}\| \leq \alpha\|\bar{u} - \bar{v}\|$$

which is a contradiction, unless $\|\bar{u} - \bar{v}\| = 0$, giving $\bar{u} = \bar{v}$. Hence, the solution is unique.

The theorem not only presents an existence and uniqueness result but also gives an algorithm to obtain the solution by an iterative procedure. For this reason Eq. (5-25) is known as the *method of successive approximations*.

The contraction mapping theorem (or the method of successive approximations) can be used in proofs of existence and in obtaining approximate solutions of algebraic, differential, and integral equations. Some illustrative examples are presented here.

Example 5-5 (Solution of systems of algebraic equations) Consider the n-dimensional euclidean space $\mathbb{R}^n$ of elements of the type $\mathbf{x} = (x_1, x_2, \ldots, x_n)$.

Define the metric (norm) in $\mathbb{R}^n$ by

$$\|\mathbf{x} - \mathbf{y}\| = \max_i |x_i - y_i|$$

The space $\mathbb{R}^n$ is complete with respect to the max-norm defined above. We consider in $\mathbb{R}^n$ the operator equation

$$\mathbf{x} = A\mathbf{x}$$

defined by

$$x_i = \sum_{j=1}^{n} a_{ij} x_j + b_i \qquad i = 1, 2, \ldots, n$$

where $\mathbf{b}$ is a fixed element in $\mathbb{R}^n$. Consider

$$\begin{aligned} \|A\mathbf{x} - A\mathbf{y}\| &= \max_i \left| \sum_{j=1}^{n} a_{ij} (x_j - y_j) \right| \\ &\leq \max_i \sum_{j=1}^{n} |a_{ij}| \, |x_j - y_j| \\ &\leq \max_i \sum_{j=1}^{n} |a_{ij}| \cdot \max_j |x_j - y_j| \\ &= \max_i \sum_{j=1}^{n} |a_{ij}| \, \|\mathbf{x} - \mathbf{y}\| \end{aligned}$$

Thus the operator A is contractive if

$$\alpha = \max_i \sum_{j=1}^{n} |a_{ij}| < 1$$

and consequently the contractive mapping theorem applies and operator A has a unique fixed point.

As a specific example, consider the set of algebraic equations

$$\begin{aligned} \tfrac{4}{5}x_1 - \tfrac{1}{2}x_2 - \tfrac{1}{4}x_3 &= -1 \\ -\tfrac{1}{3}x_1 + \tfrac{4}{3}x_2 + \tfrac{1}{4}x_3 &= 2 \\ -\tfrac{1}{4}x_1 - \tfrac{2}{15}x_2 + \tfrac{3}{4}x_3 &= 3 \end{aligned}$$

To apply the contractive mapping theorem, we first rewrite the above equations in the form $\mathbf{x} = A\mathbf{x} + \mathbf{b}$, and then prove that $\max_i \sum_{j=1}^{n} |a_{ij}| < 1$. Thus the representation $\mathbf{x} = A\mathbf{x} + \mathbf{b}$ should be guided by the requirement $\max_i \sum_{j=1}^{n} |a_{ij}| < 1$. One such representation is given by

$$\begin{aligned} x_1 &= \tfrac{1}{5}x_1 + \tfrac{1}{2}x_2 + \tfrac{1}{4}x_3 - 1 \\ x_2 &= \tfrac{1}{3}x_1 - \tfrac{1}{3}x_2 - \tfrac{1}{4}x_3 + 2 \\ x_3 &= \tfrac{1}{4}x_1 + \tfrac{2}{15}x_2 + \tfrac{1}{4}x_3 + 3 \end{aligned}$$

$$\text{or} \qquad \begin{Bmatrix} x_1 \\ x_2 \\ x_3 \end{Bmatrix} = \begin{bmatrix} \frac{1}{5} & \frac{1}{2} & \frac{1}{4} \\ \frac{1}{3} & -\frac{1}{3} & -\frac{1}{4} \\ \frac{1}{4} & \frac{2}{15} & \frac{1}{4} \end{bmatrix} \begin{Bmatrix} x_1 \\ x_2 \\ x_3 \end{Bmatrix} + \begin{Bmatrix} -1 \\ 2 \\ 3 \end{Bmatrix}$$

Clearly, the condition $\max_i \sum_{j=1}^{n} |a_{ij}| < 1$ is met.

The iterative scheme can be used to obtain the solution. For example, the initial guess $\mathbf{x} = (0, 0, 0)$ gives

$$\mathbf{x}^{(1)} = (-1, 2, 3)$$

$$\mathbf{x}^{(5)} = (0.507535, 0.942640, 4.303294)$$

$$\mathbf{x}^{(10)} = (0.641659, 0.835535, 4.360547)$$

$$\mathbf{x}^{(19)} = (0.639519, 0.841889, 4.362843)$$

$$\mathbf{x}^{(20)} = (0.639559, 0.841833, 4.362843)$$

Of course, any other initial guess can be used. The number of iterations required to achieve convergence depends on the initial guess. The exact solution is given by

$$\mathbf{x} = (0.639548, 0.841853, 4.362845)$$

The next example deals with the existence of solutions to a nonlinear algebraic equation.

Example 5-6 Consider the nonlinear algebraic equation $x^3 - x - 1 = 0$. The equation has three roots. It has a root between 1 and 2. There are several ways of putting the equation in the form $x = Tx$:

$$Tx = x^3 - 1 \qquad Tx = (1 + x)^{1/3} \qquad Tx = \frac{1}{x^2 - 1}$$

On $S = [1, 2]$ only T defined by $Tx = (1 + x)^{1/3}$ is contractive and the other two operators are not contractive.

We can use the method of successive approximations to determine the root (unique) in S. We begin with $x_0 = 1$. We obtain $x_1 = 1.2599$, $x_2 = 1.3123$, $x_3 = 1.3224$, $x_4 = 1.3243$, $x_5 = 1.3246$, $x_6 = 1.3247$, $x_7 = 1.3247$. Thus, the root, to four significant figures, is obtained at the end of the seventh iteration.

Since every differential equation can be transformed to an equivalent integral equation, we now turn our attention to the application of the contraction mapping theorem to integral equations.

Example 5-7 Consider the Sturm–Liouville equation

$$\frac{d^2u}{dx^2} + \lambda f(x, u) = 0 \qquad u(0) = u(1) = 0 \tag{5-27}$$

which arises in connection with the problem of an elastic rod (where $f(x, u) = \sin u$) under compression with its ends fixed; here u denotes the axial displacement and λ is the axial load. For small values of λ, the solution of the equation is $u = 0$ (unless λ corresponds to an eigenvalue), which corresponds to the rod being straight. But as the load λ is increased, the rod will buckle, giving rise to a nonzero solution u.

The equation can be cast as an integral equation

$$u(x) = \lambda \int_0^1 G(x, y) f(y, u(y))\, dy \tag{5-28}$$

where G is Green's function

$$G(x, y) = \begin{cases} x(1 - y) & \text{for } x \le y \\ y(1 - x) & \text{for } y \le x \end{cases} \tag{5-29}$$

We have $u = Tu$, with T given by

$$Tu = \lambda \int_0^1 G(x, y) f(y, u(y))\, dy \tag{5-30}$$

To insure that T is a contraction in $L_2[0, 1]$, we assume that $\partial f/\partial u$ is bounded,

$$\left| \frac{\partial f}{\partial u}(x, u) \right| \le M \qquad \text{for } 0 \le x \le 1 \text{ and all } u \in L_2[0, 1]$$

By the Mean Value theorem, we have $[K(x, y, u) = G(x, y) f(y, u)]$

$$\begin{aligned} |K(x, y, u_1) - K(x, y, u_2)| &= |G(x, y)| \; \frac{\partial f}{\partial u}(y, u) \;\; |u_1 - u_2| \\ &\le M\, |G(x, y)|\, |u_1 - u_2| \end{aligned}$$

for some u between u_1 and u_2. Let

$$\mu^2 = M^2 \int_0^1 \int_0^1 |G(x, y)|^2\, dx\, dy = \frac{M^2}{90} \tag{5-31}$$

We now show that T is a contraction for a fixed value of λ. For any u_1, $u_2 \in L_2[0, 1]$ we have

$$\begin{aligned} |Tu_1 - Tu_2| &= \left| \lambda \int_0^1 [K(x, y, u_1(y)) - K(x, y, u_2(y))]\, dy \right| \\ &\le |\lambda| \int_0^1 M\, |G(x, y)|\, |u_1(y) - u_2(y)|\, dy \\ &\le |\lambda| \left\{ \int_0^1 M^2 |G(x, y)|^2\, dy \cdot \int_0^1 |u_1(y) - u_2(y)|^2\, dy \right\}^{1/2} \end{aligned}$$

where, in arriving at the last step, we used the Cauchy–Schwarz inequality.

$$\|Tu_1 - Tu_2\| \le |\lambda| \left\| \left\{ \int_0^1 M^2 |G(x, y)|^2 \, dy \|u_1 - u_2\|^2 \right\}^{1/2} \right\|$$

$$= |\lambda| \mu \|u_1 - u_2\|$$

Now let $\alpha = \mu|\lambda| > 0$. Note that $\alpha < 1$ only if $|\lambda| < 1/\mu$. Thus T is a contraction in $L_2[0, 1]$ provided

(i) $f \in L_2[0, 1]$ and $\dfrac{\partial f}{\partial u} \le M$ for $0 \le x \le 1$ and all $u \in L_2[0, 1]$

$$\text{(ii)} \qquad |\lambda| < 1/\mu, \; \mu^2 = M^2 \int_0^1 \int_0^1 |G(x, y)|^2 \, dx \, dy \text{ exists.} \tag{5-32}$$

Under these conditions the equation $u = Tu$ has a unique solution in $L_2[0, 1]$.

More specifically, for the problem of buckling of a rod we have

$$f(x, u) = \sin u \qquad \left|\frac{\partial f}{\partial u}\right| = |\cos u| \le 1 \quad (M = 1)$$

$$\lambda < 3\sqrt{10}$$

Hence for loads up to $3\sqrt{10}$, the column problem has a unique solution. Since $u = 0$ satisfies the equation, $u = 0$ is the unique solution (i.e., the rod will not buckle for loads up to $3\sqrt{10}$). The nonzero solutions exist for $\lambda > 3\sqrt{10}$. This completes the example.

PROBLEMS 5-1

For the sets of linear algebraic equations in Probs. 5-1-1 to 5-1-6, check whether the solvability condition is satisfied. Also, discuss whether the system has solutions, and if so give the linearly independent solution.

5-1-1 $x_1 + x_2 + x_3 = 1 \qquad 2x_1 - x_2 + x_3 = 2 \qquad x_1 - 2x_2 + 3x_3 = 3.$

5-1-2 $x_1 + 2x_2 + x_3 = 1 \qquad x_1 - x_2 - 2x_3 = 3 \qquad x_1 + x_2 = 1.$

5-1-3 $3x_1 - x_2 + 2x_3 = 2 \qquad 2x_1 + x_2 + x_3 = -1 \qquad x_1 + 3x_3 = 2.$

5-1-4 $10x_1 + 10x_2 + 9x_3 = 25 \qquad 10x_1 + 9x_2 + 8x_3 = 24 \qquad 9x_1 + 8x_2 + 7x_3 = 23.$

5-1-5 $2x_1 + 3x_2 - x_3 = 4 \qquad x_1 - x_2 + 2x_3 = 2 \qquad x_1 + 2x_2 - x_3 = 1.$

5-1-6 $-x_1 + 5x_2 + 4x_3 = 0 \qquad x_1 - x_2 + 2x_3 = 2 \qquad x_1 + 2x_2 - x_3 = 1.$

5-1-7 Show that the system of equations in Prob. 5-1-4 is ill-conditioned. Specifically, show that there are at least two solutions to the system that differ only in the second decimal point.

Use the Contraction Mapping Theorem 5-3 to determine the solution of the linear equations in Probs. 5-1-8 to 5-1-11.

5-1-8

$$x_1 = \tfrac{1}{5}x_1 + \tfrac{1}{5}x_2 - \tfrac{1}{5}x_3 + 1$$
$$x_2 = \tfrac{1}{4}x_1 - \tfrac{1}{4}x_2 + \tfrac{1}{4}x_3 + \tfrac{3}{2}$$
$$x_3 = -\tfrac{1}{3}x_1 - \tfrac{1}{3}x_2 + \tfrac{1}{6}x_3 + \tfrac{7}{2}$$

5-1-9 $x_1 = \tfrac{1}{3}x_1 + \tfrac{2}{5}x_2 + \tfrac{1}{6}x_3 - \tfrac{1}{2}$
$x_2 = \tfrac{1}{6}x_1 - \tfrac{1}{5}x_2 + \tfrac{1}{3}x_3 + \tfrac{29}{6}$
$x_3 = \tfrac{1}{5}x_1 + \tfrac{1}{3}x_2 - \tfrac{1}{6}x_3 + \tfrac{1}{15}$

5-1-10 $x_1 = \tfrac{1}{5}x_1 + \tfrac{1}{2}x_2 - \tfrac{1}{3}x_3 - 6$
$x_2 = -\tfrac{1}{10}x_1 + \tfrac{1}{4}x_2 + \tfrac{1}{3}x_3 + 2$
$x_3 = \tfrac{1}{5}x_1 - \tfrac{1}{4}x_2 - \tfrac{1}{6}x_3 - 2$

5-1-11 $x_1 = \tfrac{1}{4}x_1 + \tfrac{1}{8}x_2 + \tfrac{1}{2}x_3 - 1$
$x_2 = -\tfrac{1}{8}x_1 + \tfrac{1}{2}x_2 + \tfrac{1}{4}x_3$
$x_3 = \tfrac{1}{8}x_1 - \tfrac{1}{8}x_2 - \tfrac{1}{4}x_3 - 1$

5-1-12 Define $T: \mathbb{R}^2 \to \mathbb{R}^2$ by $T(x_1, x_2) = (x_2^{1/3}, x_1^{1/3})$. What are the fixed points of T? In what quadrant of the x_1, x_2-plane is T a contraction?

5-4 VARIATIONAL BOUNDARY-VALUE PROBLEMS

5-4-1 Introduction

In Sec. 5-2 we studied conditions for the existence and uniqueness of solutions of systems of algebraic equations. Such sets of equations are encountered in problems with a finite number of degrees of freedom. For example, all approximate methods lead ultimately (as will be seen in Chaps. 6 and 7) to the solution of a system of algebraic equations. For linear operator equations, existence and uniqueness of solutions are governed by conditions discussed in Sec. 5-3.

This section deals with the study of existence and uniqueness of the weak solutions of continuum problems described by differential operator equations,

$$Au = f \text{ in } \Omega \tag{5-33}$$

where A is a linear differential operator which maps a linear set $\mathscr{D}_A$, dense in a Hilbert space H, into H, and $f \in L_2(\Omega)$. Equation (5-33) is to be solved in conjunction with appropriate boundary conditions on the traces of u and its derivatives on the boundary Γ of Ω.

The weak formulation of Eq. (5-33) (and its boundary conditions) has the general form (see Chap. 4): find $u \in H$ such that

$$B(v, u) = l(v) \qquad \text{holds for every } v \in H \tag{5-34}$$

where $B(\cdot, \cdot)$ is a bilinear form on $H \times H$ and $l(\cdot)$ is a linear form on H. We will see in the forthcoming discussions that $B(v, u)$ for certain problems is defined on $H_1 \times H_2$ where $H_1 \subset H_2$. Such problems include nonhomogeneous boundary value problems and Neumann problems.

We state and prove the well-known Lax–Milgram theorem, which provides the sufficient conditions for existence and uniqueness of solutions of problems with symmetric and positive-definite bilinear forms. The theorem can also be used for problems defined on finite-dimensional spaces, and therefore it is useful in the approximation theory. In the latter case, the Lax–Milgram theorem should, as one might suspect, require the same conditions as Theorem 5-1. We also study more general existence theorems than the Lax–Milgram theorem, and establish the existence and uniqueness results for a number of problems in engineering.

5-4-2 Regularity of the Solution

Before we embark on the theory of the existence and uniqueness of solutions to the weak (or variational) problem (5-34) associated with the (differential) operator equation (5-33), we must impose certain smoothness conditions on the data f, domain Ω, and the operator A so that if the weak solution to Eq. (5-34) exists it is also the classical solution [i.e., the solution of (5-33)]. Here we shall not attempt to study the regularity theory but only point out to the reader that certain smoothness conditions are assumed in presenting the existence results (to be given shortly). Interested readers can consult Lions and Magenes (1972) and Necas (1967) for details of the regularity theory.

Let A be the $2m$th order differential operator,

$$Au \equiv \sum_{0 \leq |\boldsymbol{\alpha}|,\, |\boldsymbol{\beta}| \leq m} (-1)^{|\boldsymbol{\alpha}|} D^{\boldsymbol{\alpha}}[a_{\boldsymbol{\alpha\beta}}(\mathbf{x}) D^{\boldsymbol{\beta}} u(\mathbf{x})] \tag{5-35}$$

where $\boldsymbol{\alpha} = (\alpha_1, \alpha_2, \ldots, \alpha_N)$, N is the dimension of the domain Ω, α_i are non-negative integers and

$$D^{\boldsymbol{\alpha}} \equiv \frac{\partial^{|\boldsymbol{\alpha}|}}{\partial x_1^{\alpha_1}\, \partial x_2^{\alpha_2} \cdots \partial x_N^{\alpha_N}} \qquad |\boldsymbol{\alpha}| = \sum_{i=1}^{N} \alpha_i \tag{5-36}$$

In constructing the weak form of Eq. (5-33), where A is given by Eq. (5-35), we made use of Green's theorem to obtain the weak form in Eq. (5-34). However, Green's theorem requires that the function $u(\mathbf{x})$, the coefficients $a_{\boldsymbol{\alpha\beta}}(\mathbf{x})$ of the operator A, and the function $f(\mathbf{x})$ are sufficiently smooth in the domain Ω. The same is assumed in proving that the weak solution of Eq. (5-34) is indeed the classical solution of Eq. (5-33). In other words, if the given data (the coefficients $a_{\boldsymbol{\alpha\beta}}$, f, etc.) of the problem are sufficiently smooth, and if the weak solution is sufficiently smooth, then the weak solution is the classical solution of the problem. The Sobolev embedding theorems assert that the smoothness of the weak solution is a consequence of the smoothness of the data of the given problem.

The following theorem gives the regularity (i.e., smoothness) conditions on the data in order for the weak solution to be the classical solution. For a proof of the theorem, see Necas (1967).

Theorem 5-4 Let Ω be a domain with a Lipschitz boundary and A a differential operator of order $2m$ with the bilinear form

$$B(v, u) = \sum_{|\boldsymbol{\alpha}|,\, |\boldsymbol{\beta}| \leq m} \int_{\Omega} a_{\boldsymbol{\alpha\beta}} D^{\boldsymbol{\alpha}} v D^{\boldsymbol{\beta}} u \, d\mathbf{x} \tag{5-37}$$

which is $H_0^m(\Omega)$-elliptic, i.e., there exists a positive constant α such that

$$|B(v, v)| \geq \alpha \|v\|_m^2 \qquad \text{for } v \in H_0^m(\Omega) \tag{5-38}$$

Let s be a positive integer, $s > m + N/2$, $a_{\boldsymbol{\alpha\beta}}(\mathbf{x}) \in C^{(\alpha_i)}(\Omega)$, where $\alpha_i = s - m + |\boldsymbol{\alpha}|$ and $f \in H^{(s-m)}(\Omega)$. Then the weak solution $u(\mathbf{x})$ of Eq. (5-33) is its classical solution in Ω.

5-4-3 The Lax–Milgram Theorem

The Lax–Milgram theorem is given here for an abstract variational (or weak) problem in Eq. (5-34) so that it holds for any problem that can be cast into the general form. To fix the ideas, consider the variational problem of finding u in a Hilbert space H such that Eq. (5-34) holds:

$$B(u, v) = l(v) \qquad \text{for all } v \text{ in } H$$

We now give the conditions under which the variational equation (5-34) has a unique solution. The theorem is valid only for symmetric bilinear forms.

Theorem 5-5 (The Lax–Milgram theorem) Let H be a Hilbert space, and let $B(\cdot, \cdot)\colon H \times H \to \mathbb{R}$ be a bilinear form on $H \times H$, with the following properties:

(*a*) Continuity of $B(\cdot, \cdot)$: $|B(u, v)| \leq M\|u\|\,\|v\| \qquad 0 < M < \infty$
(*b*) Positive-definiteness of $B(\cdot, \cdot)$: $|B(u, u)| \geq \alpha\|u\|^2 \qquad \alpha > 0$

(5-39)

for all $u, v \in H$. Then for any continuous linear functional $l : H \to \mathbb{R}$ on H, there exists a unique vector u_0 in H such that

$$B(u_0, v) = l(v) \qquad \text{for every } v \in H \tag{5-40}$$

PROOF OF EXISTENCE For each $u \in H$, $B(u, v)$ defines a linear form $l_u(v)$ on H (that is, l_u is an element of the dual space H'),

$$l_u(v) = B(u, v) \qquad v \in H$$

Using the continuity of $B(\cdot, \cdot)$, we have

$$\|l_u\|_{H'} = \sup_{v \in H} \frac{|l_u(v)|}{\|v\|} = \sup_{v \in H} \frac{|B(u, v)|}{\|v\|} \leq M\|u\|$$

By the Reisz representation theorem, Theorem 4-1, there exists a unique element $u_l = Au$, such that

$$B(u, v) = l_u(v) = (u_l, v) = (Au, v) \tag{5-41}$$

where $(\cdot, \cdot)$ is the inner product on H, and A is a linear operator, $A : H \to H$. The operator A is continuous,

$$\|Au\| = \|l_u\|_{H'} \leq M\|u\|$$

and bounded below,

$$(Au, u) = B(u, u) \geq \alpha\|u\|^2$$

or

$$\|Au\| \geq \alpha\|u\| \qquad \text{for all } u \in H \tag{5-42}$$

Since A is bounded below, A has a continuous inverse (see Theorem 3-5) $A^{-1} : \mathscr{R}(A) \to H$,

$$\|A^{-1}u\| \leq \frac{1}{\alpha} \|u\| \tag{5-43}$$

Further, the range of A, $\mathscr{R}(A)$ is the whole space, H. To prove this, first we show that $\mathscr{R}(A)$ is a closed subspace of H. Let $\{Au_n\}$ be a Cauchy sequence in H. Since A is linear and bounded below, we have

$$0 = \lim_{m, n \to \infty} \|Au_n - Au_m\| = \lim_{m, n \to \infty} \|A(u_n - u_m)\|$$
$$\geq \alpha \lim_{m, n \to \infty} \|u_m - u_n\|$$

which shows that $\{u_n\}$ is a Cauchy sequence. Since A is continuous, $\mathscr{R}(A)$ is closed in H.

Now suppose that $\mathscr{R}(A) \neq H$. Then, since $\mathscr{R}(A)$ is a closed subspace of H, there exists a nonzero element $u_0 \in \mathscr{R}(A)^{\perp}$ [the orthogonal complement of $\mathscr{R}(A)$] such that

$$(Au, u_0) = 0 \qquad \text{for every } u \in H$$

In particular, we have

$$(Au_0, u_0) = B(u_0, u_0) = 0 \qquad u_0 \neq 0$$

which contradicts the assumption (*b*) of Eq. (5-39). Hence, $\mathscr{R}(A) = H$.

Now let u^* be the unique element (by Reisz representation theorem 4-1) in H corresponding to $l(u)$,

$$l(u) = (u^*, u) \qquad \|l\| = \|u^*\| \tag{5-44}$$

Hence,

$$0 = B(u, v) - l(v) = (Au, v) - (u^*, v) = (Au - u^*, v)$$

for every $v \in H$. Hence a solution to Eq. (5-40) does exist, and is given by

$$u_0 = A^{-1}u^*$$

We also have, for $H = H^m(\Omega)$,

$$\alpha\|u_0\|_m^2 \leq |B(u_0, u_0)| = |(u_0, f)_0| \leq \|u_0\|_m \|f\|_0$$

from which it follows that

$$\|u_0\|_m \leq \frac{\|f\|_0}{\alpha} = \beta\|f\|_0 \tag{5-45}$$

This inequality expresses the *continuous dependence of the weak solution on the given data* of the problem. If the function f changes "slightly" in the norm of $L_2(\Omega)$ (e.g., due to numerical approximation) to $\tilde{f}$, then the weak solution u also changes "slightly" in the norm of $H^m(\Omega)$ to $\tilde{u}$. Then the linearity of the problem and Eq. (5-45) imply

$$\|u - \tilde{u}\|_m \leq \beta\|f - \tilde{f}\|_0$$

PROOF OF UNIQUENESS Suppose that Eq. (5-34) has two solutions, u_1 and u_2. This implies that

$$0 = B(u_1 - u_2, v) = (A(u_1 - u_2), v) \qquad \text{for every } v \in H$$

By choosing $v = A(u_1 - u_2)$, we obtain $\|A(u_1 - u_2)\| = 0$. This implies that $u_1 = u_2$, hence the solution is unique.

Note that the variational problem (5-40) can be written in terms of the operator A as,

$$0 = B(u_0, v) - l(v) = (Au_0, v) - l(v)$$

If the linear functional l is of the form

$$l(v) = (f, v) \tag{5-46}$$

where f is a fixed element in H, known as the *data*, we have

$$(Au_0 - f, v) = 0 \qquad \text{for every } v \text{ in } H \tag{5-47}$$

This implies that

$$Au_0 = f \text{ in } H \tag{5-48}$$

Thus Eq. (5-47) is the variational (or weak) problem associated with the operator equation (5-48).

We now consider applications of the Lax–Milgram theorem to several problems.

Example 5-8 Consider a boundary-value problem described by the operator equation,

$$Tu \equiv -\nabla^2 u + c_0 u = f \text{ in } \Omega \subset \mathbb{R}^2 \tag{5-49}$$

$$u = 0 \text{ on } \Gamma$$

where Γ is the boundary of the two-dimensional region Ω, and $c_0 > 0$ is a positive function of $\mathbf{x}$. The solution to the above problem belongs to the space,

$$H = H_0^1(\Omega) \cap H^2(\Omega)$$

$$(u, v)_1 = \int_\Omega \left(uv + \frac{\partial u}{\partial x}\frac{\partial v}{\partial x} + \frac{\partial u}{\partial y}\frac{\partial v}{\partial y} \right) dx_1\, dx_2 \tag{5-50}$$

$$B(v, u) \equiv (v, Tu)_0 = \int_\Omega v(-\nabla^2 u + c_0 u)\, dx_1\, dx_2$$

$$= \int_\Omega (\nabla v \cdot \nabla u + c_0 vu)\, dx_1\, dx_2 \tag{5-51}$$

$$l(v) = \int_\Omega fv\, dx_1\, dx_2 \qquad u, v \in H_0^1(\Omega)$$

A quadratic form (or functional) for the problem can be defined on $H_0^1(\Omega)$ by

$$I(u) \equiv \frac{1}{2} B(u, u) - l(u) = \int_\Omega \left\{ \frac{1}{2} \left[\left(\frac{\partial u}{\partial x_1} \right)^2 + \left(\frac{\partial u}{\partial x_2} \right)^2 \right] - fu \right\} dx_1 \, dx_2 \quad (5\text{-}52)$$

Note that $B(\cdot, \cdot)$ is symmetric, continuous, and positive-definite. Indeed, the continuity follows from the Schwarz inequality,

$$\begin{aligned} |B(v, u)| \le & \left[\int_\Omega \left(\left| \frac{\partial u}{\partial x_1} \right|^2 + \left| \frac{\partial u}{\partial x_2} \right|^2 \right) dx_1 \, dx_2 \right]^{1/2} \\ & \cdot \left[\int_\Omega \left(\left| \frac{\partial v}{\partial x_1} \right|^2 + \left| \frac{\partial v}{\partial x_2} \right|^2 \right) dx_1 \, dx_2 \right]^{1/2} \\ & + c_0 \left(\int_\Omega |u|^2 \, dx_1 \, dx_2 \right)^{1/2} \left(\int_\Omega |v|^2 \, dx_1 \, dx_2 \right)^{1/2} \\ \le & \, M \|v\|_1 \, \|u\|_1 \qquad M = \max(1, c_0) \end{aligned} \quad (5\text{-}53)$$

The positive-definiteness follows from (5-51) by setting $u = v$:

$$\begin{aligned} B(u, u) &= \int_\Omega (|\nabla u|^2 + c_0 u^2) \, dx_1 \, dx_2 \ge \alpha \int_\Omega (|\nabla u|^2 + |u|^2) \, dx_1 \, dx_2 \\ &= \alpha \|u\|_1^2 \qquad \alpha = \min(1, c_0) \end{aligned} \quad (5\text{-}54)$$

Since the linear form $l(v)$ is continuous, it follows from the Lax–Milgram theorem that the problem

$$B(v, u) = l(v) \quad (5\text{-}55)$$

with $B(v, u)$ and $l(v)$ given by Eq. (5-51), has a unique solution in $H_0^1(\Omega)$.

Note that the positive-definiteness of $B(u, u)$ for the case $c_0 = 0$ can be established for only Dirichlet boundary-value problems and mixed boundary-value problems. For $c_0 = 0$, $B(u, u) = 0$ implies that

$$0 = B(u, u) = \int_\Omega \left(\left| \frac{\partial u}{\partial x_1} \right|^2 + \left| \frac{\partial u}{\partial x_2} \right|^2 \right) dx_1 \, dx_2$$

which in turn gives

$$\left| \frac{\partial u}{\partial x_1} \right|^2 = 0 \quad \text{and} \quad \left| \frac{\partial u}{\partial x_2} \right|^2 = 0$$

These conditions imply that u is a constant. If $u = 0$ on Γ (or on a portion of Γ), this constant is necessarily zero. By the equivalence of norm and semi-norm on $H_0^1(\Omega)$ [see Eq. (4-27)], we have

$$c_2^2 \|u\|_1^2 \le |B(u, u)| \le c_1^2 \|u\|_1^2$$

Now suppose that we wish to find an approximate solution to the variational problem in (5-55). Let S be a finite-dimensional subspace of $H_0^1(\Omega)$, and let $\{\phi_i\}_{i=1}^N$ be the basis of S. Then a typical element u_N in S is of the form

$$u_N = \sum_{i=1}^{N} \phi_i \alpha_i \approx u \tag{5-56}$$

where α_i, $i = 1, 2, \ldots, N$, are (components of u) constants to be determined such that the problem (5-55) holds on the finite-dimensional space. Thus the approximate problem associated with (5-55) is to find u_N in S such that

$$B(u_N, v) = l(v) \qquad \text{for every } v \text{ in } S \tag{5-57}$$

In view of (5-56), the problem can be stated as one of determining α_i such that

$$\sum_{i=1}^{N} B(\phi_i, \phi_k)\alpha_i = l(\phi_k) \qquad (k = 1, 2, \ldots, N) \tag{5-58}$$

Note that Eq. (5-58) is of the same form as (5-1):

$$a_{ik} = B(\phi_i, \phi_k) \qquad l(\phi_k) = f_k$$

Thus, the discrete problem involves solving a system of algebraic equations. The existence and uniqueness of solutions to the approximate problem (5-57) is guaranteed by the conditions of Theorem 5-1. Theorems 5-1 and 5-5 are the same for finite-dimensional spaces. The positive-definiteness condition (5-39) on $B(\cdot, \cdot)$ is equivalent to the positive-definiteness of the matrix $[a_{ik}]$, and the condition (*a*) in Eq. (5-39) is satisfied by all bilinear forms on finite-dimensional spaces.

Example 5-9 (Existence and uniqueness of plane elasticity problems) Consider the bilinear form associated with plane elasticity problems associated with isotropic bodies (see Sec. 2-6-2):

$$\begin{aligned} B(\bar{\Lambda}, \Lambda) = \int_\Omega \bigg\{ &\lambda\left(\frac{\partial \bar{u}}{\partial x} + \frac{\partial \bar{v}}{\partial y}\right)\left(\frac{\partial u}{\partial x} + \frac{\partial v}{\partial y}\right) + \mu\bigg[\left(\frac{\partial \bar{u}}{\partial y} + \frac{\partial \bar{v}}{\partial x}\right)\left(\frac{\partial u}{\partial y} + \frac{\partial v}{\partial x}\right) \\ &+ 2\frac{\partial \bar{u}}{\partial x}\frac{\partial u}{\partial x} + 2\frac{\partial \bar{v}}{\partial y}\frac{\partial v}{\partial y}\bigg]\bigg\}\, dx\, dy \end{aligned} \tag{5-59}$$

$$l(\bar{\Lambda}) = \int_\Omega (f_x \bar{u} + f_y \bar{v})\, dx\, dy + \int_{\Gamma_2} (\hat{t}_x \bar{u} + \hat{t}_y \bar{v})\, ds$$

where u and v are the displacements, f_x and f_y are the body forces, $\hat{t}_x$ and $\hat{t}_y$ are specified tractions on the portion Γ_2 of the boundary Γ, and $\mu, \lambda > 0$ are the *Lamé constants*. We assume that the boundary Γ is *Lipschitz continuous* [in order to apply the *Korn's inequality*; see Hlavacek and Necas (1970)].

We introduce the following space,

$$H(\Omega) = \{\Lambda = (u, v) : (u, v) \in H^1(\Omega) \times H^1(\Omega),\ u = 0,\ v = 0 \text{ on } \Gamma_1\} \tag{5-60}$$

equipped with the norm,

$$\|\Lambda\|_{H(\Omega)} = (\|u\|_{1,\Omega}^2 + \|v\|_{1,\Omega}^2)^{1/2} \tag{5-61}$$

It can be shown that $H(\Omega)$ is a closed subspace of $H^1(\Omega) \times H^1(\Omega)$. We note that the bilinear form $B(\cdot, \cdot)$ is symmetric on $H(\Omega)$. The quadratic form $\Pi(\Lambda) \equiv \frac{1}{2}B(\Lambda, \Lambda) - l(\Lambda)$ represents the *total potential energy* of the elastic continuum Ω (assumed to be isotropic). We have

$$\Pi(\Lambda) = \int_\Omega \left\{ \frac{\lambda}{2}\left(\frac{\partial u}{\partial x} + \frac{\partial v}{\partial y}\right)^2 + \mu\left[\frac{1}{2}\left(\frac{\partial u}{\partial y} + \frac{\partial v}{\partial x}\right)^2 + \left(\frac{\partial u}{\partial x}\right)^2 + \left(\frac{\partial v}{\partial y}\right)^2\right] - f_x u - f_y v \right\} dx\, dy - \int_{\Gamma_2} (u\hat{t}_x + v\hat{t}_y)\, ds \tag{5-62}$$

The problem of minimizing the total potential energy in (5-62) is equivalent to seeking the solution $(u, v) = \Lambda \in H(\Omega)$ to the weak problem

$$B(\Lambda, \bar{\Lambda}) = l(\bar{\Lambda}) \qquad \text{for every } \bar{\Lambda} \in H(\Omega) \tag{5-63}$$

In order to show that there exists a unique solution to the problem (5-63), it is sufficient to verify the conditions of Theorem 5-5.

Continuity: Using the *Hölder's inequality* for sums

$$\sum_{i=1}^{n} |x_i y_i| \le \left(\sum_{i=1}^{n} |x_i|^p\right)^{1/p} \left(\sum_{i=1}^{n} |y_i|^q\right)^{1/q} \qquad 1/p + 1/q = 1 \tag{5-64}$$

we write

$$|B(\Lambda, \bar{\Lambda})| \le M |\Lambda|_{H(\Omega)} |\bar{\Lambda}|_{H(\Omega)} \le M \|\Lambda\|_{H(\Omega)} \|\bar{\Lambda}\|_{H(\Omega)}$$

where $M = \max(\lambda, 2\mu)$. Continuity of $l(\bar{\Lambda})$ is obvious. Note that for incompressible elastic materials M is not finite because $\lambda = \infty$. Hence the present discussion is valid only for compressible materials (Poisson's ratio, $\nu \neq 0.5$).

$H(\Omega)$-ellipticity: Here we make use of the Korn's inequality: For $\Lambda = (u, v) \in H(\Omega)$, and Γ Lipschitz continuous, there exists a constant $C(\Omega)$ such that

$$\|\Lambda\|_{H(\Omega)} \le C(\Omega) \left\{ \int_\Omega \left[2\left(\frac{\partial u}{\partial x}\right)^2 + 2\left(\frac{\partial v}{\partial y}\right)^2 + \frac{1}{2}\left(\frac{\partial u}{\partial y} + \frac{\partial v}{\partial x}\right)^2 \right] dx\, dy \right\}^{1/2} \tag{5-65}$$

Then we have H-ellipticity of $B(\cdot, \cdot)$,

$$B(\Lambda, \Lambda) = \int_\Omega \left\{ \lambda\left(\frac{\partial u}{\partial x} + \frac{\partial v}{\partial y}\right)^2 + \mu\left[\left(\frac{\partial u}{\partial y} + \frac{\partial v}{\partial x}\right)^2 + 2\left(\frac{\partial u}{\partial x}\right)^2 + 2\left(\frac{\partial v}{\partial y}\right)^2\right]\right\} dx\, dy$$

$$\ge [\alpha/C^2(\Omega)] \|\Lambda\|_{H(\Omega)}^2 = \alpha_0 \|\Lambda\|_{H(\Omega)}^2 \tag{5-66}$$

Therefore, the variational problem (5-63) has a unique solution. This completes the example.

The next example deals with the Stokes equations [see Reddy (1978)].

Example 5-10 (Stationary Stokes equations) The Stokes equations are the linear version of the Navier–Stokes equations governing the steady, laminar flow of an incompressible fluid [see Sec. 2-6-3]. The Stokes equations are given by [the linear version of Eq. (2-100)]

$$\left.\begin{aligned} -\mu\nabla^2\mathbf{u} + \text{grad } P - \mathbf{f} &= \mathbf{0} \\ \text{div } \mathbf{u} &= 0 \end{aligned}\right\} \quad \text{in } \Omega \tag{5-67}$$

$$\mathbf{u} = \mathbf{0} \quad \text{on } \Gamma \tag{5-68}$$

where $\mathbf{f} \in \mathbf{L}_2(\Omega)$ is the body force vector, $\mathbf{u} = (u_1, u_2, u_3)$ is the velocity vector, P is the pressure, and μ is the viscosity.

We introduce the following spaces:

$$\mathscr{D} = \{\mathbf{u} \in C_0^\infty(\Omega), \text{ div } \mathbf{u} = 0\}$$

$$\mathbf{V} = \{\mathbf{u} \in H_0^1(\Omega) \times H_0^1(\Omega) \text{: div } \mathbf{u} = 0\}$$

$$Q = \left\{P \in L_2(\Omega)\text{: } \int_\Omega P\, d\mathbf{x} = 0\right\} \tag{5-69}$$

The space V is equipped with the inner product

$$(\mathbf{v}, \mathbf{u})_\mathbf{V} = \sum_{i=1}^N \int_\Omega \text{grad } v_i \cdot \text{grad } u_i \, d\mathbf{x} \tag{5-70}$$

where N is the dimension of the domain $\Omega \subset \mathbb{R}^N$.

The weak formulation of Eqs. (5-67) and (5-68) is obtained using the familiar procedure (i.e., multiply each equation with a test function and integrate by parts to distribute the differentiation to the test function). We obtain for $\mathbf{v} \in \mathscr{D}$,

$$\begin{aligned} 0 &= \int_\Omega \mathbf{v}\cdot(-\mu\nabla^2\mathbf{u} + \text{grad } P - \mathbf{f})\, d\mathbf{x} \\ &= \int_\Omega (\mu \text{ grad } \mathbf{v}\text{: grad } \mathbf{u} - P \text{ div } \mathbf{v} - \mathbf{v}\cdot\mathbf{f})\, d\mathbf{x} - \oint_\Gamma \mathbf{v}\cdot(\mu\hat{\mathbf{n}}\cdot\text{grad } \mathbf{u} - P\hat{\mathbf{n}})\, ds \end{aligned}$$

Since $\mathbf{v} \in \mathscr{D}$ we have div $\mathbf{v} = 0$ and $\mathbf{v} = 0$ on Γ, giving

$$B(\mathbf{v}, \mathbf{u}) = (\mathbf{v}, \mathbf{f})_0 \qquad \text{for every } \mathbf{v} \in \mathscr{D}$$

where

$$\begin{aligned} B(\mathbf{v}, \mathbf{u}) &= \int_\Omega \mu \text{ grad } \mathbf{v}\text{: grad } \mathbf{u}\, d\mathbf{x} \\ &= \sum_{i=1}^N \int_\Omega \mu \text{ grad } v_i \cdot \text{grad } u_i\, d\mathbf{x} \end{aligned} \tag{5-71}$$

Now we have the following weak problem: find $\mathbf{u} \in \mathbf{V}$ such that

$$B(\mathbf{v}, \mathbf{u}) = (\mathbf{v}, \mathbf{f})_0 \tag{5-72}$$

holds for every $\mathbf{v} \in \mathbf{V}$.

The proof that the weak solution of Eq. (5-72) is the classical solution of Eqs. (5-67) and (5-68) follows from the argument [see Temam (1977)]

$$(\mathbf{v}, -\mu\nabla^2\mathbf{u} - \mathbf{f})_0 = \mathbf{0} \qquad \text{for every } \mathbf{v} \in \mathbf{V} \tag{5-73}$$

This *does not* imply that $-\mu\nabla^2\mathbf{u} - \mathbf{f} = \mathbf{0}$ because $\mathbf{v}$ is subjected to the constraint (because $\mathbf{v} \in \mathbf{V}$) div $\mathbf{v} = 0$. Instead, Eq. (5-73) implies that

$$-\mu\nabla^2\mathbf{u} - \mathbf{f} = -\text{grad } P$$

because (necessary and sufficient)

$$(\mathbf{v}, \text{grad } P)_0 = (\text{div } \mathbf{v}, P)_0 = 0 \qquad \text{for every } P \in Q \tag{5-74}$$

The bilinear form in Eq. (5-71) satisfies the conditions of Theorem 5-5 with $H = \mathbf{V}$. The continuity follows from Eq. (5-71) using the Schwarz inequality,

$$\begin{aligned}
|B(\mathbf{v}, \mathbf{u})| &= \left| \sum_{i=1}^{N} \int_\Omega \mu \text{ grad } v_i \cdot \text{grad } u_i \, d\mathbf{x} \right| \\
&\le \mu \left[\sum_{i=1}^{N} \int_\Omega |\text{grad } v_i|^2 \, d\mathbf{x} \right]^{1/2} \left[\sum_{i=1}^{N} \int_\Omega |\text{grad } u_i|^2 \, d\mathbf{x} \right]^{1/2} \\
&= \mu \|\mathbf{v}\|_{\mathbf{V}} \|\mathbf{u}\|_{\mathbf{V}}
\end{aligned}$$

The $\mathbf{V}$-ellipticity of $B(\mathbf{v}, \mathbf{v})$ follows from

$$\begin{aligned}
|B(\mathbf{v}, \mathbf{v})| &= \mu \sum_{i=1}^{N} \int_\Omega \text{grad } v_i \cdot \text{grad } v_i \, d\mathbf{x} \\
&= \mu \|\mathbf{v}\|_{\mathbf{V}} \ge \alpha \|\mathbf{v}\|_{\mathbf{V}}
\end{aligned}$$

for $\alpha \le \mu$. Thus, the variational problem (5-72) has a unique solution in $\mathbf{V}$.

The following regularity result can be established for the Stokes problem [see Ladyzhenskaya (1969)]: If $\mathbf{f} \in \mathbf{H}^s(\Omega)$, $s \ge 0$, the solution $(\mathbf{u}, P)$ of the problem (5-67) and (5-68) has the smoothness properties

$$\begin{gathered}
\mathbf{u} \in \mathbf{H}^{s+2}(\Omega), \text{ grad } P \in H^s(\Omega) \\
\|\mathbf{u}\|_{s+2} + \|\text{grad } P\|_s \le c\|\mathbf{f}\|_s
\end{gathered} \tag{5-75}$$

where $\|\cdot\|_s$ denotes the $\mathbf{H}^s(\Omega)$-norm. This completes the example.

5-4-4 Nonhomogeneous Boundary Conditions

In the previous discussions we assumed that the essential boundary conditions of the given problem are assumed to be zero. This enabled us to identify the solution space H to be a linear vector space which coincides with the space V of admissible variations (or weight functions). When the specified essential boundary conditions are nonhomogeneous, we have $V \neq H$. In this case the Lax–Milgram theorem 5-5 is not valid, and we need to extend the theorem to a more general case in which $V \neq H$.

Here we consider a general boundary-value problem with nonhomogeneous boundary conditions, its variational formulation, and the existence and uniqueness of its solution. We begin with the operator equation (5-33), where

$$A = \sum_{|\alpha|, |\beta| \leq m} (-1)^{|\alpha|} D^{\alpha}(a_{\alpha\beta} D^{\beta})$$

$$\begin{aligned} \Omega &= \text{open bounded domain with a Lipschitz boundary } \Gamma \text{ (i.e.,} \\ &\quad \text{has the outward normal } \mathbf{n} \text{ defined almost everywhere)} \end{aligned} \tag{5-76}$$

$$f \in L_2(\Omega)$$

$$a_{\alpha\beta}(\mathbf{x}) - \text{bounded measurable coefficients}$$

Recall from previous discussions that the weak solution of the equation $Au = f$ of order $2m$ is a function from $H^m(\Omega)$. The essential (or Dirichlet) boundary conditions associated with the operator consist of the given function and its derivatives of orders at most $m - 1$:

$$u = g_0(s) \qquad \frac{\partial u}{\partial n} = g_1(s), \qquad \ldots, \qquad \frac{\partial^{m-1} u}{\partial n^{m-1}} = g_{m-1} \tag{5-77}$$

The natural (or Neumann) boundary conditions consist of derivatives of orders higher than $m - 1$:

$$\frac{\partial^m u}{\partial n^m} = g_m \qquad \frac{\partial^{m+1} u}{\partial n^{m+1}} = g_{m+1}, \qquad \ldots, \qquad \frac{\partial^{2m-1} u}{\partial n^{2m-1}} = g_{2m-1} \tag{5-78}$$

We assume that for the problem at hand, n Dirichlet boundary conditions are specified with $m - n$ Neumann boundary conditions. Thus only n of the essential boundary conditions in Eq. (5-77) and $m - n$ of the natural boundary conditions in Eq. (5-78) are specified. Let us denote the specified essential and natural boundary conditions by

$$\begin{aligned} &B_1 u = g_1 \qquad B_2 u = g_2, \ldots, B_n u = g_n \ (n \leq m) \text{ on } \Gamma_1 \\ &C_1 u = h_1 \qquad C_2 u = h_2, \ldots, C_{m-n} u = h_{m-n} \text{ on } \Gamma_2 \end{aligned} \tag{5-79}$$

where the traces $g_1, g_2, \ldots, g_n$ belong to $L_2(\Gamma)$ and Γ_1 and Γ_2 are disjoint portions such that $\Gamma_1 \cup \Gamma_2 = \Gamma$. We define the space

$$V = \{v : v \in H^m(\Omega), B_i v = 0 \text{ on } \Gamma_1, i = 1, 2, \ldots, n\} \tag{5-80}$$

The function $u \in H^m(\Omega)$ is called the *weak solution* of the boundary-value problem $Au = f$ defined by Eq. (5-76) if

$$u - w \in V$$

$$B(v, u) = \int_\Omega vf \, d\mathbf{x} + \sum_{i=1}^{m-n} \int_{\Gamma_2} B_i v \, h_i \, ds \tag{5-81}$$

where
$$B(v, u) = \sum_{|\alpha|, |\beta| \le m} \int_\Omega a_{\alpha\beta} D^\alpha v D^\beta u \, d\mathbf{x} \tag{5-82}$$

and $w \in H^m(\Omega)$ such that

$$B_i w = g_i \text{ on } \Gamma_1 \qquad i = 1, 2, \ldots, n \tag{5-83}$$

Recall that a bilinear form $B(v, u)$ is called *V-elliptic* if there exists a constant $\alpha > 0$ such that

$$B(v, v) \ge \alpha \|v\|_V^2 \qquad \text{for every } v \in V \tag{5-84}$$

Note that, in general, u does not belong to V.

The existence and uniqueness of solution to the weak problem (5-81) are not governed by Theorem 5-5 because the bilinear form $B(v, u)$ is not defined on $H \times H$. It is defined on $V \times H$, where V is the space of admissible variations and H is the solution space. The following theorem provides sufficient conditions for the existence and uniqueness of the weak solution of a boundary-value problem.

Theorem 5-6 The weak form of a boundary-value problem defined by Eq. (5-81) with the data in Eqs. (5-76), (5-79), and (5-80) has a unique solution $u \in H^m(\Omega)$ if the bilinear form $B(v, u)$ is V-elliptic. Further, there exists a positive constant c, independent of f, w, and h_i, such that

$$\|u\|_m \le \beta \left[\|f\|_0 + \|w\|_m + \sum_{i=1}^{m-n} \|h_i\|_{L_2(\Gamma)} \right] \tag{5-85}$$

holds.

PROOF In proving the existence we shall make use of Theorem 5-5, with $H = V$. Note that $V \subset H^m(\Omega)$ and the norm in V is the same as that in $H^m(\Omega)$, that is, $\| \ \|_V = \| \ \|_m$. From the continuity and V-ellipticity of $B(v, u)$ it follows that

$$|B(v, z)| \le M\|v\|_V \|z\|_V \tag{5-86}$$

$$B(v, v) \ge \alpha \|v\|_V^2$$

which are similar to those in the hypothesis of Theorem 5-5. Consider the linear functional in V

$$l(v) = \int_\Omega fv \, d\mathbf{x} + \sum_{i=1}^{m-n} \int_{\Gamma_2} B_i v \, h_i \, ds - B(v, w) \tag{5-87}$$

where $f \in L_2(\Omega)$ and $h_i \in L_2(\Gamma)$ and $w \in H^m(\Omega)$ are given functions in Eqs. (5-76), (5-79), and (5-83). If we can prove that $l(v)$ is bounded in V, then it follows from the Lax–Milgram theorem 5-5 that there exists exactly one $z \in V$ such that

$$B(v, z) = l(v) \qquad \text{for every } v \in V \tag{5-88}$$

which is the same as ($z = u - w$) the weak problem in Eq. (5-81).

Thus, we need to prove that $l(v)$ in Eq. (5-87) is bounded in V. We prove the boundedness of each of the three terms in Eq. (5-87). Consider the first of the three terms. By the Schwarz inequality, we have

$$\left| \int_\Omega vf \, d\mathbf{x} \right| \leq \|v\|_0 \, \|f\|_0 = c\|v\|_0 \leq c\|v\|_m$$

Next consider the third term,

$$|-B(v, w)| \leq K\|v\|_m \, \|w\|_m = K\|v\|_V \, \|w\|_V$$

We now turn to the second term. Recall from Eqs. (5-77) and (5-79) that B_i involves the normal derivative of order equal to or greater than n:

$$B_i = \frac{\partial^i}{\partial n^i} \qquad i = 1, 2, \ldots, n$$

By the Extension Theorem 4-2, there exists a bounded operator T which assigns to every function $v(\mathbf{x}) \in H^1(\Omega)$ its trace $v(s) \in L_2(\Gamma)$

$$\|v(s)\|_{L_2(\Gamma)} \leq \|T\| \, \|v(\mathbf{x})\|_1$$

where $\|T\|$ is the norm of the operator, which depends only on the domain Ω and its boundary Γ. In the present case we have

$$\left(\int_{\Gamma_2} v^2(s) \, ds \right)^{1/2} \leq \|T\| \, \|v\|_1$$

$$\left\| \frac{\partial v}{\partial x_j} \right\|_{L_2(\Omega)} \leq \|T\| \left\| \frac{\partial v}{\partial x_j} \right\|_1 \leq \|T\| \, \|v\|_2 \qquad j = 1, 2, \ldots, n$$

Since

$$\frac{\partial v}{\partial n} = \sum_{j=1}^{N} \frac{\partial v}{\partial x_j} n_j$$

and n_j are bounded functions on Γ, we have for every $v \in H^2(\Omega)$

$$\left\| \frac{\partial v}{\partial n} \right\|_{L_2(\Gamma)} \leq c_1 \|v\|_2 = c_1 \|v\|_V \tag{5-89}$$

By similar arguments, we obtain

$$\left[\int_{\Gamma_2} \left(\frac{\partial^{m-1} v}{\partial n^{m-1}} \right)^2 ds \right]^{1/2} \leq c_{m-1} \|v\|_m = c_{m-1} \|v\|_V \tag{5-90}$$

Letting

$$c = \max_i \{c_i\} \qquad \left[\int_{\Gamma_2} h_i^2 \, ds\right]^{1/2} = \|h_i\|_{L_2(\Gamma_2)}$$

we obtain

$$\left|\int_{\Gamma_2} B_i v \, h_i \, ds\right| \le c \sum_{i=1}^{m-n} \|h_i\|_{L_2(\Gamma_2)} \|v\|_V \tag{5-91}$$

Thus, the linear form $l(v)$ is bounded

$$|l(v)| \le M \|v\|_V \tag{5-92a}$$

where

$$M = \|f\|_0 + K\|w\|_m + c \sum_{i=1}^{m-n} \|h_i\|_{L_2(\Gamma)} \tag{5-92b}$$

The uniqueness of the solution can be proved by contradiction. Let $u_1 - w \in V$ and $u_2 - w \in V$ be the solutions of Eq. (5-81). We have

$$B(v, u_1) = \int_\Omega vf \, d\mathbf{x} + \sum_{i=1}^{m-n} \int_{\Gamma_2} B_i v \, h_i \, ds \qquad \text{for every } v \in V$$

$$B(v, u_2) = \int_\Omega vf \, d\mathbf{x} + \sum_{i=1}^{m-n} \int_{\Gamma_2} B_i v \, h_i \, ds \qquad \text{for every } v \in V$$

Subtracting the first equation from the second, we obtain

$$B(v, u_2 - u_1) = 0 \qquad \text{for every } v \in V$$

Note that $u_2 - u_1 = (u_2 - w) - (u_1 - w) \in V$. Hence, we can take $v = u_2 - u_1 \in V$ and obtain

$$0 = B(u_2 - u_1, u_2 - u_1) \ge \alpha \|u_2 - u_1\|_V^2$$

This implies that

$$\|u_2 - u_1\|_m^2 = 0$$

from which we obtain that $u_2 = u_1$ in $H^m(\Omega)$ which was to be proved.

To prove the inequality (5-85), we first note from Eq. (5-92a) that

$$\|l\|_V \le M$$

and from Eq. (5-45) (with u_0 replaced by $u - w$)

$$\|u - w\|_V \le \frac{\|l\|}{\alpha} \le \frac{M}{\alpha}$$

Next consider

$$\|u\|_V = \|(u - w) + w\|_V \le \|u - w\|_V + \|w\|_V$$

$$\le \frac{M}{\alpha} + \|w\|_m$$

Substituting for M from Eq. (5-92b), we obtain

$$\|u\|_V \leq \frac{1}{\alpha}\left(\|f\|_0 + K\|w\|_m + c\sum_{i=1}^{m-n}\|h_i\|_{L_2(\Gamma)}\right) + \|w\|_m$$

$$\leq \beta\left[\|f\|_0 + \|w\|_m + \sum_{i=1}^{m-n}\|h_i\|_{L_2(\Gamma)}\right]$$

where
$$\beta = \max\left(\frac{1}{\alpha}, \frac{K}{\alpha} + 1, \frac{c}{\alpha}\right) \tag{5-93}$$

Note that the function $w(\mathbf{x})$ characterizes the function that satisfies the essential boundary conditions of the problem and belongs to $H^m(\Omega)$, and the functions $h_i(s)$ characterize the data in the natural boundary conditions. A slight change in the data (i.e., in f, w, and h_i) causes a corresponding slight change in the weak solution. The change in the weak solution is bounded by the changes in the data according to

$$\|u - \tilde{u}\|_m \leq \beta\left[\|f - \tilde{f}\|_0 + \|w - \tilde{w}\|_m + \sum_{i=1}^{m-n}\|h_i - \tilde{h}_i\|_{L_2(\Gamma_2)}\right] \tag{5-94}$$

where $\tilde{u}, \tilde{f}, \tilde{w}$, and $\tilde{h}_i$ denote the changed solution and data.

Inequalities of type (5-94) that show the dependence of the weak solution on the slight changes in the coefficients $a_{\alpha\beta}(\mathbf{x})$ and/or the domain $\bar{\Omega}$ are of considerable interest in the numerical analysis of boundary-value problems. We shall return to these questions later in this book.

We now consider some examples of boundary-value problems with nonhomogeneous boundary conditions and their weak formulations.

Example 5-11 Consider the second-order equation

$$-\frac{\partial}{\partial x_1}\left(a_{11}\frac{\partial u}{\partial x_1}\right) - \frac{\partial}{\partial x_2}\left(a_{22}\frac{\partial u}{\partial x_2}\right) - f = 0 \text{ in } \Omega \qquad (a_{11}, a_{22} > 0) \tag{5-95}$$

subjected to various types of boundary conditions.

1. *The Dirichlet problem.* We consider Eq. (5-95) with the boundary condition

$$u = g(s) \text{ on } \Gamma \tag{5-96}$$

For $g(s) \in C(\Gamma)$ and a_{11}, a_{22}, $f \in C(\bar{\Omega})$, the classical solution u belongs to $C^2(\bar{\Omega})$. The operator of this equation is a special case of A in Eq. (5-76). We have $N = 2$ (two-dimensional problem), $\boldsymbol{\alpha} = (\alpha_1, \alpha_2)$, $m = 1$ (second-order equation) and

$$a_{(0,0)(0,0)} = 0 \qquad a_{(1,0)(1,0)} = a_{11} \qquad a_{(0,1)(0,1)} = a_{22}$$

$$a_{(1,0)(0,1)} \equiv a_{12} = 0 \qquad a_{(0,1)(1,0)} \equiv a_{21} = 0, \text{ etc.}$$

and the summation becomes

$$\sum_{|\alpha|,\,|\beta|\leq 1} D^{\alpha}(a_{\alpha\beta}D^{\beta}u) = \frac{\partial}{\partial x_1}\left[a_{(1,\,0)(0,\,0)}u + a_{(1,\,0)(1,\,0)}\frac{\partial u}{\partial x_1} + a_{(1,\,0)(0,\,1)}\frac{\partial u}{\partial x_2}\right]$$

$$+\frac{\partial}{\partial x_2}\left[a_{(0,\,1)(0,\,0)}u + a_{(0,\,1)(1,\,0)}\frac{\partial u}{\partial x_1} + a_{(0,\,1)(0,\,1)}\frac{\partial u}{\partial x_2}\right]$$

$$+ a_{(0,\,0)(0,\,0)}u + a_{(0,\,0)(1,\,0)}\frac{\partial u}{\partial x_1} + a_{(0,\,0)(0,\,1)}\frac{\partial u}{\partial x_2}$$

$$= \frac{\partial}{\partial x_1}\left(a_{11}\frac{\partial u}{\partial x_1}\right) + \frac{\partial}{\partial x_2}\left(a_{22}\frac{\partial u}{\partial x_2}\right)$$

We now transform the boundary-value problem (5-95) and (5-96) to an equivalent weak form, as described in Section 4-2. Take an arbitrary $v \in V \equiv H_0^1(\Omega)$, and multiply Eq. (5.95) and integrate the result over the domain Ω. We obtain

$$0 = \int_{\Omega} v\left[-\frac{\partial}{\partial x_1}\left(a_{11}\frac{\partial u}{\partial x_1}\right) - \frac{\partial}{\partial x_2}\left(a_{22}\frac{\partial u}{\partial x_2}\right) - f\right] dx_1\, dx_2$$

Since $V \in H_0^1(\Omega)$, we know that $\partial v/\partial x_1, \partial v/\partial x_2 \in L_2(\Omega)$. Using the divergence theorem we can write

$$0 = \int_{\Omega}\left(a_{11}\frac{\partial v}{\partial x_1}\frac{\partial u}{\partial x_1} + a_{22}\frac{\partial v}{\partial x_2}\frac{\partial u}{\partial x_2} - vf\right) dx_1\, dx_2$$

$$-\oint_{\Gamma} v\left(a_{11}\frac{\partial u}{\partial x_1}n_1 + a_{22}\frac{\partial u}{\partial x_2}n_2\right) ds \tag{5-97}$$

The traces $(\partial u/\partial x_1)(s)$ and $(\partial u/\partial x_2)(s)$ belong to $L_2(\Gamma)$. Since $v \in H_0^1(\Omega)$, we have $v = 0$ on Γ. Consequently, we have

$$\int_{\Omega}\left(a_{11}\frac{\partial v}{\partial x_1}\frac{\partial u}{\partial x_1} + a_{22}\frac{\partial v}{\partial x_2}\frac{\partial u}{\partial x_2}\right) dx_1\, dx_2 = \int_{\Omega} vf\, dx_1\, dx_2 \tag{5-98}$$

To begin with we assumed that a_{11}, a_{22}, $f \in C(\bar{\Omega})$ and $u \in C^2(\Omega)$. But Eq. (5-98) also makes sense under weaker continuity conditions: a_{11}, a_{22}, $f \in L_2(\Omega)$ and $u \in H^1(\Omega)$. Hence the name weak form is appropriate for Eq. (5-98). We have the following weak formulation of Eqs. (5-95) and (5-96).

Let $g(s)$ be the trace of the function $w \in H^1(\Omega)$ such that $w = g(s)$ on Γ, and let $f \in L_2(\Omega)$. Define

$$V = \{v : v \in H^1(\Omega),\ v = 0 \text{ on } \Gamma\} \equiv H_0^1(\Omega)$$

Then the function $u \in H^1(\Omega)$ is called the weak solution of the problem (5-95) and (5-96) if $u - w \in V$ and Eq. (5-98) holds for every $v \in V$.

The bilinear form

$$B(v, u) = \int_{\Omega}\left(a_{11}\frac{\partial v}{\partial x_1}\frac{\partial u}{\partial x_1} + a_{22}\frac{\partial v}{\partial x_2}\frac{\partial u}{\partial x_2}\right) dx_1\, dx_2$$

is V-elliptic because

$$B(v, v) = \int_\Omega \left(a_{11} \left| \frac{\partial v}{\partial x_1} \right|^2 + a_{22} \left| \frac{\partial v}{\partial x_2} \right|^2 \right) dx_1\, dx_2$$

$$\geq \alpha \|v\|_1^2 \qquad \text{for all } v \in H_0^1(\Omega)$$

with $\alpha = c_2 \min(a_{11}, a_{22})$, and $c_2 = c_2(\Omega)$ is as defined in Eq. (4-25). The linear form is obviously bounded. Therefore, by Theorem 5-6 a unique solution $u - w \in H_0^1(\Omega)$ to Eq. (5-98) exists.

2. *The mixed problem.* Next suppose that we have the mixed boundary conditions

$$u = g(s) \text{ on } \Gamma_1$$

$$\left(a_{11} \frac{\partial u}{\partial x_1} n_1 + a_{22} \frac{\partial u}{\partial x_2} n_2 \right) = h(s) \text{ on } \Gamma_2 \tag{5-99}$$

in place of Eq. (5-96). The portions Γ_1 and Γ_2 of the boundary Γ are disjoint but their union is equal to Γ. For this case we define

$$V = \{v\colon\ v \in H^1(\Omega),\ v = 0 \text{ on } \Gamma_1\}$$

The weak form associated with Eqs. (5-95) and (5-99) can be obtained (noting that $v = 0$ on Γ_1 and $v \neq 0$ on Γ_2) from Eq. (5-97): find $u \in H^1(\Omega)$ such that $u - w \in V$ and

$$\int_\Omega \left(a_{11} \frac{\partial v}{\partial x_1} \frac{\partial u}{\partial x_1} + a_{22} \frac{\partial v}{\partial x_2} \frac{\partial u}{\partial x_2} \right) dx_1\, dx_2 = \int_\Omega vf\, dx_1\, dx_2 + \int_{\Gamma_2} vh\, ds \tag{5-100}$$

for all $v \in V$. Here w is the function in $H^1(\Omega)$ that satisfies the essential boundary condition $w = g(s)$ on Γ_1. The V-ellipticity of the bilinear form can be proved using the inequality (4-22*a*). We have

$$B(v, v) = \int_\Omega \left[a_{11} \left(\frac{\partial v}{\partial x_1} \right)^2 + a_{22} \left(\frac{\partial v}{\partial x_2} \right)^2 \right] dx_1\, dx_2$$

$$\geq c_0 \int_\Omega \left[\left(\frac{\partial v}{\partial x_1} \right)^2 + \left(\frac{\partial v}{\partial x_2} \right)^2 \right] dx_1\, dx_2$$

where $c_0 = \min\{a_{11}(\mathbf{x}), a_{22}(\mathbf{x})\}$. Using Eq. (4-22*a*) we have

$$B(v, v) \geq c_0 c_2 \|v\|_1^2 = \alpha \|v\|_V^2$$

Thus, $B(v, v)$ is V-elliptic, and hence by Theorem 5-6, the weak form (5-100) of the mixed boundary-value problem (5-95) and (5-98) has a unique solution in V.

3. *The Neumann problem.* Now consider the Neumann problem for the Poisson equation (5-95):

$$\left(a_{11} \frac{\partial u}{\partial x_1} n_1 + a_{22} \frac{\partial u}{\partial x_2} n_2 \right) = h(s) \text{ on } \Gamma \tag{5-101}$$

The space V is given by

$$V = H^1(\Omega)$$

and the weak problem involves finding $u \in H^1(\Omega)$ such that

$$\int_\Omega \left(a_{11} \frac{\partial v}{\partial x_1} \frac{\partial u}{\partial x_1} + a_{22} \frac{\partial v}{\partial x_2} \frac{\partial u}{\partial x_2} \right) dx_1 \, dx_2 = \int_\Omega vf \, dx_1 \, dx_2 + \oint_\Gamma vh(s) \, ds \quad \text{(5-102)}$$

for every $v \in H^1(\Omega)$. The bilinear form is not elliptic on $H^1(\Omega)$, because for $v =$ constant, we have $B(v, v) = 0$ while $v \in V$. In other words, the inequality (5-84) cannot be satisfied for any nonzero constant element in V. Therefore, Theorem 5-6 does not apply. We shall consider the question of existence and uniqueness of solutions to Neumann boundary-value problems in the next section (Sec. 5-4-5).

4. *The Newton problem.* The last case of this example is concerned with the *Newton problem* for the Poisson equation (5-95) with the boundary condition

$$\left(a_{11} \frac{\partial u}{\partial x_1} n_1 + a_{22} \frac{\partial u}{\partial x_2} n_2 + a_{00} u \right) = 0 \text{ on } \Gamma \quad \text{(5-103)}$$

In this case we have $V = H^1(\Omega)$, and the weak problem involves seeking $u \in H^1(\Omega)$ such that

$$\int_\Omega \left(a_{11} \frac{\partial v}{\partial x_1} \frac{\partial u}{\partial x_1} + a_{22} \frac{\partial v}{\partial x_2} \frac{\partial u}{\partial x_2} \right) dx_1 \, dx_2 + \oint_\Gamma a_{00} vu \, ds = \int_\Omega vf \, dx_1 \, dx_2 \quad \text{(5-104)}$$

for all $v \in H^1(\Omega)$. The V-ellipticity of the bilinear form can be established as follows.

Let $a_{00}(s) \geq c_0 =$ constant > 0 on Γ, and let $c = \min(a_{11}, a_{22}, c_0)$. Then

$$B(v, v) = \int_\Omega \left[a_{11} \left(\frac{\partial v}{\partial x_1} \right)^2 + a_{22} \left(\frac{\partial v}{\partial x_2} \right)^2 \right] dx_1 \, dx_2 + \int_\Gamma a_{00} v^2 \, ds$$

$$\geq c \left\{ \int_\Omega \left[\left(\frac{\partial v}{\partial x_1} \right)^2 + \left(\frac{\partial v}{\partial x_2} \right)^2 \right] dx_1 \, dx_2 + \oint_\Gamma v^2 \, ds \right\}$$

Using the Friedrichs inequality (4-22*a*), we obtain

$$B(v, v) \geq \frac{c}{c_2} \|v\|_1^2 = \alpha \|v\|_V$$

Thus, $B(v, v)$ is V-elliptic and hence the weak form (5-104) associated with the Newton problem has a unique solution. This completes the example.

The next example deals with the existence of solutions of the biharmonic equation

$$\frac{\partial^2}{\partial x_1^2}\left(\frac{\partial^2 w}{\partial x_1^2} + \nu \frac{\partial^2 w}{\partial x_2^2}\right) + 2 \frac{\partial^2}{\partial x_1 \, \partial x_2}\left[(1-\nu)\frac{\partial^2 w}{\partial x_1 \, \partial x_2}\right] + \frac{\partial^2}{\partial x_2^2}\left(\nu \frac{\partial^2 w}{\partial x_1^2} + \frac{\partial^2 w}{\partial x_2^2}\right) = \frac{q}{D} \tag{5-105}$$

which arises in connection with the bending of thin, elastic, isotropic plates. Here D denotes the flexural rigidity ($D > 0$), $\nu(0 \le \nu \le 0.5)$ the Poisson ratio, and q the transverse distributed load. When simplified, Eq. (5-105) reduces to $\nabla^4 w = q/D$. The particular form in Eq. (5-105) allows us to include the Poisson effect in the analysis. The biharmonic equation $\nu\nabla^4\psi = f$ also arises in connection with the stationary Stokes problem, where ψ denotes the *stream function* that is selected to satisfy the continuity equation $\partial u_1/\partial x_1 + \partial u_2/\partial x_2 = 0$ by setting $\partial\psi/\partial x_2 = u_1$ and $\partial\psi/\partial x_1 = -u_2$, ν is the kinematic viscosity, and f is a function related to the body forces of the fluid medium.

Example 5-12 First we consider the boundary conditions

$$B_1 w \equiv w = g(s) \text{ on } \Gamma$$
$$B_2 w \equiv \nu\nabla^2 w + (1-\nu)\frac{\partial^2 w}{\partial n^2} = h(s) \text{ on } \Gamma \tag{5-106}$$

These boundary conditions correspond to a plate with specified deflection of the supports and specified moments along the boundary. The space V is given by

$$V = \{v\colon\ v \in H^2(\Omega),\ v = 0 \text{ on } \Gamma\} \tag{5-107}$$

For $h(s) \in L_2(\Gamma)$ and $f \equiv q/D \in L_2(\Omega)$, the weak solution of Eqs. (5-105) and (5-106) is the solution $w \in H^2(\Omega)$ such that

$$w - w_0 \in V$$

and

$$B(v, w) = \int_\Omega vf \, d\mathbf{x} + \oint_\Gamma \frac{\partial v}{\partial n} h(s) \, ds \tag{5-108}$$

holds for every $v \in V$. Here w_0 denotes a function in $H^2(\Omega)$ such that $w_0 = g(s)$ on Γ, and $B(v, w)$ is the bilinear form

$$B(v, w) = \int_\Omega \left[\left(\frac{\partial^2 v}{\partial x_1^2} + \nu\frac{\partial^2 v}{\partial x_2^2}\right)\frac{\partial^2 w}{\partial x_1^2} + 2(1-\nu)\frac{\partial^2 v}{\partial x_1 \, \partial x_2}\frac{\partial^2 w}{\partial x_1 \, \partial x_2} + \left(\frac{\partial^2 v}{\partial x_2^2} + \nu\frac{\partial^2 v}{\partial x_1^2}\right)\frac{\partial^2 w}{\partial x_2^2}\right] d\mathbf{x} \tag{5-109}$$

The bilinear form $B(v, w)$ is V-elliptic for $0 \leq \nu < 1$:

$$B(v, v) = \int_\Omega \left\{ \nu \left(\frac{\partial^2 v}{\partial x_1^2} + \frac{\partial^2 v}{\partial x_2^2} \right)^2 + (1 - \nu) \left[\left(\frac{\partial^2 v}{\partial x_1^2} \right)^2 + 2 \left(\frac{\partial^2 v}{\partial x_1 \, \partial x_2} \right)^2 + \left(\frac{\partial^2 v}{\partial x_2^2} \right)^2 \right] \right\} d\mathbf{x}$$

$$\geq \frac{(1 - \nu)}{c_3} \|v\|_2^2 = \alpha \|v\|_V^2$$

where inequality (4-22*b*) (with $v = 0$ on Γ) is used in arriving at the last step. Thus $B(v, w)$ is V-elliptic for $\nu < 1$, and hence by Theorem 5-6 the weak problem (5-108) has a unique solution w in $H^2(\Omega)$.

The expression $B(v, v)$ denotes two times the elastic strain energy due to bending of the plate. We have proved that the strain energy of the plate is positive definite.

Next we consider the boundary conditions

$$w = g_1(s) \text{ on } \Gamma \qquad \frac{\partial w}{\partial n} = g_2(s) \text{ on } \Gamma \tag{5-110}$$

which correspond to a plate with specified deflections and slopes of the supported edges. For $g_1 = g_2 = 0$, the boundary conditions correspond to a clamped plate. In this case, the space V is given by

$$V = \left\{ v\colon v \in H^2(\Omega),\ v = 0 \text{ and } \frac{\partial v}{\partial n} = 0 \text{ on } \Gamma \right\} \equiv H_0^2(\Omega) \tag{5-111}$$

The weak form is given by Eq. (5-108) with the boundary integral equal to zero (because $\partial v/\partial n = 0$ for $v \in V$). The bilinear form $B(v, w)$ is again V-elliptic and hence the weak problem associated with Eqs. (5-105) and (5-110) has a unique solution (in fact Theorem 5-5 applies for this case). For the regularity results for this case, see Kondratev (1967).

The Neumann problem for Eq. (5-105) involves specifying

$$B_1 w = h_1(s) \text{ on } \Gamma \qquad B_2 w = h_2(s) \text{ on } \Gamma \tag{5-112}$$

where B_2 is defined by Eq. (5-106), and B_1 is defined by

$$B_1 w \equiv -\frac{\partial}{\partial n}(\nabla^2 w) + \frac{\partial M_s}{\partial s}$$
$$M_s = (1 - \nu)\left[\frac{\partial^2 w}{\partial x_1^2} n_1 n_2 - \frac{\partial^2 w}{\partial x_1 \, \partial x_2}(n_1^2 - n_2^2) - \frac{\partial^2 w}{\partial x_2^2} n_1 n_2 \right] \tag{5-113}$$

For this case we have $V = H^2(\Omega)$, and the V-ellipticity of $B(v, w)$ *cannot* be established. This completes the example.

5-4-5 The Neumann Boundary-Value Problems

As discussed in the last two examples, the existence and uniqueness results for the Neumann problems for the Poisson equation and the biharmonic equation are not governed by Theorems 5-5 and 5-6. In general, the problems are not solvable (unless the equations have terms that make the bilinear forms positive-definite; see Probs. 5-2-5 to 5-2-7). To see this, assume that $u(\mathbf{x})$ is the solution of Eq. (5-102) in $H^1(\Omega)$ for *any* arbitrary element $v \in V = H^1(\Omega)$. For $v = c$, where c is an arbitrary real number, we have

$$B(c, u) = \int_\Omega cf \, d\mathbf{x} + \oint_\Gamma ch \, ds$$

Since $B(c, u) = 0$ and c is arbitrary (but $c \neq 0$), we have

$$\int_\Omega f \, d\mathbf{x} + \oint_\Gamma h \, ds = 0 \tag{5-114}$$

Thus, if $u \in H^1(\Omega)$ is the solution of Eq. (5-102) then the data (f, h) should be such that Eq. (5-114) is necessarily satisfied. It also follows from Eqs. (5-102) and (5-114) that if $u \in H^1(\Omega)$ is the solution of Eq. (5-102), then $u(\mathbf{x}) + c$, where c is an arbitrary nonzero constant, is also a solution. Thus, if Eq. (5-102) is solvable, then there are an infinite number of solutions.

The difficulty (i.e., nonuniqueness of solutions of a Neumann problem) can be overcome by seeking solutions in an appropriate space. Here we describe the approach for the Neumann problem for the Poisson equation. The discussion is applicable almost immediately to a more general case.

Solvability conditions for second-order equations From the above discussion it is clear that the weak form (5-102) of the Neumann problem has a solution only if Eq. (5-114) is satisfied. Even then the solution can be determined only within an arbitrary constant. In the following we show that Eq. (5-114) is not only necessary but it is also sufficient for the existence of a weak solution to the Neumann problem, and that the solution is unique in the orthogonal complement of the space of functions constant in Ω with respect to $H^1(\Omega)$.

Let $\mathscr{P}_0$ be the linear subspace of $H^1(\Omega)$ whose elements are all functions constant in Ω. The space $\mathscr{P}_0$ contains elements

$$u_1(\mathbf{x}) = k_1, \; u_2(\mathbf{x}) = k_2, \ldots$$

It can be shown that $\mathscr{P}_0$ is indeed complete [hence it qualifies as a subspace of $H^1(\Omega)$]. The set Q of all elements $w \in H^1(\Omega)$ orthogonal to the subspace is itself a linear subspace. Since $Q \perp \mathscr{P}_0$ and $Q \cap \mathscr{P}_0 = \{0\}$, from Theorem 3-11 we have

$$H^1(\Omega) = Q \oplus \mathscr{P}_0 \tag{5-115}$$

and every element $v \in H^1(\Omega)$ can be uniquely decomposed as $v(\mathbf{x}) = v_1(\mathbf{x}) + v_2(\mathbf{x})$, $v_1(\mathbf{x}) \in Q$ and $v_2(\mathbf{x}) \in \mathscr{P}_0$. The space Q can be characterized by

$$Q = \left\{ v\colon\ v \in H^1(\Omega),\ \int_\Omega v(\mathbf{x})\, d\mathbf{x} = 0 \right\} \tag{5-116}$$

Now consider the weak problem of finding $u \in Q$ such that

$$\int_\Omega \left(a_{11} \frac{\partial v}{\partial x_1} \frac{\partial u}{\partial x_1} + a_{22} \frac{\partial v}{\partial x_2} \frac{\partial u}{\partial x_2} \right) d\mathbf{x} = \int_\Omega vf\, d\mathbf{x} + \oint_\Gamma vh\, ds \tag{5-117}$$

for all $v \in Q$. We show that the bilinear form in Eq. (5-117) is Q-elliptic so that by the Lax–Milgram theorem 5-5 the weak solution $u(\mathbf{x})$ exists in Q and is unique. Using the Poincaré inequality (4-23*b*), we have for every $v \in Q$

$$\begin{aligned} B(v, v) &= \int_\Omega \left[a_{11} \left| \frac{\partial v}{\partial x_1} \right|^2 + a_{22} \left| \frac{\partial v}{\partial x_2} \right|^2 \right] d\mathbf{x} \\ &\geq c_0 \int_\Omega \left(\left| \frac{\partial v}{\partial x_1} \right|^2 + \left| \frac{\partial v}{\partial x_2} \right|^2 \right) d\mathbf{x} \\ &\geq \frac{c_0}{c_3} \|v\|_1^2 = c\|v\|_Q^2 \end{aligned} \tag{5-118}$$

where $c_0 = \min(a_{11}, a_{22})$. Thus $B(v, v)$ is Q-elliptic.

Next we prove that the necessary and sufficient condition for the weak problem (5-117) to have a solution is that condition (5-114) is satisfied. The necessity was already established. To show the sufficiency assume that Eq. (5-114) is satisfied. Replacing v by $w - k$ in Eq. (5-117), where $w \in H^1(\Omega)$ and $k \in \mathscr{P}_0$, we obtain

$$\begin{aligned} \int_\Omega \left(a_{11} \frac{\partial w}{\partial x_1} \frac{\partial u}{\partial x_1} + a_{22} \frac{\partial w}{\partial x_2} \frac{\partial u}{\partial x_2} \right) d\mathbf{x} &= \int_\Omega wf\, d\mathbf{x} + \oint_\Gamma wh\, ds \\ &\quad - k\left[\int_\Omega f\, d\mathbf{x} + \oint_\Gamma h\, ds \right] \\ &= \int_\Omega wf\, d\mathbf{x} + \oint_\Gamma wh\, ds \end{aligned} \tag{5-119}$$

for all $w \in H^1(\Omega)$. Since $w \in H^1(\Omega)$ is arbitrary, Eq. (5-119) holds for $w = v \in Q$, giving Eq. (5-117). Equation (5-114) is known as the *solvability* or *compatibility condition.*

We summarize the above discussion in the following theorem.

Theorem 5-7 The necessary and sufficient condition for the existence of a solution to the weak problem (5-102) is that the compatibility condition

$$\int_\Omega f\, d\mathbf{x} + \oint_\Gamma h\, ds = 0 \tag{5-120}$$

is satisfied. If this condition is satisfied, then there exist an infinite number of weak solutions in $H^1(\Omega)$, each differing from the others by a constant.

By requiring $u \in Q$, where Q is defined by Eq. (5-116), the weak solution is uniquely determined by solving Eq. (5-117). Also, the following estimate holds:

$$\|u\|_Q = \|u\|_1 \le c(\|f\|_0 + \|h\|_{L_2(\Gamma)}) \tag{5-121}$$

Solvability conditions for higher-order equations A generalization of the results presented in Theorem 5-7 to higher-order equations is straightforward. Consider the problem of finding the weak solution $u \in H^m(\Omega)$ (of a $2m$th order partial differential equation with Neumann boundary conditions) such that

$$B(v, v) = \int_\Omega vf\, dx + \sum_{i=0}^{m-1} \oint_\Gamma \frac{\partial^i v}{\partial n^i} h_{i+1}\, ds \tag{5-122}$$

for all $v \in H^m(\Omega)$. Here $B(v, v)$ is a continuous bilinear form associated with the $2m$th order differential operator:

$$B(v, u) = \sum_{|\alpha|,\, |\beta| = m} \int_\Omega a_{\alpha\beta} D^\alpha v D^\alpha u\, d\mathbf{x} \tag{5-123}$$

We introduce the following new inner product in $H^m(\Omega)$:

$$((v, u))_m = \int_\Omega vu\, d\mathbf{x} + \sum_{|\alpha| = m} \int_\Omega D^\alpha v D^\alpha u\, d\mathbf{x} \tag{5-124}$$

This inner product differs from the usual inner product in $H^m(\Omega)$ in that the derivatives of only the highest order are included in (5-124). We shall denote the space of functions from $H^m(\Omega)$ that are equipped with the inner product (5-124) by the symbol $\bar{H}^m(\Omega)$. The natural norm induced by the inner product in Eq. (5-124) is given by

$$|||u|||_m^2 = \|u\|_0^2 + |u|_m^2 \tag{5-125}$$

where $|\cdot|_m$ denotes the seminorm in $H^m(\Omega)$. It can be shown [see Necas (1967)] that the norm $|||\cdot|||_m$ is equivalent to the norm $\|\ \|_m$,

$$\|u\|_m^2 \le \gamma |||u|||_m^2 \le \|u\|_m^2 \qquad \text{for } u \in H^m(\Omega) \tag{5-126}$$

Therefore, the completeness of the space $\bar{H}^m(\Omega)$ follows from the completeness of the space $H^m(\Omega)$.

Let $\mathscr{P}_{m-1}$ be the linear space of polynomials of degree less than or equal to $(m - 1)$. We use in $\mathscr{P}_{m-1}$ the same norm as in $\bar{H}^m(\Omega)$. The space $\mathscr{P}_{m-1}$ is a complete space for $m \ge 1$, and hence it is a linear subspace of $\bar{H}^m(\Omega)$.

Next consider the linear subspace $\bar{Q}$ of those functions from $\bar{H}^m(\Omega)$ that are orthogonal, with respect to the inner product in Eq. (5-124), to the space $\mathscr{P}_{m-1}$. Hence $\bar{Q} \perp \mathscr{P}_{m-1}$ and

$$\bar{H}^m(\Omega) = \bar{Q} \oplus \mathscr{P}_{m-1} \tag{5-127}$$

For $p \in \mathscr{P}_{m-1}$ we have $D^\alpha p = 0$ for $|\alpha| = m$. Hence the condition

$$((p, v))_m = 0$$

implies, in view of Eq. (5-124), that $\bar{Q}$ is orthogonal to $\mathscr{P}_{m-1}$ with respect to the L_2-inner product,

$$(p, v)_0 = 0$$

for every $v \in \bar{H}^m(\Omega)$. Thus, $v \in \bar{H}^m(\Omega)$ belongs to $\bar{Q}$ if and only if v satisfies the equation

$$\int_\Omega pv \, d\mathbf{x} = 0 \qquad \text{for every } p \in \mathscr{P}_{m-1} \tag{5-128}$$

Returning to the question of the solvability of the problem (5-122), we note that, for $v, u \in H^m(\Omega)$ and $p \in \mathscr{P}_{m-1}$, we have

$$\begin{aligned} B(v + p, u) &= B(v, u) \\ B(v, u + p) &= B(v, u) \end{aligned} \tag{5-129}$$

Then if the weak solution exists, the condition

$$\int_\Omega pf \, d\mathbf{x} + \sum_{i=0}^{m-1} \oint_\Gamma \frac{\partial^i p}{\partial n^i} h_{i+1} \, ds = 0 \tag{5-130}$$

for every $p \in \mathscr{P}_{m-1}$ must be necessarily satisfied. From the second equation in (5-129) it follows that if $u(\mathbf{x})$ is a weak solution of Eq. (5-122) then $u(\mathbf{x}) + p(\mathbf{x})$, where $p \in \mathscr{P}_{m-1}$, is also a solution of Eq. (5-122).

The compatibility condition (5-130), which is necessary for the existence of a weak solution of Eq. (5-122), is also sufficient. Analogous to the development presented for the second-order problem, it is sufficient to consider the following weak problem: find $u \in \bar{Q}$ such that

$$B(v, u) = \int_\Omega vf \, d\mathbf{x} + \sum_{i=0}^{m-1} \oint_\Gamma \frac{\partial^i v}{\partial n^i} h_{i+1} \, ds \tag{5-131}$$

holds for all $v \in \bar{Q}$.

Next we show the $\bar{Q}$-ellipticity of $B(v, u)$. We have for $v \in \bar{Q}$

$$\begin{aligned} B(v, v) &= \sum_{|\alpha|\, |\beta| = m} \int_\Omega a_{\alpha\beta} D^\alpha v D^\beta v \, d\mathbf{x} \\ &\geq c_0 \sum_{|\alpha| = m} \int_\Omega |D^\alpha v|^2 \, d\mathbf{x} \end{aligned} \tag{5-132}$$

For arbitrary $v \in \bar{Q}$ consider the class of functions of the form $v(\mathbf{x}) + p(\mathbf{x})$, where $p \in \mathscr{P}_{m-1}$. For each $v(\mathbf{x})$, it is possible to uniquely determine a $p_1 \in \mathscr{P}_{m-1}$ such that

$$\int_\Omega D^{\boldsymbol{\alpha}}(v + p_1)\, d\mathbf{x} = 0 \tag{5-133}$$

holds for each $\boldsymbol{\alpha}$ for which $|\boldsymbol{\alpha}| < m$. Then by the Poincaré inequality we have

$$\begin{aligned} \|v + p_1\|_m^2 &\leq c_3 \sum_{|\boldsymbol{\alpha}|=m} \int_\Omega [D^{\boldsymbol{\alpha}}(v + p_1)]^2\, d\mathbf{x} \\ &= c_3 \sum_{|\boldsymbol{\alpha}|=m} \int_\Omega (D^{\boldsymbol{\alpha}} v)^2\, d\mathbf{x} \end{aligned} \tag{5-134}$$

because $D^{\boldsymbol{\alpha}} p_1 = 0$ for $|\boldsymbol{\alpha}| = m$. Since v and p_1 are orthogonal in $\bar{H}^m(\Omega)$, we obtain

$$\begin{aligned} |||v + p_1|||_m^2 &= |||v|||_m^2 + |||p_1|||_m^2 \\ &= |||v|||_m^2 + \int_\Omega p_1^2\, d\mathbf{x} \end{aligned} \tag{5-135}$$

From Eqs. (5-134) and (5-135) it follows that

$$|||v|||_m^2 \leq c_4 \sum_{|\boldsymbol{\alpha}|=m} \int_\Omega (D^{\boldsymbol{\alpha}} v)^2\, d\mathbf{x} \tag{5-136}$$

From Eqs. (5-132) and (5-136) we have the $\bar{Q}$-ellipticity of $B(v, v)$,

$$B(v, v) \geq \frac{c_0}{c_4} |||v|||_m^2 = \alpha |||v|||_{\bar{Q}}^2 \tag{5-137}$$

for every $v \in \bar{Q}$.

We summarize the above discussions in the following theorem.

Theorem 5-8 Let $u \in H^m(\Omega)$ be the weak solution of the problem

$$B(v, u) = \int_\Omega vf\, d\mathbf{x} + \sum_{i=0}^{m-1} \int_\Omega \frac{\partial^i v}{\partial n^i} h_{i+1}\, ds \tag{5-138}$$

where $f \in L_2(\Omega)$ and $h_i \in L_2(\Gamma)$ $(i = 1, 2, \ldots, m)$. Let $B(v, u)$ satisfy the conditions

$$\begin{aligned} &\text{(i)}\quad |B(v, u)| \leq M \|v\|_m \|u\|_m \\ &\text{(ii)}\quad B(v + p, u) = B(v, u) \quad \text{and} \quad B(v, u + p) = B(v, u) \\ &\text{(iii)}\quad B(v, v) \geq \alpha \sum_{|\boldsymbol{\alpha}|=m} \int_\Omega (D^{\boldsymbol{\alpha}} v)^2\, d\mathbf{x} \end{aligned} \tag{5-139}$$

for every $u, v \in H^m(\Omega)$, where M and α are constants independent of u and v.

Then the necessary and sufficient condition for the existence of a weak solution of (5-138) is

$$\int_{\Omega} pf \, d\mathbf{x} + \sum_{i=0}^{m-1} \oint_{\Gamma} \frac{\partial^i p}{\partial n^i} h_{i+1} \, ds = 0 \qquad \text{for every } p \in \mathscr{P}_{m-1} \tag{5-140}$$

If this condition is satisfied, then there exists an infinite number of weak solutions of (5-138), and each pair of solutions differ by $p \in \mathscr{P}_{m-1}$.

If Q is the subspace of functions $v(\mathbf{x})$ from the space $H^m(\Omega)$ which are orthogonal in $L_2(\Omega)$ to the space $\mathscr{P}_{m-1}$,

$$\int_{\Omega} pv \, d\mathbf{x} = 0 \qquad \text{for every } p \in \mathscr{P}_{m-1} \tag{5-141}$$

then the requirement $u \in Q$ uniquely determines the weak solution of (5-138) and the following estimate

$$\|u\|_Q = \|u\|_2 \leq c\left[\|f\|_0 + \sum_{i=1}^{m} \|h_i\|_{L_2(\Gamma)}\right] \tag{5-142}$$

holds.

We close this section with an example.

Example 5-13 Consider the biharmonic equation (5-105) with the Neumann boundary conditions on Γ:

$$B_1 w \equiv -\frac{\partial}{\partial n}(\nabla^2 w) + (1 - \nu)\frac{\partial}{\partial s}$$

$$\cdot \left[\frac{\partial^2 w}{\partial x_1^2} n_1 n_2 - \frac{\partial^2 w}{\partial x_1 \, \partial x_2}(n_1^2 - n_2^2) - \frac{\partial^2 w}{\partial x_2^2} n_1 n_2\right] = h_1 \tag{5-143}$$

$$B_2 w \equiv \nu \nabla^2 w + (1 + \nu)\frac{\partial^2 w}{\partial n^2} = h_2$$

where $h_1, h_2 \in L_2(\Gamma)$. Consider the weak formulation of the problem: find $w \in H^2(\Omega)$ such that

$$B(v, u) = \int_{\Omega} vf \, d\mathbf{x} + \oint_{\Gamma}\left(vh_1 + \frac{\partial v}{\partial n} h_2\right) ds$$

holds for every $v \in H^2(\Omega)$. The bilinear form $B(v, u)$ is given by Eq. (5-109). We have $n = 2$, $m = 2$, and $\mathscr{P}_{m-1} = \mathscr{P}_1$. A typical element in $\mathscr{P}_1$ is of the form

$$p(x_1, x_2) = a + bx_1 + cx_2$$

The elements of Q are those functions from $H^2(\Omega)$ which satisfy the condition

$$\int_{\Omega} pv \, d\mathbf{x} = 0 \qquad \text{for every } p \in \mathscr{P}_1$$

This is equivalent to the three conditions

$$\int_{\Omega} v \, d\mathbf{x} = 0 \qquad \int_{\Omega} x_1 v \, d\mathbf{x} = 0 \qquad \int_{\Omega} x_2 v \, d\mathbf{x} = 0$$

The bilinear form $B(v, u)$ of the problem satisfies the conditions (5-139):

$$B(v, v) \geq (1 - \nu) \sum_{|\alpha| = 2} \int_{\Omega} (D^{\alpha} v)^2 \, d\mathbf{x}$$

Thus by Theorem 5-8, the necessary and sufficient conditions for the existence of a weak solution are given by [set $p = 1$, $p = x_1$, $p = x_2$ in Eq. (5-140)]

$$\int_{\Omega} f \, d\mathbf{x} + \oint_{\Gamma} h_1 \, ds = 0$$

$$\int_{\Omega} x_1 f \, d\mathbf{x} + \oint_{\Gamma} x_1 h_1 \, ds + \oint_{\Gamma} \frac{\partial x_1}{\partial n} h_2 \, ds = 0$$

$$\int_{\Omega} x_2 f \, d\mathbf{x} + \oint_{\Gamma} x_2 h_1 \, ds + \oint_{\Gamma} \frac{\partial x_2}{\partial n} h_2 \, ds = 0$$

where $\partial x_1/\partial n = n_1$ and $\partial x_2/\partial n = n_2$.

If the above conditions are satisfied, then the problem has an infinite number of weak solutions. By the requirement $u \in Q$, we mean

$$\int_{\Omega} u \, d\mathbf{x} = 0 \qquad \int_{\Omega} x_1 u \, d\mathbf{x} = 0 \qquad \int_{\Omega} x_2 u \, d\mathbf{x} = 0$$

the weak solution $u(\mathbf{x})$ is uniquely determined. All other solutions have the form $u(\mathbf{x}) + p(\mathbf{x})$ where $p(\mathbf{x}) \in \mathscr{P}_1$. The solution $u(\mathbf{x})$ depends continuously on the data

$$\|u\|_2 \leq c[\|f\|_0 + \|h_1\|_{L_2(\Gamma)} + \|h_2\|_{L_2(\Gamma)}]$$

This completes the example.

PROBLEMS 5-2

5-2-1 Convert the following boundary-value problem with nonhomogeneous boundary conditions to one with homogeneous boundary conditions:

$$-\frac{d}{dx}\left[(1 + 2x^2)\frac{du}{dx}\right] + u = x^2 \qquad 0 < x < 1$$

$$u(0) = 1 \qquad \frac{du}{dx}(1) = 2$$

5-2-2 Consider the first-order ordinary differential equation $c_1 \, du/dx + c_0 u = f$, $a < x < b$ where c_1 and c_0 are constants. Give the least squares formulation (see the next exercise) of the equations and investigate the existence of the solution in an appropriate space.

5-2-3 (*Least squares formulation of the Poisson equation*) Consider the equation $-\nabla^2 u + c_0 u = f$ in $\Omega \subset \mathbb{R}^2$, $\partial u/\partial n + \beta u = g$ on Γ. The least squares formulation of the equation is given by

$$\delta\left\{\int_\Omega (-\nabla^2 u + c_0 u - f)^2 \, d\mathbf{x} + \oint_\Gamma \left(\frac{\partial u}{\partial n} + \beta u - g\right)^2 ds\right\} = 0$$

Equivalently, we have $B(v, u) = l(v)$, where

$$B(v, u) = \int_\Omega (-\nabla^2 v + c_0 v)(-\nabla^2 u + c_0 u) \, d\mathbf{x} + \oint_\Gamma \left(\frac{\partial v}{\partial n} + \beta v\right)\left(\frac{\partial u}{\partial n} + \beta u\right) ds$$

$$l(v) = \int_\Omega (-\nabla^2 v + c_0 v) f \, d\mathbf{x} + \oint_\Gamma \left(\frac{\partial v}{\partial n} + \beta v\right) g \, ds$$

Define

$$H = \{u \in C_0^\infty(\Omega): \|u\|_H^2 = \|-\nabla^2 u + c_0 u\|_{L_2(\Omega)}^2 + \|u\|_{H^{3/2}(\Omega)}^2\}$$

Show that $B(v, u)$ is continuous and H-elliptic.

5-2-4 Consider the second-order ordinary differential equation

$$-\frac{d}{dx}\left[p(x)\frac{du}{dx}\right] + q(x)u = f(x) \qquad a < x < b$$

Subject to the boundary conditions $u(a) = 0$, $u(b) = 0$. Give the weak formulation of the problem, identify the space V, and show the V-ellipticity, for $p(x) \geq p_0 > 0$ and $q(x) \geq 0$ in $[a, b]$, of the bilinear form. Estimate the constants M and α of Eq. (5-39) in terms of p_0, a, and b.

5-2-5 Repeat Prob. 5-2-4 for the Neumann boundary conditions $(du/dx)(a) = 0$ and $(du/dx)(b) = 0$. Assume that $q(x) \geq q_0 > 0$ in $[a, b]$. If $q(x) = 0$, does the V-ellipticity of the bilinear form hold?

5-2-6 Prove that the Neumann problem for the partial differential equation,

$$-\sum_{i,j=1}^{n} \frac{\partial}{\partial x_i}\left(a_{ij}\frac{\partial u}{\partial x_j}\right) + a_0 u = f \text{ in } \Omega \subset \mathbb{R}^n$$

$$\sum_{i,j=1}^{n} a_{ij}\frac{\partial u}{\partial x_j} n_i = g \text{ on } \Gamma$$

where $a_{ij}(\mathbf{x}) \geq \mu_0 > 0$ and $a_0(\mathbf{x}) \geq \mu_1 > 0$, has a unique solution. Use Theorem 5-5 (*not* Theorem 5-7) in your proof.

5-2-7 Consider the problem of an elastic plate on an elastic foundation, with foundation modulus $k > 0$. The governing equation of equilibrium is given by Eq. (5-105) with the term kw added to the left side of the equality. Investigate the existence and uniqueness of the solution to the problem in an appropriate space.

5-2-8 Show that the compatibility conditions for the Neumann problem for the equations of linear elasticity are given by $\int_\Omega \mathbf{f} \, d\mathbf{x} + \oint_\Gamma \hat{\mathbf{t}} \, ds = 0$, $\int_\Omega \mathbf{r} \times \mathbf{f} \, d\mathbf{x} + \oint_\Gamma \mathbf{r} \times \hat{\mathbf{t}} \, ds = 0$, where $\mathbf{f}$ is the body force vector, $\hat{\mathbf{t}}$ the specified traction vector and $\mathbf{r}$ is the position vector of a typical point $\mathbf{x}$ in the body.

5-2-9 (*Classification of partial differential operators*) The classification of partial differential equations of second order is based on the analogy of classifying the equation of a conic section:

$$ax^2 + bxy + cy^2 + dx + ey + f = 0. \tag{a}$$

The equation represents

1. an ellipse if $b^2 - 4ac < 0$,
2. a parabola if $b^2 - 4ac = 0$, and
3. a hyperbola if $b^2 - 4ac > 0$.

(*b*)

We now consider the general partial differential equation of second order in two-dimensions (note the analogy between this and the equation of a conic section given above):

$$a \frac{\partial^2 u}{\partial x^2} + b \frac{\partial^2 u}{\partial x \, \partial y} + c \frac{\partial^2 u}{\partial y^2} + d \frac{\partial u}{\partial x} + e \frac{\partial u}{\partial y} + f = 0 \tag{c}$$

The partial differential equation in (*c*) is classified as elliptic, parabolic, and hyperbolic according to Eq. (*b*). Classify the following differential operators into *elliptic*, *parabolic*, or *hyperbolic*

(i) $$A = \frac{\partial^4}{\partial x^4} + \frac{\partial^4}{\partial y^4} - \frac{\partial^4}{\partial z^4} + i \frac{\partial^2}{\partial z^2}\left(\frac{\partial^2}{\partial x^2} + \frac{\partial^2}{\partial y^2}\right) \qquad i = \sqrt{-1}$$

(ii) $$A = \frac{\partial^2}{\partial x^2} - \frac{\partial^2}{\partial y^2}$$

(iii) $$A = \frac{\partial}{\partial x} + \frac{\partial}{\partial y}$$

5-2-10 Consider the following problem, which arises in heat transfer problems: $-\nabla \cdot (\mathbf{k}\nabla\theta) + c\theta = f$ in Ω, $\mathbf{n}\cdot(\mathbf{k}\nabla\theta) + \alpha\theta = g$ on Γ, where $\mathbf{k}$, c, f, α, and g are given functions of position and θ is the temperature. Construct the variational formulation of the problem and show that the bilinear and linear forms are continuous on $H^1(\Omega)$ and the bilinear form is $H^1(\Omega)$-elliptic. Assume that $f \in L_2(\Omega)$, $g \in L_2(\Gamma)$, $c(\mathbf{x}) \in L_\infty(\Omega)$ with $c(x) \geq c_0 > 0$, where c_0 is a constant, $\alpha(s) \in L_\infty(\Gamma)$, $\alpha(s) \geq \alpha_0 > 0$, and that the conductivity tensor $\mathbf{k}$ satisfies the condition

$$\mathbf{k}(\mathbf{x})\boldsymbol{\xi}\cdot\boldsymbol{\xi} \geq \mu|\boldsymbol{\xi}|^2 \qquad \text{for every } \boldsymbol{\xi} = (\xi_1, \xi_2, \ldots, \xi_n) \in \mathbb{R}^n$$

where $\mu > 0$ is a constant and $|\boldsymbol{\xi}|$ denotes the euclidean norm of $\boldsymbol{\xi}$.

5-2-11 Consider the Dirichlet problem for the biharmonic operator

$$\nabla^4\psi = f \text{ in } \Omega \subset \mathbb{R}^2,\ \psi = 0 \text{ and } -\nabla^2\psi = \mu \text{ on } \Gamma \tag{a}$$

Here ψ denotes the stream function in the context of fluid mechanics, and the transverse deflection of a thin plate in the context of solid mechanics. Let

$$V = H^2(\Omega) \cap H_0^1(\Omega) \tag{b}$$

Show that for $\lambda \in H^{-1/2}(\Gamma)$ the problem in Eq. (*a*) has a weak unique solution in V. Note that the weak problem is given by

$$\int_\Omega \nabla^2 v \nabla^2 \psi \, dx_1 \, dx_2 = -\oint_\Gamma \frac{\partial v}{\partial n} \lambda \, ds \qquad \text{for all } v \in V \tag{c}$$

Hint: Show that $\|\nabla^2 v\|_{L_2(\Omega)}$ defines on V a norm equivalent to the norm on $H^2(\Omega)$ and therefore show that the bilinear form is continuous and V-elliptic. The continuity of the linear functional [from $H^2(\Omega)$ to $H^{1/2}(\Gamma)$] can be proved using the trace properties.

5-5 BOUNDARY-VALUE PROBLEMS WITH EQUALITY CONSTRAINTS

5-5-1 Introduction

The results presented in the preceding sections of this chapter can be extended to variational formulations based on the Lagrange multiplier and penalty function techniques. In this section we will consider specific model problems to investigate the existence and uniqueness questions. We begin with the Lagrange multiplier formulation of the Dirichlet problem for the Poisson equation.

5-5-2 The Dirichlet Problem for a Second-Order Differential Equation

Consider the model problem

$$-\nabla^2 u + u = f \text{ in } \Omega \subset \mathbb{R}^N \qquad (N = 2 \text{ or } 3) \tag{5-144a}$$

with the boundary condition

$$u = g \text{ on } \Gamma \tag{5-144b}$$

where Ω is assumed to be a bounded domain and that its boundary Γ is sufficiently smooth.

As discussed in Sec. 4-4, the usual (i.e. classical) variational method is to minimize the functional

$$J(u) = \frac{1}{2}\int_\Omega \left[\sum_{i=1}^{N}\left(\frac{\partial u}{\partial x_i}\right)^2 + u^2\right] d\mathbf{x} - \int_\Omega fu\, d\mathbf{x} \tag{5-145}$$

over $H_0^1(\Omega)$. Existence and uniqueness of solutions to this variational problem were discussed in Sec. 5-4 (see Example 5-8). Here we consider the alternative formulations discussed in Sec. 4-4: the Lagrange multiplier and penalty function formulations.

The Lagrange multiplier formulation The method of Lagrange multipliers seeks the stationary point (u, λ) of the augmented functional

$$L(u, \lambda) = J(u) + \oint_\Gamma \lambda(u - g)\, ds \tag{5-146}$$

where λ is the Lagrange multiplier. The bilinear and linear forms associated with this functional are given by

$$\begin{aligned} B(u, \lambda; v, \mu) &= \int_\Omega \left[\sum_{i=1}^{3} \frac{\partial v}{\partial x_i}\frac{\partial u}{\partial x_i} + vu\right] d\mathbf{x} + \oint_\Gamma (\lambda v + \mu u)\, ds \\ l(v, \mu) &= \int_\Omega vf\, d\mathbf{x} + \oint_\Gamma \mu g\, ds \end{aligned} \tag{5-147}$$

The weak problem involves seeking $(u, \lambda) \in \mathbf{H} \equiv H^1(\Omega) \times H^{-1/2}(\Gamma)$ such that

$$B(u, \lambda; v, \mu) = l(v, \mu) \tag{5-148}$$

for all $(v, \mu) \in \mathbf{H}$. The space $\mathbf{H}$ is a Hilbert space with the norm

$$\|(u, \lambda)\|_{\mathbf{H}}^2 = \|u\|_{H^1(\Omega)}^2 + \|\lambda\|_{H^{-1/2}(\Gamma)}^2 \tag{5-149a}$$

where

$$\|\lambda\|_{H^{-1/2}(\Gamma)} = \sup_{v \in H^{1/2}(\Gamma)} \frac{|\oint_\Gamma \lambda v\, ds|}{\|v\|_{H^{1/2}(\Gamma)}} \tag{5-149b}$$

We now use Theorem 5-5 to prove the existence and uniqueness of solutions to the weak problem (5-148). To this end we must show that the bilinear form in Eq. (5-147) satisfies the conditions in Eq. (5-39) [see Babuska (1973*a*)]. Let us first verify condition (5-39*a*). Using the Trace Theorem 4-3 we have for $u \in H^m(\Omega)$,

$$\|u\|_{H^{m-1/2}(\Gamma)} \le c\|u\|_{H^m(\Omega)} \qquad m > \tfrac{1}{2} \tag{5-150}$$

Hence, using the Hölder inequality for finite sums, we obtain,

$$\begin{aligned} |B(u, \lambda; v, \mu)| &\le \|v\|_{H^1(\Omega)}\|u\|_{H^1(\Omega)} \\ &\quad + c[\|\lambda\|_{H^{-1/2}(\Gamma)}\|v\|_{H^1(\Omega)} + \|\mu\|_{H^{-1/2}(\Gamma)}\|u\|_{H^1(\Omega)}] \\ &\le M\|(u, \lambda)\|_{\mathbf{H}}\|(v, \mu)\|_{\mathbf{H}} \end{aligned} \tag{5-151}$$

Next we prove Eq. (5-39*b*). Let $(u, \lambda) \in \mathbf{H}$ and $w \in H^1(\Omega)$ be the solution of the Neumann problem for the differential equation

$$\begin{aligned} -\nabla^2 w + w &= 0 \text{ in } \Omega \\ \frac{\partial w}{\partial n} &= -\lambda \text{ on } \Gamma \end{aligned} \tag{5-152}$$

Then by Theorem 5-7, the function $w \in H^1(\Omega)$ exists and the following estimate [more precise than that given in Eq. (5-121)] holds

$$\|w\|_{H^1(\Omega)} \le c\|\lambda\|_{H^{-1/2}(\Gamma)} \tag{5-153}$$

Now let $\hat{w}$ be the solution of the Dirichlet problem

$$\begin{aligned} -\nabla^2 \hat{w} + \hat{w} &= 0 \text{ in } \Omega \\ \hat{w} &= v \text{ on } \Gamma \end{aligned} \tag{5-154}$$

From the weak form of Eq. (5-152) and the inequality $\|\hat{w}\|_{H^1(\Omega)} \le c_1\|v\|_{H^{1/2}(\Gamma)}$, we obtain

$$\begin{aligned} \frac{|\oint_\Gamma \lambda v\, ds|}{\|v\|_{H^{1/2}(\Gamma)}} &= \frac{|\int_\Omega (\text{grad } \hat{w} \cdot \text{grad } w + \hat{w}w)\, d\mathbf{x}|}{\|v\|_{H^{1/2}(\Gamma)}} \\ &\le c\,\frac{\|\hat{w}\|_{H^1(\Omega)}\|w\|_{H^1(\Omega)}}{\|\hat{w}\|_{H^1(\Omega)}} \\ &\le c\|w\|_{H^1(\Omega)} = c\left[-\oint_\Gamma \lambda w\, ds\right]^{1/2} \end{aligned}$$

where in arriving at the last equality, the weak form of Eq. (5-152) is used. Thus, in view of Eq. (5-149*b*), we have

$$\|\lambda\|^2_{H^{-1/2}(\Gamma)} \le -c\oint_\Gamma \lambda w\, ds \tag{5-155}$$

For the choice of $v = u - w$ and $\mu = -2\lambda$, we have

$$\|(v, \mu)\|_{\mathbf{H}} \leq c_1 \|(u, \lambda)\|_{\mathbf{H}}$$

Finally, consider the bilinear form (with $v = u - w$ and $\mu = -2\lambda$),

$$B(u, \lambda; v, \mu) = \int_\Omega \left[\sum_{i=1}^n \left(\frac{\partial u}{\partial x_i} \right)^2 + u^2 \right] d\mathbf{x} - \int_\Omega \left[\sum_{i=1}^n \frac{\partial u}{\partial x_i} \frac{\partial w}{\partial x_i} + uw \right] d\mathbf{x}$$

$$+ \oint_\Gamma (\lambda u + \mu u)\, ds - \oint_\Gamma \lambda w \, ds$$

Noting that the expression under the second integral is equal to [by the weak problem of Eq. (5-152)] $-\oint_\Gamma \lambda u \, ds$, we obtain

$$B(u, \lambda; v, \mu) = \|u\|^2_{H^1(\Omega)} + \oint_\Gamma (2\lambda + \mu) u \, ds - \oint_\Gamma \lambda w \, ds$$

The second integral is identically zero (because $\mu = -2\lambda$) and the third integral is bounded below by $\|\lambda\|^2_{H^{-1/2}(\Gamma)}$:

$$B(u, \lambda; v, \mu) \geq \|u\|^2_{H^1(\Omega)} + c^{-1} \|\lambda\|^2_{H^{-1/2}(\Gamma)}$$

$$\geq \alpha \|(u, \lambda)\|^2_{\mathbf{H}}$$

where $\alpha = \min(1, c^{-1})$. Since the right-hand side does not depend on v and μ it follows that condition (5-39*b*) is satisfied.

The continuity of the linear form $l(v, \mu)$, for $f \in L_2(\Omega)$ and $g \in H^{1/2}(\Gamma)$, can be easily established. Hence the weak problem (5-148) has a unique solution.

The penalty formulation In the penalty function formulation we minimize the functional,

$$P_K(u) = J(u) + \frac{K}{2} \oint_\Gamma (u - g)^2 \, ds \tag{5-156}$$

on $H^1_K(\Omega)$. The weak problem associated with $P_K(u)$ involves finding $u \in H^1_K(\Omega)$ such that

$$B_K(v, u) = l_K(v) \qquad \text{for all } v \in H^1_K(\Omega) \tag{5-157a}$$

where

$$B_K(v, u) = \int_\Omega \left(\sum_{i=1}^3 \frac{\partial v}{\partial x_i} \frac{\partial u}{\partial x_i} + vu \right) d\mathbf{x} + K \oint_\Gamma vu \, ds$$

$$l_K(v) = \int_\Omega fv \, d\mathbf{x} + K \oint_\Gamma gv \, ds \tag{5-157b}$$

$$H^1_K(\Omega) = \{u \in H^1(\Omega): \|u\|^2_K = \|u\|^2_{H^1(\Omega)} + K \|u\|^2_{L_2(\Gamma)}\}$$

The space $H_K^1(\Omega)$ is a subspace of $H^1(\Omega)$, and the norm $\|u\|_K$ is equivalent to the $H^1(\Omega)$ norm (the constants appearing in the definition of equivalent norms depend on K).

We now show that the bilinear form $B_K(v, u)$ on $H_K^1(\Omega) \times H_K^1(\Omega)$ satisfies the conditions (5-39) of Theorem 5-5, with constants M and α independent of K. The continuity follows from the Cauchy–Schwarz inequality:

$$\begin{aligned} |B_K(u, v)| &\leq (\|u\|_{H^1(\Omega)}\|v\|_{H^1(\Omega)} + K\|u\|_{L_2(\Gamma)}\|v\|_{L_2(\Gamma)}) \\ &\leq M[\|u\|^2_{H^1(\Omega)} + K\|u\|^2_{L_2(\Gamma)}]^{1/2}[\|v\|^2_{H^1(\Omega)} + K\|v\|^2_{L_2(\Gamma)}]^{1/2} \\ &= M\|u\|_K\|v\|_K \end{aligned}$$

The $H_K^1(\Omega)$-ellipticity follows from

$$B_K(u, u) = \|u\|^2_{H^1(\Omega)} + K\|u\|^2_{L_2(\Gamma)} = \|u\|^2_{H_K^1(\Omega)}$$

with $0 < \alpha < 1$. Hence the weak problem (5-157) has a unique solution for any fixed $0 < K < \infty$.

We note that the solution of (5-157), which depends on K, is not the solution of the original problem (5-144) or (5-145). However, we have shown in Sec. 4-4 that the sequence of the penalty solutions $u_n \equiv u_{K_n}$, $0 < K_1 < K_2 \cdots < K_n < \cdots < \infty$, converges to the solution of the original problem.

5-5-3 Nearly Incompressible and Incompressible Elastic Solids and Viscous Fluids

The results presented in Example 5-9 for Navier's equations of plane elasticity are not valid when the Poisson ratio ν is nearly equal or equal to 1/2. This value of ν causes $\lambda = E\nu/[(1 + \nu)(1 - 2\nu)]$ to take a large unbounded value. Elastic materials for which $\nu = 1/2$ are known as *incompressible materials.* Rubber and photoelastic materials at elevated temperatures provide examples of incompressible materials.

The proof of existence and uniqueness of solutions to elasticity problems involving incompressible materials makes use of alternative formulations (to that of the total potential energy formulation) that include the constraint

$$P\gamma - \left(\frac{\partial u_1}{\partial x_1} + \frac{\partial u_2}{\partial x_2}\right) = 0 \tag{5-158}$$

where γ is a positive constant. Here we consider the Lagrange multiplier formulation [see Babuska and Aziz (1972)].

The original problem involves seeking (u_1, u_2, P) such that

$$J(u_1, u_2, P) = \int_\Omega \left\{\frac{\lambda}{2}\gamma^2 P^2 + \mu\left[\frac{1}{2}\left(\frac{\partial u_1}{\partial x_2} + \frac{\partial u_2}{\partial x_1}\right)^2 + \left(\frac{\partial u_1}{\partial x_1}\right)^2 + \left(\frac{\partial u_2}{\partial x_2}\right)^2\right] - f_1 u_1 - f_2 u_2\right\} dx_1\, dx_2 \tag{5-159}$$

subject to the constraint in Eq. (5-158) and homogeneous essential boundary conditions ($u_1 = u_2 = 0$ on Γ). Using the technique of Lagrange multipliers, we introduce the lagrangian functional

$$L(u_1, u_2, P, \xi) = J(u_1, u_2, P) + \int_\Omega \xi\left[P\gamma - \left(\frac{\partial u_1}{\partial x_1} + \frac{\partial u_2}{\partial x_2}\right)\right] dx_1\, dx_2 \quad (5\text{-}160)$$

on $\mathbf{H}$, where ξ is the Lagrange multiplier and

$$\mathbf{H} = H_0^1(\Omega) \times H_0^1(\Omega) \times \tilde{H}^0(\Omega) \times \tilde{H}^0(\Omega)$$

$$\tilde{H}^0(\Omega) = \left\{P \in L_2(\Omega)\colon \int_\Omega P\, dx_1\, dx_2 = 0\right\} \quad (5\text{-}161)$$

The weak problem corresponding to the Lagrange multiplier formulation is: find $(u_1, u_2, P, \xi) \in \mathbf{H}$ such that

$$B(u_1, u_2, P, \xi;\ v_1, v_2, Q, \eta) = l(v_1, v_2, Q, \eta) \quad (5\text{-}162)$$

holds for every $(v_1, v_2, Q, \eta) \in \mathbf{H}$. The bilinear and linear forms are given by

$$\begin{aligned} B(u_1, u_2, P, \xi;\ v_1, v_2, Q, \eta) = \int_\Omega \Bigg\{ & \lambda\gamma^2 PQ + \mu\left[\left(\frac{\partial u_1}{\partial x_1} + \frac{\partial u_2}{\partial x_2}\right)\left(\frac{\partial v_1}{\partial x_1} + \frac{\partial v_2}{\partial x_2}\right)\right. \\ & \left. + 2\frac{\partial u_1}{\partial x_1}\frac{\partial v_1}{\partial x_1} + 2\frac{\partial u_2}{\partial x_2}\frac{\partial v_2}{\partial x_2}\right] \\ & + \xi\left[Q\gamma - \left(\frac{\partial v_1}{\partial x_1} + \frac{\partial v_2}{\partial x_2}\right)\right] \\ & + \eta\left[P\gamma - \left(\frac{\partial u_1}{\partial x_1} + \frac{\partial u_2}{\partial x_2}\right)\right]\Bigg\} dx_1\, dx_2 \end{aligned}$$

$$l(v_1, v_2, Q, \eta) = \int_\Omega (v_1 f_1 + v_2 f_2)\, dx_1\, dx_2 \quad (5\text{-}163)$$

Toward proving that the bilinear and linear forms satisfy the conditions of Theorem 5-5, we first introduce the following norm in $\mathbf{H}$:

$$\|(u_1, u_2, P, \xi)\|_{\mathbf{H}}^2 = \|u_1\|_{H^1(\Omega)}^2 + \|u_2\|_{H^1(\Omega)}^2 + \sqrt{\lambda}\gamma\|P\|_{L_2(\Omega)}^2 + \|\xi\|_{L_2(\Omega)}^2 \quad (5\text{-}164)$$

The bilinear form satisfies the continuity condition in Eq. (5-39*a*) with $M = c \max\{\mu, \sqrt{\lambda}\gamma, \gamma^{-1/2}\lambda^{-1/4}\}$, and $c > 0$ is independent of μ, λ, and γ. To prove the $\mathbf{H}$-ellipticity of the bilinear form, we need the following two basic results.

Theorem 5-9 If $Q \in \tilde{H}^0(\Omega)$, then there exist $v_1, v_2 \in H_0^1(\Omega)$ such that

$$-\left(\frac{\partial v_1}{\partial x_1} + \frac{\partial v_2}{\partial x_2}\right) = Q$$

$$\|v_1\|_{H^1(\Omega)}^2 + \|v_2\|_{H^1(\Omega)}^2 \le c\|Q\|_{L_2(\Omega)}^2 \quad (5\text{-}165)$$

PROOF Since $Q \in \tilde{H}^0(\Omega)$, by Theorem 5-7 there exists a $w \in H^1(\Omega)$ such that

$$\int_\Omega \text{grad } v \cdot \text{grad } w \, dx_1 \, dx_2 = \int_\Omega Qv \, dx_1 \, dx_2 \qquad \text{for all } v \in \tilde{H}^0(\Omega)$$

$$\|w\|_{H^1(\Omega)} \le c\|Q\|_{L_2(\Omega)}$$

By the regularity of solution, we have $w \in H^2(\Omega)$ and $\|w\|_{H^2(\Omega)} \le c\|Q\|_{L_2(\Omega)}$. If we choose $u_1 = (\partial w/\partial x_1) \in H^1(\Omega)$ and $u_2 = (\partial w/\partial x_2) \in H^1(\Omega)$, we have

$$\|u_1\|_{H^1(\Omega)} + \|u_2\|_{H^1(\Omega)} \le c\|Q\|_{L_2(\Omega)}$$

and
$$-\left(\frac{\partial u_1}{\partial x_1} + \frac{\partial u_2}{\partial x_2}\right) = -\nabla^2 w = Q$$

Clearly, $u_1, u_2 \in H_0^1(\Omega)$.

Let $U \in H^2(\Omega)$ and define

$$v_1 = u_1 + \frac{\partial U}{\partial x_2} = \frac{\partial w}{\partial x_1} + \frac{\partial U}{\partial x_2}$$

$$v_2 = u_2 - \frac{\partial U}{\partial x_1} = \frac{\partial w}{\partial x_2} - \frac{\partial U}{\partial x_1}$$

so that $-(\partial v_1/\partial x_1 + \partial v_2/\partial x_2) = -\nabla^2 w = Q$. We wish to show that $v_1 = v_2 = 0$ on Γ to complete the proof. Suppose that $\partial w/\partial x_1 + \partial U/\partial x_2 = 0$ and $\partial w/\partial x_2 - \partial U/\partial x_1 = 0$ on Γ. Since $\partial w/\partial n = 0$ on Γ, we have

$$0 = \frac{\partial w_1}{\partial x_1} n_1 + \frac{\partial w_2}{\partial x_2} n_2 = -\frac{\partial U}{\partial x_2} n_1 + \frac{\partial U}{\partial x_1} n_2 \equiv -\frac{\partial U}{\partial s} \text{ on } \Gamma$$

Thus, if $U \in H^2(\Omega)$ is the weak solution of a plate (see Example 5-12) with the boundary conditions $U = 0$ (equivalent to $\partial U/\partial s = 0$) and

$$\frac{\partial U}{\partial n} = \frac{\partial w}{\partial x_2} n_1 - \frac{\partial w}{\partial x_1} n_2 \equiv g \text{ on } \Gamma$$

we have
$$\|U\|_{H^2(\Omega)} \le \|g\|_{L_2(\Gamma)}$$

and
$$\|g\|_{L_2(\Gamma)} = \left\|\left(\frac{\partial w}{\partial x_2} n_1 - \frac{\partial w}{\partial x_1} n_2\right)\right\|_{L_2(\Gamma)}$$
$$\le c\|w\|_{H^2(\Omega)} = c\|Q\|_{L_2(\Omega)}$$

In summary, we have $v_1 = v_2 = 0$ on Γ [hence $v_1, v_2 \in H_0^1(\Omega)$] and

$$\|v_1\|_{H^1(\Omega)}^2 + \|v_2\|_{H^1(\Omega)}^2 = \left\|\left(\frac{\partial w}{\partial x_1} + \frac{\partial U}{\partial x_2}\right)\right\|_{H^1(\Omega)}^2 + \left\|\left(\frac{\partial w}{\partial x_2} - \frac{\partial U}{\partial x_1}\right)\right\|_{H^1(\Omega)}^2$$
$$\le c\|w\|_{H^2(\Omega)}$$
$$= c\|Q\|_{L_2(\Omega)}$$

This completes the proof.

Theorem 5-10 If $Q \in \tilde{H}^0(\Omega)$ and $u_1, u_2 \in H_0^1(\Omega)$ such that

$$\int_\Omega \left[\frac{\partial u_1}{\partial x_1}\frac{\partial v_1}{\partial x_1} + \frac{\partial u_2}{\partial x_2}\frac{\partial v_2}{\partial x_2} + \frac{1}{2}\left(\frac{\partial u_1}{\partial x_2} + \frac{\partial u_2}{\partial x_1}\right)\left(\frac{\partial v_1}{\partial x_2} + \frac{\partial v_2}{\partial x_1}\right)\right] dx_1\, dx_2 = -\int_\Omega Q\left(\frac{\partial v_1}{\partial x_1} + \frac{\partial v_2}{\partial x_2}\right) dx_1\, dx_2$$

or

$$B_0(u_1, u_2\,;\, v_1, v_2) = l(v_1, v_2) \tag{5-166}$$

holds for every $v_1, v_2 \in H_0^1(\Omega)$. Then

$$-\int_\Omega Q\left(\frac{\partial u_1}{\partial x_1} + \frac{\partial u_2}{\partial x_2}\right) dx_1\, dx_2 \geq \hat{c}\|Q\|^2_{L_2(\Omega)} \tag{5-167a}$$

$$\|u_1\|^2_{H^1(\Omega)} + \|u_2\|^2_{H^1(\Omega)} \leq c\|Q\|^2_{L_2(\Omega)} \tag{5-167b}$$

PROOF The bilinear form in Eq. (5-166) satisfies the conditions of Theorem 5-5 (as was shown in Example 5-9). Therefore, Eq. (5-166) has a unique solution $(u_1, u_2) \in H_0^1(\Omega) \times H_0^1(\Omega)$ and Eq. (5-167*b*) holds.

By Theorem 5.9 and Eq. (5-166) we have

$$\|Q\|^2_{L_2(\Omega)} = B_0(u_1, u_2\,;\, v_1, v_2)$$

where $v_1, v_2 \in H_0^1(\Omega)$ are the functions appearing in Theorem 5-9. From Example 5-9 and the Schwarz inequality we derive

$$\begin{aligned}\|Q\|^2_{L_2(\Omega)} &\leq c[B_0(u_1, u_2; u_1, u_2)]^{1/2}[B_0(v_1, v_2; v_1, v_2)]^{1/2} \\ &\leq c[l(u_1, u_2)]^{1/2}[\|v_1\|^2_{H^1(\Omega)} + \|v_2\|^2_{H^1(\Omega)}]^{1/2} \\ &\leq c_1[l(u_1, u_2)]^{1/2}\|Q\|_{L_2(\Omega)}\end{aligned}$$

from which we obtain Eq. (5-167*a*).

Returning to the **H**-ellipticity of the bilinear form, let $v_1 = u_1 + z_1$ and $v_2 = u_2 + z_2$ in $B(u_1, u_2, P, \xi; v_1, v_2, Q, \eta)$ of Eq. (5-163). We obtain

$$\begin{aligned}B(u_1, u_2, P, \xi;\, v_1, v_2, Q, \eta) = \int_\Omega \Bigg\{ &\lambda\gamma^2 PQ + 2\mu\left[\left(\frac{\partial u_1}{\partial x_1}\right)^2 + \left(\frac{\partial u_2}{\partial x_2}\right)^2 + \frac{1}{2}\left(\frac{\partial u_1}{\partial x_2} + \frac{\partial u_2}{\partial x_1}\right)^2\right] + 2\mu\left[\frac{\partial u_1}{\partial x_1}\frac{\partial z_1}{\partial x_1}\right. \\ &\left. + \frac{\partial u_2}{\partial x_2}\frac{\partial z_2}{\partial x_2} + \frac{1}{2}\left(\frac{\partial u_1}{\partial x_2} + \frac{\partial u_2}{\partial x_1}\right)\left(\frac{\partial z_1}{\partial x_2} + \frac{\partial z_2}{\partial x_1}\right)\right] \\ &+ \xi\left[Q\gamma - \left(\frac{\partial u_1}{\partial x_1} + \frac{\partial u_2}{\partial x_2}\right) - \left(\frac{\partial z_1}{\partial x_1} + \frac{\partial z_2}{\partial x_2}\right)\right] \\ &+ \eta\left[P\gamma - \left(\frac{\partial u_1}{\partial x_1} + \frac{\partial u_2}{\partial x_2}\right)\right]\Bigg\} dx_1\, dx_2\end{aligned}$$

Now let z_1 and z_2 be such that Eq. (5-166) holds with $u_1 = z_1$, $u_2 = z_2$, $v_1 = u_1$, $v_2 = u_2$, and $Q = \xi$ in Theorem 5-10. We obtain

$$B(u_1, u_2, P, \xi;\ v_1, v_2, Q, \eta) = \int_\Omega \left\{2\mu\left[\left(\frac{\partial u_1}{\partial x_1}\right)^2 + \left(\frac{\partial u_2}{\partial x_2}\right)^2 + \frac{1}{2}\left(\frac{\partial u_1}{\partial x_2} + \frac{\partial u_2}{\partial x_1}\right)^2\right]\right.$$
$$- [(2\mu + 1)\xi + \eta]\left(\frac{\partial u_1}{\partial x_1} + \frac{\partial u_2}{\partial x_2}\right) - \xi\left(\frac{\partial z_1}{\partial x_1} + \frac{\partial z_2}{\partial x_2}\right)$$
$$\left. + \gamma(Q\xi + P\eta + \lambda\gamma PQ)\right\} dx_1\, dx_2$$

Taking $\eta = -(2\mu + 1)\xi$ and $Q = (2\mu + 1)P$ in the above expression, and using Eq. (5-167a), we obtain

$$B(u_1, u_2, P, \xi;\ v_1, v_2, Q, \eta) \geq \int_\Omega \left\{2\mu\left[\left(\frac{\partial u_1}{\partial x_1}\right)^2 + \left(\frac{\partial u_2}{\partial x_2}\right)^2 + \frac{1}{2}\left(\frac{\partial u_1}{\partial x_2} + \frac{\partial u_2}{\partial x_1}\right)^2\right]\right.$$
$$\left. + (2\mu + 1)\lambda\gamma^2 P^2\right\} dx_1\, dx_2 + \hat{c}\|\xi\|^2_{L_2(\Omega)}$$
$$\geq c[2\mu(\|u_1\|^2_{H^1(\Omega)} + \|u_2\|^2_{H^1(\Omega)}) + \hat{c}\|\xi\|^2_{L_2(\Omega)}$$
$$+ (2\mu + 1)\lambda\gamma^2\|P\|^2_{L_2(\Omega)}]$$
$$\geq \alpha\|(u_1, u_2, P, \xi)\|^2_{\mathbf{H}} \tag{5-168}$$

We also have $\|(v_1, v_2, Q, \eta)\|_{\mathbf{H}} \leq \|(u_1, u_2, P, \xi)\|_{\mathbf{H}}$ so that

$$B(u_1, u_2, P, \xi;\ v_1, v_2, Q, \eta) \geq \alpha\|(u_1, u_2, P, \xi)\|_{\mathbf{H}}\,\|(v_1, v_2, Q, \eta)\|_{\mathbf{H}} \tag{5-169}$$

Thus, $B(\cdot,\ \cdot)$ satisfies the conditions of Theorem 5-5. The continuity of the linear functional can be easily established. Therefore the weak problem in Eq. (5-162) has a unique solution.

The Euler equations of the functional (5-160), or equivalently, the weak problem (5-162) are given by

$$\mu\left[2\frac{\partial^2 u_1}{\partial x_1^2} + \frac{\partial}{\partial x_2}\left(\frac{\partial u_1}{\partial x_2} + \frac{\partial u_2}{\partial x_1}\right)\right] = f_1 + \frac{\partial \xi}{\partial x_1}$$
$$\mu\left[\frac{\partial}{\partial x_1}\left(\frac{\partial u_1}{\partial x_2} + \frac{\partial u_2}{\partial x_1}\right) + 2\frac{\partial^2 u_2}{\partial x_2^2}\right] = f_2 + \frac{\partial \xi}{\partial x_2} \tag{5-170}$$
$$\lambda\gamma^2 P + \gamma\xi = 0$$
$$P\gamma - \left(\frac{\partial u_1}{\partial x_1} + \frac{\partial u_2}{\partial x_2}\right) = 0$$

For the case $\lambda < \infty$ (i.e., compressible and nearly incompressible elasticity), we can solve the last two equations for ξ,

$$\xi = -\lambda\left(\frac{\partial u_1}{\partial x_1} + \frac{\partial u_2}{\partial x_2}\right)$$

and substitute into the first two equations in Eq. (5-170) to obtain the Navier equations.

Now consider the incompressible case, $\lambda \to \infty$. In this case, the third and fourth equations give, respectively, $P = 0$ and

$$\frac{\partial u_1}{\partial x_1} + \frac{\partial u_2}{\partial x_2} = 0 \tag{5-171}$$

Thus, the first two equations of (5-170) and Eq. (5-171) describe the equilibrium of incompressible elastic solids.

It should also be noted that the equations of the stationary Stokes problem are precisely the same as those of incompressible elasticity with the following changes in the meaning of the variables:

$$u_1, u_2 = \text{velocity components}$$

$$\xi = \text{hydrostatic pressure}$$

$$\mu = \text{viscosity}$$

Therefore, the existence results presented here for incompressible elasticity are immediately applicable to the Lagrange multiplier formulation, also known as the *mixed formulation*, presented in Example 4-12.

5-5-4 The Penalty Formulation of Incompressible Fluids

Of the two formulations we discussed for the Stokes problem (see Examples 4-12 and 4-17), the Lagrange multiplier formulation is already dealt with in the last section. Here we consider the penalty function formulation [see Reddy (1979)].

Recall the weak problem of the penalty function formulation from Example 4-17: find $\mathbf{u} \in \mathbf{H}_K^1 \subset [H_0^1(\Omega)]^N$ such that

$$B_K(\mathbf{v}, \mathbf{u}) = (\mathbf{f}, \mathbf{v})_0 \tag{5-172}$$

for all $\mathbf{v} \in \mathbf{H}_K^1$. The space $\mathbf{H}_k^1(\Omega)$ and the bilinear form are defined in Eqs. (4-157) and (4-158). The space $\mathbf{H}_K^1$ is the same as $\mathbf{H}_0^1(\Omega)$ except for the norm; however, the norms in the two spaces can be shown to be equivalent.

The continuity and $\mathbf{H}_K^1$-ellipticity of $B_K(\mathbf{v}, \mathbf{u})$ can be easily established (see Sec. 5-5-2). The constants M and α in the inequalities of Eq. (5-39) for the present problem are independent of the penalty parameter K. Hence, the weak problem (5-172) has a unique solution.

The convergence of the penalty solution to the solution of the original problem is governed by Theorem 4-8. We can derive the a-priori error estimate [see Eq. (4-151)] for the problem. To this end we consider the *mixed* version of the penalty function formulation: find $(\mathbf{u}_K, P_K) \in H_K^1 \times \tilde{H}^0$ such that

$$\begin{aligned} B(\mathbf{v}, \mathbf{u}_K) - (P_K, \operatorname{div} \mathbf{v})_0 &= (\mathbf{f}, \mathbf{v})_0 \\ -(\operatorname{div} \mathbf{u}_K, Q)_0 &= \frac{1}{K}(P_K, Q)_0 \end{aligned} \tag{5-173}$$

for every $(\mathbf{v}, Q) \in H_K^1 \times \tilde{H}^0$. Here $\tilde{H}^0$ denotes the subspace of $L_2(\Omega)$ that is orthogonal to the space of constants, and $B(\mathbf{v}, \mathbf{u})$ is the bilinear form defined in Eq. (4-98). Subtracting Eq. (5-173) from the corresponding equations of the mixed formulation, we obtain

$$B(\mathbf{v}, \mathbf{u} - \mathbf{u}_K) - (P - P_K, \operatorname{div} \mathbf{v})_0 = 0$$

$$-(\operatorname{div}(\mathbf{u} - \mathbf{u}_K), Q)_0 = \frac{1}{K}(P - P_K, Q)_0 - \frac{1}{K}(P, Q)_0$$

Letting $\mathbf{v} = \mathbf{u} - \mathbf{u}_K$ and $Q = P - P_K$, and using the second equation in the first, we obtain

$$\|\mathbf{u} - \mathbf{u}_K\|^2_{H^1(\Omega)} + \frac{1}{K}\|P - P_K\|^2_{L_2(\Omega)} = \frac{1}{K}(P, P - P_K)_0$$

$$\leq \frac{1}{K}\|P\|_{L_2(\Omega)}\|P - P_K\|_{L_2(\Omega)}$$

Now using the *elementary inequality*

$$|a|\,|b| \leq \frac{1}{2\alpha}|a|^2 + \frac{\alpha}{2}|b|^2 \qquad \text{for any } \alpha > 0$$

we obtain (with $\alpha = 1$)

$$\|\mathbf{u} - \mathbf{u}_K\|^2_{H^1(\Omega)} + \frac{1}{2K}\|P - P_K\|^2_{L_2(\Omega)} \leq \frac{1}{2K}\|P\|^2_{L_2(\Omega)} \tag{5-174}$$

This estimate also shows the convergence of $\mathbf{u} \to \mathbf{u}_K$ as $K \to \infty$.

We close this section with an example of application of the Lagrange multiplier and penalty methods to a constrained discrete problem.

Example 5-14 Let $\mathbf{B}$ be a matrix operator mapping $\mathbb{R}^n$ into $\mathbb{R}^m$ (i.e., $\mathbf{B}$ is a $m \times n$ matrix). The range of $\mathbf{B}$ is defined by

$$\mathscr{R}(\mathbf{B}) = \{\mathbf{b} \in \mathbb{R}^m\colon\ \mathbf{b} = \mathbf{B}\mathbf{v} \text{ for every } \mathbf{v} \in \mathbb{R}^n\}$$

and the set $\mathbf{V}$ is defined by

$$\mathbf{V} = \{\mathbf{v} \in \mathbb{R}^n\colon \mathbf{B}\mathbf{v} = \mathbf{c}\} \tag{5-175}$$

Clearly $\mathbf{c} \in \mathbb{R}^m$, hence $\mathbf{V}$ is not empty.

Consider the variational problem: find $\mathbf{u} \in \mathbf{V}$ such that

$$(\mathbf{A}\mathbf{u}, \mathbf{v} - \mathbf{u}) = (\mathbf{b}, \mathbf{v} - \mathbf{u}) \qquad \text{for every } \mathbf{v} \in \mathbf{V} \tag{5-176}$$

The problem has a unique solution since we can apply Theorem 5-7 with $H = \mathbb{R}^n$.

If $\mathbf{A} = \mathbf{A}^T$ (transpose of $\mathbf{A}$), then the variational (inequality) problem is equivalent to the minimization problem:

Find $\mathbf{u} \in \mathbf{V}$ such that

$$J(\mathbf{u}) \leq J(\mathbf{v}) \qquad \text{for every } \mathbf{v} \in \mathbf{V} \tag{5-177}$$

where

$$J(\mathbf{v}) = \tfrac{1}{2}(\mathbf{A}\mathbf{v}, \mathbf{v}) - (\mathbf{b}, \mathbf{v}) \tag{5-178}$$

The Euler equations of the variational problem (5-176) can be derived as follows. Let $\mathbf{u}$ be the solution of (5-176). Then $\mathbf{u} + \mathbf{w} \in \mathbf{V}$ for every $\mathbf{w} \in \ker(\mathbf{B})$, where $\ker(\mathbf{B}) = \{\mathbf{v} \in \mathbb{R}^n\colon \mathbf{B}\mathbf{v} = 0\}$. Taking $\mathbf{v} = \mathbf{u} + \mathbf{w}$ in (5-176), we obtain

$$(\mathbf{A}\mathbf{u} - \mathbf{b}, \mathbf{w}) = 0 \qquad \text{for every } \mathbf{w} \in \ker(\mathbf{B})$$

Hence $\mathbf{A}\mathbf{u} - \mathbf{b} \in \ker(\mathbf{B})^\perp$, where $\ker(\mathbf{B})^\perp$ stands for the orthogonal complement of $\ker(\mathbf{B})$. It is well known from linear algebra that

$$\ker(\mathbf{B})^\perp = \mathscr{R}(\mathbf{B}^T)$$

Therefore, there exists an element $-P \in \mathbb{R}^m$ such that (the negative sign is used to cast the equation in a standard form)

$$\mathbf{A}\mathbf{u} - \mathbf{b} = -\mathbf{B}^T P \qquad \text{or} \qquad \mathbf{A}\mathbf{u} + \mathbf{B}^T P = \mathbf{b} \tag{5-179}$$

The fact that $\mathbf{u} \in \mathbf{V}$ implies that

$$\mathbf{B}\mathbf{u} = \mathbf{c} \tag{5-180}$$

Thus the variational problem (5-176) is characterized by the matrix equations (5-179) and (5-180).

Conversely, Eq. (5-179) implies Eq. (5-176) for $\mathbf{v}, \mathbf{u} \in \mathbf{V}$, $\mathbf{u} - \mathbf{v} \in \ker(\mathbf{B})$. Taking the inner product of Eq. (5-179) with $\mathbf{u} - \mathbf{v}$, we obtain

$$\begin{aligned}(\mathbf{A}\mathbf{u}, \mathbf{u} - \mathbf{v}) - (\mathbf{b}, \mathbf{u} - \mathbf{v}) &= -(\mathbf{B}^T\mathbf{p}, \mathbf{u} - \mathbf{v}) \qquad \text{for all } \mathbf{u} - \mathbf{v} \in \ker(\mathbf{B}) \\ &= -(\mathbf{P}, \mathbf{B}(\mathbf{u} - \mathbf{v})) \\ &= 0\end{aligned}$$

Equations (5-179) and (5-180) define a constrained variational problem. For $\mathbf{A} = \mathbf{A}^T$, $\mathbf{P}$ is the Lagrange multiplier associated with the linear equality constraint (5-180), which defines $\mathbf{V}$. The unique solution $(\mathbf{u}, \mathbf{P})$ of the Eqs. (5-179) and (5-180) is in $\mathbb{R}^n \times \mathscr{R}(\mathbf{B})$, and all other solutions of it can be expressed as $(\mathbf{u}, \mathbf{P} + \mathbf{q})$ where $\mathbf{q} \in \ker(\mathbf{B}^T)$.

Since the problem described by Eqs. (5-179) and (5-180) is a constrained problem, we can apply the penalty function method to formulate the problem. Define $G\colon \mathbb{R}^n \to \mathbb{R}$ by

$$G(\mathbf{v}) = \tfrac{1}{2}|\mathbf{B}\mathbf{v} - \mathbf{c}|^2 \tag{5-181}$$

where $|\cdot|$ denotes the euclidean norm of $\mathbb{R}^m$. The Gateaux derivative of G is given by

$$\delta G(\mathbf{v}) = \mathbf{B}^T(\mathbf{B}\mathbf{v} - \mathbf{c})$$

Now we can define the penalty function formulation for Eqs. (5-175) and (5-176):

Find $\mathbf{u}_k \in \mathbb{R}^n$ such that

$$(\mathbf{A}\mathbf{u}_k, \mathbf{v} - \mathbf{u}_k) + K[G(\mathbf{v}) - G(\mathbf{u}_k)] = (\mathbf{b}, \mathbf{v} - \mathbf{u}_k) \tag{5-182}$$

for all $\mathbf{v} \in \mathbb{R}^n$, where $K > 0$ is the penalty parameter. The problem (5-182) is equivalent to the algebraic equations

$$(\mathbf{A} + K\mathbf{B}^T\mathbf{B})\mathbf{u}_k = K\mathbf{B}^T\mathbf{c} + \mathbf{b} \tag{5-183}$$

The Lagrange multiplier $\mathbf{P}_k \in \mathbb{R}^m$ is given by

$$\mathbf{P}_k = K(\mathbf{B}\mathbf{u}_k - \mathbf{c}) \tag{5-184}$$

In view of Eqs. (5-183) and (5-184), Eq. (5-182) is equivalent to the system [note the form of the equations, which resemble Eq. (5-173)]

$$\begin{aligned} \mathbf{A}\mathbf{u}_k + \mathbf{B}^T\mathbf{P}_k &= \mathbf{b} \\ -\mathbf{B}\mathbf{u}_k + \frac{1}{K}\mathbf{P}_k &= -\mathbf{c} \end{aligned} \tag{5-185a}$$

or

$$\begin{bmatrix} \mathbf{A} & \mathbf{B}^T \\ -\mathbf{B} & \dfrac{1}{K}\mathbf{I} \end{bmatrix} \begin{Bmatrix} \mathbf{u}_k \\ \mathbf{P}_k \end{Bmatrix} = \begin{Bmatrix} \mathbf{b} \\ -\mathbf{c} \end{Bmatrix} \tag{5-185b}$$

The coefficient matrix in Eq. (5-185*b*) is an $n + m$ by $n + m$ positive-definite matrix.

The penalty solution $(\mathbf{u}_k, \mathbf{P}_k)$ will be a good approximation of the true solution $(\mathbf{u}, \mathbf{P})$, provided that K is *sufficiently large*. On the other hand, the *condition number* of the matrix $\mathbf{A}_k \equiv \mathbf{A} + K\mathbf{B}^T\mathbf{B}$ will be large, spoiling the *condition* of the matrix $\mathbf{A}_k$ (i.e., for large condition numbers of $\mathbf{A}_k$, $\mathbf{A}_k$ is *ill-conditioned*). The condition number $n(\mathbf{A}_k)$ for a symmetric matrix $\mathbf{A}_k$ is defined by

$$n(\mathbf{A}_k) = \|\mathbf{A}_k^{-1}\| \, \|\mathbf{A}_k\| = K \frac{\rho(\mathbf{B}^T\mathbf{B})}{\alpha} [1 + \beta(K)] \tag{5-186}$$

where $\lim_{k \to \infty} \beta(K) = 0$, $\rho(\mathbf{B}^T\mathbf{B})$ is the *spectral radius* of $\mathbf{B}^T\mathbf{B}$ (i.e., the largest eigenvalue of $\mathbf{B}^T\mathbf{B}$) and α is defined by

$$\alpha = \inf_{\mathbf{v} \in \ker(\mathbf{B})} \frac{(\mathbf{A}\mathbf{v}, \mathbf{v})}{\|\mathbf{v}\|^2} \qquad \mathbf{v} \neq \mathbf{0} \tag{5-187}$$

From Eq. (5-186) it is clear that $n(\mathbf{A}_k)$ increases with increasing value of K, and $\mathbf{A}_k$ is ill-conditioned for large K. The ill-conditioning property of the finite-dimensional version of the penalty function formulation of a continuous problem is the main drawback of the penalty method.

The drawback of the penalty method is overcome in practice by using the *augmented lagrangian method*. In the augmented lagrangian method a combination of penalty and Lagrange multiplier methods is employed to allow smaller K, because the method produces the exact solution (by virtue of the Lagrange multiplier method). As applied to the system (5-179) and (5-180), the augmented lagrangian functional is given by (assume $\mathbf{A} = \mathbf{A}^T$),

$$\mathscr{L}_K(\mathbf{u}, \mathbf{P}) = \tfrac{1}{2}(\mathbf{Au}, \mathbf{u}) - (\mathbf{b}, \mathbf{u}) + (\mathbf{P}, \mathbf{Bu} - \mathbf{c}) + \frac{K}{2} \,|\, \mathbf{Bu} - \mathbf{c} \,|^2 \qquad (5\text{-}188)$$

The stationary points $(\mathbf{u}, \mathbf{P})$ of $\mathscr{L}_K(\mathbf{u}, \mathbf{P})$ are characterized by the system of equations

$$\begin{aligned} \mathbf{Au} + K\mathbf{B}^T(\mathbf{Bu} - \mathbf{c}) + \mathbf{B}^T\mathbf{P} &= \mathbf{b} \\ \mathbf{Bu} &= \mathbf{c} \end{aligned} \qquad (5\text{-}189)$$

Note that we *do not* use the second equation in the first to simplify the equation [in which case it reduces to Eq. (5-179)]. The point $(\mathbf{u}, \mathbf{P})$ is said to be a *saddle point* over $\mathbb{R}^n \times \mathbb{R}^m$ of the augmented lagrangian functional $\mathscr{L}_K$ if

$$\mathscr{L}_K(\mathbf{u}, \mathbf{q}) \le \mathscr{L}_K(\mathbf{u}, \mathbf{P}) \le \mathscr{L}_K(\mathbf{v}, \mathbf{P}) \qquad (5\text{-}190)$$

for all $(\mathbf{v}, \mathbf{q})$ in $\mathbb{R}^n \times \mathbb{R}^m$.

Computationally, Eqs. (5-189) are solved in two steps. For example, the so-called *Uzawa algorithm* for Eq. (5-189) is as follows. For $n \ge 0$ (iteration number), define $\mathbf{u}^n$ and $\mathbf{P}^{n+1}$ by the equations

$$\begin{aligned} (\mathbf{A} + K\mathbf{B}^T\mathbf{B})\mathbf{u}^n &= \mathbf{b} + K\mathbf{B}^T\mathbf{c} - \mathbf{B}^T\mathbf{P}^n \\ \mathbf{P}^{n+1} &= \mathbf{P}^n + \rho(\mathbf{Bu}^n - \mathbf{c}) \qquad \rho > 0 \end{aligned} \qquad (5\text{-}191)$$

where $\mathbf{b}$ and $\mathbf{c}$ are given vectors and ρ is a convergence parameter. The initial value $\mathbf{P}^0 \in \mathbb{R}^m$ is assumed to be known.

5-6 EIGENVALUE PROBLEMS

5-6-1 Introduction

In the preceding sections, we briefly discussed what an eigenvalue problem is. A more formal treatment of this topic forms the core of the discussion in this section.

Recall from Sec. 5-4 that the weak form of the operator equation

$$Au = f \text{ in } \Omega \qquad (5\text{-}192)$$

where A is the linear operator, with homogeneous essential boundary conditions, is given by

$$B(v, u) = (v, f)_0 \qquad \text{for all } v \in V \qquad (5\text{-}193)$$

Here V denotes the subspace of those functions from $H^m(\Omega)$ which satisfy the homogeneous essential boundary conditions, where $2m$ denotes the order of the differential operator A. If $B(v, u)$ is V-elliptic, we know that the weak problem (5-193) has a unique solution $u(\mathbf{x})$ and that there exists a constant $c > 0$ independent of u and $f \in L_2(\Omega)$ such that

$$\|u\|_V \leq c\|f\|_0 \tag{5-194}$$

holds. Thus, to every $f \in L_2(\Omega)$ there corresponds one weak solution $u \in V$. This defines a linear bounded operator $T: L_2(\Omega) \to V$ such that

$$u = Tf \tag{5-195}$$

Using Eqs. (5-193) and (5-194), we have

$$\begin{gathered} B(v, Tf) = (v, f)_0 \qquad \text{for every } f \in L_2(\Omega) \qquad v \in V \\ \|Tf\| \leq c\|f\|_0 \end{gathered} \tag{5-196}$$

In this section we study the properties of the operator T. In particular we study the properties of T when $T: V \subset L_2(\Omega) \to V$, because the eigenvalue problem is defined by operator equations of the form

$$Tu = \mu u, \qquad \mu = 1/\lambda$$

where λ is an eigenvalue.

5-6-2 Existence and Uniqueness Results

Consider the following problem associated with Eq. (5-192),

$$Au - \lambda u = f \tag{5-197}$$

with homogeneous boundary conditions. The weak problem associated with Eq. (5-197) consists of finding $u \in V$ such that

$$B(v, u) - \lambda(v, u)_0 = (v, f)_0 \tag{5-198}$$

holds for every $v \in V$. We assume that $B(v, u)$ is symmetric and V-elliptic. The problem of finding a number λ and a nonzero function $u \in V$ such that

$$B(v, u) - \lambda(v, u)_0 = 0 \qquad \lambda \neq 0 \tag{5-199}$$

holds for every $v \in V$ is called the *eigenvalue problem* of the bilinear form $B(v, u)$.

The following theorem relates Eq. (5-198) with Eq. (5-199).

Theorem 5-11 The function $u \in V$ is the weak solution of the problem (5-198) if and only if

$$u - \lambda Tu = Tf \tag{5-200}$$

holds in V. The number λ is an eigenvalue of the bilinear form $B(v, u)$ and the function $u(\mathbf{x})$ is the corresponding eigenfunction if and only if

$$u - \lambda Tu = 0 \qquad u \neq 0 \tag{5-201}$$

holds in V.

PROOF If $u(\mathbf{x})$ is the weak solution of the problem (5-198),

$$B(v, u) = (v, f + \lambda u)_0$$

then by the definition of operator T [see Eqs. (5-193) and (5-195)] we have

$$u = T(f + \lambda u) \qquad \text{or} \qquad u - \lambda T u = Tf$$

The converse follows very easily.

If the problem (5-199) has a nontrivial solution, it follows from the definition of T that

$$B(v, u) = (v, \lambda u)_0 \rightarrow u = T(\lambda u) = \lambda T u$$

i.e., Eq. (5-201) holds, and vice versa.

Because of Theorem 5-11 it is possible to investigate the question of the solvability of Eq. (5-198) by considering Eqs. (5-200) and (5-201). The advantage of working with Eqs. (5-200) and (5-201) is that they involve the completely continuous operator T. A linear operator $T: U \rightarrow V$, where U and V are Banach spaces, is said to be *completely continuous* (also called *compact*) if T maps every bounded set in U into a compact subset of V. Every bounded linear operator in a finite-dimensional space is completely continuous, because bounded sets in finite-dimensional spaces are compact. In an infinite-dimensional space, any bounded operator whose range is finite-dimensional is compact for the same reason. However, not every bounded operator from an infinite-dimensional space into another infinite-dimensional space is compact. For example, the identity operator is bounded but not compact in an infinite-dimensional space because a unit sphere is bounded but not compact. The proof that T is compact is given in Rektorys (1980).

The operator T is also self-adjoint and positive in H_A, the energy space. Indeed, from Eq. (5-196) we have for $f = u$,

$$(v, Tu)_A = (v, u)_0$$
$$= (u, v)_0 = (u, Tv)_A$$

for all $u, v \in H_A$. Further,

$$(Tu, u)_A = (u, Tu)_A = (u, u)_0 \geq 0$$

for all $u \in H_A$.

We now state, without proof, the properties of the eigenvalue problem (5-201) associated with a self-adjoint compact operator. For additional details see Rektorys (1980) and Mikhlin (1964).

Theorem 5-12 Consider the problem of finding a number λ and a function $u(\mathbf{x}) \in V$ such that Eq. (5-198) holds for every $v \in V$. Let $f \in L_2(\Omega)$ and let the form $B(v, u)$ be symmetric and V-elliptic. Then the form $B(v, u)$ has a countable set of eigenvalues, each of them being positive,

$$\lambda_1 \leq \lambda_2 \leq \lambda_3 \leq, \cdots, \lim_{n \rightarrow \infty} \lambda_n = \infty \tag{5-202}$$

The corresponding orthogonal system of eigenfunctions $\{\phi_i\}$ constitutes a basis in the space H_A (hence in V). Further, we have

$$\lambda_1 = \min_{v \in V, v \neq 0} \frac{(v, v)_A}{\|v\|_0^2} = \frac{(u_1, u_1)_A}{(u_1, u_1)_0}, \text{ etc.} \tag{5-203}$$

For $\lambda \neq \lambda_n$, $n = 1, 2, 3, \ldots$, the problem (5-198) has exactly one weak solution for every $f \in L_2(\Omega)$. In particular, for $f = 0$ in $L_2(\Omega)$ it has only the null solution. If $\lambda = \lambda_n$, then the problem (5-198) is solvable, but not uniquely, if and only if the function $f(\mathbf{x})$ is orthogonal in $L_2(\Omega)$ to every eigenfunction corresponding to this eigenvalue.

From Theorem 5-12, the so-called *Fredholm alternative* follows: either the problem (5-198) is uniquely solvable for every $f \in L_2(\Omega)$, or the corresponding homogeneous problem (5-199) has a nonzero solution.

Since the space V is dense in $L_2(\Omega)$ the set of eigenfunctions $\{\phi_i(\mathbf{x})\}$ is complete and orthogonal in $L_2(\Omega)$. Therefore every function $v \in L_2(\Omega)$ can be represented by

$$v = \sum_{j=1}^{\infty} (v, \phi_j)\phi_j \tag{5-204}$$

A number μ is called a *regular value* of the operator T in a Hilbert space H if the equation

$$Tu - \mu u = f$$

has exactly one solution $u = (T - \mu I)^{-1}f$, where I is the identity operator, and the solution depends continuously on f. The set of all $\lambda = 1/\mu$ such that the range of the transformation $(\lambda I - T)$ is dense in V and such that $(\lambda I - T)$ has a continuous inverse defined on its range is said to be the *resolvent set* of T. The set of all numbers that are not in the resolvent set is termed the *spectrum* of T. If there exists a countable set of positive eigenvalues with the property (5-202) and, in addition, the corresponding set of eigenfunctions is complete, then the spectrum is termed *discrete*.

If the eigenfunctions $\phi_i(\mathbf{x})$ belong to the domain $\mathscr{D}_A$ of the operator A, then the relation

$$(A\phi_j, v) = B(\phi_j, v) = \lambda_j(\phi_j, v) \qquad \text{for every } v \in V$$

gives the classical eigenvalue problem

$$A\phi_j = \lambda_j \phi_j \tag{5-205}$$

If we choose $\{\phi_j\}$ as the basis for H_A, then for the N-parameter Ritz solution of $Au = f$ we write,

$$u_N(\mathbf{x}) = \sum_{j=1}^{N} c_j \phi_j$$

and c_j are given by $c_j = (f, \phi_j)$. Thus the eigenfunctions associated with the eigenvalue problem $Au = \lambda u$ can be used to advantage in the Ritz method (see Chap. 6) to find the solution of the equation $Au = f$.

PROBLEMS 5-3

5-3-1 Consider the fourth-order ordinary differential equation,

$$\frac{d^2}{dx^2}\left(EI\frac{d^2w}{dx^2}\right) = f \qquad 0 < x < L$$

$$w = \frac{dw}{dx} = 0 \text{ at } x = 0 \qquad EI\frac{d^2w}{dx^2} = \frac{d}{dx}\left(EI\frac{d^2w}{dx^2}\right) = 0 \text{ at } x = L$$

where $EI > 0$ for $0 < x < L$. The equations describe the bending of a cantilever beam of length L, flexural rigidity EI, and subjected to distributed load. The mixed formulation of this problem is to consider the alternative pair of second-order equations

$$\frac{d^2w}{dx^2} - \frac{M}{EI} = 0 \text{ in } (0, L) \qquad w = \frac{dw}{dx} = 0 \text{ at } x = 0$$

$$\frac{d^2M}{dx^2} = f \text{ in } (0, L) \qquad M = \frac{dM}{dx} = 0 \text{ at } x = L$$

Construct a weak problem of these equations on the space $H = H^1_*(0, L) \times H^1_+(0, L)$, where

$$H^1_*(0, L) = \{w \in H^1(0, L)\colon w(0) = 0\}$$

$$H^1_+(0, L) = \{M \in H^1(0, L)\colon M(L) = 0\}$$

and investigate the existence of solutions.

5-3-2 The Lagrange multiplier formulation of the fourth-order equation in Prob. 5-3-1 involves seeking the minimum of the functional,

$$J(\psi, w) = \int_0^L \left[\frac{EI}{2}\left(\frac{d\psi}{dx}\right)^2 - fw\right] dx$$

subject to the constraint $\psi + dw/dx = 0$. Give the weak problem associated with the formulation and investigate the question of existence of solution.

5-3-3 Repeat Prob. 5-3-2 using the penalty function formulation.

5-3-4 Consider the second-order elliptic problem $-\nabla^2 u = f$ in Ω, $u = 0$ on Γ. A mixed formulation of the equation is to seek $(u, \mathbf{q})$ such that $-\text{div}\, \mathbf{q} = f$ in Ω, $\mathbf{q} - \text{grad}\, u = 0$ in Ω, and $u = 0$ on Γ. Introduce the bilinear forms $B(\mathbf{P}, \mathbf{q}) = \int_\Omega \mathbf{P}\cdot\mathbf{q}\, d\mathbf{x}$ and the function space $H(\text{div}; \Omega) = \{\mathbf{q}\colon \mathbf{q} \in [L_2(\Omega)]^2;\ \text{div}\, \mathbf{q} \in L_2(\Omega)\}$. Prove that the weak problem: find $(u, \mathbf{q}) \in L_2(\Omega) \times H(\text{div}; \Omega)$ such that $B(\mathbf{P}, \mathbf{q}) + \int_\Omega u \,\text{div}\, \mathbf{P}\, d\mathbf{x} = 0$, $\int_\Omega v(\text{div}\, \mathbf{q} + f)\, d\mathbf{x} = 0$ holds for all $(v, \mathbf{P}) \in L_2(\Omega) \times H(\text{div}; \Omega)$ has a unique solution.

Hint: See Raviart and Thomas (1979).

5-3-5 Prove the existence of the unique solution $(\mathbf{u}, P)$ to the Lagrange multiplier (or mixed) formulation of the Stokes problem; i.e., specialize the discussion of Sec. 5-5-3 to the Stokes problem.

5-3-6 Consider the mixed formulation of thin, elastic, and isotropic plates: find the stationary values of the functional,

$$J(w, M_1, M_2) = \int_\Omega \Bigg[\frac{1}{2}(-S_{11}M_1^2 + 2S_{12}M_1M_2 - S_{22}M_2^2)$$

$$+ \frac{\partial w}{\partial x_1}\frac{\partial M_1}{\partial x_1} + \frac{\partial w}{\partial x_2}\frac{\partial M_2}{\partial x_2} + 2D_{66}\left(\frac{\partial^2 w}{\partial x_1\, \partial x_2}\right)^2 - fw\Bigg] dx_1\, dx_2$$

on $[H^2(\Omega)]^3$, where S_{11}, S_{12}, S_{22}, and D_{66} are positive constants. Derive the associated weak problem, and prove the continuity and the V-ellipticity of the bilinear form for an appropriate V.

5-3-7 Consider the Lagrange multiplier formulation for thin plates (also, see Example 4-14): Given $q \in L_2(\Omega)$, find $\Lambda \equiv (w, \psi_1, \psi_2, \lambda_1, \lambda_2) \in V$ such that [see Eq. (4-125)], $B(\Lambda, \bar{\Lambda}) + M(\lambda_1, \lambda_2; \bar{\Lambda}) = (q, \bar{w})_0$, $M(\Lambda; \bar{\lambda}_1, \bar{\lambda}_2) = 0$ for all $\bar{\Lambda} = (\bar{w}, \bar{\psi}_1, \bar{\psi}_2, \bar{\lambda}_1, \bar{\lambda}_2) \in V$, where $V = [H_0^1(\Omega)]^3 \times [L_2(\Omega)]^2$ with $\|\Lambda\|_V^2 = \|w\|_1^2 + \|\psi_1\|_1^2 + \|\psi_2\|_1^2 + \|\lambda_1\|_0^2 + \|\lambda_2\|_0^2$. Prove the existence of the unique solution to the problem.

5-3-8 Consider an open bounded domain Ω with portions Ω_1 and Ω_2, with the property $\Omega_1 \cap \Omega_2$ is empty and $\bar{\Omega}_1 \cup \bar{\Omega}_2 = \bar{\Omega}$. Let us denote the intersection of $\bar{\Omega}_1 \cap \bar{\Omega}_2$ by Γ_{12} (the interface of $\bar{\Omega}_1$ and $\bar{\Omega}_2$), and the boundary of Ω by Γ. Introduce the bilinear form,

$$B(v, w) = \sum_{i=1}^{2} \int_{\Omega_i} \left[\frac{\partial^2 v}{\partial x_1^2} \frac{\partial^2 w}{\partial x_1^2} + 2 \frac{\partial^2 v}{\partial x_1 \partial x_2} \cdot \frac{\partial^2 w}{\partial x_1 \partial x_2} + \frac{\partial^2 v}{\partial x_2^2} \frac{\partial^2 w}{\partial x_2^2} \right] d\mathbf{x}$$

$$+ DK \left[\int_{\Gamma_{12}} \left(\frac{\partial v}{\partial n} + \frac{\partial v}{\partial(-n)} \right) \left(\frac{\partial w}{\partial n} + \frac{\partial w}{\partial(-n)} \right) ds + \oint_{\Gamma} \frac{\partial v}{\partial n} \frac{\partial w}{\partial n} ds \right]$$

on $H = H^1(\Omega) \supset V$, where V is the space of piecewise cubic polynomials which are equal to zero on Γ, K is the penalty parameter, and $(-n)$ denotes the normal in the opposite direction to n. Show that if $B(v, v) = 0$ for $v \in V$ then $v = 0$; that is, $B(\cdot, \cdot)$ is positive-definite on V. Hence show that $B(v, v) = \|v\|_K^2$ defines a norm on $[H^2(\Omega) \cap H_0^1(\Omega)] \cup V \equiv \hat{H}$. Finally, show that for $f \in L_2(\Omega)$, the problem $B(v, w) = (v, f)_0$ for every $v \in V$ has a unique solution.

Note: This problem arises in connection with the finite-element analysis of thin plates in bending. Here Ω_1 and Ω_2 can be viewed as two plate elements and V is the space of the Hermite cubic polynomials. The constraint condition is to make the normal derivative, $\partial w/\partial n$, continuous across the interface Γ_{12} of the two elements [see Babuska and Zlamal (1973)].

5-3-9 Consider the Dirichlet problem for Poisson's equation $-\nabla^2 u = f$ in Ω, $u = 0$ on Γ. Use the penalty functional,

$$P_K(u) = \int_{\Omega} [\tfrac{1}{2}|\text{grad } u|^2 + fu] \, d\mathbf{x} + \frac{K}{2} \int_{\Omega} (-\nabla^2 u - f)^2 \, d\mathbf{x}$$

to derive a penalty formulation that enables the convergence of $-\nabla^2 u$ to f and prove the existence of the solution in appropriate space.

5-3-10 Consider the problem: suppose that $\beta(K) > \beta_0 > 0$ be given for $0 < K < \infty$. Let $H_K(\Omega)$ be the completion of $C_0^\infty(\bar{\Omega})$ with respect to the norm, $\|u\|_H^2 = \int_\Omega |\nabla^2 u|^2 \, d\mathbf{x} + \beta(K) \int_\Gamma |u|^2 \, ds$, $u \in C_0^\infty(\bar{\Omega})$. Define on $H_K(\Omega) \times H_K(\Omega)$ the bilinear form $B_K(v, u) = \int_\Omega \nabla^2 v \nabla^2 u \, d\mathbf{x} + \beta(K) \int_\Gamma vu \, ds$. Show that the weak problem $B_K(v, u) = -\int_\Omega fv \, d\mathbf{x}, f \in L_2(\Omega)$ has a unique solution.

5-3-11 Give the penalty function formulation of the constrained minimization problem of Example 4-14; i.e., minimize the functional $\Pi_0(w, \psi_1, \psi_2)$ in Eq. (4-123) subject to the constraints in Eq. (4-124). Derive the weak problem of the penalty formulation and investigate the existence of the solutions.

5-3-12 Give a penalty function formulation of the mixed boundary-value problem $-\nabla \cdot (\mathbf{k} \nabla u) + a_0 u = f$ in Ω, $u = g_0$ on Γ_0, $\mathbf{n} \cdot (\mathbf{k} \nabla u) = g_1$ on Γ_1, where $\Gamma_0 \cup \Gamma_1 = \Gamma$ and $\Gamma_0 \cap \Gamma_1 =$ empty. Treat the Dirichlet boundary condition on Γ_0 to be a constraint, and investigate the existence of solution of the weak problem.

5-3-13 *The Method of Discretization in Time* [see Rektorys (1982)]. Consider the parabolic equation

$$\frac{\partial u}{\partial t} + Au = f(\mathbf{x}) \text{ in } \Omega \times (0, T) \tag{a}$$

$$u(\mathbf{x}, 0) = u_0(\mathbf{x}) \text{ in } \Omega \tag{b}$$

with homogeneous essential boundary conditions on $\Gamma \times (0, T)$. Here A is the $2m$th order differential operator in Eq. (5-76). Divide the time interval $I = [0, T]$ into p intervals $I_j = [t_{j-1}, t_j]$, $j = 1, 2, \ldots,$

p, $t_j = j\,\Delta t$ and $\Delta t = T/p$. Replacing $\partial u/\partial t$ by $(w_j - w_{j-1})/\Delta t$ and u by w_j in Eq. (*a*) we obtain the approximate problem,

$$Aw_j + \frac{w_j}{\Delta t} = f + \frac{w_{j-1}}{\Delta t} \qquad j = 1, 2, \ldots, p \tag{c}$$

where $w_0(\mathbf{x}) = u_0(\mathbf{x})$. Thus, given $w_0(\mathbf{x}) = u_0(\mathbf{x})$, we solve Eq. (*c*) for $j = 1$ (with the boundary conditions of the problem). Then using w_1 thus obtained we solve Eq. (*c*) for $j = 2$ (with the same boundary conditions). This procedure is repeated for $j = 1, 2, \ldots, p$. In this way, the solution of the parabolic equation (*a*) is reduced to the solution of an elliptic problem (*c*). Give the weak formulation of Prob. (*c*) and show that it has a unique solution for each k (the constants in the continuity and positive-definiteness depend on Δt).

5-3-14 Prove the following error estimate for the Prob. 5-3-13:

$$\|u(\mathbf{x}, t_j) - w_j(\mathbf{x})\| \le \frac{\|Af\|\, j(\Delta t)^2}{2}$$

5-3-15 Consider the hyperbolic equation,

$$\begin{aligned} &\frac{\partial^2 u}{\partial t^2} + Au = f(\mathbf{x}) \text{ in } \Omega \times (0, T) \\ &u(\mathbf{x}, 0) = u_0(\mathbf{x}) \\ &\frac{\partial u}{\partial t}(\mathbf{x}, 0) = v_0(\mathbf{x}) \end{aligned} \tag{a}$$

with homogeneous essential boundary conditions. The operator A is given by Eq. (5-76). Replace $\partial^2 u/\partial t^2$ by $(w_j - 2w_{j-1} + w_{j-2})/(\Delta t)^2$ and u by w_j to write the approximate equation of (*a*), formulate variationally and show that it has a unique solution. Note that $w_0(\mathbf{x}) = u_0(\mathbf{x})$ and $w_{-1}(\mathbf{x}) = \frac{1}{2}(\Delta t)^2 f(\mathbf{x})$.

5-3-16 If $A: H \to H$ is a self-adjoint operator on a Hilbert space, prove that the eigenvalues of the operator A are real (i.e., not imaginary) numbers.

5-3-17 Let H be a real Hilbert space $\mathbb{R}^3$, and let $A: H \to H$ be a self-adjoint operator. Let $\{\hat{e}_1, \hat{e}_2, \hat{e}_3\}$ be any orthonormal basis in H, and let a_{ij} denote the matrix representation of A with respect to this basis (that is, $A = a_{ij}\hat{e}_i\hat{e}_j$, sum on repeated indices is implied). Show that the characteristic polynomial $p(\lambda) \equiv \det(a_{ij} - \lambda\delta_{ij})$ is given by $p(\lambda) = -\lambda^3 + I_1\lambda^2 - I_2\lambda + I_3$, where

$$I_1 = a_{11} + a_{22} + a_{33} \text{ (trace of } A) = a_{ii} \text{ (sum on } i)$$

$$I_2 = \tfrac{1}{2}(a_{ii}a_{jj} - a_{ij}a_{ji}) \text{ (sum on } i \text{ and } j)$$

$$I_3 = \det A$$

5-3-18 Find the eigenvalues and eigenvectors associated with the following matrices:

$$(a)\ \begin{bmatrix} 2 & -1 & 1 \\ -1 & 0 & 1 \\ 1 & 1 & 2 \end{bmatrix} \qquad (b)\ \begin{bmatrix} 2 & -1 & 0 \\ -1 & 2 & -1 \\ 0 & -1 & 2 \end{bmatrix}$$

CHAPTER

SIX

VARIATIONAL METHODS OF APPROXIMATION

6-1 INTRODUCTION

In this chapter we deal with approximate methods that employ the variational statements (i.e., either variational principles or weak formulations) to determine continuous solutions of problems in engineering and applied sciences. The methods to be described in this chapter seek a solution to the given problem in terms of adjustable parameters that are determined by either minimizing the functional or solving the weak form of the problem. Such solution methods are called *direct methods* because the approximate solutions are obtained directly by utilizing a variational formulation equivalent to the given problem.

The assumed solutions in the variational methods are in the form of a finite linear combination of vectors from an appropriate vector space. This amounts to seeking the solution in a finite-dimensional subspace V_N of the solution space V of the original continuous problem. Since the solution of a continuous problem in general lies in an infinite-dimensional space V and therefore cannot be represented by a finite set of functions exactly, error is introduced into the solution. The solution obtained is an *approximation* to the true solution of the equations describing a physical problem. As the number of linearly independent terms in the assumed solution is increased (i.e., as the space V_N is enlarged) the error in the approximation will be reduced, and the assumed solution converges to the desired solution of the given equations.

We assume that the space V is approximated by a finite-dimensional subspace V_N (where N denotes a parameter characteristic of the approximation such that $\lim_{N \to \infty} V_N = V$) and that the variational problems are approximated such that the ellipticity, symmetry, etc. of the original problem are preserved. These assumptions allow us to carry the existence and uniqueness results of the continuous problem to the approximate (or discrete) problem. When considering an approximate problem we not only investigate the aspects of existence and uniqueness of solutions, but we must also study the numerical stability of the approximate problem.

The variational methods of approximation to be described here include the Rayleigh–Ritz method, the Bubnov–Galerkin method, the Petrov–Galerkin method, the Kantorovitch method, and the Trefftz method. These methods differ from each other in the form of the variational statement used and hence on the space of approximation functions. Emphasis is placed on the selection of the approximation functions, and on the convergence of the approximate solution to the exact solution.

6-2 THE RITZ METHOD

6-2-1 Introduction

Consider the operator equation,

$$Au = f \text{ in } \Omega \qquad u = 0 \text{ on } \Gamma \tag{6-1}$$

where A is a linear positive-definite operator on a linear set $\mathscr{D}_A$ in a separable Hilbert space H, and $f \in H$. Let H_A be the energy (Hilbert) space introduced in Chap. 4. From the discussions of Chap. 4 we know that the generalized solution u of Eq. (6-1) is an element of H_A and minimizes the functional

$$J(u) = \tfrac{1}{2}B(u, u) - (f, u)_H \equiv \tfrac{1}{2}\|u\|_A^2 - (f, u)_H \tag{6-2a}$$

or, equivalently, satisfies the weak form

$$B(v, u) = (v, f)_H \tag{6-2b}$$

for all $v \in H_A$. Here $B(u, u) = (Au, u)_H$, $\|u\|_A$ is the norm of u in H_A, and $(\cdot, \cdot)_H$ is the inner product in H. The equivalence of solving Eq. (6-1) and minimizing the functional $J(u)$ of Eq. (6-2) on H_A is amply discussed in Chaps. 4 and 5. Of course, the equivalence holds only under certain regularity conditions on the data (e.g., on f).

Because of the equivalence of Eqs. (6-1) and (6-2), it is preferable to solve Eq. (6-2) over Eq. (6-1) for two main reasons. First, the exact solution of Eq. (6-1) is often a difficult task. Second, the variational problem of finding u such that $J(u) \leq J(v)$ for all $v \in H_A$ provides a natural means for determining an approximation to the weak solution. The Ritz method is one such means. The method was proposed by Lord Rayleigh (1842–1919), and independently and in more general form by Ritz (1878–1909).

6-2-2 Description of the Method

The space H_A is a separable Hilbert space because H is separable. In H_A we consider the basis

$$\phi_1, \phi_2, \ldots, \phi_N, \ldots \tag{6-3}$$

We choose a positive integer N and seek an approximation u_N of the solution u in the form

$$u_N = \sum_{j=1}^{N} c_j \phi_j \tag{6-4}$$

where c_j, called the *Ritz coefficients*, are as yet unknown parameters. The parameters are determined by the condition

$$J(u_N) \leq J(v_N) \tag{6-5a}$$

or, equivalently,

$$B(v_N, u_N) = (v_N, f)_H \tag{6-5b}$$

for all v_N of the form $v_N = \sum_{i=1}^{N} b_i \phi_i$, where b_i are arbitrary real numbers. This is equivalent to seeking u_N in the N-dimensional subspace S_N generated by the finite set $\{\phi_1, \phi_2, \ldots, \phi_N\}$. Since by assumption $\{\phi_1, \phi_2, \ldots, \phi_N, \ldots\}$ is a basis in H_A, the weak solution $u \in H_A$ can be approximated to arbitrary accuracy by a *suitable* linear combination of its elements. Under appropriate conditions on the set $\{\phi_i\}$, the approximate solution u_N is expected to converge to the actual solution as $N \to \infty$.

Once the set $\{\phi_j\}$ is chosen, the computation of the parameters c_j in Eq. (6-4) is simple. The functional $J(u_N)$ becomes, after carrying out the inner product operation (i.e., integrating over the domain), an ordinary function of the parameters $c_1, c_2, \ldots, c_N$. Therefore, the condition (6-5) is equivalent to

$$\frac{\partial J}{\partial c_i}(c_1, c_2, \ldots, c_N) = 0 \qquad i = 1, 2, \ldots, N$$

or

$$0 = \frac{\partial}{\partial c_i}\left[\frac{1}{2} B\left(\sum_{j=1}^{N} c_j \phi_j, \sum_{k=1}^{N} c_k \phi_k\right) - \left(f, \sum_{k=1}^{N} c_k \phi_k\right)_H\right]$$

$$= \frac{1}{2}\left[\sum_{k=1}^{N} B(\phi_i, \phi_k)c_k + \sum_{j=1}^{N} B(\phi_j, \phi_i)c_j\right] - (f, \phi_i)_H$$

where the bilinearity of $B(\cdot, \cdot)$ is used to bring the summation operation out. In view of the symmetry of $B(\cdot, \cdot)$, we can combine the first two terms to obtain

$$0 = \sum_{j=1}^{N} B(\phi_i, \phi_j)c_j - (f, \phi_i)_H \qquad i = 1, 2, \ldots, N \tag{6-6}$$

Equation (6-6) can also be obtained directly from the weak problem (6-5*b*):

$$B\left(v_N, \sum_{j=1}^{N} c_j \phi_j\right) = (v_N, f)_H$$

or

$$\sum_{j=1}^{N} B(v_N, \phi_j)c_j - (v_N, f)_H = 0$$

Since v_N is an arbitrary element of S_N, the above equation must also hold for $v_N = \phi_i$, $i = 1, 2, \ldots, N$. Thus we obtain Eq. (6-6).

Equation (6-6) contains a system of N equations in the N unknowns, $c_1, c_2, \ldots, c_N$. Since $B(\phi_i, \phi_j) = (\phi_i, \phi_j)_A$ and ϕ_i are linearly independent, the coefficient matrix

$$b_{ij} = B(\phi_i, \phi_j) = (\phi_i, \phi_j)_A \tag{6-7}$$

is nonsingular, and consequently, the system (6-6) is uniquely solvable.

We now summarize the essence of the Ritz method. Given the problem of solving an operator equation, the operator being linear and positive-definite, we determine the N-parameter Ritz solution of the form (6-4) by solving the linear algebraic equations characterized by the matrix equation

$$[B]\{c\} = \{F\} \tag{6-8}$$

where the coefficients b_{ij} of the coefficient matrix $[B]$ are defined in Eq. (6-7), and the elements f_i of the column vector $\{F\}$ are given by

$$f_i = (f, \phi_i)_H \tag{6-9}$$

When the operator A is not positive-definite but allows us to construct the weak problem of the form in Eq. (6-2*b*), with $B(\cdot, \cdot)$ not necessarily symmetric, we can still use Eq. (6-8).

We now list the properties of the approximation functions, also called the *coordinate functions*:

1. $\{\phi_i\} \subset H_A$
2. For any N, the elements $\phi_1, \phi_2, \ldots, \phi_N$ are linearly independent (6-10)
3. $\{\phi_i\}$ is complete in H_A

Note that the space H_A contains elements that are in $H^m(\Omega)$, where $2m$ is the order of the differential operator A, and satisfy the specified (assumed to be homogeneous) essential boundary conditions. If the specified essential boundary conditions are nonhomogeneous, we either transform the boundary-value problem so that the essential boundary conditions become homogeneous, or seek the Ritz approximation in the alternative form,

$$u_N = \sum_{j=1}^{N} c_j \phi_j + \phi_0 \tag{6-11}$$

where ϕ_0 is a function which satisfies the nonhomogeneous boundary conditions. In this case, Eq. (6-8) holds with f_i of Eq. (6-9) replaced by

$$f_i = (f, \phi_i)_H - B(\phi_0, \phi_i) \tag{6-12}$$

For example, consider the equation $-d^2u/dx^2 = f$ with the boundary conditions $u(0) = 0$ and $u(1) = h$. The problem can be converted to $-d^2z/dx^2 = f$ with $z(0) = 0$ and $z(1) = 0$, where $u = z + hx$. If we were to solve the original problem using the Ritz method, we would use Eq. (6-11) with $\phi_0 = hx$ and $\phi_j = x^j(1 - x)$.

Several comments are in order on the Ritz method.

1. It might appear from the description (and described as such by many authors) that the Ritz method is limited to linear problems for which it is possible to construct a quadratic functional of the form in Eq. (6-2*a*). It is possible to construct functionals for nonlinear problems [see Oden and Reddy (1983)] and most linear problems can be formulated variationally as one of solving the weak problem

$$B(v, u) = l(v)$$

where $B(v, u)$ is bilinear but not necessarily symmetric. If we have a functional or a weak formulation for any problem, then it is possible to apply the Ritz method. Thus, the Ritz method can be applied to any variational problem that includes the natural boundary conditions of the problem. In the case of nonlinear problems, the resulting algebraic equations are also nonlinear [Eq. (6-6) is not valid]. If the bilinear form is symmetric, then the coefficient matrix of the algebraic equations is symmetric, and consequently, we need to compute only upper or lower diagonal elements of the matrix.

2. When N is increased from a previous value, the previously computed coefficients of the algebraic equations remain unchanged, and one must add newly computed coefficients to the system of equations.

3. It is interesting to note that the condition (6-5) is equivalent [in view of Eq. (6-2)] to seeking the minimum of $\|u_N - u\|_A^2$. The application of the Ritz method to this problem is equivalent to the determination of the best approximation to u_0 with respect to the metric $\|u\|_A \equiv B(u, u)$.

4. If the basis $\{\phi_i\}$ is orthonormal in H_A, $(\phi_i, \phi_j)_A = \delta_{ij}$, then Eq. (6-6) takes the form

$$0 = c_i - (f, \phi_i)_H \tag{6-13}$$

or $c_i = (f, \phi_i)_H$. Thus, the coefficient matrix becomes the identity matrix, and consequently, no inversion is required to calculate the Ritz coefficients. The Ritz solution (6-4) becomes

$$u_N = \sum_{j=1}^{N} (f, \phi_j)_H \phi_j \tag{6-14}$$

Note that this expression is similar to that encountered in the Fourier series in Chap. 3. For this reason $a_j = (f, \phi_j)_H$ are also known as the *Fourier coefficients*.

5. The choice of the basis $\{\phi_i\}$ in H_A is arbitrary. While all choices of ϕ_i ensure convergence of the Ritz approximation to the exact solution in the limit, for any finite value of N (especially for small N) one choice of functions might give better accuracy over the other, or be computationally more efficient than the other.

6-2-3 Convergence and Stability

Convergence Since, by assumption, $\{\phi_1, \phi_2, \dots\}$ is a basis in H_A, we can orthonormalize the set using the Gram–Schmidt orthonormalization process (see Prob. 3-4-11). Let us denote the resulting orthonormal set by $\{\hat{\phi}_i\}$. The generalized solution $u \in H_A$ of Eq. (6-1) can be expressed by the Fourier series,

$$u = \sum_{j=1}^{\infty} c_j \phi_j = \sum_{j=1}^{\infty} \alpha_j \hat{\phi}_j \tag{6-15}$$

By orthonormalizing the vectors ϕ_j we altered the coordinates of u. The coordinates α_j with respect to the orthonormal basis $\hat{\phi}_j$ can be determined from [see Eq. (6-14)]

$$\left.\begin{aligned} \alpha_j &= (u, \hat{\phi}_j)_A \\ &= (f, \hat{\phi}_j)_H \end{aligned}\right\} \qquad j = 1, 2, \dots \tag{6-16}$$

Note that from the convergence of the series (6-15) in H_A there also follows its convergence in H. If u_N is the Nth partial sum of (6-15), it follows from the inequality

$$\|u\|_H \le c\|u\|_A$$

that

$$\|u_N - u\|_H \le c\|u_N - u\|_A \tag{6-17}$$

Since $\{\hat{\phi}_j\}$ is a basis in H_A, we have

$$\lim_{N\to\infty} u_N = u \text{ in } H_A \tag{6-18a}$$

and, in view of Eq. (6-17),

$$\lim_{N\to\infty} u_N = u \text{ in } H \tag{6-18b}$$

We summarize the above discussion in the following theorem.

Theorem 6-1 Let A be a positive operator on a linear set $\mathscr{D}_A$ which is dense in a separable Hilbert space H, and let $f \in H$. Let H_A be the completion of $\mathscr{D}_A$ with respect to the inner product induced by the bilinear form $B(\cdot, \cdot)$. Further, let $\phi_1, \phi_2, \dots$ be a basis in H_A. Then the Ritz sequence $\{u_N\}$, with the constants c_i uniquely determined for each fixed N by (6-8), converges in H_A to the generalized solution of the equation (6-1). Moreover, if $N > M$, then

$$\|u_N - u_0\|_A \le \|u_M - u_0\|_A, \tag{6-19}$$

where $\|\cdot\|_A$ is the norm in H_A.

Although $u_N \to u$ in H_A, it need not generally be true that $Au_N \to f$ in H. Therefore, the estimate in Eq. (4-46) does not generally hold. By suitable choice of the basis we can achieve the convergence of $Au_N \to f$. An alternative way of achieving the uniform convergence of $Au_N \to f$ is to treat $Au_N = f$ as a constraint (see Prob. 5-3-9). Also, note that Theorem 6-1 is valid for only linear positive operators.

Stability The inequality in Eq. (6-19) implies that for increasing N the norm $\|u_N - u_0\|_A$ does not increase, provided the resulting Ritz equations are solved exactly. However, in practice the Ritz equations are solved on a digital computer with a finite word length. Consequently, the process of increasing N can lead to a sharp growth in the numerical error for the corresponding Ritz equations, and this error can outweigh the useful effect of increasing the number of coordinate functions. Of course, such numerical instability does not occur when the approximation functions are orthonormal in the energy space. We now formulate a criterion for stability of the Ritz approximation.

Consider the equations for the N-parameter Ritz approximation,

$$[B]\{c\} = \{F\} \tag{6-20a}$$

Suppose that the coefficients b_{ij} are evaluated with small error Δb_{ij} and the right-hand side f_i with small error Δf_i. The solution of the resulting non-exact Ritz equations is $\{c + \Delta c\}$:

$$[B + \Delta B]\{c + \Delta c\} = \{F + \Delta F\} \tag{6-20b}$$

We shall say that the Ritz process is *stable* if there exist constants α, β, and γ independent of N such that for $\|[\Delta B]\| \leq \alpha$ and arbitrary $\{\Delta F\}$ the equations (6-20*b*) are solvable and the inequality

$$\|\Delta c\| \leq \beta \|[\Delta B]\| + \gamma \|\Delta F\| \tag{6-21}$$

holds, where $\|\cdot\|$ denotes the euclidean norm. This definition is based on the assumption that Eqs. (6-20*a*) and (6-20*b*) are solved exactly.

A necessary and sufficient condition for the stability of the Ritz process is given in Theorem 6-2. Before we state the theorem, we give the definition of strongly minimal systems. A countable set $\{\phi_1, \phi_2, \ldots\}$ of elements in the Hilbert space H is called *strongly minimal* in H if

$$\inf \lambda_1^{(N)} = \lim_{N \to \infty} \lambda_1^{(N)} > 0 \tag{6-22}$$

where $\lambda_1^{(N)}$ is the smallest eigenvalue of the Gram matrix,

$$[G] = \begin{bmatrix} (\phi_1, \phi_1)_H & (\phi_1, \phi_2)_H & \cdots & (\phi_1, \phi_N)_H \\ (\phi_2, \phi_1)_H & (\phi_2, \phi_2)_H & \cdots & \vdots \\ \vdots & & & \vdots \\ (\phi_N, \phi_1)_H & (\phi_N, \phi_2)_H & \cdots & (\phi_N, \phi_N)_H \end{bmatrix} \tag{6-23}$$

We now state the theorem. The proof of the theorem can be found in Mikhlin (1971).

Theorem 6-2 The necessary and sufficient condition for the Ritz process to be stable is that its generating coordinate system be strongly minimal in the corresponding energy space.

The inequality (6-21) and Theorem 6-2 also hold for the Ritz approximation; i.e., if v_N is the non-exact approximate Ritz solution and u_N is the approximate Ritz solution,

$$v_N = \sum_{j=1}^{N} (c_j + \Delta c_j)\phi_j \qquad u_N = \sum_{j=1}^{N} c_j \phi_j \tag{6-24}$$

then the Ritz approximation u_N is called *stable* if there exist constants α_1, β_1, and γ_1 such that

$$\|u_N - v_N\|_A \leq \beta_1 \|[\Delta B]\| + \gamma_1 \|\Delta F\| \text{ for } \|[\Delta B]\| \leq \alpha_1 \tag{6-25}$$

An estimate for the error in the numerical solution of the Ritz equations is given by

$$\|v_N - u_N\|_A^2 \leq \|\Delta c\| \cdot \|\Delta F - (\Delta B)(c + \Delta c)\| \tag{6-26}$$

Conditioning of the equations In all of the above discussion of the stability of Ritz approximations, it is assumed that the Ritz equations are solved exactly so that errors arise only from the errors introduced during the construction of the Ritz equations (i.e., in the evaluation of the elements of the coefficient matrices $[B]$ and $\{F\}$). However, in practice, round-off errors are introduced into the equation during their solution. Such round-off errors have a more significant effect on the solution than the errors introduced during the evaluation of the coefficient matrices. The round-off errors introduced during the solution of the equations are proportional to the so-called condition number of the coefficient matrix $[B]$. For the positive-definite matrix $[B]$, the *condition number* is characterized by

$$n(\mathbf{B}) = \frac{\lambda_N^{(N)}}{\lambda_1^{(N)}} \tag{6-27}$$

where $\lambda_1^{(N)}$ is the smallest and $\lambda_N^{(N)}$ is the largest eigenvalue of the matrix $[B]$.

In order that the Ritz equations are well-conditioned it is important to require that as N increases the condition number $n(\mathbf{B})$ remains bounded. Since, for increasing N, the smallest eigenvalue does not increase and the largest eigenvalue does not decrease, the condition number does not decrease. Therefore, a necessary and sufficient condition for the condition number to be bounded is that the eigenvalues of $[B]$ be contained between two positive numbers which are independent of N.

6-2-4 Applications

We now consider several examples of application of the Ritz method. Additional examples can be found in the works of Mikhlin (1964), Kantorovich and Krylov (1964), Rektorys (1980), Reddy and Rasmussen (1982), and Reddy (1984a,b).

Example 6-1 Consider the differential equation

$$-\frac{d^2u}{dx^2} = \cos \pi x \qquad 0 < x < 1 \qquad \left(A = -\frac{d^2}{dx^2}\right) \tag{6-28}$$

We seek the Ritz solution of the equation for various types of boundary conditions.

1. *Dirichlet boundary conditions*

$$u(0) = u(1) = 0 \tag{6-29}$$

In this case the domain $\mathscr{D}_A$ of the operator A consists of twice differentiable functions in (0, 1) which satisfy the boundary conditions in Eq. (6-29). The operator A is symmetric and positive on $\mathscr{D}_A \subset H = L_2(0, 1)$. The null space $\mathscr{N}(A^*)$ contains elements of the form $v(x) = a + bx$. Since $v \in \mathscr{D}_A$, we have $v(0) = v(1) = 0$, hence $\mathscr{N}(A^*)$ is trivial. Consequently, the solvability condition (5-19) is identically satisfied. The operator equation (6-28) with the boundary conditions (6-29) has a unique solution. The exact solution of the problem is given by

$$u_0(x) = \frac{1}{\pi^2}(\cos \pi x + 2x - 1) \tag{6-30}$$

The functional for the problem is

$$\begin{aligned} J(u) &= \frac{1}{2}(Au, u)_0 - (f, u)_0 \\ &= \frac{1}{2}\int_0^1 \left[-\frac{d^2u}{dx^2} u - 2 \cos \pi x \, u \right] dx \\ &= \frac{1}{2}\int_0^1 \left[\left(\frac{du}{dx}\right)^2 - 2 \cos \pi x \, u \right] dx \end{aligned} \tag{6-31}$$

for all $u \in \mathscr{D}_A$. The energy space is $H_A = H_0^1(0, 1) = \{u \in H^1(0, 1): u(0) = u(1) = 0\}$.

Next, we determine the Ritz approximation to the generalized solution of Eqs. (6-28) and (6-29). We choose the basis $\{\phi_i\} = \{\sin i\pi x\}$ for the N-parameter Ritz approximation:

$$u_N = \sum_{j=1}^{N} c_j \sin j\pi x \tag{6-32}$$

The set $\{\phi_i\} = \{\sin i\pi x\}$ is clearly a subset of $H_0^1(0, 1)$ and spans the subspace S_N of H_A. The set $\{\phi_i\}$ is orthogonal in $H_0^1(\Omega)$. It is the set of eigenfunctions for the operator $A = -d^2/dx^2$ (see the comment made at the end of Sec. 5-6).

The weak problem of Eqs. (6-28) and (6-29) is to find $u \in H_A$ such that

$$\int_0^1 \left(\frac{dv}{dx}\frac{du}{dx} - v \cos \pi x\right) dx = 0$$

or, equivalently,

$$B(v, u) - (v \cos \pi x)_0 = 0 \tag{6-33a}$$

holds for all $v \in H_A$. The discrete analog of Eq. (6-33*a*), when the Ritz method is used, is to find $u_N \in S_N \subset H_A$ such that

$$B(v_N, u_N) - (v_N \cos \pi x)_0 = 0 \tag{6-33b}$$

holds for all $v_N \in S_N$. Substituting Eq. (6-32) into Eq. (6-33*b*), we obtain Eq. (6-6):

$$\sum_{j=1}^{N} b_{ij} c_j - f_i = 0$$

where

$$b_{ij} = B(\phi_i, \phi_j) = \int_0^1 (i\pi) \cos i\pi x\,(j\pi) \cos j\pi x \, dx = \begin{cases} 0 & \text{if } j \neq i \\ \dfrac{(i\pi)^2}{2} & \text{if } j = i \end{cases} \tag{6-34a}$$

$$\begin{aligned} f_i &= (\phi_i, \cos \pi x)_0 \\ &= \int_0^1 \cos \pi x \sin i\pi x \, dx = \frac{1}{2}\int_0^1 [\sin \pi(i+1)x + \sin \pi(i-1)x] \, dx \\ &= -\frac{1}{2}\left[\frac{(-1)^{i+1} - 1}{\pi(i+1)} + \frac{(-1)^{i-1} - 1}{\pi(i-1)}\right] \\ &= \begin{cases} 0 & \text{if } i \text{ is odd} \\ \dfrac{2i}{\pi(i^2 - 1)} & \text{if } i \text{ is even} \end{cases} \end{aligned} \tag{6-34b}$$

Thus, $[B]$ is a diagonal matrix as expected; consequently, we can solve for c_i very easily,

$$c_i = \frac{4}{\pi^3}\frac{1}{i(i^2 - 1)} \qquad (i = 2, 4, 6, \ldots)$$

The Ritz solution is given by

$$\begin{aligned} u_N(x) &= \frac{4}{\pi^3}\sum_{i=1}^{N} \frac{\sin i\pi x}{(i^2 - 1)i} \qquad (i \text{ even}) \\ &= \frac{2}{\pi^3}\sum_{j=1}^{N} \frac{\sin 2j\pi x}{(4j^2 - 1)j} \end{aligned} \tag{6-35}$$

It is not difficult to show that the series is uniformly convergent in the interval [0, 1] to the exact solution (6-30). The error estimate (4-46) for the problem at hand becomes (with $c = 1$)

$$\|u_N - u_0\|_A \le \|Au_N - f\|_0 = \left\| \sum_{j=1}^{N} \frac{8j^2 \sin 2j\pi x}{\pi j(4j^2-1)} - \cos \pi x \right\|_0$$

$$= \left[\frac{\pi}{2} - \frac{32}{\pi} \sum_{j=1}^{N} \frac{j^2}{(4j^2-1)^2} \right]$$

For $N = 1$, 2, and 3, this error estimate gives

$$\|u_N - u_0\|_A \le \begin{cases} 0.439, & N = 1 \\ 0.258, & N = 2 \\ 0.183, & N = 3 \end{cases}$$

2. *Mixed boundary conditions*

$$u(0) = 0 \qquad u'(1) = 0 \tag{6-36}$$

The operator A is symmetric and positive-definite on $\mathscr{D}_A$, which consists of twice-differentiable functions that satisfy the boundary conditions (6-36). The null space $\mathscr{N}(A^*)$ consists of the zero element. Hence the solvability condition is trivially satisfied. Hence the solution exists and is unique. The energy space is $H_A = \{u \in H^1(0, 1): u(0) = 0\}$. The exact solution is given by

$$u_0 = \frac{1}{\pi^2}(\cos \pi x - 1) \tag{6-37}$$

The functional for the problem is given by Eq. (6-31) because $u(0) = 0$ and $u'(1) = 0$.

For the N-parameter Ritz approximation we select a basis that satisfies only the essential boundary condition, $u(0) = 0$. Obviously, $\{\sin i\pi x\}$ meets the boundary condition $u(0) = 0$. However, the set $\{\sin i\pi x\}$, being not complete, cannot form a basis for H_A. The set is incomplete because it cannot generate the functions in H_A that are nonzero at $x = 1$. If we overlook the completeness and use $\{\sin i\pi x\}$ for computing the Ritz coefficients c_i, we obtain the same Ritz solution as in part 1, which does not converge to the exact solution (6-37).

To remedy this situation, we must select a complete basis $\{\phi_i\}$ in H_A. If we add a function $\phi_0 = x$, then the set $\{x, \sin \pi x, \sin 2\pi x, \ldots\}$ will be complete in H_A. Note that $\phi_0 = 1$ is not in H_A because it does not satisfy $\phi_0(0) = 0$. Since we have already computed b_{ij}, for $i, j = 1, 2, \ldots, N$, we need only to compute b_{ij} for $i = 0, j = 0, 1, 2, \ldots, N$, and $(\phi_0, f)_0$. We obtain

$$b_{0j} = B(\phi_0, \phi_j) = \int_0^1 \phi_j' \, dx = \begin{cases} 0 & \text{if } j \ne 0 \\ 1 & \text{if } j = 0 \end{cases}$$

$$f_0 = (\phi_0, f)_0 = \int_0^1 x \cos \pi x \, dx = -\frac{2}{\pi^2} \tag{6-38}$$

The Ritz approximation of Eqs. (6-28) and (6-36) is given by

$$u_N = -\frac{2x}{\pi^2} + \frac{2}{\pi^3} \sum_{j=1}^{N} \frac{\sin 2\pi jx}{(4j^2 - 1)j} \tag{6-39}$$

It can be shown that u_N converges to

$$\left[-\frac{2x}{\pi^2} + \frac{1}{\pi^2} (\cos \pi x + 2x - 1) \right] = \frac{1}{\pi^2} (\cos \pi x - 1)$$

the exact solution.

3. *Neumann boundary conditions*

$$u'(0) = u'(1) = 0 \tag{6-40}$$

The set $\mathscr{D}_A$ consists of twice differentiable functions that satisfy the boundary conditions in Eq. (6-40). The operator A is not positive-definite on $\mathscr{D}_A$. The set $\mathscr{N}(A^*)$ consists of constant functions. Hence, the solvability condition becomes

$$(f, c)_0 \equiv c \int_0^1 \cos \pi x \, dx = 0 \qquad c \text{ is a constant}$$

which is clearly satisfied. Hence the solution exists. However, the solution is not unique in $H_A = H^1(0, 1)$, because if $u(x)$ is a solution of Eqs. (6-28) and (6-40), then $v(x) = u(x) + c$, where c is a constant, is also a solution. To make the solution unique, we impose the additional condition (see Sec. 5-4-5),

$$\int_0^1 u(x) \, dx = 0 \tag{6-41}$$

Let Q denote the space of all functions u from $H^1(0, 1)$ such that Eq. (6-41) holds. The unique solution in Q of Eqs. (6-28) and (6-40) is given by

$$u_0 = \frac{\cos \pi x}{\pi^2}$$

For the N-parameter Ritz approximation neither of the two sets $\{\sin i\pi x\}$ and $\{x, \sin \pi x, \sin 2\pi x, \ldots\}$ is complete in Q. We must add an element to the set $\{\sin i\pi x\}$ that does not vanish at $x = 0$ and $x = 1$, and at the same time $u_N(0) \neq u_N(1)$. If the last requirement is not satisfied by the new element, then the new set spans only the subspace of Q that consists of functions $u(x)$ with the property $u(0) = u(1)$, and is therefore not complete in Q.

We select $\phi_0 = x - c$, where c is a constant not equal to 1. We obtain b_{0j} and $l(\phi_0)$ as in Eq. (6-38). Hence the solution becomes

$$u_N(x) = -\frac{2}{\pi^2} (x - c) + \frac{2}{\pi^3} \sum_{j=1}^{n} \frac{\sin 2\pi jx}{(4j^2 - 1)} \tag{6-42}$$

The constant c must be selected such that the conduction (6-41) is satisfied. This gives $c = \frac{1}{2}$. Then the Ritz approximation $u_N(x)$ converges to

$$-\frac{2}{\pi^2}(x - \tfrac{1}{2}) + \frac{1}{\pi^2}(\cos \pi x + 2x - 1) = \frac{\cos \pi x}{\pi^2}$$

This completes the example.

The next example is concerned with the solution of the Dirichlet problem in Eqs. (6-28) and (6-29) using the Ritz method, but with a different basis of $H_0^1(0, 1)$.

Example 6-2 Here we consider the Ritz approximation of Eqs. (6-28) and (6-29) using algebraic basis functions. We select $\phi_i = x^i(1 - x) \in H_A$. Since ϕ_i are not orthogonal, the matrix $[b_{ij}]$ will not be diagonal. One must invert the matrix $[b_{ij}]$ to compute the Ritz coefficients c_i. We have

$$b_{ij} = \int_0^1 [ix^{i-1} - (i+1)x^i][jx^{j-1} - (j+1)x^j]\, dx$$

$$= \frac{2ij}{(i+j)[(i+j)^2 - 1]} \qquad i, j = 1, 2, \ldots \tag{6-43a}$$

$$l(\phi_i) = \int_0^1 (x^i - x^{i+1}) \cos \pi x \, dx$$

$$= \sum_{k=0}^{[(i-1)/2]} (-1)^k \frac{i!}{(i-2k-1)!} \frac{1}{\pi^{2(k+1)}} [\cos \pi x \; x^{i-2k-1}]_0^1$$

$$- \sum_{S=0}^{[i/2]} (-1)^S \frac{(i+1)!}{(i-2S)!} \frac{1}{\pi^{2(S+1)}} [\cos \pi x \; x^{i-2S}]_0^1 \tag{6-43b}$$

where $S = [i/2]$ = greatest integer less than or equal to $i/2$.

For the one-parameter Ritz approximation ($N = 1$), we obtain

$$b_{11} = \tfrac{1}{3} \qquad l(\phi_1) = \left[\frac{1}{\pi^2}(-2)\right] - \left[\frac{2}{\pi^2} \cdot (-1)\right] = 0$$

and hence $c_1 = 0$. For $N = 2$, we obtain

$$b_{11} = \tfrac{1}{3},\ b_{12} = \tfrac{1}{6},\ b_{22} = \tfrac{2}{15},\ l(\phi_1) = 0,\ l(\phi_2) = \frac{1}{\pi^2} - \frac{12}{\pi^4}$$

This yields $c_2 = -2c_1 = 20(\pi^2 - 12)/\pi^4$. The error estimate (4-46) yields

$$\|u_2 - u_0\|_A \leq 0.5875$$

The solution accuracy improves with increasing N, as can be seen from the Ritz coefficients listed in Table 6-1. One can show that $\{\phi_k\} = x^k(1 - x)$ is not strongly minimal in H_A.

Table 6-1 The Ritz solution (at $x = 0.2197$) and coefficients for the problem in Example 6-2 ($N = 1, 2, 3, 4, 5, 6$)

Solution		Coefficients					
Exact	Ritz	c_1	c_2	c_3	c_4	c_5	c_6
0.02133	0.00000	0.00000					
0.02133	0.02102	0.21870	−0.43740				
0.02133	0.02102	0.21870	−0.43740	−0.00000			
0.02133	0.02134	0.20220	−0.28910	−0.34610	0.23070		
0.02133	0.02133	0.20220	−0.28930	−0.34510	0.22930	0.0007	
0.02133	0.02133	0.20270	−0.29880	−0.28800	0.08623	0.1583	−0.06312

In the next example we illustrate the application of the Ritz method for the solution of the eigenvalue problem associated with Eqs. (6-28) and (6-29):

$$-\frac{d^2u}{dx^2} = \lambda u \qquad 0 < x < 1$$
$$u(0) = u(1) = 0 \tag{6-44}$$

Example 6-3 We now consider the question of finding the eigenvalues λ of Eq. (6-44), which exist by Theorem 5-12, using the Ritz approximation (6-32). The Ritz system associated with the eigenvalue problem is given by [cf. Eq. (5-198) for $f = 0$]

$$\sum_{j=1}^{N} [B(\phi_i, \phi_j) - \lambda(\phi_i, \phi_j)_0]c_j = 0 \tag{6-45}$$

We need to compute the coefficients m_{ij} (called the mass coefficients),

$$m_{ij} \equiv (\phi_i, \phi_j)_0$$
$$= \int_0^1 \sin i\pi x \sin j\pi x \, dx = \begin{cases} 0 & \text{if } i \neq j \\ \frac{1}{2} & \text{if } i = j \end{cases}$$

The eigenvalue problem involves solving for λ_i ($i = 1, 2, \ldots, N$) such that Eq. (6-45) holds for any *nontrivial* c_j. This requirement amounts to finding the roots of the polynomial (in λ) obtained by setting the determinant of (6-45) to zero:

$$\det(b_{ij} - \lambda m_{ij}) = 0$$

Since both b_{ij} and m_{ij} are diagonal in the present case, the polynomial is given by

$$\prod_{i}^{N} [(i\pi)^2 - \lambda] = 0$$

and the eigenvalues λ_i become

$$\lambda_i = i^2\pi^2$$

Table 6-2 The minimum and maximum eigenvalues and the condition number as a function of the number of parameters in the approximation

N	$\lambda_1^{(N)}$	$\lambda_N^{(N)}$	$n(N)$
1	10.00	10.0	1.0
2	10.00	42.0	4.2
3	9.8697	102.1	10.3445
4	9.8696	200.5	20.3147
5	9.8696	350.96	35.5597
6	9.8696	570.53	57.8068
7	9.8696	878.88	89.0492
8	9.8696	1298.1	131.5251

The values $\{\lambda\} = \{i^2\pi^2\}$ and functions $\{\phi_i\} = \{\sin i\pi x\}$ are the exact eigenvalues and eigenfunctions of the operator, $A = -d^2/dx^2$.

If we use the system $\{\phi_k\} = \{x^k(1-x)\}$, which is not minimal in H_A, we obtain

$$\det(b_{ij} - \lambda m_{ij}) = 0$$

with b_{ij} given by Eq. (6-43a) and m_{ij} given by

$$m_{ij} = \frac{1}{i+j+1} + \frac{1}{i+j+3} - \frac{2}{i+j+2}$$

The values of $\lambda_1^{(N)}$ and $\lambda_N^{(N)}$ for various values of N are listed in Table 6-2. The table also contains the condition number as a function of N. Clearly, the condition number is increasing at a rapid rate than N^2.

Next we consider a mixed boundary-value problem to illustrate the accuracy of solutions obtained using ϕ_i from $\mathscr{D}_A$ and ϕ_i from H_A.

Example 6-4 Consider the boundary-value problem

$$-\frac{d^2u}{dx^2} - u + x^2 = 0 \qquad 0 < x < 1 \tag{6-46a}$$

$$u(0) = 0 \qquad \frac{du}{dx}(1) = 1 \tag{6-46b}$$

The domain $\mathscr{D}_A$ of the operator $A = -d^2/dx^2 - 1$ contains functions that are twice-differentiable and satisfy the boundary conditions in (6-46b). Note that $\mathscr{D}_A$ is not a linear set. The energy space H_A contains elements from $H^1(0, 1)$ which vanish at $x = 0$.

We consider a three-parameter Ritz solution to the above equation. First, we choose ϕ_0 and ϕ_i from $\mathscr{D}_A$ (i.e., choose a function ϕ_0 that satisfies all of the specified boundary conditions, and choose ϕ_i that satisfy the associated homogeneous boundary conditions):

$$\phi_0(0) = 0 \qquad \phi'_0(1) = 1 \qquad \phi_i(0) = 0 \qquad \phi'_i(1) = 0 \qquad \text{for } i = 1, 2, 3$$

We confine our search for these functions among polynomials. If $\phi_0 = a + bx$, the conditions $\phi_0(0) = 0$ and $\phi'_0(1) = 1$ immediately give us $\phi_0 = x$. Next, suppose that $p(x)$ is an nth degree polynomial in x:

$$p(x) = a_0 + a_1 x + a_2 x^2 + \cdots$$

where a_i are arbitrary constants. The condition $p(0) = 0$ gives $a_0 = 0$, and the condition $p'(1) = 0$ gives

$$a_1 + 2a_2 + 3a_3 + \cdots = 0$$

Then the simplest polynomial satisfying the required conditions is obtained by setting $a_j = 0$ for $j > 2$ and $a_1 + 2a_2 = 0$ (or $a_2 = -\frac{1}{2}a_1$):

$$p_1(x) = x - \frac{x^2}{2} \equiv \phi_1$$

In order to obtain ϕ_2, we use the same procedure and set $a_j = 0$ for $j > 3$, and set either a_1 or a_2 to zero:

$$a_1 = 0 \qquad 2a_2 + 3a_3 = 0$$

$$a_2 = 0 \qquad a_1 + 3a_3 = 0$$

These give, respectively,

$$p_2 = x^2 - \tfrac{2}{3}x^3 \qquad \text{and} \qquad p_3 = x - \tfrac{1}{3}x^3 \tag{6-47a}$$

At first glance it might appear that we have three functions for a three-parameter approximation

$$u_3 = \phi_0 + c_1 p_1 + c_2 p_2 + c_3 p_3$$

If we had proceeded to compute the coefficient matrix $b_{ij} = B(\phi_i, \phi_j)$, $i, j = 1, 2, 3$, the determinant of the coefficient matrix would have been zero (i.e., matrix is singular). This is due to the fact that the set $\{p_1, p_2, p_3\}$ is *not* linearly independent. Indeed, we can express p_2 as a linear combination of p_1 and p_3:

$$p_2 = 2(p_3 - p_1)$$

However, any two of them are linearly independent. Therefore we must select one more function that is linearly independent.

We proceed as before to select a fourth-order polynomial (that is, $a_j = 0$ for $j > 4$) and set two of the three constants a_1, a_2, or a_3 to zero:

$$a_2 = a_3 = 0\colon a_4 = -\tfrac{1}{4}a_1 \qquad a_1 = a_3 = 0\colon a_4 = -\tfrac{2}{4}a_2$$

$$a_1 = a_2 = 0\colon a_4 = -\tfrac{3}{4}a_3$$

This results in the following functions

$$p_4 = x - \frac{x^4}{4} \qquad p_5 = x^2 - \frac{x^4}{2} \qquad p_6 = x^3 - \frac{3}{4}x^4 \tag{6-47b}$$

Any one of these three would be linearly independent of $\{p_1, p_2\}$, $\{p_1, p_3\}$, and $\{p_2, p_3\}$, since the fourth power of x is not represented by p_1, p_2, and p_3. We choose the following set (of course, any other combination of a linearly independent set from $\{p_i\}$, $i = 1, 2, \ldots, 6$ would give the same result):

$$\phi_1 = p_1 \qquad \phi_2 = p_3 \qquad \phi_3 = p_4$$

Substituting these functions into $B(\phi_i, \phi_j)$ and $l(\phi_i)$, we obtain

$$b_{11} = \tfrac{1}{3} - (\tfrac{1}{3} - \tfrac{1}{5}) = \tfrac{1}{5} \qquad b_{12} = \tfrac{5}{12} - \tfrac{61}{360} = \tfrac{89}{360}$$

$$b_{13} = \tfrac{9}{20} - \tfrac{31}{168} = \tfrac{223}{840} \qquad b_{22} = \tfrac{8}{15} - \tfrac{68}{315} = \tfrac{100}{315}$$

$$b_{23} = \tfrac{7}{12} - \tfrac{113}{480} = \tfrac{167}{480} \qquad b_{33} = \tfrac{9}{14} - \tfrac{37}{144} = \tfrac{389}{1008}$$

$$f_1 = \tfrac{7}{120} \qquad f_2 = \tfrac{13}{180} \qquad f_3 = \tfrac{13}{168}$$

In matrix form, we have

$$\frac{1}{10080}\begin{bmatrix} 2016 & 2492 & 2676 \\ 2492 & 3200 & 3507 \\ 2676 & 3507 & 3890 \end{bmatrix} \begin{Bmatrix} c_1 \\ c_2 \\ c_3 \end{Bmatrix} = \begin{Bmatrix} \frac{7}{120} \\ \frac{13}{180} \\ \frac{13}{168} \end{Bmatrix}$$

Note that the above matrix equation can also be used to obtain the one-parameter and the two-parameter Ritz solutions of the problem by simply deleting the rows and columns of the above matrix equation. On the other hand if a four-parameter Ritz solution is desired, one needs to compute only the entries, $b_{4j} = b_{j4}$, $j = 1, 2, 3, 4$ and f_4 associated with ϕ_4 (say, $x - x^5/5$).

The one-, two-, and three-parameter solutions of the above equations are given by

One-parameter solution ($N = 1$): $c_1 = 7/24$

$$u_1(x) = 1.29167x - 0.14583x^2$$

Two-parameter solution ($N = 2$): $c_1 = 602/2153$, $c_2 = 21/2153$

$$u_2(x) = 1.28936x - 0.13980x^2 - 0.00325x^3$$

Three-parameter solution ($N = 3$): $c_1 = -1559/379336$, $c_2 = 248892/379336$, $c_3 = -147252/379336$

$$u_3(x) = 1.26383x + 0.00206x^2 - 0.21871x^3 + 0.09705x^4$$

Next we consider a Ritz approximation by functions from H_A (that is, ϕ_i do not satisfy the natural boundary conditions). Since the essential boundary condition is homogeneous, we choose ϕ_0 to be zero. Then ϕ_i must be selected such that $\phi_i(0) = 0$ for all $i = 1, 2, \ldots, N$. Clearly, the following set meets these conditions:

$$\phi_1 = x \qquad \phi_2 = x^2, \ldots, \phi_N = x^N$$

For $N = 3$ we obtain

$$\begin{bmatrix} \frac{2}{3} & \frac{3}{4} & \frac{4}{5} \\ \frac{3}{4} & \frac{17}{15} & \frac{4}{3} \\ \frac{4}{5} & \frac{4}{3} & \frac{58}{35} \end{bmatrix} \begin{Bmatrix} c_1 \\ c_2 \\ c_3 \end{Bmatrix} = -\begin{Bmatrix} \frac{1}{4} \\ \frac{1}{5} \\ \frac{1}{6} \end{Bmatrix} + \begin{Bmatrix} 1 \\ 1 \\ 1 \end{Bmatrix}$$

Solving the above matrix equation for one-, two-, and three-parameter solutions, we obtain

One-parameter solution: $c_1 = 9/8$

$$u_1(x) = 1.125x$$

Two-parameter solution: $c_1 = 900/695$, $c_2 = -105/695$

$$u_2(x) = 1.294964x - 0.15108x^2$$

Three-parameter solution: $c_1 = 127680/99512$, $c_2 = -11368/99512$, $c_3 = -2450/99512$

$$u_3(x) = 1.283x - 0.11424x^2 - 0.02462x^3$$

These Ritz solutions are compared in Table 6-3 with the exact solution, along with the solution obtained earlier for the problem. The Ritz solution obtained with $\phi_i \in H_A$ is relatively less accurate than the Ritz solution obtained with $\phi_i \in \mathscr{D}_A$. The difference is attributable to the term x^4 that is present in the basis $\phi_i \in \mathscr{D}_A$.

Table 6-3 Comparison of the Ritz solution with the exact solution of $-d^2u/dx^2 - u + x^2 = 0$, $u(0) = 0$, $du/dx(1) = 1$

	Ritz method ($\phi_i \in H_A$)			Ritz method ($\phi_i \in \mathscr{D}_A$)			Exact
x	$N = 1$	$N = 2$	$N = 3$	$N = 1$	$N = 2$	$N = 3$	solution†
0.1	0.1125	0.1280	0.1271	0.1277	0.1275	0.1262	0.1262
0.2	0.2250	0.2529	0.2518	0.2525	0.2523	0.2513	0.2513
0.3	0.3375	0.3749	0.3740	0.3744	0.3741	0.3742	0.3742
0.4	0.4500	0.4938	0.4934	0.4933	0.4932	0.4943	0.4943
0.5	0.5625	0.6097	0.6099	0.6094	0.6093	0.6112	0.6112
0.6	0.6750	0.7226	0.7234	0.7225	0.7226	0.7244	0.7244
0.7	0.7875	0.8324	0.8337	0.8327	0.8329	0.8340	0.8340
0.8	0.9000	0.9393	0.9407	0.9400	0.9404	0.9402	0.9401
0.9	1.0125	1.0431	1.0443	1.0444	1.0448	1.0433	1.0433
1.0	1.1250	1.1439	1.1442	1.1458	1.1463	1.1442	1.1442

† $u(x) = 2(\cos x + \tan 1 \sin x - 1) - (\sin x/\cos 1) + x^2$

The next example illustrates the use of the Ritz method in the approximation of various weak solutions of fourth-order ordinary differential equations (for beams).

Example 6-5 We wish to find Ritz approximations of the transverse deflection w of the beam (see Fig. 4-5 with $F = 0$) under applied transverse uniform load of intensity f per unit length and end moment M. We consider various weak formulations of the following fourth-order equation

$$\frac{d^2}{dx^2}\left(EI\,\frac{d^2w}{dx^2}\right) = f \qquad 0 < x < L \qquad EI > 0 \tag{6-48a}$$

$$w(0) = \frac{dw}{dx}(0) = 0 \qquad \left(EI\,\frac{d^2w}{dx^2}\right)\Bigg|_{x=L} = M \qquad \left[\frac{d}{dx}\left(EI\,\frac{d^2w}{dx^2}\right)\right]\Bigg|_{x=L} = 0 \tag{6-48b}$$

Here we have $A \equiv (d^2/dx^2)[EI(d^2/dx^2)]$.

1. *Conventional formulation* The given boundary-value problem can be transformed to

$$\frac{d^2}{dx^2}\left(EI\,\frac{d^2u}{dx^2}\right) = \hat{f} \qquad 0 < x < L \tag{6-49}$$

$$u(0) = \frac{du}{dx}(0) = EI\,\frac{d^2u}{dx^2}(L) = \left[\frac{d}{dx}\left(EI\,\frac{d^2u}{dx^2}\right)\right]_{x=L} = 0$$

where $u = w - w_0$, $\hat{f} = f - (d^2/dx^2)[EI(d^2w_0/dx^2)]$, and w_0 satisfies the actual (nonhomogeneous) boundary conditions in (6-48b). We can choose w_0 to be

$$w_0 = Mx^2/2EI$$

which satisfies the required boundary conditions. This choice gives $\hat{f} = f$. The energy space H_A is given by

$$H_A = \left\{u\colon\ u \in H^2(0, L),\ u(0) = \frac{du}{dx}(L) = 0\right\} \tag{6-50}$$

It can be shown that the operator A is self-adjoint and positive-definite on H_A, and that the solvability conditions are satisfied.

The quadratic functional for the problem is given by

$$J(u) = \int_0^L \left[\frac{EI}{2}\left(\frac{d^2u}{dx^2}\right)^2 - fu\right] dx \tag{6-51}$$

The weak problem is given by $B(v, u) = l(v)$, where

$$B(v, u) = \int_0^L EI\,\frac{d^2v}{dx^2}\,\frac{d^2u}{dx^2}\,dx \qquad l(v) = \int_0^L vf\,dx \tag{6-52}$$

For the N-parameter Ritz approximation we choose $\phi_i \in H_A$ [i.e., boundary conditions $\phi_i(0) = \phi_i'(0) = 0$]. The set $\{\phi_i\} = \{x^{i+1}\}$ is a basis in H_A. We write

$$u_N(x) = \sum_{i=1}^{N} c_i \phi_i \qquad \phi_i = x^{i+1} \tag{6-53}$$

Substituting (6-53) into (6-52), we obtain

$$b_{ij} = \int_0^L EI(i+1)ix^{i-1}j(j+1)x^{j-1}\,dx$$

$$= \frac{EIij(i+1)(j+1)(L)^{i+j-1}}{(i+j-1)}$$

$$l(\phi_i) = \frac{f(L)^{i+2}}{i+2} + M(i+1)(L)^i \tag{6-54}$$

The two-parameter solution is given by $w_2 = u_2 + w_0$,

$$w_2(x) = \frac{(12M + 5fL^2)}{24EI}\,x^2 - \frac{fL}{12EI}\,x^3 \tag{6-55}$$

The three-parameter Ritz approximation yields the exact solution

$$w(x) = \left(\frac{2M + fL^2}{4EI}\right)x^2 - \frac{fL}{6EI}\,x^3 + \frac{f}{24EI}\,x^4 \tag{6-56}$$

2. *The Lagrange multiplier formulation* Equations (6-48) can be formulated alternatively as one of minimizing the functional

$$J(\psi, w) = \int_0^L \left[\frac{EI}{2}\left(\frac{d\psi}{dx}\right)^2 - fw\right] dx + M\psi(L) \tag{6-57}$$

subject to the constraint

$$G(\psi, w) \equiv \psi + \frac{dw}{dx} = 0 \tag{6-58}$$

Note that the substitution of Eq. (6-58) in (6-57) for ψ gives the total potential energy functional associated with the conventional formulation (see part 1 of this example).

The lagrangian functional is given by

$$L(\psi, w, \lambda) = \int_0^L \left[\frac{EI}{2}\left(\frac{d\psi}{dx}\right)^2 - fw\right] dx + \int_0^L \lambda\left(\psi + \frac{dw}{dx}\right) dx + M\psi(L) \tag{6-59}$$

where λ is the Lagrange multiplier, which in the present case represents the shear force. It is easily seen that the specified essential boundary conditions for the formulation are

$$\psi(0) = w(0) = 0 \tag{6-60}$$

The space H_A in the present case is the product space, $H_A = [H_*^1(0, L)]^2 \times L_2(0, L)$, where

$$H_*^1(0, L) = \{w\colon\ w \in H^1(0, L),\ w(0) = 0\} \tag{6-61}$$

The weak problem of the formulation is given by setting δL to zero:

$$0 = \int_0^L \left(EI \frac{d\psi}{dx} \frac{d\delta\psi}{dx} + \lambda\, \delta\psi \right) dx + M\, \delta\psi(L) + \int_0^L \left(\lambda \frac{d\delta w}{dx} - f\, \delta w \right) dx + \int_0^L \delta\lambda \left(\psi + \frac{dw}{dx} \right) dx \tag{6-62}$$

For a one-parameter (for each variable) Ritz solution, we have

$$\psi(x) = a_1 \phi_1^1 \qquad w(x) = b_1 \phi_1^2 \qquad \lambda(x) = c_1 \phi_1^3 \tag{6-63}$$

where ϕ_1^α ($\alpha = 1, 2, 3$) denote the approximation functions for the three variables. The element $(x, x, 1) \in H_A$ is chosen for $(\phi_1^1, \phi_1^2, \phi_1^3)$. Substituting Eq. (6-63) into (6-62) and collecting the coefficients of δa_1, δb_1, and δc_1 separately, we obtain

$$EILa_1 + \frac{L^2}{2} c_1 = -ML \qquad c_1 = \frac{fL}{2} \qquad b_1 = \frac{4ML + fL^3}{8EI} \tag{6-64}$$

Thus, the one-parameter solution becomes,

$$\psi(x) = -\left(\frac{4M + fL^2}{4EI} \right) x \qquad w(x) = \frac{(4M + fL^2)L}{8EI}\, x \qquad \lambda(x) = fL/2 \tag{6-65}$$

whereas the exact solutions for ψ and λ are

$$\psi(x) = \frac{-(2M + fL^2)x}{2EI} + \frac{fLx^2}{2EI} - \frac{fx^3}{6EI} \qquad \lambda = f(L - x) \tag{6-66}$$

and $w(x)$ is given by Eq. (6-56). The maximum values of ψ, w, and λ obtained by the Ritz method and the exact solution are compared below:

Variable	Ritz solution	Exact solution
$\psi(L)$	$-\dfrac{ML}{EI} - \dfrac{fL^3}{4EI}$	$-\dfrac{ML}{EI} - \dfrac{fL^3}{6EI}$
$w(L)$	$\dfrac{ML^2}{2EI} + \dfrac{fL^4}{8EI}$	$\dfrac{ML^2}{2EI} + \dfrac{fL^4}{8EI}$
$\lambda(0)$	fL	$fL/2$

The one-parameter solution is exact in representing the response due to M, but in error in representing the response due to f.

A two-parameter approximation of the form

$$\psi(x) = a_1 x + a_2 x^2 \qquad w(x) = b_1 x + b_2 x^2 \qquad \lambda(x) = c_1 + c_2 x \tag{6-67}$$

results in the Ritz solution

$$\begin{aligned}\psi(x) &= -\frac{12M + 5fL^2}{12EI}x + \frac{fL}{4EI}x^2 \\ w(x) &= \frac{fL^3}{24EI}x + \frac{6M + fL^2}{12EI}x^2 \\ \lambda(x) &= f(L - x)\end{aligned} \tag{6-68}$$

This solution gives the maximum values of ψ, w, and λ that coincide with the exact solution.

3. *The penalty function formulation* The augmented functional for this formulation is given by

$$P_K(\psi, w) = J(\psi, w) + \frac{K}{2}\int_0^L \left(\psi + \frac{dw}{dx}\right)^2 dx \tag{6-69}$$

where K is the penalty parameter. The energy space is given by $H_A = [H_*^1(0, L)]^2$, where $H_*^1(0, L)$ is defined in Eq. (6-61).

The weak problem for this formulation is given by $\delta P_K = 0$:

$$0 = \int_0^L \left(EI\frac{d\delta\psi}{dx}\frac{d\psi}{dx} - \delta w f\right) dx + M\,\delta\psi(L) + K\int_0^L \left(\delta\psi + \frac{d\delta w}{dx}\right)\left(\psi + \frac{dw}{dx}\right) dx \tag{6-70}$$

First we consider a one-parameter Ritz approximation for each variable:

$$\psi = a_1\phi_1^1 = a_1 x \qquad w = b_1\phi_1^2 = b_1 x \tag{6-71}$$

Substituting Eq. (6-71) for ψ and w into Eq. (6-70) and collecting the coefficients of δa_1 and δb_1 separately, we obtain

$$\begin{aligned}EILa_1 + K\left(\frac{L^3}{3}a_1 + \frac{L^2}{2}b_1\right) &= -ML \\ K\left(\frac{L^2}{2}a_1 + Lb_1\right) &= \frac{fL^2}{2}\end{aligned} \tag{6-72}$$

From the second equation we obtain

$$b_1 = \frac{fL}{2K} - \frac{a_1 L}{2} \tag{6-73a}$$

Substituting into the first equation and solving for a_1, we obtain

$$a_1 = \frac{-ML + (-fL^3/4)}{(EIL + KL^3/12)} \tag{6-73b}$$

Thus, in the limit $K \to \infty$ we obtain the trivial solution: $a_1 = b_1 = 0$.

Next we consider a one-parameter approximation for ψ and a two-parameter approximation for w:

$$\psi = a_1 x \qquad w = b_1 x + b_2 x^2 \tag{6-74}$$

This results in the equations

$$EILa_1 + K\left(\frac{L^3}{3} a_1 + \frac{L^2}{2} b_1 + \frac{2L^3}{3} b_2\right) = -ML$$

$$K\left(\frac{L^2}{2} a_1 + Lb_1 + L^2 b_2\right) = \frac{fL^2}{2}$$

$$2K\left(\frac{L^3}{3} a_1 + \frac{L^2}{2} b_1 + \frac{2L^3}{3} b_2\right) = \frac{fL^3}{3}$$

The approximate solution is given by

$$\begin{aligned} \psi &= \frac{-6M - fL^2}{6EI} x \\ w &= -\frac{f}{K} x + \frac{6M + fL^2}{12EI} x^2 \end{aligned} \tag{6-75}$$

This is definitely an improvement over the one-parameter solution. However, the solution for w is still in considerable error compared to the exact solution.

4. *The mixed formulation* The mixed method involves rewriting a given higher order equation as a pair of lower order equations by introducing secondary dependent variables. This decomposition of a higher order equation into a pair of lower order equations enables one to seek approximation in lower order Hilbert spaces. Equation (6-48) can be decomposed into the equations,

$$\frac{M(x)}{EI} = \frac{d^2 w}{dx^2}; \qquad \frac{d^2 M(x)}{dx^2} = f \quad \text{in } 0 < x < L \tag{6-76}$$

The functional corresponding to Eq. (6-76) is given by

$$I(w, M) = \int_0^L \left(\frac{dw}{dx}\frac{dM}{dx} + \frac{M^2}{2EI} + fw\right) dx \tag{6-77}$$

The specified essential boundary conditions for the formulation are

$$w(0) = 0 \qquad M(L) = \hat{M}$$

The space H_A in this case is given by $H_A = H^1_*(0, L) \times H^1_+(0, L)$, where

$$\begin{aligned} H^1_*(0, L) &= \{w \in H^1(0, L)\colon\ w(0) = 0\} \\ H^1_+(0, L) &= \{M \in H^1(0, L)\colon\ M(L) = 0\} \end{aligned} \tag{6-78}$$

We seek a two-parameter approximation for both w and M:

$$w(x) = \phi_0 + a_1\phi_1 + a_2\phi_2 \qquad M(x) = \psi_0 + b_1\psi_1 + b_2\psi_2 \tag{6-79}$$

where $\phi_0 = 0$, $\phi_1 = x$, $\phi_2 = x^2$, $\psi_0 = \hat{M}$, $\psi_1 = x - L$, and $\psi_2 = (x - L)^2$.
Substituting into the weak form associated with (6-77), we obtain

$$b_1 = 0 \qquad b_2 = f/2 \qquad a_1 = \frac{11fL^3}{120EI} \qquad a_2 = \frac{\hat{M}}{2EI} + \frac{3fL^3}{40EI} \tag{6-80}$$

The Ritz solution becomes

$$\begin{aligned} w(x) &= \frac{11fL^3}{120EI}\,x + \left(\frac{\hat{M}}{2EI} + \frac{3fL^2}{40EI}\right)x^2 \\ M(x) &= \frac{f}{2}\,(x - L)^2 + \hat{M} \end{aligned} \tag{6-81}$$

The Ritz solution for $M(x)$ coincides with the exact solution. The Ritz solution and exact solutions for the maximum deflection are

$$w(L) = \frac{3\hat{M}L^2 + fL^4}{6EI} \qquad w_{\text{exact}}(L) = \frac{4\hat{M}L^2 + fL^4}{8EI}$$

This completes the example.

The next example deals with the Dirichlet and Neumann boundary-value problems associated with the Poisson equation in two dimensions.

Example 6-6 Consider the Poisson equation

$$-\nabla^2 u = f \text{ in } \Omega \subset \mathbb{R}^2 \tag{6-82}$$

where Ω is the unit square. The origin of the coordinate system is taken at the lower left corner of the square. We shall consider the Dirichlet and Neumann boundary conditions for the problem.

1. *Dirichlet boundary conditions* (see Fig. 6-1*a*)

$$u = 0 \text{ on } \Gamma \tag{6-83}$$

The domain of the operator, $A \equiv -\nabla^2$, in this case is $\mathscr{D}_A = H^2(\Omega) \cap H^1_0(\Omega)$. Recall that the bilinear form for the operator is given by

$$B(v, u) = \int_0^1 \int_0^1 \left(\frac{\partial v}{\partial x}\frac{\partial u}{\partial x} + \frac{\partial v}{\partial y}\frac{\partial u}{\partial y}\right) dx\, dy$$

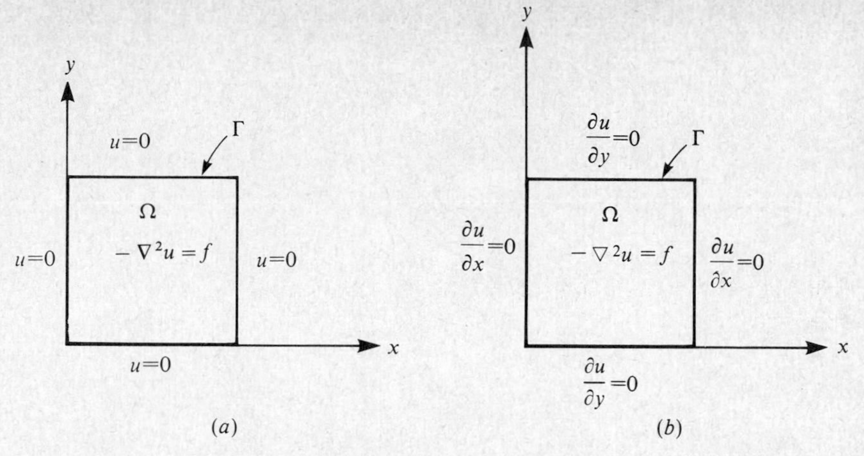

Figure 6-1 A square domain with (*a*) Dirichlet boundary conditions and (*b*) Neumann boundary conditions.

The energy space H_A is identified with $H_0^1(\Omega)$. The energy norm $\|u\|_A = \sqrt{B(u, u)}$ is equal to the seminorm $|u|_{1,\Omega}$ on $H_0^1(\Omega)$, which is equivalent to the norm $\|u\|_{1,\Omega}$ on $H_0^1(\Omega)$ [see Eq. (4-27)]. The operator A is self-adjoint and positive-definite on $\mathscr{D}_A$ (and also in H_A). Hence $\mathscr{N}(A^*)$ contains only the zero element and therefore the solvability condition (5-19) is identically satisfied. Hence there exists a unique solution to the problem for any $f \in L_2(\Omega)$.

For the N-parameter Ritz approximation we choose

$$u_N = \sum_{n=1}^{N} \sum_{m=1}^{N} c_{mn} \phi_{mn} \qquad \phi_{mn} = \sin m\pi x \sin n\pi y \tag{6-84}$$

The coefficient matrix $[B] \equiv [B(\phi_{ij}, \phi_{mn})]$ can be computed to be

$$B(\phi_{ij}, \phi_{mn}) = \begin{cases} \dfrac{\pi^2}{4}(i^2 + j^2) & \text{if } m = i \text{ and } n = j \\ 0 & \text{otherwise} \end{cases} \tag{6-85a}$$

and for $f = f_0 =$ constant, the right-hand side is given by

$$l(\phi_{ij}) = \begin{cases} \dfrac{4f_0}{\pi^2 ij} & \text{if both } i \text{ and } j \text{ odd} \\ 0 & \text{otherwise} \end{cases} \tag{6-85b}$$

The N-parameter Ritz solution becomes

$$u_N = \frac{16f_0}{\pi^4} \sum_{m,n=1,3,\ldots}^{N} \frac{\sin m\pi x \sin n\pi y}{mn[m^2 + n^2]} \tag{6-86}$$

which coincides with the exact solution.

For the case $f = \cos \pi x$, the coefficient matrix is as defined in Eq. (6-85a), but the right hand side becomes

$$l(\phi_{ij}) = \begin{cases} \dfrac{8}{\pi^2} \dfrac{1}{j(i^2 - 1)} & \text{if } j \text{ odd and } i \text{ even} \\ 0 & \text{otherwise} \end{cases} \tag{6-87}$$

and the Ritz solution becomes

$$u_N = \frac{32}{\pi^4} \sum_{j=1,3,\ldots}^{N} \sum_{i=2,4,\ldots}^{N} \frac{\sin i\pi x \sin j\pi y}{j(i^2 - 1)(i^2 + j^2)} \tag{6-88}$$

2. *Neumann boundary conditions* (see Fig. 6-1b)

$$\frac{\partial u}{\partial n} = 0 \left(\frac{\partial u}{\partial x} = 0 \text{ at } x = 0, 1; \frac{\partial u}{\partial y} = 0 \text{ at } y = 0, 1 \right) \tag{6-89}$$

The domain of the operator is $\mathscr{D}_A$, which contains functions from $H^2(\Omega)$ whose normal derivatives vanish on Γ. The energy space is $H_A = H^1(\Omega)$, but $\sqrt{B(u, u)}$ does not qualify as a norm because the null space $\mathscr{N}(A^*)$ contains functions that are constant in Ω. We define

$$Q = \left\{ u \in H^1(\Omega): \int_\Omega u(x, y)\, dx\, dy = 0 \right\} \tag{6-90}$$

Then for $u \in Q$ we have

$$\int_\Omega \left| \frac{\partial u}{\partial x} \right|^2 dx\, dy = 0 \qquad \int_\Omega \left| \frac{\partial u}{\partial y} \right|^2 dx\, dy = 0 \text{ implies } u = \text{constant}$$

and, since $u \in Q$ the constant must be zero. Consequently, $\sqrt{B(u, u)} = \|u\|_A$ defines a norm on Q. Then the Neumann problem has a unique solution in Q for all $f \in L_2(\Omega)$ such that (solvability condition)

$$\int_\Omega f(x, y)\, dx\, dy = 0 \tag{6-91}$$

holds. Clearly, in the present case f cannot be a nonzero constant; $f = \cos \pi x$ satisfies the solvability condition.

For an N-parameter Ritz approximation, we choose

$$u_N = \sum_{i=0}^{N} \sum_{j=0}^{N} c_{ij} \cos i\pi x \cos j\pi y \tag{6-92}$$

Upon substituting into the variational problem $B(u, v) = l(v)$, we obtain the system of equations

$$\begin{aligned} \frac{\pi^2}{4} (j^2 + i^2) c_{ij} &= \int_0^1 \int_0^1 f \cos i\pi x \cos j\pi y\, dx\, dy, \; i, j \neq 0 \\ B_{ij} c_{ij} &= \begin{cases} \frac{1}{2} & \text{for } i = 1, j = 0 \\ 0 & \text{otherwise} \end{cases} \end{aligned} \tag{6-93}$$

and the solution becomes [because $B(\phi_{10}, \phi_{10}) = \pi^2/2$]

$$u_N = \frac{1}{\pi^2} \cos \pi x \qquad 0 < x < 1 \tag{6-94}$$

which coincides with the classical solution of the Neumann problem for $f = \cos \pi x$.

The last example of this section deals with the solution of the Poisson equation with mixed boundary conditions.

Example 6-7 Consider the Poisson equation in a square

$$\begin{aligned} -\nabla^2 u &= f_0 \text{ in } \Omega = \{(x, y)\colon 0 < (x, y) < 1\} \\ u &= 0 \text{ on sides } x = 1 \text{ and } y = 1 \\ \frac{\partial u}{\partial n} &= 0 \text{ on sides } x = 0 \text{ and } y = 0 \end{aligned} \tag{6-95}$$

The domain $\mathscr{D}_A$ of the operator $A = -\nabla^2$ is the linear space of functions from $H^2(\Omega)$ which satisfy the boundary conditions in (6-95). The null space $\mathscr{N}(A^*)$ consists of functions of the form

$$v = a + bx + cy$$

Since $v \in \mathscr{D}_A$, we have $v(1, y) = v(x, 1) = \partial v/\partial x(0, y) = \partial v/\partial y(x, 0) = 0$. These conditions imply $v = 0$. Thus, $\mathscr{N}(A^*)$ is trivial, and therefore the equation (6-95) has a unique solution. The energy space is $H_A = \{u\colon u \in H^1(\Omega), u(1, y) = u(x, 1) = 0\}$.

We select the following Ritz approximation,

$$u_N = \sum_i^N c_i \phi_i \qquad \phi_i = \cos \frac{(2i-1)\pi x}{2} \cos \frac{(2i-1)\pi y}{2} \qquad i = 1, 2, \ldots, N \tag{6-96}$$

Incidentally, the ϕ_i also satisfy the natural boundary conditions of the problem. While the choice $\phi_i = \sin i\pi x \cdot \sin i\pi y$ meets the essential boundary conditions, it is not complete because it cannot be used to generate the solution that does not vanish on the sides $x = 0$ and $y = 0$.

The coefficient matrix b_{ij} and source vector f_i can be computed using ϕ_i of (6-96):

$$b_{ij} = -\int_0^1 \int_0^1 \alpha_i \alpha_j (f_i f_j \cdot g_i g_j + p_i p_j \cdot q_i q_j)\, dx\, dy,$$

where $f_i = \sin \alpha_i x$, $g_i = \cos \alpha_i y$, $p_i = \cos \alpha_i x$, $q_i = \sin \alpha_i y$, and $\alpha_i = [(2i - 1)\pi/2]$. Carrying out the integration we obtain,

$$b_{ij} = \begin{cases} \alpha_i^2/2 & \text{if } i = j \\ \dfrac{\alpha_i^2 \alpha_j^2}{2(\alpha_i^2 - \alpha_j^2)^2} & \text{if } i \neq j \end{cases} \tag{6-97a}$$

$$f_i = l(\phi_i) = \frac{f_0}{\alpha_i^2} \tag{6-97b}$$

For the one-parameter approximation, we have

$$b_{11} = \frac{\alpha_1^2}{2} \qquad l(\phi_1) = \frac{f_0}{\alpha_1^2} \qquad \text{and} \qquad c_1 = \frac{32 f_0}{\pi^4}$$

Hence, the solution is given by

$$u_1 = \frac{32 f_0}{\pi^4} \cos \frac{\pi x}{2} \cos \frac{\pi y}{2} \tag{6-98}$$

The exact solution to this problem is given by

$$u(x, y) = \frac{f_0}{2} \left\{ (1 - y^2) + \frac{32}{\pi^3} \sum_{n=0}^{\infty} \frac{(-1)^{n+1} \cos [(2n + 1)\pi y/2] \cosh [(2n + 1)\pi x/2]}{(2n + 1)^3 \cosh (2n + 1)\pi/2} \right\} \tag{6-99}$$

The one-parameter Ritz solution at the center of the domain is $0.1643 f_0$ whereas the exact solution is $0.1811 f_0$ (9.3 percent error). The two-parameter solution is given by

$$u_2 = f_0 \left(0.3283988 \cos \frac{\pi x}{2} \cos \frac{\pi y}{2} + 0.001976 \cos \frac{3\pi x}{2} \cos \frac{3\pi y}{2} \right) \tag{6-100}$$

which differs from the exact solution at the center of the domain only by 0.54 percent. Table 6-4 shows the Ritz coefficients for $N = 1, 2, \ldots, 6$.

If algebraic polynomials are to be used, one can choose $\phi_1 = (1 - x)(1 - y)$ or $\phi_1 = (1 - x^2)(1 - y^2)$, both of which satisfy the homogeneous essen-

Table 6-4 Ritz coefficients for the N-parameter Ritz approximation (6-96) of Eq. (6-95)

N	c_1/f_0	c_2/f_0	c_3/f_0	c_4/f_0	c_5/f_0	c_6/f_0
1	0.32851					
2	0.328399	0.001976				
3	0.328395	0.001966	0.000267			
4	0.328394	0.001965	0.000264	0.000070		
5	0.328394	0.001965	0.000264	0.000069	0.000026	
6	0.328394	0.001965	0.000264	0.000069	0.000025	0.000012

tial boundary conditions. Obviously, the choice $\phi_1 = (1 - x)(1 - y)$ does not meet the natural boundary conditions of the problem whereas $\phi_1 = (1 - x^2)(1 - y^2)$ does.

For the one-parameter approximation with $\phi_1 = (1 - x^2)(1 - y^2) \in \mathscr{D}_A$ we obtain

$$u_1(x, y) = \frac{5f_0}{16}(1 - x^2)(1 - y^2)$$

This gives a value of $0.17578 f_0$ for u at the center of the domain. This is in an error of 2.94 percent when compared to the exact solution.

For the one-parameter approximation with $\phi_1 = (1 - x)(1 - y) \in H_A$, we obtain

$$u_1(x, y) = \frac{3f_0}{8}(1 - x)(1 - y)$$

When compared with the exact solution at the center of the domain, this is in error of 48.23 percent! Of course, the solution would improve with a larger number of parameters of the type $\phi_i = (1 - x^i)(1 - y^i)$.

PROBLEMS 6-1

6-1-1 Consider the differential equation $-(d^2u/dx^2) = 2x - 1$, $0 < x < 1$ with the three types of boundary conditions.

(a) $u(0) = u(1) = 0$
(b) $u'(0) = u'(1) = 0$
(c) $u(0) = u'(1) = 0$

For each pair of boundary conditions, give the domain and energy spaces and two-parameter Ritz approximation. Compare your solution with the exact solution. Use algebraic basis functions. For case (a) compute the error estimate in Eq. (4-46).

6-1-2 Consider the differential equation $-(d^2u/dx^2) = 1/(1 + x)$, $0 < x < 1$, subjected to the boundary conditions: $u(0) = u'(1) = 0$. Use an N-parameter Ritz approximation with algebraic polynomials, and solve the resulting equations for $N = 1, 2, \ldots, 8$ and compare the Ritz solution with the exact solution, $u(x) = x(1 + \ln 2) - (1 + x)\ln(1 + x)$.

6-1-3 Solve the differential equation $-(d/dx)[(2 + x)\, du/dx] = 1$, $-1 < x < 1$, $u(-1) = u(1) = 0$ using a 2-parameter Ritz approximation with polynomial basis functions.

6-1-4 Find the first two eigenvalues associated with Prob. 6-1-3.

6-1-5 Consider the self-adjoint differential equation of the second order

$$-\frac{d}{dx}\left[p(x)\frac{du}{dx}\right] + q(x)u = f \qquad a < x < b$$

$$u(a) = u_0 \qquad u(b) = u_1$$

Reduce the equation to one with homogeneous boundary conditions, and derive the algebraic form of the N-parameter Ritz equations. Specialize the equations for (a) $\phi_i = \sin i\pi x$ and (b) $\phi_i = x^i(1 - x)$. Take $p(x) = 1$, $q(x) = -1$, $f = x$, and $u_0 = u_1 = 0$, $a = 0$, and $b = 1$. Compare the two approximations, for various values of x, for $N = 1, 2, \ldots, 8$ with the exact solution $u = (\sin x/\sin 1) - x$.

6-1-6 Give a one-parameter Ritz solution of the equation in Prob. 6-1-5 for $p(x) = x$, $q(x) = (1 - x^2)/x$, $f(x) = x^2$, $a = 1$, $b = 2$, $u_0 = u_1 = 0$. Use a polynomial basis function.

6-1-7 Consider the problem $-(d/dr)[a(r)\, du/dr] = 0$, $r_i < r < r_0$, $u(r_i) = u_i$, $u(r_0) = u_0$ which arises in connection with the axisymmetric potential flow (e.g., heat transfer, electrostatics, etc.) problems. For the following data

$$a(r) = r \qquad r_i = 20 \qquad r_0 = 50 \qquad u_i = 100 \qquad u_0 = 0$$

obtain a one-parameter Ritz solution.

6-1-8 Find the first eigenvalue of the equation

$$-\frac{1}{r}\frac{d}{dr}\left(r\frac{du}{dr}\right) = \lambda u \qquad 0 < r < a$$

$$u(a) = 0 \qquad \frac{du}{dr}(0) = 0$$

using one- and two-parameter Ritz approximations with trigonometric basis functions.

6-1-9 Repeat Prob. 6-1-3 with trigonometric basis functions.

6-1-10 Find the two-parameter Ritz approximation, using algebraic polynomials, of the problem $-(1 + x)\, d^2u/dx^2 - du/dx = x$, $0 < x < 1$, $u(0) = u(1) = 0$ and compare the result with the exact solution,

$$u(x) = -\frac{1}{4}\left[\frac{\log(1 + x)}{\log 2} + x^2 - 2x\right]$$

6-1-11 Derive a three parameter Ritz approximation of the problem $-(d^2u/dx^2) = x^2$, $0 < x < 1$, $u(0) = 0$, $(du/dx)(1) + 2u(1) = 1$.

6-1-12 Find the N-parameter Ritz solution of the eigenvalue problem $-(d^2u/dx^2) = \lambda u$, $0 \le x \le 1$, $u'(0) = \alpha u(0)$, $u(1) = 0$. Use an approximation of the form

$$u = \sum_{n=1}^{N} c_n \cos\left(\frac{2n - 1}{2}\right)\pi x$$

and determine λ_n ($n = 1, 2, \ldots, N$) by minimizing the functional (check that the eigenvalue problem is equivalent to finding the minimum of the functional)

$$J(u) = \int_0^1 \left[\left(\frac{du}{dx}\right)^2 - \lambda u^2\right] dx + \alpha[u(0)]^2$$

Use $\alpha = 0.5$.

6-1-13 Consider the differential equation

$$\frac{d^2}{dx^2}\left[b(x)\frac{d^2w}{dx^2}\right] + c(x)w = q_0 \qquad 0 < x < 1$$

$$w(0) = w(1) = 0 \qquad w''(0) = w''(1) = 0$$

associated with a simply-supported beam on an elastic foundation. Show that the N-parameter Ritz approximation is given by

$$w_N = \sum_{n=1}^{N} a_n \sin\frac{n\pi x}{L}$$

where
$$a_n = -\frac{4q_0 L^4}{n\pi(n^4\pi^4 EI + KL^4)} \qquad \text{for odd values of } n$$

$$a_n = 0 \qquad \text{for even values of } n$$

6-1-14 For a clamped-hinged beam of length L, constant flexural rigidity EI, and mass per unit length m, the functional associated with its free vibration (i.e., the eigenvalue problem) is given by

$$I(w) = \frac{1}{2}\int_0^L [EI(w'')^2 - \lambda w^2]\, dx$$

where $\lambda = m\omega^2$, ω being the natural frequency of vibration. Using a two-parameter Ritz approximation with algebraic functions, solve for the first two eigenvalues in terms of EI and L. The specified boundary conditions are $w = 0$ at $x = 0, L$; $dw/dx = 0$ at $x = 0$ and $d^2w/dx^2 = 0$ at $x = L$.

6-1-15 Consider the mixed boundary value problem for the Poisson equation $-\nabla^2 u = 1$ in $\Omega \subset \mathbb{R}^2$ (Ω is the unit square), $u = 0$ along $x = 0$ and $x = 1$, $\partial u/\partial n = 0$ along $y = 0$ and $y = 1$. Find the N-parameter Ritz solution in the form

$$u_N = \sum_{i=1}^{N}\sum_{j=0}^{N} c_{ij} \sin i\pi x \cos j\pi y.$$

6-1-16 Consider the Poisson equation $-\nabla^2 u = 2(x + y) - 4$ in $\Omega =$ unit square with the boundary conditions $u(0, y) = y^2$, $u(x, 0) = x^2$, $\partial u/\partial x(1, y) = 2 - 2y - y^2$, $\partial u/\partial y(x, 1) = 2 - 2x - x^2$. Find a two-parameter approximation using algebraic polynomials.

6-1-17 Consider the Poisson equation $-\nabla^2 u = 1$ over the triangular domain $\Omega = \{(x, y)\colon x \geq 0, y \geq 0, x + y - 1 \leq 0\}$ with the boundary conditions $u = 0$ on $x = 0$ and $y = 0$, $\partial u/\partial n = 0$ on $x + y - 1 = 0$. Find a three-parameter approximation of the form $u = c_1 xy + c_2(x^2 y + xy^2) + c_3 x^2 y^2$ by minimizing the functional

$$J(u) = \int_0^1 \left\{ \int_0^x \left[\left(\frac{\partial u}{\partial x}\right)^2 + \left(\frac{\partial u}{\partial y}\right)^2 - 2u \right] dy \right\} dx$$

on H, $H = \{u \in H^1(\Omega)\colon u(x, 0) = u(0, y) = 0\}$.

6-1-18 Find the first two eigenvalues of Prob. 6-1-15.

6-1-19 Find the first three eigenvalues of Prob. 6-1-17.

6-1-20 A square plate of constant thickness is clamped (i.e., $w = \partial w/\partial n = 0$) along all of its four sides and subjected to the uniformly distributed load f_0 in the transverse (to the plane of the plate) direction. The functional associated with the bending of a thin elastic plate is given by

$$J(w) = \frac{1}{2}\int_{-a}^{a}\int_{-a}^{a} [(\nabla^2 w)^2 - 2f_0 w]\, dx\, dy$$

Assuming a three-parameter Ritz approximation of the form,

$$w_3 = (x^2 - a^2)^2(y^2 - a^2)^2(c_1 + c_2 x^2 + c_3 y^2)$$

determine the constants c_1, c_2, and c_3 such that $J(w)$ is a minimum.

6-3 THE WEIGHTED-RESIDUAL METHOD

6-3-1 Introduction

Consider a separable Hilbert space H and let S be a dense subspace of H. If for some element $u \in H$

$$(u, v) = 0 \text{ holds for every } v \in S$$

where $(\cdot, \cdot)$ is an inner product in H, then it follows that $u = 0$ in H. If $\{\psi_i\}$ is a basis in H, then

$$(u, \psi_k) = 0 \qquad \text{for all } k \text{ implies } u = 0 \text{ in } H$$

Now consider the operator equation, $A: \mathscr{D}_A \subset H \to H$,

$$Au = f \text{ in } \Omega \tag{6-101}$$

with appropriate homogeneous boundary conditions. By definition, the elements of $\mathscr{D}_A$ satisfy these boundary conditions. If $u \in \mathscr{D}_A$ is such that

$$(Au - f, \psi_k) = 0 \qquad \text{for every } k = 1, 2, \ldots \tag{6-102}$$

where $\{\psi_k\}$ is a basis in H, then we have $Au - f = 0$ in H, that is, u is the solution of (6-101) in H. In other words, finding the solution of (6-101) is equivalent to finding the solution of (6-102). This equivalence forms the basis of the weighted-residual method.

Note that u is not necessarily represented with respect to the basis $\{\psi_i\}$. Any basis $\{\phi_i\}$ in $\mathscr{D}_A$ can be used to represent u, while the residual, R_N,

$$R_N \equiv Au_N - f \tag{6-103}$$

is made orthogonal to the subspace spanned by $\{\psi_k\}$:

$$(Au_N - f, \psi_k) = 0 \qquad k = 1, 2, \ldots, N \tag{6-104}$$

Equation (6-104) is known in the literature by different names for different choices of ψ_k. All these methods, which will be outlined shortly, are grouped into a class of methods, called the *weighted-residual methods.* The general case in which $\psi_k \neq \phi_k$ is known as the *Petrov–Galerkin method.*

In the weighted-residual method, we look for an approximate solution u_N of equation (6-101) in the form,

$$u_N = \sum_{i=1}^{N} c_i \phi_i \tag{6-105}$$

where N is an arbitrary but fixed positive integer and c_i are constants to be determined using (6-104). This gives N equations for the N unknown constants $c_1, c_2, \ldots, c_N$. If the operator A is linear, then Eq. (6-104) becomes

$$\sum_{i=1}^{N} (A\phi_i, \psi_k)c_i = (f, \psi_k) \qquad k = 1, 2, \ldots, N \tag{6-106}$$

Note that the requirement $\phi_i \in \mathscr{D}_A$ means that ϕ_i are differentiable $2m$ times if A is a differential operator of order $2m$, and satisfy the specified boundary conditions (assumed to be homogeneous).

We immediately note that the weighted-residual method is applicable to a much larger class of operator equations than those allow the incorporation of natural boundary conditions into the weak formulation, as required by the Ritz method. In the weighted-residual method, the most general form of which is given by (6-104), the operator A is not restricted to be positive-definite, or even linear. Due to this general nature of the weighted-residual method, the questions of solvability, existence, and uniqueness of solutions, as one might suspect, are in general more difficult. We now consider various special cases of the weighted residual method.

6-3-2 The Bubnov–Galerkin Method

A generalization of the Ritz method, which is applicable to positive-definite (bounded below) operators that admit a functional formulation or operator equations that permit a weak formulation which includes the natural boundary conditions to general operator equations is known as the Bubnov–Galerkin method. The method seeks an approximate solution to Eq. (6-101) in the form of a linear combination (6-105), and determines the coefficients c_i from the condition that the residual R_N is orthogonal to the basis functions $\phi_1, \phi_2, \ldots, \phi_N$ (which satisfy all of the boundary conditions of the problem):

$$(R_N, \phi_k) = 0 \qquad k = 1, 2, \ldots, N \tag{6-107}$$

If A is linear, this leads to the following system of linear algebraic equations in the unknowns c_i:

$$\sum_{i=1}^{N} (A\phi_i, \phi_k)c_i = (f, \phi_k) \qquad k = 1, 2, \ldots, N \tag{6-108}$$

for every $\phi_k \in \mathscr{D}_A$.

Note that if A is positive-definite, we can write $A = T^*T$, where T is a linear operator and T^* is its adjoint. We have $(A\phi_i, \phi_k) = (T\phi_i, T\phi_k) = B(\phi_i, \phi_k)$, and Eq. (6-108) becomes identical to Eq. (6-6) in $\mathscr{D}_A$. Hence it follows that the Bubnov–Galerkin and Ritz methods are identical if the operator A is linear and positive-definite on $\mathscr{D}_A$. In this case, the Bubnov–Galerkin method can be formulated using ϕ_i from H_A, which is a much larger space than $\mathscr{D}_A$ because the elements of H_A satisfy only the homogeneous form of the specified essential boundary conditions and the functions need to have only the mth order generalized derivatives for a $2m$th order differential operator A. For $A = T^*T$ and ϕ_i from H_A, Eq. (6-108) becomes [see Sec. 5-4-4]

$$\sum_{i=1}^{N} (T\phi_i, T\phi_k)c_i = (f, \phi_k) + \sum_{j=1}^{m-n} (h_j, B_j\phi_k) \tag{6-109}$$

where B_j are the boundary operators appearing in the essential boundary conditions, h_j are the specified values appearing in the natural boundary conditions, $2m$ is the order of the differential operator A, and n is the number of specified essential boundary conditions. Of course, if all of the specified boundary conditions are of the essential type, the last term in Eq. (6-109) drops out and Eqs. (6-108) and (6-109) become identical. Thus, in general, the Bubnov–Galerkin and Ritz methods give the same equations for all problems, whether linear or not, provided that the natural boundary conditions can be included in the variational statement of the problem.

Although we have not considered many examples of integral equations in the present study, the operator A can be an integral operator, in which case Eq. (6-108) is still valid but it is not meaningful to consider Eq. (6-109).

The following theorem [see Mikhlin (1964) for a proof] gives sufficient conditions for the convergence of the Bubnov–Galerkin method.

Theorem 6-3 Let the operator A in Eq. (6-101) be linear and have the form $A = A_0 + A_1$, Eq. (6-101) has no more than one weak solution in H_0, and the operator $T = A_0^{-1}A_1$ is completely continuous in H_0 (here H_0 denotes the energy space associated with the operator A_0). Then the approximate solutions of Eq. (6-101) constructed by the Bubnov–Galerkin method converge, in the energy norm of H_0, to the exact solution of this equation.

The operator A_0 is assumed to be positive-definite (so that H_0 can be constructed) and the set $\{\phi_i\}$ is assumed to be complete in H_0. An example of the decomposition of A into $A_0 + A_1$ is given by the operator

$$\begin{aligned} Au &\equiv \sum_{i,j=1}^{m} \frac{\partial}{\partial x_i}\left(a_{ij}\frac{\partial u}{\partial x_j}\right) + \sum_{i=1}^{m} b_i \frac{\partial u}{\partial x_i} + cu \\ &= A_0 u + A_1 u \end{aligned}$$

where $$A_0 u = \sum_{i,j=1}^{m} \frac{\partial}{\partial x_i}\left(a_{ij}\frac{\partial u}{\partial x_j}\right) \qquad A_1 u = \sum_{i=1}^{m} b_i \frac{\partial u}{\partial x_i} + cu$$

We consider several examples of application of the Bubnov–Galerkin method. The first example is concerned with the torsion of a prismatic member of rectangular section Ω.

Example 6-8 Consider the torsion (see Fig. 6-2) of a rod of rectangular section $\Omega = \{(x, y)\colon -a < x < a,\ -b < y < b\}$. The problem reduces to the

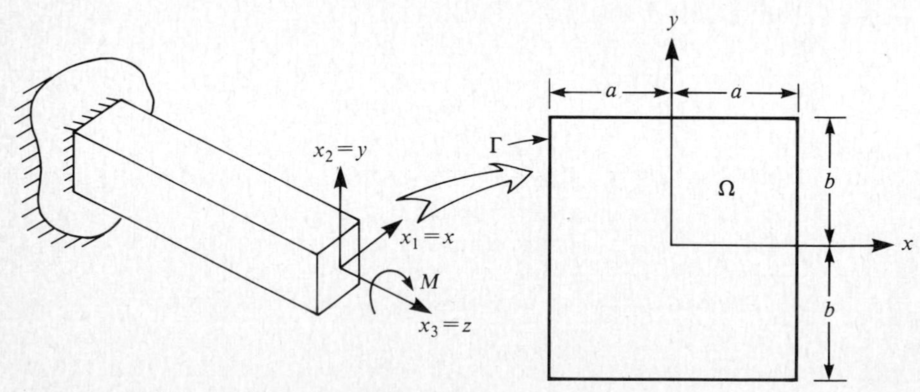

Figure 6-2 Torsion of a rectangular section member.

solution of a Dirichlet boundary-value problem for the Poisson equation (see Example 6-6),

$$-\nabla^2 u = 2 \text{ in } \Omega \qquad u = 0 \text{ on } \Gamma \tag{6-110}$$

where u is the Prandtl stress function.

The domain $\mathscr{D}_A \subset H = L_2(\Omega)$ of the operator $A = -\nabla^2$ consists of functions that are twice-differentiable with respect to x and y in Ω and vanish on Γ. Clearly, $\phi_{mn} = \cos m\pi x/2a \cos n\pi y/2b$ (m, $n = 1, 3, 5, \ldots$) are in $\mathscr{D}_A$ and form a complete orthogonal basis for $\mathscr{D}_A$. Hence, we seek the N-parameter Bubnov–Galerkin solution in the form

$$u_N(x, y) = \sum_{m,\, n=1,\, 3,\, 5,\, \ldots}^{N} c_{mn} \cos \frac{m\pi x}{2a} \sin \frac{n\pi y}{2b} \tag{6-111}$$

The residual R_N becomes

$$R_N = Au_N - 2 = -\nabla^2\left(\sum_{m,\, n=1}^{N} c_{mn}\phi_{mn}\right) - 2$$

$$= \sum_{m,\, n=1}^{N} \left[\left(\frac{m\pi}{2a}\right)^2 + \left(\frac{n\pi}{2b}\right)^2\right] c_{mn}\phi_{mn} - 2$$

Substituting this expression into Eq. (6-107) with ϕ_k replaced by ϕ_{kl}, we obtain

$$0 = \int_{-a}^{a}\int_{-b}^{b} R_N \cos\frac{k\pi x}{2a} \cos\frac{l\pi y}{2b}\, dx\, dy$$

$$= ab\left(\frac{k^2\pi^2}{4a^2} + \frac{l^2\pi^2}{4b^2}\right)c_{kl} - 2\cdot\frac{16ab}{\pi^2 kl}(-1)^{(k+l)/2-1}$$

or
$$c_{kl} = \frac{128a^2b^2(-1)^{(k+l)/2-1}}{\pi^4 kl(b^2k^2 + a^2l^2)}$$

The solution (6-111) becomes

$$u_N(x, y) = \frac{128a^2b^2}{\pi^4} \sum_{m,\, n=1,\, 3,\, 5,\, \ldots}^{N} \frac{(-1)^{(m+n)/2-1} \cos m\pi x/2a \ \cos n\pi y/2b}{mn(b^2m^2 + a^2n^2)} \tag{6-112}$$

This solution coincides, for $N \to \infty$, with that obtained by means of the double Fourier series. Note that Eq. (6-86) gives Eq. (6-112) for $a = b = \frac{1}{2}$, when x and y in Eq. (6-86) are replaced by $\frac{1}{2}[(x/a) + 1]$ and $\frac{1}{2}[(y/b) + 1]$, respectively. This proves that the Ritz and Bubnov–Galerkin methods, for the same choice of approximation functions, give the same solutions.

Next suppose that we wish to seek the solution to Eq. (6-110) in the form of algebraic polynomials. We choose the solution in the form,

$$u_N(x, y) = (a^2 - x^2)(b^2 - y^2)(c_1 + c_2x^2 + c_3y^2 + c_4x^2y^2 + \cdots) \tag{6-113}$$

For $N = 1$, we obtain

$$0 = \int_{-a}^{a}\int_{-b}^{b} [2(b^2 - y^2)c_1 + 2(a^2 - x^2)c_1 - 2](a^2 - x^2)(b^2 - y^2)\,dx\,dy$$

$$= \tfrac{128}{45}a^3b^3(a^2 + b^2)c_1 - \tfrac{32}{9}a^3b^3$$

and the one-parameter solution becomes

$$u_1 = \frac{5}{4}\frac{(a^2 - x^2)(b^2 - y^2)}{a^2 + b^2}$$

The twisting moment M (see Fig. 6-2) is related to the Prandtl stress function u by

$$M = 2G\theta \int_{-a}^{a}\int_{-b}^{b} u(x, y)\,dx\,dy$$

$$= \frac{40}{9}G\theta\frac{a^3b^3}{a^2 + b^2}$$

where G is the shear modulus and θ is the twist per unit length of the rod. For a square section ($b = a$) member the one-parameter approximation gives $M = \frac{20}{9}G\theta a^4 = 2.2208G\theta a^4$ compared to the exact value of $2.2496G\theta a^4$. The shear stresses, σ_{xz} and σ_{yz} (z being taken along the axis of the member; i.e., perpendicular to Ω) will be in even greater error because they are computed from the derivatives of u. The solution can be improved by considering additional terms in Eq. (6-113).

The next example is concerned with the solution of an eigenvalue problem.

Example 6-9 Consider the problem of finding the eigenvalues of a membrane having the form of a circle of radius a. The membrane is fixed at its edge. The governing equation is given by

$$-\nabla^2 u = \lambda u \text{ in } \Omega \qquad u = 0 \text{ on } \Gamma$$

For the circular membrane, which is axisymmetric, u depends only on the radial coordinate r. Hence, the above equation can be transformed to the polar coordinate r,

$$-\frac{1}{r}\frac{d}{dr}\left(r\frac{du}{dr}\right) = \lambda u$$

For an N-parameter approximation, we choose the following basis

$$\phi_1 = \cos\frac{\pi r}{2a},\ \phi_2 = \cos\frac{3\pi r}{2a},\ \ldots,\ \phi_N = \cos(2N - 1)\frac{\pi r}{2a} \tag{a}$$

We consider the case $N = 1$. Substituting into the Bubnov–Galerkin integral,

$$2\pi\int_0^a (-\nabla^2 u - \lambda u)\phi_i r\,dr = 0 \tag{b}$$

we obtain

$$0 = \int_0^a \left[\frac{1}{r} \frac{d}{dr} \left(-\frac{\pi r}{2a} \sin \frac{\pi r}{2a} \right) c_1 + \lambda \cos \frac{\pi r}{2a} c_1 \right] \cos \frac{\pi r}{2a} r \, dr$$

or

$$\pi c_1 \left[\frac{\pi^2}{4} \left(\frac{1}{2} + \frac{2}{\pi^2} \right) - \lambda a^2 \left(\frac{1}{2} - \frac{2}{\pi^2} \right) \right] = 0$$

For nonzero c_1, we obtain

$$\lambda_1 = \frac{5.832}{a^2}$$

This value is less than 1 percent in error compared to the exact value, $5.779/a^2$.

Next consider the case $N = 2$. Substituting Eq. (*a*) into the integral (*b*) and evaluating it, we obtain

$$\left(\begin{bmatrix} 1.7337 & -1.5 \\ -1.5 & 11.603 \end{bmatrix} - \lambda \begin{bmatrix} 0.29736a^2 & -0.20264a^2 \\ -0.20264a^2 & 0.47748a^2 \end{bmatrix} \right) \begin{Bmatrix} c_1 \\ c_2 \end{Bmatrix} = \begin{Bmatrix} 0 \\ 0 \end{Bmatrix}$$

The characteristic polynomial of this eigenvalue problem is

$$(0.10092a^4)\lambda^2 - (3.6701a^2)\lambda + 17.866 = 0$$

The smallest root of this equation is

$$\lambda_1 = \frac{5.792}{a^2}$$

which is only 0.23 percent in error.

The next example illustrates the use of the Bubnov–Galerkin and Ritz methods in the solution of a nonlinear differential equation.

Example 6-10 Consider the nonlinear differential equation

$$u \frac{d^2u}{dx^2} + \left(\frac{du}{dx} \right)^2 = 1 \qquad 0 < x < 1$$
$$u'(0) = 0 \qquad u(1) = \sqrt{2} \tag{6-114a}$$

The domain of the operator $\bar{A}u \equiv u(d^2u/dx^2) + (du/dx)^2$ is the set $\bar{\mathscr{D}}_A$ of twice-differentiable functions that satisfy the boundary conditions in (6-114*a*). The set is not linear because of the nonhomogeneous boundary condition. We can transform the equation to one with homogeneous boundary conditions. Let $u = w + \sqrt{2}$ in (6-114*a*) to obtain

$$(w + \sqrt{2}) \frac{d^2w}{dx^2} + \left(\frac{dw}{dx} \right)^2 = 1 \qquad 0 < x < 1$$
$$w'(0) = 0 \qquad w(1) = 0 \tag{6-114b}$$

The domain of the new operator defined by (6-114*b*) is the set $\mathscr{D}_A$ of twice-differentiable functions which satisfy the homogeneous boundary conditions in (6-114*b*). The set constitutes a linear subspace of $H^2(0, 1)$. The energy space H_A contains elements from $H^1(0, 1)$ that vanish at $x = 1$.

In the present study we did not develop any theoretical results to establish that a given nonlinear problem has a solution, and if it has one, it may not be unique. However, the present problem has a unique solution

$$u_e(x) = \sqrt{1 + x^2} \tag{6-115}$$

As will be seen, the Ritz and Bubnov–Galerkin methods yield quadratic algebraic equations relating the parameters, c_i. We select one set of the two values using the requirement that $\|R_N\|$ be a minimum.

For this mixed boundary-value problem the Bubnov–Galerkin method can be applied either in its original form, i.e., Eq. (6-108), or in the weak form, i.e., Eq. (6-109) [actually we use Eq. (6-107) and its weak form for this nonlinear problem]. The main difference is that the form (6-108) requires the use of ϕ_i from the domain of the operator $\mathscr{D}_A$ whereas Eq. (6-109) requires the use of ϕ_i from H_A. The latter also forms the basis for the Ritz method. Both equations, Eqs. (6-108) and (6-109), give the same result if ϕ_i from $\mathscr{D}_A$ are used. With these comments we now seek solutions to Eq. (6-114*b*) using equations equivalent to (6-108) and (6-109).

First we note that Eq. (6-114*b*) can be recast in the adjoint form ($A = T^*T$),

$$-\frac{d}{dx}\left[(w + \sqrt{2})\frac{dw}{dx}\right] = -1$$

$$w'(0) = 0 \qquad w(1) = 0$$

Equation (6-107) for the problem at hand has the explicit form

$$\int_0^1 \left\{-\frac{d}{dx}\left[(w_N + \sqrt{2})\frac{dw_N}{dx}\right] + 1\right\}\phi_k\, dx = 0 \qquad k = 1, 2, \ldots, N \tag{6-116}$$

for each $\phi_k \in \mathscr{D}_A$. The weak form is obtained by integrating the first term in the brackets by parts:

$$\int_0^1 \left[(w_N + \sqrt{2})\frac{dw_N}{dx}\frac{d\phi_k}{dx} + \phi_k\right] dx = 0 \qquad k = 1, 2, \ldots, N \tag{6-117}$$

for each $\phi_k \in H_A$. Note that the boundary term resulting from the integration by parts is zero because $w'(0) = 0$ and $\phi_k(1) = 0$.

To use Eq. (6-116) we must select the coordinate functions ϕ_i from $\mathscr{D}_A$. Suppose that we choose algebraic polynomials for the coordinate functions:

$$w_N(x) = \sum_{i=1}^{N} c_i\phi_i \qquad \phi_i = 1 - x^{i+1} \tag{6-118}$$

In the interest of algebraic simplicity, we seek only the one-parameter solution:

$$A(w_1) = -[c_1(1 - x^2) + \sqrt{2}](-2c_1) - (2xc_1)^2$$
$$= \sqrt{2}\,2c_1 + 2c_1^2 - 6x^2c_1^2$$
$$[A(w_1), \phi_1]_0 = \int_0^1 A(w_1)(1 - x^2)\,dx$$
$$= \tfrac{4}{3}\sqrt{2}\,c_1 + \tfrac{8}{15}c_1^2$$
$$(f, \phi_1)_0 = \int_0^1 (-1)(1 - x^2)\,dx = -\tfrac{2}{3}$$

Thus we have a quadratic expression in c_1. Solving for c_1, we obtain

$$c_1^{(1)} = \frac{-5\sqrt{2} + \sqrt{30}}{4} \approx -0.39846$$
$$c_1^{(2)} = \frac{-5\sqrt{2} - \sqrt{30}}{4} \approx -3.13707 \tag{6-119}$$

We must use an appropriate criterion to discard one of the two values of c_1 calculated above. Here we use the minimum residual criterion. We compute the L_2-norm of the residual, $R_1 = A(w_1) - f$ using $c_1^{(1)}$ and $c_1^{(2)}$. The smaller value of R_1 is provided by $c_1^{(1)}$. Consequently, the one-parameter solution to (6-114*a*) becomes

$$u_1 \equiv w_1 + \sqrt{2} = 1.01575 + 0.39846x^2 \tag{6-120}$$

The one-parameter solution is in fairly good agreement with the first two terms of the exact solution $\sqrt{1 + x^2} \simeq 1 + 0.5x^2 + \cdots$

If we solve the differential equation (6-114*a*) with nonhomogeneous boundary conditions, we select the approximate solution u_N in the form

$$u_N = \phi_0 + \sum_{i=1}^{N} c_i\phi_i \tag{6-121}$$

where ϕ_0 must be selected to satisfy the specified boundary conditions. For the problem at hand, we can choose $\phi_0 = \sqrt{2}$ so that $\phi_0(1) = \sqrt{2}$ and $\phi_0'(0) = 0$. For $N = 1$ once again we obtain the solution in (6-120).

Next we seek a solution using Eq. (6-117). Since $\phi_i \in H_A = \{u \in H^1(0, 1): u(1) = 0\}$, we choose $\phi_i = 1 - x^i$. Let

$$w_N(x) = \sum_{i=1}^{N} c_i\phi_i \qquad \phi_i = 1 - x^i \tag{6-122}$$

Substituting Eq. (6-122) into Eq. (6-117) and carrying out the indicated integration, we obtain

$$c^2 + 2\sqrt{2}\,c + 1 = 0$$

$$c_1^{(1)} = 1 - \sqrt{2} \qquad c_1^{(2)} = -1 - \sqrt{2}$$

Using the same criterion as before (i.e., minimum residual criteria in the L_2-sense), we select $c_1^{(1)}$. The solution becomes

$$u_1 \equiv w_1 + \sqrt{2} = (1 - x) + \sqrt{2}\,x \tag{6-123}$$

The one-parameter solution is not very good (especially for larger values of x, $0 < x < 1$), but one can get better accuracy by taking more parameters in the approximation.

For $N \geq 2$, the selection of the right parameters from $c_i^{(1)}$ and $c_i^{(2)}$ becomes tedious. One must use numerical methods, such as Newton's method, to solve the nonlinear algebraic equations.

Next we consider the solution of Eq. (6-114*a*) by the Petrov–Galerkin method. In the Petrov–Galerkin method we select the "weight" function ψ_k in $(A(w_N) - f, \psi_k) = 0$ to be different from the coordinate functions ϕ_k. In the present case we select $\psi_1 \in \mathscr{D}_A$ [this is not a requirement; ψ_k should be in $H = L_2(0, 1)$]

$$\psi_1 = \cos\frac{\pi x}{2} \text{ so that } \psi_1(1) = 0 \qquad \psi_1'(0) = 0 \tag{6-124}$$

We have (for $\phi_i = 1 - x^{i+1}$)

$$(A(w_1), \psi_1) = \frac{1}{\pi}\left\{4\sqrt{2}\,c_1 + \left[-8 + \frac{96}{\pi^2}\right]c_1^2\right\}$$

$$(f, \psi_1) = -\frac{2}{\pi}$$

Solving for c_1, we obtain

$$c_1^{(1,\,2)} = \frac{-2\sqrt{2} \pm \sqrt{8 - 2[(96/\pi^2) - 8]}}{(96/\pi^2) - 8}$$

$$c_1^{(1)} = -0.40317 \qquad c_2^{(1)} = -2.87268$$

Once again we use the "minimum residual" criterion, and select $c_1^{(1)}$. The solution is given by

$$u_1 \equiv w_1 + \sqrt{2} = 1.01104 + 0.40317x^2 \tag{6-125}$$

which is a slight improvement over the Bubnov–Galerkin solution (also see Example 6-12).

We close this section with a remark concerning the Bubnov–Galerkin method. For differential equations with even order highest derivatives, the Ritz and Bubnov–Galerkin methods give the same solutions for $\phi_i \in H_A$ (i.e., the weak form of the equation is used). However, for first-order equations, the Bubnov–Galerkin method is a natural choice over the Ritz method because we cannot construct a weak form for it. Also, we can relax the conditions on ϕ_i by not requiring them to satisfy the boundary conditions of the problem; the boundary conditions can also be approximated in the Bubnov–Galerkin sense (sometimes, this might be the only way to solve the problem). This is illustrated in Example 6-17.

6-3-3 The Method of Least Squares

Consider the operator equation,

$$Au = f \text{ in } \Omega \tag{6-126}$$

where $A: \mathscr{D}_A \to H$ is not necessarily symmetric. The linear set $\mathscr{D}_A$ is assumed to be dense in the Hilbert space H. In the least squares method, the solution of Eq. (6-126) is constructed as the function $u_N \in \mathscr{D}_A$,

$$u_N = \sum_{i=1}^{N} c_i \phi_i$$

which minimizes the residual in the H-norm,

$$\|R_N\|_H = \|Au_N - f\|_H = \left\| \sum_{i=1}^{N} A(c_i \phi_i) - f \right\|_H \tag{6-127}$$

This requirement leads to a system of equations in the unknowns $c_1, c_2, \ldots, c_N$:

$$0 = \frac{\partial}{\partial c_k} \|R_N\|_H = \frac{\partial}{\partial c_k} \sqrt{(R_N, R_N)_H} = \left(R_N, \frac{\partial R_N}{\partial c_k} \right)_H \frac{1}{\|R_N\|_H}$$

Since $R_N \neq 0$, it follows that Eq. (6-127) is equivalent to

$$\left(R_N, \frac{\partial R_N}{\partial c_k} \right)_H = 0 \tag{6-128}$$

Thus, the least squares method is a special case of the weighted residual method for the weight function

$$\psi_k = \frac{\partial R_N}{\partial c_k} \tag{6-129}$$

Equation (6-128) has the form

$$\left(Au_N - f, \frac{\partial}{\partial c_k} (Au_N) \right)_H = 0 \tag{6-130}$$

If A is linear, we have

$$\sum_{i=1}^{N} (A\phi_i, A\phi_k)_H c_i = (f, A\phi_k)_H \tag{6-131}$$

Note that the coefficient matrix, $G_{ij} = (A\phi_i, A\phi_j)_H$, is symmetric for any operator A. The matrix $[G]$ is invertible if and only if the set $\{A\phi_i\}$ is linearly independent. But the set $\{A\phi_i\}$ is always linearly independent if A is linear and $\{\phi_i\}$ is linearly independent. To prove this statement, assume that $\{A\phi_i\}$ is not linearly independent. Then it is possible to find constants α_i, not all of them zero, such that

$$\alpha_1 A\phi_1 + \alpha_2 A\phi_2 + \cdots + \alpha_N A\phi_N = A\left(\sum_{i=1}^{N} \alpha_i \phi_i \right) = 0$$

for all $\phi_i \in \mathscr{D}_A$. This implies that $\sum_{i=1}^{N} \alpha_i \phi_i = 0$. Since ϕ_i are linearly independent, $\alpha_i = 0$ for all $i = 1, 2, \ldots, N$. This in turn proves that $\{A\phi_i\}$ is a linearly independent set.

We now show that the sequence $u_N = \sum_{i=1}^{N} c_i \phi_i$, with c_i determined by Eq. (6-131), converges in the energy space H_A associated with the positive-definite operator A to the generalized solution u of the equation (6-126). Since the operator A is positive-definite on the set $\mathscr{D}_A$, there exists a constant $\mu > 0$ such that

$$(Au, u) = \|u\|_A^2 \geq \mu \|u\|_H^2 \qquad \text{for every } u \in H_A$$

By Eq. (4-46), every approximation $u_N \in \mathscr{D}_A$ of the generalized solution u of the equation (6-126) satisfies the inequality

$$\|u_N - u\|_A \leq \frac{1}{\sqrt{\mu}} \|Au_N - f\|_H \tag{6-132}$$

If we can prove that the expression $\|Au_N - f\|_H$ can be made arbitrarily small for N sufficiently large, then the convergence of u_N to u follows. Since $\{A\phi_i\}$ constitutes a base in H, it is possible to find, for every given $f \in H$ and $\eta > 0$, a positive integer M and constants $b_1, b_2, \ldots, b_M$ such that

$$\left\| \sum_{i=1}^{M} A(b_i \phi_i) - f \right\|_H < \eta \tag{6-133}$$

The constants b_i are determined subject to the requirement

$$\left\| \sum_{i=1}^{M} A(b_i \phi_i) - f \right\|_H = \text{minimum}$$

However, if $N > M$ and if the constants c_i in Eq. (6-127) are determined so that $\|R_N\|_H$ is minimized, then

$$\left\| \sum_{i=1}^{N} A(c_i \phi_i) - f \right\|_H < \eta \tag{6-134}$$

also holds. If we take $c_i = b_i$ for $i = 1, 2, \ldots, M$, and $c_{M+1}, c_{M+2}, \ldots, c_N$ are determined such that the quantity in Eq. (6-127) is minimized, the norm in Eq. (6-134) can only decrease. Hence

$$\|Au_N - f\|_H < \eta$$

holds for every $N > M$. Now take $\eta = \sqrt{\mu\varepsilon}$ and use the inequality (6-132) to obtain

$$\|u_N - u\|_A < \varepsilon \qquad \text{for } N > M$$

which we started to prove. We summarize the above discussion in the following theorem.

Theorem 6-4 Let A be a positive-definite operator on a linear set $\mathscr{D}_A$ which is dense in a separable Hilbert space H, and let $f \in H$. If $\{A\phi_i\}$ constitutes a base in H, then the sequence of elements $u_N = \sum_{i=1}^{N} c_i \phi_i$ with constants c_i uniquely determined by the requirement

$$\|Au_N - f\|_H \text{ be a minimum}$$

converges in H_A, and thus also in H, to the generalized solution u of the equation $Au = f$. Moreover, we have

$$\lim_{N\to\infty} Au_N \to f \text{ in } H \tag{6-135}$$

The fact that $Au_N \to f$ in H for the least squares method allows us to estimate the error using Eq. (4-46):

$$\|u_N - u\|_A \leq \frac{1}{\sqrt{\mu}} \|Au_N - f\|_H \tag{6-136}$$

Note that the Ritz method does not guarantee the convergence of Au_N to f while the least squares method does. Of course, in both methods the sequence u_N converges in H_A to the weak solution of the equation $Au = f$. It would be nice to know which method gives better convergence in H_A. To this end let u_N denote the sequence obtained by the least squares method and let v_N be the sequence obtained by the Ritz method, both using the same basis (from $\mathscr{D}_A$). Since v_N is obtained by minimizing the functional $J(v) = \frac{1}{2}B(v, v) - (f, v) = \frac{1}{2}(\|v - u\|_A^2 - \|u\|_A^2)$, where u is the generalized solution of $Au = f$, it follows that

$$J(v_N) \leq J(u_N)$$

or

$$\|v_N - u\|_A \leq \|u_N - u\|_A \tag{6-137}$$

Thus, under the same choice of the basis the Ritz sequence converges in H_A to the generalized solution u of the equation $Au = f$ at least as well as or better than the least squares sequence. Moreover, in the Ritz method the elements of the basis need not be chosen from $\mathscr{D}_A$.

Note that if the operator A and basis $\{\phi_i\}$ are such that $A\phi_k = \alpha_k \phi_k$ (no sum on k), then Eq. (6-131) yields

$$\sum_{i=1}^{N} (A\phi_i, \phi_k)c_i = (f, \phi_k)_H \qquad k = 1, 2, \ldots, N$$

which is identical to that obtained in the Bubnov–Galerkin method. For this case, the Ritz, Bubnov–Galerkin and least squares methods give the same solution.

A combination of the Ritz method and least squares method was proposed by Courant (1943). The *Courant method* seeks a solution $u_N = \sum_{i=1}^{N} c_i \phi_i$, $\phi_i \in \mathscr{D}_A$, by minimizing the augmented functional,

$$J_c(u) = J(u) + \| Au - f \|_H^2 \qquad u \in \mathscr{D}_A \tag{6-138}$$

where $J(u)$ is the functional used in the Ritz method. This method allows convergence of $u_N \to u$ in H_A as well as $Au_N \to f$ in H. However, the method entails considerable numerical calculations.

We consider several examples of application of the least squares method.

Example 6-11 Consider the torsion problem of Example 6-8. Using the basis of trigonometric functions, we write

$$u_N(x, y) = \sum_{m,\, n=1,\, 3,\, 5,\, \ldots}^{N} c_{mn} \cos \frac{m\pi x}{2a} \cos \frac{n\pi y}{2b}$$

$$A\phi_{mn} = -\left[\left(\frac{m\pi}{2a}\right)^2 + \left(\frac{n\pi}{2b}\right)^2\right]\phi_{mn}$$

Hence Eq. (6-131) becomes [here $H = L_2(\Omega)$],

$$\sum_{m,\, n=1,\, 3,\, 5,\, \ldots}^{N} (A\phi_{mn}, A\phi_{kl})_0\, c_{mn} = (2, A\phi_{kl})_0$$

Since $A\phi_{kl} = \alpha_{kl}\,\phi_{kl}$ (no sum on k and l), we obtain

$$\sum_{m,\, n=1,\, 3,\, 5,\, \ldots}^{N} (A\phi_{mn}, \phi_{kl}) c_{mn} = (2, \phi_{kl})_0$$

which is precisely the same equation as that in the Bubnov–Galerkin method, and therefore we obtain the same result for c_{mn} as in Example 6-8:

$$c_{mn} = \frac{128a^2b^2(-1)^{(m+n)/2-1}}{\pi^4 mn(b^2m^2 + a^2n^2)}$$

Now if we use the approximation,

$$u_N(x, y) = \sum_{m,\, n=1,\, 3,\, 5,\, \ldots}^{N} (a^2 - x^2)(b^2 - y^2)(c_1 + c_2 x^2 + c_3 y^2 + \cdots)$$

we do not expect to get the same result as in Example 6-8. Let us consider the one-parameter case. We have

$$A\phi_1 = 2(b^2 - y^2) + 2(a^2 - x^2)$$

$$(A\phi_1, A\phi_1)_0\, c_1 = (2, A\phi_1)_0$$

or

$$\int_{-a}^{a}\int_{-b}^{b}[2(b^2-y^2)+2(a^2-x^2)]^2 c_1\,dx\,dy$$

$$=\int_0^a\int_0^b 2[2(b^2-y^2)+2(a^2-x^2)]c_1\,dx\,dy$$

$$(\tfrac{32}{15}ab^5+\tfrac{32}{9}a^3b^3+\tfrac{32}{15}a^5b)c_1=\tfrac{8}{3}(ba^3+b^3a)$$

$$c_1=\frac{15}{4}\frac{(a^2+b^2)}{(3a^4+5a^2b^2+3b^4)}$$

so that

$$u_1(x,y)=\frac{15}{4}\frac{a^2+b^2}{(3a^4+5a^2b^2+3b^4)}(a^2-x^2)(b^2-y^2) \tag{6-139}$$

The approximate value for the moment is given by

$$M=2G\theta\int_{-a}^{a}\int_{-b}^{b}u(x,y)\,dx\,dy=\frac{15}{4}\frac{a^2+b^2}{(3a^4+5a^2b^2+3b^4)}\frac{16}{9}a^3b^3\cdot 2G\theta$$

For the case of a square section, $b=a$, we obtain $M=\frac{80}{33}G\theta a^4=2.424G\theta a^4$, which is an overestimation of the exact value, $2.2496G\theta a^4$.

In the next example we use the least squares method to solve the nonlinear problem of Example 6-10.

Example 6-12 Consider the nonlinear differential equation (6-114*a*) of Example 6-10. We use the approximation

$$u_1(x)=\phi_0+c_1\phi_1=\sqrt{2}\,x^2+c_1(1-x^2) \tag{6-140}$$

and compute the approximate solution using both the Bubnov–Galerkin and least squares methods.

For the Bubnov–Galerkin method, we obtain

$$4c^2+2\sqrt{2}\,c-7=0$$

which gives the two values

$$c_1^{(1)}=\frac{1}{2\sqrt{2}}(-1+\sqrt{15})=1.01575$$

$$c_1^{(2)}=\frac{1}{2\sqrt{2}}(-1-\sqrt{15})=-1.72289$$

Of the two values, $c_1^{(1)}$ is found to give the smaller residual. Hence the solution becomes

$$u_1=1.01575+0.39846x^2 \tag{6-141}$$

which is the same as in Eq. (6-120).

Table 6-5 Comparison of the one-term variational solutions with the exact solution of the nonlinear equation (6-114*a*)

x	Exact	Two-term exact	RBG-1	RBG-2	LS	PG
0.0	1.0000	1.0000	1.0000	1.0158	1.0885	1.0110
0.1	1.0050	1.0050	1.0414	1.0197	1.0918	1.0150
0.2	1.0198	1.0200	1.0828	1.0317	1.1015	1.0272
0.3	1.0440	1.0450	1.1243	1.0516	1.1178	1.0473
0.4	1.0770	1.0800	1.1657	1.0795	1.1406	1.0755
0.5	1.1180	1.1250	1.2071	1.1154	1.1699	1.1118
0.6	1.6620	1.1800	1.2485	1.1592	1.2058	1.1562
0.7	1.2206	1.2450	1.2899	1.2110	1.2481	1.2086
0.8	1.2806	1.3200	1.3314	1.2708	1.2969	1.2691
0.9	1.3454	1.4050	1.3728	1.3385	1.3523	1.3376
1.0	1.4142	1.5000	1.4142	1.4142	1.4142	1.4142

RBG = Ritz–Bubnov–Galerkin methods [RBG-1 refers to Eq. (6-123) and RBG-2 to Eq. (6-120)].
LS = Least squares method, Eq. (6-142).
PG = Petrov–Galerkin method, Eq. (6-143).

Using the approximation (6-140) in the least squares method, we obtain a cubic equation for c_1:

$$c_1^3 - 3\sqrt{2}\,c_1^2 + 7.35c_1 - \tfrac{47}{16}\sqrt{2} = 0$$

Of the three roots, only $c_1^{(1)} = 1.0256$ is found to be real, the other two being complex. The least squares solution becomes

$$u_1 = 1.0256 + 0.3257x^2 \tag{6-142}$$

The Petrov–Galerkin method, with $\psi_1 = \cos(\pi x/2) + \sqrt{2}$ and the same choice of approximation as in (6-140), gives (see Example 6-10)

$$u_1 = 1.01104 + 0.40317x^2 \tag{6-143}$$

The exact and the two-term "exact" solutions to the nonlinear equation are given by

$$u_e = \sqrt{1 + x^2} \qquad u_{2e} = 1 + 0.5x^2 \tag{6-144}$$

The variational solutions are compared with the exact solutions in Table 6-5.

6-3-4 Collocation and Subdomain Methods

Collocation method In the collocation method the parameters c_i in $u_N = \sum_{i=1}^{N} c_i \phi_i$ are determined by forcing the residual in the approximation of the governing equation to vanish at N selected points $\mathbf{x}^k$ ($k = 1, 2, \ldots, N$) in the domain Ω:

$$R_N(\mathbf{x}^k, c_j, \phi_j, f) = 0 \qquad k = 1, 2, \ldots, N \tag{6-145}$$

For a linear operator equation $Au = f$, this yields

$$\sum_{j=1}^{N} A[\phi_j(\mathbf{x}^k)]c_j - f(\mathbf{x}^k) = 0 \tag{6-146}$$

Equation (6-146) can be recast in a form that resembles the weighted-residual method. Let $\delta(\mathbf{x} - \mathbf{x}^k)$ denote the *Dirac delta function* [see Eq. (4-12)]

$$\int_{\Omega} \delta(\mathbf{x} - \mathbf{x}^k) f(\mathbf{x})\, d\mathbf{x} \equiv f(\mathbf{x}^k) \tag{6-147}$$

Then Eq. (6-146) can be expressed in the alternative form

$$\sum_{i=1}^{N} c_i \left[\int_{\Omega} A[\phi_i(\mathbf{x})]\, \delta(\mathbf{x} - \mathbf{x}^k)\, d\mathbf{x} - \int_{\Omega} f(\mathbf{x})\, \delta(\mathbf{x} - \mathbf{x}^k)\, d\mathbf{x} = 0 \right. \tag{6-148}$$

A comparison of Eq. (6-148) with Eq. (6-106) shows [for $H = L_2(\Omega)$] that the weight functions ψ_k are given by

$$\psi_k(\mathbf{x}) = \delta(\mathbf{x} - \mathbf{x}^k) \tag{6-149}$$

Thus, the collocation method is a special case of the weighted-residual method.

The selection of the *collocation points* $\mathbf{x}^k$ is crucial in obtaining a well-conditioned system of equations and a convergent solution. The collocation points should be located as evenly as possible to avoid ill-conditioning of the resulting equations. We now consider an example.

Example 6-13 Consider a simply-supported beam under a uniformly distributed load f_0. The governing differential equation is given by Eq. (6-48*a*), and the boundary conditions are

$$w(0) = w(L) = 0 \qquad \frac{d^2w}{dx^2}(0) = \frac{d^2w}{dx^2}(L) = 0 \tag{6-150}$$

We consider a two-parameter approximation,

$$w_2 = c_1 \sin\frac{\pi x}{L} + c_2 \sin\frac{3\pi x}{L} \tag{6-151}$$

with the collocation points at $x = L/4$ and $x = L/2$. We obtain

$$\begin{aligned} EI\left[c_1\left(\frac{\pi}{L}\right)^4 \sin\frac{\pi}{4} + c_2\left(\frac{3\pi}{L}\right)^4 \sin\frac{3\pi}{4}\right] &= f_0 \\ EI\left[c_1\left(\frac{\pi}{L}\right)^4 \sin\frac{\pi}{2} + c_2\left(\frac{3\pi}{L}\right)^4 \sin\frac{3\pi}{2}\right] &= f_0 \end{aligned} \tag{6-152}$$

which yields

$$\begin{aligned} c_1 &= \frac{(1+\sqrt{2})f_0 L^4}{2EI\pi^4} \qquad c_2 = \frac{(\sqrt{2}-1)f_0 L^4}{162EI\pi^4} \\ w_2(x) &= \frac{f_0 L^4}{162EI\pi^4}\left(195.55 \sin\frac{\pi x}{L} + 0.414 \sin\frac{3\pi x}{L}\right) \end{aligned} \tag{6-153}$$

Note that if we had used the points $x = L/3$ and $x = 2L/3$ as collocation points, it would not have been possible to compute c_2. Because of the symmetry, the use of points $x = L/4$ and $x = L/2$ in the half beam is sufficient. The maximum deflection, $w_2(L/2) = 1.205 f_0 L^4/EI\pi^4 = f_0 L^4/80.87EI$ is 5 percent in error compared to the exact value, $f_0 L^4/76.8EI$.

Subdomain method In the subdomain method, the domain of the problem is subdivided into as many subdomains as there are adjustable parameters and then the parameters are determined by making the residual orthogonal to a constant (say unity) in each domain:

$$\int_{\Omega_i} R_N(c_j, \phi_j, f)\, d\mathbf{x} = 0 \qquad i = 1, 2, \ldots, N \tag{6-154}$$

where Ω_i denotes the ith subdomain. Equation (6-154) states that the average value of the residual in each subdomain is zero. Obviously, in this method negative errors can cancel positive errors to give least net error. This is not always desirable because errors are errors whether they are positive or negative. Once again we can interpret the subdomain method as a special case of the weighted-residual method for the weight function

$$\psi_i(\mathbf{x}) = \begin{cases} 1 & \text{if } \mathbf{x} \text{ is in } \Omega_i \\ 0 & \text{otherwise} \end{cases} \tag{6-155}$$

We consider the simply supported beam of Example 6-13.

Example 6-14 Reconsider the simply supported beam of Example 6-13. Due to the symmetry, we can consider two subdomains in the first half of the beam: $\Omega_1 = (0, L/4)$, $\Omega_2 = (L/4, L/2)$. For the same choice of coordinate functions as in Example 6-13, we obtain

$$\int_0^{L/4} \left[c_1 EI\left(\frac{\pi}{L}\right)^4 \sin\frac{\pi x}{L} + c_2 EI\left(\frac{3\pi}{L}\right)^4 \sin\frac{3\pi x}{L} - f_0 \right] dx = 0$$

$$\int_{L/4}^{L/2} \left[c_1 EI\left(\frac{\pi}{L}\right)^4 \sin\frac{\pi x}{L} + c_2 EI\left(\frac{3\pi}{L}\right)^4 \sin\frac{3\pi x}{L} - f_0 \right] dx = 0$$

Carrying out the integration, we obtain

$$EI\left(\frac{\pi}{L}\right)^4 \frac{L}{\pi}\left(1 - \frac{1}{\sqrt{2}}\right)c_1 + EI\left(\frac{3\pi}{L}\right)^4 \frac{L}{3\pi}\left(1 + \frac{1}{\sqrt{2}}\right)c_2 - \frac{f_0 L}{4} = 0$$

$$EI\left(\frac{\pi}{L}\right)^4 \frac{L}{\pi}\frac{1}{\sqrt{2}}c_1 + EI\left(\frac{3\pi}{L}\right)^4 \frac{L}{3\pi}\left(-\frac{1}{\sqrt{2}}\right)c_2 - \frac{f_0 L}{4} = 0$$

whose solution is

$$c_1 = \frac{(1+\sqrt{2}) f_0 L^4}{4\sqrt{2}\, EI\pi^3} \qquad c_2 = \frac{(\sqrt{2}-1) f_0 L^4}{108\sqrt{2}\, EI\pi^3}$$

$$w_2(x) = \frac{f_0 L^4}{108\sqrt{2}\, EI\pi^3}\left(65.184 \sin\frac{\pi x}{L} + 0.414 \sin\frac{3\pi x}{L}\right) \tag{6-156}$$

The center deflection obtained in the subdomain method, $w_2(L/2) = f_0 L^4/73.12EI$, is -5 percent in error compared to the exact value.

PROBLEMS 6-2

6-2-1 Consider the problem of a vertically loaded clamped circular plate of radius a with a central hole of radius b. The governing differential equation is given by

$$\left(\frac{d^2}{dr^2} + \frac{1}{r}\frac{d}{dr}\right)\left(\frac{d^2 w}{dr^2} + \frac{1}{r}\frac{dw}{dr}\right) = 1 \qquad b < r < a \qquad a = 2 \qquad b = 1$$

where w is the transverse deflection, and q is the uniformly distributed loading. Obtain a two-parameter Bubnov–Galerkin solution by choosing

$$\phi_1 = (r^2 - b^2)^2(a^2 - r^2)^2 \qquad \phi_2 = r^2(r^2 - b^2)^2(a^2 - r^2)^2$$

These functions satisfy the essential boundary conditions $w = dw/dr = 0$ at the boundary.

6-2-2 Solve Prob. 6-1-8 using the Bubnov–Galerkin method.

6-2-3 Solve Prob. 6-1-10 using the Bubnov–Galerkin method.

6-2-4 Solve Prob. 6-1-10 using the least squares method.

6-2-5 Solve Prob. 6-1-10 using the collocation method.

6-2-6 Solve Prob. 6-1-10 using the Petrov–Galerkin method with $\psi_1 = 1$ and $\psi_2 = x$.

6-2-7 Solve Prob. 6-1-11 using the Bubnov–Galerkin method.

6-2-8 Solve Prob. 6-1-11 using the least squares method.

6-2-9 Solve Prob. 6-1-11 using the collocation method.

6-2-10 Solve Prob. 6-1-11 using the subdomain method.

6-2-11 Solve the problem $-d^2u/dx^2 - u = x$, $0 < x < 1$, $u(0) = u(1) = 0$, using (*a*) the collocation method and (*b*) the subdomain method. Use two parameters and algebraic polynomials.

6-2-12 Solve the problem $-d^2u/dx^2 + 0.1u = 1$, $0 \le x \le 10$, $u'(0) = 0$, $u(10) = 0$, using (*a*) the Bubnov–Galerkin method, (*b*) the least squares method, (*c*) the collocation method with $x^i = \frac{10}{6} + \frac{10}{3}(i-1)$, and (*d*) the subdomain method with $\Omega^i = \{x: \frac{10}{3}(i-1) < x \le \frac{10}{3}i\}$. Use a three-parameter approximation of the form $u(x) = c_1 \cos \pi x/20 + c_2 \cos 3\pi x/20 + c_3 \cos 5\pi x/20$.

6-2-13 Consider the two-point boundary-value problem $-d^2u/dx^2 = \lambda u$, $u(0) = 0$, $u(1) + (du/dx)(1) = 0$. Determine the first eigenvalue using a two-parameter approximation by (*a*) the Bubnov–Galerkin method, (*b*) the least squares method, and (*c*) collocation at $x = \frac{1}{3}$ and $\frac{2}{3}$.

6-2-14 Solve Prob. 6-1-2 using a one-parameter least squares approximation.

6-2-15 Solve Prob. 6-1-3 using a one-parameter least squares approximation.

6-2-16 Solve Prob. 6-1-3 using a three-point collocation, $x = -\frac{1}{3}, 0, \frac{1}{3}$.

6-2-17 Consider the partial differential equation $-\nabla^2 u = 1$ in $\Omega \subset \mathbb{R}^2$, $u = 0$ on Γ, where the domain Ω is an elliptical region, that is, Γ is the arc of an ellipse $x^2/a^2 + y^2/b^2 = 1$. Find a one-parameter Bubnov–Galerkin solution of the equation.

6-2-18 Consider the torsion of a prismatic member of square cross-section Ω. The problem reduces to the solution of (see Example 6-8) the Poisson equation $-\nabla^2 u = 2$ in Ω and $u = 0$ on Γ. Take the side of the square to be $2a$, the origin being at the center of the square. Using an approximation of the form $u(x, y) = (a^2 - x^2)(a^2 - y^2)[c_1 + c_2(x^2 + y^2)]$ determine the two-parameter Bubnov–Galerkin solution.

6-2-19 Solve Prob. 6-2-18 using the collocation method. Take $x = y = 0$ and $x = y = a/2$ as the collocation points.

6-2-20 Consider the Neumann problem for the Poisson equation over a rectangle Ω of sides $2a$ and $2b$, with the origin being at the center of the rectangle: $-\nabla^2 u = -y$ in Ω, $\partial u/\partial n = 0$ on Γ. Assume an N-parameter approximation of the form (odd function of y and even function of x)

$$U_N(x, y) = A_1 + A_2 y + A_3 x^2 + A_4 y^3 + A_5 x^2 y + A_6 x^4 + A_7 x^4 y + A_8 x^2 y^3 + A_9 y^5 + \cdots$$

and determine the relationship between $A_i (i = 1, 2, \ldots, 9)$ by requiring that $\partial u/\partial n = 0$ on Γ:

$$\left.\frac{\partial u_N}{\partial x}\right|_{x=\pm a} = 0 \qquad \left.\frac{\partial u_N}{\partial y}\right|_{y=\pm b} = 0$$

To eliminate the arbitrary constant, set $A_1 = 0$. Show that the resulting approximation is ($A_6 = c_1$, $A_4 = c_2$, $A_9 = c_3$)

$$u_3(x, y) = c_1(x^4 - 2a^2x^2) + c_2(y^3 - 3b^2y) + c_3(y^5 - 5b^4y)$$

Use the Bubnov–Galerkin method to determine the constants c_i ($i = 1, 2, 3$) and compare with the exact solution $u(x, y) = \frac{1}{6}(y^3 - 3b^2y)$.

6-2-21 Find the first eigenvalue associated with the Laplace operator $-\nabla^2$ on the domain Ω lying in the first quadrant bounded by x and y axes and by the arc of the ellipse $x^2/a^2 + y^2/b^2 - 1 = 0$. Assume $u = 0$ on the boundary, and use a one-parameter Bubnov–Galerkin approximation of the form $u_1 = c_1 xy(1 - x^2/a^2 - y^2/b^2)$ to obtain $\lambda_1 = 15/a^2 + 15/b^2$.

6-2-22 Recall from Chaps. 4 and 5 that the problem of solving the equation $\nabla^2 u = 0$ in Ω, $u = \hat{u}$ on Γ is equivalent to finding u such that $u = \hat{u}$ on Γ and the functional

$$I(u) = \int_\Omega \frac{1}{2}\left[\left(\frac{\partial u}{\partial x}\right)^2 + \left(\frac{\partial u}{\partial y}\right)^2\right] dx\, dy$$

attains its minimum. Show that

(a) $I(v) = I(u) + I(u - v)$ for any v such that $v = \hat{u}$ on Γ

(b) $I(w) = I(w) - I(u - w)$ for any w such that $\nabla^2 w = 0$ on Ω and $\oint_s (w - \hat{u})\, \partial w/\partial n\, ds = 0$.

6-2-23 Show that the conditions in part (b) of Prob. 6-2-22 are equivalent to the requirement that $I(u - w)$ be a minimum.

6-2-24 Determine a one-parameter Bubnov–Galerkin solution that satisfies the boundary conditions $u(0) = 1$, $u(1) = 0$, and minimizes the integral

$$I(u) = \int_0^1 \left[\frac{1}{48}\left(\frac{du}{dx}\right)^4 + \left(\frac{du}{dx}\right)^2 + u^6 - 6u\right] dx$$

Compare the solution with the exact solution, $u = 1 - x^2$.

6-4 THE KANTOROVICH AND TREFFTZ METHODS

6-4-1 The Kantorovich Method

In the Ritz method, the solution of an ordinary or partial differential equation is reduced to the solution of a set of algebraic equations in terms of undetermined parameters. In the Kantorovich method, the problem of solving partial differential equations is reduced to the solution of ordinary differential equations in terms of undetermined functions. It is hoped that it is possible to solve the ordinary differential equations more accurately, possibly in exact form, to achieve better accuracy than the Ritz approximation. Thus, the Kantorovich method is appropriate for problems with two or more independent variables. We describe the gist of the method using a partial differential equation in two dimensions and the ideas introduced in connection with the Ritz method.

Consider the problem of determining the solution $u \in V$ to the variational problem

$$B(v, u) = l(v) \qquad \text{for any } v \in V \tag{6-157a}$$

where $B(v, u)$ and $l(v)$ are the bilinear and linear forms on the vector space V:

$$B(v, u) = (Tv, Tu) \qquad l(v) = (v, f) + (v, g)_{\Gamma_2} \tag{6-157b}$$

where T and T^* are linear operators such that $T^*T = A$ (see Sec. 6-3-2) and g is the specified value of Bu on the boundary Γ_2. In the Ritz method, we seek an N-parameter approximation of u in the form (for the case of nonhomogeneous essential boundary conditions)

$$u_N = \sum_{j=1}^{N} c_j \phi_j(x, y) + \phi_0(x, y) \tag{6-158}$$

where ϕ_0 is selected to satisfy the specified nonzero essential boundary conditions, and ϕ_j satisfy the conditions specified in Eq. (6-10), by requiring that Eq. (6-157a) be satisfied for every $v \equiv \phi_i$, $i = 1, 2, \ldots, N$. Substitution of Eq. (6-158) into Eq. (6-157a) gives

$$B\left(\phi_i, \sum_{j=1}^{N} c_j \phi_j + \phi_0\right) = l(\phi_i)$$

$$\sum_{j=1}^{N} B(\phi_i, \phi_j)c_j = l(\phi_i) - B(\phi_i, \phi_0)$$

or

$$\sum_{j=1}^{N} b_{ij} c_j = f_i \tag{6-159}$$

where

$$b_{ij} = B(\phi_i, \phi_j) \qquad f_i = l(\phi_i) - B(\phi_i, \phi_0) \tag{6-160}$$

In the Kantorovich method, we seek an N-parameter solution of Eq. (6-157) in the form

$$u_N = \sum_{j=1}^{N} c_j(x)\phi_j(x, y) + \phi_0(x, y) \tag{6-161}$$

where now c_j are undetermined *functions* of the coordinate x, and ϕ_0 satisfies the specified essential boundary condition

$$\phi_0 = \hat{u} \text{ on } \Gamma_1 \tag{6-162a}$$

and

$$\text{either } \phi_j = 0 \text{ on } \Gamma_1 \text{ or } c_j = 0 \text{ on } \Gamma_1 \tag{6-162b}$$

so that $c_j \phi_j = 0$ on Γ_1 for any $j = 1, 2, \ldots, N$. Thus, we have some flexibility in choosing the functions ϕ_j. Substitution of Eq. (6-161) for u and $\delta u = \sum \delta c_i \phi_i$ for v into Eq. (6-157a) gives

$$\sum_{i=1}^{N} \delta c_i [B(\phi_i, u_N) - l(\delta c_i \phi_i)] = 0 \tag{6-163}$$

where δc_i is the variation of c_i; δc_i is arbitrary except for that it should vanish on Γ_1. Since this is the weak form of the Euler equation $Au = f$, we can integrate it by parts to obtain

$$\sum_{i=1}^{N} \left\{ \iint_{\Omega} \delta c_i \phi_i [A(u_N) - f] \, dx \, dy + \int_{\Gamma_1} \delta c_i \phi_i B(u_N) \, ds + \int_{\Gamma_2} \delta c_i \phi_i [B(u_N) - \hat{g}] \, ds \right\} = 0 \tag{6-164}$$

The boundary terms drop out because $\delta c_i = 0$ on Γ_1 and $B(u_N) = \hat{g}$ on Γ_2.

Next we assume that the region Ω is such that we can separate the integration with respect to x and y (see Fig. 6-3):

$$\sum_{i=1}^{N} \int_a^b \delta c_i \left\{ \int_{y=P(x)}^{y=Q(x)} [A(u_N) - f] \phi_i(x, y) \, dy \right\} dx = 0 \tag{6-165}$$

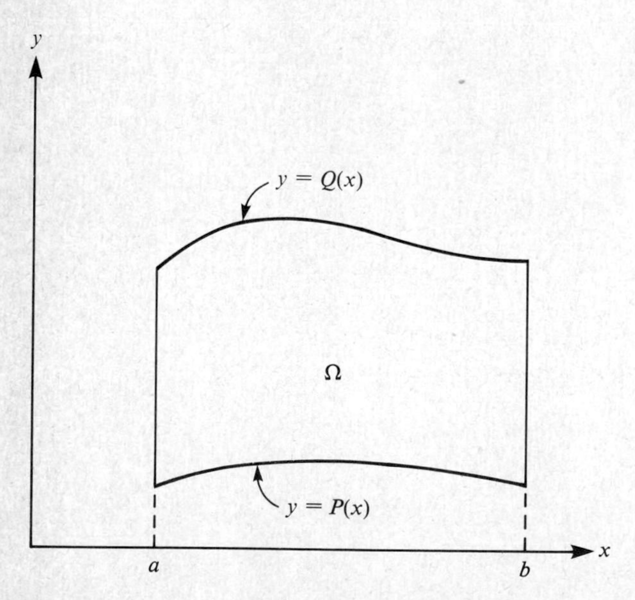

Figure 6-3 A typical two-dimensional domain in the Kantorovich method.

Since $\delta c_i(\mathbf{x})$ are arbitrary, it follows from Eq. (6-165) that

$$\int_P^Q [A(u_N) - f]\phi_i \, dy = 0 \tag{6-166}$$

or

$$\sum_{j=1}^{N} \left[\int_P^Q \phi_i A(c_j \phi_j) \, dy \right] = \int_P^Q \phi_i [f - A(\phi_0)] \, dy \tag{6-167}$$

which gives a system of N ordinary differential equations in c_j, which can be solved either exactly or by a second approximation.

We illustrate the method via the problem of the torsion of a square cross-section member.

Example 6-15 Consider the torsion problem considered in Example 6-8. First we seek a one-parameter Kantorovich approximation of the form

$$u_1 = c_1(x)\phi_1(y) = c_1(x)(y^2 - b^2) \tag{6-168}$$

Since $u = 0$ on $x = \pm a$ and on $y = \pm b$, it follows that we must determine $c_1(x)$ such that it satisfies the boundary conditions

$$c_1(-a) = c_1(a) = 0 \tag{6-169}$$

and the ordinary differential equation resulting from Eq. (6-166),

$$-\int_{-b}^{b} \left[(y^2 - b^2) \frac{d^2 c_1}{dx^2} + 2c_1 + 2 \right] (y^2 - b^2) \, dy = 0$$

Performing the integration and dividing throughout by the coefficient of d^2c_1/dx^2 we obtain

$$\frac{d^2 c_1}{dx^2} - \frac{5}{2b^2} c_1 - \frac{5}{2b^2} = 0 \tag{6-170}$$

This completes the application of the Kantorovich method.

We can solve the ordinary differential equation (6-170) either exactly or by an approximate method. In the interest of improving the accuracy (over the Bubnov–Galerkin solution) here we consider the exact solution of Eq. (6-170) subjected to the boundary conditions in Eq. (6-169).

The exact solution of Eq. (6-170) is given by

$$c_1(x) = A \cosh kx + B \sinh kx - 1 \qquad k = \frac{1}{b}\sqrt{\frac{5}{2}} \tag{6-171}$$

Using the conditions in Eq. (6-169), the constants A and B can be evaluated,

$$A = \frac{1}{\cosh ka} \qquad B = 0$$

The one-parameter solution becomes

$$u_1 = (b^2 - y^2)\left[1 - \frac{\cosh kx}{\cosh ka}\right] \tag{6-172}$$

The moment is given by

$$M = 2G\theta \int_{-a}^{a} \left(1 - \frac{\cosh kx}{\cosh ka}\right) dx \int_{-b}^{b} (b^2 - y^2)\, dy$$

$$= \frac{16}{3} b^3 a\left[1 - \sqrt{\frac{2}{5}}\frac{b}{a} \tanh \sqrt{\frac{5}{2}}\frac{a}{b}\right] G\theta \tag{6-173}$$

For the case of square section ($b = a$), the one-parameter Kantorovich method gives $M = 2.234G\theta a^4$, which is an improvement over the Bubnov–Galerkin solution.

Next, consider a two-parameter Kantorovich approximation of the form

$$u_2(x, y) = (y^2 - b^2)[c_1(x) + c_2(x)y^2] \tag{6-174a}$$

This gives [see Reddy (1984b)]:

$$c_1(x) = -1 + 1.0312 \frac{\cosh k_1 x}{\cosh k_1 a} - 0.0312 \frac{\cosh k_2 x}{\cosh k_2 a}$$

$$c_2(x) = -0.2276 \frac{\cosh k_1 x}{\cosh k_1 a} + 0.2276 \frac{\cosh k_2 x}{\cosh k_2 a} \tag{6-174b}$$

Table 6-6 Comparison of the solutions $\bar{u}(x, 0)$ of Eq. (6-110) ($-\nabla^2 u = f_0$ over a square with $u = 0$ on the boundary) obtained by the Ritz, Bubnov–Galerkin, and Kantorovich methods with the series solution.

($\bar{u} = u/f_0 a^2$, $\bar{\sigma}_{yz} = \sigma_{yz}/f_0$, $f_0 = 2$)

		The Ritz–Bubnov–Galerkin method		The Kantorovich method	
x ($y = 0$)	Series solution Eq. (6-112)	One-parameter	Two-parameter	One-parameter	Two-parameter
0.0	0.29445	0.31250	0.29219	0.30261	0.29473
0.10000	0.29194	0.30937	0.28986	0.30014	0.29222
0.20000	0.28437	0.30000	0.28278	0.29266	0.28462
0.30000	0.27158	0.28437	0.27075	0.27999	0.27177
0.40000	0.25328	0.26250	0.25340	0.26180	0.25339
0.50000	0.22909	0.23437	0.23025	0.23765	0.22909
0.60000	0.19854	0.20000	0.20065	0.20693	0.19839
0.70000	0.16104	0.15937	0.16382	0.16886	0.16072
0.80000	0.11591	0.11250	0.11884	0.12250	0.11546
0.90000	0.06239	0.05937	0.06463	0.06667	0.06203
1.00000	−0.00038	0.0	0.0	0.00000	0.00000
		shear stress, $\bar{\sigma}_{yz}(a, 0)$			
1.00000	0.67500	0.62500	0.70284	0.72636	0.66416

where $b^2k_1^2 = 14 - \sqrt{133}$ and $b^2k_2^2 = 14 + \sqrt{133}$ ($k_1 = 1.5708/b$ and $k_2 = 5.0530/b$). Equations (6-174a) and (6-174b) define the two-parameter Kantorovich solution of the torsion problem.

A comparison of the solutions obtained using the Bubnov–Galerkin method and Kantorovich method (with one and two parameters) with the series solution for the case of square section are presented in Table 6-6. The table also includes the shearing stress, $\sigma_{yz} = -\partial u/\partial x$, at $x = a$ and $y = 0$. The exact value of the maximum shear stress, which occurs at $x = a$ and $y = 0$, is 1.35. It is clear from the results that the two-parameter solutions are more accurate than the one-parameter solutions, and that the Kantorovich method gives slightly more accurate solutions than those obtained using the Bubnov–Galerkin method (which are the same as those obtained by the Ritz method for this problem).

The next example deals with the application of the Kantorovich method to an eigenvalue problem.

Example 6-16 Consider the problem of finding the eigenvalues associated with the Laplace operator over the rectangle $\Omega = \{(x, y)\colon -a < x < a, -b < y < b\}$,

$$-\nabla^2 u = \lambda u \text{ in } \Omega \qquad u = 0 \text{ on } \Gamma$$

We seek a solution in the form

$$u_N = \sum_{i=1}^{N} c_i(x)\phi_i(x, y) + \phi_0(x, y)$$

For $N = 1$, we choose $\phi_0 = 0$ and $\phi_1 = (y^2 - b^2)$,

$$u_1 = (y^2 - b^2)c_1(x)$$

Equation (6-166) takes the form

$$\int_{-b}^{b} \left[2c_1 + (y^2 - b^2)\frac{d^2c_1}{dx^2} + \lambda(y^2 - b^2)c_1 \right](y^2 - b^2)\, dy = 0$$

or

$$\frac{16}{15} b^5 \frac{d^2c_1}{dx^2} + \left(\frac{16}{15} b^5 \lambda - \frac{8}{3} b^3 \right) c_1 = 0$$

subject to the end conditions $c_1(a) = c_1(-a) = 0$. The general solution of the second-order equation is

$$c_1(x) = A \sin \alpha x + B \cos \alpha x \qquad \alpha = \sqrt{\lambda - \frac{5}{2b^2}}$$

The symmetry and condition $c_1(a) = 0$ together give $A = 0$ and

$$B \cos \alpha a = 0$$

This implies, for the nontrivial solution, the value

$$\sqrt{\lambda - \frac{5}{2b^2}}\, a = (2n-1)\frac{\pi}{2}$$

or

$$\lambda_n = \left[(2n-1)\frac{\pi}{2a}\right]^2 + \frac{5}{2b^2}$$

The first eigenvalue is given by

$$\lambda_1 = \frac{\pi^2}{(2a)^2} + \frac{10}{(2b)^2}$$

which differs by less than 1.3 percent compared to the exact value

$$\lambda_1 = \frac{\pi^2}{(2a)^2} + \frac{\pi^2}{(2b)^2}$$

We remark that the Kantorovich method gives, in general, more accurate solutions provided one can solve the resulting differential equations exactly. The method can be used to reduce partial differential equations of time and space to ordinary differential equations in time, which then can be solved using either the Laplace transforms or numerical (e.g., finite difference) methods. In the next example we illustrate the use of the Kantorovich method for time-dependent problems [see Reddy (1984a,b)].

Example 6-17 Consider the following nondimensionalized partial differential equation,

$$\frac{\partial u}{\partial t} - \frac{\partial^2 u}{\partial x^2} = 0 \qquad 0 < x < 1 \tag{6-175}$$

$$u(0, t) = \frac{\partial u}{\partial x}(1, t) = 0 \text{ for } t > 0 \tag{6-176}$$

$$u(x, 0) = 1.0 \text{ for } 0 < x < 1 \tag{6-177}$$

We consider a one-parameter Kantorovich approximation of the form

$$u_1(x, t) = c_1(t)\phi_1(x) = c_1(t)(2x - x^2) \tag{6-178}$$

The function $\phi_1(x)$, which is a function of x only, satisfies the boundary conditions in Eq. (6-176). We should determine $c_1(t)$ such that the initial condition (6-177) is satisfied. This requires

$$c_1(0) = 1 \tag{6-179}$$

Substituting Eq. (6-178) into Eq. (6-166), we obtain

$$\int_0^1 \left[\frac{dc_1}{dt}(2x - x^2) - c_1(-2)\right](2x - x^2)\,dx = 0$$

or

$$\frac{dc_1}{dt} + \frac{5}{2}c_1 = 0 \tag{6-180}$$

The exact solution of Eq. (6-180) is given by

$$c_1(t) = Ae^{-(5/2)t}$$

Using the condition (6-179), we determine A to be 1. The one-parameter solution becomes

$$u_1(x, t) = e^{-2.5t}(2x - x^2) \tag{6-181}$$

The exact solution of Eqs. (6-175)–(6-177) is given by

$$u(x, t) = 2\sum_{n=0}^{\infty} \frac{\exp(-\lambda_n^2 t)\sin\lambda_n x}{\lambda_n} \qquad \lambda_n = (2n+1)\pi/2 \tag{6-182}$$

Next we consider a two-parameter approximation. We use the weak form

$$\begin{aligned} 0 &= \int_0^1 \phi_i\left(\frac{\partial u}{\partial t} - \frac{\partial^2 u}{\partial x^2}\right)dx \\ &= \int_0^1 \left(\phi_i\frac{\partial u}{\partial t} + \frac{\partial \phi_i}{\partial x}\frac{\partial u}{\partial x}\right)dx \end{aligned} \tag{6-183}$$

where $\phi_i \in H_A$ $(A = -d^2/dx^2)$ satisfy only the essential boundary conditions. Note that such a weak formulation is not always possible with Eq. (6-166). It is possible only if ϕ_i is not a function of t. Note that the spatial approximation used here is the same for the Kantorovich method, the Bubnov–Galerkin method and the Ritz method. For the Kantorovich and Bubnov–Galerkin methods we employed the weak form of the equation [see Eq. (6-109)].

For the problem at hand, we seek a solution in the form,

$$u_2(x, t) = c_1(t)\phi_1 + c_2(t)\phi_2 = c_1(t)x + c_2(t)x^2 \tag{6-184}$$

where ϕ_1 and ϕ_2 satisfy the essential boundary condition, $u(0) = 0$. Substitution of Eq. (6-184) into Eq. (6-183) gives the ordinary differential equations in time,

$$\begin{aligned} \frac{1}{3}\frac{dc_1}{dt} + \frac{1}{4}\frac{dc_2}{dt} + c_1 + c_2 &= 0 \\ \frac{1}{4}\frac{dc_1}{dt} + \frac{1}{5}\frac{dc_2}{dt} + c_1 + \frac{4}{3}c_2 &= 0 \end{aligned} \tag{6-185}$$

which must be solved subject to the initial condition in Eq. (6-177). From Eq. (6-184) it is clear that there is no way the initial condition will be satisfied exactly by c_i $(i = 1, 2)$. Therefore, we satisfy the initial condition in the Bubnov–Galerkin sense:

$$\int_0^1 [u(x, 0) - 1]\phi_i \, dx = 0 \qquad i = 1, 2$$

We obtain

$$\tfrac{1}{3}c_1(0) + \tfrac{1}{4}c_2(0) = \tfrac{1}{2} \qquad \tfrac{1}{4}c_1(0) + \tfrac{1}{5}c_2(0) = \tfrac{1}{3} \tag{6-186}$$

Now we can use the Laplace transform method to solve Eqs. (6-185) and (6-186). We obtain [see Reddy (1984a)]

$$\begin{aligned} c_1(t) &= 1.6408e^{-32.1807t} + 2.3592e^{-2.486t} \\ c_2(t) &= -(2.265e^{-32.1807t} + 1.068e^{-2.486t}) \end{aligned} \tag{6-187}$$

Equations (6-184) and (6-187) together give the two-parameter solution.

The one- and two-parameter solutions in Eq. (6-181) and (6-184) are compared, for various values of t and x, with the series solution (6-182) in Table 6-7. Clearly, the two-parameter solution is more accurate than the one-parameter solution, and agrees more closely with the series solution for large values of time.

Table 6-7 Comparison of the variational solution with the series solution of the transient motion of a cable subject to initial displacement $u(x, 0) = 1$ [see Eqs. (6-175) to (6-177)]

		Series solution	Variational solution				Series solution	Variational solution	
t	x	Eq. (6-182)	Eq. (6-181)	Eq. (6-184)	t	x	Eq. (6-182)	Eq. (6-181)	Eq. (6-184)
	0.2	0.4727	0.3177	0.4265		0.2	0.0617	0.0552	0.0665
	0.4	0.7938	0.5648	0.7413		0.4	0.1174	0.0981	0.1198
0.05	0.6	0.9418	0.7413	0.9443	0.75	0.6	0.1616	0.1288	0.1598
	0.8	0.9880	0.8472	1.0357		0.8	0.1906	0.1472	0.1866
	1.0	0.9965	0.8825	1.0154		1.0	0.1997	0.1534	0.2001
	0.2	0.2135	0.1927	0.2306		0.2	0.0333	0.0296	0.0357
	0.4	0.4052	0.3426	0.4152		0.4	0.0633	0.0525	0.0643
0.25	0.6	0.5560	0.4496	0.5538	1.0	0.6	0.0872	0.0690	0.0858
	0.8	0.6520	0.5139	0.6466		0.8	0.1025	0.0788	0.1002
	1.0	0.6848	0.5353	0.6934		1.0	0.1077	0.0821	0.1075
	0.2	0.1145	0.1031	0.1238		0.2	0.0097	0.0085	0.0103
	0.4	0.2177	0.1834	0.2230		0.4	0.0184	0.0151	0.0186
0.50	0.6	0.2996	0.2407	0.2975	1.50	0.6	0.0254	0.0198	0.0248
	0.8	0.3522	0.2751	0.3473		0.8	0.0298	0.0226	0.0289
	1.0	0.3703	0.2865	0.3725		1.0	0.0313	0.0235	0.0310

The above example illustrates the use of variational methods for the solution of time-dependent problems. The approximation is selected such that the parameters c_i are functions of time, while ϕ_i are functions of spatial coordinates. The procedure used in Example 6-17 to obtain ordinary differential equations in time is known as the *semidiscrete* approximation. In Sec. 6-5 we consider time-dependent problems in more detail.

6-4-2 The Trefftz Method

In all variational methods discussed in this chapter, the approximation functions were selected such that they satisfy the boundary conditions (most often, only the essential boundary conditions) of the problem, and the parameters were determined using a variational procedure, such as the minimization of a quadratic functional or setting the weighted-residual to zero. An alternative approach is to select approximation functions that satisfy the governing differential equation but not the boundary conditions, and then determine the parameters in the approximation by a variational procedure that requires the satisfaction of the boundary conditions. The Trefftz method is one such method. We now describe the essence of the method.

Consider the problem of finding a function u which is harmonic (i.e., $\nabla^2 u = 0$) in the domain Ω and which satisfies the boundary condition

$$u = \hat{u} \text{ on } \Gamma \tag{6-188}$$

where $\hat{u}(s)$ is a given continuous function of position on the boundary Γ. From the previous discussions we know that this problem is equivalent to finding the minimum of the quadratic functional

$$J(u) = \int_\Omega |\operatorname{grad} u|^2 \, d\mathbf{x} \tag{6-189}$$

Let $\{\phi_k\}$ be a sequence of linearly independent functions which are harmonic in Ω and complete in the following sense: for any function v which is harmonic and together with its derivatives is quadratically summable in Ω, it is possible to find for a given $\varepsilon > 0$ an integer N and constants c_i such that

$$J\left(v - \sum_{i=1}^{N} c_i \phi_i\right) = \int_\Omega \left|\operatorname{grad}\left(v - \sum_{i=1}^{N} c_i \phi_i\right)\right|^2 d\mathbf{x} < \varepsilon \tag{6-190}$$

We seek an approximate solution of the problem in the usual form

$$u_N = \sum_{i=1}^{N} c_i \phi_i$$

where N is an arbitrarily chosen number, and determine the coefficients c_i from the requirement that

$$J(u - u_N) \text{ is a minimum}$$

where u is the required solution of the problem. By setting $\partial J/\partial c_k$ to zero, we obtain the system of algebraic equations,

$$\delta J(u - u_N, \phi_k) = B(u - u_N, \phi_k) = 0 \qquad k = 1, 2, \ldots, N \tag{6-191}$$

Using the Green–Gauss theorem, we write

$$B(u - u_N, \phi_k) = -\int_\Omega (u - u_N)\nabla^2 \phi_k \, d\mathbf{x} + \oint_\Gamma (u - u_N) \frac{\partial \phi_k}{\partial n} \, ds$$

where $\partial/\partial n$ is the normal derivative. Since the ϕ_k are harmonic, the first integral is zero, and since $u = \hat{u}$ on Γ we have

$$\oint_\Gamma (\hat{u} - u_N) \frac{\partial \phi_k}{\partial n} \, ds = 0 \qquad k = 1, 2, \ldots, N \tag{6-192}$$

Equation (6-192) is the statement of the Trefftz method. It can also be viewed as the weighted residual method because the residual $R_N = u - u_N$ is made orthogonal to trace $\partial \phi_k/\partial n$ of ϕ_k on the boundary [i.e., Eq. (6-188) is solved by the method of weighted residuals]:

$$\sum_{i=1}^{N} \left(\oint_\Gamma \frac{\partial \phi_k}{\partial n} \phi_i \, ds \right) c_i = \oint_\Gamma \hat{u} \frac{\partial \phi_k}{\partial n} \, ds \qquad k = 1, 2, \ldots, N \tag{6-193}$$

We note that u_N does not change if u is changed by a constant because the functional (6-189) does not change when u is changed from u to u + constant. Consequently, the Trefftz solution u_N can be determined only within an arbitrary constant. To eliminate this constant we assume that the mean values of the harmonic functions u and ϕ_k over the boundary Γ are equal to zero:

$$\oint_\Gamma u \, ds = 0 \qquad \oint_\Gamma \phi_k \, ds = 0$$

The existence of a unique solution to Eq. (6-193) is to be established. Assume the converse. Then the homogeneous equation corresponding to Eq. (6-193) has a nontrivial solution v_N:

$$\oint_\Gamma v_N \frac{\partial \phi_k}{\partial n} \, ds = 0 \qquad k = 1, 2, \ldots, N$$

where $v_N = \sum_{i=1}^{N} b_i \phi_i$. Multiplying the above equation with b_k and summing, we obtain

$$\oint_\Gamma v_N \frac{\partial v_N}{\partial n} \, ds = 0$$

Using the Green–Gauss theorem, we write

$$0 = \oint_\Gamma v_N \frac{\partial v_N}{\partial n} \, ds = \int_\Omega |\text{grad } v_N|^2 \, d\mathbf{x} - \int_\Omega v_N \nabla^2 v_N \, d\mathbf{x}$$

Since $\nabla^2 v_N = 0$, we obtain grad $v_N = 0$ or $v_N =$ constant in Ω. Since the mean value of $v_N|_\Gamma$ equals zero, this constant must be zero on Γ giving $v_N = 0$ in Ω, which is contrary to our assumption. Hence Eq. (6-193) has a solution. The uniqueness follows from a similar argument (use $u_N^{(1)} - u_N^{(2)}$ for v_N in the above discussion).

The Trefftz solution also provides a lower bound because

$$J(u_N) \leq J(v_N)$$

for any $v_N \in Q$, where

$$Q = \left\{ v\colon \nabla^2 v = 0 \text{ and } \oint_\Gamma v\, ds = 0 \right\} \tag{6-194}$$

Next we prove the convergence of the Trefftz solution. It can be shown (see Sec. 5-5-2) that the operator $\partial/\partial n$ is a positive-bounded-below operator in the space Q. We assume that the elements of Q are representable through the Green's function in terms of their limiting values. Since $\partial/\partial n$ is bounded below, there exists a constant μ such that

$$\oint_\Gamma v \frac{\partial v}{\partial n}\, ds \geq \mu \oint_\Gamma v^2\, ds \tag{6-195}$$

holds for all $v \in Q$. By the assumption of completeness of the sequence $\{\phi_i\}$, we can find for given $\varepsilon > 0$ a number M and constants $c_1, c_2, \ldots, c_M$ such that

$$J\left(u - \sum_{i=1}^{M} c_i \phi_i\right) < \varepsilon$$

Hence for $N \geq M$,

$$J(u - u_N) \leq J(u - u_M) < \varepsilon$$

where u_M and u_N are the approximate solutions constructed by the Trefftz method. Since the function $u - u_N$ is harmonic, we have

$$J(u - u_N) = \oint_\Gamma (u - u_N) \frac{\partial(u - u_N)}{\partial n}\, ds < \varepsilon$$

Using Eq. (6-195), we obtain

$$\oint_\Gamma (u - u_N)^2\, ds < \frac{\varepsilon}{\mu} \tag{6-196}$$

If $G(p, q)$ is Green's function in the domain Ω, then

$$u(p) - u_N(p) = \oint_\Gamma [u(q) - u_N(q)] \frac{\partial G(p, q)}{\partial n}\, ds$$

We suppose that the point p lies within a closed domain which lies entirely within Ω. Then $\partial G(p, q)/\partial n$ and all its derivatives with respect to the coordinates of point

p are bounded. Using the Cauchy–Schwarz inequality and inequality (6-196), we obtain

$$\|u(p) - u_N(p)\| \leq \frac{c\varepsilon}{\mu}$$

and therefore $u \to u_N$ uniformly in any closed domain lying entirely within Ω.

All of our discussion thus far used the Laplace equation, $\nabla^2 u = 0$, as the model problem. The discussion can be generalized to the operator equation $Au = f$.

Consider the operator equation

$$Au - f = 0 \text{ in } \Omega \tag{6-197}$$

$$Bu - \hat{u} = 0 \text{ on } \Gamma \tag{6-198}$$

where A and B are linear operators and f and $\hat{u}$ are given continuous functions of position. In the Trefftz method we seek an approximate solution of the form in Eq. (6-158), with ϕ_0 and ϕ_i satisfying the following properties:

ϕ_0 is a particular solution of Eq. (6-197)

ϕ_i are (nonconstant) solutions of the homogeneous equation,

$$Au = 0 \tag{6-199}$$

The parameters c_i are determined by setting the weighted-residual in the boundary condition (6-198) to zero:

$$\oint_\Gamma [B(u_N) - \hat{u}] \frac{\partial \phi_k}{\partial n} \, ds = 0 \qquad k = 1, 2, \ldots, N \tag{6-200}$$

or

$$\sum_{i=1}^{N} \left[\oint_\Gamma \frac{\partial \phi_i}{\partial n} B(\phi_k) \, ds \right] c_k = \oint_\Gamma \frac{\partial \phi_k}{\partial n} \hat{u} \, ds \tag{6-201}$$

Equation (6-201) gives N linear algebraic equations for the parameters c_k, $k = 1, 2, \ldots, N$. Note that the coefficient matrix is not symmetric. We illustrate the application of the method via an example.

Example 6-18 Consider the torsion problem considered in Example 6-8. Put $u = \Psi - x^2$ in Eq. (6-110) to obtain

$$\begin{aligned} -\nabla^2 \Psi &= 0 \text{ in } \Omega \\ \Psi &= x^2 \text{ on } \Gamma \end{aligned} \tag{6-202}$$

We consider a one-parameter Trefftz solution of the form

$$\Psi_1 = c_1 \cos \frac{\pi x}{2a} \cosh \frac{\pi y}{2a} \tag{6-203}$$

Equation (6-201) reduces to (here $B = 1$)

$$c_1 \oint_\Gamma \frac{\partial \phi_1}{\partial n} \phi_1 \, ds = \oint_\Gamma x^2 \frac{\partial \phi_1}{\partial n} \, ds$$

where Γ is the boundary of the rectangle. We have

$$2c_1 \left[\int_{-b}^{b} \left(-\frac{\pi}{2a} \sin \frac{\pi x}{2a} \cosh \frac{\pi y}{2a} \cdot \cos \frac{\pi x}{2a} \cosh \frac{\pi y}{2a} \right) \Bigg|_{x=a} dy \right.$$

$$\left. + \int_{a}^{-a} \left(\frac{\pi}{2a} \cos \frac{\pi x}{2a} \sinh \frac{\pi y}{2a} \cos \frac{\pi x}{2a} \cosh \frac{\pi y}{2a} \right) \Bigg|_{y=b} (-dx) \right]$$

$$= 2 \int_{-b}^{b} \left(-\frac{\pi}{2a} x^2 \sin \frac{\pi x}{2a} \cosh \frac{\pi y}{2a} \right) \Bigg|_{x=a} dy$$

$$+ 2 \int_{a}^{-a} \left(\frac{\pi}{2a} x^2 \cos \frac{\pi x}{2a} \sinh \frac{\pi y}{2a} \right) \Bigg|_{y=b} (-dx)$$

$$c_1 \left(\frac{\pi}{2a} \cosh \frac{\pi b}{2a} \sinh \frac{\pi b}{2a} \right) a = \left(-2a^2 \sinh \frac{\pi b}{2a} \right)$$

$$+ \left(2a^2 \sinh \frac{\pi b}{2a} - \frac{16a^2}{\pi^2} \sinh \frac{\pi b}{2a} \right)$$

or

$$c_1 = -\frac{32a^3}{\pi^3} \frac{1}{\cosh \pi b/2a}$$

The one-parameter approximation, within a constant, equals

$$u_1 = \Psi_1 - x^2 = -x^2 - \frac{32a^3}{\pi^3} \frac{\cosh \pi y/2a}{\cosh \pi b/2a} \cos \frac{\pi x}{2a} \tag{6-204}$$

PROBLEMS 6-3

6-3-1 Use the Kantorovich method to find a two-parameter approximation of the form $u(x, y) = c_1(y) \cos \pi x/2a + c_2(y) \cos 3\pi x/2a$ to determine an approximate solution of the torsion problem in Example 6-8.

6-3-2 Consider the problem of the deflection of a rectangular plate clamped along the sides $y = \pm b$ and simply supported along $x = \pm a$. Assume an approximation of the form $u(x, y) = c_1(x)(y^2 - b^2)^2$, $\phi_1 = (y^2 - b^2)^2$, and determine $c_1(x)$.

6-3-3 Find the first eigenvalue of the equation $-\nabla^2 u = \lambda u$ in Ω, $u = 0$ on Γ, using the Kantorovich method. Use a trigonometric basis function. See Example 6-16.

6-3-4 Use a variational method in space and the Laplace transform method in time to find a one-parameter approximation of the form (see Example 6-17) $w(x, t) = c_1(t)(1 - \cos 2\pi x)$ of the equations $\partial^4 w/\partial x^4 + \partial^2 w/\partial t^2 = 0$, $0 < x < 1$, $t > 0$; $w = \partial w/\partial x = 0$ at $x = 0, 1$, and $t > 0$; $w = \sin \pi x - \pi x(1 - x)$, $\partial w/\partial t = 0$ at $t = 0$, $0 < x < 1$. Use the Galerkin method to satisfy the initial conditions. Show that the solution is given by $w(x, t) = 0.1107 \cos 22.7929t(1 - \cos 2\pi x)$.

6-3-5 Consider the Poisson equation $-\nabla^2 u = 0, 0 < x < 1, 0 < y < \infty, u(0, y) = u(1, y) = 0$ for $y > 0$, $u(x, 0) = x(1 - x)$, $u(x, \infty) = 0$, $-1 \le x \le 1$. Assuming an approximation of the form $u(x, y) = c_1(y)x(1 - x)$ find the differential equation for $c_1(y)$ and solve it exactly.

6-3-6 Use the Trefftz method to solve the torsion problem by substituting $u = \Psi - \frac{1}{2}(x^2 + y^2)$, and solving the differential equation $\nabla^2\Psi = 0$ in $-a < x < a$, $-b < y < b$; $\Psi = \frac{1}{2}(x^2 + y^2)$ on the boundary. Use a three-parameter approximation of the form

$$\Psi = c_1(x^2 - y^2) + c_2(x^4 - 6x^2y^2 + y^4) + c_3(x^6 - 15x^4y^2 + 15x^2y^4 - y^6)$$

6-3-7 In the Trefftz method, show that if one of ϕ_1 is a constant, the corresponding c_i are indeterminate.

6-3-8 Find the first eigenvalue associated with the Laplace operator on a rectangle using the two-parameter Kantorovich approximation (see Example 6-16), $u^2 = (y^2 - b^2)(c_1 + c_2 y^2)$. The requirement $-\nabla^2 u - \lambda u = 0$ for $y = \pm b$ gives $c_1 = -5b^2 c_2$. Hence u_2 becomes $u_2 = (y^2 - b^2)(y^2 - 5b^2)c_2$. Show that the first eigenvalue is given by

$$\lambda_1 = \frac{\pi^2}{(2a)^2} + \frac{(306/31)}{(2b)^2}$$

6-5 TIME-DEPENDENT PROBLEMS

6-5-1 Introduction

In the variational solution of time-dependent problems, the undetermined parameters c_i in (6-105) are assumed to be functions of time while ϕ_i are assumed to depend on spatial coordinates. This leads to two stages of approximation, both of which could employ variational methods, as was illustrated in Example 6-17. In the solution of time-dependent problems, the spatial approximation is considered first, and the time approximation next. Such a procedure is commonly known as *semidiscrete approximation*. Semidiscrete variational approximation in space, as discussed in the preceding sections, results in a set of ordinary differential equations in time, which must be further approximated to obtain a set of algebraic equations. The spatial approximation of time-dependent problems leads, in place of Eq. (6-8), to a matrix differential equation in time of the form (irrespective of the space dimension)

$$[M]\left\{\frac{\partial c}{\partial t}\right\} + [B]\{c\} = \{F\} \tag{6-205}$$

for equations involving first-order time derivatives, and

$$[M]\left\{\frac{\partial^2 c}{\partial t^2}\right\} + [B]\{c\} = \{F\} \tag{6-206}$$

for equations containing second-order time derivatives. Here $[B]$ and $\{F\}$ are the matrices defined in Eq. (6-8), and $[M]$ is the *mass* or *gram* matrix, whose coefficients are given by

$$m_{ij} = \int_\Omega \phi_i \phi_j \, d\mathbf{x} \tag{6-207}$$

In the following sections we discuss approximation schemes for first-order and second-order time derivatives.

We shall use the following notation. We use Δt to denote a *time step*, and u^n to denote the approximation of u at time $t = n\,\Delta t$ ($n = 0, 1, 2, \ldots$). Here we describe time-approximation schemes to compute u^{n+1} if u^n (and u^{n-1}) are known.

We recall the notion of *numerical stability* of approximation. We say that a *numerical process is stable* if the error in the approximation $e = u - u_N$, measured in an appropriate norm, does not become unbounded as N is increased [see Eq. (6-21)]. Analogously, a *time approximation scheme is said to be stable* if for a fixed N the error in the approximation measured in an appropriate norm does not become unbounded as time increases. For a parabolic equation of the type

$$\left\{\frac{du}{dt}\right\} = [A]\{u\} \tag{6-208}$$

with initial condition $\{u\}^0 = \{u_0\}$, where $[A]$ is independent of t and $\{u\}$, the error e in the approximation due to the error $\{e_0\}$ in the initial values is governed by the equation

$$\left\{\frac{de}{dt}\right\} = [A]\{e_0\} \tag{6-209}$$

The solution of this equation is given by

$$\{e(t)\} = [\exp\,([A]t)]\{e_0\} \tag{6-210}$$

If the error is to be bounded in the norm $\|\cdot\|$, we must have

$$\|\{e(t)\}\| \leq \|[\exp\,([A]t)]\| \cdot \|\{e_0\}\| \tag{6-211}$$

For stability, the *amplification matrix* $[\exp\,([A]t)]$ should be bounded as t is increased. A sufficient condition is that

$$\|[\exp\,([A]t)]\| \leq M < \infty \qquad \text{for all } t > 0 \tag{6-212}$$

In practice, when Eq. (6-205) is discretized, we obtain an equation of the form

$$[A^1]^n\{c\}^{n+1} = [A^2]^n\{c\}^n + \{F\} \tag{6-213}$$

where $[A^i]^n$ are matrices which depend on $[B]$, $[M]$, n, and Δt. A time integration scheme is said to be *unconditionally stable* if the error in the solution does not grow without bound with time t, for any n and time step Δt. The scheme is *conditionally stable* if the error is bounded for certain fixed values of n and Δt. Of course, a scheme that is not unconditionally or conditionally stable is *unstable*.

A numerical scheme (i.e., time-approximation method) is termed *implicit* if the application of the scheme to the equation

$$\left\{\frac{dc}{dt}\right\} = [B]\{c\} + \{F\}$$

results in Eq. (6-213) with $[A^1]^n$ nondiagonal. The scheme is termed *explicit* if the above equation results in Eq. (6-213) with $[A^1]^n$ diagonal.

We say a *numerical process is implicit* if it results in an equation of the form (6-213) in which $[A^1]^n$ is *not* diagonal. The process is called *explicit* if the coefficient matrix $[A^1]^n$ is diagonal or an identity matrix. When a process is explicit and the equations are linear, Eq. (6-213) can be solved by simply marching forward in time, which involves matrix multiplication of $[A^2]^n$ with the column vector $\{c\}^n$. For an implicit process the solution of Eq. (6-213) requires the inversion of the matrix $[A^1]^n$ at each time step. Whether a given numerical process is implicit or explicit depends not only on the method of approximating the time derivatives but also on the method of spatial discretization (and on the basis used in the spatial approximation). An implicit numerical scheme always results in an implicit numerical process, but an explicit numerical scheme does not always result in an explicit numerical process.

6-5-2 Parabolic Equations

Consider a matrix differential equation of the form

$$[M]\{\dot{c}\} + [B]\{c\} = \{F\} \qquad 0 < t \leq T \tag{6-214}$$

where $[M]$, $[B]$, and $\{F\}$ are known matrices, and $\{c\}$ is the column matrix of the undetermined parameters. The superposed dot on $\{c\}$ indicates differentiation with respect to time, t. Equation (6-214) is valid for any time $t > 0$.

We introduce a *θ-family of approximation* which approximates a weighted average of the time derivative of a dependent variable at two consecutive time steps by linear interpolation of the values of the variable at the two time steps:

$$(\theta\{\dot{c}\}^{n+1} + (1-\theta)\{\dot{c}\}^n) = (\{c\}^{n+1} - \{c\}^n)/\Delta t \qquad 0 \leq \theta \leq 1 \tag{6-215}$$

where $\{\cdot\}^n$ refers to the value of the enclosed quantity at time $t = t_n = n\,\Delta t$. From Eq. (6-215) we can obtain a number of well-known numerical schemes for parabolic equations:

1. *Euler's (forward difference) scheme:* $\theta = 0$; explicit and conditionally stable; order of accuracy $= O(\Delta t)$.
2. *Backward Euler's scheme:* $\theta = 1$; implicit and unconditionally stable; order of accuracy $= O(\Delta t)$.
3. *The Crank–Nicholson scheme:* $\theta = 1/2$; implicit and unconditionally stable; order of accuracy $= O((\Delta t)^2)$.
4. *The Galerkin scheme:* $\theta = 2/3$; implicit and unconditionally stable; order of accuracy $= O((\Delta t)^2)$. (6-216)

Using the approximation (6-215) for time t_n and t_{n+1} in (6-214), we obtain

$$[M]\{c\}^{n+1} = [M]\{c\}^n + \theta\,\Delta t(\{F\}^{n+1} - [B]\{c\}^{n+1}) + (1-\theta)\,\Delta t(\{F\}^n - [B]\{c\}^n)$$

Rearranging the terms, we obtain

$$([M] + \theta\,\Delta t[B])\{c\}^{n+1} = ([M] - (1-\theta)\,\Delta t[B])\{c\}^n + \Delta t(\theta\{F\}^{n+1} + (1-\theta)\{F\}^n)$$

or

$$[\hat{M}]\{c\}^{n+1} = [\hat{B}]\{c\}^n + \{F\} \equiv \{\hat{F}\} \tag{6-217}$$

where

$$[\hat{M}] = [M] + \theta\,\Delta t[B] \qquad [\hat{B}] = [M] - (1-\theta)\,\Delta t[B] \tag{6-218}$$

The solution at time $t = t_{n+1}$ is obtained in terms of the solution at time t_n by inverting the matrix $[\hat{M}]$. At $t = 0$, the solution is known from the *initial* conditions of the problem, and therefore, Eq. (6-217) can be used to obtain the solution at $t = \Delta t_1$. Since the column vector $\{F\}$ is known at all times, $\{\hat{F}\}$ is known in advance.

It should be noted that one can expect better results if smaller time steps are used. In practice, however, one wishes to take as large a time step as possible to cut down the computational expense. Larger time steps, in addition to decreasing the accuracy of the solution, can introduce some unwanted, numerically induced oscillations into the solution. Thus an estimate of an upper bound on the time step, even for unconditionally stable schemes, proves to be very useful. A stability analysis shows [see Eqs. (6-209) to (6-212)] that the numerical scheme (6-217) is stable (i.e., no unbounded oscillations) if the minimum eigenvalue λ of the equation

$$\det([\hat{B}] - \lambda[\hat{M}]) = 0 \tag{6-219}$$

is nonnegative. More specifically we have

$$\begin{aligned} 0 < \lambda < 1&: \text{ stable without oscillations} \\ -1 < \lambda < 0&: \text{ stable with oscillations} \\ \lambda < -1&: \text{ unstable} \end{aligned} \tag{6-220}$$

Example 6-19 Consider the one-dimensional partial differential equation (see Example 6-17)

$$\frac{\partial u}{\partial t} - \frac{\partial^2 u}{\partial x^2} = 0 \qquad 0 < x < 1 \tag{6-221}$$

with the boundary conditions

$$u(0, t) = 0 \qquad \frac{\partial u}{\partial x}(1, t) = 0 \tag{6-222}$$

and the initial condition

$$u(x, 0) = 1.0 \tag{6-223}$$

where u is the dependent variable, t is the time, and x is the independent coordinate.

First, we write Eq. (6-221) in a weak form. We have

$$0 = \int_0^1 \left(v \frac{\partial u}{\partial t} + \frac{\partial v}{\partial x} \frac{\partial u}{\partial x} \right) dx \tag{6-224}$$

The boundary term vanishes in view of the homogeneous natural boundary conditions at $x = 1$ and the specified essential boundary condition at $x = 0$.

Next, we consider a two-parameter (semidiscrete) Ritz approximation of the form

$$u(x, t) = \phi_0(x) + \sum_{i=1}^{2} c_i(t)\phi_i(x) \tag{6-225}$$

with $\phi_0 = 0$, $\phi_1 = x$, and $\phi_2 = x^2$. Substituting $v = \phi_i$ and (6-225) into (6-224), we obtain [cf. Eq. (6-185)]

$$\frac{1}{60}\begin{bmatrix} 20 & 15 \\ 15 & 12 \end{bmatrix}\begin{Bmatrix} \dot{c}_1 \\ \dot{c}_2 \end{Bmatrix} + \frac{1}{3}\begin{bmatrix} 3 & 3 \\ 3 & 4 \end{bmatrix}\begin{Bmatrix} c_1 \\ c_2 \end{Bmatrix} = \begin{Bmatrix} 0 \\ 0 \end{Bmatrix} \tag{6-226}$$

This completes the semidiscretization in space.

Equations (6-226) can be solved either exactly (see Example 6-17) or approximately. Since the initial condition $u(x, 0) = 1$ cannot be satisfied exactly by the present approximation we find the initial values $c_i(0)$ by the Bubnov–Galerkin method [see Eq. (6-186)]. We obtain

$$c_1(0) = 4 \qquad c_2(0) = -10/3 \tag{6-227}$$

We now solve the ordinary differential equations in (6-226), subject to the initial conditions (6-227), by the theta-family of approximation. We obtain

$$\begin{bmatrix} (\frac{1}{3} + \theta\, \Delta t) & (\frac{1}{4} + \theta\, \Delta t) \\ (\frac{1}{4} + \theta\, \Delta t) & \left(\frac{1}{5} + \dfrac{4\theta\, \Delta t}{3}\right) \end{bmatrix}\begin{Bmatrix} c_1 \\ c_2 \end{Bmatrix}^{n+1} = \begin{bmatrix} \frac{1}{3} - (1-\theta)\, \Delta t & \frac{1}{4} - (1-\theta)\, \Delta t \\ \frac{1}{4} - (1-\theta)\, \Delta t & \frac{1}{5} - \dfrac{4(1-\theta)\, \Delta t}{3} \end{bmatrix}\begin{Bmatrix} c_1 \\ c_2 \end{Bmatrix}^{n} \tag{6-228}$$

Upon selecting the values of θ and Δt, the solution c_i at $t = (n + 1)\, \Delta t$ can be obtained in terms of c_i at $t = n\, \Delta t$; the initial conditions are given by (6-227).

The stability criterion in Eq. (6-220) for the problem (after lengthy algebra) gives

$$-\frac{1}{(240)^2} + \frac{13}{21600}(1-\theta)\,\Delta t - \frac{1}{720}(1-\theta)^2(\Delta t)^2 - \frac{1}{7200}\theta\,\Delta t$$
$$+\frac{13}{2700}(1-\theta)\theta(\Delta t)^2 - \frac{1}{90}\theta(1-\theta)^2(\Delta t)^3 > 0 \qquad (6\text{-}229a)$$

for stability without oscillations and

$$\text{LHS} > -1 \qquad (6\text{-}229b)$$

for stability with oscillations, where LHS stands for the "left-hand side" of Eq. (6-229*a*). Of course, Eq. (6-229*b*) is satisfied for *any* Δt when $\theta = 1/2$ and 2/3. Equation (6-229*a*) is satisfied for $\Delta t < 0.1$ when $\theta = 0.5$ but is not satisfied for any Δt when $\theta = 2/3$.

Table 6-8 shows a comparison of the solution obtained from Eq. (6-228) and the two-parameter solution obtained in Example 6-17 using the variational method and Laplace transforms [see Eqs. (6-184) and (6-187)] with the analytical solution in Eq. (6-182). Both the Laplace transform and Crank–Nicholson methods give more accurate solutions than the Galerkin scheme (note that the spatial discretization is common for columns 4, 5, and 6).

Table 6-8 Comparison of the approximate solutions with the analytical solution ($\Delta t = 0.05$) of Examples 6-17 and 6-19

				Eq. (6-228)	
t	x	Analytical Eq. (6-182)	Eq. (6-184)	$\theta = 2/3$	$\theta = 1/2$
0.2	0.5	0.5532	0.5555	0.5610	0.5548
	1.0	0.7723	0.7844	0.7913	0.7848
0.4	0.5	0.3356	0.3376	0.3441	0.3372
	1.0	0.4745	0.4777	0.4868	0.4771
0.6	0.5	0.2049	0.2054	0.2113	0.2045
	1.0	0.2897	0.2906	0.2989	0.2900
0.8	0.5	0.1251	0.1249	0.1297	0.1246
	1.0	0.1769	0.1767	0.1836	0.1763
1.0	0.5	0.0764	0.0760	0.0797	0.0757
	1.0	0.1080	0.1075	0.1127	0.1071
1.2	0.5	0.0466	0.0463	0.0489	0.0460
	1.0	0.0659	0.0654	0.0692	0.0652
1.4	0.5	0.0285	0.0281	0.0300	0.0396
	1.0	0.0402	0.0398	0.0425	0.0406

6-5-3 Hyperbolic Equations

In structural dynamics problems we encounter the second-order time derivatives of the dependent variable(s). The semidiscrete (spatial) approximation of the equations results in a matrix differential equation of the form

$$[M]\{\ddot{c}\} + [B]\{c\} = \{F\} \qquad 0 < t < T \tag{6-230}$$

The most commonly used numerical schemes are listed below [see Bathe and Wilson (1973)].

The central difference scheme

$$\begin{aligned} \{\ddot{c}\}^n &= \frac{1}{(\Delta t)^2}(\{c\}^{n-1} - 2\{c\}^n + \{c\}^{n+1}) \\ \{\dot{c}\}^n &= \frac{1}{2\,\Delta t}(\{c\}^{n+1} - \{c\}^{n-1}) \end{aligned} \tag{6-231}$$

The scheme is explicit and conditionally stable, and the order of accuracy of the approximation is $(\Delta t)^2$. Applied to Eq. (6-230) at $t = t_n$, the central difference scheme results in

$$[\hat{M}]\{c\}^{n+1} = [\hat{B}]\{c\}^n - [\hat{M}]\{c\}^{n-1} + \{F\} \tag{6-232}$$

where
$$[\hat{M}] = \frac{1}{(\Delta t)^2}[M] \qquad [\hat{B}] = -[B] + \frac{2}{(\Delta t)^2}[M] \tag{6-233}$$

Observe that the calculation of $\{c\}^{n+1}$ involves $\{c\}^{n-1}$. Therefore a special starting procedure is required to calculate $\{c\}$ at $t = \Delta t$. Equation (6-230) can be used to compute $\{\ddot{c}\}^0 = -[M]^{-1}[B]\{c\}^0$ (assuming $\{F\} = \{0\}$ initially). Using $\{c\}^0$, $\{\dot{c}\}^0$, and $\{\ddot{c}\}^0$ in Eq. (6-231), we obtain

$$\{c\}^{(-1)} = \{c\}^0 - \Delta t\{\dot{c}\}^0 + \frac{(\Delta t)^2}{2}\{\ddot{c}\}^0 \tag{6-234}$$

The critical time step for the central difference method is

$$\Delta t \le \Delta t_{cr} = \frac{T_N}{\pi} \tag{6-235}$$

where T_N is the smallest period (i.e., reciprocal of the frequency ω) of the N-parameter variational solution of the equation $([B] - \omega^2[M])\{c\} = \{0\}$.

The Hubolt method

$$\begin{aligned} \{\ddot{c}\}^{n+1} &= \frac{1}{(\Delta t)^2}(2\{c\}^{n+1} - 4\{c\}^n + 4\{c\}^{n-1} - \{c\}^{n-2}) \\ \{\dot{c}\}^{n+1} &= \frac{1}{6\,\Delta t}(11\{c\}^{n+1} - 18\{c\}^n + 9\{c\}^{n-1} - 2\{c\}^{n-2}) \end{aligned} \tag{6-236}$$

The scheme is implicit and unconditionally stable, and the order of accuracy is $(\Delta t)^2$. Using Eq. (6-236) in Eq. (6-230) at $t = t_{n+1}$, we obtain

$$[\hat{M}]\{c\}^{n+1} = \frac{5}{(\Delta t)^2}[M]\{c\}^n - \frac{4}{(\Delta t)^2}[M]\{c\}^{n-1} + \frac{1}{(\Delta t)^2}[M]\{c\}^{n-2} \quad (6\text{-}237)$$

where
$$[\hat{M}] = \frac{1}{(\Delta t)^2}[M] + [B] \quad (6\text{-}238)$$

To use the Hubolt method it is suggested that one obtain $\{c\}^1$ and $\{c\}^2$ using a different scheme, such as the central difference scheme, for the first two time steps. Although there is no critical time step for the Hubolt method, Eq. (6-235) gives an order of magnitude of the time step.

The Newmark scheme In the Newmark scheme the first time derivative $\{\dot{c}\}$ and the function $\{c\}$ itself are approximated at the $(n + 1)$th time step by the following expressions:

$$\begin{aligned} \{\dot{c}\}^{n+1} &= \{\dot{c}\}^n + [(1-\alpha)\{\ddot{c}\}^n + \alpha\{\ddot{c}\}^{n+1}]\,\Delta t \\ \{c\}^{n+1} &= \{c\}^n + \{\dot{c}\}^n\,\Delta t + [(\tfrac{1}{2} - \beta)\{\ddot{c}\}^n + \beta\{\ddot{c}\}^{n+1}](\Delta t)^2 \end{aligned} \quad (6\text{-}239)$$

where α and β are parameters that control the accuracy and stability of the scheme. The choice $\alpha = 1/2$, and $\beta = 1/4$ is known to give the *constant-average-acceleration method* (also called the *trapezoidal rule*). The case $\alpha = 1/2$ and $\beta = 1/6$ corresponds to the *linear acceleration method*. Both schemes are implicit and the first method is unconditionally stable.

Rearranging Eqs. (6-239) and (6-230), we arrive at

$$[\hat{M}]\{c\}^{n+1} = \{\hat{F}\} \quad (6\text{-}240)$$

where

$$[\hat{M}] = [B] + a_0[M],\ \{\hat{F}\} = \{F\}^{n+1} + [M](a_0\{c\}^n + a_1\{\dot{c}\}^n + a_2\{\ddot{c}\}^n) \quad (6\text{-}241)$$

Once the solution $\{c\}$ is known at $t_{n+1} = (n + 1)\,\Delta t$, the first and second derivatives (velocity and acceleration) of $\{c\}$ at t_{n+1} can be computed from [rearranging the expressions in Eq. (6-240)]

$$\begin{aligned} \{\ddot{c}\}^{n+1} &= a_0(\{c\}^{n+1} - \{c\}^n) - a_1\{\dot{c}\}^n - a_2\{\ddot{c}\}^n \\ \{\dot{c}\}^{n+1} &= \{\dot{c}\}^n + a_3\{\ddot{c}\}^n + a_4\{\ddot{c}\}^{n+1} \\ a_0 = [\beta(\Delta t^2)]^{-1} \qquad & a_1 = a_0\,\Delta t \qquad a_2 = \frac{1}{2\beta} - 1 \\ a_3 = (1-\alpha)\,\Delta t \qquad & a_4 = \Delta t \end{aligned} \quad (6\text{-}242)$$

For a given set of initial conditions $\{c\}_0$, $\{\dot{c}\}_0$, and $\{\ddot{c}\}_0$ [obtained from Eq. (6-230)], we can solve Eq. (6-240) repeatedly, marching forward in time, for the column vector $\{c\}$ and its time derivatives at any time $t > 0$.

Example 6-20 Consider the motion (see Prob. 6-3-4) of a uniform cross-section beam clamped at the ends. The equation of motion, in non-dimensional form, can be written as

$$\frac{\partial^4 w}{\partial x^4} + \frac{\partial^2 w}{\partial t^2} = 0 \qquad 0 < x < 1 \qquad t > 0 \tag{6-243}$$

subject to the boundary conditions

$$w = \frac{\partial w}{\partial x} = 0 \qquad \text{at } x = 0, 1 \text{ and } t > 0 \tag{6-244}$$

and initial conditions

$$w = \sin \pi x - \pi x(1 - x) \qquad \frac{\partial w}{\partial t} = 0 \text{ at } t = 0 \qquad \text{for } 0 < x < 1 \tag{6-245}$$

Note that all boundary conditions in the present problem are of the essential type. Hence, the selection criteria for the approximation functions in the Bubnov–Galerkin method and the Ritz method coincide. The following approximation functions meet the boundary conditions:

$$\phi_1 = 1 - \cos 2\pi x \qquad \phi_2 = 1 - \cos 4\pi x, \ldots, \phi_N = 1 - \cos 2N\pi x \tag{6-246}$$

The semidiscrete approximation results in

$$[M]\{\ddot{c}\} + [B]\{c\} = 0 \tag{6-247}$$

where
$$M_{ij} = \int_0^1 \phi_i \phi_j \, dx \qquad B_{ij} = \int_0^1 \frac{d^2\phi_i}{dx^2} \frac{d^2\phi_j}{dx^2} \, dx \tag{6-248}$$

We consider the one-parameter ($N = 1$) approximation. We have

$$M_{11} = \tfrac{3}{2} \qquad B_{11} = \frac{(2\pi)^4}{2}$$

and Eq. (6-247) becomes

$$\frac{d^2 c_1}{dt^2} + k^2 c_1 = 0 \qquad k = \frac{(2\pi)^2}{\sqrt{3}} = 22.7929 \tag{6-249}$$

Application of the Newmark scheme to Eq. (6-249), with the initial conditions in (6-245), results in

$$\hat{M}_{11} = \frac{(2\pi)^4}{2} + \frac{3a_0}{2} \qquad \hat{F}_1(0) = 0.1661 a_0 - 57.5105 a_2 \tag{6-250}$$

and for $\alpha = \frac{1}{2}$ and $\beta = \frac{1}{4}$ we have $a_0 = 4/(\Delta t)^2$. Then we must solve Eq. (6-240) repeatedly, marching forward in time.

The numerical results for the center deflection (as a function of time) as computed in Prob. 6-3-4 and the Newmark method are compared in Table 6-9 for various values of the time step.

Table 6-9 Comparison of solutions of the time dependent problem of Example 6-20

	Galerkin in space/Newmark's integration in time			Galerkin in space/exact in time
t	$\Delta t = 0.01$	$\Delta t = 0.005$	$\Delta t = 0.0025$	(Prob. 6-3-4)
0.02	0.19898	0.19884	0.19877	0.19879
0.04	0.13626	0.13574	0.13550	0.13558
0.06	0.04595	0.04498	0.04468	0.04455
0.08	−0.05367	−0.05495	−0.05550	−0.05534
0.10	−0.14242	−0.14368	−0.14421	−0.14406
0.12	−0.20233	−0.20312	−0.20347	−0.20336
0.14	−0.22126	−0.22116	−0.22113	−0.22113
0.16	−0.19537	−0.19411	−0.19355	−0.19374
0.18	−0.12992	−0.12750	−0.12640	−0.12677
0.20	−0.03816	−0.03490	−0.03343	−0.03392
0.22	0.06133	0.06482	0.06637	0.06586
0.24	0.14840	0.15132	0.15261	0.15219
0.26	0.20542	0.20699	0.20767	0.20744
0.28	0.22083	0.22046	0.22028	0.22032

6-5-4 The Method of Discretization in Time

Here we describe the so-called method of discretization in time [see Rektorys (1982)]. The method is described for both parabolic and hyperbolic initial value problems.

Parabolic equations Consider the problem:

$$\frac{\partial u}{\partial t} + Au = f \text{ in } \Omega \times (0, T) \tag{6-251a}$$

$$u(\mathbf{x}, 0) = u_0(\mathbf{x}) \text{ in } \Omega \tag{6-251b}$$

$$\begin{aligned} B_i u &= g_i \text{ in } \Gamma_1 \times (0, T) \qquad i = 1, 2, \ldots, k \\ C_i u &= h_i \text{ in } \Gamma_2 \times (0, T) \qquad i = 1, 2, \ldots, 2m - k \end{aligned} \tag{6-251c}$$

where A is a linear, positive-definite operator, B_i and C_i are the boundary operators associated with the operator A (see Sec. 5-4-4), and T is the total time.

We divide the time interval $[0, T]$ into p subintervals $I_1, I_2, \ldots, I_p$ of length Δt, such that $I_j = [t_{j-1}, t_j]$ with $t_0 = 0$. The method of discretization involves seeking the functions $z_j, j = 1, 2, \ldots, p$, such that

$$Az_j + \frac{1}{\Delta t} z_j = \frac{1}{\Delta t} z_{j-1} + f \tag{6-252}$$

holds. It is clear that the quotient $(z_j - z_{j-1})/\Delta t$ replaced $\partial u/\partial t$ at $t = t_j$. Equation (6-252) is to be solved for $z_1, z_2, \ldots, z_p$. For $j = 1$, we have

$$Az_1 + \frac{1}{\Delta t} z_1 = \frac{1}{\Delta t} z_0 + f$$

where $z_0 = u_0$. The solution z_1 is used to determine z_2, and so on.

The method of discretization in time allows us to find approximate solutions $z_j(\mathbf{x})$, $j = 1, 2, \ldots, p$ of the given problem at discrete values of time t only [i.e., z_j is the approximation of $u(\mathbf{x}, t_j)$]. To get an approximation in the whole domain $Q = \Omega \times (0, T)$, we construct a function $\bar{u}(\mathbf{x}, t)$, defined in the jth interval by

$$\bar{u}(\mathbf{x}, t) = z_{j-1}(\mathbf{x}) + \frac{t - t_{j-1}}{\Delta t} [z_j(\mathbf{x}) - z_{j-1}(\mathbf{x})] \tag{6-253}$$

Existence and uniqueness Let A be the $2m$th order differential operator in Eq. (5-76), with its bilinear form $B(v, u)$ in Eq. (5-82). Let B_i $(i = 1, 2, \ldots, k \leq m)$ be the boundary operators appearing in the essential boundary conditions [see Eq. (5-79)]. Define V by

$$V = \{u\colon\ u \in H^m(\Omega),\ B_1 u = 0,\ B_2 u = 0, \ldots, B_k u = 0 \quad (k \leq m) \text{ on } \Gamma_1\} \tag{6-254a}$$

Suppose that $w \in H^m(\Omega)$ exists such that $B_i w = g_i$ $(i = 1, 2, \ldots, k)$ on Γ_1. Then by Theorem 5-6 the weak problem in Eq. (5-81) has a unique solution $u \in H^m(\Omega)$ provided $B(\cdot, \cdot)$ is V-elliptic.

In the case of the problem defined by Eqs. (6-252) and (6-251c), the operator A is replaced by the operator

$$\bar{A} = A + \frac{1}{\Delta t} \tag{6-254b}$$

Hence the associated bilinear form is given by

$$\bar{B}(v, u) = B(v, u) + \frac{1}{\Delta t} (v, u)_0 \tag{6-255}$$

where $B(v, u)$ is the bilinear form associated with the operator A [see Eq. (5-82)]. For fixed j $(1 \leq j \leq p)$, let $f + z_{j-1}/\Delta t \in L_2(\Omega)$. Then the weak problem associated with Eqs. (6-252) and (6-251c) involves seeking the function $z_j(\mathbf{x})$ in $H^m(\Omega)$ such that $z_j - u(\mathbf{x}, t_j) \in V$ and

$$\bar{B}(v, z_j) = \int_\Omega v\left(f + \frac{z_{j-1}}{\Delta t}\right) d\mathbf{x} + \sum_{i=1}^{m-k} \int_{\Gamma_2} B_i v h_i \, ds \tag{6-256}$$

holds. The following theorem gives sufficient conditions for the existence of the unique solution to Eqs. (6-252) and (6-251c).

Theorem 6-5 Let j be fixed $(1 \leq j \leq p)$, and

$$\begin{aligned} &\text{(i)}\ \ \bar{B}(v, u) \text{ be continuous and } V\text{-elliptic} \\ &\text{(ii)}\ f + (1/\Delta t) z_{j-1} \in L_2(\Omega) \end{aligned} \tag{6-257}$$

Then there exists only one solution $z_j \in H^m(\Omega)$ such that $z_j - u(\mathbf{x}, t_j) \in V$ and Eq. (6-256) is satisfied.

The proof of the theorem follows along the same lines as that given for Theorem 5-6. For V-symmetric bilinear forms, Theorem 6-5 reduces to one equivalent to Theorem 5-5. The following lemma establishes that if the bilinear form $B(v, u)$ is continuous and V-elliptic then the same holds for the bilinear form $\bar{B}(v, u)$. This result implies that if the weak problem associated with the steady (or static) problem has a unique solution then the corresponding unsteady (or transient) problem also has a unique solution with $\Delta t > 0$ fixed.

Lemma 6-1 Let the bilinear form $B(v, u)$ be continuous and V-elliptic. Then the same holds for the bilinear form $\bar{B}(v, u)$ in Eq. (6-255) for fixed $\Delta t > 0$.

PROOF By the Schwarz inequality we have

$$|(v, u)_0| \leq \|v\|_0 \|u\|_0 \leq \|v\|_m \|u\|_m$$

and, by assumption,

$$|B(v, u)| \leq M \|v\|_m \|u\|_m$$

for all $v, u \in H^m(\Omega)$. Then it follows that

$$|\bar{B}(v, u)| \leq |B(v, u)| + \frac{1}{\Delta t}|(v, u)_0|$$

$$\leq \left(M + \frac{1}{\Delta t}\right)\|v\|_m \|u\|_m$$

for all $v, u \in H^m(\Omega)$, i.e., $\bar{B}(v, u)$ is continuous.

The V-ellipticity of $\bar{B}(v, u)$ follows easily from the V-ellipticity of $B(v, u)$ [the converse is not true, in general]:

$$\bar{B}(v, u) = B(v, u) + \frac{1}{\Delta t}(v, v)_0$$

$$\geq \alpha \|v\|_V^2 + \frac{1}{\Delta t}\|v\|_0^2$$

$$\geq \alpha \|v\|_V^2$$

because $\Delta t > 0$. This completes the proof of the lemma.

Theorem 6-5 and Lemma 6-1 state that if the weak problem of the operator equation

$$Au = f$$

satisfies the conditions of Theorem 5-5 (that is, $B(v, u)$ is continuous and V-elliptic), hence has a unique solution, then the solution of the corresponding parabolic equation

$$\frac{\partial u}{\partial t} + Au = f$$

can be *approximately* determined by Eq. (6-253), wherein each $z_j(\mathbf{x})$ is the unique solution of the weak problem (6-256) for $j = 1, 2, \ldots, p$. The error in the time approximation for the case $u(\mathbf{x}, 0) = 0$ is given by [see Rektorys (1982)]

$$\|u(\mathbf{x}, t_j) - z_j(\mathbf{x})\|_0 \leq \frac{Mj(\Delta t)^2}{2} \qquad M = \|Af\|_0 \tag{6-258}$$

which is based on the estimate in Eq. (4-46). Further, if the coefficient α in the positive-definiteness of $B(v, u)$ can be found, the following estimate can be used

$$\|u(\mathbf{x}, t_j) - z_j(\mathbf{x})\|_0 \leq \frac{M(\Delta t)}{2\alpha^2}(1 - e^{-\alpha^2 j \Delta t}) \tag{6-259}$$

Note that estimates in Eqs. (6-258) and (6-259) hold only for the case $u_0(\mathbf{x}) = 0$.

For a discussion of the method of discretization in time for more general operators $A = A(t)$, see Rektorys (1982).

Variational approximation The weak problem in Eq. (6-256) can be spatially approximated, for each fixed j, by the Ritz and Galerkin methods. Let $\{\phi_i\}_{i=1}^N$ be the first N vectors of a base in V, and let

$$z_j^N(\mathbf{x}) = \sum_{r=1}^N c_r^j \phi_r(\mathbf{x}) \qquad j = 1, 2, \ldots, p \tag{6-260}$$

be the N-parameter variational approximation of $z_j(\mathbf{x})$. Substituting Eq. (6-260) into Eq. (6-256), we obtain for the jth problem

$$\sum_{r=1}^N \bar{B}(\phi_i, \phi_r)c_r^j = \int_\Omega \phi_i\left(f + \frac{1}{\Delta t} z_{j-1}^N\right) d\mathbf{x} + \sum_{s=1}^{m-k} \int_{\Gamma_2} B_s \phi_i h_s \, ds \tag{6-261}$$

where z_{j-1}^N is known from the solution of Eq. (6-261) for $(j-1)$th problem. For $j = 1$, we take $z_0^N(\mathbf{x}) = u_0(\mathbf{x})$.

The ideas presented thus far in this section are illustrated below through an example problem.

Example 6-21 Consider the problem

$$\begin{gathered} \frac{\partial u}{\partial t} - \frac{\partial^2 u}{\partial x^2} = \sin x \text{ in } Q = (0, \pi) \times (0, 1) \\ u(x, 0) = 0, \qquad u(0, t) = u(\pi, t) = 0 \end{gathered} \tag{6-262}$$

The operator A is given by $A = -d^2/dx^2$. Equation (6-252) takes the form

$$-\frac{d^2 z_j}{dx^2} + \frac{1}{\Delta t} z_j = \sin x + \frac{1}{\Delta t} z_{j-1} \qquad z_j(0) = z_j(\pi) = 0 \tag{6-263}$$

with $z_0 = 0$. Since A is positive-definite on the set of twice differentiable functions u which vanish at $x = 0, \pi$, the bilinear form

$$\bar{B}(v, u) = \int_0^\pi \frac{dv}{dx}\frac{du}{dx}\,dx + \frac{1}{\Delta t}\int_0^\pi vu\,dx$$

satisfies the conditions of Theorem 6-5. Therefore the variational problem

$$\bar{B}(v, z_j) = \int_0^\pi v\left(\sin x + \frac{1}{\Delta t} z_{j-1}\right) dx \tag{6-264}$$

has a unique solution in $H_0^1(0, \pi)$ for each j ($j = 1, 2, \ldots, p$) and fixed $\Delta t > 0$.

Equation (6-263) can be solved exactly. For $j = 1$, we have

$$-\frac{d^2 z_1}{dx^2} + \frac{1}{\Delta t} z_1 = \sin x \qquad z_1(0) = z_1(\pi) = 0$$

The solution is of the form

$$z_1(x) = A_1 \sin x$$

Substituting this into the equation, we obtain

$$A_1 = \frac{\Delta t}{1 + \Delta t} = 1 - \frac{1}{1 + \Delta t}$$

For $j = 2$, we solve the equation

$$-\frac{d^2 z_2}{dx^2} + \frac{1}{\Delta t} z_2 = \sin x + \frac{1}{(1 + \Delta t)} \sin x = \left(\frac{2 + \Delta t}{1 + \Delta t}\right) \sin x$$

Assuming a solution in the form

$$z_2(x) = A_2 \sin x$$

and substituting into the differential equation, we obtain

$$A_2 = \frac{(2 + \Delta t)\,\Delta t}{(1 + \Delta t)^2} = 1 - \frac{1}{(1 + \Delta t)^2}$$

Continuing this way, we can show that the solution of Eq. (6-263) is given by

$$z_j(x) = \left[1 - \frac{1}{(1 + \Delta t)^j}\right] \sin x \qquad j = 1, 2, \ldots, p \tag{6-265}$$

Table 6-10 Comparison of the approximate solution with the exact solution of the problem in Example 6-21

Time	u_{exact}	$u_{\text{approx.}}$	Error
	$p = 10, \Delta t = 0.1$		
0.2000	0.1813	0.1736	0.0077
0.4000	0.3297	0.3170	0.0127
0.6000	0.4512	0.4355	0.0157
0.8000	0.5507	0.5335	0.0172
1.0000	0.6321	0.6145	0.0177
	$p = 50, \Delta t = 0.02$		
0.2000	0.1813	0.1797	0.0016
0.4000	0.3297	0.3270	0.0027
0.6000	0.4512	0.4479	0.0033
0.8000	0.5507	0.5471	0.0036
1.0000	0.6321	0.6285	0.0036
	$p = 100, \Delta t = 0.01$		
0.2000	0.1813	0.1805	0.0008
0.4000	0.3297	0.3283	0.0013
0.6000	0.4512	0.4496	0.0016
0.8000	0.5507	0.5489	0.0018
1.0000	0.6321	0.6303	0.0018
	$p = 200, \Delta t = 0.005$		
0.2000	0.1813	0.1809	0.0004
0.4000	0.3297	0.3290	0.0007
0.6000	0.4512	0.4504	0.0008
0.8000	0.5507	0.5498	0.0009
1.0000	0.6321	0.6312	0.0009

The exact solution of Eq. (6-262) is given by

$$u(x, t) = (1 - e^{-t}) \sin x \tag{6-266}$$

In Table 6-10 a comparison of the exact solution with the solutions obtained using the method of discretization in time for $p = 10$, 50, 100, and 200 (equivalently, $\Delta t = 0.1$, 0.02, 0.01, and 0.005) is presented. It is clear that the approximate solution converges to the exact solution as the time step is reduced (or p is increased: $\Delta t = 1/p$).

The next example deals with a parabolic problem with nonzero initial condition. Also, we use a variational method to solve the discretized problem.

Example 6-22 Consider the problem

$$\frac{\partial u}{\partial t} - \frac{\partial^2 u}{\partial x^2} = 0 \text{ in } Q = (0, \pi) \times (0, 1)$$
$$u(x, 0) = 1 \qquad u(0, t) = u(\pi, t) = 0 \tag{6-267}$$

The problem has the exact solution

$$u(x, t) = \frac{4}{\pi} \sum_{n=1,3,5} \frac{\exp(-n^2 t)}{n} \sin nx \tag{6-268}$$

For the problem at hand, Eq. (6-252) takes the form

$$-\frac{d^2 z_j}{dx^2} + \frac{1}{\Delta t} z_j = \frac{1}{\Delta t} z_{j-1} \qquad z_0 = 1 \qquad z_j(0) = z_j(\pi) = 0 \tag{6-269}$$

For $j = 1$, we have

$$-\frac{d^2 z_1}{dx^2} + \frac{1}{\Delta t} z_1 = \frac{1}{\Delta t}$$

The general solution of this equation is

$$z_1(x) = 1 + c_1 \sinh \frac{x}{\sqrt{\Delta t}} + c_2 \cosh \frac{x}{\sqrt{\Delta t}}$$

The constants c_1 and c_2 can be evaluated using the boundary conditions $z_1(0) = z_1(\pi) = 0$. However, with increasing j, the solution of Eq. (6-269) becomes more difficult. Hence we choose to solve Eq. (6-269) using the Ritz method.

Let

$$z_j^N(x) = \sum_{n=1}^{N} c_n^j \phi_n^j(x) \qquad \phi_n^j(x) = \sin nx \tag{6-270}$$

Substituting this expression into the weak problem

$$\bar{B}(v, z_j) = \frac{1}{\Delta t} \int_0^\pi v z_{j-1}\, dx$$

we obtain

$$\sum_{n=1}^{N} \bar{B}(\phi_i, \phi_n) c_n^j = \frac{1}{\Delta t} \int_0^\pi \phi_i z_{j-1}\, dx \tag{6-271}$$

where

$$\bar{B}(\phi_i, \phi_n) = \int_0^\pi \left(\frac{d\phi_i}{dx}\frac{d\phi_n}{dx} + \frac{1}{\Delta t} \phi_i \phi_n \right) dx$$

$$= \begin{cases} 0 & i \neq n \\ \dfrac{\pi}{2}\left(n^2 + \dfrac{1}{\Delta t}\right) & i = n \end{cases} \tag{6-272}$$

For $j = 1$, we have

$$l(\phi_i) = \frac{1}{\Delta t} \int_0^\pi \phi_i z_0\, dx = \begin{cases} 0 & i = \text{even} \\ \dfrac{2}{i} & i = \text{odd} \end{cases}$$

Table 6-11 Comparison of the approximate solution with the exact solution of the problem in Example 6-22

Time	u_{exact}	$u_{approx.}$	Error
	$p = 10$, $\Delta t = 0.1$		
0.2000	0.9740	0.9513	0.0227
0.4000	0.8419	0.8386	0.0033
0.6000	0.6969	0.7098	−0.0130
0.8000	0.5718	0.5915	−0.0197
1.0000	0.4683	0.4902	−0.0219
	$p = 50$, $\Delta t = 0.02$		
0.2000	0.9740	0.9676	0.0064
0.4000	0.8419	0.8414	0.0005
0.6000	0.6969	0.7000	−0.0031
0.8000	0.5718	0.5761	−0.0043
1.0000	0.4683	0.4729	−0.0046
	$p = 100$, $\Delta t = 0.01$		
0.2000	0.9740	0.9706	0.0034
0.4000	0.8419	0.8417	0.0002
0.6000	0.6969	0.6984	−0.0016
0.8000	0.5718	0.5740	−0.0022
1.0000	0.4683	0.4707	−0.0023
	$p = 200$, $\Delta t = 0.005$		
0.2000	0.9740	0.9723	0.0017
0.4000	0.8419	0.8418	0.0001
0.6000	0.6969	0.6977	−0.0008
0.8000	0.5718	0.5729	−0.0011
1.0000	0.4683	0.4695	−0.0012

Thus,

$$z_1^N(x) = \frac{4}{\pi} \sum_{n=1,3,5}^{N} \frac{1}{n(n^2\,\Delta t + 1)} \sin nx$$

For $j = 2$, we have

$$l(\phi_i) = \frac{1}{\Delta t} \cdot \frac{4}{\pi} \sum_{n=1,3,5}^{N} \frac{1}{n(n^2\,\Delta t + 1)} \int_0^{\pi} \sin n\pi \sin ix\, dx$$

$$= \frac{1}{\Delta t}\frac{4}{\pi}\frac{1}{i(i^2\,\Delta t + 1)}\frac{\pi}{2} \qquad i \text{ odd}$$

and

$$z_2^N(x) = \frac{4}{\pi} \sum_{n=1,3,5}^{N} \frac{1}{n(n^2\,\Delta t + 1)^2} \sin nx$$

Proceeding in this manner, we obtain

$$z_j(x) = \frac{4}{\pi} \sum_{n=1,3,5}^{N} \frac{1}{n(n^2 \Delta t + 1)^j} \sin nx$$
$$= \frac{4}{\pi} \left[\frac{1}{(1 + \Delta t)^j} \sin x + \frac{1}{3(1 + 9 \Delta t)^j} \sin 3x + \frac{1}{5(1 + 25 \Delta t)^j} \sin 5x + \cdots \right] \tag{6-273}$$

for $j = 1, 2, \ldots, p$.

A comparison of the solution (6-273) with the exact solution (6-268) is presented in Table 6-11. Once again we see the convergence of the approximate solution to the exact solution.

Hyperbolic Equations Consider the problem

$$\frac{\partial^2 u}{\partial t^2} + Au = f \text{ in } Q = \Omega \times (0, T)$$
$$u(\mathbf{x}, 0) = u_0 \qquad \frac{\partial u}{\partial t}(\mathbf{x}, 0) = v_0 \text{ in } \Omega \tag{6-274}$$
$$B_i u = g_i \qquad i = 1, 2, \ldots, k \text{ on } \Gamma_1 \times (0, T)$$
$$C_i u = h_i \qquad i = 1, 2, \ldots, 2m - k \text{ on } \Gamma_2 \times (0, T)$$

where A, f, u_0, etc. is as defined earlier.

As applied to Eq. (6-274), the method of discretization in time involves successively solving the equation (for $j = 1, 2, \ldots, p$)

$$Az_j + \frac{z_j - 2z_{j-1} + z_{j-2}}{(\Delta t)^2} = f$$
$$B_i z_j = g_i \text{ on } \Gamma_1 \qquad c_i z_j = h_i \text{ on } \Gamma_2 \tag{6-275}$$

with $z_0 = u_0$. Note that $\partial^2 u/\partial t^2$ is replaced by (central difference approximation)

$$\frac{\partial^2 u}{\partial t^2} \approx \frac{z_j - 2z_{j-1} + z_{j-2}}{(\Delta t)^2}$$

Equation (6-275) requires the knowledge of the solution at two previous time steps. For example, for $j = 1$ we have

$$Az_1 + \frac{1}{(\Delta t)^2} z_1 = f + \frac{1}{(\Delta t)^2}(2z_0 - z_{-1}) \tag{6-276}$$

We use

$$z_0 = u_0 \qquad z_{-1} = (\Delta t)^2 f + 2z_0 \tag{6-277}$$

Existence and uniqueness The weak problem associated with Eq. (6-275) can be shown to have a unique solution if the bilinear form associated with the operator A is continuous and V-elliptic [see Eq. (6-254a) for the definition of V]. The operator defined by Eq. (6-275) is given by

$$A' = A + \frac{1}{(\Delta t)^2} \tag{6-278}$$

The weak problem corresponding to Eq. (6-275) involves finding $z_j(\mathbf{x}) \in H^m(\Omega)$ such that $z_j - w(\mathbf{x}, t_j) \in V$ and

$$B'(v, z_j) = \int_\Omega v\left[f + \frac{1}{(\Delta t)^2}(2z_{j-1} - z_{j-2})\right] d\mathbf{x} + \sum_{i=1}^{m-k} \int_{\Gamma_2} B_i v h_i \, ds \tag{6-279}$$

holds for all $v \in V$. Here $B'(\cdot, \cdot)$ is the bilinear form

$$B'(v, z_j) = B(v, z_j) + \frac{1}{(\Delta t)^2}(v, z_j)_0 \tag{6-280}$$

The existence and uniqueness of the solution to Eq. (6-279) follows from Theorem 6-5 provided

$$\left[f + \frac{1}{(\Delta t)^2}(2z_{j-1} - z_{j-2})\right] \in L_2(\Omega) \text{ and } \Delta t > 0$$

The approximation in the whole domain $Q = \Omega \times (0, T)$ can be constructed according to the equation

$$u'(\mathbf{x}, t) = z_{j-1} + \frac{1}{\Delta t}(z_j - z_{j-1})(t - t_{j-1}) \text{ in } [t_{j-1}, t_j] \tag{6-281}$$

The error estimate is given by [see Rektorys (1982)]

$$\|u(\mathbf{x}, t_j) - z_j(\mathbf{x})\|_0 \leq \tfrac{3}{2} \cdot j(j+1)(\Delta t)^3 \|f\|_V \tag{6-282}$$

Variational approximation For fixed j ($j = 1, 2, \ldots, p$) and $\Delta t > 0$, the Ritz approximation of Eq. (6-279) is given by

$$\sum_{r=1}^{N} B'(\phi_i, \phi_r)c_r^j = \int_\Omega \phi_i\left[f + \frac{1}{(\Delta t)^2}(2z_{j-1} - z_{j-2})\right] d\mathbf{x} + \sum_{s=1}^{m-k} \int_{\Gamma_2} B_s \phi_i h_s \, ds \tag{6-283}$$

where the initial functions z_{j-1} and z_{j-2} are known from the previous time step solutions. For $j = 1$, we use Eq. (6-277).

An illustrative example of the procedure for a hyperbolic equation is presented below.

Example 6-23 Consider the problem

$$\frac{\partial^2 u}{\partial t^2} - \frac{\partial^2 u}{\partial x^2} = \sin x \text{ in } Q = (0, \pi) \times (0, 1)$$

$$u(x, 0) = 0 \qquad \frac{\partial u}{\partial t}(x, 0) = 0 \tag{6-284}$$

$$u(0, t) = u(\pi, t) = 0$$

The problem has the exact solution

$$u(x, t) = (1 - \cos t) \sin x$$

The approximate problem involves solving a sequence of problems successively:

$$-\frac{d^2 z_j}{dx^2} + \frac{1}{(\Delta t)^2} z_j = \sin x + (2z_{j-1} - z_{j-2})/(\Delta t)^2 \tag{6-285}$$

with $z_0 = 0$ and $z_{-1} = \frac{1}{2}(\Delta t)^2 f = \frac{1}{2}(\Delta t)^2 \sin x$.

For $j = 1$, we have

$$-\frac{d^2 z_1}{dx^2} + \frac{1}{(\Delta t)^2} z_1 = \frac{\sin x}{2} \qquad z_1(0) = z_1(\pi) = 0$$

Assuming the solution in the form

$$z_1(x) = A_1 \sin x$$

we substitute into the differential equation for $j = 1$ and obtain

$$A_1 = \frac{(\Delta t)^2}{2[1 + (\Delta t)^2]}$$

In general, the solution of Eq. (6-285) can be assumed in the form

$$z_j(x) = A_j \sin x \tag{6-286a}$$

Substituting into Eq. (6-285), we obtain

$$A_j\left[1 + \frac{1}{(\Delta t)^2}\right] = 1 + \frac{2A_{j-1} - A_{j-2}}{(\Delta t)^2} \qquad j = 1, 2, \ldots, p$$

or

$$A_j = \frac{1}{1 + (\Delta t)^2}[2A_{j-1} - A_{j-2} + (\Delta t)^2] \tag{6-286b}$$

with $A_0 = 0$ and $A_{-1} = (\Delta t)^2/2$.

A comparison of the exact solution with the present solution is presented in Table 6-12. The error between the two solutions decreases as p is increased (or as Δt is decreased).

Table 6-12 A comparison of the approximate solution with the exact solution of the problem in 6-23

Time	u_{exact}	$u_{\text{approx.}}$	Error
	$p = 10$, $\Delta t = 0.1$		
0.2000	0.0199	0.0197	0.0002
0.4000	0.0789	0.0776	0.0014
0.6000	0.1747	0.1705	0.0041
0.8000	0.3033	0.2943	0.0090
1.0000	0.4597	0.4433	0.0164
	$p = 50$, $\Delta t = 0.02$		
0.2000	0.0199	0.0199	0.0000
0.4000	0.0789	0.0787	0.0002
0.6000	0.1747	0.1739	0.0007
0.8000	0.3033	0.3017	0.0016
1.0000	0.4597	0.4566	0.0031
	$p = 100$, $\Delta t = 0.01$		
0.2000	0.0199	0.0199	0.0000
0.4000	0.0789	0.0788	0.0001
0.6000	0.1747	0.1743	0.0004
0.8000	0.3033	0.3025	0.0008
1.0000	0.4597	0.4582	0.0015
	$p = 200$, $\Delta t = 0.005$		
0.2000	0.0199	0.0199	0.0000
0.4000	0.0789	0.0789	0.0001
0.6000	0.1747	0.1745	0.0002
0.8000	0.3033	0.3029	0.0004
1.0000	0.4597	0.4589	0.0008

In summary, the method of discretization in time allows us to reduce an initial boundary-value problem to a boundary-value problem which can be solved either exactly or approximately. The idea, in spirit, is similar to that of the Kantorovich method. The method of discretization in time allows us to use the spatial methods of discretization (e.g., the Ritz and Galerkin methods and the finite element method).

PROBLEMS 6-4

6-4-1 Solve the equation $\partial u/\partial t - \partial^2 u/\partial x^2 = \pi x - x^2$, $0 < x < \pi$; $u(x, 0) = 0$, $u(0, t) = u(\pi, t) = 0$, on the rectangle $\Omega_T = (0, \pi) \times (0, 1)$ using a three-parameter Ritz approximation of the form $u_3(x, t) = c_1(t) \sin x + c_2(t) \sin 2x + c_3(t) \sin 3x$. Solve the resulting ordinary differential equations exactly.

6-4-2 Repeat Prob. 6-4-1 but solve the resulting differential equations using the Crank–Nicholson method.

6-4-3 Consider the nondimensionalized heat conduction equation, $\partial u/\partial t - \partial^2 u/\partial x^2 = 0$, $0 < x < 1$, subjected to the boundary conditions $u(0, t) - (\partial u/\partial x)(0, t) = 0$, $(\partial u/\partial x)(1, t) = 0$, and initial condition $u(x, 0) = 1$. Find a two-parameter Ritz–Laplace transform solution.

6-4-4 Consider the partial differential equation $\partial^2 u/\partial x^2 - \partial^2 u/\partial t^2 = 0$, $0 < x < 1$, subjected to the boundary conditions $u(0, t) = u(1, t) = 0$, and initial conditions $u(x, 0) = x(1 - x)$, $(\partial u/\partial t)(x, 0) = 0$. Find the one-parameter solution using (*a*) the Bubnov–Galerkin method and (*b*) collocation at $x = \frac{1}{2}$. Choose the approximation in the form $u(x, t) = c_1(t)x(1 - x)$. Solve the resulting ordinary differential equation for c_1 exactly.

6-4-5 Repeat Prob. 6-4-4 for a two-parameter Bubnov–Galerkin solution in space and Laplace transforms in time.

6-4-6 Show that Eqs. (6-230) and (6-239) can be expressed in the alternative form [to Eq. (6-240)] $[H]\{\ddot{c}\}^{n+1} = \{F\} - [B]\{b\}^n$ where $[H] = \beta(\Delta t)^2[B] + [M]$ and $\{b\}^n = \{c\}^n + \Delta t\{\dot{c}\}^n + (\frac{1}{2} - \beta) \times (\Delta t)^2\{\ddot{c}\}^n$.

6-4-7 Using the Newmark integration scheme (6-239), express the equation,

$$[M]\{\ddot{c}\} + [C]\{\dot{c}\} + [K]\{c\} = \{F\}$$

in the form $[M]\{c\}^{n+1} = \{P\}$, where

$$[M] = [K] + a_0[M] + a_5[C] \qquad a_5 = \frac{\alpha}{\beta\,\Delta t}$$

$$\{P\} = \{F\}^{n+1} + [M](a_0\{c\}^n + a_1\{\dot{c}\}^n + a_2\{\ddot{c}\}^n) + [C](a_5\{c\}^n + a_6\{\dot{c}\}^n + a_7\{\ddot{c}\}^n)$$

$$a_6 = \frac{\alpha}{\beta} - 1 \qquad a_7 = \frac{\Delta t}{2}\left(\frac{\alpha}{\beta} - 2\right)$$

6-4-8 Repeat Prob. 6-4-4 with the Bubnov–Galerkin method in space and the Newmark scheme in time.

6-4-9 Use the central difference method to approximate the hyperbolic equations in Prob. 6-4-7.

6-4-10 Reduce the following integral equation to a set of N algebraic equations by using the Petrov–Galerkin method: $u(x) = \alpha \int_{-1}^{1} K(x, y)u(y)\,dy$. Use the approximation of the form

$$u_N(x) = \sum_{i=1}^{N} c_i \phi_i(x)$$

$$\phi_i(x) = \begin{cases} 1 & 2\left(\dfrac{i-1}{N}\right) - 1 < x < \dfrac{2i}{N} - 1 \\ 0 & \text{otherwise} \end{cases}$$

and the weight functions $\psi_j(x)$,

$$\psi_j(x) = \delta(x - x^j)$$

$$x^j = \frac{1}{N} + \frac{2}{N}(j - 1) - 1$$

where $\delta(x)$ is the Dirac delta function (distribution).

6-4-11 Consider the problem $\partial u/\partial t - \partial^2 u/\partial x^2 = \sin 2x$ in $(0, \pi) \times (0, 1)$, $u(x, 0) = 0$, $u(0, t) = u(\pi, t) = 0$. Solve the problem using a one-parameter Ritz or Galerkin approximation in space and exact in time. The exact solution is given by $u(x, t) = \frac{1}{4}(1 - e^{-4t}) \sin 2x$.

6-4-12 Solve the problem $\partial u/\partial t - (1 + t)(\partial^2 u/\partial x^2) = (1 + t) \sin x$ in $(0, \pi) \times (0, 1)$, $u(x, 0) = 0$, $u(0, t) = u(\pi, t) = 0$, using the Galerkin approximation in space. Use one parameter. The exact solution is given by $u(x, t) = \{1 - \exp[-(t^2 + 2t)/2]\} \sin x$.

6-4-13 Consider the problem $\partial^2 u/\partial t^2 - \partial^2 u/\partial x^2 = \sin x$ in $(0, \pi) \times (0, 1)$, $u(x, 0) = (\partial u/\partial t)(x, 0) = 0$, $u(0, t) = u(\pi, t) = 0$. Solve the problem using the Ritz approximation in space and exact in time. Use one parameter.

6-4-14 Repeat Prob. 6-4-13 with the central difference approximation in time.

Use the method of discretization to solve Probs. 6-4-15 to 6-4-21. Solve the spatial equations exactly if possible, or by the Ritz method.

6-4-15 $\partial u/\partial t - \partial^2 u/\partial x^2 = \sin nx$ in $Q = (0, \pi) \times (0, 1)$, $u(x, 0) = 0$, $u(0, t) = u(\pi, t) = 0$, where $n > 1$ is an integer.

6-4-16 $\partial u/\partial t - \partial^2 u/\partial x^2 = 0$ in $Q = (0, \pi) \times (0, 1)$, $u(x, 0) = \sin x$, $u(0, t) = u(\pi, t) = 0$. The exact solution is $u_{\text{exact}} = e^{-t} \sin x$.

6-4-17 $$\frac{\partial u}{\partial t} - \left\{\frac{\partial}{\partial x}\left[\left(4 + \frac{y^2}{8}\right)\frac{\partial u}{\partial x}\right] + \frac{\partial}{\partial y}\left[\left(4 + \frac{x^2}{25}\right)\frac{\partial u}{\partial y}\right]\right\} = 1 - \frac{x^2}{16} - \frac{y^2}{4} \text{ in } Q = \Omega \times (0, 1)$$

$u(x, y, 0) = 0$ in Ω, $u = 0$ on $\Gamma \times (0, 1)$, where Ω is an ellipse with half axes $a = 4$ and $b = 2$ along the coordinate axes x and y.

6-4-18 $$\frac{\partial^2 u}{\partial t^2} - \left\{\frac{\partial}{\partial x}\left[a(x, y)\frac{\partial u}{\partial x}\right] + \frac{\partial}{\partial y}\left[a(x, y)\frac{\partial u}{\partial y}\right]\right\} = \sin x \sin y$$

$$\left.\begin{aligned} u(x, y, 0) &= 0 \\ \frac{\partial u}{\partial t}(x, y, 0) &= 0 \end{aligned}\right\} \text{ in } \Omega = \text{square} = (0, \pi) \times (0, \pi)$$

$$u = 0 \text{ on } \Gamma \times (0, 1)$$

where $a(x, y) = 4 + \sin x \sin y$.

CHAPTER

SEVEN

THE FINITE-ELEMENT METHOD

7-1 SOME GENERAL PROPERTIES OF THE METHOD

7-1-1 Introduction

The traditional variational methods presented in Chap. 6 are ineffective in solving problems that are geometrically complex, have discontinuous loads, or involve discontinuous material or geometric properties. In such cases the selection of the approximating functions is a formidable task. Even in cases where the approximating functions are available, the computation of associated coefficient matrices cannot be automatized because the approximating functions are problem dependent. The variational methods would be effective if we could construct the approximating functions for arbitrary domains and variationally consistent boundary conditions for a given class of problems.

The finite-element method is a variational procedure in which the approximating functions are systematically derived by representing the given domain as a collection of simple subdomains. The method differs in two ways from the traditional variational methods in generating the algebraic equations of the problem. First, the approximating functions are often algebraic polynomials that are developed using ideas from the interpolation theory; second, the approximating functions are developed for subdomains into which a given domain is divided. The subdomains, called *finite elements*, are geometrically simple shapes that permit a systematic construction of the approximating functions. Since the approximating functions are algebraic polynomials, the computation of the coefficient matrices of the algebraic equations can be automatized on the computer. As

Table 7-1 The basic steps in the finite-element analysis of a typical problem

1. *Division of whole into parts.* Represent the given domain (i.e., given system) as a collection of a finite number of simple subdomains, called *finite elements.* The number, shape and type of element depend on the domain and differential equation being solved. The principal parts of this step include:
   (*a*) Number the nodes (see Step 2 below) and elements of the collection, called the *finite-element mesh.*
   (*b*) Generate the coordinates of the nodes in the mesh and array, called the *connectivity matrix,* that indicate the relative position of each element in the mesh.

2a. *Derivation of the approximating functions.* For each element in the mesh, derive the approximation functions needed in the variational method. These functions are generally algebraic polynomials generated by interpolating the unknown function in terms of its values (which are unknown) at preselected points, called *nodes*, of the element.

2b. *Variational approximation of the equation.* Using the functions derived in Step 2a and any appropriate variational method (see Chap. 6), derive the algebraic equations among the nodal values of an element.

3. *Connectivity (or assembly) of elements.* Combine the algebraic equations of all elements in the mesh by imposing the continuity of the nodal variables (i.e., the values of the nodal variables at a node shared by two or more elements are the same). This can be viewed as putting the elements (which were isolated in steps 2a and 2b from the mesh to derive the algebraic equations) back into their original places. This gives the algebraic equations governing the whole problem.

4. *Imposition of boundary conditions.* Impose the boundary conditions, both essential and natural, on the assembled equations.

5. *Solution of equations.* Solve the equations for the unknown nodal values.

6. *Computation of additional quantities.* Using the nodal values, compute the solution and its derivatives at other points (other than the nodes) of the domain.

will be seen later, the construction of the approximating functions employs concepts from interpolation theory, and the process is independent of the specific boundary conditions and the problem data; the undetermined parameters represent the values of the dependent variables appearing in the essential boundary conditions at a finite number of preselected points (whose number and location dictates the degree and form of the approximating functions) in the element. The method is well-suited for electronic computation and the development of general purpose computer programs. The basic steps involved in the finite-element method are outlined in Table 7-1.

7-1-2 Division of Whole into Parts (Partitioning of Domain)

One of the underlying features of the finite-element method that distinguishes it from the traditional variational methods and other numerical methods is the representation of the given domain by a collection of "simple" subdomains. This representation is necessitated by two reasons: First, the derivation of the approximation functions using algebraic interpolation is possible only for certain well-defined (or simple) geometries, such as lines, triangles, rectangles, cubes, etc.;

second, since the approximating functions are defined element-wise, the accuracy of the approximation can be improved by increasing the number of elements (i.e., refining the mesh). Note that in general a given domain can be represented accurately by fewer elements than necessary for an accurate representation of the solution. The representation of a domain by a collection of elements is called *mesh generation* and the collection is called the *finite-element mesh* Ω_h. A further subdivision of a finite element mesh is called the *mesh refinement*.

In mathematical terms all of the above discussion can be expressed as follows. The domain $\bar{\Omega} = \Omega \cup \Gamma$ is divided into a finite number N of subsets $\bar{\Omega}^e$, called *finite elements*, in such a way that the following properties are satisfied:

1. Each $\bar{\Omega}^e$ is closed and nonempty.
2. The boundary Γ^e of each Ω^e is Lipschitz-continuous.
3. The intersection of any two distinct elements is empty, i.e.,

$$\Omega^e \cap \Omega^f = \varnothing, \qquad e \neq f$$

4. The union $\bar{\Omega}_h$ of all elements $\bar{\Omega}^e$ is equal (at least approximately) to the total domain $\bar{\Omega}$:

$$\bar{\Omega} \approx \bar{\Omega}_h = \sum_{e=1}^{N} \bar{\Omega}^e$$

As an example, consider the mixed boundary-value problem for the Poisson equation on the domain shown in Fig. 7-1*a*. The domain can be represented exactly either by a quadrilateral and a rectangle or by a triangle and three rectangles. As will be seen later, quadrilaterals, rectangles, and triangles are simple geometric shapes that allow the derivation of the approximation functions. Suppose that the solution $u(x, 0)$ varies nonlinearly along the x-axis, as shown in Fig. 7-1*a*. If linear interpolation functions $\psi^e(x, y) \in H^1(\Omega)$ are used to represent u within each element, the finite-element solution $U(x, 0)$ will be a piecewise linear function. The accuracy of U would increase if we subdivide the original mesh as shown in Fig. 7-1*b*.

In the above example it is possible to represent the domain exactly by a finite-element mesh. This is not always the case. An example of a domain that cannot be represented exactly by a finite number of elements is shown in Fig. 7-2. In such cases, the finite-element solution will have additional error, in addition to the approximation and computational errors, due to the inexact representation of the domain.

7-1-3 Finite-Element Interpolation

As mentioned earlier, in the finite-element method the approximating functions are developed for each element separately. If we wish to generate the approximating functions for the Ritz finite-element approximation, we must require them to satisfy the essential boundary conditions of the problem. Since the

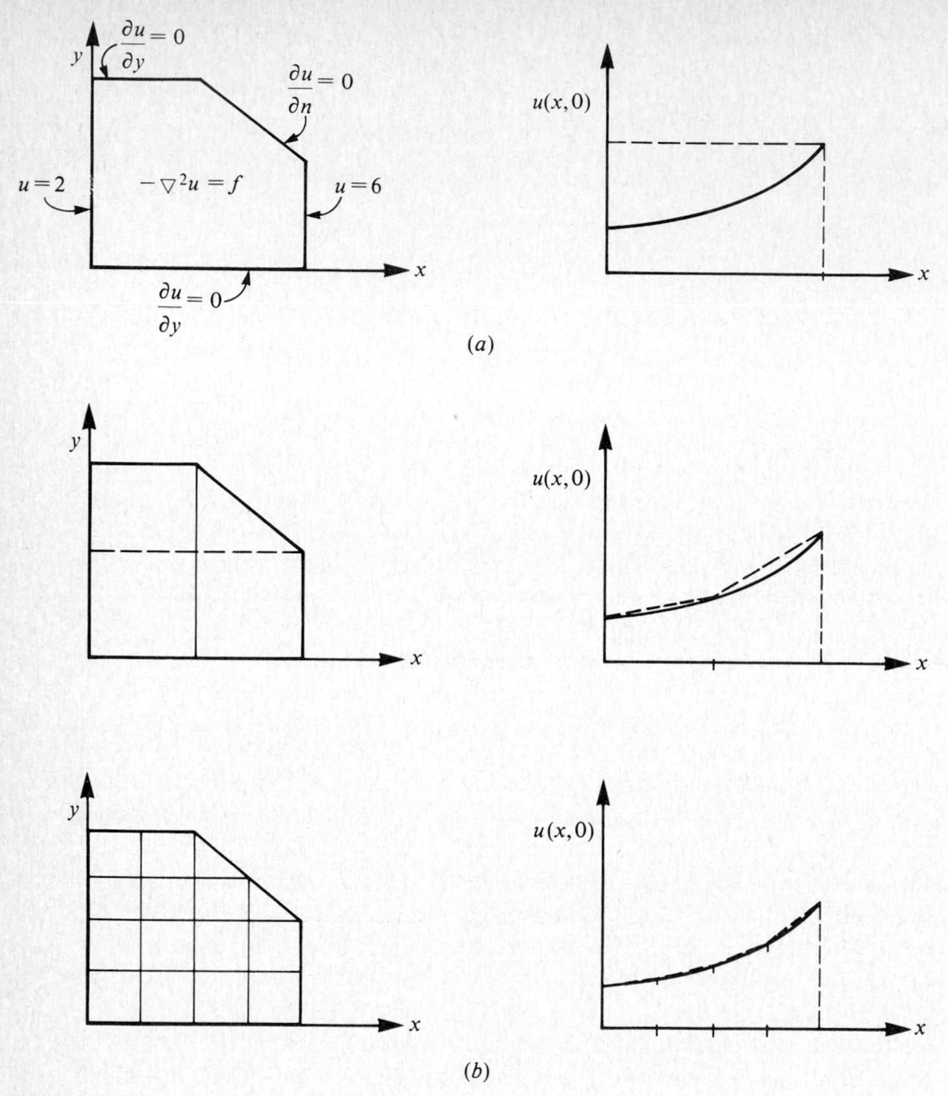

Figure 7-1 Representation of a domain and solution in the finite-element method.

boundary of the domain is shared in general by more than one element, the satisfaction of the essential boundary conditions of the problem by the approximating functions of an element cannot be achieved. To circumvent this difficulty, we assume that each element itself is a possible representation of the entire domain with the general boundary conditions of the differential equation. In other words, we assume that the given differential equation is to be formulated and approximated over each element using any one of the variational methods. To account for both types of boundary conditions in the problem, we include, in the Ritz finite-element model, the essential boundary conditions by interpolation and the

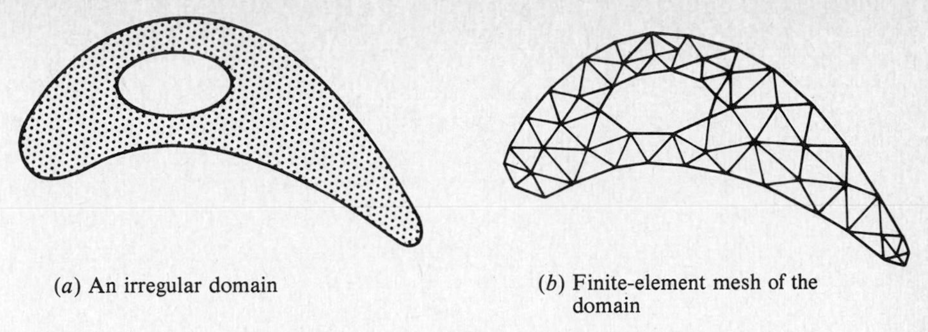

Figure 7-2 Finite-element representation of an irregular domain by triangular elements.

natural boundary conditions via the variational form of the equation over the element. After assembling the elements, the boundary values on portions of the boundaries of elements that share the boundary of the given domain are replaced by the actual specified values (*imposition of boundary conditions*).

In the finite-element method the approximating functions are conventionally algebraic polynomials, which facilitate (1) accurate numerical evaluation of the coefficient matrices on a digital computer, and (2) the proof of convergence of the finite-element approximation. The minimum degree of the polynomial used for approximation functions depends on the order of the differential equation being solved, and the degree of the polynomial in turn dictates the number of interpolation points, called *nodes*, to be identified in the element. The number of nodes also depends on the geometric shape of the element. In general, the number n of interpolation functions and the number m of nodes in an element are not the same ($n \geq m$).

The approximation functions are also called *interpolation functions* because they are derived by interpolating the function and possibly its derivatives at the nodes of the element. The nodes are placed on the boundary of the element such that they uniquely define the element geometry. Any additional nodes required to define the interpolation functions can be placed at other points, either in the interior or on the boundary. The boundary nodes also enable us to connect elements together by requiring the primary degrees of freedom (i.e., variables that appear in essential boundary conditions) to be the same at the nodes common to any two elements.

For each Ω^e, let P_e denote the finite-dimensional spaces spanned by linearly independent *local* interpolation functions $\{\psi_i^e\}_{i=1}^n$ of the nodal points. Over each element $\bar{\Omega}^e \subset \bar{\Omega}_h$, we approximate the restriction u_e of u to $\bar{\Omega}^e$ by a linear combination of the form

$$u_e \approx u_h^e = \sum_{j=1}^{n} u_j^e \psi_j^e \tag{7-1}$$

where the coefficients u_j^e are usually taken to be the values of u and possibly its derivatives at the preselected nodes $\{\mathbf{x}_i^e\}_{i=1}^m$ in the element $\bar{\Omega}^e$. The values u_i^e are called the *element* (or local) *degrees of freedom*. In general, we require that P_e contains $\mathscr{P}_k(\bar{\Omega}_e)$, the space of polynomials in $\mathbf{x} \in \bar{\Omega}^e$ of degree $\leq k$. Stating in other terms, the interpolation functions ψ_i^e, $1 \leq i \leq n$, and nodal points $\mathbf{x}_i^e$, $1 \leq i \leq m$, are selected in such a way that the linear combination (7-1) is a polynomial of degree $\leq k$ on $\bar{\Omega}^e$.

Now suppose that we wish to derive the linear interpolation functions, which when approximating the unknown function u should satisfy the essential boundary conditions at the boundary points of the element. For example, for the weak fomulation of Eq. (4-34a) over an element, the essential boundary conditions are given by

$$u(x_e) = u_1^e \qquad u(x_{e+1}) = u_2^e$$

where x_e denotes the x-coordinate of the global node e, which is the same as the node 1 of element Ω^e. A linear function u_e that satisfies the above conditions is given by

$$u_e = \frac{x_{e+1} - x}{x_{e+1} - x_e} u_1^e + \frac{x - x_e}{x_{e+1} - x_e} u_2^e$$

$$\equiv \psi_1^e u_1^e + \psi_2^e u_2^e = \sum_{i=1}^{n=2} u_i^e \psi_i^e$$

Clearly, $u_e \in \mathscr{P}_1(\Omega_e) \subset P_e$, where P_e is the space spanned by $\{\psi_1^e, \psi_2^e\}$.

In order to derive algebraic equations relating u_i^e $(i = 1, 2, \ldots, n)$, we first obtain the variational formulation of the given equation *over the element*, Ω^e. The variational problem over Ω^e involves seeking $u_e \in P_e$ such that (see Chap. 6)

$$B_e(v_e, u_e) = l_e(v_e) \tag{7-2}$$

holds for every $v_e \in P_e$, where B_e and l_e denote the bilinear and linear forms associated with the given equation over the element Ω^e. Substituting Eq. (7-1) for u_e and $v_e = \psi_i^e$ into Eq. (7-2), we obtain the element equations

$$[K^e]\{u^e\} = \{F^e\} \tag{7-3a}$$

where

$$K_{ij}^e = B_e(\psi_i^e, \psi_j^e) \qquad F_i^e = l_e(\psi_i^e) \tag{7-3b}$$

The coefficient matrix $[K^e]$ is known as the *stiffness matrix* and the column vector $\{F^e\}$ is known as the *force vector* in solid mechanics applications. Since ψ_i^e are algebraic polynomials, the coefficients K_{ij}^e and F_i^e can be numerically evaluated on a digital computer quite easily. It should be noted that in general some of the F_i^e are not known because l_e contains unknown internal fluxes, and therefore Eq. (7-3a) cannot be solved at the element level.

7-1-4 Connectivity (or Assembly) of Elements

As mentioned earlier, all elements contain boundary nodes which define their geometry. These boundary nodes allow us to connect the element to its neighboring elements. The connectivity of elements is accomplished by requiring that at a node common to adjacent elements, the nodal values from the elements are the same. This allows us to define continuous, linearly independent, global inter-

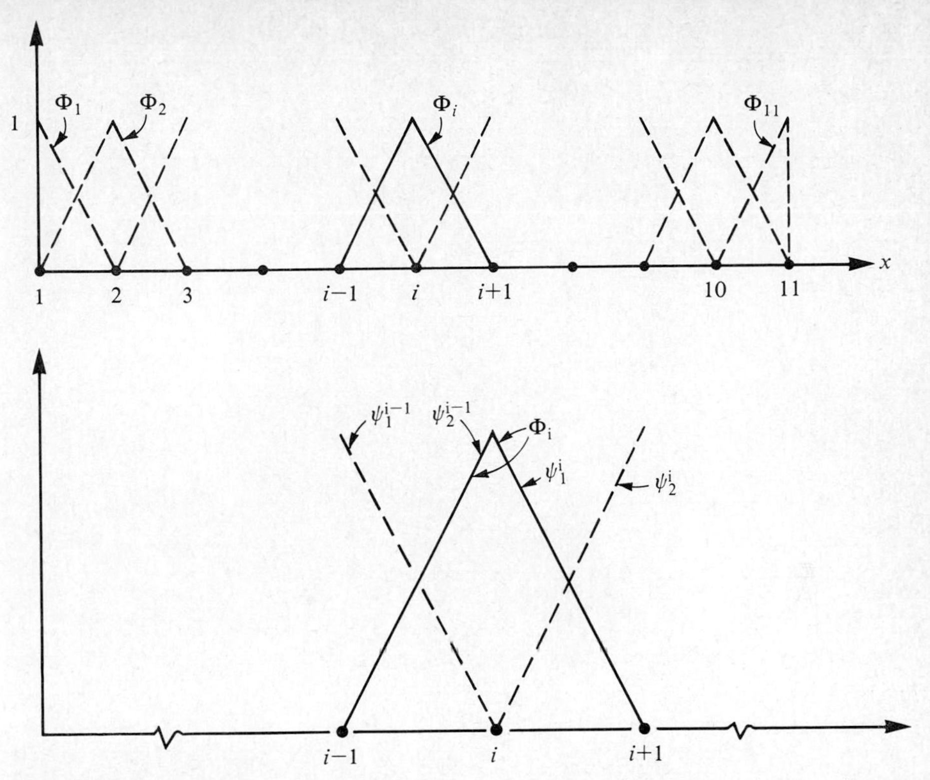

Figure 7-3 Local and global linear basis functions in the finite-element approximation of a second-order equation in one dimension.

polation functions $\{\Phi_J\}_{J=1}^{N}$ with compact support (i.e., the functions are nonzero only within a small region of the domain $\bar{\Omega}$),

$$u \approx u_h = \sum_{J=1}^{N} U_J \Phi_J \tag{7-4}$$

where U_J denotes the *global degree of freedom* at the global node J. For example, if the domain $\Omega = [0, L]$ is divided into ten subintervals of length $h = L/10$, the functions Φ_I are given by (see Fig. 7-3)

$$\Phi_I(x) = \begin{cases} 0 & 0 \leq x \leq (I-2)\dfrac{L}{10} \\[2ex] \dfrac{10x}{L} - (I-2) & (I-2)\dfrac{L}{10} \leq x \leq (I-1)\dfrac{L}{10} \\[2ex] I - \dfrac{10x}{L} & (I-1)\dfrac{L}{10} \leq x \leq \dfrac{IL}{10} \\[2ex] 0 & \dfrac{IL}{10} \leq x \leq L \end{cases}$$

Note that $\Phi_I(x)$ is defined by two disjoint local functions, $\psi_2^{I-1}(x)$ and $\psi_1^I(x)$.

The space spanned by the linearly independent set $\{\Phi_I\}_{I=1}^N$ is denoted by $S^h(\Omega_h)$, called the *finite-element space.*

7-1-5 Existence and Convergence of Finite-Element Solutions

In brief, the finite-element method seeks an approximate solution of Eq. (5-34) in the form of linear combinations (7-4), which define families of finite-dimensional (hence, closed) subspaces S^h of H. The subscript h is a geometric parameter associated with the finite element mesh. In the Ritz finite-element method we seek $u_h \in S^h$ such that

$$B(v_h, u_h) = l(v_h) \qquad \text{for all } v_h \in S^h$$

The approximate problem has a unique solution if $B(\cdot, \cdot)$ satisfies the conditions of Theorem 5-5 on $S^h \times S^h$. Further, the error $\varepsilon = u - u_h$ is orthogonal to the space S^h in the sense that

$$B(\varepsilon, v_h) = 0 \qquad \text{for all } v_h \in S^h$$

and satisfies the inequality [see Babuska and Aziz (1972)]

$$\|\varepsilon\|_H \leq \left(1 + \frac{M}{\alpha_h}\right) \|u - v_h\|_H \qquad \text{for all } v_h \in S^h \tag{7-5}$$

where M and α_h are the constants in Eq. (5-39) when H is replaced by S^h, and u is the exact solution.

Inequality (7-5) is the key to the proof of the convergence of the finite-element method. It indicates that the finite-element approximation error is bounded above by the term $\|u - v_h\|_H$. The magnitude of this term depends on how close the space S^h is to the space H. If S^h is the space spanned by functions that are polynomial interpolants of u, then the magnitude of $\|u - v_h\|_H$ and the rate of convergence can be estimated in terms of the degree of the polynomials. Also, note that the error depends inversely on the ellipticity constant α_h, which in turn depends on the mesh parameter. The rate of convergence as well as the stability of the approximation depends on how α_h varies as a function of h.

With this general introduction, we now turn to several model problems to illustrate the application of the finite-element method. Many books that are devoted to the finite-element method are available in the literature, and the reader is asked to consult the bibliography at the end of the book [see Reddy (1984a), and references therein].

7-2 ONE-DIMENSIONAL SECOND-ORDER EQUATIONS

7-2-1 The Model Problem

We use the following model equation to describe the finite-element method:

$$-\frac{d}{dx}\left(a\frac{du}{dx}\right) = f \qquad 0 < x < L \tag{7-6}$$

where $a = a(x)$ and $f = f(x)$ are given functions of x. We are interested in determining an approximation of the true solution u that satisfies Eq. (7-6) and appropriate boundary conditions. Equation (7-6) arises in one-dimensional representation of many physical problems (see Table 4-1).

In order to account for all possible cases of Eq. (7-6), we allow the data (a and f) to take zero, continuous, or discontinuous values. For example, a stepped shaft has discontinuities in its area of cross-section at a finite number of points along the length,

$$a \equiv EA = \begin{cases} a_1(x) & 0 = x_1 < x < x_2 \\ a_2(x) & x_2 < x < x_3 \\ \vdots & \\ a_N(x) & x_N < x < x_{N+1} = L \end{cases} \tag{7-7}$$

Also, we must allow for various combinations of the two types of boundary conditions we derived for Eq. (7-6) in Chap. 4,

$$\textit{Essential}: \text{Specify } u \qquad \textit{Natural}: \text{Specify } a\frac{du}{dx} \tag{7-8}$$

Thus, in a general case, we must seek an approximate solution to Eq. (7-6) in each of the subintervals of (0, L) in which the equation has continuous coefficients.

7-2-2 The Ritz Finite-Element Model

A step-by-step procedure for the finite-element analysis of Eq. (7-6) is given below.

1 Discretization of the domain The domain, $\Omega \equiv (0, L)$, of the problem is divided into a set of "line elements." The mesh shown in Fig. 7-4 is a *nonuniform mesh* because the elements are not of equal length. The minimum number of elements is equal to the number of subdivisions created by the discontinuities in the data of the problem. The element Ω^e is between global nodes e and $e + 1$. The coordinate of the eth global node is denoted by x_e. Therefore, $\bar{\Omega}_e = [x_e, x_{e+1}]$. The two end points of $\bar{\Omega}_e$ are the element nodes. The length of element $\bar{\Omega}_e$ is denoted by $h_e = x_{e+1} - x_e$.

2 Derivation of the element equations Consider an arbitrary element, $\Omega^e = (x_e, x_{e+1})$, located between points $x = x_e$ and $x = x_{e+1}$ in the domain (see Fig. 7-5a). We isolate the element and study a variational approximation of Eq. (7-6) in Ω^e using the Ritz method. The objective here is to find algebraic equations relating the nodal values of the primary variable u to the nodal values of the secondary variable, $P \equiv a(du/dx)$. The *primary variables* are those involved in the specification of the essential boundary conditions and the *secondary variables* are those involved in the specification of the natural boundary conditions. The following three steps describe the finite-element formulation of Eq. (7-6) over a typical element.

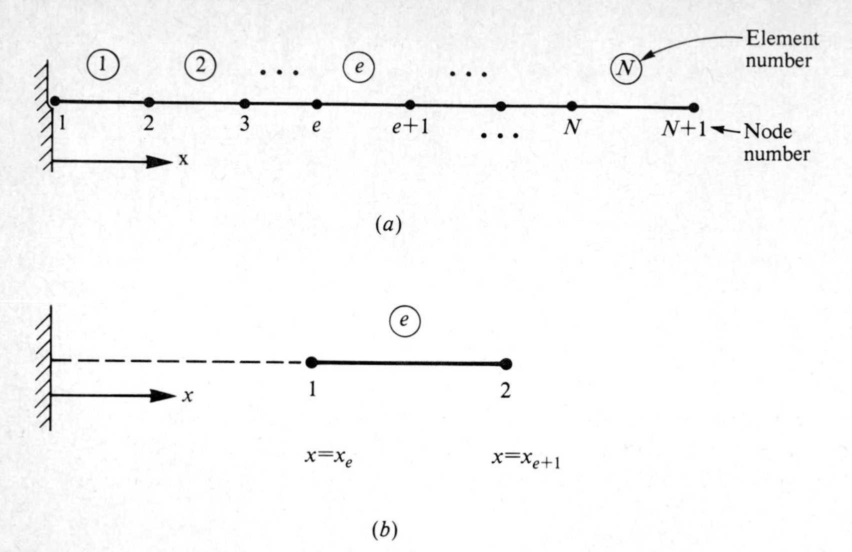

Figure 7-4 Finite-element representation of a line (one-dimensional domain) by line elements.

(i) ***Variational formulation*** Since Eq. (7-6) is valid over the domain (0, L), it is valid, in particular, over the element Ω^e. Following the procedure described in Chap. 6, we can construct the variational formulation of Eq. (7-6) over the element:

$$0 = \int_{x_e}^{x_{e+1}} v\left[-\frac{d}{dx}\left(a_e\frac{du}{dx}\right) - f_e\right] dx$$
$$= \int_{x_e}^{x_{e+1}} \left(a_e\frac{dv}{dx}\frac{du}{dx} - vf_e\right) dx + \left[v\left(-a_e\frac{du}{dx}\right)\right]_{x_e}^{x_{e+1}} \tag{7-9}$$

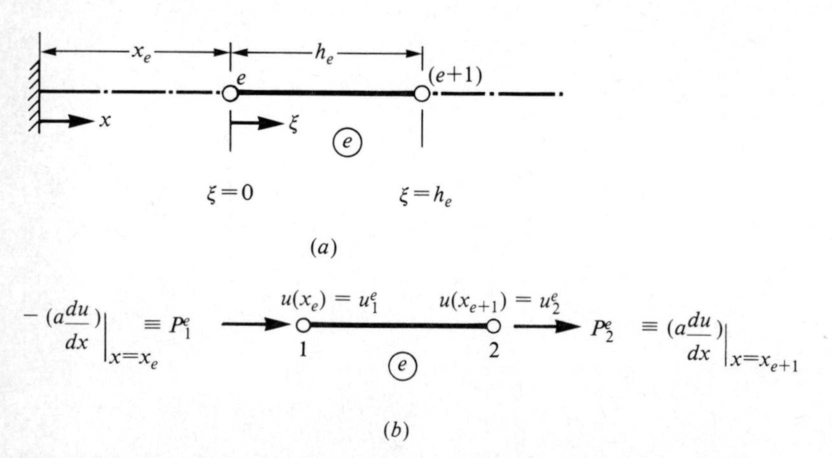

Figure 7-5 Finite-element discretization of a one-dimensional domain.

where $v \in P_e$ denotes an arbitrary continuous function. From the boundary term in the above expression, we immediately note that the specification of u at $x = x_e$ and x_{e+1} constitutes the essential boundary conditions, and the specification of $(-a(du/dx))$ at $x = x_e$ and x_{e+1} constitutes the natural boundary conditions for the element. Thus the basic unknowns at the element nodes are the primary variables (u), and secondary variable ($a(du/dx)$). Let

$$u(x_e) \equiv u_1^e \qquad u(x_{e+1}) \equiv u_2^e \tag{7-10a}$$

$$\left(-a_e \frac{du}{dx}\right)\bigg|_{x_e} \equiv P_1^e \qquad \left(-a_e \frac{du}{dx}\right)\bigg|_{x_{e+1}} \equiv -P_2^e \tag{7-10b}$$

These boundary conditions of an element are shown on the typical element in Fig. 7-5*b*. Students of engineering will recognize that Fig. 7-5*b* is the so-called free-body diagram of the typical element. With the notation in (7-10*b*), the variational form becomes

$$0 = \int_{x_e}^{x_{e+1}} \left[a_e \frac{dv}{dx}\frac{du}{dx} - vf_e \right] dx - P_1^e v(x_e) - P_2^e v(x_{e+1}) \tag{7-11}$$

or $$0 = B_e(v, u) - l_e(v)$$

where the bilinear form and linear form are given by

$$B_e(v, u) = \int_{x_e}^{x_{e+1}} a_e \frac{dv}{dx}\frac{du}{dx}\, dx \qquad l_e(v) = \int_{x_e}^{x_{e+1}} vf_e\, dx + v(x_e)P_1^e + v(x_{e+1})P_2^e \tag{7-12}$$

(ii) ***Variational approximation*** Now suppose that we wish to find an approximate solution of the variational problem in (7-11) with boundary conditions in (7-10*a*) using the Ritz method. Let the Ritz approximation of u on the element be given by

$$u_e(x) = \sum_{j=1}^{n} u_j^e \psi_j^e(x) \tag{7-13}$$

where u_j^e are the parameters to be determined, and $\psi_j^e(x)$ are the approximation functions. Substituting (7-13) for u and $v = \psi_i^e$ into (7-11), we obtain

$$0 = \sum_{j=1}^{n} \left(\int_{x_e}^{x_{e+1}} a_e \frac{d\psi_i^e}{dx}\frac{d\psi_j^e}{dx}\, dx \right) u_j^e - \left[\int_{x_e}^{x_{e+1}} \psi_i^e f_e\, dx + P_1^e \psi_i^e(x_e) + P_2^e \psi_i^e(x_{e+1}) \right]$$

$$= \sum_{j=1}^{n} K_{ij}^e u_j^e - F_i^e \qquad (i = 1, 2, \ldots, n)$$

or $$[K^e]\{u^e\} = \{F^e\} \tag{7-14}$$

where the coefficient matrix K_{ij}^e and the column vector F_i^e are given by

$$K_{ij}^e = \int_{x_e}^{x_{e+1}} a_e \frac{d\psi_i^e}{dx}\frac{d\psi_j^e}{dx}\,dx$$

$$F_i^e = \int_{x_e}^{x_{e+1}} \psi_i^e f_e\,dx + P_1^e \psi_i^e(x_e) + P_2^e \psi_i^e(x_{e+1}) \tag{7-15}$$

(iii) ***Derivation of the interpolation functions*** We must select $\psi_i^e \in P_e$ such that: (a) u_e is differentiable at least once with respect to x and satisfies the essential boundary conditions (7-10a), (b) $\{\psi_i^e\}$ are linearly independent, and (c) $\{\psi_i^e\}$ are complete. These conditions are met if we choose at least a linear approximation of the form

$$u_e(x) = c_1^e + c_2^e x \tag{7-16}$$

The continuity is obviously satisfied, $\{1, x\}$ is linearly independent and complete. In addition, we require u_e to satisfy the essential boundary conditions of the element, Eq. (7-10a):

$$u(x_e) \equiv u_1^e = c_1^e + c_2^e x_e$$

$$u(x_{e+1}) \equiv u_2^e = c_1^e + c_2^e x_{e+1}$$

In matrix form we have

$$\begin{Bmatrix} u_1^e \\ u_2^e \end{Bmatrix} = \begin{bmatrix} 1 & x_e \\ 1 & x_{e+1} \end{bmatrix} \begin{Bmatrix} c_1^e \\ c_2^e \end{Bmatrix} \tag{7-17}$$

Solving for c_1^e and c_2^e in terms of u_1^e and u_2^e, we obtain

$$c_1^e = \frac{u_1^e x_{e+1} - u_2^e x_e}{x_{e+1} - x_e} \qquad c_2^e = \frac{u_2^e - u_1^e}{x_{e+1} - x_e} \tag{7-18}$$

Substituting (7-18) for c_i^e in (7-16) and collecting the coefficients of u_i^e we obtain

$$\begin{aligned} u_e(x) &= \frac{u_1^e x_{e+1} - u_2^e x_e}{x_{e+1} - x_e} + \frac{u_2^e - u_1^e}{x_{e+1} - x_e}\,x \\ &= \left(\frac{x_{e+1} - x}{x_{e+1} - x_e}\right) u_1^e + \left(\frac{x - x_e}{x_{e+1} - x_e}\right) u_2^e \\ &= \sum_{i=1}^{2} u_i^e \psi_i^e \end{aligned} \tag{7-19a}$$

where $$\psi_1^e = \frac{x_{e+1} - x}{x_{e+1} - x_e} \qquad \psi_2^e = \frac{x - x_e}{x_{e+1} - x_e} \qquad x_e \le x \le x_{e+1} \tag{7-19b}$$

The expression in (7-19) satisfies the essential boundary conditions of the element, and $\{\psi_i^e\}$ are continuous, linearly independent and complete over the element. Comparing (7-13) with (7-19a), we immediately note that $n = 2$.

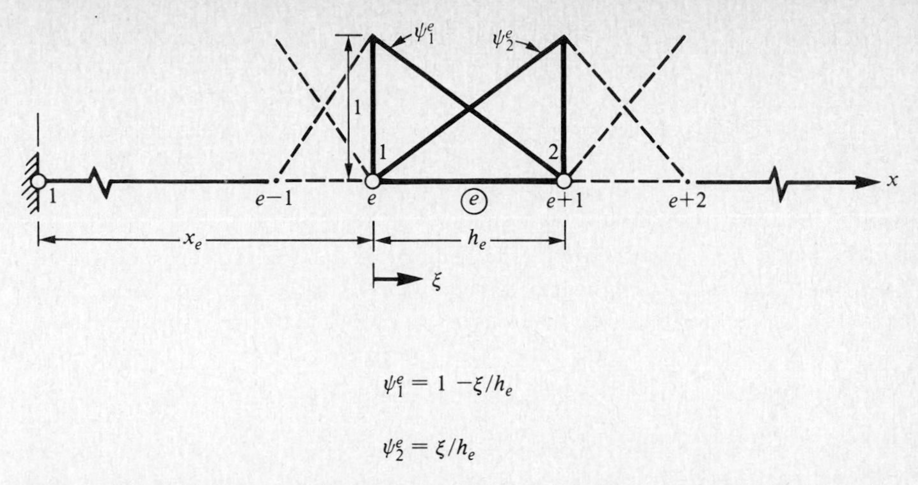

Figure 7-6 Local interpolation functions (linear) for a one-dimensional (Lagrange) element.

Since the approximation functions are derived in such a way that $u_e(x)$ is equal to u_1^e at node 1 and u_2^e at node 2, i.e., interpolated, they are also called a *Lagrange family of interpolation functions.* The interpolation functions have the following properties, in addition to the property that $\psi_i^e = 0$ outside element $\bar{\Omega}^e$ (see Fig. 7-6)

$$\text{(i)} \quad \psi_i^e(x_j) = \begin{cases} 0 & \text{if } i \neq j \\ 1 & \text{if } i = j \end{cases} \qquad (x_1 = x_e,\, x_2 = x_{e+1})$$

$$\text{(ii)} \quad \sum_i^2 \psi_i^e(x) = 1 \tag{7-20}$$

We note that the derivation of interpolation function does not depend on the specific problem being modeled. They depend only on the form of the essential boundary conditions and the type of element (geometry and number of nodes). Thus, the linear interpolation functions derived above are not only useful in the finite-element approximation of the problem at hand, they are also useful in all problems that admit linear interpolation of the variables.

In view of property (i) in Eq. (7-20), the coefficient F_i^e of Eq. (7-15) becomes

$$F_i^e = \int_{x_e}^{x_{e+1}} \psi_i^e f_e \, dx + P_i^e \tag{7-21}$$

Note that the approximation chosen for u_e in Eq. (7-16) satisfies the minimum continuity requirements. One can choose a quadratic, cubic, etc. approximation by adding various powers of x (not leaving out any lower-order terms, otherwise the completeness condition will be violated). For higher-order (i.e., quadratic, cubic, etc.) elements, the steps involved in the derivation of the approximation functions remain the same as those described in Eqs. (7-16) to (7-19). For higher-order elements, Eq. (7-16) requires identification of additional nodes (in addition to the nodes at $x = x_e$ and $x = x_{e+1}$) in the element. In other words, a quadratic approximation (which has three constants, c_1^e, c_2^e, and c_3^e), requires identification of three nodes (the third one, say, at the center of the element). Thus, there is a relationship between the order of the approximation used for the dependent variable u_e and the number of nodes in the element.

A local coordinate system (i.e., a coordinate system in the element) proves to be more convenient in the derivation of the interpolation functions and in the computation of K_{ij}^e. Suppose that ξ is the local or element coordinate with origin at the node 1 (left end) of the element (see Fig. 7-6). The local coordinate ξ is related to the global coordinate x [which is used to describe the problem in Eq. (7-6)] according to the linear "translation" transformation

$$x = \xi + x_e \tag{7-22}$$

In terms of the local coordinate ξ, $\psi_i^e(\xi)$ are given by

$$\psi_1^e(\xi) = 1 - \frac{\xi}{h_e} \qquad \psi_2^e(\xi) = \frac{\xi}{h_e} \qquad 0 < \xi < h_e \tag{7-23}$$

Note that mathematically and geometrically Eqs. (7-19*b*) and (7-23) are the same. The coefficient matrix K_{ij}^e and column vector F_i^e in (7-15) can be written in the local coordinate system as

$$K_{ij}^e = \int_0^{h_e} \hat{a}_e \frac{d\psi_i^e}{d\xi}\frac{d\psi_j^e}{d\xi}\, d\xi \qquad F_i^e = \int_0^{h_e} \hat{f}_e \psi_i^e \, d\xi + P_i^e \tag{7-24}$$

where $\qquad \hat{a}_e \equiv a_e$ evaluated at $(x_e + \xi) = a(x_e + \xi)$

$$\hat{f}_e \equiv f_e(x_e + \xi)$$

When the linear interpolation functions are used to approximate the dependent variable of the present problem and a_e and f_e are element-wise constant, Eq. (7-24) has the explicit form:

$$[K^e] = \frac{a_e}{h_e}\begin{bmatrix} 1 & -1 \\ -1 & 1 \end{bmatrix} \qquad \{F^e\} = \frac{f_e h_e}{2}\begin{Bmatrix} 1 \\ 1 \end{Bmatrix} + \begin{Bmatrix} P_1^e \\ P_2^e \end{Bmatrix} \tag{7-25}$$

3 Assembly (or connectivity) of element equations Since Eq. (7-14) is derived for an arbitrary typical element, it holds for any element from the finite-element mesh. For the sake of discussion, suppose that the domain of the problem, $\Omega = (0, L)$, is divided into three elements of possibly unequal lengths. Since these elements are connected at global nodes 2 and 3 and u is continuous, u_2^e of element Ω^e should be the same as u_1^e of element Ω^{e+1} for $e = 1, 2$. To express this correspondence mathematically, we label the value of u_e at global node I by U_I, $I = 1, 2, \ldots, N$, where N is the total number of global nodes. Then we have the following correspondence between the local (element) nodal values and global nodal values (see Fig. 7-7*a*).

$$u_1^1 = U_1 \quad u_2^1 = U_2 = u_1^2 \quad u_2^2 = U_3 = u_1^3 \quad u_2^3 = U_4 \tag{7-26}$$

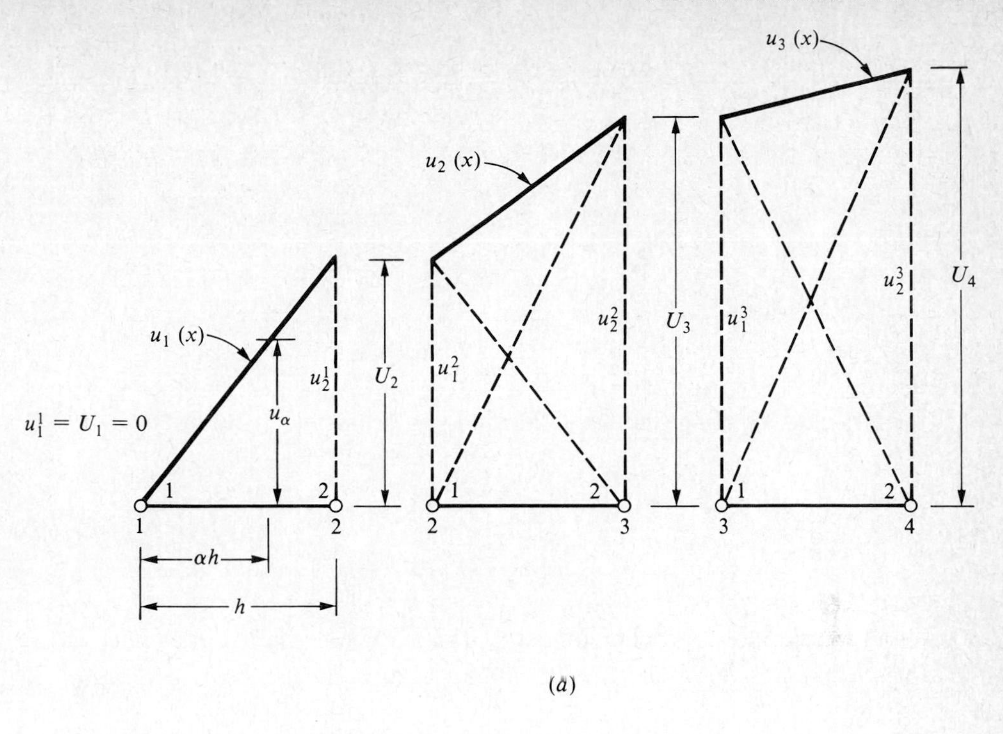

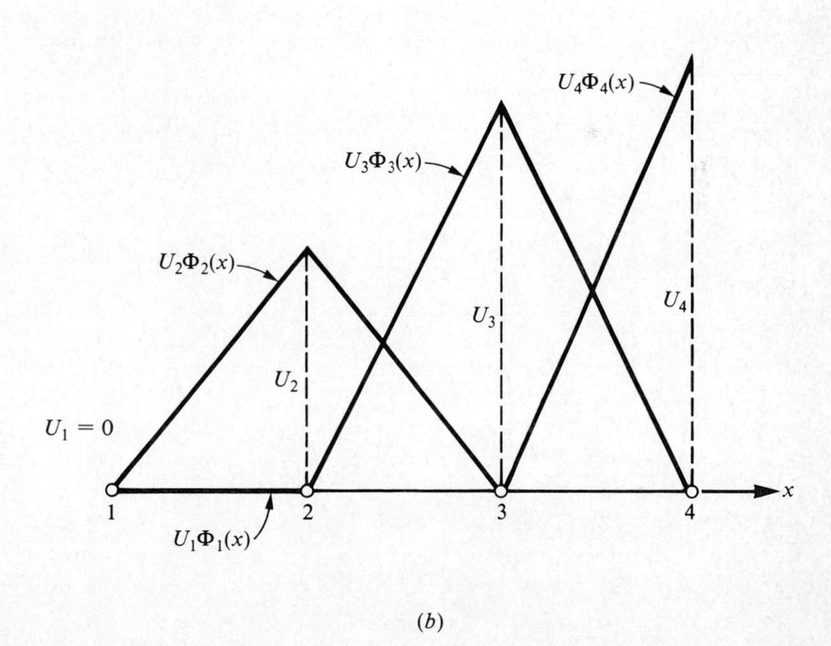

Figure 7-7 Correspondence of local and global nodal values and representation of the finite-element solution by global (linear) interpolation functions for the model problem.

We shall call these relations *interelement continuity conditions*. The global finite-element solution $u_h(x)$ is given by

$$u_h(x) = \begin{cases} \sum_{j=1}^{2} u_j^1 \psi_j^1 & x_1 \le x \le x_2 \\ \sum_{j=1}^{2} u_j^2 \psi_j^2 & x_2 \le x \le x_3 \\ \sum_{j=1}^{2} u_j^3 \psi_j^3 & x_3 \le x \le x_4 \end{cases} \tag{7-27}$$

In view of the conditions (7-26), $u_h(x) \in S^h(\Omega_h) \subset H^1(\Omega)$ can be expressed in terms of N linearly independent global basis functions $\{\Phi_I\}_{I=1}^N$ (see Fig. 7-7*b*)

$$u_h(x) = \sum_{I=1}^{N} U_I \Phi_I(x) \tag{7-28}$$

where

$$\Phi_I(x) = \begin{cases} \psi_2^{(I-1)}(x) & x_{I-1} < x < x_I \\ \psi_1^{(I)}(x) & x_I < x < x_{I+1} \end{cases} \tag{7-29}$$

The connectivity of elements is based on the fact that the global variational statement is the sum of local variational statements:

$$B(u_h, v_h) - l(v_h) = \sum_{e=1}^{E} [B_e(u_e, v_e) - l_e(v_e)] = 0 \tag{7-30}$$

where E denotes the total number of elements in the mesh. Substituting

$$u_e = \sum_{j=1}^{n} u_j^e \psi_j^e \qquad \text{and} \qquad v_e = \sum_{i=1}^{n} v_j^e \psi_j^e$$

into Eq. (7-30) and using Eq. (7-26) for both u_j^e and v_j^e (replace U_i by V_i for v_j^e), we obtain

$$\begin{aligned} 0 &= \sum_{e=1}^{E} \left[\sum_{i=1}^{2} \sum_{j=1}^{2} v_i^e K_{ij}^e u_j^e - \sum_{i=1}^{2} v_i^e F_i^e \right] \\ &= v_1^1(K_{11}^1 u_1^1 + K_{12}^1 u_2^1 - F_1^1) + v_2^1(K_{21}^1 u_1^1 + K_{22}^1 u_2^1 - F_2^1) \\ &\quad + v_1^2(K_{11}^2 u_1^2 + K_{12}^2 u_2^2 - F_1^2) + v_2^2(K_{21}^2 u_1^2 + K_{22}^2 u_2^2 - F_2^2) \\ &\quad + v_1^3(K_{11}^3 u_1^3 + K_{12}^3 u_2^3 - F_1^3) + v_2^3(K_{21}^3 u_1^3 + K_{22}^3 u_2^3 - F_2^3) \\ &= V_1[K_{11}^1 U_1 + K_{12}^1 U_2 - F_1^1] \\ &\quad + V_2[K_{21}^1 U_1 + (K_{22}^1 + K_{11}^2) U_2 + K_{12}^2 U_3 - (F_2^1 + F_1^2)] \\ &\quad + V_3[K_{21}^2 U_2 + (K_{22}^2 + K_{11}^3) U_3 + K_{12}^3 U_4 - (F_2^2 + F_1^3)] \\ &\quad + V_4[K_{21}^3 U_3 + K_{22}^3 U_4 - F_2^3] \end{aligned} \tag{7-31}$$

Since Eq. (7-30) holds for any arbitrary $v_h = \sum_{I=1}^{N} V_I \Phi_I$, Eq. (7-31) must hold for any V_I, $I = 1, 2, 3, 4$. This implies that the coefficients of each V_I should be equal to zero. The resulting four equations, expressed in matrix form, are the finite-element equations of the assembly $\bar{\Omega}_h = \bigcup_{e=1}^{E} \bar{\Omega}_e$:

$$\begin{bmatrix} K_{11}^1 & K_{12}^1 & 0 & 0 \\ K_{21}^1 & K_{22}^1 + K_{11}^2 & K_{12}^2 & 0 \\ 0 & K_{21}^2 & K_{22}^2 + K_{11}^3 & K_{12}^3 \\ 0 & 0 & K_{21}^3 & K_{22}^3 \end{bmatrix} \begin{Bmatrix} U_1 \\ U_2 \\ U_3 \\ U_4 \end{Bmatrix} = \begin{Bmatrix} F_1^1 \\ F_2^1 + F_1^2 \\ F_2^2 + F_1^3 \\ F_2^3 \end{Bmatrix} \tag{7-32}$$

4 Imposition of boundary conditions Equation (7-32) is valid for any problem described by the differential equation in (7-6), irrespective of the boundary conditions. The coefficient matrix in (7-32) is singular prior to the imposition of the (essential) boundary conditions on the primary nodal degrees of freedom. Also, the right-hand side of Eq. (7-32) is not completely known prior to the imposition of the "equilibrium" conditions among the secondary degrees of freedom. Recall that each F_i^e contains the contribution from the source term, f, in the differential equation (7-6) and the secondary degree of freedom, P_i^e. The individual P_i^e are unknown (so-called reactive forces in solid mechanics problems) but the sums of the P_i^e's at a given node are known if the primary degree of freedom at the node is unknown and vice versa. Recalling that $P_1^e = -a_e(du_e/dx)$ at $x = x_e$ and $P_2^e = a_e(du_e/dx)$ at $x = x_{e+1}$, we require the sum of P_2^e and P_1^{e+1} at global node $(e + 1)$ to equal the specified value of $a_e(du_e/dx)$ at that node, which is equal to zero in the present problem:

$$\left(a_e \frac{du}{dx}\right)^+ \Bigg|_{x=x_{e+1}} - \left(a_e \frac{du}{dx}\right)^- \Bigg|_{x=x_{e+1}} \equiv P_2^e + P_1^{e+1} = 0 \tag{7-33}$$

The difference accounts for any point source applied at the point $x = x_{e+1}$. If no point source is applied the difference is zero there.

5 Solution of equations The global finite-element equations (7-32) can be partitioned conveniently into the following form

$$\begin{bmatrix} [K^{11}] & [K^{12}] \\ [K^{21}] & [K^{22}] \end{bmatrix} \begin{Bmatrix} \{\Delta^1\} \\ \{\Delta^2\} \end{Bmatrix} = \begin{Bmatrix} \{F^1\} \\ \{F^2\} \end{Bmatrix} \tag{7-34}$$

where $\{\Delta^1\}$ is the column of known primary degrees of freedom, $\{\Delta^2\}$ is the column of the unknown primary degrees of freedom, $\{F^1\}$ is the column of unknown secondary degrees of freedom, and $\{F^2\}$ is the column of the known secondary degrees of freedom. Equation (7-34) can be written as a pair of matrix equations

$$[K^{11}]\{\Delta^1\} + [K^{12}]\{\Delta^2\} = \{F^1\} \tag{7-35a}$$

$$[K^{21}]\{\Delta^1\} + [K^{22}]\{\Delta^2\} = \{F^2\} \tag{7-35b}$$

From Eq. (7-35*b*), we have

$$\{\Delta^2\} = [K^{22}]^{-1}(\{F^2\} - [K^{21}]\{\Delta^1\}) \tag{7-36}$$

Once $\{\Delta^2\}$ is known, $\{F^1\}$ can be computed from Eq. (7-35*a*).

6 Postprocessing of the solution The solution of finite-element equations (7-36) gives only the nodal values of the unknown function. Often one is interested in finding the solution and possibly its derivatives, which are needed in the determination of the secondary variables, at points other than the nodes. The finite-element solution u_h at any point x can be computed using Eq. (7-27). For example, for $x = \alpha h$, $0 < \alpha < 1$, we have (see Fig. 7-7a)

$$\begin{aligned} u(\alpha h) &= \frac{h - \alpha h}{h} u_1^{(1)} + \frac{\alpha h - 0}{h} u_2^{(1)} \\ &= (1 - \alpha) u_1^{(1)} + \alpha u_2^{(1)} \end{aligned} \tag{7-37}$$

That is, u at $x = \alpha h$ is a weighted average of its values at the nodes on either side of the point $x = \alpha h$. Similarly, for $x_0 = h + \alpha h$, we obtain from the second line of (7-27) the value $u(x_0) = (1 - \alpha)u_1^{(2)} + \alpha u_2^{(2)}$.

Similarly, the derivative of the solution can be computed from Eq. (7-27) by differentiation,

$$\begin{aligned} \frac{du_h}{dx} &= \sum_{j=1}^{2} u_j^e \frac{d\psi_j^e}{dx} \\ &= \frac{u_2^e - u_1^e}{h_e} \qquad \text{for} \qquad x_e < x < x_{e+1} \end{aligned} \tag{7-38}$$

Thus the derivative of the finite-element solution is element-wise constant for a linear element. Also du_h/dx at a node is discontinuous, for any degree of approximation, in going from one element to another because we did not strictly impose the continuity of the derivatives at the nodes.

This completes the finite-element analysis of the model problem for a second-order equation in one dimension. We shall consider some specific examples of the application of the method in Sec. 7-2-3.

Some remarks on the method We make some remarks concerning the finite-element procedure for second-order problems.

1. The element equations are derived for the linear operator

   $$T \equiv -(d/dx)(a(d/dx)).$$

   Hence, they are valid for any physical problem that is described by the operator T (one needs only to interpret the quantities). Examples of problems described by this operator are listed in Table 4-1. Thus, a computer program written for the finite-element analysis of Eq. (7-6) can be used to analyze any of the problems in Table 4-1. Also, note that the property $a = a(x)$ (which depends on the geometry as well as the material) can be different in each of the elements.
2. An interpretation of the equations (7-14) and (7-25) (for $f = 0$) can be given in terms of the finite difference approximation. The axial force at any point x is

given by $P(x) = a(du/dx)$. Using the forward difference approximation, we can write

$$-P_1^{(e)} \equiv P(x)\Big|_{x=x_e} = a_e[u(x_{e+1}) - u(x_e)]/h_e$$

and using the backward difference approximation, we write

$$P_2^{(e)} \equiv P(x)\Big|_{x=x_{e+1}} = a_e[u(x_{e+1}) - u(x_e)]/h_e$$

These two equations give Eq. (7-14) with $[K^e]$ and $\{F^e\}$ given by Eq. (7-25).

3. For the model problem considered here, the element matrices $[K^e]$ in (7-14) are symmetric, $(K_{ij}^e = K_{ji}^e)$. In all problems involving the symmetric operator T, we have symmetry of the coefficient matrix. This enables one to compute K_{ij}^e, $i = 1, 2, \ldots, n$, for $j \leq i$ only. In other words, one needs to compute only the diagonal terms and upper (or lower) diagonal terms. Because of the symmetry of the element matrices, the assembled global matrix will also be symmetric. Thus, one needs to store only the upper triangle (including the diagonal) of the assembled matrix in a finite-element program. Another property that is characteristic of the finite-element method is the *sparseness* of the assembled matrix. Since the global interpolation function Φ_I is nonzero only on two neighboring elements (see Fig. 7-7*b*) at node I, its contribution to nodes $I - 2$, $I - 3$, etc., and nodes $I + 2$, $I + 3$, etc. is zero; that is, entries K_{13}, K_{24}, etc. in the global coefficient matrix are zero. In general, $K_{IJ} = 0$ if nodes I and J do not belong to the same element. Thus, if the global nodes are numbered sequentially, the contribution to entries two or more places away on either side of the main diagonal of the global coefficient matrix is zero, and the matrix is *banded*. When a matrix is banded and symmetric, one needs to store only the entries in the upper (or lower) band of the matrix. Equation solvers which are written for the solution of banded symmetric equations are available for use in such cases. While symmetry of the coefficient matrix is a property transferred from the variational formulation of the equation being modeled, the sparseness of the matrix is a result of the finite-element interpolation functions, which have *local support*.
4. Equilibrium of the secondary variables P_i at the interelement boundaries is expressed by Eq. (7-33). This amounts to imposing the condition that the secondary variable $(a(du/dx))$ at a node be continuous. Although Eq. (7-33) is imposed in solving the global equations, the continuity of $(a(du/dx))$ will not be satisfied because the finite-element solution is defined only element-wise.
5. Equations (7-32), after imposing the equilibrium conditions on the P_i's, can be obtained directly by considering the global variational problem. The variational form of (7-6) with $u(0) = 0$ and $(a(du/dx))|_{x=L} = P$ is given by

$$0 = \int_0^L \left[a \frac{dv}{dx}\frac{du}{dx} - vf \right] dx - Pv(L)$$

When u is approximated by functions that are defined only on a local interval (which is the case in the finite-element method), use of the above variational form implies the omission of the sum of the interelement contributions,

$$\left[\sum_{e=1}^{N}\left(\sum_{i=1}^{2} P_i^e u_i^e\right)\right]$$

Substituting (7-28) for u and $v = \Phi_I$ into the variational form above, we obtain

$$0 = \int_0^L \left[a \frac{d\Phi_I}{dx}\left(\sum_{J=1}^{4} U_J \frac{d\Phi_J}{dx}\right) - \Phi_I f \right] dx - \Phi_I(L)P$$

Since Φ_I is nonzero between x_{I-1} and x_{I+1}, the integral becomes

$$0 = \int_{x_{I-1}}^{x_{I-1}} \left[a \frac{d\Phi_I}{dx}\left(U_{I-1}\frac{d\Phi_{I-1}}{dx} + U_I \frac{d\Phi_I}{dx} + U_{I+1}\frac{d\Phi_{I+1}}{dx}\right) - \Phi_I f \right] dx$$
$$- \Phi_I(L)P$$

for $I = 1, 2, 3, 4$. We have

$$I = 1:\ \ 0 = \int_{x1=0}^{x2} \left[a \frac{d\Phi_1}{dx}\left(U_1 \frac{d\Phi_1}{dx} + U_2 \frac{d\Phi_2}{dx}\right) - \Phi_1 f \right] - \Phi_1(L)P$$

$$I = 2:\ \ 0 = \int_{x1=0}^{x3} \left[a \frac{d\Phi_2}{dx}\left(U_1 \frac{d\Phi_1}{dx} + U_2 \frac{d\Phi_2}{dx} + U_3 \frac{d\Phi_3}{dx}\right) - \Phi_2 f \right] - \Phi_2(L)P$$

$$I = 3:\ \ 0 = \int_{x2}^{x4} \left[a \frac{d\Phi_3}{dx}\left(U_2 \frac{d\Phi_2}{dx} + U_3 \frac{d\Phi_3}{dx} + U_4 \frac{d\Phi_4}{dx}\right) - \Phi_3 f \right] dx - \Phi_3(L)P$$

$$I = 4:\ \ 0 = \int_{x3}^{x4=L} \left[a \frac{d\Phi_4}{dx}\left(U_3 \frac{d\Phi_3}{dx} + U_4 \frac{d\Phi_4}{dx}\right) - \Phi_4 f \right] dx - \Phi_4(L)P$$

These equations, upon performing the integration, yield (7-32) with the column of P's in $\{F\}$ replaced by

$$\begin{Bmatrix} 0 \\ 0 \\ 0 \\ P \end{Bmatrix}$$

Although this procedure gives the assembled equations directly, the procedure is algebraically complicated (especially for two-dimensional problems) and not amenable to simple computer implementation. In addition, it does not provide the means to calculate the P_i^e's (the internal reactions).

6. Recall that in assembling the elements, we used the continuity of the primary nodal variables. This means that we cannot accurately represent primary variables that are discontinuous [but still belong to $L_2(\Omega)$], such as the one shown in Fig. 7-8*a*. Such problems arise, for example, in the study of compressible fluid flows where shock waves contain discontinuities. Thus such functions

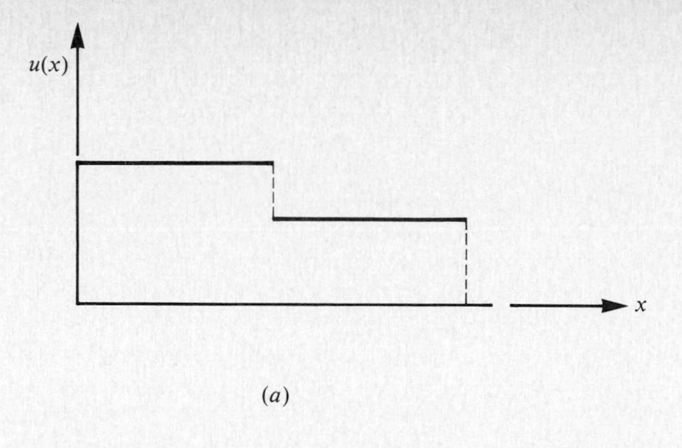

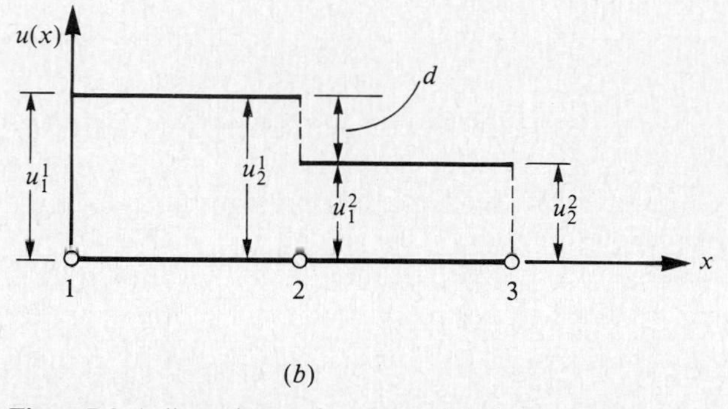

Figure 7-8 A discontinuous function and its finite-element approximation with linear elements.

should not be used as primary variables in the finite-element model, or special procedures must be used in the assembly procedure.

For example, consider a two-element approximation of the function in Fig. 7-8*b*. Before we assemble the element equations, we note that

$$u_1^1 = U_1 \qquad u_2^1 = U_2 \qquad u_1^2 = u_2^1 - d = U_2 - d \qquad u_2^2 = U_3$$

Then the assembled equations are given by

$$\begin{bmatrix} K_{11}^1 & K_{12}^1 & \\ K_{21}^1 & K_{22}^1 + K_{11}^2 & K_{12}^2 \\ & K_{21}^2 & K_{22}^2 \end{bmatrix} \begin{Bmatrix} U_1 \\ U_2 \\ U_3 \end{Bmatrix} = \begin{Bmatrix} F_1^1 \\ F_2^1 + F_1^2 \\ F_2^2 \end{Bmatrix} + \begin{Bmatrix} 0 \\ K_{11}^2 d \\ K_{21}^2 d \end{Bmatrix}$$

Obviously, we need an additional condition to determine the discontinuity, d (see Fig. 7-8*b*).

The so-called mixed and hybrid finite-element methods utilize variational formulations that contain variables (e.g., stresses or moments) that can be discontinuous in some problems. In utilizing such elements caution should be exercised to see if the problem being modeled contains any discontinuities in the primary variables in the interior of the domain.

7-2-3 Applications of the Ritz Finite-Element Model

Here we consider some examples of applications of the finite-element method. The first example deals with the finite-element analysis of a radially symmetric problem.

Example 7-1 As noted earlier (see Table 4-2), the Poisson equation arises in connection with the solution of many field problems, including heat conduction, electrostatics, etc. When the domain of these field problems is circular (e.g., heat transfer in a long cylindrical member, electric field in coaxial cylindrical cables) and the source is either zero or distributed axisymmetrically, the Poisson equation in cylindrical coordinates reduces to the ordinary differential equation

$$-\frac{d}{dr}\left[a(r)\frac{du}{dr}\right] = 0 \tag{a}$$

In heat transfer problems, u denotes the temperature and $a = 2\pi rk$, k being the thermal conductivity (which can be a function of radial distance r) of the medium. In electrostatics, u denotes the electrostatic potential and $a = 2\pi r\varepsilon$, ε being the permittivity of the medium.

Equation (a) is in the same form as the model equation (7-6). Therefore, Eqs. (7-14) and (7-15) describe the finite-element model of Eq. (a). For the choice of linear interpolation functions, we have

$$K_{ij}^e = \int_{r_e}^{r_{e+1}} 2\pi k_e r \frac{d\psi_i^e}{dr}\frac{d\psi_j^e}{dr}\, dr$$

where $\quad \psi_1^e = \dfrac{r_{e+1} - r}{r_{e+1} - r_e} \qquad \psi_2^e = \dfrac{r - r_e}{r_{e+1} - r_e} \qquad h_e = r_{e+1} - r_e$

For example, K_{11}^e is given by

$$\begin{aligned} K_{11}^e &= 2\pi k_e \int_{r_e}^{r_{e+1}} r\left(-\frac{1}{h_e}\right)^2 dr \\ &= \frac{\pi k_e}{h_e}(r_{e+1} + r_e) \end{aligned}$$

We have

$$\frac{\pi k_e}{h_e}(r_{e+1} + r_e)\begin{bmatrix} 1 & -1 \\ -1 & 1 \end{bmatrix}\begin{Bmatrix} u_1^e \\ u_2^e \end{Bmatrix} = \begin{Bmatrix} P_1^e \\ P_2^e \end{Bmatrix}$$

where P_i^e denote the internal fluxes

$$P_1^e = -2\pi k_e \left(\frac{du}{dr}\right)\bigg|_{r=r_e} \qquad P_2^e = 2\pi k_e \left(\frac{du}{dr}\right)\bigg|_{r=r_{e+1}}$$

The assembled equations for an N-element case are given by

$$\pi \begin{bmatrix} \frac{\bar{k}_1}{h_1} & -\frac{\bar{k}_1}{h_1} & 0 & \cdot & \cdot & \cdot \\ -\frac{\bar{k}_1}{h_1} & \frac{\bar{k}_1}{h_1}+\frac{\bar{k}_2}{h_2} & -\frac{\bar{k}_2}{h_2} & \cdot & \cdot & \cdot \\ 0 & -\frac{\bar{k}_2}{h_2} & \frac{\bar{k}_2}{h_2}+\frac{\bar{k}_3}{h_3} & \cdot & \cdot & \cdot \\ \cdot & \cdot & \cdot & \cdot & -\frac{\bar{k}_N}{h_N} & 0 \\ \cdot & \cdot & \cdot & -\frac{\bar{k}_N}{h_N} & \frac{\bar{k}_N}{h_N}+\frac{\bar{k}_{N+1}}{h_{N+1}} & -\frac{\bar{k}_{N+1}}{h_{N+1}} \\ \cdot & \cdot & \cdot & 0 & -\frac{\bar{k}_{N+1}}{h_{N+1}} & \frac{\bar{k}_{N+1}}{h_{N+1}} \end{bmatrix} \begin{Bmatrix} U_1 \\ U_2 \\ U_3 \\ \vdots \\ U_N \\ U_{N+1} \end{Bmatrix} = \begin{Bmatrix} P_1^1 \\ P_2^1 + P_1^2 \\ P_2^2 + P_1^3 \\ \vdots \\ P_2^{N-1} + P_1^N \\ P_2^N \end{Bmatrix}$$

where $\bar{k}_e = k_e(r_e + r_{e+1})$.

We now impose the boundary conditions of the problem. Suppose that the domain is the cross-section of a coaxial cylinder with two dielectrics (or a hollow cylinder made of two materials with different thermal conductivities), as shown in Fig. 7-9. Let the internal and external radii be $r_1 = 20$ mm and

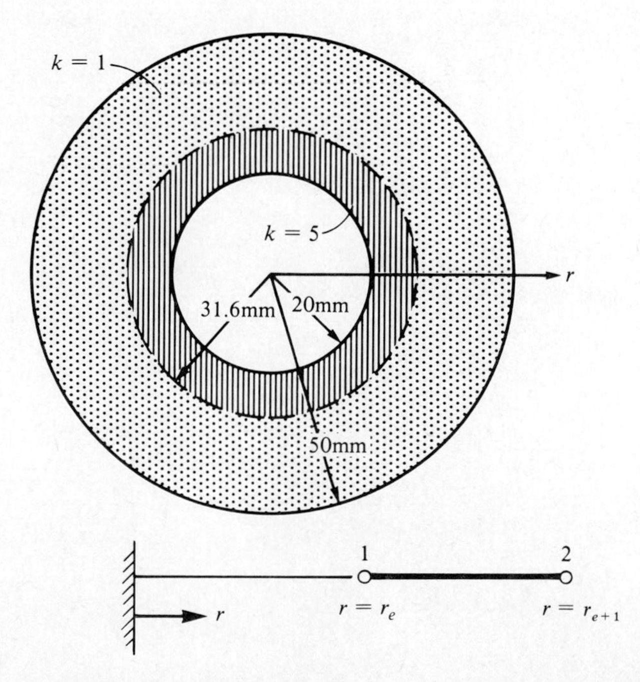

Figure 7-9 A coaxial cylinder made of two materials and a typical finite element in the mesh.

$r_{N+1} = 50$ mm, and let the thickness of the first material be 11.6 mm and that of the second material be 18.4 mm, and the associated material constants be 5 and 1. We assume that $u(20) = 100$ and $u(50) = 0.0$. Thus the boundary conditions are

$$U_1 = 100.0 \qquad U_{N+1} = 0.0$$

$$P_2^1 + P_1^2 = 0, \ldots, \qquad P_2^{N-1} + P_1^N = 0$$

For a nonuniform mesh of four elements, the equations for element 1 are given by ($h_1 = 5.1$, $h_2 = 6.5$, $h_3 = 8.2$, $h_4 = 10.2$; equivalently, $r_1 = 20$, $r_2 = 25.1$, $r_3 = 31.6$, $r_4 = 39.8$ and $r_5 = 50.0$)

$$[K^1] = 2\pi \cdot \frac{5(20 + 25.1)}{2 \times 5.1} \begin{bmatrix} 1 & -1 \\ -1 & 1 \end{bmatrix} = 2\pi \begin{bmatrix} 22.108 & -22.108 \\ -22.108 & 22.108 \end{bmatrix}$$

The assembled equations are given by

$$2\pi \begin{bmatrix} 22.108 & -22.108 & 0 & 0 & 0 \\ -22.108 & 43.916 & -21.808 & 0 & 0 \\ 0 & -21.808 & 26.162 & -4.354 & 0 \\ 0 & 0 & -4.354 & 8.756 & -4.402 \\ 0 & 0 & 0 & -4.402 & 4.402 \end{bmatrix} \begin{Bmatrix} U_1 \\ U_2 \\ U_3 \\ U_4 \\ U_5 \end{Bmatrix} = \begin{Bmatrix} P_1^1 \\ P_2^1 + P_1^2 \\ P_2^2 + P_1^3 \\ P_2^3 + P_1^4 \\ P_2^4 \end{Bmatrix}$$

The boundary conditions are

$$U_1 = 100.0 \qquad U_5 = 0.0 \qquad P_2^1 + P_1^2 = 0 \qquad P_2^2 + P_1^3 = 0 \qquad P_2^3 = 0$$

The solution for U_2, U_3, and U_4 is obtained by solving Eqs. 2, 3, and 4:

$$\begin{bmatrix} 43.916 & -21.808 & 0 \\ -21.808 & 26.162 & -4.354 \\ 0 & -4.354 & 8.756 \end{bmatrix} \begin{Bmatrix} U_2 \\ U_3 \\ U_4 \end{Bmatrix} = \begin{Bmatrix} 22.108\, U_1 \\ 0 \\ 0 \end{Bmatrix}$$

or,

$$U_2 = 91.745, \; U_3 = 83.377, \; U_4 = 41.458$$

Table 7-2 Comparison of the finite-element solution with the analytical solution of $-d/dr(a(du/dr)) = 0$, $u(20) = 100$, $u(50) = 0$

r	Two elements	Four elements	Eight elements	Analytical solution
20.0	100.000	100.000	100.000	100.000
22.6	—	—	95.559	95.559
25.1	—	91.745	91.746	91.746
28.4	—	—	87.258	87.257
31.6	83.375	83.377	83.377	83.377
35.7	—	—	61.213	61.210
39.8	—	41.458	41.457	41.457
44.9	—	—	19.551	19.549
50.0	0.000	0.000	0.000	0.000

Table 7-2 contains a comparison of the finite-element solutions obtained by three different, nonuniform meshes with the analytical solution. The numerical convergence and accuracy of the finite element solution is apparent from the results.

Example 7-2 Consider the problem in Example 6-1. Equation (6-28) is a special case of the model Eq. (7-6) with $L = 1$, $a = 1$ and $f = \cos \pi x$ for $0 < x < 1$. For a linear element, the element coefficient matrix $[K^e]$ is given by Eq. (7-25) with $a_e = 1$ for all e. The column vector $\{F^e\}$ needs to be computed using Eq. (7-21):

$$F_i^e = \int_{x_e}^{x_{e+1}} \cos \pi x \, \psi_i^e \, dx + P_i^e = f_i^e + P_i^e$$

We have

$$\begin{aligned}
f_1^e &= \int_{x_e}^{x_{e+1}} \cos \pi x \left(\frac{x_{e+1} - x}{h_e} \right) dx \\
&= \left(\frac{x_{e+1} - x}{h_e} \frac{\sin \pi x}{\pi} - \frac{\cos \pi x}{\pi^2 h_e} \right) \Bigg|_{x_e}^{x_{e+1}} \\
&= -\frac{\sin \pi x_e}{\pi} - \frac{(\cos \pi x_{e+1} - \cos \pi x_e)}{\pi^2 h_e} \\
f_2^e &= \int_{x_e}^{x_{e+1}} \cos \pi x \left(\frac{x - x_e}{h_e} \right) dx \\
&= \left(\frac{x - x_e}{h_e} \frac{\sin \pi x}{\pi} + \frac{\cos \pi x}{\pi^2 h_e} \right) \Bigg|_{x_e}^{x_{e+1}} \\
&= \frac{\sin \pi x_{e+1}}{\pi} + \frac{\cos \pi x_{e+1} - \cos \pi x_e}{\pi^2 h_e}
\end{aligned} \tag{a}$$

For a mesh of four equal elements, we have $h = 0.25$ and

$$f_1^1 = 0.1187 \qquad f_2^1 = 0.1064$$

The assembled equations for the four-element case are given by

$$\frac{1}{h} \begin{bmatrix} 1 & -1 & & & \\ -1 & 2 & -1 & & \\ & -1 & 2 & -1 & \\ & & -1 & 2 & -1 \\ & & & -1 & 1 \end{bmatrix} \begin{Bmatrix} U_1 \\ U_2 \\ U_3 \\ U_4 \\ U_5 \end{Bmatrix} = \begin{Bmatrix} f_1^1 \\ f_2^1 + f_1^2 \\ f_2^2 + f_1^3 \\ f_2^3 + f_1^4 \\ f_2^4 \end{Bmatrix} + \begin{Bmatrix} P_1^1 \\ P_2^1 + P_1^2 \\ P_2^2 + P_1^3 \\ P_2^3 + P_1^4 \\ P_2^4 \end{Bmatrix} \tag{b}$$

The above equation is valid for any set of admissible boundary conditions. We now consider three types of boundary conditions.

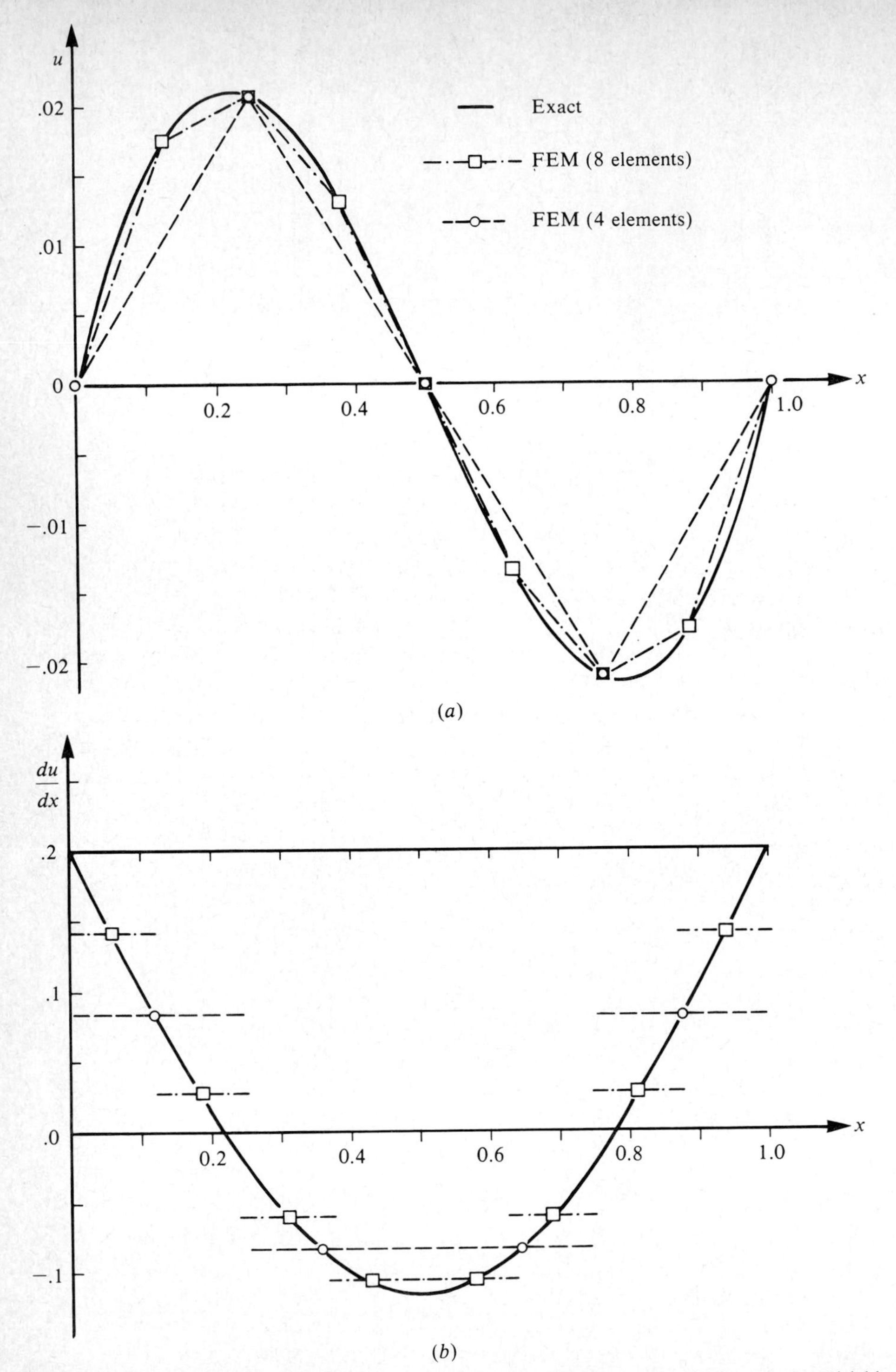

Figure 7-10 A comparison of the finite-element and exact solutions of Eq. (6-28) with the Dirichlet boundary conditions [$u(0) = u(1) = 0$].

1. *Dirichlet boundary conditions.* $u(0) = u(1) = 0$. These boundary conditions imply that $U_1 = U_5 = 0$. In addition, the following equilibrium conditions hold:

$$P_2^1 + P_1^2 = 0 \qquad P_2^2 + P_1^3 = 0 \qquad P_2^3 + P_1^4 = 0 \tag{c}$$

Hence, we must solve Eq. (*b*) for the vector $\{\Delta^2\} = \{U_2\, U_3\, U_4\}^T$:

$$\begin{bmatrix} 8 & -4 & 0 \\ -4 & 8 & -4 \\ 0 & -4 & 8 \end{bmatrix} \begin{Bmatrix} U_2 \\ U_3 \\ U_4 \end{Bmatrix} = \begin{Bmatrix} f_2^1 + f_1^2 \\ f_2^2 + f_1^3 \\ f_2^3 + f_1^4 \end{Bmatrix}$$

The solution is given by

$$U_2 = -U_4 = 0.02098 \qquad U_3 \approx 0.0$$

which coincide with the values of the exact solution at $x = 0.25, 0.75$, and 0.5, respectively. Figure 7-10*a* shows a comparison of the four-element and eight-element finite-element solutions with the exact solution of the Dirichlet problem associated with Eq. (6-28). The corresponding first derivatives of the solutions are compared in Fig. 7-10*b*. Although the nodal values obtained in the finite-element method are accurate, the solution is in considerable error at other points of the domain. The accuracy of the solution is improved when the mesh is refined. Note that the derivatives obtained by the finite-element method at the midpoints of each element coincide with the exact values.

2. *Mixed boundary conditions.* $u(0) = 0$, $u'(1) = 0$. These boundary conditions imply that $U_1 = 0$ and $P_2^4 = 0$; in addition we have the conditions in Eq. (*c*) for the four-element case. We solve the equations,

$$\begin{bmatrix} 8 & -4 & 0 & 0 \\ -4 & 8 & -4 & 0 \\ 0 & -4 & 8 & -4 \\ 0 & 0 & -4 & 4 \end{bmatrix} \begin{Bmatrix} U_2 \\ U_3 \\ U_4 \\ U_5 \end{Bmatrix} = \begin{Bmatrix} f_2^1 + f_1^2 \\ f_2^2 + f_1^3 \\ f_2^3 + f_1^4 \\ f_2^4 \end{Bmatrix}$$

The solution of these equations is given by

$$U_2 = -0.02968 \quad U_3 = -0.10132 \quad U_4 = -0.17297 \quad U_5 = -0.20264$$

which coincide with the values of the exact solution at the nodes. The four-element and eight-element solutions are compared with the exact solution for u and du/dx in Fig. 7-11. In this case, even the four-element solution is accurate at all points of the domain.

3. *Neumann boundary conditions.* $u'(0) = u'(1) = 0$. These boundary conditions imply that $P_1^1 = P_2^4 = 0$ in Eq. (*b*). In addition we have the equilibrium conditions of Eq. (*c*). Equation (*b*), with the column of P's being zero, cannot be inverted because the coefficient matrix is singular. To eliminate the arbitrary constant, we set $U_3 = 0$ (in the present case, we happened to know this) and solve the equations. We obtain

$$U_1 = -U_5 = 0.10132 \qquad U_2 = -U_4 = 0.071645$$

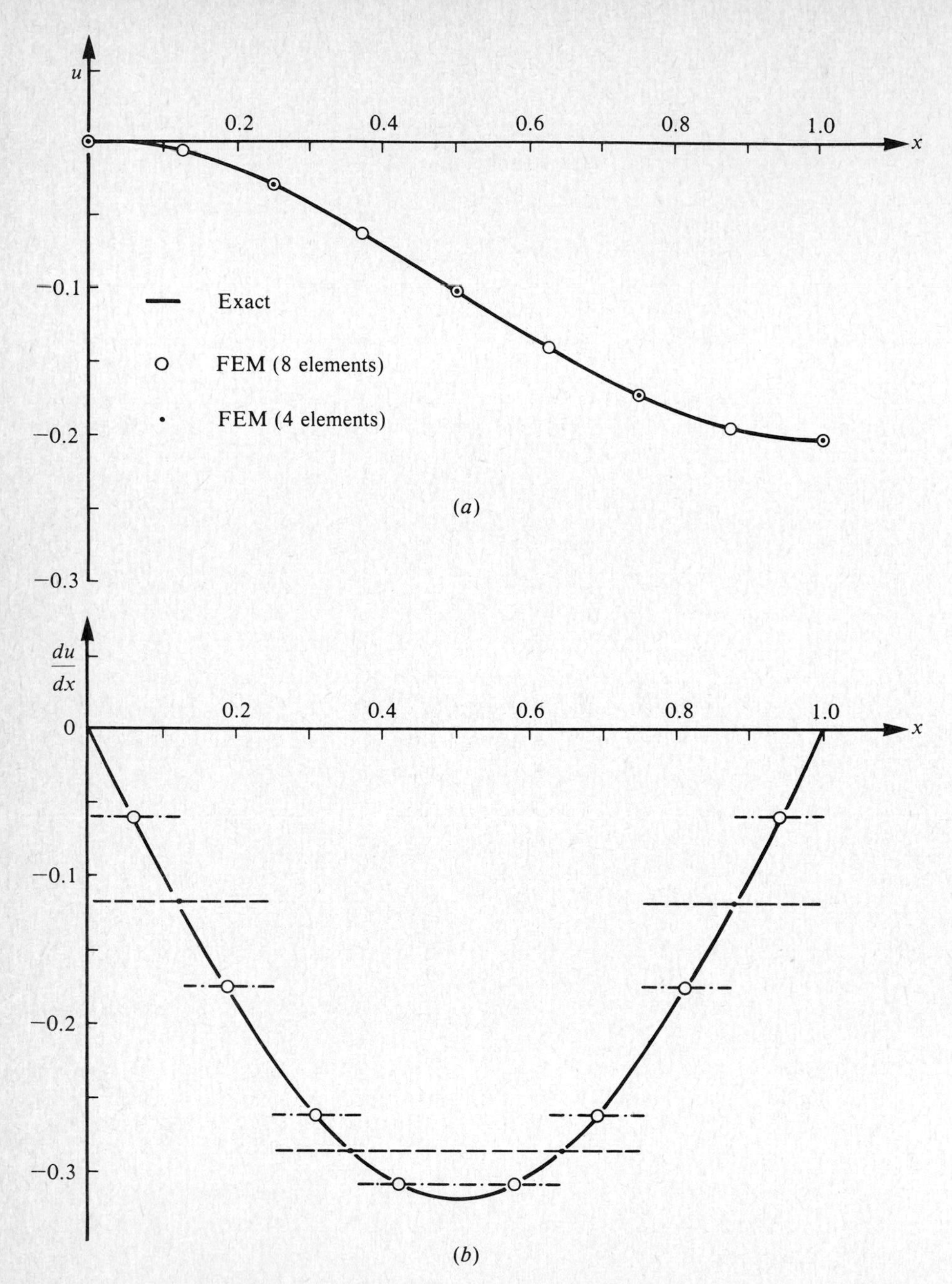

Figure 7-11 A comparison of the finite-element and exact solutions of Eq. (6-28) with mixed boundary conditions $[u(0) = u'(1) = 0]$.

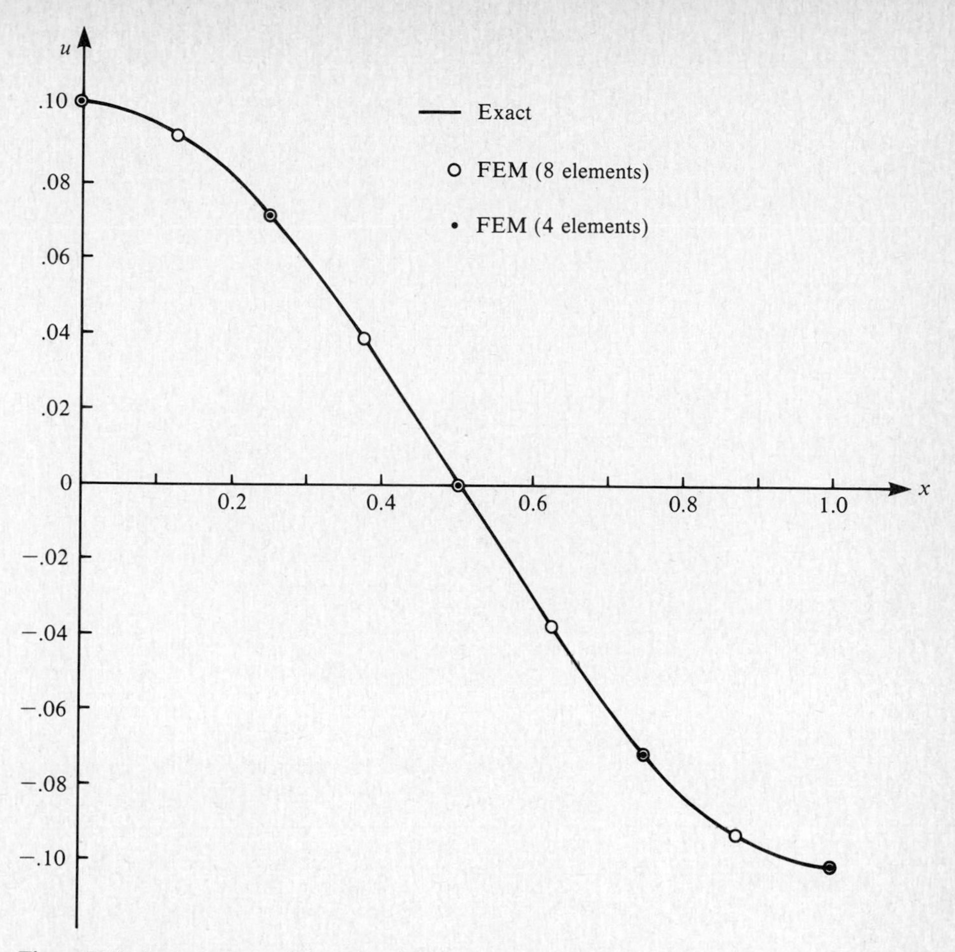

Figure 7-12 A comparison of the finite-element and exact solutions of Eq. (6-28) with the Neumann boundary conditions $[u'(0) = u'(1) = 0]$.

Once again these values coincide with the values of the exact solution at the nodes. The four-element and eight-element solutions are compared with the exact solution in Fig. 7-12. The derivative in the present case is the same as that in Fig. 7-11*b*.

From the results of the three problems, it is clear that the finite-element solution converges, with the refinement of the mesh, to the true solutions of the problems. The next example deals with a mixed boundary-value problem that is not a special case of Eq. (7-6).

Example 7-3 Consider the differential equation (6-46) in Example 6-4. This equation is of the same form as Eq. (7-6) with $a = 1$, $f = -x^2$ and $L = 1$, except that Eq. (6-46) has an additional term, $-u$. This is reflected in the bilinear form

$$B(v, u) = \int_0^1 \left(\frac{dv}{dx} \frac{du}{dx} - vu \right) dx$$

Hence, the element matrix $[K^e]$ becomes

$$K_{ij}^e = \int_{x_e}^{x_{e+1}} \left(\frac{d\psi_i^e}{dx} \frac{d\psi_j^e}{dx} - \psi_i^e \psi_j^e \right) dx \tag{7-39}$$

For the choice of linear interpolation functions, the coefficient matrix is given by

$$[K^e] = \frac{1}{h_e} \begin{bmatrix} 1 & -1 \\ -1 & 1 \end{bmatrix} - \frac{h_e}{6} \begin{bmatrix} 2 & 1 \\ 1 & 2 \end{bmatrix}$$

The column vector is given by ($F_i^e = f_i^e + P_i^e$),

$$f_i^e = \int_{x_e}^{x_{e+1}} (-x^2) \psi_i^e \, dx$$

$$= \begin{cases} -\dfrac{1}{h^e} \left[\dfrac{x_{e+1}}{3} (x_{e+1}^3 - x_e^3) - \dfrac{1}{4} (x_{e+1}^4 - x_e^4) \right] & i = 1 \\ -\dfrac{1}{h_e} \left[\dfrac{1}{4} (x_{e+1}^4 - x_e^4) - \dfrac{x_e}{3} (x_{e+1}^3 - x_e^3) \right] & i = 2 \end{cases}$$

For two elements, the boundary conditions are

$$U_1 = 0 \qquad P_2^1 + P_1^2 = 0 \qquad P_2^2 = 1.0$$

The solution of the finite-element equations is

$$U_2 = 0.6075 \qquad U_3 = 1.1392$$

whereas the exact solution at the nodes is given by 0.6112 and 1.1442. The four-element and eight-element solutions at these points are given by (0.6102, 1.1429) and (0.6109, 1.1439), respectively.

As mentioned in earlier discussions, the accuracy of finite-element approximations can be improved either by refining the mesh or by using higher-order interpolation. To this end, let us consider the quadratic element for Eq. (7-6). Let

$$u_e(x) = c_1^e + c_2^e x + c_3^e x^2 \tag{7-40}$$

In order to express this approximation in terms of the values of u_e at selected points of the element $\Omega^e = (x_e, x_{e+1})$, we now identify three nodes: the two end nodes (which allow us to connect the element to the adjacent elements) and a node in the interior of the element, say in the middle (see Fig. 7-13*a*).

We can follow the procedure used in Eqs. (7-17) to (7-19) to obtain the three interpolation functions. This procedure requires the inversion of a 3 × 3 matrix.

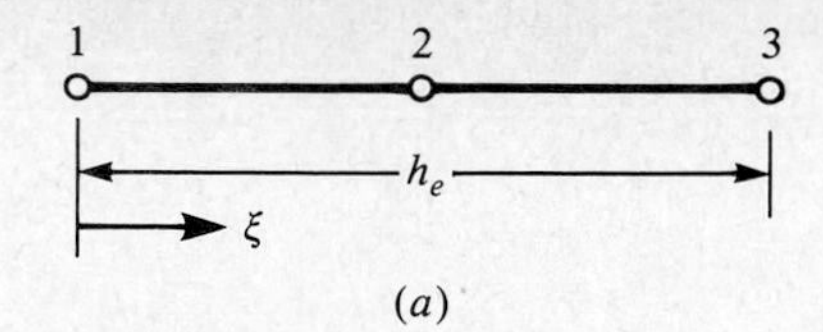

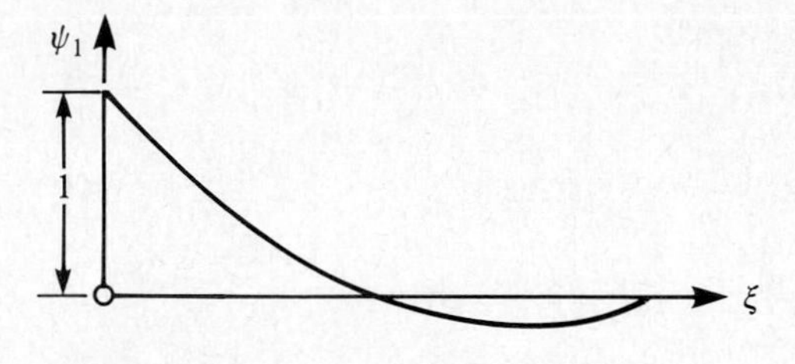

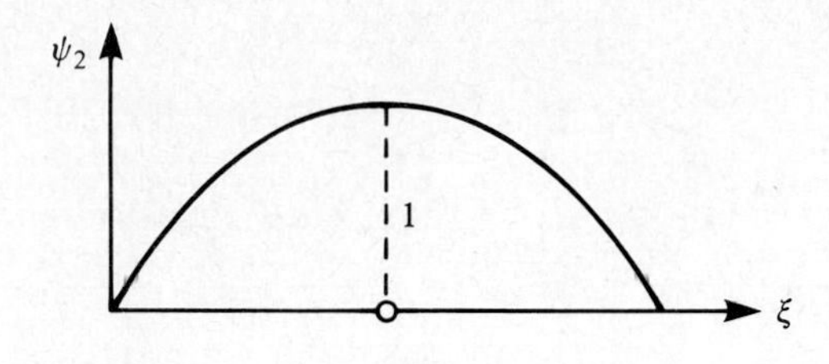

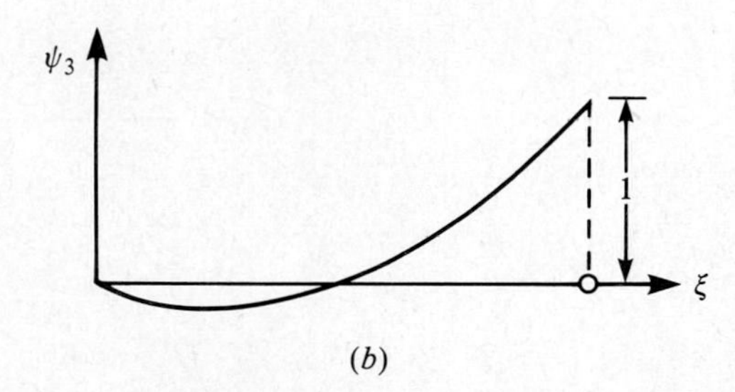

Figure 7-13 The Lagrange quadratic element and associated interpolation functions.

Here we present an alternative procedure to derive the interpolation functions in the local (i.e., element) coordinate ξ. The procedure is based on the interpolation property of Eq. (7-20):

$$\psi_i^e(\xi_j) = \delta_{ij} \qquad (\xi_1 = 0 \qquad \xi_2 = h_e/2 \qquad \xi_3 = h_e) \tag{7-41}$$

For instance, $\psi_1^e = 1$ at node 1 and is zero at the other two nodes. A quadratic polynomial satisfying these conditions has the form

$$\psi_1^e(\xi) = \alpha_1(\xi_2 - \xi)(\xi_3 - \xi)$$

The constant α_1 can be evaluated by requiring $\psi_1^e(0) = 1$, which gives $\alpha_1 = 2/h_e^2$. Similarly, $\psi_2^e(\xi)$ and $\psi_3^e(\xi)$ can be determined from

$$\psi_2^e(\xi) = \alpha_2(\xi_1 - \xi)(\xi_3 - \xi)$$
$$\psi_3^e(\xi) = \alpha_3(\xi_1 - \xi)(\xi_2 - \xi)$$

After determining the constants, we obtain

$$\begin{aligned} \psi_1^e(\xi) &= \left(1 - \frac{2\xi}{h_e}\right)\left(1 - \frac{\xi}{h_e}\right) \\ \psi_2^e(\xi) &= 4\frac{\xi}{h_e}\left(1 - \frac{\xi}{h_e}\right) \\ \psi_3^e(\xi) &= -\frac{\xi}{h_e}\left(1 - \frac{2\xi}{h_e}\right) \end{aligned} \tag{7-42}$$

These functions also satisfy the second property of Eq. (7-20). The functions are geometrically represented in Fig. 7-13*b*.

Note that one quadratic element is as good as, and often better than, two linear elements because the quadratic element approximates a function quadratically between its nodes. The following example supports this statement.

Example 7-4 Consider the second-order differential equation of Example 7-3. The element coefficient matrix in Eq. (7-39) for the choice of quadratic interpolation functions is given by

$$[K^e] = \frac{1}{3h_e}\begin{bmatrix} 7 & -8 & 1 \\ -8 & 16 & -8 \\ 1 & -8 & 7 \end{bmatrix} - \frac{h_e}{30}\begin{bmatrix} 4 & 2 & -1 \\ 2 & 16 & 2 \\ -1 & 2 & 4 \end{bmatrix}$$

For the one-element mesh ($h_e = 1$), we have

$$\{f^e\} = \{0.01667, 0.2, -0.15\}^T$$

At the internal nodes, such as node 2, the P_i's are equal to the point source specified. In the present case, we have $P_2^1 = 0$. The boundary conditions are $U_1 = 0$, $P_3^1 = 1.0$. The solution is given by

$$U_2 = 0.6097 \qquad U_3 = 1.1439$$

which is more accurate than that predicted by a mesh of two linear elements.

Table 7-3 contains a comparison of the linear and quadratic finite-element solutions with the exact solution. It is clear that the finite-element solution predicted by the quadratic element converges to the exact solution faster, for equal number of global nodes, than that predicted by the linear element.

Table 7-3 A comparison of the finite-element solution with the exact solution of the problem in Example 7-4

		Finite-element solution					
		Linear elements			Quadratic elements		
x	Exact solution	2	4	8	1	2	4
0.125	0.1576	0.1519	0.1563	0.1575*	0.1595	0.1583	0.1576*
0.250	0.3130	0.3038	0.3125*	0.3129*	0.3143	0.3130*	0.3130*
0.375	0.4646	0.4556	0.4614	0.4644*	0.4644	0.4639	0.4646*
0.500	0.6112	0.6075*	0.6102*	0.6109*	0.6097*	0.6111*	0.6112*
0.625	0.7521	0.7404	0.7482	0.7518*	0.7503	0.7517	0.7521*
0.750	0.8875	0.8734	0.8863*	0.8872*	0.8862	0.8874*	0.8875*
0.875	1.0178	1.0063	1.0146	1.0175*	1.0174	1.0182	1.0178*
1.000	1.1442	1.1392*	1.1429*	1.1439*	1.1439*	1.1442*	1.1442*

* Nodal values; all other values are computed by interpolation.

Thus far we have focused our attention on second-order equations. Of course, we have not discussed the finite-element formulation of the fourth-order equation governing bending of beams. We can solve a pair of second-order equations [see Eqs. (6-76)], or use the moment-deflection equations with the bending moment known throughout the length of the beam. In that case, the equation to

Table 7-4 Comparison of the finite-element solution with the exact solution of a simply supported beam (see Example 7-5)

		Finite-element solution		
x	Exact‡ $\bar{w}$	2	4	8
		Deflection, $\bar{w}$		
30	0.0685	0.0555	0.0659	0.0685*
60	0.1319	0.1111	0.1319*	0.1319*
90	0.1849	0.1667	0.1771	0.1849*
120	0.2222	0.2222*	0.2222*	0.2222*
180	0.2396	0.1667	0.2083	0.2396*
240	0.1944	0.1111	0.1944*	0.1944*
300	0.1076	0.0555	0.0972	0.1076*
		Slope, $\bar{\theta}$‡‡		
30	2.2280	1.8519	2.1991	2.2859
60	1.9676	—	—	2.1123
90	1.5336	—	1.5046	1.7650
120	0.9259	—	—	1.2442
180	−0.2893	−0.9295	−0.2315	−0.2893
240	−1.1574	—	—	−0.7523
300	−1.6782	—	−1.6204	−1.4468
360	−1.8519	—	—	−1.7940

‡ $\bar{w} = (-w \times 10^3)/P$, $\bar{\theta} = (-dw/dx)10^6/P$.

* Nodal values.

‡‡ Values are constant within each element.

be solved is

$$-EI\,\frac{d^2w}{dx^2} = -M(x) \tag{7-43}$$

which is a special case of Eq. (7-6). Here w denotes the transverse deflection, M the bending moment, and $a = EI$ the flexural rigidity (assumed to be a constant). Statically determinate problems (otherwise, M will not be known) with constant flexural rigidity can be modeled by Eq. (7-43). The next example deals with one such problem.

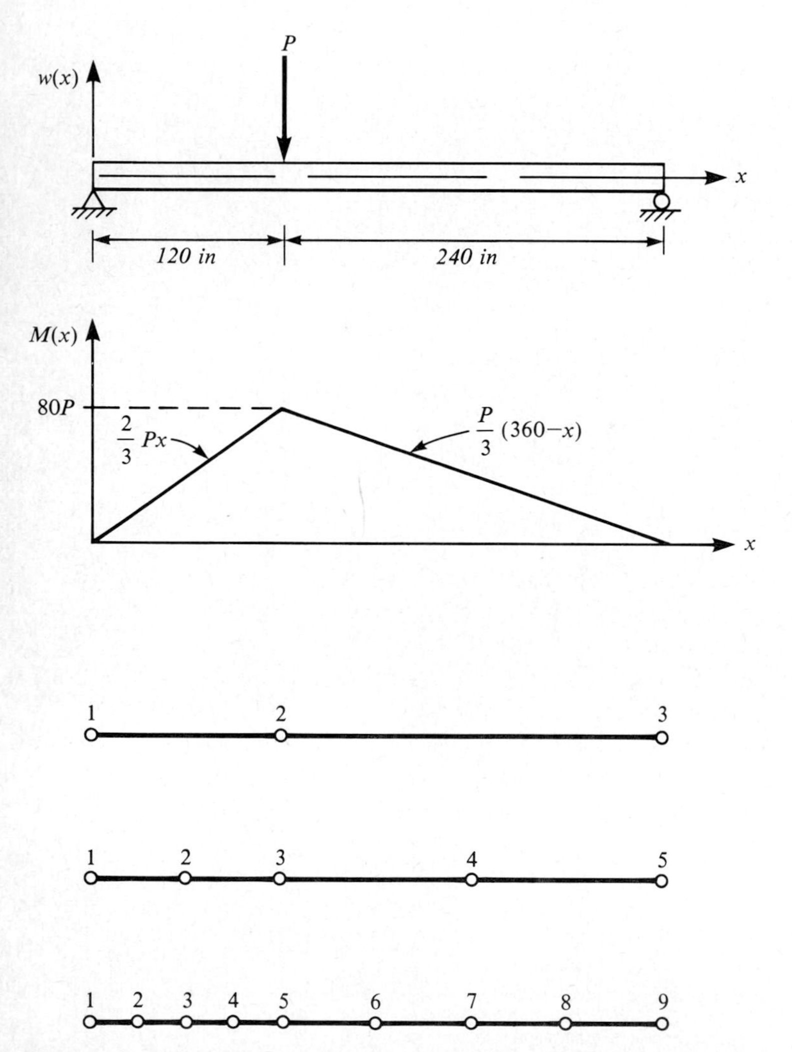

Figure 7-14 Finite-element analysis of a simply supported beam $[w(0) = w(360) = 0]$ using linear Lagrange elements.

Example 7-5 Consider the simply supported beam shown in Fig. 7-14. The variation of the bending moment and the three, nonuniform, finite-element meshes used are also shown in Fig. 7-14. Compared to Eq. (7-6), we have $f = -M(x)$. The solution for various meshes is shown in Table 7-4. Note that the derivatives of w are not continuous from element to element. Hence, the solution would not be as accurate as the solution obtained using the beam element discussed in Section 7-3.

In the next example we consider the finite-element modeling of the pair (6-76) of second-order equations involving the transverse deflection and bending moment.

Example 7-6 Consider the fourth-order Eq. (6-48a) governing the bending of beams. Here we study the finite-element model associated with the mixed formulation discussed in Example 6-5. Toward developing the finite-element model of Eqs. (6-76), we first develop the variational form over an element, $\Omega^e = (x_e, x_{e+1})$.

Consider the first of the two equations in (6-76). Multiplying with a test function v_1^e and integrating by parts over Ω^e we obtain

$$0 = \int_{x_e}^{x_{e+1}} \left(\frac{dv_1^e}{dx} \frac{dw_e}{dx} + \frac{1}{b_e} v_1^e M_e \right) dx - P_1^e v_1^e(x_e) - P_2^e v_1^e(x_{e+1}) \tag{7-44}$$

where $b_e = E_e I_e$, $P_1^e = (-dw_e/dx)(x_e)$ and $P_2^e = (dw_e/dx)(x_{e+1})$. Similarly, from the second equation we have

$$0 = \int_{x_e}^{x_{e+1}} \left(\frac{dv_2^e}{dx} \frac{dM_e}{dx} + v_2^e f_e \right) dx - Q_1^e v_2^e(x_e) - Q_2^e v_2^e(x_{e+1}) \tag{7-45}$$

where v_2^e is a test function, and $Q_1^e = (-dM_e/dx)(x_e)$ and $Q_2^e = (dM_e/dx)(x_{e+1})$. Note that Eq. (7-44) is coupled to Eq. (7-45).

The finite-element models of Eqs. (7-44) and (7-45) are obtained by selecting appropriate interpolations of w_e and M_e over Ω^e. Comparison of the weak forms (7-44) and (7-45) with those obtained from the functional in Eq. (6-77) show that v_1^e should be interpolated by the same functions as those used for M_e and v_2^e should be interpolated by the same functions as those used for w_e. Substituting

$$\begin{gathered} w_e = \sum_{j=1}^{n} w_j^e \psi_j^{1e} \qquad M_e = \sum_{j=1}^{m} M_j^e \psi_j^{2e} \\ v_1^e = \psi_i^{2e} \qquad \text{and} \qquad v_2^e = \psi_i^{1e} \end{gathered} \tag{7-46}$$

into Eq. (7-44) and (7-45), we obtain

$$\begin{gathered} [A^e]\{w^e\} + [B^e]\{M^e\} = \{P^e\} \\ [A^e]^T\{M^e\} = \{f^e\} + \{Q^e\} \end{gathered} \tag{7-47}$$

$$\text{where} \qquad A_{ij}^e = \int_{x_e}^{x_{e+1}} \frac{d\psi_i^{2e}}{dx} \frac{d\psi_j^{1e}}{dx} \, dx \qquad B_{ij}^e = \int_{x_e}^{x_{e+1}} \frac{1}{b_e} \psi_i^{2e} \psi_j^{2e} \, dx$$

$$f_i^e = -\int_{x_e}^{x_{e+1}} f_e \psi_i^{1e} \, dx \tag{7-48}$$

Note that $[A^e]$ is not a square matrix.

For the choice of linear interpolation functions, $\psi_i^{1e} = \psi_i^{2e} = \psi_i^e$, we obtain

$$[A^e] = \frac{1}{h_e} \begin{bmatrix} 1 & -1 \\ -1 & 1 \end{bmatrix} \qquad [B^e] = \frac{h_e}{6b_e} \begin{bmatrix} 2 & 1 \\ 1 & 2 \end{bmatrix}$$

The element equation (7-47) becomes ($\alpha_e = h_e^2/2b_e$)

$$\frac{1}{h_e} \begin{bmatrix} 0 & 1 & 0 & -1 \\ 1 & 2\alpha_e & -1 & \alpha_e \\ 0 & -1 & 0 & 1 \\ -1 & \alpha_e & 1 & 2\alpha_e \end{bmatrix} \begin{Bmatrix} w_1^e \\ M_1^e \\ w_2^e \\ M_2^e \end{Bmatrix} = \begin{Bmatrix} f_1^e \\ 0 \\ f_2^e \\ 0 \end{Bmatrix} + \begin{Bmatrix} Q_1^e \\ P_1^e \\ Q_2^e \\ P_2^e \end{Bmatrix} \tag{7-49}$$

The remaining steps of the finite-element procedure are routine by now. Note that the P's denote the slopes and the Q's denote the shear forces. The main advantages of the mixed model are that a linear interpolation of the variables can be used, and moments can be computed at the nodes. A specific example of application of the mixed model is given in the finite-element text by the author (1984a, p. 402).

7-2-4 Weighted-Residual Finite-Element Models

The finite-element model presented in Sec. 7-2-2 is based on the Ritz method. Here we describe finite-element models based on weighted-residual methods. Recall from Chap. 6 that the Ritz method is based on a weak formulation (i.e., formulation in which the solution lies in an energy space H_A instead of in the domain space $\mathscr{D}_A$; equivalently, in the weak formulation the generalized integration by parts is used to include the natural boundary conditions into the variational problem). The weighted-residual methods are based on an integral form of the given operator equation and therefore require the approximations to be chosen from the domain space $\mathscr{D}_A$. Of course, for first-order equations we do not have the energy space (because there is no weak formulation for them) and we must work with the domain space, $\mathscr{D}_A$. In other words, weighted-residual methods are the natural (and only) choice of solving first-order equations by the finite-element method. For second-order equations we have a choice between the weak (or Ritz) formulation and the weighted residual formulation. We begin with the description of weighted-residual finite-element models of a first-order equation.

First-order equations Consider the first-order equation

$$a \frac{du}{dx} + cu = f \qquad 0 < x < L \tag{7-50}$$

where a, c, and f are given functions of x. An example of the situation in which the above equation arises is given by Newton's law of cooling of a body with temperature u in an environment at temperature u_0. The law states that the time rate of cooling is proportional to the difference between the body and the environment temperatures,

$$\frac{du}{dt} + k(u - u_0) = 0$$

which is the same as Eq. (7-50) with $x = t$, $c = k$, $a = 1$, and $f = ku_0$.

Since the equation under consideration is first order, there is no advantage of transferring the derivative to the test function. Hence the variational form to be used here naturally falls under the weighted residual method. Over an element, we have

$$0 = \int_{x_e}^{x_{e+1}} v_e \left(a_e \frac{du_e}{dx} + c_e u_e - f_e \right) dx \tag{7-51}$$

Various choices of v_e dictate various special methods (see Chap. 6 for additional details). Note that there are no "flux" terms in the equation.

The Bubnov–Galerkin model Here we use $u_e = \sum_{j=1}^{n} u_j^e \psi_j^e$ and $v_e = \psi_i^e$ to obtain the finite-element model

$$[K^e]\{u^e\} = \{f^e\} \tag{7-52}$$

where $$K_{ij}^e = \int_{x_e}^{x_{e+1}} \left(a_e \psi_i^e \frac{d\psi_j^e}{dx} + c_e \psi_i^e \psi_j^e \right) dx \qquad f_i^e = \int_{x_e}^{x_{e+1}} f_e \psi_i^e \, dx \tag{7-53}$$

Collocation model In this case we have $v_e = \delta(x - x_i)$ $(i = 1, 2, \ldots, n)$, the Dirac delta function. The collocation points x_i can be chosen arbitrarily, usually the quadrature points in $\bar{\Omega}^e = [x_e, x_{e+1}]$. We have

$$K_{ij}^e = a_e(x_i) \frac{d\psi_j^e}{dx}(x_i) + c_e(x_i)\psi_j^e(x_i) \qquad f_i^e = f_e(x_i) \tag{7-54}$$

Subdomain model In this model the element domain Ω^e is subdivided into n subdomains. The weight function for the ith subdomain Ω_i^e is unity over the subdomain and zero outside the subdomain

$$v_e = w_i^e(x) = \begin{cases} 1 & x \in \Omega_i^e \\ 0 & x \notin \Omega_i^e \end{cases}$$

We obtain

$$K_{ij}^e = \int_{\Omega_i^e} \left(a_e \frac{d\psi_j^e}{dx} + c_e \psi_j^e \right) dx \qquad f_i^e = \int_{\Omega_i^e} f_e \, dx \tag{7-55}$$

Least-squares model Recall from Chap. 6 that the weight function in this case is $v_e = a_e\,(d\psi_i^e/dx) + c_e\,\psi_i^e$. Consequently, we have

$$K_{ij}^e = \int_{x_e}^{x_{e+1}} \left(a_e \frac{d\psi_i^e}{dx} + c_e\,\psi_i^e\right)\left(a_e \frac{d\psi_j^e}{dx} + c_e\,\psi_j^e\right) dx$$
$$f_i^e = \int_{x_e}^{x_{e+1}} \left(a_e \frac{d\psi_i^e}{dx} + c_e\,\psi_i^e\right) f_e\, dx \qquad (7\text{-}56)$$

Note that only the least-squares method gives a symmetric coefficient matrix.

Example 7-7 Let us solve Eq. (7-50) for the data $a = 1$, $c = 2$, $f = 1$ and $L = 1$, and initial condition $u(0) = 1$. We shall use two linear elements in the domain.

For the Bubnov–Galerkin model we have

$$[K^e] = \frac{a_e}{2}\begin{bmatrix} -1 & 1 \\ -1 & 1 \end{bmatrix} + \frac{c_e h_e}{6}\begin{bmatrix} 2 & 1 \\ 1 & 2 \end{bmatrix} \qquad \{f^e\} = \frac{h_e}{2}\begin{Bmatrix} 1 \\ 1 \end{Bmatrix}$$

The assembled equations are given by

$$\left(\frac{1}{2}\begin{bmatrix} -1 & 1 & 0 \\ -1 & 0 & 1 \\ 0 & -1 & 1 \end{bmatrix} + \frac{1}{6}\begin{bmatrix} 2 & 1 & 0 \\ 1 & 4 & 1 \\ 0 & 1 & 2 \end{bmatrix}\right)\begin{Bmatrix} U_1 \\ U_2 \\ U_3 \end{Bmatrix} = \frac{1}{4}\begin{Bmatrix} 1 \\ 2 \\ 1 \end{Bmatrix}$$

Using the initial condition $U_1 = 1$, we obtain from the last two equations

$$\frac{1}{2}\begin{bmatrix} \frac{4}{3} & \frac{4}{3} \\ -\frac{2}{3} & \frac{5}{3} \end{bmatrix}\begin{Bmatrix} U_2 \\ U_3 \end{Bmatrix} = \frac{1}{6}\begin{Bmatrix} 5 \\ 1.5 \end{Bmatrix} \qquad \text{and } U_2 = 0.6786 \qquad U_3 = 0.5714$$

The exact solution is given by

$$u(x) = u(0)e^{-(c/a)x} - \frac{f}{c}\left(1 - e^{-(c/a)x}\right)$$

The exact values of u at $x = \frac{1}{2}$ and 1 are given by 0.6839 and 0.5677, respectively.

For the collocation model, we choose $x_1 = h_e/3$ and $x_2 = 2h_e/3$ and obtain

$$[K^e] = \frac{a_e}{h_e}\begin{bmatrix} -1 & 1 \\ -1 & 1 \end{bmatrix} + \frac{c_e}{3}\begin{bmatrix} 2 & 1 \\ 1 & 2 \end{bmatrix} \qquad \{f^e\} = \begin{Bmatrix} 1 \\ 1 \end{Bmatrix}$$

Clearly, the element equations obtained using the collocation points $x_1 = h_e/3$ and $x_2 = 2h_e/3$ are the same as those obtained in the Bubnov–Galerkin method (actually a multiple of $h_e/2$). Hence we will get the same solution as given above.

In the case of the subdomain model, we use $\Omega_1^e = (0, h_e/2)$, $\Omega_2^e = (h_e/2, h_e)$. For this choice, we obtain

$$[K^e] = \frac{a_e}{2}\begin{bmatrix} -1 & 1 \\ -1 & 1 \end{bmatrix} + \frac{c_e h_e}{8}\begin{bmatrix} 3 & 1 \\ 1 & 3 \end{bmatrix} \qquad \{f^e\} = \frac{3}{8} h_e \begin{Bmatrix} 1 \\ 1 \end{Bmatrix}$$

The assembled equations become

$$\frac{1}{8}\begin{bmatrix} -1 & 5 & 0 \\ -3 & 6 & 5 \\ 0 & -3 & 7 \end{bmatrix}\begin{Bmatrix} U_1 \\ U_2 \\ U_3 \end{Bmatrix} = \frac{3}{16}\begin{Bmatrix} 1 \\ 2 \\ 1 \end{Bmatrix}$$

The solution, from the last two equations, is given by $U_2 = 0.6053$ and $U_3 = 0.4737$.

Lastly, consider the least squares model. We have

$$[K^e] = \frac{a_e^2}{h_e}\begin{bmatrix} 1 & -1 \\ -1 & 1 \end{bmatrix} + \frac{c_e^2 h_e}{6}\begin{bmatrix} 2 & 1 \\ 1 & 2 \end{bmatrix} + \frac{a_e c_e}{2}\begin{bmatrix} -2 & 0 \\ 0 & 2 \end{bmatrix}$$

$$\{f^e\} = a^e \begin{Bmatrix} -1 \\ 1 \end{Bmatrix} + \frac{c_e h_e}{2}\begin{Bmatrix} 1 \\ 1 \end{Bmatrix}$$

$$\frac{1}{3}\begin{bmatrix} -2.5 & -0.5 & 0.0 \\ -0.5 & 7.0 & -0.5 \\ 0.0 & -0.5 & 9.5 \end{bmatrix}\begin{Bmatrix} U_1 \\ U_2 \\ U_3 \end{Bmatrix} = \begin{Bmatrix} -0.5 \\ 1.0 \\ 1.5 \end{Bmatrix}$$

The solution of the last two equations is given by

$$U_2 = 0.6793 \qquad \text{and} \qquad U_3 = 0.5094$$

This completes the example.

Second-order equations Use of weighted-residual finite-element models in the solution of second-order equations is a bit more involved. We describe the finite-element models for the model equation in Eq. (7-6). For the weak formulation of this equation the admissible functions are in $H^1(0, L)$. Therefore, piecewise-continuous functions are adequate for the Ritz finite-element model; i.e., the Ritz finite-element model requires the interpolation functions to be continuous across an interface between elements, or equivalently, the basis functions $\{\psi_i^e\}$ span a subspace of $C^0(0, L)$.

In weighted-residual finite-element models, we must require that the approximating functions be in $\mathscr{D}_A \subset H^2(0, L)$. In other words, the basis functions must span a subspace of $C^1(0, L)$. This requirement is the result of not including the natural boundary conditions in the variational equation. For the second-order equation under consideration, the natural boundary condition involves specifying $a(du/dx)$ at the element boundaries. Therefore, the interpolation functions must be selected such that $a(du/dx)$ is continuous across an interface between elements. This in turn implies, if a is continuous, that du/dx is continuous throughout the domain $\Omega = (0, L)$. Hence a C^1-approximation of u is required.

Because u and du/dx are required to be continuous across an interface between elements and a typical element (in one dimension) has two such interfaces, the polynomial approximation of u must involve four parameters, i.e., a cubic polynomial. Thus, the finite element is a line element with two nodes and two degrees of freedom, u and du/dx, at each node. The element is different from the Lagrange cubic element, which has four nodes with one degree of freedom per node. The Lagrange interpolation functions (of any order) do not satisfy the continuity of u' across element interfaces and therefore do not belong to $C^1(0, L)$. The two-node element, with continuous u and du/dx at element interfaces, is called a *Hermite cubic element* and the associated interpolation functions are called *Hermite cubics*. We now set out to derive them.

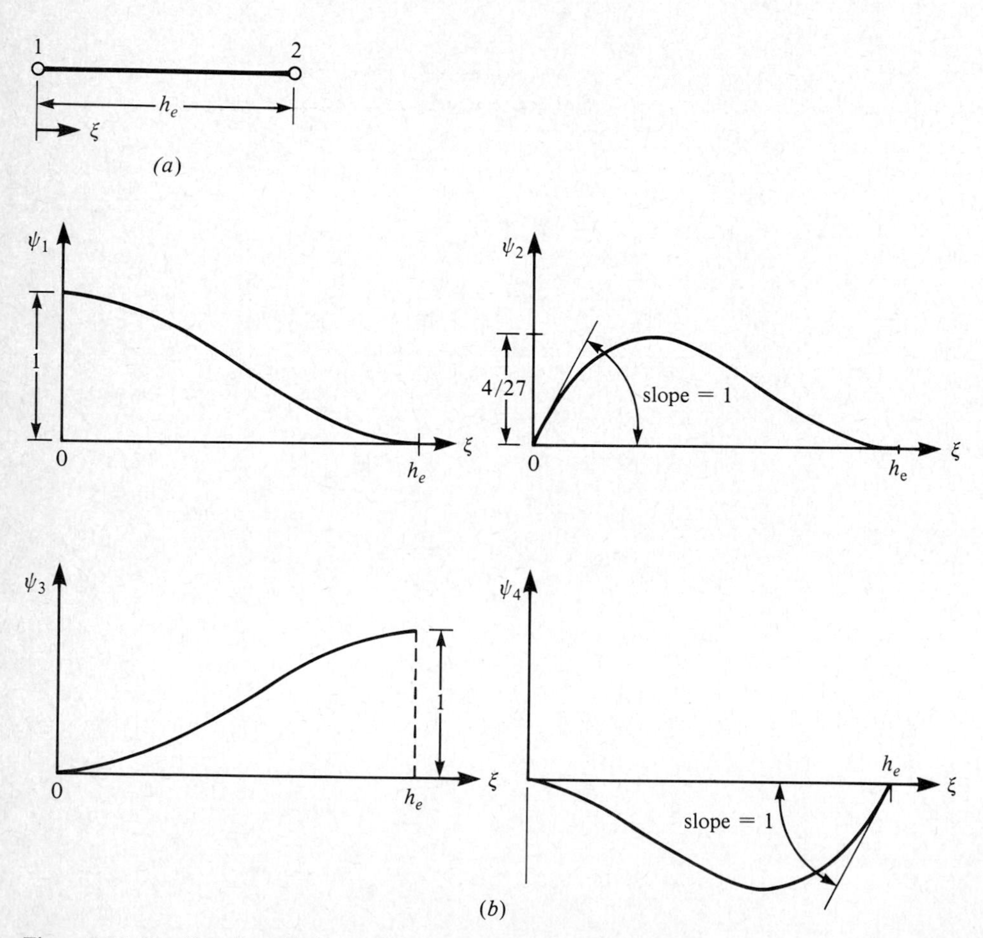

Figure 7-15 The Hermite cubic interpolation functions in one dimension.

Consider the two-node element shown in Fig. 7-15*a*. Let u_e be of the form

$$u_e = \alpha_0 + \alpha_1 \xi + \alpha_2 \xi^2 + \alpha_3 \xi^3 \tag{7-57}$$

where ξ is the local coordinate. The parameters α_i are now evaluated in terms of the nodal values of u_e as well as $du_e/d\xi$. We have

$$\begin{aligned} u_e(0) &\equiv u_1^e = \alpha_0 \qquad \frac{du_e}{d\xi}(0) \equiv u_2^e = \alpha_1 \\ u_e(h_e) &\equiv u_3^e = \alpha_0 + \alpha_1 h_e + \alpha_2 h_e^2 + \alpha_3 h_e^3 \\ \frac{du_e}{d\xi}(h_e) &\equiv u_4^e = \alpha_1 + 2\alpha_2 h_e + 3\alpha_3 h_e^2 \end{aligned} \tag{7-58}$$

where h_e is the length of the element. Solving these equations for α_0, α_1, α_2, and α_3 in terms of u_1^e, u_2^e, u_3^e, and u_4^e, and substituting the result into Eq. (7-57), we obtain

$$u_e(\xi) = \sum_{j=1}^{4} u_j^e \psi_j^e(\xi) \tag{7-59}$$

where u_1^e and u_3^e are the nodal values of u_e, u_2^e, and u_4^e are the nodal values of $du_e/d\xi$, and ψ_j^e are the Hermite cubics,

$$\begin{aligned} \psi_1^e(\xi) &= 1 - 3\left(\frac{\xi}{h_e}\right)^2 + 2\left(\frac{\xi}{h_e}\right)^3 \\ \psi_2^e(\xi) &= \xi\left[1 - \left(\frac{\xi}{h_e}\right)\right]^2 \\ \psi_3^e(\xi) &= 3\left(\frac{\xi}{h_e}\right)^2 - 2\left(\frac{\xi}{h_e}\right)^3 \\ \psi_4^e(\xi) &= \xi\left[\left(\frac{\xi}{h_e}\right)^2 - \left(\frac{\xi}{h_e}\right)\right] \end{aligned} \tag{7-60}$$

These functions are geometrically depicted in Fig. 7-15b.

Next, we consider various weighted-residual finite-element models of Eq. (7-6). The weighted-residual form of Eq. (7-6) over $\Omega^e = (x_e, x_{e+1})$ is

$$0 = \int_{x_e}^{x_{e+1}} v_e\left[-\frac{d}{dx}\left(a\frac{du_e}{dx}\right) - f_e\right] dx$$

Substituting ϕ_i^e for the weight function v_e and Eq. (7-59) for u_e we obtain

$$0 = \sum_{j=1}^{4} K_{ij}^e u_j^e - f_i^e \tag{7-61a}$$

where $$K_{ij}^e = \int_{x_e}^{x_{e+1}} -\phi_i^e \frac{d}{dx}\left(a_e \frac{d\psi_j^e}{dx}\right) dx \qquad f_i^e = \int_{x_e}^{x_{e+1}} \phi_i^e f_e \, dx \tag{7-61b}$$

For different choices of ϕ_i^e we get different finite-element models.

The Bubnov–Galerkin model For $\phi_i^e = \psi_i^e$ Eq. (7-61*b*) becomes

$$K_{ij}^e = \int_{x_e}^{x_{e+1}} -\psi_i^e \frac{d}{dx}\left(a_e \frac{d\psi_j^e}{dx}\right) dx \qquad f_i^e = \int_{x_e}^{x_{e+1}} \psi_i^e f_e \, dx \tag{7-62}$$

Least-squares model For $\phi_i^e = A(\psi_i^e) \equiv -d/dx(a_e(d\psi_i^e/dx))$, we have

$$K_{ij}^e = \int_{x_e}^{x_{e+1}} \frac{d}{dx}\left(a_e \frac{d\psi_i^e}{dx}\right) \frac{d}{dx}\left(a_e \frac{d\psi_j^e}{dx}\right) dx \qquad f_i^e = -\int_{x_e}^{x_{e+1}} \frac{d}{dx}\left(a_e \frac{d\psi_i^e}{dx}\right) f_e \, dx \tag{7-63}$$

Collocation model For $\phi_i^e = \delta(x - x_i)$, Eq. (7-61*b*) takes the form

$$K_{ij}^e = -\left\{\frac{d}{dx}\left[a_e(x) \frac{d\psi_j^e}{dx}(x)\right]\right\}_{x=x_i} \qquad f_i^e = f_e(x_i) \tag{7-64}$$

where x_i are the collocation points (should not be placed at the nodes).

While the Bubnov–Galerkin and least-squares models have the same form as the Ritz model, the collocation model does not. In the collocation model the coefficient matrices and column vector are simply evaluated at the collocation points (instead of integrating the expressions). The number of collocation points should be equal to the number of unknowns, after imposing the boundary conditions, of the problem. For second-order equations, we have two boundary conditions and $2(N + 1)$ number of nodal degrees of freedom for an N-element model. Hence, a total of $2N$ collocation points, two per element, are needed. Also, note that the coefficient matrix in Eq. (7-64) is of order 2×4 ($i = 1, 2$) and there is no overlap of element matrices because there is no summation of equations over the number of elements [see Eq. (7-31)]. However the continuity conditions [of the type in Eq. (7-26)] are imposed.

Example 7-8 Consider the mixed boundary-value problem,

$$\begin{gathered} -\frac{d}{dx}\left[(1 + x)\frac{du}{dx}\right] = 0 \qquad 0 < x < 1 \\ u(0) = 0 \qquad \left[(1 + x)\frac{du}{dx}\right]_{x=1} = 1 \end{gathered} \tag{7-65}$$

The exact solution of this problem is $u = \log_e(1 + x)$.

Consider a two-element discretization of the problem. There are three nodes and six degrees of freedom. The known degrees of freedom for the weighted-residual models are

$$U_1 = 0 \qquad U_6 = 0.5$$

whereas in the four-element Ritz model they are

$$U_1 = 0 \qquad U_5 = 1.00$$

Table 7-5 A comparison of weighted-residual finite-element solutions† with the exact solution of the problem in Eq. (7-65)

x		Exact	Collocation	Bubnov–Galerkin	Least squares	Ritz
0.00	u	0.00000	0.00000	0.00000	0.00000	0.00000
	u'	1.00000	0.99390	0.99604	0.99902	0.99612
			0.99967	0.99930	0.99993	0.99930
0.25	u	0.22314	—	—	—	—
			0.22326	0.22313	0.22315	0.22313
	u'	0.80000	—	—	—	—
			0.80028	0.80004	0.79997	0.80004
0.50	u	0.40547	0.40487	0.40537	0.40554	0.40538
			0.40562	0.40546	0.40547	0.40546
	u'	0.66667	0.66299	0.66707	0.66645	0.66728
			0.66646	0.66673	0.66665	0.66674
0.75	u	0.55962	—	—	—	—
			0.55975	0.55961	0.55962	0.55961
	u'	0.57143	—	—	—	—
			0.57123	0.57146	0.57142	0.57148
1.00	u	0.69315	0.69202	0.69309	0.69324	0.69315
			0.69325	0.69314	0.69315	0.69315
	u'	0.50000	0.50000	0.50000	0.50000	0.50102
			0.50000	0.50000	0.50000	0.50010

† The first line corresponds to two elements and the second line corresponds to four elements.

For the collocation model, the two-point Gauss quadrature points are used as the collocation points. The two-element and four-element finite-element solutions obtained by various finite-element models, all using the Hermite cubics, are compared with the exact solution in Table 7-5.

7-3 ONE-DIMENSIONAL FOURTH-ORDER EQUATIONS

7-3-1 Model Problem

Here we consider the finite-element formulation of the one-dimensional fourth-order differential equation that arises in the elastic bending of beams (see Example 4-6). The finite-element formulation of the fourth-order equation involves the same steps as described in Sec. 7-2 for second-order equations, but the mathematical details are somewhat different, especially in the finite-element formulation of the equation.

Consider the fourth-order differential equation

$$\frac{d^2}{dx^2}\left(b\,\frac{d^2w}{dx^2}\right) = f \qquad 0 < x < L \tag{7-66}$$

where $b = b(x)$ and $f = f(x)$ are given functions of x (i.e., data), and w is the dependent variable. In the case of bending of beams, b denotes the product of modulus of elasticity E and moment of inertia I of the beam, f denotes the distributed transverse load (positive upward), and w is the transverse deflection (positive upward) of the beam (see Fig. 7-16*a*). In addition to satisfying the differential equation (7-66), w must also satisfy appropriate boundary conditions: since the equation is fourth-order, four boundary conditions are needed to solve the equation. We now give a step-by-step procedure for the finite-element analysis of Eq. (7-66).

7-3-2 The Ritz Finite-Element Model

1 Discretization of the domain The domain is divided into a set (say N) of line elements, each element having at least the two end nodes. Although the element, geometrically, is the same as that used in the second-order equations, the number and form of the primary unknowns at each node are dictated by the variational formulation of the differential equation (7-66). Due to the physical background of the equation (7-66) the element is referred to as the *beam element*.

2 Derivation of element equations In this step we isolate a typical element, $\Omega^e = (x_e, x_{e+1})$ (see Fig. 7-16*b*) and construct the variational form of Eq. (7-66) over the element. The variational formulation will allow us to identify the primary and secondary variables of the problem. Then suitable approximations for the primary variables are developed and the element equations are derived.

(i) *Variational formulation.* Following the procedure used earlier, we write

$$\begin{aligned} 0 &= \int_{x_e}^{x_{e+1}} v\left[\frac{d^2}{dx^2}\left(b\,\frac{d^2w}{dx^2}\right) - f\right]dx \\ &= \int_{x_e}^{x_{e+1}} \left\{-\frac{dv}{dx}\frac{d}{dx}\left[b\,\frac{d^2w}{dx^2}\right] - vf\right\}dx + \left\{v\,\frac{d}{dx}\left[b\,\frac{d^2w}{dx^2}\right]\right\}_{x=x_e}^{x=x_{e+1}} \\ &= \int_{x_e}^{x_{e+1}} \left[b\,\frac{d^2v}{dx^2}\frac{d^2w}{dx^2} - vf\right]dx + \left\{v\,\frac{d}{dx}\left[b\,\frac{d^2w}{dx^2}\right] - \frac{dv}{dx}\,b\,\frac{d^2w}{dx^2}\right\}_{x=x_e}^{x=x_{e+1}} \end{aligned} \tag{7-67}$$

where $v \in H^2(\Omega^e)$ is a test function. An examination of the boundary terms indicates that the essential boundary conditions involve the specification of w and dw/dx, and the natural boundary conditions involve the specification of $b(d^2w/dx^2)$ and $d/dx(b(d^2w/dx^2))$ at the end points of the element (see Fig. 7-16*b*). Thus there are two essential boundary conditions and two natural boundary conditions. Therefore, we must identify w and dw/dx as the primary variables at each node so that the essential boundary conditions are included in the interpolation.

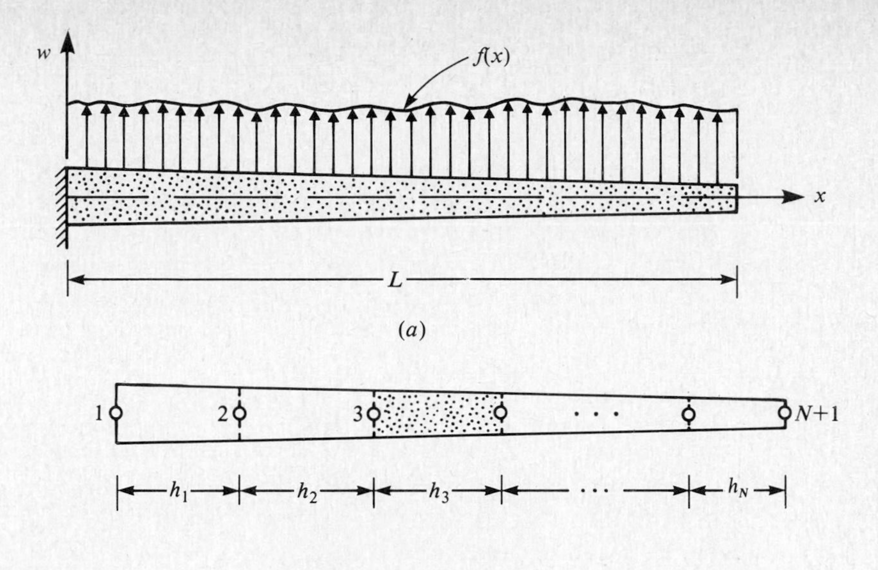

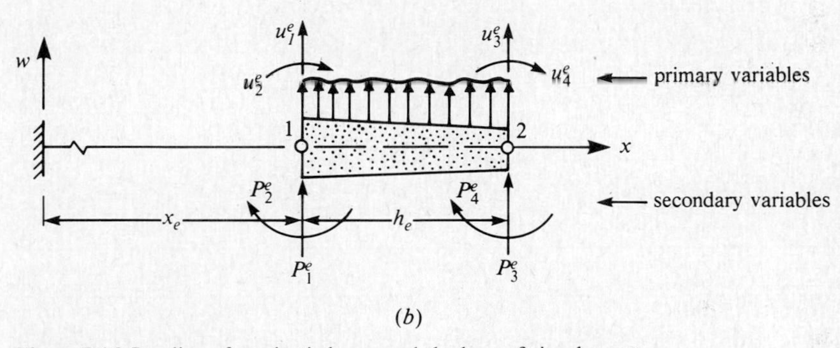

Figure 7-16 Bending of an elastic beam and the beam finite element.

The natural boundary conditions always remain in the variational form and end up on the right-hand side (i.e., source vector) of the matrix equation. Thus we have

Essential:

$$w(x_e) \equiv u_1^e \qquad \frac{dw}{dx}(x_e) \equiv -u_2^e \qquad w(x_{e+1}) \equiv u_3^e \qquad \frac{dw}{dx}(x_{e+1}) \equiv -u_4^e \tag{7-68a}$$

Natural:

$$\begin{aligned} \left[\frac{d}{dx}\left(b\,\frac{d^2w}{dx^2}\right)\right]_{x=x_e} &\equiv P_1^e \qquad \left(b\,\frac{d^2w}{dx^2}\right)\bigg|_{x=x_e} \equiv P_2^e \\ \left[\frac{d}{dx}\left(b\,\frac{d^2w}{dx^2}\right)\right]_{x=x_{e+1}} &\equiv -P_3^e \qquad \left(b\,\frac{d^2w}{dx^2}\right)\bigg|_{x=x_{e+1}} \equiv -P_4^e \end{aligned} \tag{7-68b}$$

With this notation, the variational form (7-67) becomes

$$0 = \int_{x_e}^{x_{e+1}} \left[b \frac{d^2v}{dx^2}\frac{d^2w}{dx^2} - vf \right] dx - v(x_e)P_1^e - \left[-\frac{dv}{dx}(x_e) \right] P_2^e$$

$$- v(x_{e+1})P_3^e - \left[-\frac{dv}{dx}(x_{e+1}) \right] P_4^e$$

$$\equiv B(v, w) - l(v) \tag{7-69a}$$

where the bilinear form and linear forms are given by

$$B(v, w) = \int_{x_e}^{x_{e+1}} b \frac{d^2v}{dx^2}\frac{d^2w}{dx^2}\, dx$$

$$l(v) = \int_{x_e}^{x_{e+1}} fv\, dx + v(x_e)P_1^e + \left(-\frac{dv}{dx} \right)\bigg|_{x_e} P_2^e + v(x_{e+1})P_3^e + \left(-\frac{dv}{dx} \right)\bigg|_{x_{e+1}} P_4^e \tag{7-69b}$$

(*ii*) *Interpolation functions.* To derive the interpolation functions for the finite-element approximation of the function w_e, we must select a polynomial that is continuous and complete up to the degree required. It is clear from the variational problem (7-69) that the function we select should be an element from $H^2(\Omega^e)$. In order to satisfy all four essential boundary conditions in Eq. (7-68*a*), we need to use a cubic polynomial,

$$w_e = \alpha_0 + \alpha_1 \xi + \alpha_2 \xi^2 + \alpha_3 \xi^3$$

where ξ denotes the local coordinate. Next, select the parameters α_0, α_1, α_2, and α_3 such that w_e satisfies the essential boundary conditions Eq. (7-68*a*). This leads to the Hermite cubic interpolation functions of Eq. (7-60), with $\psi_2^e \to -\psi_2^e$ and $\psi_4^e \to -\psi_4^e$.

We note the following properties of ψ_i^e (see Fig. 7-15*b*):

$$\psi_{2i-1}^e(\xi_j) = \delta_{ij} \qquad \psi_{2i}^e(\xi_j) = 0 \qquad \sum_{i=1}^{2} \psi_{2i-1}^e = 1$$

$$\frac{d\psi_{2i-1}^e}{d\xi}(\xi_j) = 0 \qquad -\frac{d\psi_{2i}^e}{d\xi}(\xi_j) = \delta_{ij} \tag{7-70}$$

where $\xi_1 = 0$ and $\xi_2 = h_e$ are the local coordinates of nodes 1 and 2 of element $\Omega^e = (x_e, x_{e+1})$, and x_e is the global coordinate of global node e. The Lagrange cubic element is not admissible here because it does not contain the derivatives as the nodal values (hence we cannot impose the continuity of the derivatives).

Substituting Eq. (7-59) into Eq. (7-69*a*), we obtain

$$\sum_{j=1}^{4} K_{ij}^e u_j^e = F_i^e \tag{7-71}$$

where $[K^e]$ is the stiffness matrix and $\{F^e\}$ is the force vector

$$K_{ij}^e = \int_{x_e}^{x_{e+1}} b_e \frac{d^2\psi_i^e}{dx^2}\frac{d^2\psi_j^e}{dx^2}\,dx \qquad F_i^e = \int_{x_e}^{x_{e+1}} f_e \psi_i^e\,dx + P_i^e \tag{7-72a}$$

In Eq. (7-72*a*), it is understood that ψ_i^e are expressed in terms of the global coordinate x. In the element coordinate ξ, Eq. (7-72*a*) takes the simple form

$$K_{ij}^e = \int_0^{h_e} \bar{b}_e \frac{d^2\psi_i^e}{d\xi^2}\frac{d^2\psi_j^e}{d\xi^2}\,d\xi \qquad F_i^e = \int_0^{h_e} \bar{f}_e \psi_i^e\,d\xi + P_i^e \tag{7-72b}$$

where $\bar{b}_e$ and $\bar{f}_e$ are the transformed functions

$$\bar{b}_e = b_e(\xi) \qquad \bar{f}_e = f_e(\xi)$$

For the case in which b_e and f_e are constant over the element Ω^e, the stiffness matrix $[K^e]$ and force vector $\{F^e\}$ are given by

$$[K^e] = \frac{2b_e}{h_e^3}\begin{bmatrix} 6 & -3h_e & -6 & -3h_e \\ -3h_e & 2h_e^2 & 3h_e & h_e^2 \\ -6 & 3h_e & 6 & 3h_e \\ -3h_e & h_e^2 & 3h_e & 2h_e^2 \end{bmatrix}, \quad \{F^e\} = \frac{f_e h_e}{12}\begin{Bmatrix} 6 \\ -h_e \\ 6 \\ h_e \end{Bmatrix} + \begin{Bmatrix} P_1^e \\ P_2^e \\ P_3^e \\ P_4^e \end{Bmatrix} \tag{7-73}$$

One can show that the first column of the force vector represents the "statically equivalent" forces and moments at the nodes due to uniformly distributed load f_e over the elements.

3 Assembly of the element equations The assembly procedure for the beam element is analogous to that described for the element for a second-order equation. The difference is that in the bar element there are two degrees of freedom per node. Consequently, at every global node that is shared by two elements the elements in the last two rows and columns of the Ith element overlap with the elements in the first two rows and columns of the $(I + 1)$th element. For a two element case the assembled stiffness matrix and force column are shown below:

$$[K^e] = \begin{bmatrix} K_{11}^1 & K_{12}^1 & K_{13}^1 & K_{14}^1 & & \\ K_{21}^1 & K_{22}^1 & K_{23}^1 & K_{24}^1 & & \\ K_{31}^1 & K_{32}^1 & K_{33}^1 + K_{11}^2 & K_{34}^1 + K_{12}^2 & K_{13}^2 & K_{14}^2 \\ K_{41}^1 & K_{42}^1 & K_{43}^1 + K_{21}^2 & K_{44}^1 + K_{22}^2 & K_{23}^2 & K_{24}^2 \\ & & K_{31}^2 & K_{32}^2 & K_{33}^2 & K_{34}^2 \\ & & K_{41}^2 & K_{42}^2 & K_{43}^2 & K_{44}^2 \end{bmatrix}$$

$$\{F\} = \begin{Bmatrix} f_1^1 \\ f_2^1 \\ f_3^1 + f_1^2 \\ f_4^1 + f_2^2 \\ f_3^2 \\ f_4^2 \end{Bmatrix} + \begin{Bmatrix} P_1^1 \\ P_2^1 \\ P_3^1 + P_1^2 \\ P_4^1 + P_2^2 \\ P_3^2 \\ P_4^2 \end{Bmatrix} \tag{7-74}$$

4 Imposition of boundary conditions The boundary conditions on the displacements and forces can be imposed in the manner described earlier. The boundary (or equilibrium) conditions on forces are imposed by modifying the second column of the assembled force [see Eq. (7-74)]. If, for example, the force at the Ith global node is specified to be F_0, and the moment at the Kth global node is specified to be M_0, then the following assembled coefficients get modified:

$$P_3^{I-1} + P_1^I = F_0 \qquad P_4^{K-1} + P_2^K = M_0$$

If no force or moment is specified at a nodal point that is unconstrained, it is understood that the force and moment are specified to be zero there.

The solution and postcomputation of the displacements and their derivatives at various points of the beam follow along the same lines as described in Sec. 7-2 for second-order problems. We now consider two examples to illustrate the steps involved in the finite-element analysis of beams.

7-3-3 Applications

The first example is concerned with the simply supported beam of Example 7-5.

Example 7-9 Consider the simply supported beam shown in Fig. 7-14. The minimum number of elements we must use is two. The element matrices are given by Eq. (7-73) with $h_1 = 120$ in., $h_2 = 240$ in., $b_1 = b_2 = 144 \times 24 \times 10^6$ lb·in.^{2}, and $f_1 = f_2 = 0$. The assembled stiffness matrix and force vectors have the same form as those in Eq. (7-74).

The boundary conditions are

$$w(0) = 0 \rightarrow U_1 = 0 \qquad w(360) = 0 \rightarrow U_5 = 0$$

$$P_3^1 + P_1^2 = -P \qquad P_4^1 + P_2^2 = 0 \text{ (zero bending moment)}$$

Solving the resulting four equations, we obtain

$$U_2 = 2.3148 \times 10^{-6} P \text{ (rad)} \qquad U_3 = -0.2222 \times 10^{-3} P \text{ (in)}$$

$$U_4 = 0.9259 \times 10^{-6} P \text{ (rad)} \qquad U_6 = -1.8519 \times 10^{-6} P \text{ (rad)}$$

Note that the rotations, dw/dx, obtained using the beam element are more accurate than those computed using the Lagrange linear elements in Example 7-5. For this problem, the element solutions for w and dw/dx coincide with the exact solutions at all points of the domain.

The next example deals with an indeterminate beam subjected to a linearly varying transverse load.

Example 7-10 Consider a beam of length $L = 180$ in, moment of inertia $I = 723$ in^4, modulus of elasticity, $E = 29 \times 10^6$ psi, and subjected to the linearly varying force, $f = -q_0 x/L$. The beam is assumed to be simply supported at the left end [$w(0) = 0$] and clamped at the right end

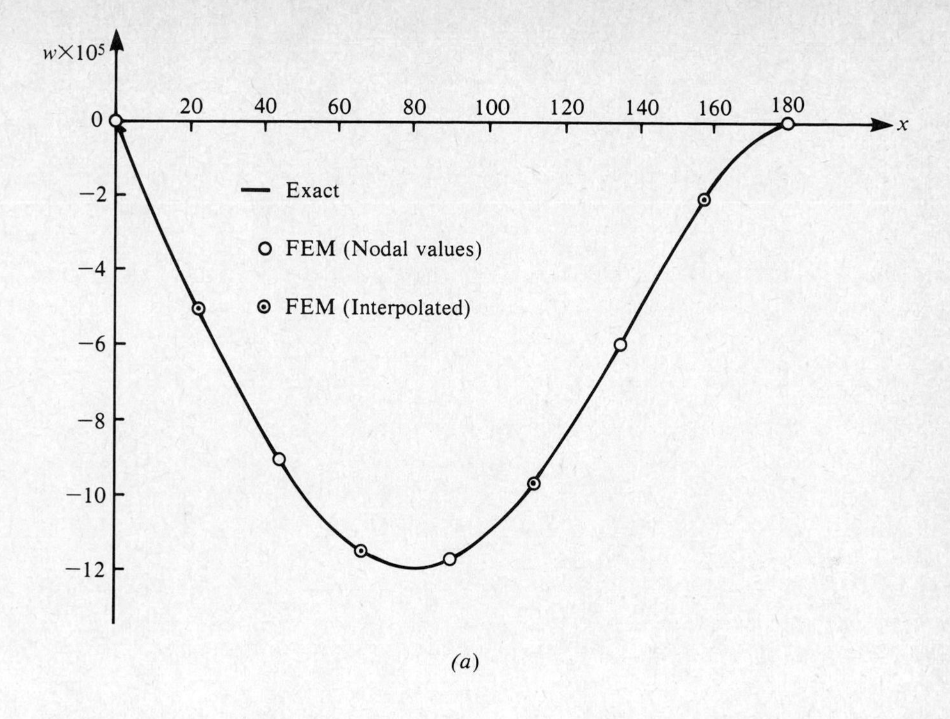

(a)

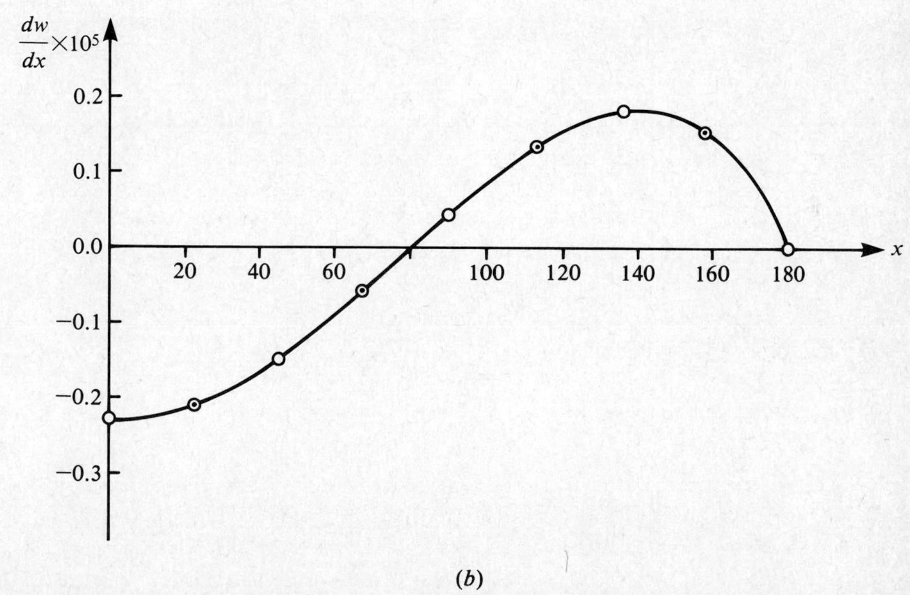

(b)

Figure 7-17 A comparison of the finite-element solution with the exact solution of the beam problem discussed in Example 7-10.

$[w(L) = dw/dx(L) = 0]$. We will consider a two-element and a four-element uniform mesh to determine the displacements.

Since $EI = 20.967 \times 10^6$ ksi is a constant, the element matrices are given by Eq. (7-73). The force vector should be calculated using Eq. (7-72*a*),

$$f_i^e = -\int_{x_e}^{x_{e+1}} \left(q_0 \frac{x}{L}\right) \psi_i^e \, dx \qquad i = 1, 2, 3, 4$$

For the two element case ($h_1 = h_2 = 90$ in), the force vector for element 1 is given by

$$\{f^1\} = \{-6.75, 135, -15.75, -202.5\}^T q_0$$

The boundary conditions are

$$U_1 = 0 \qquad U_5 = U_6 = 0 \qquad P_2^1 = 0 \qquad P_3^1 + P_1^2 = 0 \qquad P_4^1 + P_2^2 = 0$$

The solution of the resulting three equations for U_2, U_3, and U_4 coincides with the exact solution at the nodes. The two-element solution for w is not exact at other points other than the nodes. This is because the exact solution for w is a fifth-order polynomial in x.

A comparison of the finite-element solutions obtained using four elements with the exact solution is presented in Fig. 7-17. Clearly, the four-element solution is in excellent agreement with the exact solution.

7-4 TIME-DEPENDENT PROBLEMS IN ONE DIMENSION

7-4-1 Introduction

Like in the solution of time-dependent problems by traditional variational methods (see Sec. 6-5), in the finite-element solution of time-dependent problems we assume that the nodal variables are functions of time. This implies that we considered all time-dependent problems whose solutions can be represented as a product of a time function and a spatial function: $u(x, t) = T(t)U(x)$. Here we deal with the finite-element analysis of such problems.

Consider the model equation

$$c_1 \frac{\partial u}{\partial t} + c_2 \frac{\partial^2 u}{\partial t^2} - \frac{\partial}{\partial x}\left(a \frac{\partial u}{\partial x}\right) + \frac{\partial^2}{\partial x^2}\left(b \frac{\partial^2 u}{\partial x^2}\right) = f \qquad 0 < x < L \qquad (7\text{-}75)$$

which contains as special cases the second-order (for $c_2 = b = 0$ or $c_1 = b = 0$) and fourth-order (for $c_1 = a = 0$) time-dependent problems. The second-order problems, for example, involve finding the transverse motion of a cable ($c_1 = b = 0$), longitudinal motion of a rod ($c_1 = b = 0$), and temperature transients in a fin ($c_2 = b = 0$). The fourth-order problem involves finding the transverse motion of a beam ($a = c_1 = 0$). Following this introduction, we discuss the semidiscrete finite-element formulation of (7-75) and the fully discrete formulation of the form discussed in Sec. 6-5.

7-4-2 Semidiscrete Finite-Element Models

The semidiscrete formulation involves the approximation of the spatial variation of the dependent variable, which follows the same steps as described in the previous sections with f replaced by $f - c_1(\partial u/\partial t) - c_2(\partial^2 u/\partial t^2)$. For the equatio in (7-75), the typical element is either the Lagrange element ($b = 0$) or the Hermite cubic element of Sec. 7-2.

We assume that u_e over an element is interpolated by an expression of the form

$$u = \sum_{j=1}^{r} u_j^e(t)\psi_j^e(x) \tag{7-76}$$

where ψ_j^e are the Lagrange interpolation functions if $b = 0$ or the Hermite interpolation functions when $b \neq 0$. The approximation (7-76) implies that at any arbitrarily fixed time $t > 0$, the function u_e can be approximated by a linear combination of ψ_j^e with u_j^e, u_j being the value of u_e at time t at the jth node of the element; in other words, the time and spatial variations of u_e are separable. Obviously, one cannot use such approximations for wave propagation problems in which the time and spatial variations of u_e cannot be separated. Substituting for $v = \psi_i^e$ and Eq. (7-76) for u_e into the variational formulation of Eq. (7-75), we obtain

$$0 = \int_{x_e}^{x_{e+1}} \left[c_1^e \psi_i^e \sum_{j=1}^{r} \frac{du_j^e}{dt} \psi_j^e + c_2^e \psi_i^e \sum_{j=1}^{r} \frac{d^2 u_j^e}{dt^2} \psi_j^e + a_e \frac{d\psi_i^e}{dx} \sum_{j=1}^{r} u_j^e \frac{d\psi_j^e}{dx} \right.$$
$$\left. + b_e \frac{d^2\psi_i^e}{dx^2} \sum_{j=1}^{r} u_j^e \frac{d^2\psi_j^e}{dx^2} - \psi_i^e f_e \right] dx - P_i^e$$

or

$$[M^{1e}]\{\dot{u}\} + [M^{2e}]\{\ddot{u}^e\} + ([K^{1e}] + [K^{2e}])\{u^e\} = \{F^e\} \tag{7-77}$$

where

$$\dot{u}_i^e = \frac{du_i^e}{dt} \qquad \ddot{u}_i^e = \frac{d^2 u_i^e}{dt^2}$$

$$M_{ij}^{1e} = \int_{x_e}^{x_{e+1}} c_1^e \psi_i^e \psi_j^e \, dx \qquad M_{ij}^{2e} = \int_{x_e}^{x_{e+1}} c_2^e \psi_i^e \psi_j^e \, dx$$

$$K_{ij}^{1e} = \int_{x_e}^{x_{e+1}} a_e \frac{d\psi_i^e}{dx} \frac{d\psi_j^e}{dx} \, dx \qquad K_{ij}^{2e} = \int_{x_e}^{x_{e+1}} b_e \frac{d^2\psi_i^e}{dx^2} \frac{d^2\psi_j^e}{dx^2} \, dx \tag{7-78}$$

$$F_i^e = \int_{x_e}^{x_{e+1}} f_e \psi_i^e \, dx + P_i^e$$

This completes the semidiscrete finite-element formulation of (7-75) over an element.

7-4-3 Time Approximations

As special cases, Eq. (7-77) contains equations of the form in (6-205) and (6-206). Therefore, the time approximation of (7-77) for the two cases follows along the lines indicated in Secs. 6-5-2 and 6-5-3.

Parabolic equations From Eq. (7-77) we have for $c_2^e = 0$

$$[M^{1e}]\{\dot{u}^e\} + [K^e]\{u^e\} = \{F\}$$

where $[K^e] = [K^{1e}] + [K^{2e}]$. We have from Eq. (6-217)

$$[\hat{K}^{1e}]\{u^e\}^{n+1} = \{\hat{F}^{1e}\} \tag{7-79a}$$

where

$$[\hat{K}^{1e}] = [M^{1e}] + \theta\, \Delta t[K^e]$$

$$\{\hat{F}^{1e}\} = ([M^{1e}] - (1-\theta)\, \Delta t[K^e])\{u^e\}^n + \Delta t(\theta\{F^e\}^{n+1} + (1-\theta)\{F^e\}^n) \tag{7-79b}$$

and Δt is the time step. Equation (7-79*a*) is valid for a typical element. The assembly, imposition of boundary conditions, and solution of the assembled equations are the same as described before.

Hyperbolic equations When $c_1^e = 0$ equation (7-77) takes the form

$$[M^{2e}]\{\ddot{u}^e\} + [K^e]\{u^e\} = \{F^e\}$$

Here we use the Newmark direct integration method and obtain [see Eq. (6-240)]

$$[\hat{K}^{2e}]\{u^e\}^{n+1} = \{\hat{F}^{2e}\} \tag{7-80a}$$

where

$$[\hat{K}^{2e}] = [K^e] + a_0[M^{2e}]$$

$$\{\hat{F}^{2e}\} = \{F^e\}^{n+1} + [M^{2e}](a_0\{u^e\}^n + a_1\{\dot{u}^e\}^n + a_2\{\ddot{u}^e\}^n) \tag{7-80b}$$

and a_0, a_1, etc. are the parameters defined in (6-242). Again (7-80*a*) is valid over a typical element. Note that the calculation of $[\hat{K}^{2e}]$ and $\{\hat{F}^{2e}\}$ requires knowledge of the initial conditions $\{u^e\}^0$, $\{\dot{u}^e\}^0$, and $\{\ddot{u}^e\}^0$. In practice, one does not know $\{\ddot{u}^e\}^0$; it must be calculated from Eq. (7-77):

$$\{\ddot{u}^e\}^0 = [M^{2e}]^{-1}(\{F^e\}^0 - [K^e]\{u^e\}^0) \tag{7-81}$$

We now consider two examples of the time-dependent problems.

Example 7-11 Consider the transient heat conduction problem of Example 6-19. We have $a = 1$, $b = 0$, $c_1 = 1$, $c_2 = 0$, and $f = 0$. For the linear element ($r = 2$), Eq. (7-77) takes the form

$$\frac{h_e}{6}\begin{bmatrix} 2 & 1 \\ 1 & 2 \end{bmatrix}\begin{Bmatrix} \dot{u}_1^e \\ \dot{u}_2^e \end{Bmatrix} + \frac{1}{h_e}\begin{bmatrix} 1 & -1 \\ -1 & 1 \end{bmatrix}\begin{Bmatrix} u_1^e \\ u_2^e \end{Bmatrix} = \begin{Bmatrix} P_1^e \\ P_2^e \end{Bmatrix}$$

where h_e is the length of the element. The following nonuniform meshes of both linear and quadratic elements are used to compute the solution [see Reddy (1984*a*)]:

Mesh	x_1	x_2	x_3	x_4	x_5	x_6	x_7	x_8	x_9
2(1)	0.0	0.2	1.0						
4(2)	0.0	0.2	0.5	0.75	1.0				
6(3)	0.0	0.1	0.2	0.35	0.5	0.75	1.0		
8(4)	0.0	0.1	0.2	0.35	0.5	0.6	0.75	0.9	1.0

The first column indicates the number of linear (quadratic) elements in the mesh. The results for $u(1, t)$ for two different time steps and parameter θ are presented in Table 7-6. The best results, when compared to the analytical solution (6-182), are given when $\theta = 0.5$ and $\Delta t = 0.025$.

Table 7-6 The effect of time step, mesh, degree of approximation, and parameter θ on the accuracy of the solution of Example 7-11

		Analytical	Linear element				Quadratic element			
Case†	t	$u(1, t)$	2	4	6	8	1	2	3	4
	0.2	0.7723	0.8128	0.7614	0.7738	0.7714	0.7679	0.7632	0.7734	0.7732
	0.4	0.4745	0.4620	0.4648	0.4736	0.4725	0.4668	0.4710	0.4746	0.4741
1	0.6	0.2897	0.2626	0.2822	0.2867	0.2873	0.2837	0.2883	0.2873	0.2890
	0.8	0.1769	0.1493	0.1711	0.1735	0.1750	0.1725	0.1761	0.1743	0.1764
	1.0	0.1080	0.0848	0.1037	0.1051	0.1066	0.1048	0.1074	0.1060	0.1076
	0.2	0.7723	0.8129	0.7674	0.7753	0.7716	0.7680	0.7729	0.7710	0.7728
	0.4	0.4745	0.4627	0.4665	0.4727	0.4725	0.4672	0.4739	0.4740	0.4741
2	0.6	0.2897	0.2632	0.2828	0.2870	0.2878	0.2841	0.2892	0.2895	0.2894
	0.8	0.1769	0.1497	0.1715	0.1742	0.1754	0.1728	0.1765	0.1767	0.1767
	1.0	0.1080	0.0851	0.1040	0.1057	0.1068	0.1051	0.1077	0.1079	0.1079
	0.2	0.7723	0.8214	0.7723	0.7805	0.7764	0.7744	0.7768	0.7773	0.7772
	0.4	0.4745	0.4730	0.4747	0.4811	0.4806	0.4754	0.4820	0.4659	0.4824
3	0.6	0.2897	0.2720	0.2904	0.2946	0.2954	0.2916	0.2947	0.2970	0.2971
	0.8	0.1769	0.1564	0.1776	0.1804	0.1815	0.1789	0.1826	0.1828	0.1829
	1.0	0.1080	0.0899	0.1086	0.1104	0.1116	0.1097	0.1124	0.1126	0.1126
	0.2	0.7723	0.8170	0.8797	0.7775	0.7735	0.7712	0.7747	0.7744	0.7744
	0.4	0.4745	0.4782	0.4708	0.4772	0.4767	0.4715	0.4782	0.4785	0.4786
4	0.6	0.2897	0.2677	0.2868	0.2910	0.2918	0.2881	0.2932	0.2935	0.2935
	0.8	0.1769	0.1533	0.1747	0.1775	0.1786	0.1760	0.1797	0.1799	0.1800
	1.0	0.1080	0.0877	0.1064	0.1082	0.1093	0.1075	0.1102	0.1103	0.1104

† Case 1: $\theta = 0.5$, $\Delta t = 0.05$
Case 2: $\theta = 0.5$, $\Delta t = 0.025$
Case 3: $\theta = 0.65$, $\Delta t = 0.05$
Case 4: $\theta = 0.65$, $\Delta t = 0.025$

Table 7-7 Effect of time step and mesh on the accuracy of the center deflection of a clamped beam of Example 7-12

	$\Delta t = 0.005$		$\Delta t = 0.0025$			Galerkin in space and exact in time
t	2†	4	2	4	6	
0.01	0.2098	0.2097	0.2098	0.2097	0.2097	0.2157
0.02	0.1950	0.1951	0.1950	0.1951	0.1951	0.1988
0.03	0.1698	0.1698	0.1695	0.1696	0.1696	0.1716
0.04	0.1347	0.1349	0.1345	0.1348	0.1348	0.1356
0.05	0.0931	0.0935	0.0930	0.0932	0.0933	0.0925
0.06	0.0482	0.0483	0.0480	0.0483	0.0483	0.0447
0.07	0.0014	0.0017	0.0014	0.0016	0.0016	−0.0055
0.08	−0.0460	−0.0455	−0.0464	−0.0458	−0.0458	−0.0553
0.09	−0.0923	−0.0916	−0.0928	−0.0921	−0.0920	−0.1023
0.10	−0.1341	−0.1335	−0.1346	−0.1341	−0.1341	−0.1441
0.11	−0.1684	−0.1682	−0.1685	−0.1682	−0.1681	−0.1783
0.12	−0.1932	−0.1931	−0.1932	−0.1932	−0.1932	−0.2034
0.13	−0.2088	−0.2087	−0.2089	−0.2086	−0.2086	−0.2179
0.14	−0.2150	−0.2148	−0.2154	−0.2152	−0.2151	−0.2211
0.15	−0.2112	−0.2110	−0.2113	−0.2112	−0.2112	−0.2129

† Number of elements in half beam.

The next example deals with an hyperbolic equation.

Example 7-12 Consider the transverse motion of a clamped-clamped beam (see Example 6-20). We have $a = 0$, $b = 1$, $c_1 = 0$, and $f = 0$. For a typical beam element we have

$$\frac{h_e}{420}\begin{bmatrix} 156 & -22h_e & 54 & 13h_e \\ -22h_e & 4h_e^2 & -13h_e & -3h_e^2 \\ 54 & -13h_e & 156 & 22h_e \\ 13h_e & -3h_e^2 & 22h_e & 4h_e^2 \end{bmatrix}\begin{Bmatrix} \ddot{u}_1^e \\ \ddot{u}_2^e \\ \ddot{u}_3^e \\ \ddot{u}_4^e \end{Bmatrix} + \frac{2}{h_e^3}\begin{bmatrix} 6 & -3h_e & -6 & -3h_e \\ -3h_e & 2h_e^2 & 3h_e & h_e^2 \\ -6 & 3h_e & 6 & 3h_e \\ -3h_e & h_e^2 & 3h_e & 2h_e^2 \end{bmatrix}\begin{Bmatrix} u_1^e \\ u_2^e \\ u_3^e \\ u_4^e \end{Bmatrix} = \begin{Bmatrix} P_1^e \\ P_2^e \\ P_3^e \\ P_4^e \end{Bmatrix}$$

A comparison of the finite-element solution obtained using two different time steps and a different number of Hermite cubic elements (in the half beam) with the Galerkin solution [see Reddy (1984a)] is presented in Table 7-7. The solutions are in very good agreement with each other.

7-4-4 The Method of Discretization in Time

Here we describe the application of the method of discretization in time and the finite-element method in the solution of time-dependent problems. As a model equation, consider the problem

$$\frac{\partial u}{\partial t} - \frac{\partial}{\partial x}\left(a \frac{\partial u}{\partial x}\right) = f \qquad \text{in } (0, L) \times (0, T) \tag{7-82a}$$

$$u(x, 0) = u_0(x) \qquad \text{in } (0, L) \tag{7-82b}$$

$$u(0, t) = 0 \qquad u(L, t) = 0$$

The method of discretization in time gives (see Sec. 6-5-4)

$$-\frac{d}{dx}\left[a(x, t_j)\frac{dz_j}{dx}\right] + \frac{1}{\Delta t} z_j = f(x, t_j) + \frac{1}{\Delta t} z_{j-1}$$
$$z_j(0) = \frac{dz_j}{dx}(0) = 0 \tag{7-83}$$

The variational problem associated with Eq. (7-82a) over an element (x_e, x_{e+1}) is given by

$$\int_{x_e}^{x_{e+1}} \left[a(x, t_j)\frac{dv}{dx}\frac{dz_j}{dx} + \frac{1}{\Delta t} v z_j\right] dx = \int_{x_e}^{x_{e+1}} v\left[f(x, t_j) + \frac{1}{\Delta t} z_{j-1}\right] dx$$
$$+ P_{1j}^e v(x_e) + P_{2j}^e v(x_{e+1}) \tag{7-84}$$

for $j = 1, 2, \ldots, p$.

Assuming the finite-element interpolation of the form

$$z_j(x) = \sum_{k=1}^{n} u_k^j \psi_k^e(x) \tag{7-85}$$

for $z_j(x)$ and $v = \psi_i^e$ in Eq. (7-84), we obtain the finite-element model

$$\sum_{k=1}^{n}\left(K_{ik}^e + \frac{1}{\Delta t} M_{ik}^e\right)u_k^j = F_i^{ej} \tag{7-86}$$

where

$$K_{ik}^e = \int_{x_e}^{x_{e+1}} a(x, t_j)\frac{d\psi_i^e}{dx}\frac{d\psi_k^e}{dx}\,dx \qquad M_{ik}^e = \int_{x_e}^{x_{e+1}} \psi_i^e \psi_k^e\,dx$$
$$F_i^{ej} = \int_{x_e}^{x_{e+1}} \psi_i^e\left[f(x, t_j) + \frac{1}{\Delta t} z_{j-1}\right] dx + P_i^{ej} \tag{7-87}$$

Equation (7-86) can be expressed in alternative form by using the finite-element interpolation of the form (7-85) for z_{j-1}:

$$[\hat{K}^e]\{u^j\} = \frac{1}{\Delta t}[M^e]\{u^{j-1}\} + \{\hat{F}^{ej}\} \qquad j = 1, 2, \ldots, p \tag{7-88}$$

where $\quad [\hat{K}^e] = [K^e] + \dfrac{1}{\Delta t}[M^e] \qquad \hat{F}_i^{ej} = \displaystyle\int_{x_e}^{x_{e+1}} \psi_i^e f(x, t_j)\,dx + P_i^{ej}$

Table 7-8 Comparison of the finite-element solutions with the exact solution of the problem in Example 6-22

	2 elements			4 elements			8 elements			
Time	$p = 10$, $\Delta t = 0.1$	$p = 100$, $\Delta t = 0.01$	$p = 200$, $\Delta t = 0.005$	$p = 10$, $\Delta t = 0.1$	$p = 100$, $\Delta t = 0.01$	$p = 200$, $\Delta t = 0.005$	$p = 10$, $\Delta t = 0.1$	$p = 100$, $\Delta t = 0.01$	$p = 200$, $\Delta t = 0.005$	Exact
0.2	0.9485	0.9606	0.9614	0.9516	0.9731	0.9749	0.9515	0.9713	0.9730	0.9740
0.4	0.8013	0.7925	0.7918	0.8319	0.8318	0.8317	0.8371	0.8395	0.8395	0.8419
0.6	0.6608	0.6440	0.6429	0.6988	0.6854	0.6845	0.7072	0.6952	0.6944	0.6969
0.8	0.5418	0.5224	0.5213	0.5794	0.5610	0.5599	0.5885	0.5707	0.5697	0.5718
1.0	0.4438	0.4237	0.4226	0.4786	0.4587	0.4576	0.4873	0.4677	0.4665	0.4683

Clearly, Eq. (7-88) is the same as that can be obtained from Eq. (7-77) using the backward difference approximation,

$$\frac{\partial u_i}{\partial t} = \frac{u_i^j - u_i^{j-1}}{\Delta t}$$

Thus, the combination of the method of discretization in time and the finite-element method in space leads to the same finite difference and finite-element discrete models.

As an example, the parabolic problem in Example 6-22 is solved using Eq. (7-88) for $p = 10$, 100, and 200 ($\Delta t = 0.1$, 0.01, and 0.005) and 2, 4, and 8 linear elements in the half domain. The results are compared with the exact solution in Table 7-8.

7-5 APPROXIMATION ERRORS IN THE FINITE-ELEMENT METHOD

7-5-1 Introduction

The finite-element solution of a given differential equation is affected, in general, by three types of errors:

1. *Boundary error.* Error due to the inexact representation of the domain by the finite-element mesh ($\bar{\Omega} \neq \bar{\Omega}_h$).
2. *Quadrature and finite arithmetic errors.* Errors due to numerical evaluation of the coefficient matrices and column vectors and numerical computation on a computer.
3. *Approximation error.* Error due to the approximation of the solution

$$u \approx u_h = \sum_{I=1}^{N} U_I \Phi_I$$

In one-dimensional problems discussed thus far, the domains considered were straight lines. Therefore, no approximation of the domain was necessary. In two-dimensional problems posed on nonpolygonal domains $\bar{\Omega}$, domain approximation errors are introduced into the finite-element analysis. In general, these errors can be interpreted as errors in the specification of the data of the problem because we are now solving the given differential equation on a modified domain, $\bar{\Omega}_h \neq \bar{\Omega}$. As we refine the mesh, the domain is more accurately represented and therefore the boundary approximation errors are expected to approach zero.

When finite-element computations are performed on a computer, round-off errors and errors due to the numerical evaluation of integrals are introduced into the solution. In most linear problems with a reasonably small number of total degrees of freedom in the system, these errors are expected to be small compared to approximation errors.

7-5-2 Convergent Approximations

The error introduced into the finite-element solution (u_h) because of the approximation of the dependent variable (u) is inherent to any problem:

$$u \approx u_h = \sum_{e=1}^{E} \sum_{i=1}^{n} u_i^e \psi_i^e$$
$$= \sum_{I=1}^{N} U_I \Phi_I$$

where E is the number of elements in the mesh, N is the total number of global nodes, and n is the number of nodes in an element. We wish to know how the error $\varepsilon = u - u_h$, measured in a meaningful way, behaves as the number of elements in the mesh is increased.

In Chap. 6 and in the present chapter we dealt with the approximation of the solution u of the variational problem: find $u \in H$ such that

$$B(v, u) = l(v) \tag{7-89}$$

holds for all $v \in H$. The bilinear form $B(v, u)$ and linear form $l(v)$ are assumed to satisfy the assumptions of the Lax–Milgram theorem (Theorem 5-5) or its modifications (Theorems 5-6 and 5-7). In the finite-element method we consider a family of subspace S^h of the space H, where the parameter h defines the family of subspaces. We associate with each finite-element subspace S^h the discrete problem of finding the finite-element solution $u_h \in S^h$ such that

$$B_h(v_h, u_h) = l(v_h) \tag{7-90}$$

for every $v_h \in S^h$. We say that the family of discrete problems (7-90) is *convergent*, if for any problem (7-89), we have

$$\lim_{h \to 0} \|u - u_h\|_H = 0 \tag{7-91}$$

The finite-element solution u_h is said to *converge in the energy norm* (i.e., in H_A) to the true solution u if

$$\|u - u_h\|_A \leq ch^p \qquad p > 0 \tag{7-92}$$

where c is a constant independent of u and u_h, and h is the characteristic length of an element. The constant p is called the *rate of convergence*.

7-5-3 Accuracy of the Solution

For mathematical as well as practical reasons, it is worth examining the questions of convergence and estimation of error for a given finite-element model and finite-element mesh. Establishing inequalities of the type (7-92) is crucial because the inequality implies convergence and gives the rate of convergence. In the following paragraphs, we shall study the question of establishing an inequality of type (7-92) for one-dimensional equations.

Suppose that the finite-element interpolation functions Φ_I ($I = 1, 2, \ldots, N$) are complete polynomials of degree k. Then the error in the energy norm, for $H_A = H^m(\Omega)$, can be shown to satisfy the inequality [see Ciarlet (1978)]

$$\|\varepsilon\|_m \equiv \|u - u_h\|_m \leq ch^p \qquad p = k + 1 - m \tag{7-93}$$

where c is a constant. This estimate implies that the error goes to zero at the rate of p, as h is decreased (or, the number of elements is increased). In other words, log of the error in the energy norm versus log h is a straight line whose slope is $(k + 1 - m)$. The greater the degree of the interpolation functions, the faster the rate of convergence. Note also that the error in energy goes to zero at the rate of $(k + 1 - m)$; error in the L_2-norm will be even faster, namely $(k + 1)$ (in other words, derivatives converge slower than the solution itself).

Error estimates of the type in Eq. (7-93) are very useful because they give an idea of the accuracy of the approximate solution, whether we know the true solution or not. While the estimate gives an idea of how rapidly the finite-element solution converges to the true solution, it does not tell us when to stop refining the mesh. This decision rests with the analyst because only he or she knows what a reasonable tolerance is for the problem he or she is solving.

As a special case of Eq. (7-93), we now derive the error estimate for second-order problems $[a(du^2/dx^2) = f(x),\ a = \text{constant}]$ using the linear Lagrange element. Let u_h be the (global) finite-element solution, and let $\varepsilon = u - u_h$ denote the error in the approximation. Since u_h is an interpolant of u, the error vanishes at the nodes of the mesh (see the discussion at the end of this section). We can expand the restriction ε^e of ε to element $\Omega^e = (x_e, x_{e+1})$ in a Fourier series by

$$\varepsilon^e(x) = \sum_{n=1}^{\infty} \alpha_n^e \sin\left(\frac{n\pi x}{h_e}\right) \tag{7-94}$$

where h_e denotes the length of the element Ω^e. Differentiating this series and using Parseval's identity (3-119) (for nonorthogonal systems), we obtain

$$\int_{x_e}^{x_{e+1}} \left(\frac{d\varepsilon^e}{dx}\right)^2 dx = \frac{h_e}{2} \sum_{n=1}^{\infty} (\alpha_n^e)^2 \left(\frac{n\pi}{h_e}\right)^2 \tag{7-95a}$$

$$\int_{x_e}^{x_{e+1}} \left(\frac{d^2\varepsilon^e}{dx^2}\right)^2 dx = \frac{h_e}{2} \sum_{n=1}^{\infty} (\alpha_n^e)^2 \left(\frac{n\pi}{h_e}\right)^4 \tag{7-95b}$$

Note that $x_e = (e - 1)h_e$, $x_{e+1} = eh_e$, and

$$\left(\frac{n\pi}{h_e}\right)^2 \leq \frac{h^2}{\pi^2} \left(\frac{n\pi}{h_e}\right)^4 \qquad \text{for } n \geq 1$$

$$\frac{d^2 u_h}{dx^2} = \sum_{e=1}^{E} \sum_{j=1}^{n} u_j^e \frac{d^2\psi_j^e}{dx^2} = 0$$

the last equation being the result of the fact that the ψ_j^e are linear. Therefore we have from Eqs. (7-95) the inequality

$$\int_{(e-1)h_e}^{eh_e} \left(\frac{d\varepsilon^e}{dx}\right)^2 dx \le \frac{h_e^2}{\pi^2} \int_{(e-1)h_e}^{eh_e} \left(\frac{d^2\varepsilon^e}{dx^2}\right) dx$$
$$= \frac{h_e^2}{\pi^2} \int_{(e-1)h_e}^{eh_e} \left(\frac{d^2u}{dx^2}\right)^2 dx$$

Summing over the number of elements, we obtain

$$\left\|\frac{du}{dx} - \frac{du_h}{dx}\right\|_0^2 = \int_0^L \left(\frac{d\varepsilon}{dx}\right)^2 dx \le \frac{h^2}{\pi^2}\int_0^L \left(\frac{d^2u}{dx^2}\right)^2 dx \le ch^2 \|f\|_0^2$$

where
$$h = \max_{1 \le e \le E} h_e$$

Similarly, we can establish the inequality

$$\|u - u_h\|_0^2 \le \frac{h^4}{\pi^4}\left\|\frac{d^2u}{dx^2}\right\|_0 \le ch^4 \|f\|_0^2$$

Thus, we have

$$\begin{aligned} \|u - u_h\|_0 &\le c_1 h^2 \\ \|u - u_h\|_1 &\le c_2 h \end{aligned} \tag{7-96}$$

where c_1 and c_2 are constants that depend on $\Omega = (0, L)$. Clearly, Eq. (7-96) is a special case of Eq. (7-93) for $m = 1$ and $k = 1$. An example that verifies the above error estimate is presented next.

Example 7-13 Here we consider a computational example to verify the error estimate in Eq. (7-96). Consider the differential equation

$$-\frac{d^2u}{dx^2} = 2 \qquad 0 < x < 1 \qquad u(0) = u(1) = 0$$

The exact and finite-element solutions are given by

$$u(x) = x(1-x)$$

$$N = 2: \qquad u_h = \begin{cases} h^2\left(\dfrac{x}{h}\right) & 0 \le x \le h \\ h^2\left(2 - \dfrac{x}{h}\right) & h \le x \le 2h \end{cases}$$

$$N = 3: \qquad u_h = \begin{cases} 2h^2\left(\dfrac{x}{h}\right) & 0 \le x \le h \\ 2h^2\left(2 - \dfrac{x}{h}\right) + 2h^2\left(\dfrac{x}{h} - 1\right) & h \le x \le 2h \\ 2h^2\left(3 - \dfrac{x}{h}\right) & 2h \le x \le 3h \end{cases}$$

$$N=4: \qquad u_h = \begin{cases} 3h^2\left(\dfrac{x}{h}\right) & 0 \le x \le h \\ 3h^2\left(2-\dfrac{x}{h}\right)+4h^2\left(\dfrac{x}{h}-1\right) & h \le x \le 2h \\ 4h^2\left(3-\dfrac{x}{h}\right)+3h^2\left(\dfrac{x}{h}-2\right) & 2h \le x \le 3h \\ 3h^2\left(4-\dfrac{x}{h}\right) & 3h \le x \le 4h \end{cases}$$

For the two-element case, the errors are given by ($h = 0.5$)

$$\|u-u_h\|_0^2 = \int_0^h (x-x^2-hx)^2\,dx + \int_h^{2h} (x-x^2-2h^2+xh)^2\,dx = 0.002083$$

$$\left\|\frac{du}{dx}-\frac{du_h}{dx}\right\|_0^2 = \int_0^h (1-2x-h)^2\,dx + \int_h^{2h} (1-2x+h)^2\,dx = 0.08333$$

Similar calculations can be made for $N = 3$ and $N = 4$. The table below contains the errors for $N = 2$, 3, and 4.

$h(N)$	$\log_{10} h$	$\|\varepsilon\|_0$	$\log_{10} \|\varepsilon\|_0$	$\|\varepsilon\|_1$	$\log_{10} \|\varepsilon\|_1$
$\frac{1}{2}(2)$	−0.301	0.04564	−1.341	0.2887	−0.5396
$\frac{1}{3}(3)$	−0.477	0.02038	−1.693	0.1925	−0.7157
$\frac{1}{4}(4)$	−0.601	0.01141	−1.943	0.1443	−0.8406

A plot of $\log \|\varepsilon\|_0$ and $\log \|\varepsilon\|_1$ versus $\log h$ will show that

$$\log \|\varepsilon\|_0 = 2 \log h + c_1$$

$$\log \|\varepsilon\|_1 = \log h + c_2$$

In other words, the rate of convergence of the finite-element solution is 2 in the L_2-norm and 1 in the energy norm, verifying the estimates in Eq. (7-96).

In the case of both second-order equations and fourth-order equations with constant coefficients, we observed in our numerical examples that the error between the exact solution and the finite-element solution at the nodes is zero. This is not accidental. We can prove that, when the coefficients a and b are constant, the finite-element solution coincides with the exact solution at the nodes. The proof is presented below.

Consider the second-order equation,

$$-a\frac{d^2u}{dx^2} = f \qquad 0 < x < L$$

$$u(0) = 0 \qquad u(L) = 0$$

The global finite-element solution is given by ($U_1 = U_N = 0$),

$$u_h = \sum_{I=2}^{N-1} U_I \Phi_I$$

where Φ_I are the linear global interpolation functions shown in Fig. 7-3. By definition of the variational problem, we have

$$\int_0^L \left(\frac{dv_h}{dx}\frac{du_h}{dx} - v_h \hat{f} \right) dx = 0 \qquad \hat{f} = f/a$$

for any $v_h \in S^h \subset H^1(\Omega)$. For $v_h = \Phi_I$, we have the finite-element equation

$$\int_0^L \left(\frac{d\Phi_I}{dx}\frac{du_h}{dx} - \Phi_I \hat{f} \right) dx = 0 \qquad \text{for each } I = 2, \ldots, N-1$$

The exact solution also satisfies the equality

$$\int_0^L \left(\frac{d\Phi_I}{dx}\frac{du}{dx} - \Phi_I \hat{f} \right) dx = 0$$

Subtracting the finite-element equation from the last equation, we obtain

$$\int_0^L \left(\frac{du}{dx} - \frac{du_h}{dx} \right) \frac{d\Phi_I}{dx}\, dx = 0 \qquad I = 2, \ldots, N-1$$

Since $\Phi_I = 0$ for $x \geq (I+1)h$ and $x \leq (I-1)h$, and $d\Phi_I/dx = 1/h$ for $(I-1)h \leq x \leq Ih$ and $d\Phi_I/dx = -1/h$ for $Ih \leq x \leq (I+1)h$, it follows that

$$\int_{(I-1)h}^{Ih} \left(\frac{du}{dx} - \frac{du_h}{dx} \right)\left(\frac{1}{h} \right) dx + \int_{Ih}^{(I+1)h} \left(\frac{du}{dx} - \frac{du_h}{dx} \right)\left(-\frac{1}{h} \right) dx = 0$$

$$I = 2, 3, \ldots, N-1$$

Denoting $\varepsilon = u - u_h$, we have

$$\frac{1}{h}(\varepsilon_I - \varepsilon_{I-1}) + \left(-\frac{1}{h} \right)(\varepsilon_{I+1} - \varepsilon_I) = 0$$

or

$$\frac{1}{h}(-\varepsilon_{I-1} + 2\varepsilon_I - \varepsilon_{I+1}) = 0 \qquad I = 2, 3, \ldots, N-1$$

where $\varepsilon_I = \varepsilon(Ih)$. Since $\varepsilon_0 = \varepsilon_N = 0$ (because both u and u_h satisfy the essential boundary conditions), it follows from the above equation (which resembles a finite-difference equation) that the solution is $\varepsilon_1 = \varepsilon_2 = \cdots = \varepsilon_{N-1} = 0$. Thus, the finite-element solution coincides with the exact solution at the nodes.

PROBLEMS 7-1

Some of the following problems might require the use of a computer for their solution.

7-1-1 The differential equation governing (radially) symmetric two-dimensional field problems is given by (see Example 7-1) $-(d/dr)(rk(du/dr)) = f(r)$, $0 < r < L$, where r is the radial coordinate. Equations of this type arise in all fields listed in Table 4-1 when the domain is circular (or annular) and radial symmetry exists. Derive the Galerkin finite element model of the above equation. What type of interpolation functions can we use?

7-1-2 Find the first three eigenvalues λ asociated with the equation $-(d^2u/dx^2) - \lambda u = 0, 0 < x < L$; $u(0) = u(L) = 0$. (*a*) Use four Lagrange linear elements and (*b*) two Lagrange quadratic elements.

7-1-3 Find the first two eigenvalues λ of the equation $-(d^2u/dx^2) - \lambda u = 0,\ 0 < x < 1;\ u(0) = 0$, $u(1) + (du/dx)(1) = 0$, using two Lagrange linear elements.

7-1-4 Consider the differential equation $-(d^2u/dx^2) + x(1 - x)u = \sin \pi x, 0 < x < 1$, with the boundary conditions $u(0) = u(1) = 0$. Solve the problem using three Lagrange linear elements.

7-1-5 Derive the Lagrange interpolation functions of third order (i.e., cubic element). Use the alternative procedure outlined for the quadratic element.

7-1-6 Solve the first-order equation in Example 7-7 using three elements. Use (*a*) Bubnov–Galerkin, (*b*) collocation, (*c*) subdomain, and (*d*) least squares methods.

7-1-7 Solve the following differential equation using three Lagrange linear elements: $-(d^2u/dx^2) - (1 + x^2)u = 1, -1 < x < 1; u(-1) = u(1) = 0$.

7-1-8 Solve the problem in Prob. 7-1-7 using two Hermite cubic elements and (*a*) the Bubnov–Galerkin model, (*b*) least squares model, and (*c*) collocation model.

7-1-9 to 7-1-13 Solve the beam problems shown in Figs. P9–P13 using the minimum possible number of beam elements.

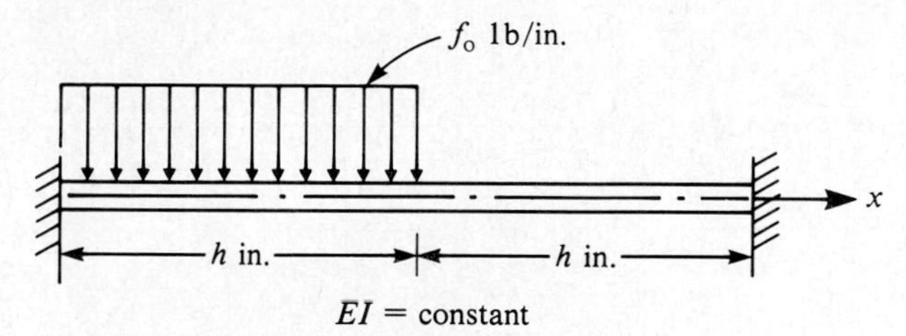

Figure P9

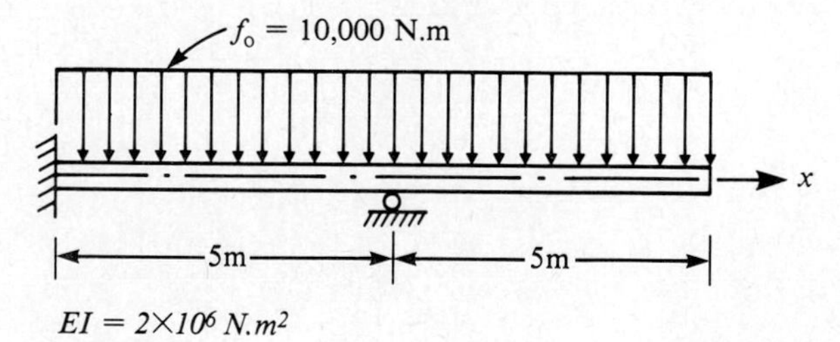

Figure P10

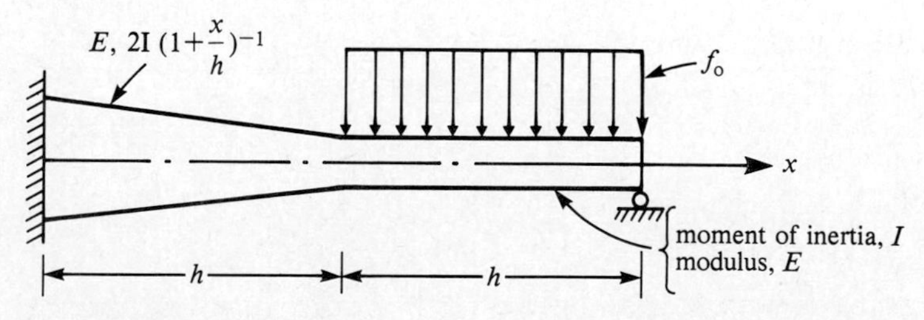

Figure P11

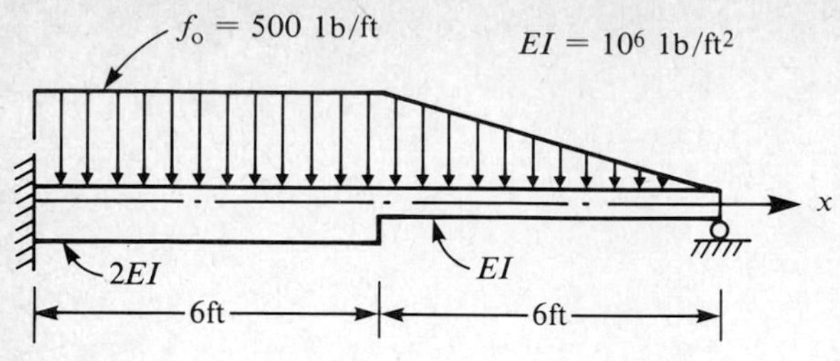

Figure P12

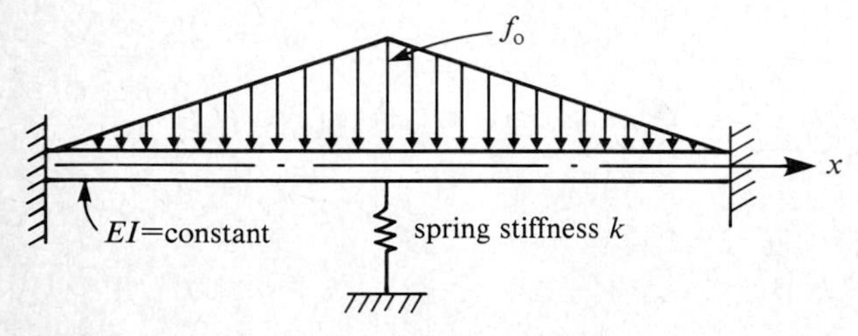

Figure P13

7-1-14 Construct the Ritz finite-element model of the equation

$$\left(\frac{d^2}{dr^2} + \frac{1}{r}\frac{d}{dr}\right)\left(\frac{d^2w}{dr^2} + \frac{1}{r}\frac{dw}{dr}\right) = f(r)$$

which arises in connection with the axisymmetric bending of circular plates.

7-1-15 Solve the following parabolic equation using (*a*) two, and (*b*) four linear elements and the Crank–Nicholson scheme: $\partial u/\partial t - \partial^2 u/\partial x^2 = 0$ in $\Omega = (0, \pi) \times (0, 1)$; $u(x, 0) = \sin x$; $u(0, t) = u(\pi, t) = 0$. The exact solution is given by $u(x, t) = e^{-t} \sin x$.

7-1-16 Repeat Prob. 7-1-15 with the following initial and boundary conditions: $u(x, 0) = 1$, $u(0, t) = u(\pi, t) = 0$. The exact solution is given by

$$u(x, t) = \frac{4}{\pi} \sum_{n=1, 3, 5, \ldots} \frac{\exp(-n^2 t)}{n} \sin nx$$

7-1-17 Solve the following hyperbolic equation using (*a*) two, and (*b*) four linear elements and the Newmark scheme: $\partial^2 u/\partial t^2 - \partial^2 u/\partial x^2 = \sin x$ in $\Omega = (0, \pi) \times (0, 1)$; $u(x, 0) = 0$, $(\partial u/\partial t)(x, 0) = 0$; $u(0, t) = u(\pi, t) = 0$. The exact solution is given by $u(x, t) = (1 - \cos t) \sin x$.

7-1-18 Functions defined over the range $[0, h]$ can be converted to $[-1, 1]$ by changing the variable ξ to $\bar{\xi} = 2(\xi/h) - 1$. The coordinate $\bar{\xi}$ is called the *natural coordinate*. The origin of the coordinate is located at the center of the range $[0, h]$ and it varies between -1 and 1. Express the linear, quadratic, and cubic Lagrange interpolation functions in terms of the natural coordinate.

7-1-19 Repeat Prob. 7-1-18 for the Hermite cubic interpolation functions.

7-1-20 Formulate the differential equation $a(du/dt) + bu = f$, $0 < t < T$; $u(0) = u_0$ using the Galerkin finite-element approximation. For the choice of linear interpolation functions

$$\psi_1^e(t) = \frac{t_{e+1} - t}{\Delta t} \qquad \psi_2^e(t) = \frac{t - t_e}{\Delta t}$$

compute the algebraic relations among u_i^e $(i = 1, 2)$. Comment on the relationship of these equations to the θ-family of approximations.

7-1-21 Compute the L_2-norms of the error in the finite element approximations of Example 7-8 and plot the $\log_e \varepsilon$ versus $\log_e h$ and determine the rates of convergence.

7-1-22 Solve the second-order equation $-(d^2u/dx^2) - u + x^2 = 0$, $0 < x < 1$; $u(0) = 0$, $(du/dx)(1) = 1$. Use (*a*) equispaced, and (*b*) Gauss point collocation points.

7-1-23 to 7-1-25 Solve Probs. 7-1-15 to 7-1-17 using the method of discretization in time and the finite-element method.

7-1-26 Show that the error estimate for the fourth-order equation in Eq. (7-66) is given by ($m = 2$) $\|w - w_h\|_2 \le ch^2$ where c is constant, w_h is the finite-element solution obtained by using the Hermite cubic interpolation, and w is the exact solution of the problem.

7-6 TWO-DIMENSIONAL SECOND-ORDER EQUATIONS

7-6-1 Model Equation

Here we consider the finite-element analysis of second-order differential equations in two dimensions. As a model equation we consider the following general second-order equation:

$$-\frac{\partial}{\partial x}\left(a_{11}\frac{\partial u}{\partial x} + a_{12}\frac{\partial u}{\partial y}\right) - \frac{\partial}{\partial y}\left(a_{21}\frac{\partial u}{\partial x} + a_{22}\frac{\partial u}{\partial y}\right) + a_0 u = f \text{ in } \Omega \quad (7\text{-}97)$$

The coefficients a_{ij} ($i, j = 1, 2$) and a_0, and the source term f are known functions of position (x, y) in the domain Ω. Equation (7-97) arises in the study of a number of engineering problems, including torsion of constant cross-section members, heat transfer, irrotational flow of a fluid, transverse deflection of a membrane, and so on (see Table 4-2). The plane elasticity and the Stokes flow problems are also described by a pair of equations of the same form as the model equation. Thus, the finite-element procedure to be described for Eq. (7-97) is applicable to *any* problem that can be formulated as one of solving equations of the form (7-97).

Our objective here is to extend the basic steps discussed earlier for the finite-element analysis of one-dimensional problems. While the basic ideas are the same, the mathematical complexity increases because we are dealing with partial differential equations on domains with possibly curved boundaries. We not only approximate the solution of a partial differential equation, but we also seek an approximation of the given domain by a suitable finite-element mesh (see Fig. 7-18). This latter property is what made the finite-element method a more attractive practical analysis tool over other competing methods.

7-6-2 The Ritz Finite-Element Model

A step by step procedure for the Ritz finite-element modeling of Eq. (7-97) is presented in the following paragraphs.

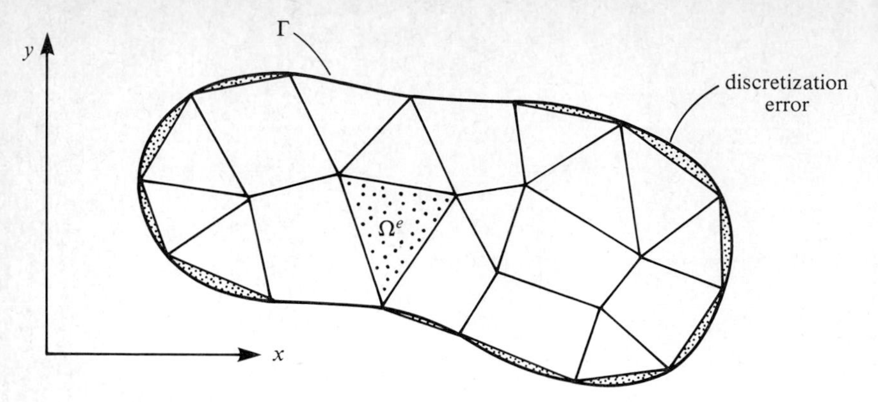

Figure 7-18 A finite-element discretization of a domain by triangular and quadrilateral elements.

1 Discretization of a domain Two-dimensional domains can be represented by more than one type of geometric shape. For example, a plane curved domain can be represented by triangular elements and/or quadrilateral elements as shown in Fig. 7-18. Without reference to a specific geometric shape, we simply denote a typical element by Ω^e and proceed to discuss the approximation of Eq. (7-97).

The choice of the finite-element mesh depends both on the element characteristics (convergence, computational simplicity, etc.) and the ability to represent the domain accurately. The concept of isoparametric elements allows us to represent the element geometry also by the same interpolation as that used in the approximation of the dependent variables. Thus, by identifying nodes on the boundary of the domain we can approximate the domain by suitable collection of elements to a desired accuracy. Due to the introductory nature of the present study of the finite-element method, we will not consider advanced topics such as the isoparametric elements, computer implementation, etc. Interested readers can consult the finite-element book by the author (1984a) and references therein.

2 Element equations Consider a typical finite element Ω^e. Let ψ_i^e $(i = 1, 2, \ldots, n)$ denote the interpolation functions for Ω^e. We do not need to know the specific form of ψ_i^e until after the formulation is complete and we are ready to compute the coefficient matrices and column vectors. Thus, the finite-element formulation of a problem can be carried in algebraic terms without identifying the specific element. This allows us to develop general equations that are valid for *any* admissible element.

Variational formulation Following the procedure of Chap. 5, we develop the variational form of Eq. (7-97) over the element Ω^e. We multiply Eq. (7-97) with a test function $v_e \in S^h \subset H^1(\Omega^e)$ and integrate over the element domain Ω^e to obtain

$$0 = \int_{\Omega^e} \left[\frac{\partial v_e}{\partial x} \left(a_{11}^e \frac{\partial u_e}{\partial x} + a_{12}^e \frac{\partial u_e}{\partial y} \right) + \frac{\partial v_e}{\partial y} \left(a_{21}^e \frac{\partial u_e}{\partial x} + a_{22}^e \frac{\partial u_e}{\partial y} \right) + a_0^e v_e u_e - v_e f_e \right] dx\, dy - \oint_{\Gamma^e} v_e \left[n_x \left(a_{11}^e \frac{\partial u_e}{\partial x} + a_{12}^e \frac{\partial u_e}{\partial y} \right) + n_y \left(a_{21}^e \frac{\partial u_e}{\partial x} + a_{22}^e \frac{\partial u_e}{\partial y} \right) \right] ds \tag{7-98}$$

where n_x and n_y are the components (i.e., direction cosines) of the unit normal $\hat{\mathbf{n}}$

$$\hat{\mathbf{n}} = n_x \hat{i} + n_y \hat{j} = \cos \alpha\, \hat{i} + \sin \alpha\, \hat{j} \tag{7-99}$$

on the boundary Γ_e, and ds is the elemental arc length along the boundary of the element. From an inspection of the boundary term in Eq. (7-98), we note that the specification of u constitutes the essential boundary condition (hence u is the primary variable) and specification of

$$q_n^e \equiv n_x \left(a_{11}^e \frac{\partial u_e}{\partial x} + a_{12}^e \frac{\partial u_e}{\partial y} \right) + n_y \left(a_{21}^e \frac{\partial u_e}{\partial x} + a_{22}^e \frac{\partial u_e}{\partial y} \right) \tag{7-100}$$

constitutes the natural boundary condition (hence q_n is the secondary variable) of the formulation. The secondary variable q_n is of physical interest in most problems. For example, in the case of the heat transfer through an anisotropic medium (where a_{ij} denote the conductivities of the medium), q_n denotes the heat flux across the boundary of the element.

The variational form in Eq. (7-98) now becomes

$$0 = \int_{\Omega^e} \left[\frac{\partial v_e}{\partial x} \left(a_{11}^e \frac{\partial u_e}{\partial x} + a_{12}^e \frac{\partial u_e}{\partial y} \right) + \frac{\partial v_e}{\partial y} \left(a_{21}^e \frac{\partial u_e}{\partial x} + a_{22}^e \frac{\partial u_e}{\partial y} \right) + a_0^e v_e u_e - v_e f_e \right] dx\, dy - \oint_{\Gamma^e} v_e q_n^e\, ds \tag{7-101}$$

This variational equation forms the basis of the finite-element model.

Finite-element formulation The variational form in (7-101) indicates that the approximation chosen for u_e should be at least bilinear in x and y so that $u_e \in H^1(\Omega^e)$. Suppose that u_e is approximated by the expression

$$u_e = \sum_{j=1}^{n} u_j^e \psi_j^e \tag{7-102}$$

where u_j^e are the values of u_e at the point (x_j, y_j) in Ω^e and ψ_j^e are the interpolation functions with the property

$$\psi_i^e(x_j, y_j) = \delta_{ij}$$

The specific form of ψ_i^e will be derived for linear triangular and rectangular elements in Sec. 7-6-3.

Substituting (7-102) for u_e and ψ_i^e for v_e into the variational form (7-101), we obtain

$$\sum_{j=1}^{n} K_{ij}^e u_j^e = F_i^e \tag{7-103}$$

where

$$K_{ij}^e = \int_{\Omega^e} \left[\frac{\partial \psi_i^e}{\partial x} \left(a_{11}^e \frac{\partial \psi_j^e}{\partial x} + a_{12}^e \frac{\partial \psi_j^e}{\partial y} \right) + \frac{\partial \psi_i^e}{\partial y} \left(a_{21}^e \frac{\partial \psi_j^e}{\partial x} + a_{22}^e \frac{\partial \psi_j^e}{\partial y} \right) + a_0^e \psi_i^e \psi_j^e \right] dx\, dy \tag{7-104}$$

$$F_i^e = \int_{\Omega^e} f_e \psi_i^e \, dx\, dy + \oint_{\Gamma^e} q_n^e \psi_i^e \, ds \equiv f_i^e + P_i^e$$

Note that $K_{ij}^e = K_{ji}^e$ (i.e., $[K^e]$ is symmetric) only when $a_{12}^e = a_{21}^e$. Equation (7-103) is called the finite-element model of Eq. (7-97).

3 Assembly of element matrices The assembly of finite-element equations is based on the same principle as that employed in one-dimensional problems. We illustrate the procedure by considering a finite-element mesh consisting of two triangular elements (see Fig. 7-19). Let K_{ij}^e and K_{ij}^f $(i, j = 1, 2, 3)$ denote the coefficient matrices and $\{F^e\}$ and $\{F^f\}$ denote the column vectors of three-node triangular elements Ω^e and Ω^f. From the finite-element mesh shown in Fig. 7-19, we note the following correspondence between the global and element nodal values:

$$U_1 = u_1^e \qquad U_2 = u_2^e = u_1^f \qquad U_3 = u_3^e = u_3^f \qquad U_4 = u_2^f \tag{7-105}$$

Note that the continuity of the nodal values at the interelement nodes guarantees the continuity of the primary variable along the *entire* interelement boundary. To see this, consider two linear triangular elements (see Fig. 7-19). The finite-element solution is linear along the boundaries of the elements. The inter-

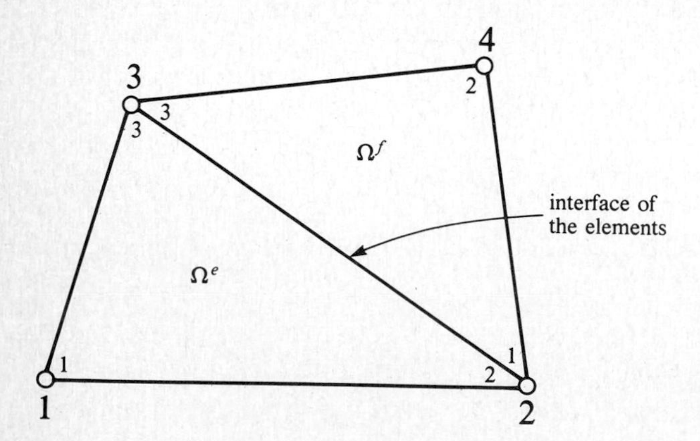

Figure 7-19 Connectivity of two linear triangular elements.

element boundary is along the line connecting global nodes 2 and 3. Since u_e is linear along side 2-3 of element Ω^e, it is uniquely determined by the two values u_2^e and u_3^e. Similarly, u_f is uniquely determined along side 1-3 of element Ω^f by the two values u_1^f and u_3^f. Since $u_2^e = u_1^f$ and $u_3^e = u_3^f$ it follows that $u_e = u_f$ along the interface. Similar arguments can be presented for higher-order Lagrange elements.

We can use the continuity conditions (7-105) and Eq. (7-30) to assemble the element equations. An alternative procedure that is considerably simpler and conceptually easy to understand is presented here. First we note that the coefficient K_{ij}^e is a representation of a physical property (e.g., in the case of the torsion of a prismatic member the property denotes the stiffness) of node i with respect to node j of element Ω^e. The assembled coefficient matrix also represents the same property among the global nodes. But the global property comes from the element nodes shared by the global nodes. For example, the coefficient K_{23} of the global coefficient matrix is the sum of the contributions from nodes 2 and 3 of Ω^e and nodes 1 and 3 of Ω^f (see Fig. 7-19):

$$K_{23} = K_{23}^e + K_{13}^f \qquad K_{32} = K_{32}^e + K_{31}^f$$

Similarly,

$$K_{22} = K_{22}^e + K_{11}^f \qquad K_{33} = K_{33}^e + K_{33}^f, \text{ etc.}$$

If the global nodes I and J do not correspond to nodes in the same element, then $K_{IJ} = 0$. The column vectors can be assembled using the same logic:

$$F_2 = F_2^e + F_1^f \qquad F_3 = F_3^e + F_3^f, \text{ etc.}$$

The complete assembled equations for the two-element mesh is given by

$$\begin{bmatrix} K_{11}^e & K_{12}^e & K_{13}^e & 0 \\ K_{21}^e & K_{22}^e + K_{11}^f & K_{23}^e + K_{13}^f & K_{12}^f \\ K_{31}^e & K_{32}^e + K_{31}^f & K_{33}^e + K_{33}^f & K_{32}^f \\ 0 & K_{21}^f & K_{23}^f & K_{22}^f \end{bmatrix} \begin{Bmatrix} U_1 \\ U_2 \\ U_3 \\ U_4 \end{Bmatrix} = \begin{Bmatrix} F_1^e \\ F_2^e + F_1^f \\ F_3^e + F_3^f \\ F_2^f \end{Bmatrix} \tag{7-106}$$

4 Imposition of boundary conditions The boundary conditions on the primary and secondary variables are imposed on the assembled equations in the same way as in the one-dimensional problems. To understand the physical significance of the P_i's [see Eq. (7-104)], let us take a closer look at the definition,

$$P_i^e \equiv \oint_{\Gamma^e} q_n^e \psi_i^e(s)\, ds \tag{7-107}$$

where $\psi_i^e(s)$ is the value of $\psi_i^e(x, y)$ on the boundary Γ^e. The quantity q_n^e [see Eq. (7-100)] is an unknown when Ω^e is an interior element of the mesh (see Fig. 7-20*a*). However, when we assemble the element equations the contribution of q_n^e to the nodes (namely, P_i^e) of Ω^e get cancelled by similar contributions from the adjoining elements (see Fig. 7-20*b*). If the element Ω^r has any of its sides on the boundary Γ of the domain Ω (see Fig. 7-20*c*), then on that side q_n^r is either

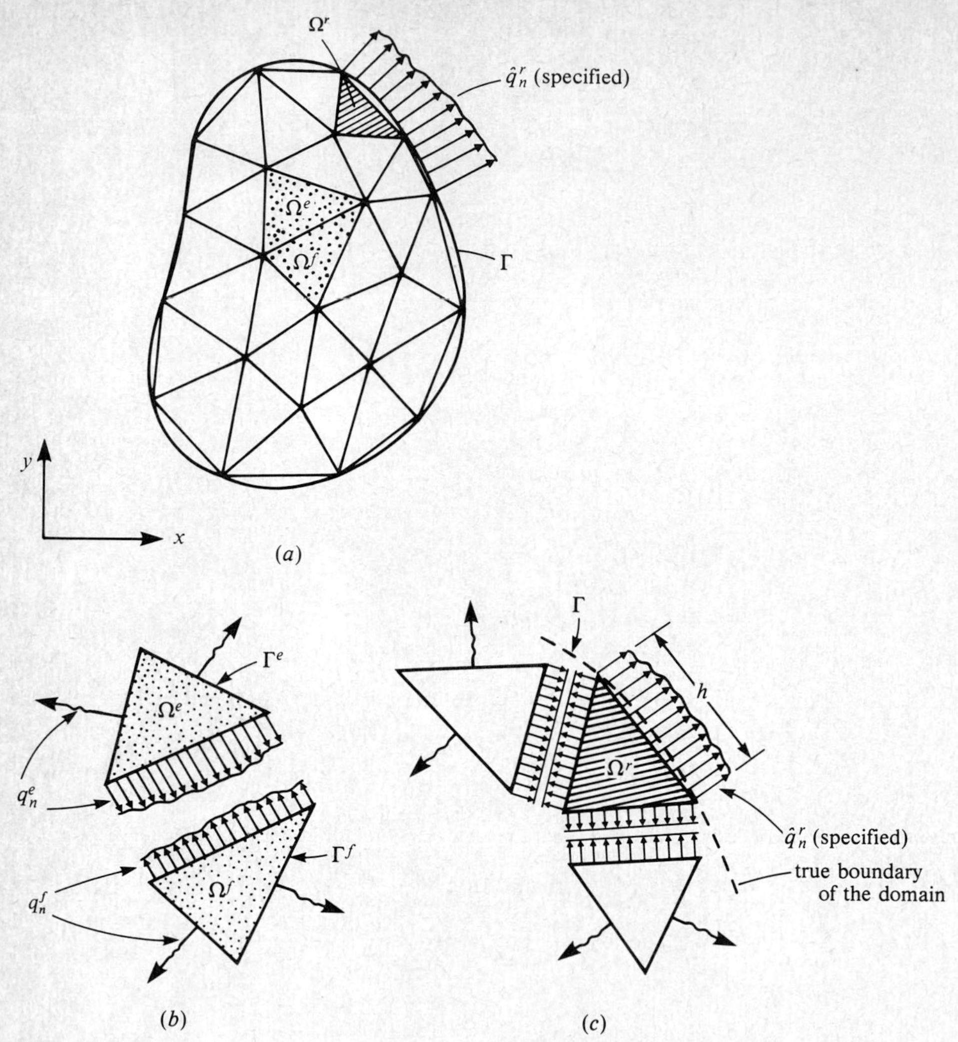

Figure 7-20 Computation of boundary forces and equilibrium of secondary variables at interelement boundaries.

specified or unspecified. If q_n^r is specified, then the P_i^r on that side can be computed using Eq. (7-107). If q_n^r is not specified, then the primary variable u_r is known on that portion of the boundary.

The remaining steps of the analysis do not differ from those of one-dimensional problems.

7-6-3 Interpolation Functions

Linear triangular element The simplest finite element in two dimensions is the triangular element. Since a triangle is defined uniquely by three points that form its vertices, we choose the vertex points as the nodes (see Fig. 7-21*a*). These nodes will be connected to the nodes of adjoining elements in a finite-element mesh.

A polynomial in x and y that is uniquely defined by three constants is of the form $p(x, y) = c_0 + c_1 x + c_2 y$. Hence, we assume an approximation of u_e in the form,

$$u_e = c_0^e + c_1^e x + c_2^e y \tag{7-108}$$

Proceeding as in the case of one-dimensional elements, we define

$$u_i^e \equiv u_e(x_i, y_i) = c_0^e + c_1^e x_i + c_2^e y_i \qquad i = 1, 2, 3$$

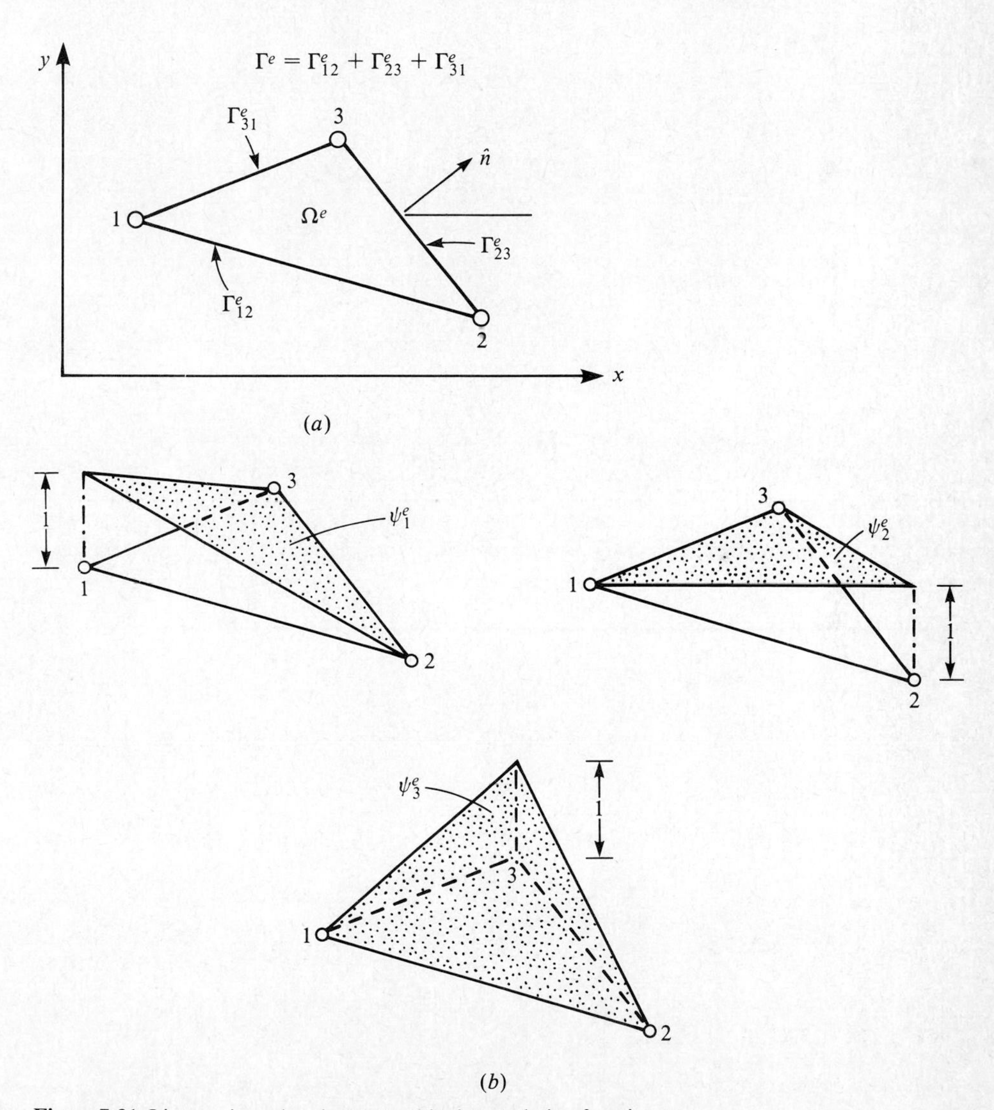

Figure 7-21 Linear triangular element and its interpolation functions.

where (x_i, y_i) denote the global coordinates of the node i in Ω^e. In explicit form this equation becomes

$$\begin{Bmatrix} u_1^e \\ u_2^e \\ u_3^e \end{Bmatrix} = \begin{bmatrix} 1 & x_1 & y_1 \\ 1 & x_2 & y_2 \\ 1 & x_3 & y_3 \end{bmatrix} \begin{Bmatrix} c_0^e \\ c_1^e \\ c_2^e \end{Bmatrix} \tag{7-109}$$

Note that the element nodes are numbered counterclockwise. Upon solving for the c's and substituting back into Eq. (7-108), we obtain

$$u_e = \sum_{i=1}^{3} u_i^e \psi_i^e(x, y) \tag{7-110}$$

$$\psi_i^e = \frac{1}{2A_e}(\alpha_i^e + \beta_i^e x + \gamma_i^e y) \tag{7.111}$$

where $2A_e$ represents the value of the determinant of the matrix in Eq. (7-109) (or A_e represents the area of the triangle), and

$$\alpha_i^e = x_j y_k - x_k y_j \qquad \beta_i^e = y_j - y_k \qquad \gamma_i^e = x_k - x_j \qquad i \neq j \neq k,$$
$$i, j, k = 1, 2, 3, \tag{7-112}$$

and the indices on α_i^e, β_i^e, and γ_i^e permute in a natural order. For example, α_1^e is given by setting $i = 1, j = 2$, and $k = 3$:

$$\alpha_1^e = x_2 y_3 - x_3 y_2$$

The sign of the determinant changes if the node numbering is changed to clockwise. The interpolation functions ψ_i^e satisfy the properties listed in Eq. (7-20). The shape of these functions is shown in Fig. 7-21*b*.

Note that the derivative of ψ_i^e with respect to x or y is a constant. Hence, the derivatives of the solution evaluated in the postcomputation would be elementwise constant. Also, the coefficient matrix

$$K_{ij}^e = \int_{\Omega^e} \left(\frac{\partial \psi_i^e}{\partial x} \frac{\partial \psi_j^e}{\partial x} + \frac{\partial \psi_i^e}{\partial y} \frac{\partial \psi_j^e}{\partial y} \right) dx\, dy$$

can be easily evaluated for the linear interpolation functions. We have

$$\frac{\partial \psi_i^e}{\partial x} = \frac{\beta_i^e}{2A_e} \qquad \frac{\partial \psi_i^e}{\partial y} = \frac{\gamma_i^e}{2A_e} \tag{7-113}$$

and

$$K_{ij}^e = \frac{1}{4A_e^2}(\beta_i^e \beta_j^e + \gamma_i^e \gamma_j^e)\left(\int_{\Omega^e} dx\, dy \right)$$
$$= \frac{1}{4A_e}(\beta_i^e \beta_j^e + \gamma_i^e \gamma_j^e) \tag{7-114}$$

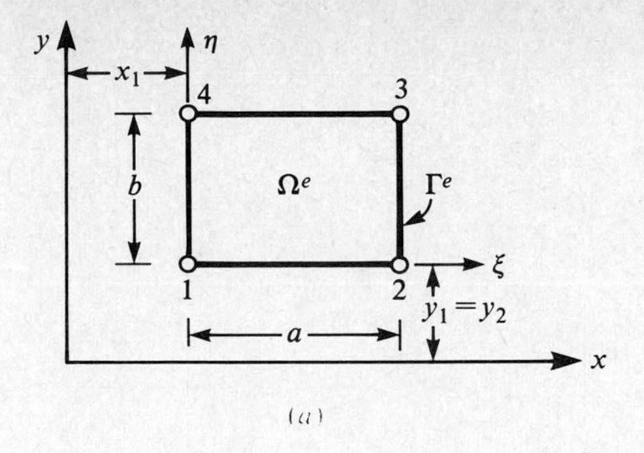

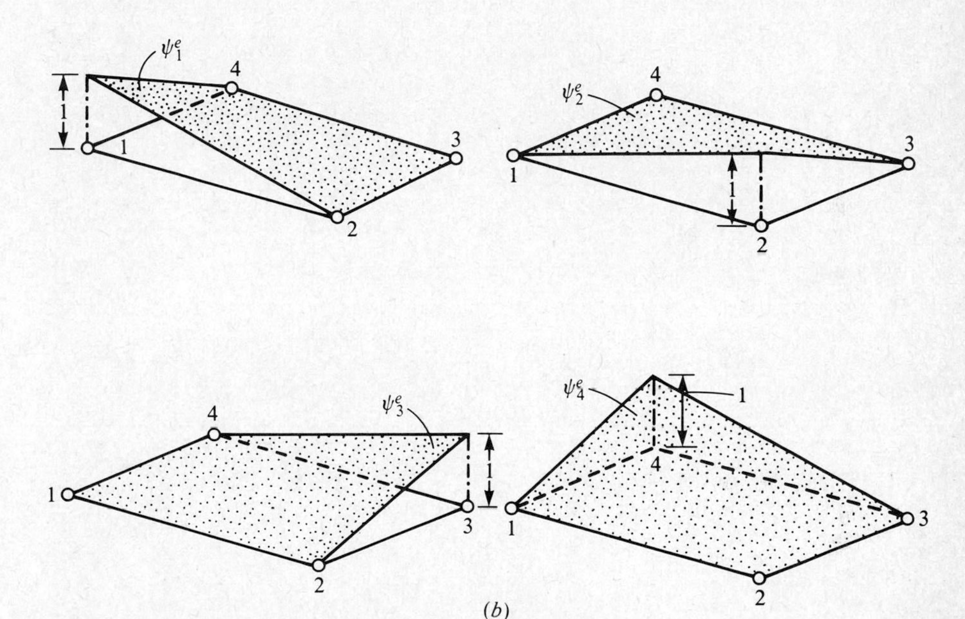

Figure 7-22 Linear rectangular element and its interpolation functions.

Linear rectangular element A rectangular element is uniquely defined by the four corner points (see Fig. 7-22*a*). Therefore, we can use the four-term polynomial to derive the interpolation functions. We have

$$u_e = c_0^e + c_1^e x + c_2^e y + c_3^e xy \tag{7-115}$$

and

$$\begin{Bmatrix} u_1^e \\ u_2^e \\ u_3^e \\ u_4^e \end{Bmatrix} = \begin{bmatrix} 1 & x_1 & y_1 & x_1 y_1 \\ 1 & x_2 & y_2 & x_2 y_2 \\ 1 & x_3 & y_3 & x_3 y_3 \\ 1 & x_4 & y_4 & x_4 y_4 \end{bmatrix} \begin{Bmatrix} c_0^e \\ c_1^e \\ c_2^e \\ c_3^e \end{Bmatrix}$$

By inverting the equations for the c's and substituting into Eq. (7-115), we obtain

$$\psi_1^e = \left(1 - \frac{\xi}{a}\right)\left(1 - \frac{\eta}{b}\right) \qquad \psi_2^e = \frac{\xi}{a}\left(1 - \frac{\eta}{b}\right)$$

$$\psi_3^e = \frac{\xi}{a}\frac{\eta}{b} \qquad \psi_4^e = \left(1 - \frac{\xi}{a}\right)\frac{\eta}{b} \tag{7-116}$$

$$\xi = x - x_1 \qquad \eta = y - y_1$$

The functions are geometrically represented in Fig. 7-22*b*. In calculating element matrices one finds that the use of the local coordinate system (ξ, η) is more convenient than using the global coordinates (x, y).

For the linear rectangular element, the derivatives of the shape functions are not constant within the element. We have

$$\frac{\partial \psi_i^e}{\partial x} = \text{linear in } y \qquad \frac{\partial \psi_i^e}{\partial y} = \text{linear in } x$$

The integration of polynomial expressions over a rectangular element is made simple by the fact that $\Omega^e = (0, a) \times (0, b)$. In other words, we have

$$\int_{\Omega^e} f(x, y)\, dx\, dy = \int_0^a \int_0^b f(x, y)\, dx\, dy$$

The following integrals can be easily evaluated over a linear rectangular element:

$$S_{ij}^{xe} = \int_{\Omega^e} \frac{\partial \psi_i^e}{\partial x}\frac{\partial \psi_j^e}{\partial x}\, dx\, dy \qquad S_{ij}^{ye} = \int_{\Omega^e} \frac{\partial \psi_i^e}{\partial y}\frac{\partial \psi_j^e}{\partial y}\, dx\, dy$$

$$S_{ij}^{xye} = \int_{\Omega^e} \frac{\partial \psi_i^e}{\partial x}\frac{\partial \psi_j^e}{\partial y}\, dx\, dy \qquad S_{ij}^e = \int_{\Omega^e} \psi_i^e \psi_j^e\, dx\, dy \tag{7-117}$$

We obtain

$$[S^{xe}] = \frac{b}{6a}\begin{bmatrix} 2 & -2 & -1 & 1 \\ -2 & 2 & 1 & -1 \\ -1 & 1 & 2 & -2 \\ 1 & -1 & -2 & 2 \end{bmatrix}, \qquad [S^{ye}] = \frac{a}{6b}\begin{bmatrix} 2 & 1 & -1 & -2 \\ 1 & 2 & -2 & -1 \\ -1 & -2 & 2 & 1 \\ -2 & -1 & 1 & 2 \end{bmatrix}$$

$$[S^{xye}] = \frac{1}{4}\begin{bmatrix} 1 & 1 & -1 & -1 \\ -1 & -1 & 1 & 1 \\ -1 & -1 & 1 & 1 \\ 1 & 1 & -1 & -1 \end{bmatrix}, \qquad [S^e] = \frac{ab}{36}\begin{bmatrix} 4 & 2 & 1 & 2 \\ 2 & 4 & 2 & 1 \\ 1 & 2 & 4 & 2 \\ 2 & 1 & 2 & 4 \end{bmatrix} \tag{7-118}$$

$$\{f^e\} = f_e \frac{ab}{4}\{1 \quad 1 \quad 1 \quad 1\}^T$$

Higher-order elements ***Triangular elements*** Higher-order triangular elements (i.e., triangular elements with interpolation functions of higher-degree) can be system-

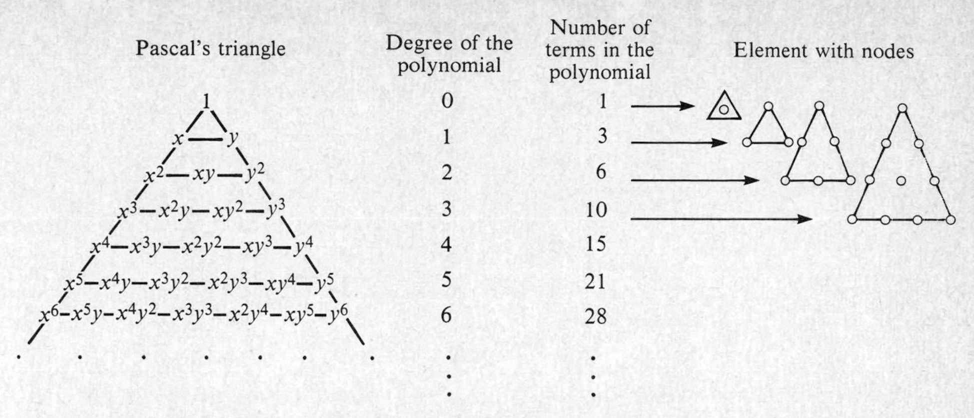

Figure 7-23 Pascal's triangle for the generation of the Lagrange family of triangular elements.

atically developed with the help of the so-called *Pascal's triangle*, which contains the terms of polynomials of various degree in the two variables x and y, as shown in Fig. 7-23. One can view the position of the terms as the nodes of the triangle with the constant term and the first and last terms of a given row being the vertices of the triangle (of course, the shape of the triangle is arbitrary—not an equilateral triangle, as might appear from the position of the terms in Pascal's triangle). For example, a triangular element of order 2 (i.e., the degree of the polynomial is 2) contains six nodes, as can be seen from the top three rows of the Pascal triangle. The positions of the six nodes in the triangle are at the three vertices and at the midpoints of the three sides. The polynomial involves six constants, which can be expressed in terms of the nodal values of the variable being interpolated:

$$u_e = \sum_{i=1}^{6} u_i^e \psi_i^e(x, y) \tag{7-119}$$

where ψ_i^e are the quadratic interpolation functions.

Rectangular elements Analogous to the Lagrange family of triangular elements, the Lagrange family of rectangular elements can be developed from the rectangular array shown in Fig. 7-24. Since a linear rectangular element has four corners (hence, four nodes), the polynomial should have the first four terms, 1, x, y, and xy (which form a rectangle in the array in Fig. 7-24). In general, a pth order Lagrange rectangular element has n nodes, with

$$n = (p + 1)^2 \qquad p = 1, 2, \ldots \tag{7-120}$$

and the associated polynomial contains terms from the pth rectangle in Fig. 7-24. The Lagrange quadratic rectangular element has nine nodes, and the associated polynomial is given by

$$u_e(x, y) = a_1 + a_2 x + a_3 y + a_4 xy + a_5 x^2 + a_6 y^2 + a_7 x^2 y + a_8 xy^2 + a_9 x^2 y^2 \tag{7-121}$$

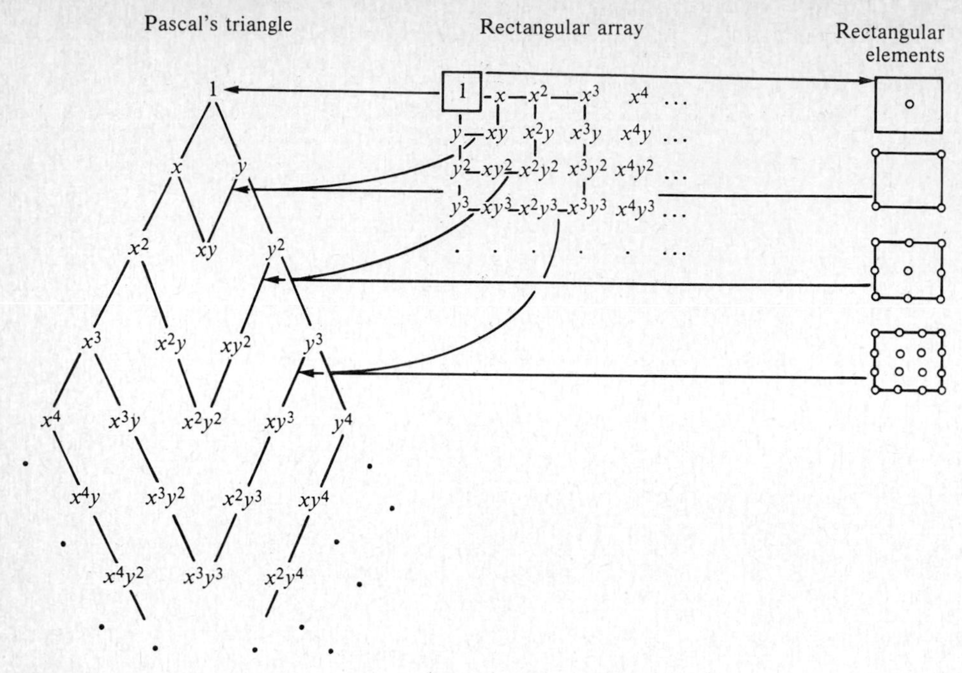

Figure 7-24 The Lagrange family of rectangular elements of various order.

The polynomial contains the complete polynomial of second-degree plus the third-degree terms x^2y and xy^2 and also the x^2y^2 term. Four of the nine nodes are placed at the four corners, four at the midpoints of the sides, and one at the center of the element. The polynomial is uniquely determined by specifying its values at each of the nine nodes.

Since the internal nodes of the higher-order elements of the Lagrange family do not contribute to the interelement connectivity, they can be condensed out at the element level so that the size of the element matrices is reduced. The *serendipity elements* are those rectangular elements which have no interior nodes. In other words, all the nodes are on the bounary of the element.

The Hermite cubic rectangular element can be constructed by taking the tensor product of the one-dimensional Hermite cubics in Eq. (7-60). The element has four corner nodes, but four degrees of freedom (u, $\partial u/\partial x$, $\partial u/\partial y$, $\partial^2 u/\partial x\, \partial y$) at each node. These elements are useful in the weighted-residual finite-element models of second-order equations (see Sec. 7-6-6).

The interpolation functions for the Lagrange linear and quadratic, and serendipity quadratic and Hermite cubic elements are listed in Table 7-9. There ξ and η denote the natural coordinates $(-1 \leq \xi \leq 1)$ and $(-1 \leq \eta \leq 1)$. For the Hermite cubic element the interpolation functions correspond to u, $\partial u/\partial \xi$, $\partial u/\partial \eta$, and $\partial^2 u/\partial \xi\, \partial \eta$ at each node. The elements discussed here are shown in Fig. 7-25.

7-6-4 Existence of Solutions and Error Estimates

Consider the Dirichlet problem (with homogeneous boundary conditions) associated with Eq. (7-97). The variational problem consists of finding $u \in H_0^1(\Omega)$ such that

$$B(v, u) = l(v) \tag{7-122}$$

for all $v \in H_0^1(\Omega)$, where

$$B(v, u) = \int_\Omega \left(a_{11} \frac{\partial v}{\partial x} \frac{\partial u}{\partial x} + a_{12} \frac{\partial v}{\partial x} \frac{\partial u}{\partial y} + a_{21} \frac{\partial v}{\partial y} \frac{\partial u}{\partial x} + a_{22} \frac{\partial v}{\partial y} \frac{\partial u}{\partial y} + a_0 vu \right) dx\, dy$$

$$l(v) = \int_\Omega fv\, dx\, dy$$

Assume that the constants $a_{ij}(\mathbf{x})$ are such that $B(\cdot, \cdot)$ satisfies the conditions of Theorem 5-5, and hence has a unique solution.

The finite-element approximation of Eq. (7-122) consists of seeking $u_h \in S_0^h(\Omega) = \{v_h \in S^h : v_h = 0 \text{ on } \Gamma\}$, S^h being the space spanned by $\{\Phi_i\}$, such that

$$B(v_h, u_h) = l(v_h) \qquad \text{for all } v_h \in S_0^h(\Omega_h) \tag{7-123}$$

This approximate problem has a unique solution in $S_0^h(\Omega)$, for every h, provided the conditions of Theorem 5-5 are satisfied on $S_0^h(\Omega)$ (with α replaced by $\alpha_h > 0$).

Table 7-9 Interpolation functions for the linear and quadratic Langrange rectangular elements, quadratic serendipity element and Hermite cubic rectangular element*

Element type	Interpolation functions	Remarks
Lagrange elements:		
Linear	$\frac{1}{4}(1 + \xi\xi_i)(1 + \eta\eta_i)$	Node i ($i = 1, 2, 3, 4$)
Quadratic	$\frac{1}{4}\xi\xi_i(1 + \xi\xi_i)\eta\eta_i(1 + \eta\eta_i)$	Corner node
	$\frac{1}{2}\eta\eta_i(1 + \eta\eta_i)(1 - \xi^2)$	Side node, $\xi_i = 0$
	$\frac{1}{2}\xi\xi_i(1 + \xi\xi_i)(1 - \eta^2)$	Side node, $\eta_i = 0$
	$(1 - \xi^2)(1 - \eta^2)$	Interior node
Serendipity element:		
Quadratic	$\frac{1}{4}(1 + \xi\xi_i)(1 + \eta\eta_i)(\xi\xi_i + \eta\eta_i - 1)$	Corner node
	$\frac{1}{2}(1 - \xi^2)(1 + \eta\eta_i)$	Side node, $\xi_i = 0$
	$\frac{1}{2}(1 + \xi\xi_i)(1 - \eta^2)$	Side node, $\eta_i = 0$
Hermite cubic element:		
Interpolation functions for		
variable, u:	$\frac{1}{16}(\xi + \xi_i)^2(\xi\xi_i - 2)(\eta + \eta_i)^2(\eta\eta_i - 2)$	For node i ($i = 1, 2, 3, 4$)
derivative, $\partial u/\partial \xi$:	$-\frac{1}{16}\xi_i(\xi + \xi_i)^2(\xi\xi_i - 1)(\eta + \eta_i)^2(\eta\eta_i - 2)$	
derivative, $\partial u/\partial \eta$:	$-\frac{1}{16}(\xi + \xi_i)^2(\xi\xi_i - 2)\eta_i(\eta + \eta_i)^2(\eta\eta_i - 1)$	
derivative, $\partial^2 u/\partial \xi\, \partial \eta$:	$\frac{1}{16}\xi_i(\xi + \xi_i)^2(\xi\xi_i - 1)\eta_i(\eta + \eta_i)^2(\eta\eta_i - 1)$	

* See Fig. 7-25 for the coordinate system; (ξ_i, η_i) denote the coordinates of the ith node of the element.

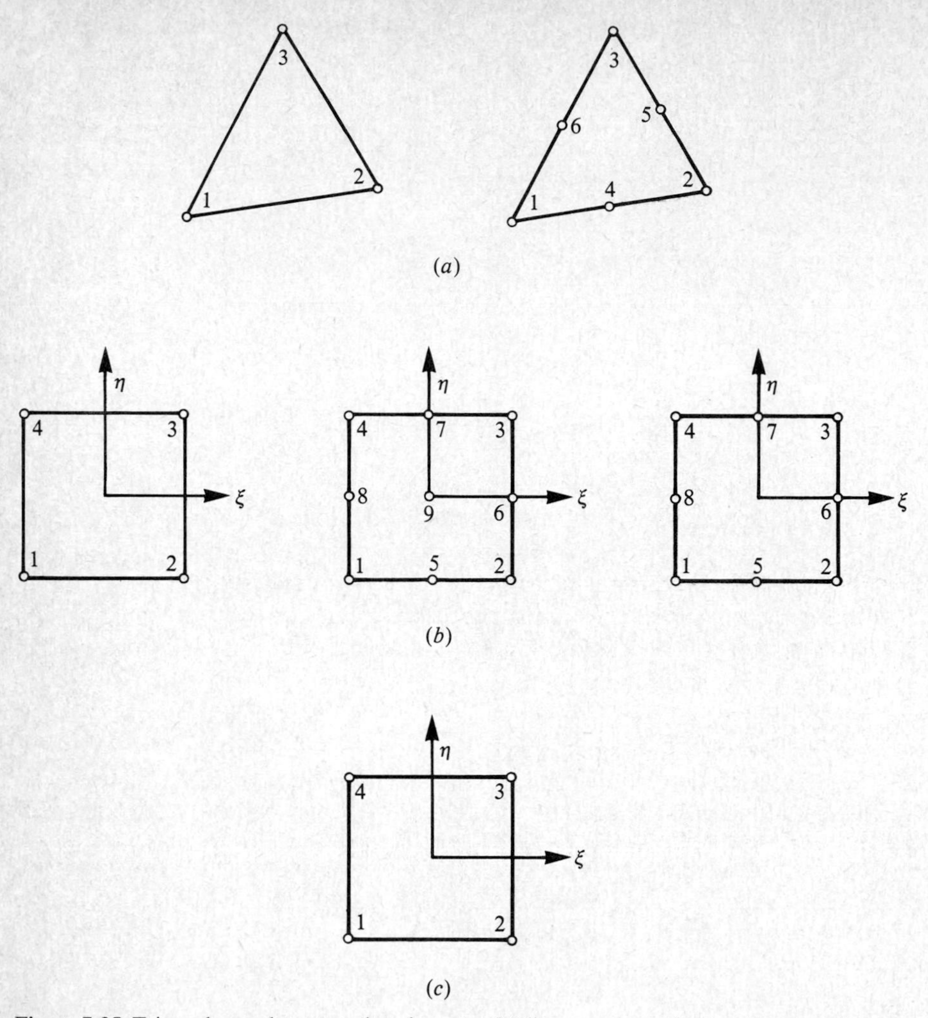

Figure 7-25 Triangular and rectangular elements: (*a*) linear and quadratic triangular elements, (*b*) linear and quadratic Lagrange elements, and (*c*) Hermite cubic element.

The following theorem gives conditions for the existence of solution and error estimates for the Dirichlet problem at hand. For a proof, see Ciarlet (1978) and Babuska and Aziz (1972).

Theorem 7-1 Let the following conditions hold:

(i) The bilinear form $B(\cdot, \cdot)$: $S_0^h(\Omega) \times S_0^h(\Omega) \rightarrow \mathbb{R}$ satisfies the conditions (5-39) of Theorem 5-5, with $\alpha = \alpha_h > 0$ and all h.

(ii) The family of subspaces $\{S_0^h(\Omega)\}$, $0 < h \leq 1$, satisfies the following conditions:

(*a*) For every h, let $\bar{\Omega}_e$, $1 \leq e \leq E$, denote a collection of finite elements whose assembly gives $\bar{\Omega}$.

(*b*) Let P_e be the polynomial space spanned by the element interpolation functions $\{\psi_i^e\}_{i=1}^n$. Upon assembly of the elements, the local interpolation functions $\{\psi_i^e\}$ give a set of global interpolation functions $\{\Phi_i\}_{i=1}^N$, which provide a basis for a finite-dimensional space, $S_0^h(\Omega)$.

(*c*) The space $S_0^h(\Omega)$ contains $\mathscr{P}_k(\Omega)$, the space of polynomials of degree $\leq k$ on Ω.

(*d*) For every $v \in H^s(\Omega)$ and every h, there exists a constant $C > 0$ and an element $\hat{v}_h \in S_0^h(\Omega)$ such that

$$\|v - \hat{v}_h\|_1 \leq Ch^p |v|_s \qquad \text{where } p = \min(k, s-1)$$

Then problem (7-123) has a unique solution. If $u \in H^s(\Omega) \subset H_0^1(\Omega)$, $s > 1$, then, as $h \to 0$, the error $\varepsilon = u - u_h$ satisfies the inequality

$$\|\varepsilon\|_1 \leq C\left(1 + \frac{M}{\alpha_h}\right) h^p |u|_s \qquad p = \min(k, s-1) \tag{7-124}$$

The inequality (7-124) states that, if α_h is uniformly bounded below by a positive constant for all h, the finite-element solution converges to the exact solution at the rate of p as $h \to 0$. When α_h depends on h, the actual rate of convergence, if it converges at all, is not given by p.

The existence and uniqueness results for the finite-element approximations of mixed and Neumann boundary-value problems can be established if the associated bilinear and linear forms satisfy the conditions of Theorems 5-6 and 5-7. Error estimates of the type (7-124) can be derived for these problems, with p depending on the spaces of exact and finite-element solutions.

7-6-5 Illustrative Examples

Here we consider a number of examples to illustrate the application of the method to two-dimensional problems.

Example 7-14 Consider the following mixed boundary-value problem for the Poisson equation

$$-\nabla^2 u = 1 \text{ in } \Omega = \{(x, y)\colon 0 < x, y < 1\}$$
$$\frac{\partial u}{\partial x}(0, y) = \frac{\partial u}{\partial y}(x, 0) = 0 \qquad u(1, y) = u(x, 1) = 0 \tag{7-125}$$

The domain and boundary conditions are shown in Fig. 7-26*a*. Note that this problem can be considered as one in Example 6-8 with $a = b = 1$ (see Fig. 6-2) and biaxial symmetry is exploited. We wish to solve the problem using the linear triangular and rectangular elements.

1. Solution by linear triangular elements. Due to the symmetry along the diagonal $x = y$, one need only model either the lower or upper triangular

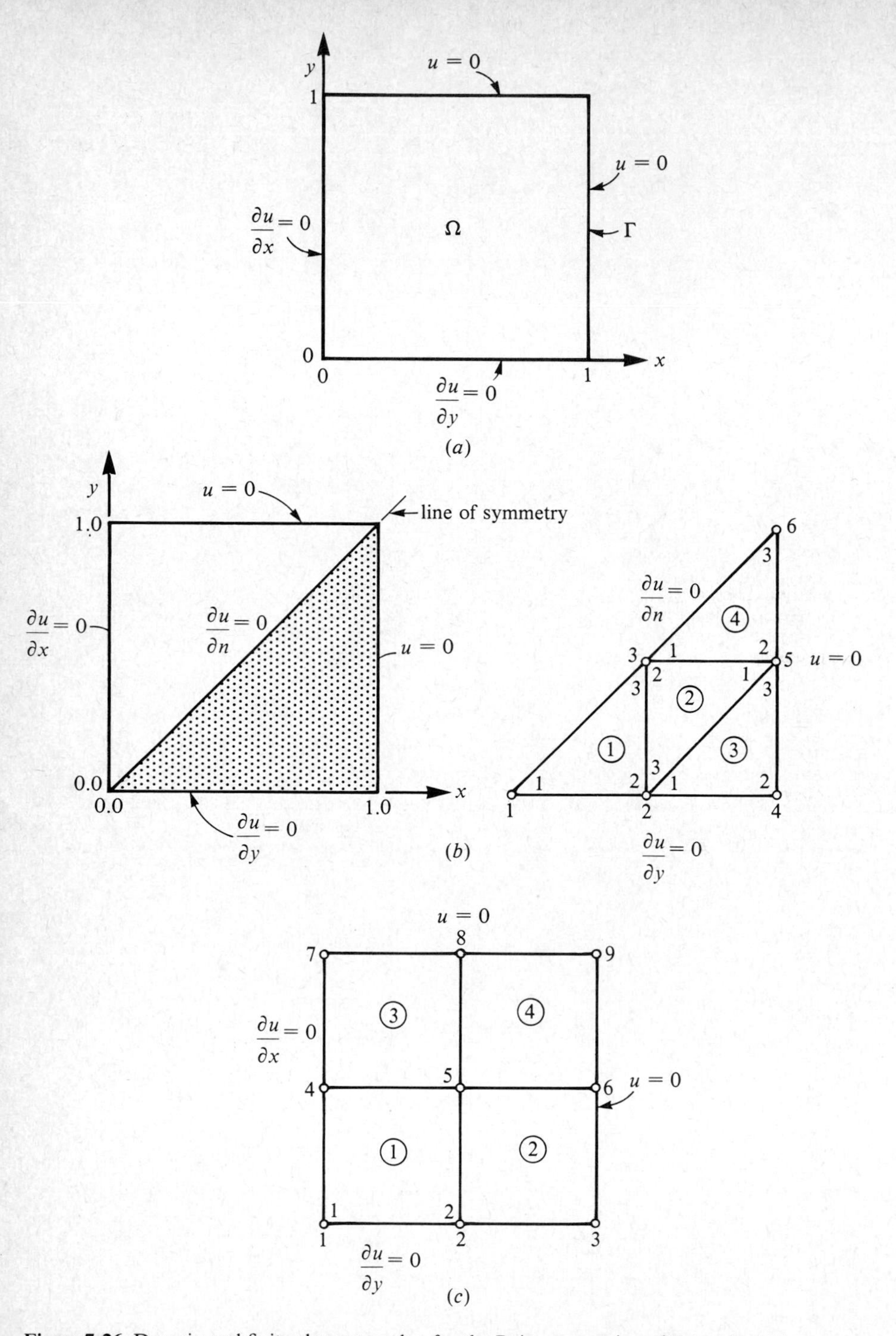

Figure 7-26 Domain and finite-element meshes for the Poisson equation of Example 7-14.

domain (see Fig. 7-26*b*). Along the line of symmetry the normal derivative of the primary variable is zero. We use a (uniform) mesh of four linear triangular elements to represent the domain. In the present case there is no domain discretization error involved in the problem. For a typical element of the mesh the element equations are given by

$$K_{ij}^e = \frac{1}{4A_e}(\beta_i^e\beta_j^e + \gamma_i^e\gamma_j^e) \qquad f_i^e = \frac{A_e}{3}$$

For example, for element 1 we have

$$[K^1] = \frac{1}{2}\begin{bmatrix} 1 & -1 & 0 \\ -1 & 2 & -1 \\ 0 & -1 & 1 \end{bmatrix} \qquad \{F^1\} = \frac{1}{24}\begin{Bmatrix} 1 \\ 1 \\ 1 \end{Bmatrix} + \begin{Bmatrix} P_1^1 \\ P_2^1 \\ P_3^1 \end{Bmatrix} \tag{7-126}$$

The assembled coefficient matrix for the finite-element mesh is 6 by 6 because there are six global nodes with one unknown per node. The assembled coefficient matrix and column vector are given by

$$[K] = \begin{bmatrix} K_{11}^1 & K_{12}^1 & K_{13}^1 & 0 & 0 & 0 \\ & K_{22}^1 + K_{33}^2 + K_{11}^3 & K_{23}^1 + K_{32}^2 & K_{12}^3 & K_{13}^3 + K_{31}^2 & 0 \\ & & K_{33}^1 + K_{22}^2 + K_{11}^4 & 0 & K_{21}^2 + K_{12}^4 & K_{13}^4 \\ & \text{symmetric} & & K_{22}^3 & K_{23}^3 & 0 \\ & & & & K_{11}^2 + K_{33}^3 + K_{22}^4 & K_{23}^4 \\ & & & & & K_{33}^4 \end{bmatrix} \tag{7-127a}$$

$$\{F\} = \begin{Bmatrix} F_1^1 \\ F_2^1 + F_3^2 + F_1^3 \\ F_3^1 + F_2^2 + F_1^4 \\ F_2^3 \\ F_1^2 + F_3^3 + F_2^4 \\ F_3^4 \end{Bmatrix} \tag{7-127b}$$

It can be shown that $[K^1] = [K^2] = [K^3] = [K^4]$ and $\{F^1\} = \{F^2\} = \{F^3\} = \{F^4\}$. We obtain from (7-126) and (7-127) the following assembled system of equations:

$$\frac{1}{2}\left[\begin{array}{ccc|ccc} 1 & -1 & 0 & 0 & 0 & 0 \\ -1 & 4 & -2 & -1 & 0 & 0 \\ 0 & -2 & 4 & 0 & -2 & 0 \\ \hline 0 & -1 & 0 & 2 & -1 & 0 \\ 0 & 0 & -2 & -1 & 4 & -1 \\ 0 & 0 & 0 & 0 & -1 & 1 \end{array}\right]\left\{\begin{array}{c} U_1 \\ U_2 \\ U_3 \\ \hline U_4 \\ U_5 \\ U_6 \end{array}\right\} = \frac{1}{24}\left\{\begin{array}{c} 1 \\ 3 \\ 3 \\ \hline 1 \\ 3 \\ 1 \end{array}\right\} + \left\{\begin{array}{c} P_1^1 \\ P_2^1 + P_3^2 + P_1^3 \\ P_3^1 + P_2^2 + P_1^4 \\ \hline P_2^3 \\ P_1^2 + P_3^3 + P_2^4 \\ P_3^4 \end{array}\right\} \tag{7-128}$$

The specified boundary conditions on the primary degrees of freedom of the problem are

$$U_4 = U_5 = U_6 = 0 \tag{7-129}$$

The specified secondary degrees of freedom (that is, q_n) and the internal equilibrium of the secondary variables are given by

$$\begin{aligned} P_1^1 &= 0 \\ P_2^1 + P_3^2 + P_1^3 &= 0 \\ P_3^1 + P_2^2 + P_1^4 &= 0 \end{aligned} \tag{7-130}$$

Thus we have the following condensed equations for the three unknown generalized displacements U_1, U_2, and U_3,

$$\begin{bmatrix} 0.5 & -0.5 & 0 \\ -0.5 & 2.0 & -1.0 \\ 0 & -1.0 & 2.0 \end{bmatrix} \begin{Bmatrix} U_1 \\ U_2 \\ U_3 \end{Bmatrix} = \frac{1}{24} \begin{Bmatrix} 1 \\ 3 \\ 3 \end{Bmatrix} \tag{7-131}$$

Solving Eq. (7-131) for U_i, $i = 1, 2, 3$ we obtain

$$\begin{Bmatrix} U_1 \\ U_2 \\ U_3 \end{Bmatrix} = \begin{Bmatrix} 0.3125 \\ 0.2292 \\ 0.1771 \end{Bmatrix} \tag{7-132}$$

2. *Solution by linear rectangular elements.* Note that we cannot exploit the symmetry along the diagonal $x = y$ to our advantage when we use rectangular elements. Therefore, we use a 2×2 uniform mesh of four linear rectangular elements (see Fig. 7-26*c*) to discretize a quadrant of the domain. Once again, no domain discretization error is introduced in the present case.

Since all elements in the mesh are identical, we will compute the element matrices for only one element, say element 1. We have

$$\psi_1 = (1 - 2x)(1 - 2y) \qquad \psi_2 = 2x(1 - 2y)$$

$$\psi_3 = 4xy \qquad \psi_4 = (1 - 2x)2y$$

$$K_{ij}^1 = \int_0^{0.5} \int_0^{0.5} \left(\frac{\partial \psi_i^1}{\partial x} \frac{\partial \psi_j^1}{\partial x} + \frac{\partial \psi_i^1}{\partial y} \frac{\partial \psi_j^1}{\partial y} \right) dx\, dy \tag{7-133}$$

$$f_i^1 = \int_0^{0.5} \int_0^{0.5} \psi_i^1 \, dx\, dy$$

Evaluating the integrals in Eq. (7-133), we obtain

$$[K^1] = \frac{1}{2} \begin{bmatrix} 4 & -1 & -2 & -1 \\ -1 & 4 & -1 & -2 \\ -2 & -1 & 4 & -1 \\ -1 & -2 & -1 & 4 \end{bmatrix} \qquad \{F^1\} = \frac{1}{16} \begin{Bmatrix} 1 \\ 1 \\ 1 \\ 1 \end{Bmatrix} + \begin{Bmatrix} P_1^1 \\ P_2^1 \\ P_3^1 \\ P_4^1 \end{Bmatrix} \tag{7-134}$$

The assembled equations are given by

$$\frac{1}{6}\begin{bmatrix} 4 & -1 & 0 & -1 & -2 & 0 & 0 & 0 & 0 \\ & 8 & -1 & -2 & -2 & -2 & 0 & 0 & 0 \\ & & 4 & 0 & -2 & -1 & 0 & 0 & 0 \\ & & & 8 & -2 & 0 & -1 & -2 & 0 \\ & & & & 16 & -2 & -2 & -2 & -2 \\ & & & & & 8 & 0 & -2 & -1 \\ & & \text{symmetric} & & & & 4 & -1 & 0 \\ & & & & & & & 8 & -1 \\ & & & & & & & & 4 \end{bmatrix}\begin{Bmatrix} U_1 \\ U_2 \\ U_3 \\ U_4 \\ U_5 \\ U_6 \\ U_7 \\ U_8 \\ U_9 \end{Bmatrix}$$

$$= \frac{1}{16}\begin{Bmatrix} 1 \\ 2 \\ 1 \\ 2 \\ 4 \\ 2 \\ 1 \\ 2 \\ 1 \end{Bmatrix} + \begin{Bmatrix} P_1^1 \\ P_2^1 + P_1^2 \\ P_2^2 \\ P_4^1 + P_1^3 \\ P_3^1 + P_4^2 + P_2^3 + P_1^4 \\ P_3^2 + P_2^4 \\ P_4^3 \\ P_3^3 + P_4^4 \\ P_3^4 \end{Bmatrix} \tag{7-135}$$

The specified primary degrees of freedom are

$$U_3 = U_6 = U_7 = U_8 = U_9 = 0$$

and the specified secondary degrees of freedom and the equilibrium of the internal generalized forces give (see Fig. 7-26*c*)

$$P_1^1 = 0 \qquad P_2^1 + P_1^2 = 0 \qquad P_4^1 + P_1^3 = 0$$

The condensed form of the matrix equation for the unknown U_i's is given by

$$\frac{1}{6}\begin{bmatrix} 4 & -1 & -1 & -2 \\ -1 & 8 & -2 & -2 \\ -1 & -2 & 8 & -2 \\ -2 & -2 & -2 & 16 \end{bmatrix}\begin{Bmatrix} U_1 \\ U_2 \\ U_4 \\ U_5 \end{Bmatrix} = \frac{1}{16}\begin{Bmatrix} 1 \\ 2 \\ 2 \\ 4 \end{Bmatrix} \tag{7-136}$$

The solution is given by ($U_2 = U_4$ by symmetry)

$$\begin{Bmatrix} U_1 \\ U_2 \\ U_5 \end{Bmatrix} = \begin{Bmatrix} 0.3107 \\ 0.2411 \\ 0.1929 \end{Bmatrix} \tag{7-137}$$

The finite-element solution $u(0, y) = u(x, 0)$ obtained by two different meshes of triangular elements and two different meshes of rectangular elements are compared in Table 7-10 with the 50-term series solution [see

Table 7-10 Comparison of the finite-element solutions $u(0, y)$ with the series solution of Eq. (7-125)

	Triangular elements		Rectangular elements		Series
y	4 elements	16 elements*	4 elements	16 elements*	solution
0.0	0.3125	0.3013	0.3107	0.3025	0.2947
0.25	0.2709†	0.2805	0.2759†	0.2857	0.2789
0.50	0.2292	0.2292	0.2411	0.2353	0.2293
0.75	0.1146†	0.1393	0.1205†	0.1429	0.1397
1.0	0.0000	0.0000	0.0000	0.0000	0.0000

* See Reddy (1984a) for the meshes.
† Interpolated values.

Eq. (6-112)]. The finite-element solution obtained by 16 triangular elements is the most accurate when compared to the series solution. However, the following calculation shows that the gradient of the solution predicted by the triangular element mesh is not as accurate as that given by the rectangular element mesh.

$$q_x(x, y) = \frac{\partial u}{\partial x} = \sum_{i=1}^{n} u_i^e \frac{\partial \psi_i^e}{\partial x}$$
$$q_y(x, y) = \frac{\partial u}{\partial y} = \sum_{i=1}^{n} u_i^e \frac{\partial \psi_i^e}{\partial y} \qquad (n = 3 \text{ or } 4) \qquad \text{(7-138)}$$

We compute the gradients for the four-element meshes.

Triangular element (4 elements)

$$q_x^1 = \frac{1}{2A} \sum_{i=1}^{3} u_i^1 \beta_i^1 = 2(U_2 - U_1) = -0.16667$$

$$q_y^1 = \frac{1}{2A} \sum_{i=1}^{3} u_i^1 \gamma_i^1 = 2(U_3 - U_2) = -0.10417$$

Rectangular element (4 elements)

$$q_x^1 = \sum_{i=1}^{4} u_i^1 \frac{\partial \psi_i^1}{\partial x} = -2U_1(1 - 2y) + 2U_2(1 - 2y) + 4yU_5 - 4yU_4$$

$$q_x^1(0.25, 0.25) = -0.11785$$

$$q_y^1 = \sum_{i=1}^{4} u_i^1 \frac{\partial \psi_i^1}{\partial y} = -2U_1(1 - 2x) + 2U_2(1 - 2x) + 4xU_5 - 4xU_4$$

$$q_y^1(0.25, 0.25) = -0.11785$$

The exact values at $x = y = 0.25$ are given by

$$q_x = q_y = -0.1193$$

The next example deals with the problem of heat flow in a square region with a heat source at the center of the plate. The point source (which is a secondary variable) can be incorporated into the problem one of the two ways: one way is to equate the heat source to the sum of the secondary variables from element nodes connecting at the global node where the heat source is specified. The alternative way is to distribute by interpolation the point source to the nodes of the element containing the source. In the first method we must have a node at the point where the source is specified and in the second method the source is located inside an element. Also, the second method gives a less accurate solution because mathematically the problem data is changed from one point source Q_0 to four point sources Q_i^e. Since the first method is the usual method we have used thus far, an explanation of the second method is given below.

Let Q_0^e be a source at point (x_0, y_0) in Ω^e. We can express Q_0^e as a function of position (x, y) by writing

$$Q_e(x, y) = Q_0^e\, \delta(x - x_0)\, \delta(y - y_0)$$

where $\delta(\cdot)$ denotes the Dirac delta function. Now using the standard expression for the computation of the contributions of a distributed source to the nodes, we obtain

$$\begin{aligned} Q_i^e &= \int_{\Omega^e} Q_e \psi_i^e \, dx\, dy \\ &= Q_0^e \psi_i^e(x_0, y_0) \end{aligned} \tag{7-139}$$

Now we consider an example.

Example 7-15 Consider the mixed boundary-value problem that arises in connection with steady heat flow in a planar medium,

$$\begin{gathered} -\left(\frac{\partial^2 u}{\partial x^2} + \frac{\partial^2 u}{\partial y^2}\right) = f \text{ in } \Omega = \{(x, y)\colon 0 < (x, y) < 2\} \\ u(x, 2) = u(0, y) = 1 \qquad \frac{\partial u}{\partial x}(2, y) = \frac{\partial u}{\partial y}(x, 0) = 0 \end{gathered} \tag{7-140}$$

where u denotes the temperature and $f = Q_0\, \delta(x - 1)\, \delta(y - 1)$ is the heat source. The analytical solution at points (1, 0), (1, 1), (2, 0), and (2, 1) is given by 0.756, 0.127, 0.719, and 0.756, respectively [see Lapidus and Pinder (1982), p. 97].

We investigate the effect of the method of incorporating the point source on the accuracy of the finite-element solution. We shall use the rectangular elements in the present example. The element calculations, assembly of elements and imposition of boundary conditions on the secondary variables were adequately described in the previous section and in the last example; therefore, we go straight to the discussion of the results.

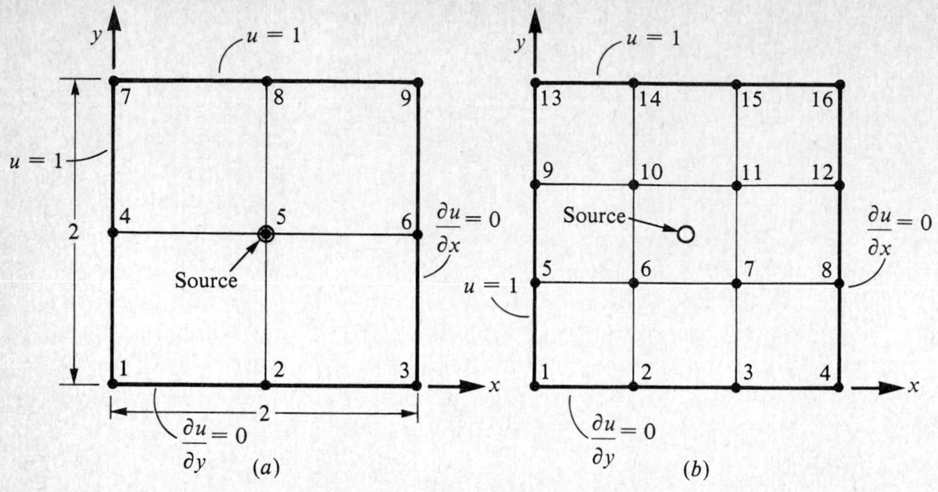

Figure 7-27 Finite-element model of a square domain with a point source (see Example 7-15).

For a 2 × 2 uniform mesh (see Fig. 7-27*a*), the specified primary degrees of freedom are

$$U_1 = U_4 = U_7 = U_8 = U_9 = 1.0$$

Note that at nodes 1 and 9 (points of mathematical singularity) we chose to impose the essential boundary conditions over the natural boundary condition. At node 5, we have $F_5 = Q_0 = -1.0$. The solution at nodes 2, 5, 3, and 6 is given by

$$0.786, 0.529, 0.657, 0.786$$

While the solution at nodes 2, 3, and 6 is in good agreement with the analytical solution, the solution at node 5 (where the source is applied) is not good.

For a 3 × 3 uniform mesh (see Fig. 7-27*b*), the source Q_0 at the center is distributed equally to the nodes 6, 7, 10, and 11. The resulting solution at node 4 ($x = 2$, $y = 0$) and the center (by interpolation) is given by

$$U_4 = 0.695 \qquad U_{\text{center}} = (U_6 + U_7 + U_{10} + U_{11})/4 = 0.708$$

Table 7-11 A comparison of the finite-element solution with the analytical solution of Eq. (7-140)

			Linear elements			Quadratic elements*		
x	y	Analytical	2	4	8	1	2	4
1	0	0.756	0.7857	0.7602	0.7579	0.7764	0.7308	0.7549
2	0	0.719	0.6571	0.7066	0.7171	0.6971	0.7065	0.7192
1	1	0.127	0.5286	0.4215	0.3120	0.7214	0.3669	0.2565
2	1	0.756	0.7857	0.7602	0.7579	0.7764	0.7308	0.7549

* 9-node rectangular elements (see Fig. 7-25*b*).

Note that the solution at the center is even less accurate than that obtained by placing the source at the node.

A refinement of the mesh should give an increasingly more accurate solution (with a node placed at the point of the source). A comparison of the finite-element solutions obtained by various meshes is presented in Table 7-11. It is clear that the convergence is slow for the solution at the center node.

The heat transfer problems in which heat is transferred between a body and the surrounding medium at the boundary are governed by a different type of natural boundary condition (i.e., the so-called Newton's type boundary condition). The following example illustrates the procedure for accounting for such boundary conditions.

Example 7-16 (*A convection heat transfer problem*) Here we consider the finite-element analysis of heat transfer problems that involve convection phenomenon (i.e., energy transfer between a solid body and a surrounding fluid medium). The governing equation is a special case of Eq. (7-97):

$$-\frac{\partial}{\partial x}\left(k_x \frac{\partial u}{\partial x}\right) - \frac{\partial}{\partial y}\left(k_y \frac{\partial u}{\partial y}\right) = f \tag{7-141}$$

where u is the temperature, k_x and k_y are conductivities (in Btu/h·ft °F or W/h·m·K) along the x- and y-directions, respectively, and f is the internal heat generation per unit volume (Btu/h·ft^3). The convective heat transfer problems are endowed with the natural boundary conditions of the Newton type,

$$\left(k_x \frac{\partial u}{\partial x} n_x + k_y \frac{\partial u}{\partial y} n_y\right) + \beta(u - u_\infty) - q_n = 0 \tag{7-142}$$

where β is the convective conductance (or convective heat transfer coefficient in Btu/h·ft^2 °F or W/h·m^2·K), and u_∞ is the temperature of the surrounding fluid medium. The first term accounts for heat transfer by conduction, the second for convection, and the third accounts for the specified heat flux (in Btu/h or W/h). It is the presence of the term $\beta(u - u_\infty)$ that requires some modification of the equations (7-101) and (7-103).

The variational form of the problem can be obtained as a special case from Eq. (7-98). The boundary integral in Eq. (7-98) should be modified using Eq. (7-142) for an element. We have, with $a_0 = a_{12} = a_{21} = 0$, $a_{11} = k_x$ and $a_{22} = k_y$,

$$0 = \int_{\Omega^e} \left\{ k_x^e \frac{\partial v_e}{\partial x} \frac{\partial u_e}{\partial x} + k_y^e \frac{\partial v_e}{\partial y} \frac{\partial u_e}{\partial y} - u_e f_e \right\} dx\, dy + \oint_{\Gamma^e} v_e[\beta^e(u_e - u_\infty^e) - q_n^e]\, ds \tag{7-143}$$

The finite-element model of Eq. (7-143) is given by

$$\sum_{j=1}^{n} [K_{ij}^e + H_{ij}^e]u_j^e = f_i^e + P_i^e + Q_i^e \tag{7-144a}$$

where
$$K_{ij}^e = \int_{\Omega^e} \left(k_x^e \frac{\partial \psi_i^e}{\partial x} \frac{\partial \psi_j^e}{\partial x} + k_y^e \frac{\partial \psi_i^e}{\partial y} \frac{\partial \psi_j^e}{\partial y} \right) dx\, dy$$

$$f_i^e = \int_{\Omega^e} f_e \psi_i^e \, dx\, dy \qquad P_i^e = \oint_{\Gamma^e} q_n^e \psi_i^e \, ds \tag{7-144b}$$

$$H_{ij}^e = \beta^e \oint_{\Gamma^e} \psi_i^e \psi_j^e \, ds \qquad Q_i^e = \beta^e \oint_{\Gamma^e} \psi_i^e u_\infty^e \, ds$$

Note that by setting $\beta^e = 0$, we get the heat transfer model that accounts for no convection.

For a linear triangular element, the coefficient matrices H_{ij}^e and P_i^e are given by

$$[H^e] = \frac{\beta_{12}^e h_{12}^e}{6} \begin{bmatrix} 2 & 1 & 0 \\ 1 & 2 & 0 \\ 0 & 0 & 0 \end{bmatrix} + \frac{\beta_{23}^e h_{23}^e}{6} \begin{bmatrix} 0 & 0 & 0 \\ 0 & 2 & 1 \\ 0 & 1 & 2 \end{bmatrix} + \frac{\beta_{31}^e h_{31}^e}{6} \begin{bmatrix} 2 & 0 & 1 \\ 0 & 0 & 0 \\ 1 & 0 & 2 \end{bmatrix}$$

$$\{Q^e\} = \frac{\beta_{12}^e u_\infty^e h_{12}^e}{2} \begin{Bmatrix} 1 \\ 1 \\ 0 \end{Bmatrix} + \frac{\beta_{23}^e u_\infty^e h_{23}^e}{2} \begin{Bmatrix} 0 \\ 1 \\ 1 \end{Bmatrix} + \frac{\beta_{31}^e u_\infty^e h_{31}^e}{2} \begin{Bmatrix} 1 \\ 0 \\ 1 \end{Bmatrix} \tag{7-145}$$

where β_{ij}^e are the convective heat transfer coefficients for side *i-j* of element Ω^e, and h_{ij}^e is the length of the side *i-j*. For a rectangular element, similar equations hold [for example, add a column and row of zeros to the first two matrices, and add two more matrices for sides 3-4 and 4-1 in Eq. (7-145)].

Note that, in actual calculations, one is not required to compute $[H^e]$ and $[Q^e]$ for all elements; the contributions from the convective terms should be calculated only for elements that have their boundaries coincide with the actual boundary that experiences convection. Even in those elements, one should compute the contributions to the nodes on the side that is subjected to convective heat transfer boundary conditions. A computational example of convective heat transfer is given in the author's text (1984a, pp. 232–235).

7-6-6 Weighted-Residual Finite-Element Models

Here we describe the Bubnov–Galerkin, least squares and collocation finite-element models of a simplified form of Eq. (7-97). We consider the Poisson equation

$$Au \equiv -\frac{\partial}{\partial x}\left(a_{11} \frac{\partial u}{\partial x}\right) - \frac{\partial}{\partial y}\left(a_{22} \frac{\partial u}{\partial y}\right) = f \text{ in } \Omega \tag{7-146}$$

In the weighted-residual finite-element model, we seek an approximate solution $u_e \in S^h \subset H^2(\Omega^e)$ of u in Ω^e in the form

$$u_e = \sum_{j=1}^{16} u_j^e \psi_j^e(x, y) \tag{7-147}$$

where $\psi_j^e(x, y)$ are the Hermite interpolation functions (see Table 7-9), and u_1^e, u_5^e, u_9^e, and u_{13}^e are the nodal values of u_e at the four nodes of the rectangular element, u_2^e, u_6^e, u_{10}^e, and u_{14}^e are the nodal values of $\partial u_e/\partial x$ at the four nodes, etc. The variational statement for an element is given by

$$0 = \int_{\Omega^e} v_e \left[-\frac{\partial}{\partial x}\left(a_{11}^e \frac{\partial u_e}{\partial x}\right) - \frac{\partial}{\partial y}\left(a_{22}^e \frac{\partial u_e}{\partial y}\right) - f_e \right] dx\, dy$$

which leads to the usual form of element equations

$$[K^e]\{u^e\} = \{f^e\} \tag{7-148}$$

where $[K^e]$ and $\{f^e\}$ are defined below for various special cases.

The Bubnov–Galerkin model For $v_e = \psi_i^e$, the coefficient matrix $[K^e]$ and column vector $\{f^e\}$ are defined by

$$\begin{aligned} K_{ij}^e &= -\int_{\Omega^e} \psi_i^e \left[\frac{\partial}{\partial x}\left(a_{11}^e \frac{\partial \psi_j^e}{\partial x}\right) + \frac{\partial}{\partial y}\left(a_{22}^e \frac{\partial \psi_j^e}{\partial y}\right) \right] dx\, dy \\ f_i^e &= \int_{\Omega^e} \psi_i^e f_e \, dx\, dy \end{aligned} \tag{7-149}$$

The least-squares model For $v_e = A(\psi_i^e)$, $[K^e]$, and $\{f^e\}$ are defined by the expressions

$$\begin{aligned} K_{ij}^e &= \int_{\Omega^e} \left[\frac{\partial}{\partial x}\left(a_{11}^e \frac{\partial \psi_i^e}{\partial x}\right) + \frac{\partial}{\partial y}\left(a_{22}^e \frac{\partial \psi_i^e}{\partial y}\right) \right] \\ &\quad \times \left[\frac{\partial}{\partial x}\left(a_{11}^e \frac{\partial \psi_j^e}{\partial x}\right) + \frac{\partial}{\partial y}\left(a_{22}^e \frac{\partial \psi_j^e}{\partial y}\right) \right] dx\, dy \\ f_i^e &= -\int_{\Omega^e} \left[\frac{\partial}{\partial x}\left(a_{11}^e \frac{\partial \psi_i^e}{\partial x}\right) + \frac{\partial}{\partial y}\left(a_{22}^e \frac{\partial \psi_i^e}{\partial y}\right) \right] f_e \, dx\, dy \end{aligned} \tag{7-150}$$

The collocation model In this model we select four collocation points per element and satisfy Eq. (7-146) exactly at those four points of each element. For the best results, the Gauss quadrature points are selected. We have

$$\begin{aligned} K_{ij}^e &= \left[-\frac{\partial}{\partial x}\left(a_{11}^e \frac{\partial \psi_j^e}{\partial x}\right) - \frac{\partial}{\partial y}\left(a_{22}^e \frac{\partial \psi_j^e}{\partial y}\right) \right]\Bigg|_{(x,\, y) = (x_i,\, y_i)} \\ f_i^e &= f_e(x_i, y_i) \end{aligned} \tag{7-151}$$

for $i = 1, 2, 3, 4$ and $j = 1, 2, \ldots, 16$. Once again (see Section 7-2-4) note that the coefficient matrix is rectangular (4×16) and that each coefficient of the global matrix and column vector are contributed to by no more than one element. After imposing the boundary conditions of the problem the number of linearly independent equations (which is equal to four times the number of elements in the finite element mesh) will be equal to the number of unknown nodal degrees of freedom.

We now consider an example of application of the weighted-residual models described above.

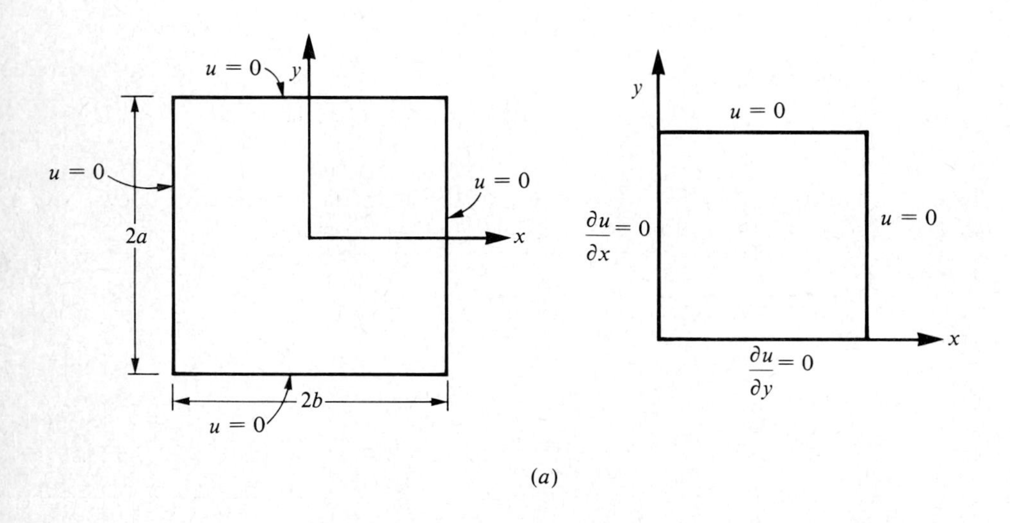

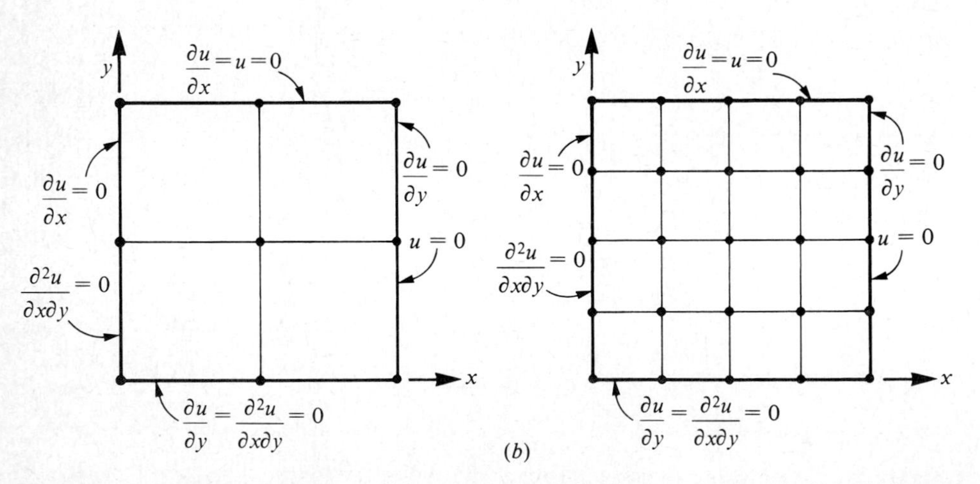

Figure 7-28 Domain, boundary conditions and the finite-element meshes of the problem in Example 7-17.

Example 7-17 Consider the Dirichlet problem for the Poisson equation (6-110),

$$-\nabla^2 u = 2 \text{ in } \Omega$$

$$u = 0 \text{ on } \Gamma$$

where Ω is a square region (see Fig. 7-28*a*). The exact solution of this problem is given by Eq. (6-112). Exploiting the biaxial symmetry, we model only a quadrant, say the region bounded by the positive axes.

Two different uniform meshes (see Fig. 7-28*b*) of Hermite rectangular elements are used to solve the problem by various finite-element models. Since the element has u, $\partial u/\partial x$, $\partial u/\partial y$, and $\partial^2 u/\partial x\, \partial y$ as nodal degrees of freedom, we must impose all known boundary values of these quantities (which are not readily known a priori). This can be considered as a drawback of the weighted-residual finite-element models (especially, the collocation finite-

Table 7-12 A comparison of the various finite-element solutions with the exact solution of the Dirichlet problem for the Poisson equation (6-110)

		$u(x, y)$				$\partial u/\partial x$			
x	y	Exact	Bubnov–Galerkin*	Least squares	Collo-cation	Exact	Bubnov–Galerkin*	Least squares	Collo-cation
0.0	0.0	0.58936	0.58903† 0.58935	0.58902 0.58935	0.58932 0.58937	0.00000	0.00000	0.00000	0.00000
0.25	0.0	0.55776	0.55774	0.55774	0.55776	0.25538	0.25568	0.25568	0.25570
0.50	0.0	0.45868	0.45837 0.45866	0.45846 0.45886	0.45862 0.45868	0.54549	0.54436 0.54541	0.54540 0.54543	0.54548 0.54555
0.75	0.0	0.27945	0.27944	0.27945	0.27946	0.90265	0.90176	0.90184	0.90192
1.00	0.00	0.00000	0.00000	0.00000	0.00000	1.3349	1.34628 1.3504	1.34912 1.35064	1.35068 1.35064
0.25	0.25	0.52830	0.52827	0.52827	0.52829	0.23827	0.2368	0.23859	0.23859
0.50	0.25	0.43560	0.43557	0.43558	0.43559	0.51168	0.51162	0.51162	0.51166
0.75	0.25	0.26665	0.26664	0.26665	0.26665	0.85513	0.85416	0.85432	0.8544
1.00	0.25	0.00000	0.00000	0.00000	0.00000	1.2819	1.29712	1.29744	1.29752
0.50	0.50	0.36230	0.36192 0.36226	0.36197 0.36226	0.36225 0.36228	0.40785	0.40744 0.40789	0.40692 0.40783	0.40724 0.40782
0.75	0.50	0.22548	0.22544	0.22545	0.22547	0.70416	0.70317	0.70314	0.70337
1.00	0.50	0.00000	0.00000	0.00000	0.00000	1.1102	1.11576 1.12448	1.1182 1.1252	1.12268 1.12552
0.75	0.75	0.14564	0.14557	0.14557	0.14563	0.42422	0.42355	0.42309	0.42315

* The Ritz finite-element solution coincides with the Galerkin solution for the same choice of the Hermite interpolation.

† The first line denotes the solution obtained by 2 × 2 mesh and the second line by 4 × 4 mesh.

element model) compared to the Ritz finite-element model, where the only boundary conditions are on u and $\partial u/\partial n$ (for the problem at hand: $u = 0$ on the $x = 1$ and $y = 1$ lines, $\partial u/\partial x = 0$ on the $x = 0$ line and $\partial u/\partial y = 0$ on the $y = 0$ line). The boundary conditions for the weighted residual models are indicated in Fig. 7-28*b*. For the 2×2 mesh, the number of known boundary conditions is twenty whereas the number of total nodal variables is thirty-six. Thus, for the collocation model we have sixteen unknowns and sixteen equations, four from each element. Similarly, for the 4×4 mesh we have thirty-six boundary conditions among one hundred nodal variables, requiring sixty-four collocation equations, which are provided by the sixteen elements.

The finite element solutions obtained by the three finite-element models for the two meshes are compared with the exact solution (6-112) in Table 7-12. The collocation finite-element solution is relatively more accurate than the other two solutions. The numerical convergence of all three models is apparent from the results.

PROBLEMS 7-2

Unless otherwise stated, all problems are to be solved using the Ritz finite-element model

7-2-1 Give the Ritz finite-element formulation of the following equation: $-\nabla^2 u + c_0 u = f$ in $\Omega \subset \mathbb{R}^2$ such that the model accounts for a boundary condition of the form $\partial u/\partial n + \beta u = g$ on Γ.

7-2-2 Give the least squares model of the above problem.

7-2-3 Evaluate the coefficients $M_{ij}^e = \int_{\Omega^e} \psi_i^e \psi_j^e \, dx \, dy$ when Ω^e is a linear triangular element.

7-2-4 Repeat Prob. 7-2-3 for the case in which Ω^e is a linear rectangle with sides a and b.

7-2-5 Evaluate $f_i^e = \int_{\Omega^e} \psi_i^e \, dx \, dy$ for a linear triangular and rectangular element.

7-2-6 Give the Ritz finite-element model of the equation governing an axisymmetric problem

$$-\frac{1}{r}\left[\frac{\partial}{\partial r}\left(r\frac{\partial u}{\partial r}\right) + \frac{\partial}{\partial z}\left(r\frac{\partial u}{\partial z}\right)\right] = f(r, z)$$

and compute the coefficients K_{ij}^e and f_i^e for a linear triangular element.

7-2-7 Derive the interpolation functions ψ_i^e $(i = 1, 2, \ldots, 9)$ of the Lagrange rectangular element (see Fig. P7) using the interpolation properties $\psi_i^e(\xi_j, \eta_j) = \delta_{ij}$.

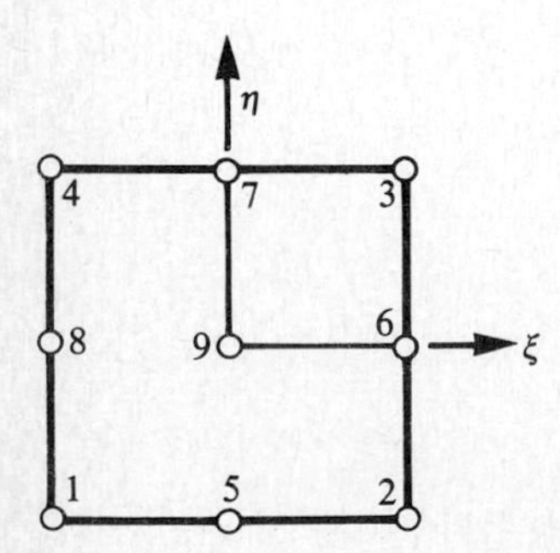

Figure P7

7-2-8 Find the contribution of a point source Q_0 located at point (2, 3) to the nodes of the triangular and rectangular elements shown in Fig. P8.

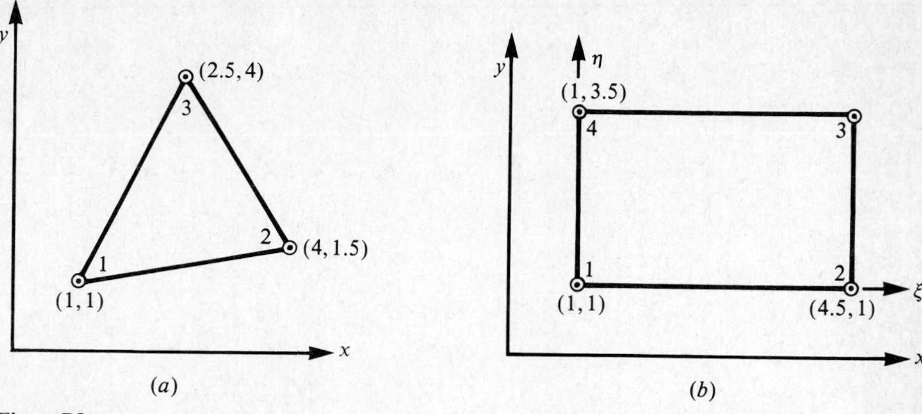

Figure P8

7-2-9 If the nodal values of the elements shown in Fig. P9 are $u_1 = 0.2645$, $u_2 = 0.2172$, $u_3 = 0.1800$ for the triangular element and $u_1 = 0.2173$, $u_2 = 0.2232$, $u_3 = 0.1870$, $u_4 = 0.2232$ for the rectangular element, compute u, $\partial u/\partial x$ and $\partial u/\partial y$ at the point $(x, y) = (0.375, 0.375)$.

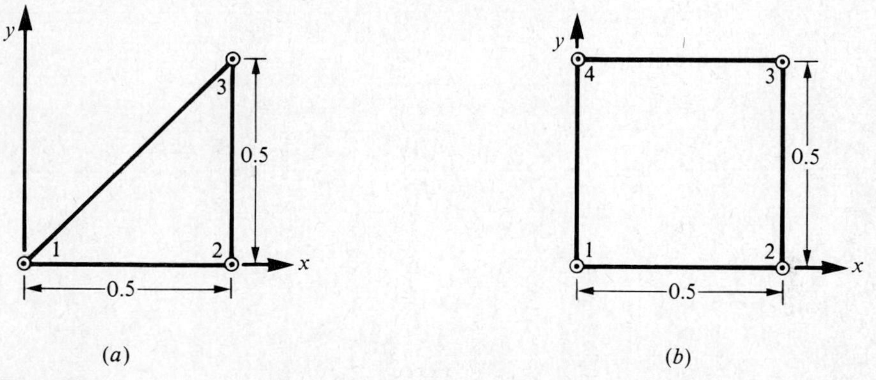

Figure P9

7-2-10 (Space-time finite elements) Consider the time-dependent problem $\partial^2 u/\partial x^2 = \partial u/\partial t$, $0 < x < 1$, $t > 0$; $u(0, t) = 0$, $(\partial u/\partial t)(1, t) = 1$, $u(x, 0) = x$. Use linear rectangular elements in the xt-plane to model the problem. Note that the finite-element model is given by $[K^e]\{u^e\} = \{P^e\}$ where

$$K_{ij}^e = \int_0^{\Delta t} \int_{x_e}^{x_{e+1}} \left(\frac{\partial \psi_i^e}{\partial x} \frac{\partial \psi_j^e}{\partial x} + \psi_i \frac{\partial \psi_j}{\partial t} \right) dx\, dt$$

$$P_1^e = - \int_0^{\Delta t} \frac{\partial u}{\partial x}\, dt \bigg|_{x = x_e} \qquad P_2^e = \int_0^{\Delta t} \frac{\partial u}{\partial x}\, dt \bigg|_{x = x_{e+1}}$$

7-2-11 Let the transformation between the global coordinates (x, y) and local normalized coordinates (ξ, η) in a Lagrange element Ω^e be

$$x_e = \sum_{i=1}^{n} x_i^e \psi_i^e(\xi, \eta) \qquad y_e = \sum_{i=1}^{n} y_i^e \psi_i^e(\xi, \eta)$$

where (x_i^e, y_i^e) denote the global coordinates of the element nodes. The differential lengths in the two coordinates are related by

$$dx_e = \frac{\partial x_e}{\partial \xi}\,d\xi + \frac{\partial x_e}{\partial \eta}\,\partial\eta, \qquad dy_e = \frac{\partial y_e}{\partial \xi}\,d\xi + \frac{\partial y_e}{\partial \eta}\,d\eta$$

or

$$\begin{Bmatrix} dx_e \\ dy_e \end{Bmatrix} = \begin{bmatrix} \dfrac{\partial x_e}{\partial \xi} & \dfrac{\partial x_e}{\partial \eta} \\ \dfrac{\partial y_e}{\partial \xi} & \dfrac{\partial y_e}{\partial \eta} \end{bmatrix} \begin{Bmatrix} d\xi \\ d\eta \end{Bmatrix} = [\mathbf{T}]\begin{Bmatrix} d\xi \\ d\eta \end{Bmatrix}$$

In the finite element literature the transpose of [T] is called the Jacobian matrix, [J]. Show that the derivatives of the interpolation functions $\psi_i^e(\xi, \eta)$ with respect to the global coordinates (x, y) are related to their derivatives with respect to the local coordinates (ξ, η) by

$$\begin{Bmatrix} \dfrac{\partial \psi_i^e}{\partial x} \\ \dfrac{\partial \psi_i^e}{\partial y} \end{Bmatrix} = [\mathbf{J}]^{-1} \begin{Bmatrix} \dfrac{\partial \psi_i^e}{\partial \xi} \\ \dfrac{\partial \psi_i^e}{\partial \eta} \end{Bmatrix}$$

and

$$\begin{Bmatrix} \dfrac{\partial^2 \psi_i^e}{\partial x^2} \\ \dfrac{\partial^2 \psi_i^e}{\partial y^2} \\ \dfrac{\partial^2 \psi_i^e}{\partial x\,\partial y} \end{Bmatrix} = \begin{bmatrix} \left(\dfrac{\partial x_e}{\partial \xi}\right)^2 & \left(\dfrac{\partial y_e}{\partial \xi}\right)^2 & 2\dfrac{\partial x_e}{\partial \xi}\dfrac{\partial y_e}{\partial \xi} \\ \left(\dfrac{\partial x_e}{\partial \eta}\right)^2 & \left(\dfrac{\partial y_e}{\partial \eta}\right)^2 & 2\dfrac{\partial x_e}{\partial \eta}\dfrac{\partial y_e}{\partial \eta} \\ \dfrac{\partial x_e}{\partial \xi}\dfrac{\partial x_e}{\partial \eta} & \dfrac{\partial y_e}{\partial \xi}\dfrac{\partial y_e}{\partial \eta} & \left(\dfrac{\partial x_e}{\partial \eta}\dfrac{\partial y_e}{\partial \xi} + \dfrac{\partial x_e}{\partial \xi}\dfrac{\partial y_e}{\partial \eta}\right) \end{bmatrix}^{-1} \left\{ \begin{Bmatrix} \dfrac{\partial^2 \psi_i^e}{\partial \xi^2} \\ \dfrac{\partial^2 \psi_i^e}{\partial \eta^2} \\ \dfrac{\partial^2 \psi_i^e}{\partial \xi\,\partial \eta} \end{Bmatrix} - \begin{bmatrix} \dfrac{\partial^2 x_e}{\partial \xi^2} & \dfrac{\partial^2 y_e}{\partial \xi^2} \\ \dfrac{\partial^2 x_e}{\partial \eta^2} & \dfrac{\partial^2 y_e}{\partial \eta^2} \\ \dfrac{\partial^2 x_e}{\partial \xi\,\partial \eta} & \dfrac{\partial^2 y_e}{\partial \xi\,\partial \eta} \end{bmatrix} \begin{Bmatrix} \dfrac{\partial \psi_i^e}{\partial x} \\ \dfrac{\partial \psi_i^e}{\partial y} \end{Bmatrix} \right\}$$

7-2-12 (continuation of Prob. 7-2-11) Show that the Jacobian can be computed from the equation

$$[\mathbf{J}] = \begin{bmatrix} \dfrac{\partial \psi_1^e}{\partial \xi} & \dfrac{\partial \psi_2^e}{\partial \xi} & \cdots & \dfrac{\partial \psi_n^e}{\partial \xi} \\ \dfrac{\partial \psi_1^e}{\partial \eta} & \dfrac{\partial \psi_2^e}{\partial \eta} & \cdots & \dfrac{\partial \psi_n^e}{\partial \eta} \end{bmatrix} \begin{bmatrix} x_1^e & y_1^e \\ x_2^e & y_2^e \\ \vdots & \vdots \\ x_n^e & y_n^e \end{bmatrix}$$

7-2-13 Consider the quadrilateral element shown in Fig. P13. Using the linear interpolation functions of a rectangular element transform the element to the local coordinate system and sketch the transformed element.

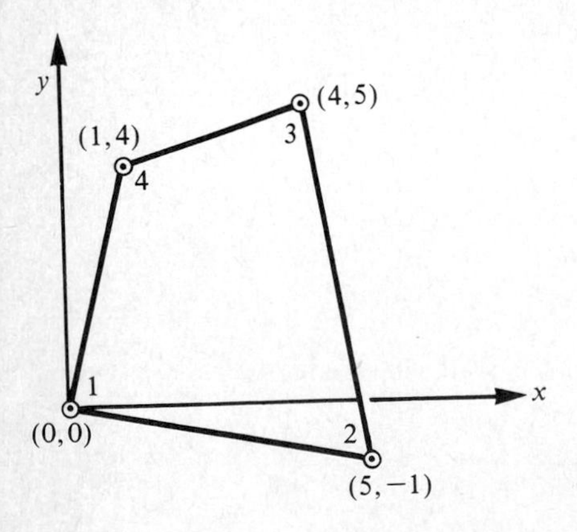

Figure P13

7-2-14 For the element in Prob. 7-2-13, express the following integral in the local coordinates,

$$K_{ij}^e = \int_{\Omega^e} \left[a\left(\frac{\partial \psi_i^e}{\partial x} \frac{\partial \psi_j^e}{\partial x} + \frac{\partial \psi_i^e}{\partial y} \frac{\partial \psi_j^e}{\partial y} \right) + b\psi_i^e \psi_j^e \right] dx\, dy$$

where Ω^e is the quadrilateral element of Prob. 7-2-13.

7-2-15 Find the Jacobian matrix for the nine-node rectangular element shown in Fig. P15. What is the determinant of the Jacobian matrix?

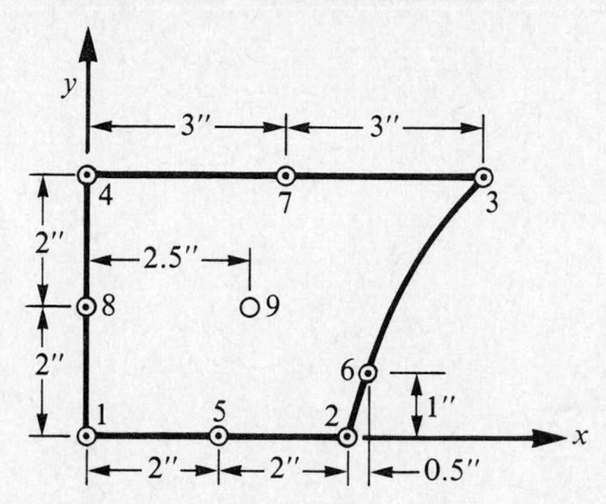

Figure P15

7-2-16 For the eight-node element shown in Fig. P16, show that the x-coordinate along the side 1-2 is related to the ξ-coordinate by the relation

$$x = -\tfrac{1}{2}\xi(1 - \xi)x_1^e + \tfrac{1}{2}\xi(1 + \xi)x_2^e + (1 - \xi^2)x_5^e$$

and that the relations

$$\xi = 2\sqrt{\frac{x}{a}} - 1, \qquad \frac{\partial x}{\partial \xi} = \sqrt{xa}$$

hold. Also, show that

$$u_e(x, 0) = -\left(2\sqrt{\frac{x}{a}} - 1\right)\left(1 - \sqrt{\frac{x}{a}}\right)u_1^e + \left(-1 + 2\sqrt{\frac{x}{a}}\right)\sqrt{\frac{x}{a}}\,u_2^e + 4\left(\sqrt{\frac{x}{a}} - \frac{x}{a}\right)u_5^e$$

$$\frac{\partial u_e(x, 0)}{\partial x} = \frac{1}{\sqrt{xa}}\left[-\frac{1}{2}\left(3 - 4\sqrt{\frac{x}{a}}\right)u_1^e + \frac{1}{2}\left(-1 + 4\sqrt{\frac{x}{a}}\right)u_2^e + 2\left(1 - 2\sqrt{\frac{x}{a}}\right)u_5^e\right]$$

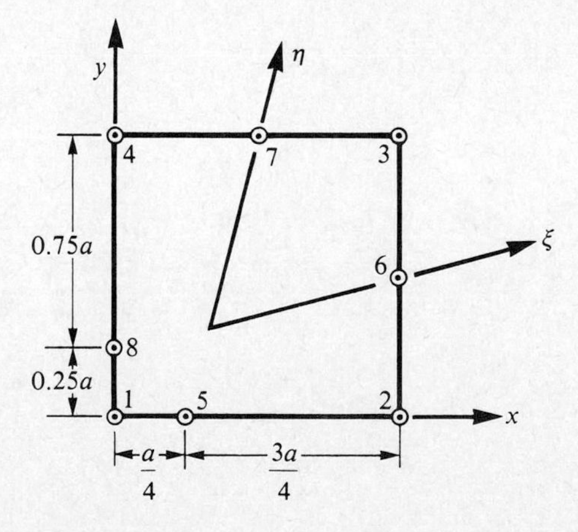

Figure P16

Thus, $\partial u_e/\partial x$ grows at a rate of $1/\sqrt{xa}$ as x approaches zero along the side 1-2. In other words, we have a $1/\sqrt{x}$ singularity at node 1.

7-2-17 Using the tensor product of the Hermite cubic interpolations functions in Eq. (7-60), obtain the Hermite cubic interpolation functions (sixteen of them) for the four-node rectangular element.

7-2-18 For the linear triangular element shown in P18, show that the element coefficient matrix associated with the Laplace operator $-\nabla^2$

$$K_{ij}^e = \int_{\Omega^e} \left(\frac{\partial \psi_i^e}{\partial x} \frac{\partial \psi_j^e}{\partial x} + \frac{\partial \psi_i^e}{\partial y} \frac{\partial \psi_j^e}{\partial y} \right) dx\, dy$$

is given by

$$[K^{(e)}] = \frac{1}{2} \begin{bmatrix} \alpha + \beta & -\alpha & -\beta \\ -\alpha & \alpha & 0 \\ -\beta & 0 & \beta \end{bmatrix} \qquad \alpha = \frac{b}{a} \quad \text{and} \quad \beta = \frac{a}{b}$$

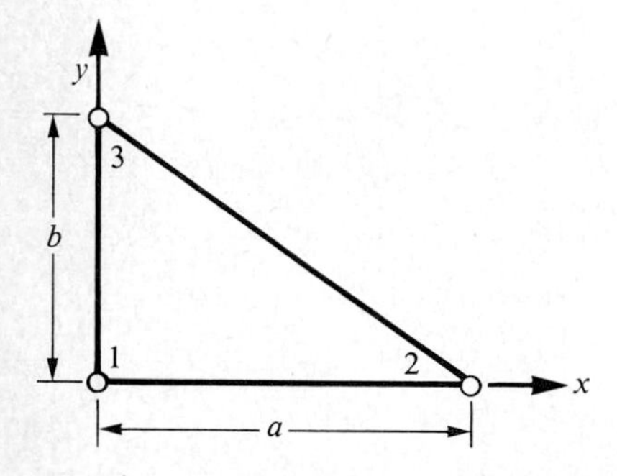

Figure P18

7-2-19 For the linear rectangular element shown in Fig. P19, show that the element coefficient matrix associated with the Laplace operator is given by

$$[K^{(e)}] = \frac{1}{6} \begin{bmatrix} 2(\alpha + \beta) & -2\alpha + \beta & -(\alpha + \beta) & \alpha - 2\beta \\ -2\alpha + \beta & 2(\alpha + \beta) & \alpha - 2\beta & -(\alpha + \beta) \\ -(\alpha + \beta) & \alpha - 2\beta & 2(\alpha + \beta) & -2\alpha + \beta \\ \alpha - 2\beta & -(\alpha + \beta) & -2\alpha + \beta & 2(\alpha + \beta) \end{bmatrix} \qquad \alpha = \frac{b}{a} \qquad \beta = \frac{a}{b}$$

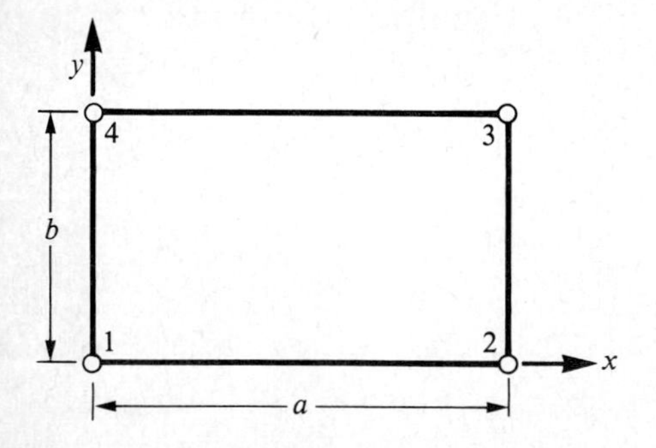

Figure P19

7-2-20 Find the coefficient matrix associated with the Laplace operator when the rectangular element in Fig. P19 is divided into two triangles by joining node 1 to node 3.

7-2-21 Solve the Laplace equation on the square domain with the mesh of triangular elements shown in Fig. P21. You are required to give the 2 × 2 matrix equations for the two unknowns U_4 and U_5.

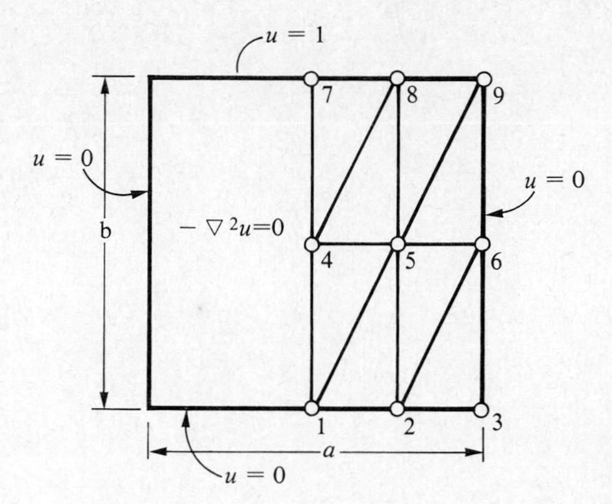

Figure P21

7-2-22 Repeat Prob. 7-2-21 using the linear rectangular elements.

7-2-23 Solve the Laplace equation for the unit square domain and boundary conditions given in Fig. P23. Use one rectangular element.

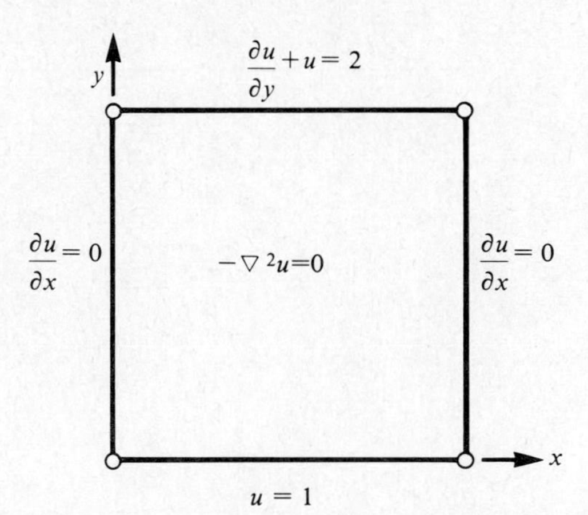

Figure P23

7-2-24 Use two triangular elements to solve the problem in Fig. P23. Use the mesh obtained by joining points (1, 0) and (0, 1).

7-2-25 Solve the Poisson equation $-\nabla^2 u = 2$ in the square whose vertices are at (0, 0), (1, 0), (1, 1), and (0, 1). The boundary conditions are: $u(0, y) = y^2$, $u(x, 0) = x^2$, $u(1, y) = 1 - y$, $u(x, 1) = 1 - x$. Use four linear rectangular elements (2 × 2 mesh).

7-2-26 Solve Prob. 7-2-25 using four triangular elements (use the mesh shown in Fig. 7-26*b*).

7-2-27 Solve Prob. 7-2-25 using the mesh of a rectangle and two triangles, as shown in Fig. P27.

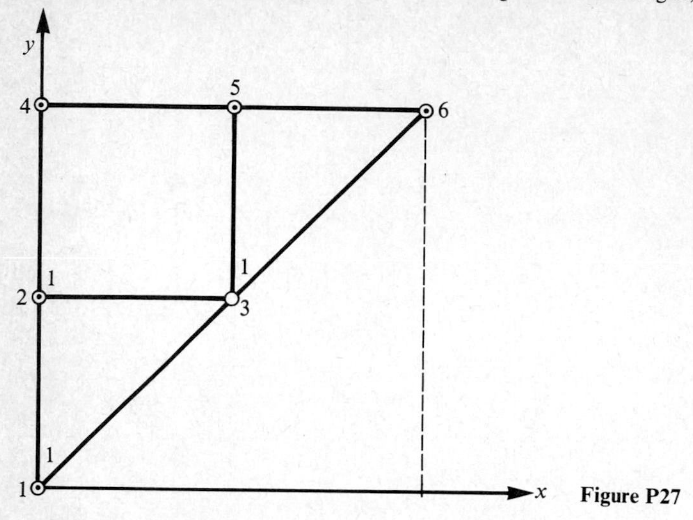

Figure P27

7-2-28 Solve the Poisson equation (torsion problem) $-\nabla^2 u = 2$ in Ω, $u = 0$ on Γ_1, $\partial u/\partial n = 0$ on Γ_2, where Ω is the first quadrant bounded by the parabola $y = 1 - x^2$ and the coordinate axes (see Fig. P28), and Γ_1 and Γ_2 are the boundaries shown in Fig. P28.

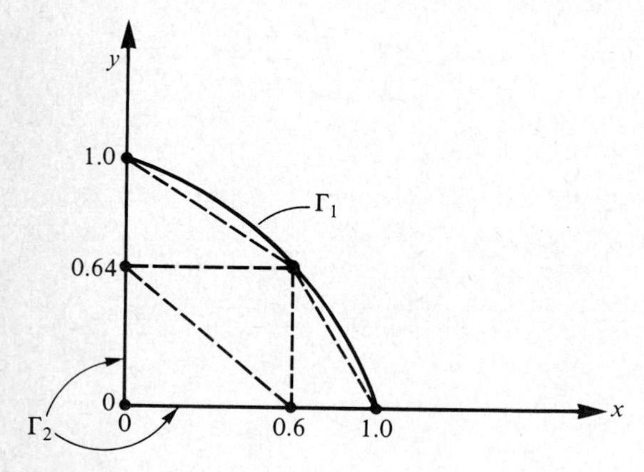

Figure P28

7-2-29 Solve Prob. 7-2-25 using the least squares method.

7-2-30 Solve Prob. 7-2-25 using the collocation method.

7-7 COUPLED SECOND-ORDER EQUATIONS

7-7-1 Introduction

In Sec. 7-6 we considered the finite-element analysis of second-order field equations that involved only one dependent unknown (hence, only one partial differential equation). Here we consider a system (often, a pair) of second-order, coupled and irreducible, partial differential equations in as many dependent variables as the number of equations. The word 'coupled' is used to imply that the same dependent variables appear in more than one equation of the set, and therefore no equation can be solved independently of the other in the set. When the equations are not coupled, the theory presented in Section 7-6 applies. Examples of two-dimensional problems in which coupled equations arise are provided by (i) plane elastic deformation of a linear elastic solid, (ii) flow of an incompressible viscous fluid, and (iii) bending of an elastic plate using a shear deformation theory, among many others.

The primary objective of this section is two-fold: first, to study the finite-element formulations of coupled partial differential equations that are irreducible (to a single equation), and second, to show how the Lagrange and Hermite finite elements developed in Sec. 7-6-3 can be used in the solution of the coupled partial differential equations in two dimensions. These objectives are accomplished via the study of the three types of problems mentioned above.

7-7-2 Plane Elasticity

Recall from Sec. 2-6-2 (also, see Example 5-9) that the equations governing the deformation of a plane elastic body are given by

$$\left.\begin{aligned} -\frac{\partial}{\partial x}\left(c_{11}\frac{\partial u}{\partial x}+c_{12}\frac{\partial v}{\partial y}\right)-c_{66}\frac{\partial}{\partial y}\left(\frac{\partial u}{\partial y}+\frac{\partial v}{\partial x}\right)=f_x \\ -c_{66}\frac{\partial}{\partial x}\left(\frac{\partial u}{\partial y}+\frac{\partial v}{\partial x}\right)-\frac{\partial}{\partial y}\left(c_{12}\frac{\partial u}{\partial x}+c_{22}\frac{\partial v}{\partial y}\right)=f_y \end{aligned}\right\}\quad \text{in } \Omega \qquad (7\text{-}152)$$

where $c_{ij} > 0$ are the material constants for an elastic body [the same as those in Eq. (2-66)], u and v are the components of the displacement vector and f_x and f_y are the components of the body force vector. Equations (7-152) are appended with the appropriate boundary conditions of the problem. The essential and natural boundary conditions associated with Eq. (7-152) are given by (see Sec. 4-3)

Essential:

$$u = \hat{u},\ v = \hat{v} \text{ on } \Gamma_1 \qquad (7\text{-}153a)$$

Natural:

$$\left.\begin{aligned} t_x \equiv \left(c_{11}\frac{\partial u}{\partial x}+c_{12}\frac{\partial v}{\partial y}\right)n_x + c_{66}\left(\frac{\partial u}{\partial y}+\frac{\partial v}{\partial x}\right)n_y = \hat{t}_x \\ t_y \equiv c_{66}\left(\frac{\partial u}{\partial y}+\frac{\partial v}{\partial x}\right)n_x + \left(c_{12}\frac{\partial u}{\partial x}+c_{22}\frac{\partial v}{\partial y}\right)n_y = \hat{t}_y \end{aligned}\right\}\quad \text{on } \Gamma_2 \qquad (7\text{-}153b)$$

We can derive the finite element model of the plane elasticity two ways: one is to make use of the quadratic functional, namely the total potential energy functional [see Eq. (5-62)] for a typical element, and the other is to make use of the weak form of Eqs. (7-152) [see Eq. (5-59)]. The latter approach is in conformity with the developments of Chaps. 4 through 7. Therefore we shall use the second approach, which begins with the variational form of the given differential equations (7-152). See the book by the author (1984a, pp. 268–270) for details of the first approach.

Variational formulation The variational form of (7-152) over an element Ω^e is given by multiplying the first equation with a test function w_1^e and the second one with a test function w_2^e and integrating the results by parts (to trade the second differential to w_i^e, $i = 1, 2$). We obtain

$$\begin{aligned} 0 &= \int_{\Omega^e} \left\{ \frac{\partial w_1^e}{\partial x}\left(c_{11}^e \frac{\partial u_e}{\partial x} + c_{12}^e \frac{\partial v_e}{\partial y}\right) + c_{66}^e \frac{\partial w_1^e}{\partial y}\left(\frac{\partial u_e}{\partial y} + \frac{\partial v_e}{\partial x}\right) - w_1^e f_x^e \right\} dx\,dy \\ &\qquad - \oint_{\Gamma^e} w_1^e t_x^e\,ds \\ 0 &= \int_{\Omega^e} \left\{ c_{66}^e \frac{\partial w_2^e}{\partial x}\left(\frac{\partial u_e}{\partial y} + \frac{\partial v_e}{\partial x}\right) + \frac{\partial w_2^e}{\partial y}\left(c_{12}^e \frac{\partial u_e}{\partial x} + c_{22}^e \frac{\partial v_e}{\partial y}\right) - w_2^e f_y^e \right\} dx\,dy \\ &\qquad - \oint_{\Gamma^e} w_2^e t_y^e\,ds \end{aligned} \tag{7-154}$$

where t_x^e and t_y^e denote the element boundary forces defined in Eq. (7-153*b*).

Finite element model Note that the primary variables of the formulation are (u, v) and the secondary variables are (t_x and t_y). Therefore, the Lagrange interpolation functions (linear, quadratic, etc.) are admissible. Using the finite-element interpolation of the form

$$u_e = \sum_{j=1}^{n} u_j^e \psi_j^e \qquad v_e = \sum_{j=1}^{n} v_j^e \psi_j^e \tag{7-155}$$

for the approximation of u_e and v_e in Ω^e, and $w_1^e = w_2^e = \psi_i^e$ in Eq. (7-154) we obtain

$$\begin{bmatrix} [K^{11e}] & [K^{12e}] \\ [K^{12e}]^T & [K^{22e}] \end{bmatrix} \begin{Bmatrix} \{u^e\} \\ \{v^e\} \end{Bmatrix} = \begin{Bmatrix} \{f^{1e}\} \\ \{f^{2e}\} \end{Bmatrix} + \begin{Bmatrix} \{P^{1e}\} \\ \{P^{2e}\} \end{Bmatrix} \tag{7-156a}$$

where

$$f_i^{1e} = \int_{\Omega^e} f_x^e \psi_i^e\,dx\,dy \qquad f_i^{2e} = \int_{\Omega^e} f_y^e \psi_i^e\,dx\,dy$$

$$P_i^{1e} = \oint_{\Gamma^e} t_x^e \psi_i^e\,ds \qquad P_i^{2e} = \oint_{\Gamma^e} t_y^e \psi_i^e\,ds$$

$$\begin{aligned} K_{ij}^{11e} &= \int_{\Omega^e} \left(c_{11}^e \frac{\partial \psi_i^e}{\partial x}\frac{\partial \psi_j^e}{\partial x} + c_{66}^e \frac{\partial \psi_i^e}{\partial y}\frac{\partial \psi_j^e}{\partial y} \right) dx\,dy \\ K_{ij}^{12e} &= \int_{\Omega^e} \left(c_{12}^e \frac{\partial \psi_i^e}{\partial x}\frac{\partial \psi_j^e}{\partial y} + c_{66}^e \frac{\partial \psi_i^e}{\partial y}\frac{\partial \psi_j^e}{\partial x} \right) dx\,dy \\ K_{ij}^{22e} &= \int_{\Omega^e} \left(c_{66}^e \frac{\partial \psi_i^e}{\partial x}\frac{\partial \psi_j^e}{\partial x} + c_{22}^e \frac{\partial \psi_i^e}{\partial y}\frac{\partial \psi_j^e}{\partial y} \right) dx\,dy \end{aligned} \tag{7-156b}$$

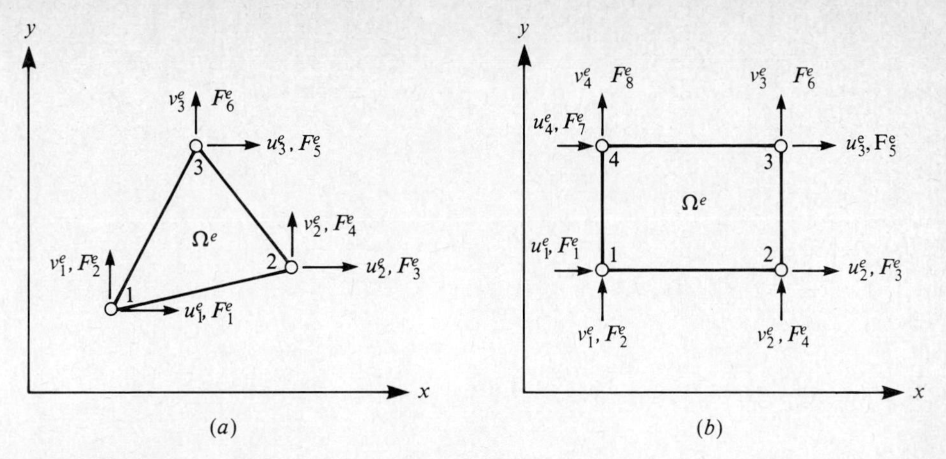

Figure 7-29 Linear triangular and rectangular elements for plane elasticity problems. (*a*) Triangular element; (*b*) rectangular element.

Note that the coefficient matrix $[K^{12e}]$, for example, corresponds to the coefficient of $\{v^e\}$ in the first equation (i.e., the first superscript corresponds to the equation number and second to the variable number).

An examination of the definition of the coefficient matrices in Eq. (7-156*b*) shows that the linear Lagrange interpolation functions are admissible. For a linear triangular or rectangular element (see Fig. 7-29), the coefficients in Eq. (7-156*b*) can be readily evaluated (for constant values of c_{ij}^e, f_x^e, and f_y^e) using Eqs. (7-113) and (7-118). We consider an example below.

Example 7-18 Consider the linear triangular and rectangular elements shown in Fig. 7-30. We wish to evaluate $[K^e]$ and $\{f^e\}$ of Eq. (7-156) for each element.

The element interpolation functions for the triangle are given by

$$\psi_1^e = \tfrac{1}{16}(40 - 3x - 4y) \qquad \psi_2^e = \tfrac{1}{16}(-16 + 4x) \qquad \psi_3^e = \tfrac{1}{16}(-8 - x + 4y)$$

we have
$$K_{ij}^{11e} = \frac{1}{4A_e}(c_{11}^e \beta_i^e \beta_j^e + c_{66}^e \gamma_i^e \gamma_j^e) \qquad f_i^{1e} = \frac{f_x^e A_e}{3}$$

$$K_{ij}^{12e} = \frac{1}{4A_e}(c_{12}^e \beta_i^e \gamma_j^e + c_{66}^e \gamma_i^e \beta_j^e) \qquad f_i^{2e} = \frac{f_y^e A_e}{3}$$

$$K_{ij}^{22e} = \frac{1}{4A_e}(c_{66}^e \beta_i^e \beta_j^e + c_{22}^e \gamma_i^e \gamma_j^e)$$

where $A_e = 16$ and

$$\beta_1^e = -3 \quad \beta_2^e = 4 \quad \beta_3^e = -1 \quad \gamma_1^e = -4 \quad \gamma_2^e = 0 \quad \gamma_3^e = 4$$

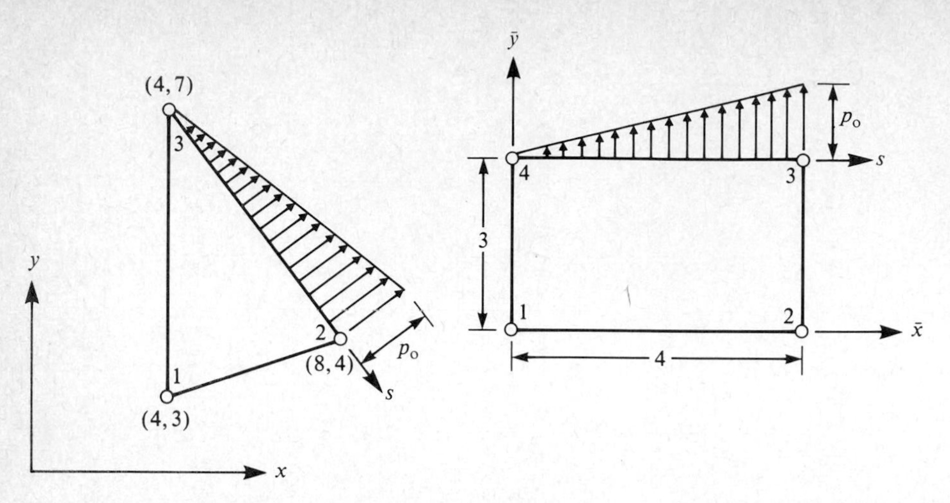

Figure 7-30 Force computation for triangular and rectangular plane elasticity elements.

The boundary forces on sides 1-2 and 3-1 are zero because $t_x^e = t_y^e = 0$ on those sides. To compute the contribution of the specified traction force,

$$t_n^e = \frac{p_0^e s}{5} \qquad 0 \le s \le 5$$

on side 2-3 contributed to both P_i^{1e} and P_i^{2e}. First we compute the contribution of t_n^e to the nodes 2 and 3 in the normal direction n of the side and then compute the components of the nodal forces along the (global) x- and y-directions. Note that $\psi_i^e(s)$ is the restriction of $\psi_i^e(x, y)$ to the boundary (i.e., sides 1-2, 2-3, and 3-1) of the element Ω^e. On side 2-3, which is of length 5, they are given by the one-dimensional linear interpolation functions,

$$\psi_3^e(s) = 1 - \frac{s}{5} \qquad \psi_2^e(s) = \frac{s}{5}$$

We obtain

$$\int_0^5 t_n^e \psi_2^e \, ds = \frac{10p_0^e}{6} \qquad \int_0^5 t_n^e \psi_3^e \, ds = \frac{5p_0^e}{6}$$

Resolving these forces along the x- and y-directions at the nodes, we have

$$P_2^{1e} = p_0^e \qquad P_2^{2e} = \tfrac{4}{3}p_0^e$$
$$P_3^{1e} = \tfrac{1}{2}p_0^e \qquad P_3^{2e} = \tfrac{2}{3}p_0^e$$

For the rectangular element with $a^e = 4$ and $b^e = 3$ we have

$$[K^{11e}] = c_{11}^e[S^{xe}] + c_{66}^e[S^{ye}] \qquad \{f^{1e}\} = \frac{f_x^e A_e}{4}$$

$$[K^{12e}] = c_{12}^e[S^{xye}] + c_{66}^e[S^{xye}]^T \qquad \{f^{2e}\} = \frac{f_y^e A_e}{4}$$

$$[K^{22e}] = c_{66}^e[S^{xe}] + c_{22}^e[S^{ye}]$$

where $A_e = a^e b^e = 12$. The contribution of t_n^e to nodes 3 and 4 is readily given by ($t_n^e = p_0^e\, s/a^e$)

$$P_3^{1e} = 0 \qquad P_3^{2e} = \int_0^{a^e} t_n^e \psi_3^e \, ds = \frac{2}{3}\left(\frac{a^e p_0^e}{2}\right)$$

$$P_4^{1e} = 0 \qquad P_4^{2e} = \int_0^{a^e} t_n^e \psi_4^e \, ds = \frac{1}{3}\left(\frac{a^e p_0^e}{2}\right)$$

The assembly, imposition of boundary conditions and solution of equations are the same as those described earlier for two-dimensional elements, with one exception. For plane elasticity elements, we have two degrees of freedom per node (u_i^e, v_i^e). If the element equations in Eq. (7-156a) are assembled, we obtain equations of the form

$$\begin{bmatrix} [K^{11}] & [K^{12}] \\ [K^{12}]^T & [K^{22}] \end{bmatrix} \begin{Bmatrix} \{U\} \\ \{V\} \end{Bmatrix} = \begin{Bmatrix} \{F^1\} \\ \{F^2\} \end{Bmatrix} \tag{7-157}$$

This form of the global equations has a greater half band width, and the band width can be reduced by rearranging the element equations (7-156a) in the form,

$$[K^e] \begin{Bmatrix} u_1^e \\ v_1^e \\ \vdots \\ u_n^e \\ v_n^e \end{Bmatrix} = \begin{Bmatrix} f_1^{1e} \\ f_1^{2e} \\ \vdots \\ f_n^{1e} \\ f_n^{2e} \end{Bmatrix} + \begin{Bmatrix} P_1^{1e} \\ P_1^{2e} \\ \vdots \\ P_n^{1e} \\ P_n^{2e} \end{Bmatrix} \tag{7-158}$$

For additional details see the author's text (1984a, pp. 337–339).

Existence and error estimates The existence of finite-element solutions to plane elasticity problems is governed by Theorem 7-1. The finite element problem consists of finding $\Lambda_h = (u_h, v_h) \in S^h \subset H$, where H is defined by Eq. (5-60), such that

$$B(\bar{\Lambda}_h, \Lambda_h) = l(\bar{\Lambda}_h) \tag{7-159}$$

holds for all $\bar{\Lambda}_h \in S^h$. The bilinear and linear forms are defined by

$$B(\bar{\Lambda}_h, \Lambda_h) = \int_{\Omega_h} \left[\frac{\partial \bar{u}_h}{\partial x}\left(c_{11}\frac{\partial u_h}{\partial x} + c_{12}\frac{\partial v_h}{\partial y}\right) + c_{66}\frac{\partial \bar{u}_h}{\partial y}\left(\frac{\partial u_h}{\partial y} + \frac{\partial v_h}{\partial x}\right) \right.$$
$$\left. + c_{66}\frac{\partial \bar{v}_h}{\partial x}\left(\frac{\partial u_h}{\partial y} + \frac{\partial v_h}{\partial x}\right) + \frac{\partial \bar{v}_h}{\partial y}\left(c_{12}\frac{\partial u_h}{\partial x} + c_{22}\frac{\partial v_h}{\partial y}\right)\right] dx\, dy \tag{7-160}$$
$$l(\bar{\Lambda}_h) = \int_{\Omega_h} (\bar{u}_h f_x + \bar{v}_h f_y)\, dx\, dy + \oint_{\Gamma_h} (\bar{u}_h t_x + \bar{v}_h t_y)\, ds$$

where Ω_h is the finite-element representation of Ω and Γ_h is the boundary of Ω_h. The error estimate is given by [see Eq. (7-124)]

$$\|\Lambda - \Lambda_h\|_1 \leq c\left(1 + \frac{M}{\alpha_n}\right) h^p |\Lambda|_s \qquad p = \min(k, s-1) \tag{7-161}$$

Next we consider several examples of application of plane elasticity elements. The computations were carried out on an IBM 370/158 computer using double precision arithmetic and the FEM2D computer program in the author's finite-element text (1984a). Both triangular and rectangular elements are used in these examples. Equivalent meshes of various elements are shown in Fig. 7-31.

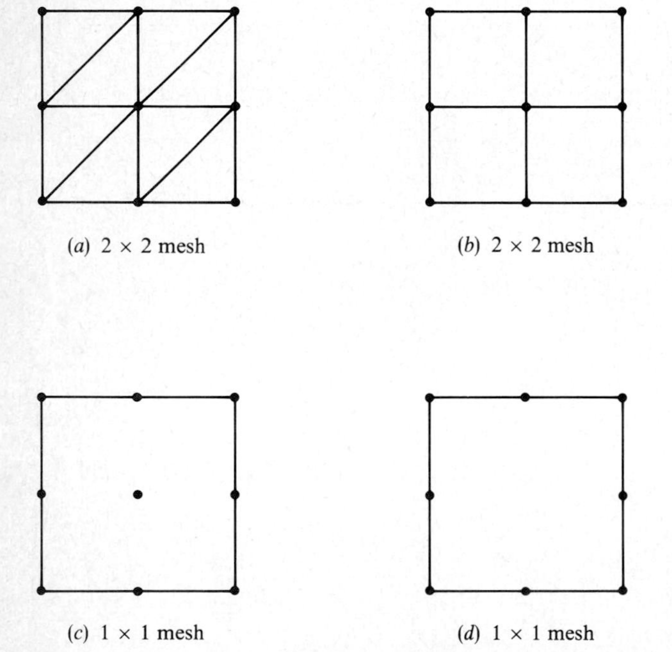

Figure 7-31 Equivalent meshes of (*a*) triangular, (*b*) linear rectangular, (*c*) nine-node (quadratic) rectangular and (*d*) eight-node (quadratic) rectangular elements.

Example 7-19 Consider the cantilever beam ($E = 30 \times 10^6$ psi, $\nu = 0.25$, $a = 10$ in., $b = c = 1$ in.) shown in Fig. 7-32*a*. We wish to determine the maximum deflection and bending stress in the beam when it is subjected to a uniformly distributed shear stress $\tau = 150$ psi. The boundary conditions of the problem are

$$u(a, y) = 0 \qquad v(a, c) = 0$$

$$t_x = t_y = 0 \text{ at } y = 0, 2c \text{ for any } x$$

$$t_x = 0 \qquad t_y = -\tau \text{ at } x = 0 \text{ for any } y$$

$$t_y = 0 \text{ at } x = a \text{ and for any } y \neq c$$

We will solve the problem using the plane stress assumption. The elastic coefficients c_{ij} for the *plane stress* case are defined (assuming that steel is isotropic) by

$$c_{11} = c_{12} = \frac{E}{1 - \nu^2} \qquad c_{12} = \frac{\nu E}{1 - \nu^2} \qquad c_{66} = \frac{E}{2(1 + \nu)} (= G) \quad (7\text{-}162)$$

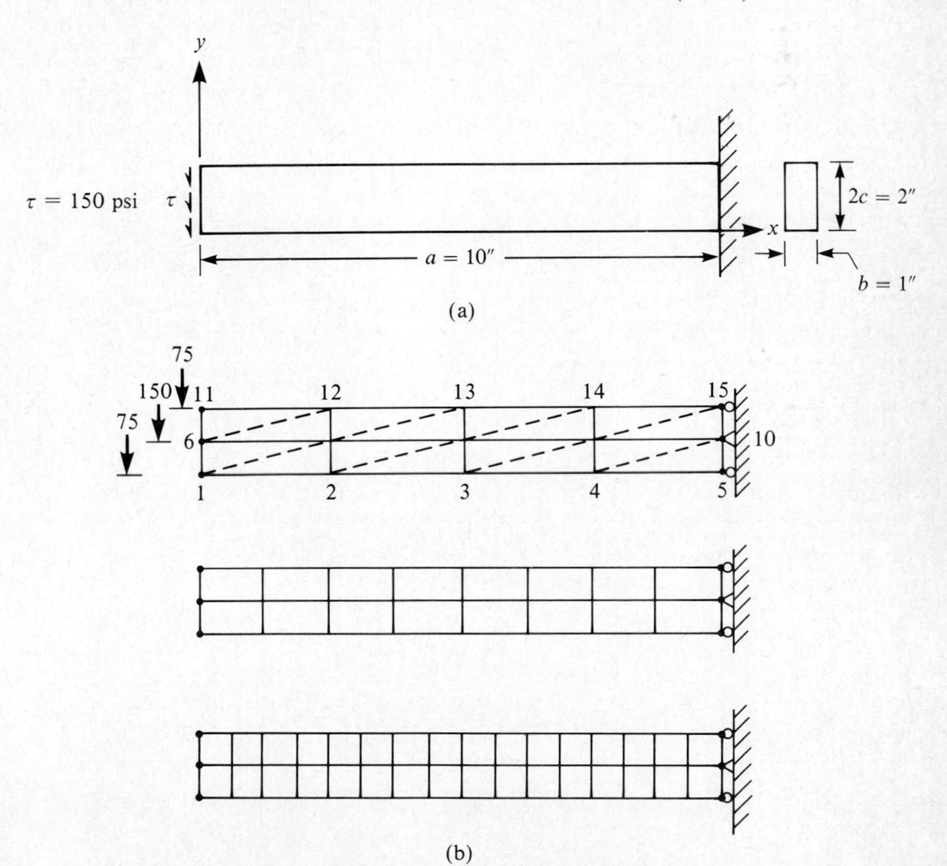

Figure 7-32 Finite-element meshes for an end-loaded cantilever beam.

Table 7-13 Comparison of the finite-element solution with the elasticity solution of a cantilever beam subjected to a uniform shear load at the free end

	Tip deflection, $-w \times 10^2$			Normal stress σ_x		
No. of Nodes	Linear triangles	Linear rectangles	Quadratic rectangles	Linear triangles	Linear rectangles	Quadratic rectangles
15	0.16112	0.31335	0.50310	1328.17 (7)*	1195.54 (8.75, 1.5)‡	2196.35 (8.943, 1.577)
27	0.26623	0.43884	0.51288	2441.67 (15)	1792.59 (9.375, 1.5)	2439.06 (9.471, 1.577)
51	0.31659	0.48779	0.51374	2978.35 (31)	2055.58 (9.6875, 1.5)	2526.21 (9.736, 1.577)
Elasticity†		0.51875			2527.95 (9.736, 1.577)	

* Element number.
† From Reddy (1984b, p. 53).
‡ Quadrature points.

Three different finite-element meshes, increasingly refined, are shown in Fig. 7-32*b*. The meshes shown are those consisting of linear rectangular elements. Equivalent triangular element meshes are obtained by joining node 1 to node 3 of each rectangular element (see the dotted lines). Equivalent meshes of nine-node quadratic Lagrange elements are obtained by considering a 2×2 mesh of linear Lagrange elements as a quadratic element.

For the Ritz finite-element model the boundary conditions on the primary and secondary variables, e.g., for the 15-node mesh, are given by

$$U_5 = U_{10} = V_{10} = U_{15} = 0.0$$

$$F_1^2 = -75.0 \qquad F_6^2 = -150.0 \qquad F_{11}^2 = -75.0$$

and all other F_i^1 and F_i^2 are zero on the boundary.

Table 7-13 contains the finite-element solutions for the tip deflection (i.e., deflection at the center node of the left end) and bending stress, σ_x, obtained using various finite elements. The numerical convergence of the deflection and stress to the elasticity solution (based on two-dimensional elasticity theory) is apparent from the results in Table 7-13. The triangular element mesh has the slowest convergence compared to the Lagrange linear and quadratic elements.

Example 7-20 Consider a long steel ($E = 30 \times 10^6$ psi, $\nu = 0.25$) cylindrical pressure vessel with internal pressure p_0. Let the internal and external radii of the cylinder be a and b, respectively. We wish to determine the state of deformation and stress in the cylinder. The problem can be solved by considering any cross-section (see Fig. 7-33*a*) of the cylinder with the *plane strain* assumption, for which case the c_{ij} are given by

$$c_{11} = c_{22} = \frac{E(1 - \nu)}{(1 + \nu)(1 - 2\nu)} \qquad c_{12} = \frac{\nu E}{(1 + \nu)(1 - 2\nu)} \qquad c_{66} = \frac{E}{2(1 + \nu)} \tag{7-163}$$

Equivalent uniform meshes of triangles, linear quadrilaterals and quadratic quadrilateral elements are used to solve the problem (see Fig. 7-33*b*). The boundary conditions of the primary degrees of freedom for the quadrant modeled are given below:

$$U_1 = U_2 = U_3 = U_4 = U_5 = V_{21} = V_{22} = V_{23} = V_{24} = V_{25} = 0$$

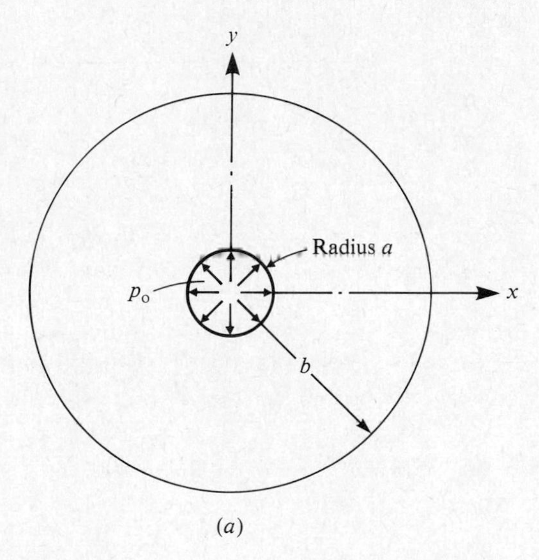

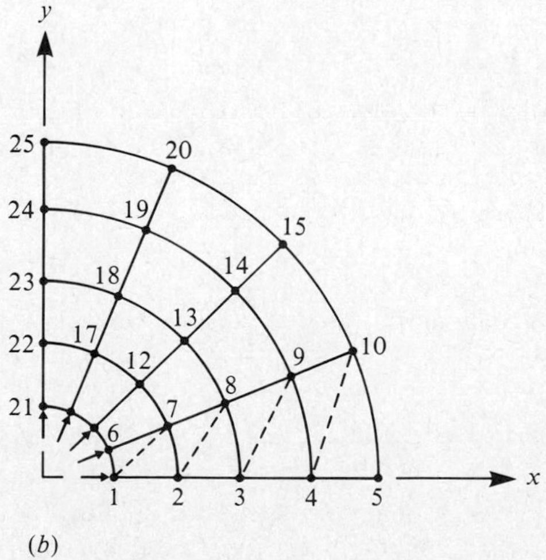

Figure 7-33 The domain and the finite-element mesh for the plane strain problem of Example 7-20.

The nonzero specified secondary degrees of freedom are at nodes 1, 6, 11, 16, and 21. For linear elements, the contribution of the pressure p_0 at the element nodes is computed by using the formula

$$\int_0^{h_e} \psi_i^e p_0 \, ds = \frac{p_0 h_e}{2}$$

where h_e is the arc length of the side of element Ω_e that is subjected to the pressure. For the mesh used here we have $h = \pi a/8$. Hence the normal (i.e., along the radial lines) forces at nodes 1, 6, 11, 16, and 21 are

$$P_1 = \frac{\pi a p_0}{16} \qquad P_6 = \frac{\pi a p_0}{8} \qquad P_{11} = \frac{\pi a p_0}{8} \qquad P_{16} = \frac{\pi a p_0}{8} \qquad P_{21} = \frac{\pi a p_0}{16}$$

The x and y components of these forces are given by

$$P_1^1 = \frac{\pi a p_0}{16} \qquad P_1^2 = 0 \qquad P_6^1 = \frac{\pi a p_0}{8} \cos 22.5 \qquad P_6^2 = \frac{\pi a p_0}{8} \sin 22.5$$

$$P_{11}^1 = \frac{\pi a p_0}{8} \cos 45 \qquad P_{11}^2 = \frac{\pi a p_0}{8} \sin 45 \qquad P_{16}^1 = P_6^2 \qquad P_{16}^2 = P_6^1$$

$$P_{21}^1 = 0 \qquad P_{21}^2 = \frac{\pi a p_0}{16}$$

For a quadratic element, the contribution of p_0 to the element nodes is given by

$$\int_0^{h_e} \psi_i^e p_0 \, ds = \begin{cases} \dfrac{p_0 h_e}{6} & i = 1 \\[2ex] \dfrac{2 p_0 h_e}{3} & i = 2 \\[2ex] \dfrac{p_0 h_e}{6} & i = 3 \end{cases}$$

where $h_e = \pi a/4$. The components of these nodal forces at nodes 1, 6, 11, 16, and 21 are given by

$$P_1^1 = \frac{\pi a p_0}{24} \qquad P_1^2 = 0 \qquad P_6^1 = \frac{\pi a p_0}{6} \cos 22.5 \qquad P_6^2 = \frac{\pi a p_0}{6} \cos 22.5$$

$$P_{11}^1 = \frac{\pi a p_0}{12} \cos 45 \qquad P_{11}^2 = \frac{\pi a p_0}{12} \sin 45 \qquad P_{16}^1 = P_6^2 \qquad P_{16}^2 = P_6^1$$

$$P_{21}^1 = 0 \qquad P_{21}^2 = \frac{\pi a p_0}{24}$$

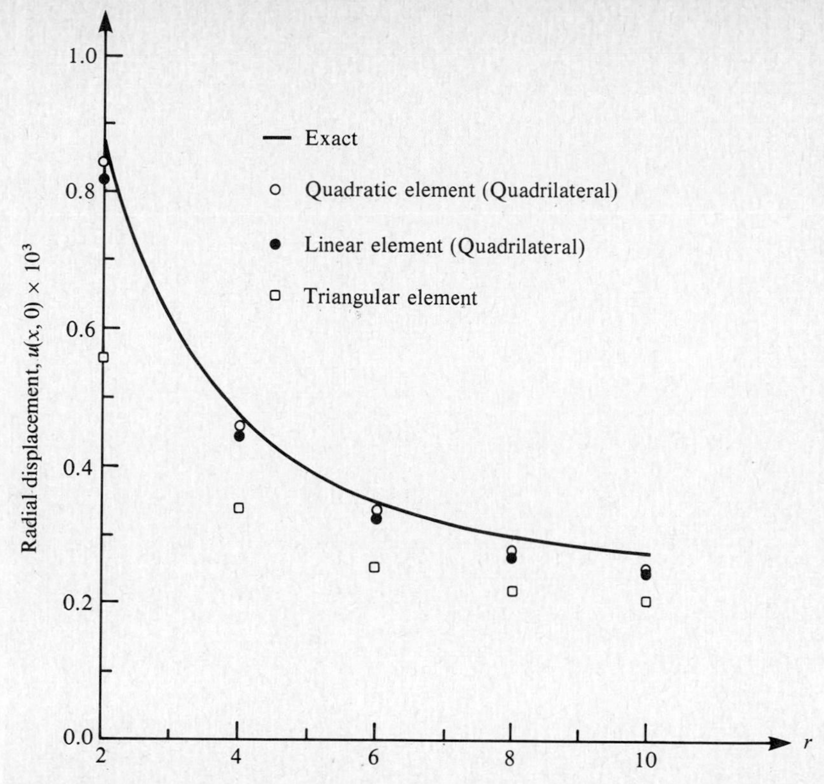

Figure 7-34 Comparison of the radial displacement as computed by various elements with the exact solution of the problem in Example 7-20.

The finite-element solution $u(x, 0)$ obtained by various elements is compared with the exact solution in Fig. 7-34. Once again we note that the triangular element mesh gives a less accurate solution than the quadrilateral element meshes. The stress values σ_x obtained by the quadrilateral element meshes are compared with the exact solution in Table 7-14. The finite-element stresses were computed at the Gauss points, which are listed in the table. This completes the example.

Table 7-14 Comparison of stress σ_x obtained by linear and quadratic quadrilateral elements with the exact solution

Quadratic element			Exact	Linear element			Exact
x	y	$-\sigma_x \times 10^{-3}$	$-\sigma_x \times 10^{-3}$	x	y	$-\sigma_x \times 10^{-3}$	$-\sigma_x \times 10^{-3}$
2.8026	0.4804	4.4360	4.4309	2.8858	0.5740	3.8784	4.0103
5.0773	0.8704	1.0020	1.0603	4.8097	0.9567	1.0858	1.1770
6.7425	1.1558	0.4141	0.4209	6.7336	1.3394	0.3505	0.3965
9.1073	1.5458	0.0510	0.0512	8.6575	1.7221	0.0511	0.0752

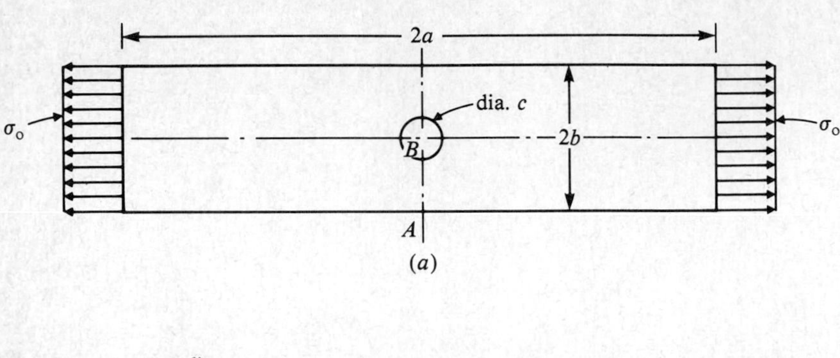

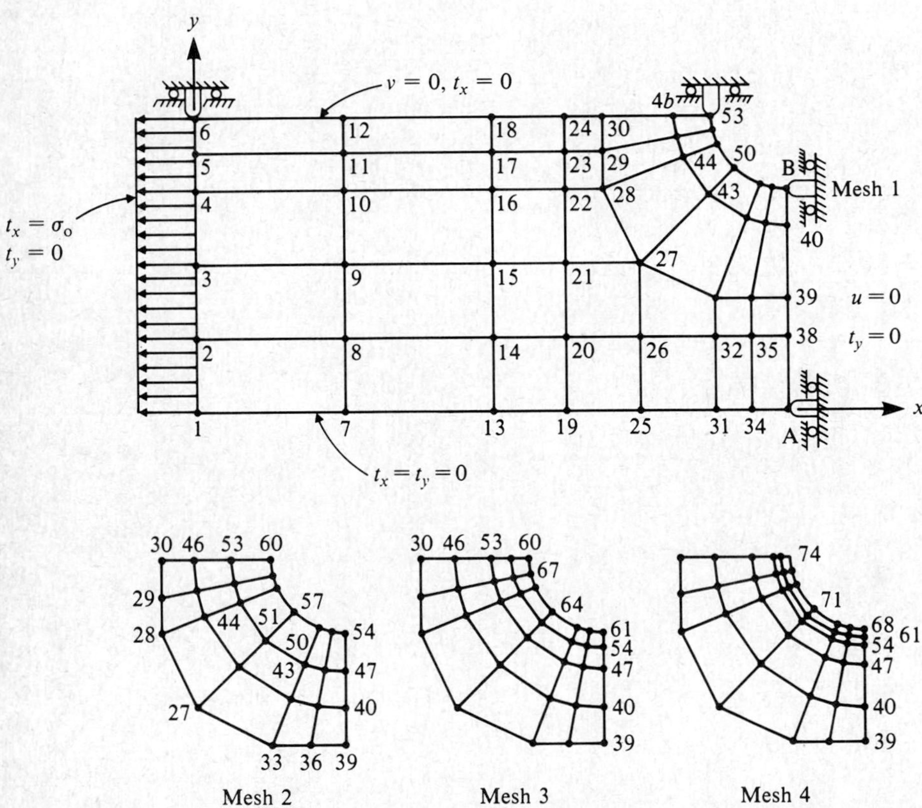

Figure 7-35 Geometry and finite-element meshes of a plate with a circular hole (*a*) geometry of the plate, (*b*) finite-element meshes.

The next example deals with the stress analysis of a rectangular plate with a circular hole and loaded along the short edges.

Example 7-21 Consider a thin steel plate with a circular hole as shown in Fig. 7-35*a*. We analyze the problem for displacements along $x = 0$ and for normal stress across the section AB. Due to the abrupt change in the cross-section along AB, a step gradient in stress and strain is expected. It is well known that the maximum normal stress σ_x along AB is three times the applied uniform stress σ_0 at the edges. To capture the maximum stress, a finer mesh near the hole is required. To illustrate the effect of mesh size, a series of gradually finer meshes near the hole are used to determine the maximum stress.

Exploiting the biaxial symmetry, only one quadrant of the plate is modeled using quadrilateral elements. The boundary conditions on the displacements are shown in Fig. 7-35*b*. The contribution of the applied distributed load along the edge $x = 0$ should be computed in each element, and should be used as the applied x-component of force at nodes 1, 2, 3, 4, 5, and 6. We have

$$P_1^1 = -\int_0^1 \sigma_0 \psi_1^1 \, dy$$

$$P_7^1 + P_1^2 = -\int_0^1 \sigma_0 \psi_2^1 \, dy - \int_0^1 \sigma_0 \psi_1^2 \, dy$$

$$P_7^2 + P_1^3 = -\int_0^1 \sigma_0 \psi_2^2 \, dy - \int_0^1 \sigma_0 \psi_1^3 \, dy$$

$$P_7^3 + P_1^4 = -\int_0^1 \sigma_0 \psi_2^3 \, dy - \int_0^{0.5} \sigma_0 \psi_1^4 \, dy$$

$$P_7^4 + P_1^5 = -\int_0^{0.5} \sigma_0 \psi_2^4 \, dy - \int_0^{0.5} \sigma_0 \psi_1^5 \, dy$$

$$P_7^5 = -\int_0^{0.5} \sigma_0 \psi_2^5 \, dy$$

where $\psi_1^e = (1 - \eta)/h_e$, $\psi_2^e = \eta/h_e$, η is the local coordinate, and h_e is the length of a typical line element along the boundary $x = 0$. The evaluation of the integrals gives

$$P_1^1 = -500 \qquad P_7^1 + P_1^2 = -1000 \qquad P_7^2 + P_1^3 = -1000$$

$$P_7^3 + P_1^4 = -750 \qquad P_7^4 + P_1^5 = -5000 \qquad P_7^5 = -250$$

Figure 7-35*b* contains the set of four meshes of quadrilateral elements used in the present study. As can be seen from Fig. 7-36, the maximum normal stress along AB varies from 1000 psi at the edge to about 3000 psi near the hole. This is consistent with the experimental and analytical results.

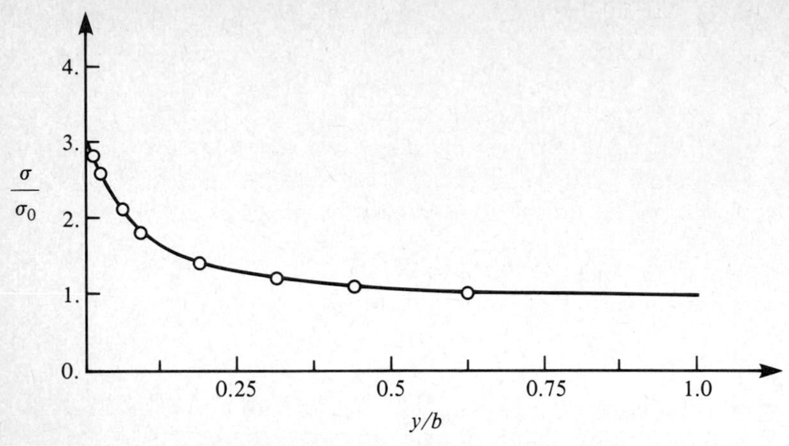

Figure 7-36 Variation of normal stress across the hole.

7-7-3 Viscous Incompressible Flows in Two Dimensions

Here we discuss finite-element models of the equations governing Stokes flow (see Sec. 2-6-3 and Examples 4-12 and 5-10). The governing equations are given by [see Eq. (4-104)]

$$\left.\begin{aligned}-\frac{\partial}{\partial x}\left(2\mu\frac{\partial u}{\partial x}\right)-\frac{\partial}{\partial y}\left[\mu\left(\frac{\partial u}{\partial y}+\frac{\partial v}{\partial x}\right)\right]+\frac{\partial P}{\partial x}=f_x\\ -\frac{\partial}{\partial x}\left[\mu\left(\frac{\partial u}{\partial y}+\frac{\partial v}{\partial x}\right)\right]-\frac{\partial}{\partial y}\left(2\mu\frac{\partial v}{\partial y}\right)+\frac{\partial P}{\partial y}=f_y\\ \frac{\partial u}{\partial x}+\frac{\partial v}{\partial y}=0\end{aligned}\right\}\text{ in }\Omega \tag{7-164}$$

where u and v are the components of the velocity vector, P is the pressure, f_x and f_y are the components of the body force vector and μ is the viscosity of the fluid. The boundary conditions are given by

Essential:

$$u=\hat{u}\text{ and }v=\hat{v}\text{ on }\Gamma_1$$

Natural:

$$\left.\begin{aligned}t_x\equiv 2\mu\frac{\partial u}{\partial x}n_x+\mu\left(\frac{\partial u}{\partial y}+\frac{\partial v}{\partial x}\right)n_y-Pn_x=\hat{t}_x\\ t_y\equiv \mu\left(\frac{\partial u}{\partial y}+\frac{\partial v}{\partial x}\right)n_x+2\mu\frac{\partial v}{\partial y}n_y-Pn_y=\hat{t}_y\end{aligned}\right\}\text{ on }\Gamma_2 \tag{7-165}$$

As discussed in Examples 4-12 and 4-17, there are two variational formulations for the Stokes problem. The finite-element model based on the Lagrange multiplier formulation is called the *mixed model* and that based on the penalty function formulation is called the *penalty model.* We shall discuss both of the models here.

Mixed model ***Variational formulation*** Recall from Sec. 4-4-2 that the quadratic functional L for the Stokes' problem is given by (4-102*b*). For Dirichlet boundary conditions on the velocity field, the space of admissible velocities is given by

$$\mathbf{H} = H_0^1(\Omega) \times H_0^1(\Omega)$$

The constraint equation is given by

$$G(\mathbf{u}) = 0$$

where $\mathbf{u} = (u, v)$,

$$G\colon \mathbf{H} \to Q \qquad G(\mathbf{u}) = \operatorname{div} \mathbf{u}$$

and Q is as defined in Eq. (5-69). The constraint set is given by

$$K = \{\mathbf{u} \in \mathbf{H}\colon\ G(\mathbf{u}) = 0\}$$

The mixed variational formulation is given by [see Eq. (4-103)]: find $(\mathbf{u}, P) \in \mathbf{H} \times Q$ such that

$$\begin{aligned} B(\bar{\mathbf{u}}, \mathbf{u}) - \int_\Omega \operatorname{div} \bar{\mathbf{u}}\, P\, d\mathbf{x} &= \int_\Omega \bar{\mathbf{u}} \cdot \mathbf{f}\, d\mathbf{x} \qquad \text{for all } \bar{\mathbf{u}} \in \mathbf{H} \\ \int_\Omega q \operatorname{div} \mathbf{u}\, d\mathbf{x} &= 0 \qquad \text{for all } q \in Q \end{aligned} \tag{7-166}$$

where $B(\cdot, \cdot)$ is the bilinear form defined in Eq. (4-98). We have discussed in Sec. 5-5-3 the existence and uniqueness of solutions to the variational problem (7-166). We now describe the finite-element model of Eq. (7-166).

The discrete form of Eq. (7-164) [i.e., explicit form of Eq. (7-166) over an element] is given by

$$\begin{aligned} 0 &= \int_{\Omega^e} \left[2\mu \frac{\partial w_1^e}{\partial x} \frac{\partial u_e}{\partial x} + \mu \frac{\partial w_1^e}{\partial y} \left(\frac{\partial u_e}{\partial y} + \frac{\partial v_e}{\partial x} \right) - \frac{\partial w_1^e}{\partial x} P_e - w_1^e f_x^e \right] dx\, dy \\ &\quad - \oint_{\Gamma^e} w_1^e t_x^e\, ds \\ 0 &= \int_{\Omega^e} \left[\mu \frac{\partial w_2^e}{\partial x} \left(\frac{\partial u_e}{\partial y} + \frac{\partial v_e}{\partial x} \right) + 2\mu \frac{\partial w_2^e}{\partial y} \frac{\partial v_e}{\partial y} - \frac{\partial w_2^e}{\partial y} P_e - w_2^e f_y^e \right] dx\, dy \\ &\quad - \oint_{\Gamma^e} w_2^e t_y^e\, ds \\ 0 &= \int_{\Omega^e} w_3^e \left(\frac{\partial u_e}{\partial x} + \frac{\partial v_e}{\partial y} \right) dx\, dy \end{aligned} \tag{7-167}$$

where $(w_1^e, w_2^e) \in \mathbf{S}^h \subset \mathbf{H}$ and $w_3^e \in Q^h \subset Q$ are appropriate test functions.

Finite-element formulation Note that the pressure P_e is a dependent variable but its specification on the boundary does not constitute an essential boundary condition because w_3^e (the variation δP_e) does not appear in the boundary integrals of the formulation; however, P_e appears in t_x^e and t_y^e. Also note that pressure appears undifferentiated in the variational equations (7-167). In other words, one can use a linear interpolation for u_e and v_e and a constant for P_e, or a quadratic approximation for u_e and v_e and a linear approximation for P_e, as implied by the spaces $\mathbf{S}^h$ and Q^h. In general, the interpolation used for P_e should be one degree less than that used for u_e and v_e (for consistency of approximation). Since P_e does not appear in the essential boundary conditions of the formulation, one need not impose the interelement continuity of P_e.

Over a typical element Ω_e, we approximate u_e, v_e, and P_e by interpolation of the form,

$$u_e = \sum_{i=1}^{r} u_i^e \psi_i^e \qquad v_e = \sum_{i=1}^{r} v_i^e \psi_i^e \qquad P_e = \sum_{I=1}^{s} P_I^e \phi_I^e \tag{7-168}$$

where ψ_i^e and ϕ_I^e are interpolation functions of degrees r and s ($r = s + 1$), respectively. Substituting Eq. (7-168) into Eq. (7-167), we obtain

$$\begin{bmatrix} [K^{11e}] & [K^{12e}] & [K^{13e}] \\ & [K^{22e}] & [K^{23e}] \\ \text{symm.} & & [K^{33e}] \end{bmatrix} \begin{Bmatrix} \{u^e\} \\ \{v^e\} \\ \{P^e\} \end{Bmatrix} = \begin{Bmatrix} \{F^{1e}\} \\ \{F^{2e}\} \\ \{0\} \end{Bmatrix} \tag{7-169}$$

where

$$[K^{11e}] = 2\mu[S^{11e}] + \mu[S^{22e}] \qquad [K^{12e}] = \mu[S^{12e}]^T$$

$$[K^{22e}] = \mu[S^{11e}] + 2\mu[S^{22e}] \qquad [K^{33e}] = [0] \text{ (zero matrix)}$$

$$S_{ij}^{11e} = \int_{\Omega^e} \frac{\partial \psi_i^e}{\partial x}\frac{\partial \psi_j^e}{\partial x}\, dx\, dy \qquad S_{ij}^{12e} = \int_{\Omega^e} \frac{\partial \psi_i^e}{\partial x}\frac{\partial \psi_j^e}{\partial y}\, dx\, dy$$

$$S_{ij}^{22e} = \int_{\Omega^e} \frac{\partial \psi_i^e}{\partial y}\frac{\partial \psi_j^e}{\partial y}\, dx\, dy \qquad K_{iJ}^{13e} = \int_{\Omega^e} \frac{\partial \psi_i^e}{\partial x}\phi_J^e\, dx\, dy \tag{7-170}$$

$$K_{iJ}^{23e} = \int_{\Omega^e} \frac{\partial \psi_i^e}{\partial y}\phi_J^e\, dx\, dy \qquad (i, j = 1, 2, \ldots, r;\ J = 1, 2, \ldots, s)$$

$$F_i^{1e} = \int_{\Omega^e} f_x^e \psi_i^e\, dx\, dy + \oint_{\Gamma^e} t_x^e \psi_i^e\, ds \qquad F_i^{2e} = \int_{\Omega^e} f_y^e \psi_i^e\, dx\, dy + \oint_{\Gamma^e} t_y^e \psi_i^e\, ds$$

Note that the three matrix equations in Eq. (7-169) correspond, respectively, to the three governing equations in Eq. (7-167).

Existence of solutions and error estimates The finite-element approximation of Eq. (7-167) involves constructing two finite-dimensional subspaces $\mathbf{S}^h \subset \mathbf{H}$ and

$Q^h \subset Q$ such that the solution $(\mathbf{u}_h, P_h) \in \mathbf{S}^h \times Q^h$ to the approximate problem

$$B(\mathbf{v}_h, \mathbf{u}_h) - \int_{\Omega_h} \operatorname{div} \mathbf{v}_h P_h \, d\mathbf{x} = \int_{\Omega_h} \mathbf{f} \cdot \mathbf{v}_h \, d\mathbf{x}$$

$$b(\mathbf{u}_h, q_h) \equiv \int_{\Omega_h} \operatorname{div} \mathbf{u}_h q_h \, d\mathbf{x} = 0 \tag{7-171}$$

exists for all $(\mathbf{v}_h, q_h) \in \mathbf{S}^h \times Q^h$. The conditions for the existence of a unique solution to Eq. (7-171) are the same as those in Theorem 5-5. In Sec. 5-5-3 it was shown that the mixed formulation has a unique solution. Alternatively, the conditions for the existence of a unique solution can be stated as follows:

(i) There exists a constant $\alpha_h > 0$ such that

$$\sup_{\mathbf{v}_h \in \ker G_h - \{0\}} \frac{|B(\mathbf{v}_h, \mathbf{u}_h)|}{\|\mathbf{v}_h\|_{\mathbf{H}}} \geq \alpha_h \|\mathbf{u}_h\|_{\mathbf{H}} \tag{7-172}$$

for every $\mathbf{v}_h \in \ker G_h$, where G_h is an approximation of the constraint operator G defined by $(G_h(\mathbf{u}_h), \mathbf{v}_h)_0 = b(\mathbf{u}_h, \mathbf{v}_h)$. The kernel of G_h is defined by

$$\ker G_h = \{\mathbf{v}_h \in \mathbf{S}^h\colon\ b(\mathbf{v}_h, q_h) = 0, \text{ for all } q_h \in Q^h\} \tag{7-173}$$

(ii) The *LBB* (*Ladyzhenskaya–Babuska–Brezzi*) *condition* There exists a constant $\beta_h > 0$ such that

$$\sup_{\mathbf{v}_h \subset S^h - \{0\}} \frac{|b(\mathbf{v}_h, q_h)|}{\|\mathbf{v}_h\|_{\mathbf{H}}} \geq \beta_h \|[q_h]\|_{\hat{Q}^h} \tag{7-174}$$

where $$\|[q_h]\|_{\hat{Q}^h} = \inf_{q_h^0 \in \ker G_h^*} \|q_h + q_h^0\|_Q \qquad q_h \in Q^h \tag{7-175a}$$

and $[q_h]$ is the set of equivalence classes,

$$[q_h] = \{\bar{q}_h \in Q^h\colon\ \bar{q}_h - q_h \in \ker G_h^*\} \tag{7-175b}$$

Here G_h^* denotes the adjoint of the operator $G_h = \operatorname{div}$,

$$(G_h \mathbf{u}_h, q_h)_0 = (\mathbf{u}_h, G_h^* q_h)_0 = -\int_{\Omega_h} \mathbf{u}_h \cdot \operatorname{grad} q_h \, d\mathbf{x} \tag{7-176}$$

and $\hat{Q}_h$ is the quotient space $\hat{Q}_h = Q_h/\ker G_h^*$, where

$$\ker G_h^* = \{q_h \in Q^h\colon\ b(\mathbf{v}_h, q_h) = 0 \text{ for all } \mathbf{v}_h \in \mathbf{S}^h\} \tag{7-177}$$

Under these conditions, a unique solution $(\mathbf{u}_h, [P_h]) \in \mathbf{S}^h \times \hat{Q}^h$ of Eq. (7-171) exists. The Lagrange multiplier P_h is unique up to the addition of any element in $\ker G_h^*$.

Error estimate Let $(\mathbf{u}, [P])$ and $(\mathbf{u}_h, [P_h])$ be the unique solutions of Eqs. (7-166) and (7-171), respectively, where $[P]$ and $[P_h]$ are elements of the quotient spaces

$\hat{Q} = Q/\ker G^*$ and $\hat{Q}^h = Q^h/\ker G_h^*$. If $\hat{Q}^h \subset \hat{Q}$, then the following error estimate can be established [see Oden and Carey (1983)]:

$$\|\mathbf{u} - \mathbf{u}_h\|_H + \|[P - P_h]\|_{\hat{Q}} \le c_h \left\{ \inf_{\mathbf{v}_h} S^h \|\mathbf{u} - \mathbf{v}_h\|_H + \inf_{q_h \in Q_h} \|[P - q_h]\|_{\hat{Q}} \right\} \quad (7\text{-}178)$$

where
$$c_h = 1 + \left[\frac{1}{\beta_h}\left(1 + \frac{M}{\alpha_h}\right) + \frac{1}{\alpha_h} \right] \max(M, m)$$

and M and m are the constants in the inequalities

$$|B(\mathbf{u}, \mathbf{v})| \le M\|\mathbf{u}\|_{\mathbf{H}}\|\mathbf{v}\|_{\mathbf{H}} \qquad |b(\mathbf{u}, q)| \le m\|\mathbf{u}\|_{\mathbf{H}}\|q\|_Q \quad (7\text{-}179)$$

Clearly, when α_h and β_h are independent of h, the parameter c_h is a constant independent of h, and hence the error estimate in Eq. (7-178) solely depends on the interpolation error.

Selection of spaces S^h and Q^h As discussed during the finite-element formulation, the interpolation for velocities and pressure should be of the following type (see Fig. 7-37):

$\mathbf{S}^h$	Q^h
Six-node, quadratic, triangular element	**Three-node, linear, triangular element**
Nine-node, quadratic, rectangular element	**Four-node, linear, rectangular element**

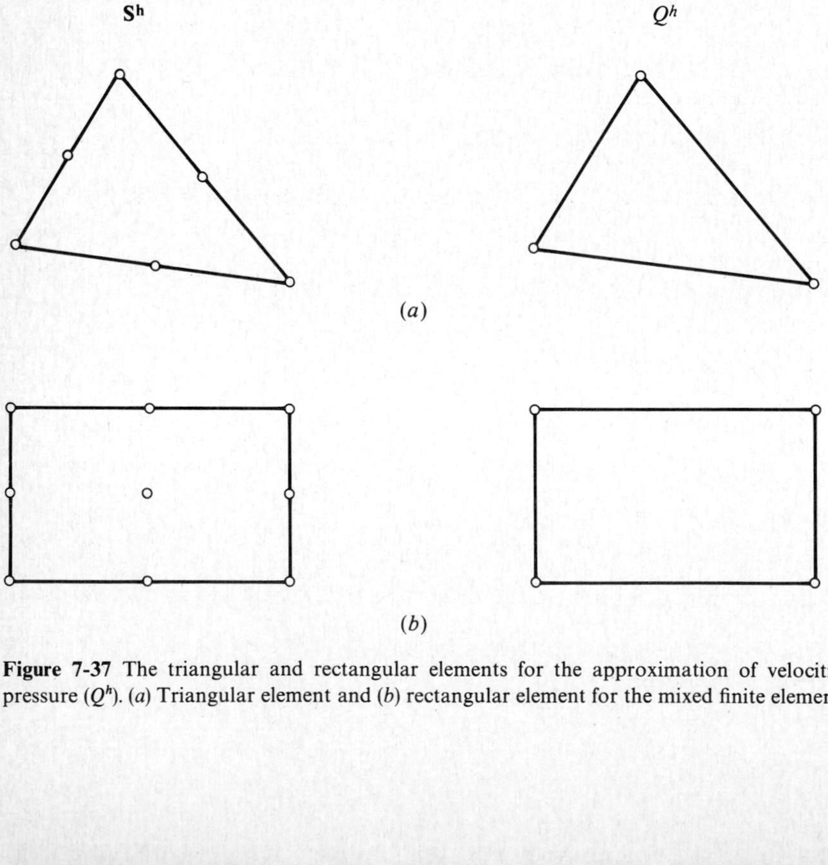

Figure 7-37 The triangular and rectangular elements for the approximation of velocities ($\mathbf{S}^h$) and pressure (Q^h). (*a*) Triangular element and (*b*) rectangular element for the mixed finite element model.

If the exact solution ($\mathbf{u}$, $[P]$) is sufficiently smooth,

$$\mathbf{u} \in (H^3(\Omega) \times H^3(\Omega)) \qquad P \in H^2(\Omega)$$

and the finite-element meshes are such that each element has at least one vertex which is not on the boundary Γ, then it can be shown that [see Bercovier and Pironneau (1977) and Oden and Carey (1983)] the error estimate in Eq. (7-178) becomes

$$\|\mathbf{u} - \mathbf{u}_h\|_1 + \|[P - P_h]\|_0 \leq ch^2(\|\mathbf{u}\|_3 + \|P\|_2) \tag{7-180}$$

Note that the rate of convergence is of order h^2 when using quadratic interpolation for the velocity field and linear interpolation for the pressure.

Penalty model ***Variational form*** Recall from Eq. (4-156) that the problem in Eq. (7-164) with homogeneous Dirichlet boundary conditions is equivalent to the variational problem: find $\mathbf{u} = (u, v) \in \mathbf{H}_k^1(\Omega)$, where $\mathbf{H}_k^1(\Omega)$ is defined by Eq. (4-157), such that Eq. (4-156) holds. The discrete form of Eq. (4-156) is given by

$$B_k(\bar{\mathbf{u}}_h, \mathbf{u}_h) = l(\bar{\mathbf{u}}_h) \qquad \text{for all } \bar{\mathbf{u}}_h \in \mathbf{S}^h$$

where $\mathbf{S}^h$ is a finite-dimensional subspace of $\mathbf{H}_k^1(\Omega)$ spanned by a set of C°-continuous basis functions $\{\Phi_I\}_{I=1}^N$ and $B_k(\cdot, \cdot)$ is given by Eq. (4-158). Locally, this is equivalent to the variational problem,

$$0 = \int_{\Omega^e} \left[2\mu \frac{\partial w_1^e}{\partial x} \frac{\partial u_e}{\partial x} + \mu \frac{\partial w_1^e}{\partial y} \left(\frac{\partial u_e}{\partial y} + \frac{\partial v_e}{\partial x} \right) + \gamma_e \frac{\partial w_1^e}{\partial x} \left(\frac{\partial u_e}{\partial x} + \frac{\partial v_e}{\partial y} \right) - w_1^e f_x^e \right] dx\, dy - \oint_{\Gamma^e} w_1^e t_x^e \, ds$$

$$0 = \int_{\Omega^e} \left[\mu \frac{\partial w_2^e}{\partial x} \left(\frac{\partial u_e}{\partial y} + \frac{\partial v_e}{\partial x} \right) + 2\mu \frac{\partial w_2^e}{\partial y} \frac{\partial v_e}{\partial y} + \gamma_e \frac{\partial w_2^e}{\partial y} \left(\frac{\partial u_e}{\partial x} + \frac{\partial v_e}{\partial y} \right) - w_2^e f_y^e \right] dx\, dy - \oint_{\Gamma^e} w_2^e t_y^e \, ds \tag{7-181}$$

where γ_e denotes the penalty parameter for element Ω^e.

Finite-element model The finite-element model associated with the penalty-function formulation (7-181) is given by substituting Eq. (7-168) for u_e and v_e:

$$\begin{bmatrix} [\bar{K}^{11e}] & [\bar{K}^{12e}] \\ [\bar{K}^{12e}]^T & [\bar{K}^{22e}] \end{bmatrix} \begin{Bmatrix} \{u^e\} \\ \{v^e\} \end{Bmatrix} = \begin{Bmatrix} \{F^{1e}\} \\ \{F^{2e}\} \end{Bmatrix} \tag{7-182}$$

where F_i^{1e} and F_i^{2e} are as defined in Eq. (7-170), and

$$[\bar{K}^{11e}] = [K^{11e}] + \gamma_e[S^{11e}] \qquad [\bar{K}^{12e}] = [K^{12e}] + \gamma_e[S^{12e}]$$
$$[\bar{K}^{22e}] = [K^{22e}] + \gamma_e[S^{22e}] \tag{7-183}$$

$K_{ij}^{\alpha\beta e}$, $S_{ij}^{\alpha\beta e}$ ($\alpha, \beta = 1, 2$) being the element matrices defined in Eq. (7-170). Globally, Eq. (7-182) is equivalent to

$$([K^1] + \gamma[K^2])\{\Delta\} = \{F\} \tag{7-184}$$

where $\{\Delta\}$ is the vector of global velocities.

Reduced integration For very large values of the penalty parameter γ_e and reasonably fine meshes, the finite-element scheme (7-184) generally does not give good solutions. This is because the discrete penalty functional

$$F(\mathbf{u}_h) = \frac{\gamma}{2} \int_{\Omega h} (\text{div } \mathbf{u}_h)^2 \, d\mathbf{x} \tag{7-185}$$

is not necessarily positive-definite [that is, $F(\mathbf{u}_h) > 0$ for div $\mathbf{u}_h \neq 0$ and $F(\mathbf{u}_h) = 0$ if div $\mathbf{u}_h = 0$] for a fixed mesh size h. Equivalently, the coefficient matrix $[K^2]$ associated with the penalty functional $F(\mathbf{u}_h)$ should be singular if div $\mathbf{u}_h = 0$. Otherwise, the assembled equations (7-184) will have a trivial solution $\mathbf{u}_h = \mathbf{0}$, which certainly satisfies the constraint condition. Thus, the discrete problem in Eq. (7-184) is overconstrained or "locked".

It is well-known in numerical analysis that the integrals appearing in finite-element coefficient matrices can be evaluated exactly, for polynomial bases, using an appropriate numerical integration rule [see Reddy (1984a)]. It is found that if the coefficients of $[K^2]$ are evaluated using a numerical integration rule of an order less than that required to integrate them exactly, the finite-element equations in Eq. (7-184) often give acceptable solutions. This technique of under integrating the penalty terms is known in the literature as *reduced integration.* For example, for a linear rectangular element, the coefficient matrix $[K^1]$ is evaluated exactly using a 2-point (in each coordinate direction) Gauss quadrature. The one-point Gauss rule is used to evaluate $[K^2]$, which would make it singular. Such selective integration gives a positive semidefinite matrix $[K^2]$ so that $[K^1] + \gamma[K^2]$ can be inverted (after imposing boundary conditions) to obtain a good finite-element approximation to the original problem.

Mathematically, the reduced integration technique is equivalent to a proper selection of the approximation subspaces $\mathbf{S}^h$ and Q^h of $\mathbf{H}_k^1(\Omega)$ and Q, respectively. Although the pressure variable P does not appear in the primary calculation, we compute it in the postcomputation using Eq. (4-159). The selection of Q^h means the selection of the integration rule to evaluate the coefficient matrix $[K^2]$ and the expression

$$P^e = -\gamma_e \text{ div } \mathbf{u}_e \tag{7-186}$$

For a given subspace $\mathbf{S}^h$, the pressure subspace Q^h is selected such that the following conditions hold [see Oden (1982)]: Let $I(f)$ denote a numerical quadrature rule for approximating the integral of a function f over the finite-element mesh

$$I(f) = \sum_{e=1}^{E} I_e(f) = \sum_{e=1}^{E} \sum_{I=1}^{G} w_I^e f(\xi_I^e) \tag{7-187}$$

where E is the total number of elements in the mesh, G the number of Gauss points, w_I^e the Gauss weights and ξ_I^e the Gauss points for the G-point Gauss rule within each element. Then

1. $$I((\text{div } \mathbf{u}_h)^2) \neq \int_{\Omega_h} (\text{div } \mathbf{u}_h)^2 \, d\mathbf{x} \qquad \text{for all } \mathbf{u}_h \in \mathbf{S}^h$$

   (i.e., the integration rule is inexact for the evaluation of the coefficient matrix $[K^2]$).
2. For $q_h \in Q^h$

   $$\int_{\Omega_h} q_h \text{ div } \bar{\mathbf{u}}_h \, d\mathbf{x} = I(q_h \text{ div } \bar{\mathbf{u}}_h) \qquad \text{for all } \bar{\mathbf{u}}_h \in \mathbf{S}^h$$

   (i.e., the integration rule is exact for the product q_h div $\bar{\mathbf{u}}_h$ when $q_h \in Q^h$ and $\bar{\mathbf{u}}_h \in \mathbf{S}^h$).
3. There is a unique $P_h \in Q^h$ such that

$$I(P_h q_h) = -\gamma I(\text{div } \mathbf{u}_h q_h) \qquad q_h \in Q^h \tag{7-188}$$

Equation (7-188) implies that the pressure is computed from the discrete relation

$$P(\xi_I^e) = -\gamma_e \text{ div } \mathbf{u}_e(\xi_I^e) \tag{7-189}$$

Figure 7-38 contains some elements and the integration points used for the nonpenalty and penalty terms. Other elements and their convergence and stability characteristics are discussed by Oden and Carey (1983).

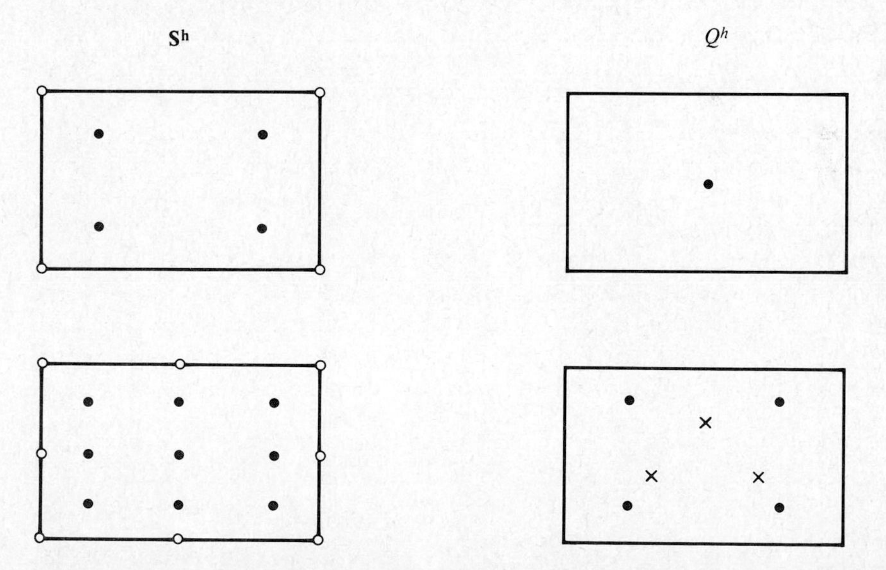

Figure 7-38 Finite elements for the approximation of velocities ($\mathbf{S}^h$) and the postcomputation of the pressure (Q^h) [open circles denote nodes and dark circles denote integration points] in the penalty finite-element model. A three-point evaluation of pressure is also suggested when velocities are approximated by quadratic polynomials.

Table 7-15 The effect of integration rule on the velocity $u(0.5, y)$ of the steady flow in a square cavity using the penalty model ($\gamma = 10^8$)

	Bilinear element		Biquadratic Lagrange element				Biquadratic serendipity element			
y	NGP = 2* LGP = 2	NGP = 2 LGP = 1	NGP = 3 LGP = 3	NGP = 3 LGP = 2	NGP = 2 LGP = 2	NGP = 3 LGP = 1	NGP = 3 LGP = 3	NGP = 3 LGP = 2	NGP = 2 LGP = 2	NGP = 3 LGP = 1
0.125	—	—	—	—	—	—	—	—	—	—
	−0.1020	−0.0558	−0.0583	−0.0600	−0.0597	−0.0694	−0.0728	−0.3255	−0.3257	−0.0670
0.250	−0.1278†	−0.0525	−0.1108	−0.0831	−0.0776	−0.0953	−0.1629	−0.2778	−0.2750	−0.0903
	−0.1410	−0.1162	−0.1105	−0.1204	−0.1226	−0.0848	−0.1158	0.2857	0.2858	−0.0964
0.375	—	—	—	—	—	—	—	—	—	—
	−0.1511	−0.1431	−0.1562	−0.1429	−0.1405	−0.1335	−0.1392	−0.2541	−0.2536	−0.1354
0.500	−0.1316	−0.3235	−0.2500	−0.2789	−0.2895	0.0066	0.1667	0.5000	0.5000	0.0182
	−0.1393	−0.2129	−0.2018	−0.2034	−0.2084	−0.0980	−0.1605	0.4893	0.4893	−0.1238
0.625	—	—	—	—	—	—	—	—	—	—
	−0.0979	−0.1804	−0.1410	−0.1583	−0.1571	−0.1068	−0.0753	−0.2505	−0.2509	−0.1170
0.750	−0.0564	−0.1240	0.0483	−0.0275	−0.0276	0.0953	−0.0038	−0.2222	−0.2250	0.0943
	0.0129	−0.0819	−0.0755	−0.0606	−0.0619	−0.1059	0.0325	0.7250	0.7250	0.0396
0.875	—	—	—	—	—	—	—	—	—	—
	0.2347	0.2903	0.3276	0.3035	0.3038	0.3097	0.2232	−0.1700	−0.1698	0.3193

* NGP = number of Gauss points used in the evaluation of nonpenalty terms, LGP = number of Gauss points used in the evaluation of penalty terms.

† The first line corresponds to 4 × 4 mesh of linear elements or 2 × 2 quadratic elements; the second line corresponds to 8 × 8 mesh of linear elements or 4 × 4 mesh of quadratic elements.

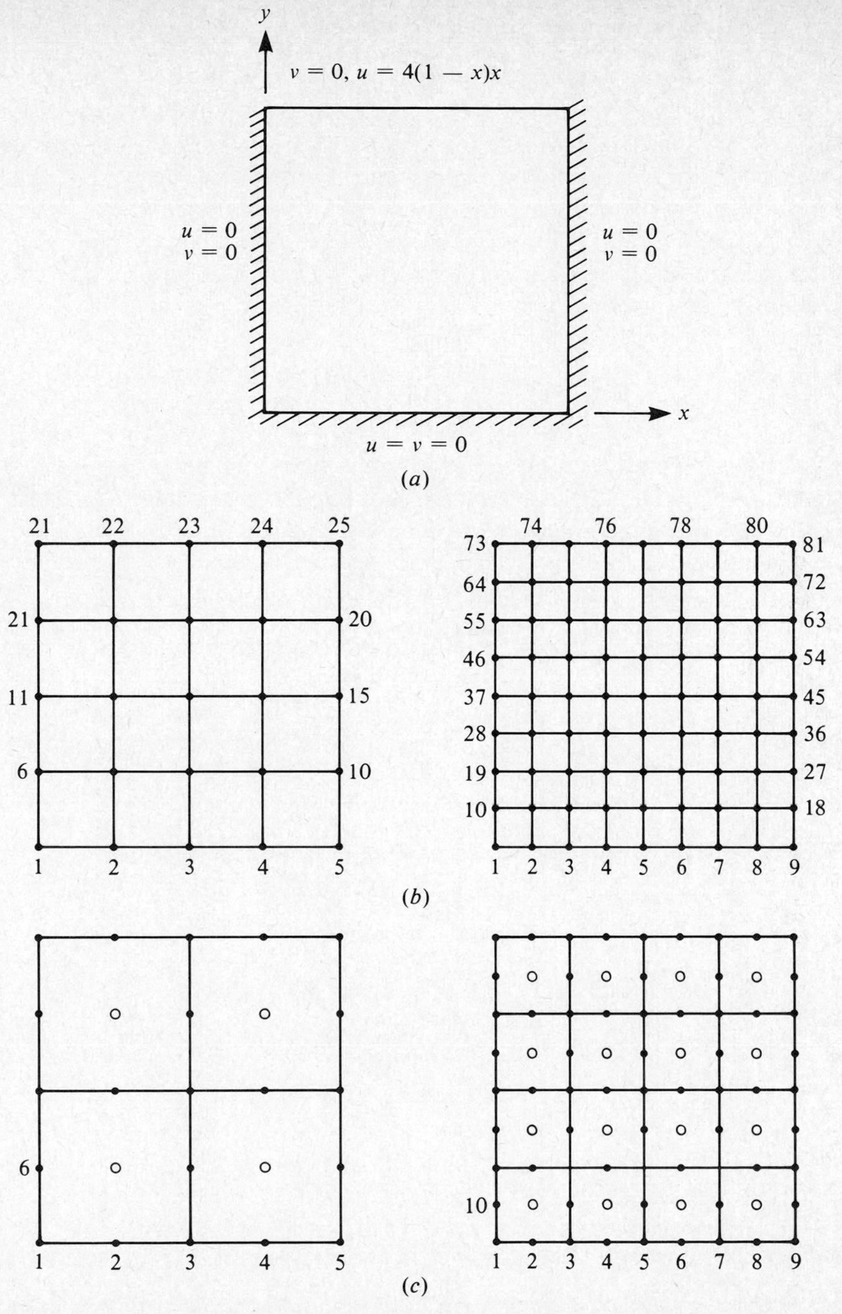

Figure 7-39 The domain and boundary conditions for the flow of a viscous fluid in a square cavity.

Applications Here we consider some applications of the mixed and penalty finite element models to incompressible fluid flow problems. The first one deals with the flow of a viscous fluid in a square cavity.

Example 7-22 Consider the flow of a viscous fluid in a square cavity (see Fig. 7-39*a*). We shall solve the problem using both the penalty model and mixed model. All of the boundary conditions are of the Dirichlet type.

Penalty model. Two different meshes of linear (4×4 and 8×8) and quadratic (2×2 and 4×4) elements are used to model the problem (see Fig. 7-39*b* and *c*). The effect of integration rule and penalty parameter on the solution, $u(0.5, y)$, is investigated (see Tables 7-15 and 7-16) using the bilinear, Lagrange biquadratic and serendipity biquadratic elements. From the results presented in Tables 7-15 and 7-16, the following observations can be made:

(i) The usual integration rule for nonpenalty terms and one order less for penalty terms gives the best results in the case of the Lagrange elements.
(ii) There exists a range of penalty parameters for which the solution is virtually unchanged (in the present example, the range is $\gamma = 10^4 - 10^8$). For values outside this range the solution gradually deteriorates.
(iii) The serendipity quadratic element gives reasonable results only when the 3×3 Gauss rule is used for nonpenalty terms and the 1×1 Gauss rule for the penalty terms. However, the accuracy, when compared to the bilinear and biquadratic Lagrange element, is very poor.

Table 7-16 The effect of the penalty parameter on the velocity $u(0.5, y)$ of the steady flow of a square cavity using the penalty model

	Linear element			Quadratic Lagrange element			Quadratic serendipity element		
y	$\gamma = 1$	$\gamma = 10^4$	$\gamma = 10^8$	$\gamma = 1$	$\gamma = 10^4$	$\gamma = 10^8$	$\gamma = 1$	$\gamma = 10^4$	$\gamma = 10^8$
0.125	—	—	—	—	—	—	—	—	—
	−0.0647*	−0.0558	−0.0558	−0.0058	−0.0600	−0.0600	−0.0057	−0.2878	−0.3255
0.250	−0.0094	−0.0526	−0.0525	−0.0009	−0.0830	−0.0831	0.0049	−0.2772	−0.2778
	−0.0051	−0.1162	−0.1162	−0.0032	−0.1203	−0.1204	−0.0021	0.2222	0.2857
0.375	—	—	—	—	—	—	—	—	—
	0.0065	−0.1431	−0.1431	0.0102	−0.1428	−0.1429	0.0102	−0.2370	−0.2541
0.500	0.0149	−0.3233	−0.3235	0.0366	−0.2786	−0.2789	0.1149	0.4988	0.5000
	0.0379	−0.2129	−0.2129	0.0439	−0.2034	−0.2036	0.0451	0.4062	0.4893
0.625	—	—	—	—	—	—	—	—	—
	0.1093	−0.1804	−0.1804	0.1178	−0.1582	−0.1583	0.1202	−0.2330	−0.2505
0.750	0.2301	−0.1238	−0.1240	0.2695	−0.0275	−0.0275	0.2360	−0.2215	−0.2222
	0.2558	−0.0819	−0.0819	0.2641	−0.0604	−0.0606	0.2637	0.6543	0.7250
0.875	—	—	—	—	—	—	—	—	—
	0.5304	0.2903	0.2903	0.5362	0.3037	0.3035	0.5354	−0.1284	−0.1700

* The first line corresponds to 4×4 mesh of linear elements or 2×2 quadratic elements, and the second line corresponds to 8×8 mesh of linear elements or 4×4 quadratic elements.

Table 7-17 Comparison of the velocity $u(0.5, y)$ obtained by various finite element models (4 × 4 mesh)

y	Penalty model 9-node*	Mixed model 9-node	Mixed model 8-node
0.125	−0.0600	−0.0613	−0.0495
0.250	−0.1204	−0.1077	−0.1240
0.375	−0.1429	−0.1500	−0.1498
0.500	−0.2034	−0.1813	−0.1502
0.625	−0.1583	−0.1729	−0.1742
0.750	−0.0606	−0.0486	−0.1104
0.875	0.3035	0.3050	0.3970

* NGP = 3 and LGP = 2, $\gamma = 10^8$.

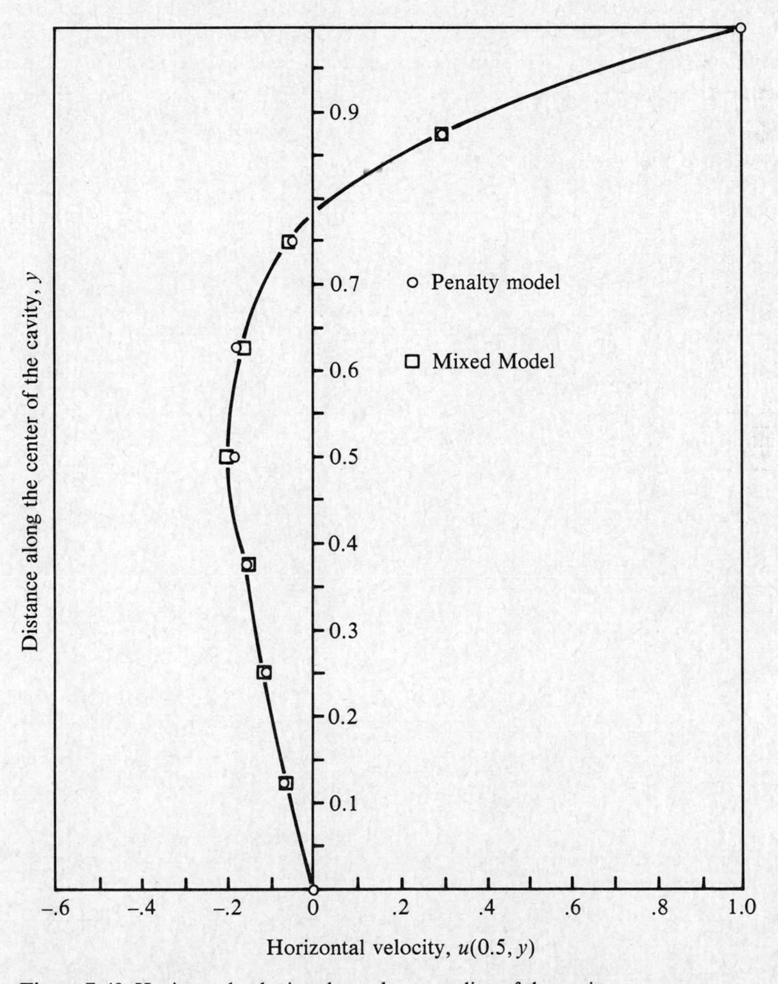

Figure 7-40 Horizontal velocity along the centerline of the cavity.

Mixed Model. The uniform 4×4 meshes of biquadratic Lagrange and serendipity elements were used to model the problem (see Fig. 7-39*c*). The pressure is specified at one point, node 9, to eliminate the constant state of pressure. The results are compared in Table 7-17 with those obtained using the penalty model (with $\gamma = 10^8$ and the three-point integration rule for non-penalty terms and the two-point-rule for penalty terms). Once again, we note that the Lagrange biquadratic element gives more accurate results over the serendipity element (also see Fig. 7-40).

The next example deals with the flow of a viscous fluid when squeezed between two parallel rigid plates [see Reddy (1984a, pp. 289–293)].

Example 7-23 Consider a viscous incompressible fluid between two long parallel plates, which are moved toward each other with a constant velocity v_0. This movement of plates causes the fluid to flow outward (parallel to the plates) from the center of the plates. Using the biaxial symmetry of the problem, we can model only a quadrant of the domain, as shown in Fig. 7-41*a*. A 5×3 nonuniform mesh of quadratic elements and the equivalent

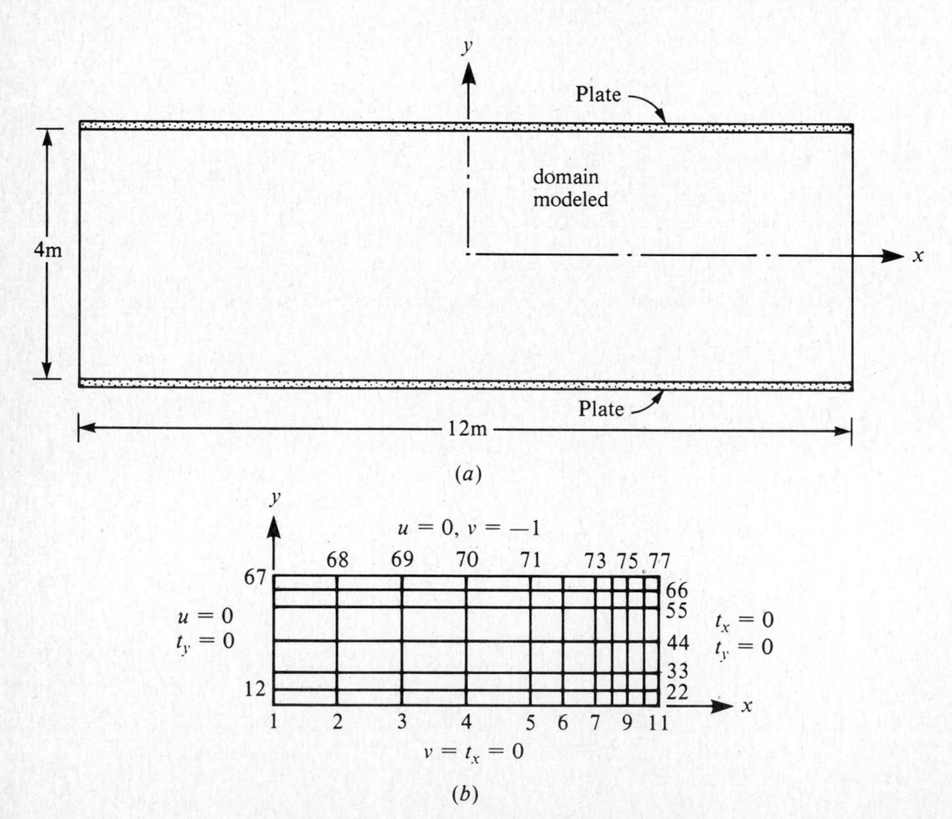

Figure 7-41 The domain and finite-element mesh for the flow of a viscous incompressible flow between two parallel plates.

Table 7-18 Comparison of the horizontal velocity $u(x, 0)$ obtained by penalty model* and mixed model with the analytical solution

	$\gamma = 1.0$		$\gamma = 100$		$\gamma = 10^8$		Mixed model	Mixed model	
x	4-node	9-node	4-node	9-node	4-node	9-node	8-node	9-node	Analytical solution
1	0.0303	0.0310	0.6563	0.6513	0.7576	0.7505	0.7496	0.7497	0.7500
2	0.0677	0.0691	1.3165	1.3062	1.5135	1.4992	1.5038	1.5031	1.5000
3	0.1213	0.1233	1.9911	1.9769	2.2756	2.2557	2.2563	2.2561	2.2500
4	0.2040	0.2061	2.6960	2.6730	3.0541	3.0238	3.0213	3.0203	3.0000
4.5	0.2611	0.2631	3.0718	3.0463	3.4648	3.4307	3.4331	3.4292	3.3750
5.0	0.3297	0.3310	3.4347	3.3956	3.8517	3.8029	3.8249	3.8165	3.7500
5.25	0.3674	0.3684	3.6120	3.5732	4.0441	3.9944	4.0074	3.9893	3.9375
5.5	0.4060	0.4064	3.7388	3.6874	4.1712	4.1085	4.1450	4.1204	4.1250
5.75	0.4438	0.4443	3.8316	3.7924	4.2654	4.2160	4.2188	4.2058	4.3125
6.0	0.4793	0.4797	3.8362	3.7862	4.2549	4.1937	4.2659	4.2364	4.5000

* Three-point Gauss rule for nonpenalty terms and two-point Gauss rule for penalty terms are used.

10×6 nonuniform mesh of linear elements are used to model the quadrant (only the 10×6 mesh of linear elements is shown in Fig. 7-41*b*).

A comparison of the horizontal velocity, $u(x, 0)$ obtained by the two finite-element models with the analytical solution is given in Table 7-18. Clearly, both models give accurate results. Figure 7-42 contains plots of pressure $P(x, 0.125)$ and velocity $u(x, y)$ for $x = 2$, 4, and 6. The pressure in the penalty model was computed using Eq. (7-189) with the one-point Gauss formula for the linear element and the two-point Gauss formula for the quadratic element. The difference between the finite-element solutions and

Table 7-19 Effect of the quadrature rule on the pressure computation in penalty elements*

Linear element		Quadratic (Lagrange) element			Location of quadrature points
1-point	2-point	1-point	2-point	3-point	
				0.2983E5 (6)	
				−0.1814E6 (7)	8• 11• 14•
	−0.4984E6 (2)		7.3238 (2)	0.8693E6 (8)	3• 5•
	0.2173E6 (3)		7.3461 (3)	0.6525E6 (9)	
7.3828 (1)†	−0.2173E6 (4)	−0.6355E6 (1)	6.8854 (4)	−0.6355E6 (10)	7• 1•10 13•
	0.4984E6 (5)		6.9304 (5)	−0.3968E6 (11)	
				0.1475E6 (12)	2• 4•
				−0.3285E6 (13)	6• 9• 12•
				0.9869E6 (14)	

* Nonuniform meshes of 10×6 linear elements and 5×3 nine-node quadratic elements with full integration on viscous terms and reduced integration on penalty terms ($\gamma = 10^8$) are used.

† The number in parentheses denotes the quadrature point number of the element (note that quadrature point locations in the two meshes are different for the same quadrature point; for example, point 2 in the mesh of linear elements has the coordinates (0.211325, 0.052831) whereas it has the coordinates (0.42265, 0.394338) in the mesh of quadratic elements).

the analytical solution is attributed partly to the singularity at point $(x, y) = (6, 2)$. Improvement is expected with further refinement of the mesh near the free surface.

If the pressure in the penalty model were computed using the full quadrature rule we would obtain erroneous values. For example, if the two-point quadrature is used for a linear element, say element 1, we obtain large values that differ in sign from one point to the next as shown in Table 7-19. Similar results are shown for the quadratic element in the same table. Thus, the same quadrature rule as that used for the evaluation of the penalty terms in the coefficient matrix must be used to evaluate the pressure.

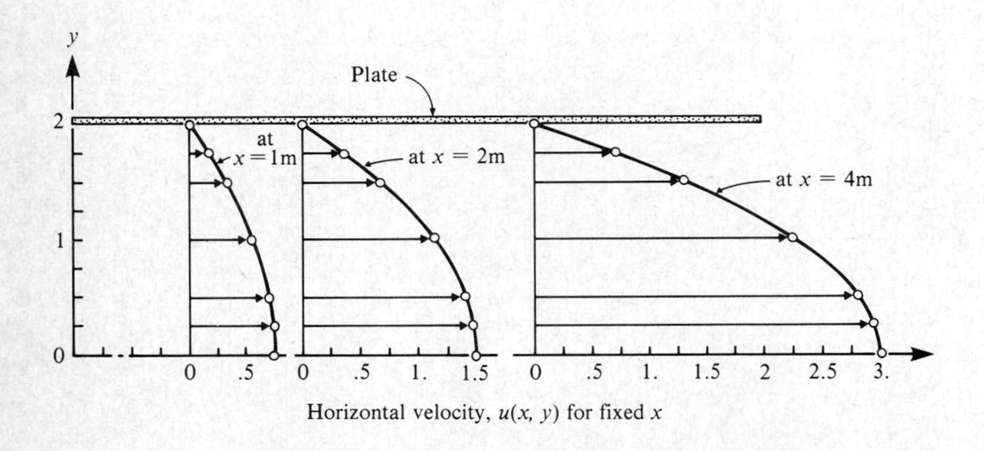

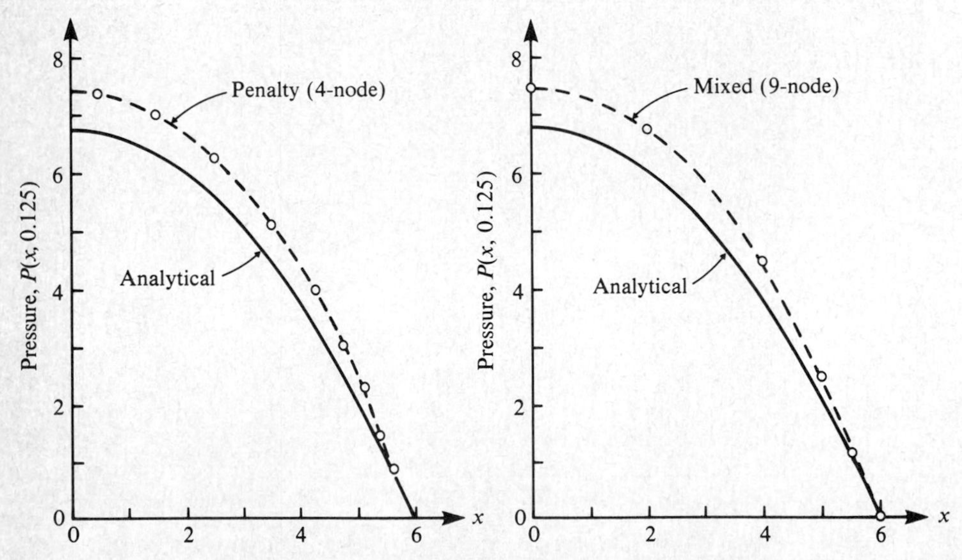

Figure 7-42 Comparison of the finite-element solutions for velocity and pressure with the analytical solutions.

The two examples presented above indicate that the bilinear and biquadratic Lagrange elements with reduced integration (i.e., one order less than usual) are the most accurate elements for the solution of Stokes (and Navier–Stokes) equations. For additional mathematical results and numerical examples the reader can consult the many references listed under Numerical Methods in the Bibliography at the end of the book.

7-7-4 Bending of Elastic Plates

There are two types of plate theories that are used to analyze plates: (i) the classical plate theory and (ii) shear deformation plate theories. The *classical plate theory* is based on the assumptions that straight lines normal to the midplane before deformation remain straight and normal to the midplane after deformation and that the transverse normals are inextensible. These assumptions lead to the neglect of transverse shear stresses. The classical plate theory is adequate for thin plates (i.e., plates with side-to-thickness ratios greater than 10). Shear deformation theories are refined theories over classical plate theory in that they account for the transverse shear stresses. In the present study we shall consider finite-element models of the classical plate theory and a first-order shear deformation theory (often called the *Reissner–Mindlin theory*).

Classical plate theory The equation governing the transverse deflection w of an orthotropic plate in the classical plate theory is given by a generalization of Eq. (5-105) [see Reddy (1984b)]:

$$\frac{\partial^2}{\partial x^2}\left(D_{11}\frac{\partial^2 w}{\partial x^2} + D_{12}\frac{\partial^2 w}{\partial y^2}\right) + 4D_{66}\frac{\partial^4 w}{\partial x^2\,\partial y^2} + \frac{\partial^2}{\partial y^2}\left(D_{12}\frac{\partial^2 w}{\partial x^2} + D_{22}\frac{\partial^2 w}{\partial y^2}\right) = f \tag{7-190}$$

where D_{11}, D_{12}, D_{22}, and D_{66} are plate rigidities,

$$D_{11} = \frac{E_1 t^3}{12(1 - \nu_{12}\nu_{21})} \qquad D_{22} = E_2 D_{11}/E_1 \qquad D_{12} = \nu_{12} D_{22}$$

$$D_{66} = G_{12} D_{11}/E_1 \tag{7-191}$$

f is the distributed transverse load, E_1 and E_2 are the Young's moduli along the x- and y-directions, respectively, ν_{12} and ν_{21} $(=\nu_{12} E_2/E_1)$ are Poisson's ratios, G_{12} is the shear modulus in the xy plane (also the material-axes) of the plate, and t is the thickness of the plate.

Equation (7-190) is an extension of Eq. (7-66) governing beams to two dimensions. Like in the case of Eq. (7-66), we can develop the conventional and mixed models of Eq. (7-190). The conventional (or displacement) model is based on the variational formulation of Eq. (7-190) and the mixed models are based on a system of second-order equations involving the deflection and bending moments of the plate. We now describe these models.

Displacement model The variational formulation of Eq. (7-190) over a typical element is given by

$$0 = \int_{\Omega^e} \left[\frac{\partial^2 v_e}{\partial x^2} \left(D_{11}^e \frac{\partial^2 w_e}{\partial x^2} + D_{12}^e \frac{\partial^2 w_e}{\partial y^2} \right) + \frac{\partial^2 v_e}{\partial y^2} \left(D_{12}^e \frac{\partial^2 w_e}{\partial x^2} + D_{22}^e \frac{\partial^2 w_e}{\partial y^2} \right) \right.$$
$$\left. + 4D_{66}^e \frac{\partial^2 v_e}{\partial x\, \partial y} \frac{\partial^2 w_e}{\partial x\, \partial y} - v f_e \right] dx\, dy$$
$$- \oint_{\Gamma^e} \left(M_n^e \frac{\partial w_e}{\partial n} + M_s^e \frac{\partial w_e}{\partial s} + Q_n^e w_e \right) ds \qquad (7\text{-}192)$$

where (see Fig. 4-11)

$$\frac{\partial w}{\partial n} = n_x \frac{\partial w}{\partial x} + n_y \frac{\partial w}{\partial y} \qquad \frac{\partial w}{\partial s} = n_x \frac{\partial w}{\partial y} - n_y \frac{\partial w}{\partial x}$$

$$M_n = M_x n_x^2 + M_y n_y^2 + 2M_{xy} n_x n_y \qquad (7\text{-}193)$$
$$M_s = (M_y - M_x) n_x n_y - M_{xy}(n_x^2 - n_y^2)$$
$$Q_n = Q_x n_x + Q_y n_y$$

and

$$M_x = -D_{11} \frac{\partial^2 w}{\partial x^2} - D_{12} \frac{\partial^2 w}{\partial y^2}$$
$$M_y = -D_{12} \frac{\partial^2 w}{\partial x^2} - D_{22} \frac{\partial^2 w}{\partial y^2} \qquad (7\text{-}194)$$
$$M_{xy} = -2D_{66} \frac{\partial^2 w}{\partial x\, \partial y}$$
$$Q_x = \frac{\partial M_x}{\partial x} + \frac{\partial M_{xy}}{\partial y} \qquad Q_y = \frac{\partial M_{xy}}{\partial x} + \frac{\partial M_y}{\partial y}$$

From Example 5-12, we recall that $w \in H^2(\Omega)$. Therefore, w_e, $\partial w_e/\partial n$ and $\partial w_e/\partial s$ must be continuous across an interface between two elements. Hence, a C^1-approximation of w_e is required. Substituting $v_e = \psi_i^e$ and $w_e = \sum_{j=1}^{16} u_j^e \psi_j^e$ into Eq. (7-192), we obtain

$$[K^e]\{u^e\} = \{f^e\} + \{P^e\} \qquad (7\text{-}195)$$

where ψ_i^e are the Hermite cubic interpolation functions, and

$$K_{ij}^e = \int_{\Omega^e} \left[\frac{\partial^2 \psi_i^e}{\partial x^2} \left(D_{11}^e \frac{\partial^2 \psi_j^e}{\partial x^2} + D_{12}^e \frac{\partial^2 \psi_j^e}{\partial y^2} \right) + \frac{\partial^2 \psi_i^e}{\partial y^2} \left(D_{12}^e \frac{\partial^2 \psi_j^e}{\partial x^2} + D_{22}^e \frac{\partial^2 \psi_j^e}{\partial y^2} \right) \right.$$
$$\left. + 4D_{66}^e \frac{\partial^2 \psi_i^e}{\partial x\, \partial y} \frac{\partial^2 \psi_j^e}{\partial x\, \partial y} \right] dx\, dy \qquad (7\text{-}196)$$

$$f_i^e = \int_{\Omega^e} f_e \psi_i^e \, dx\, dy \qquad P_i^e = \oint_{\Gamma^e} \left(M_n^e \frac{\partial \psi_i^e}{\partial n} + M_s^e \frac{\partial \psi_i^e}{\partial s} + Q_n^e \psi_i^e \right) ds$$

The stiffness (coefficient) matrix $[K^e]$ is of order 16×16 for the four-node Hermite cubic rectangular element shown in Fig. 7-25c. The primary degrees of freedom at each node are (see Fig. 7-43a)

$$w \qquad \frac{\partial w}{\partial x} \qquad \frac{\partial w}{\partial y} \qquad \text{and} \qquad \frac{\partial^2 w}{\partial x\, \partial y}$$

Note that the normal-slope continuity is satisfied by the element (hence, it is a *conforming* element).

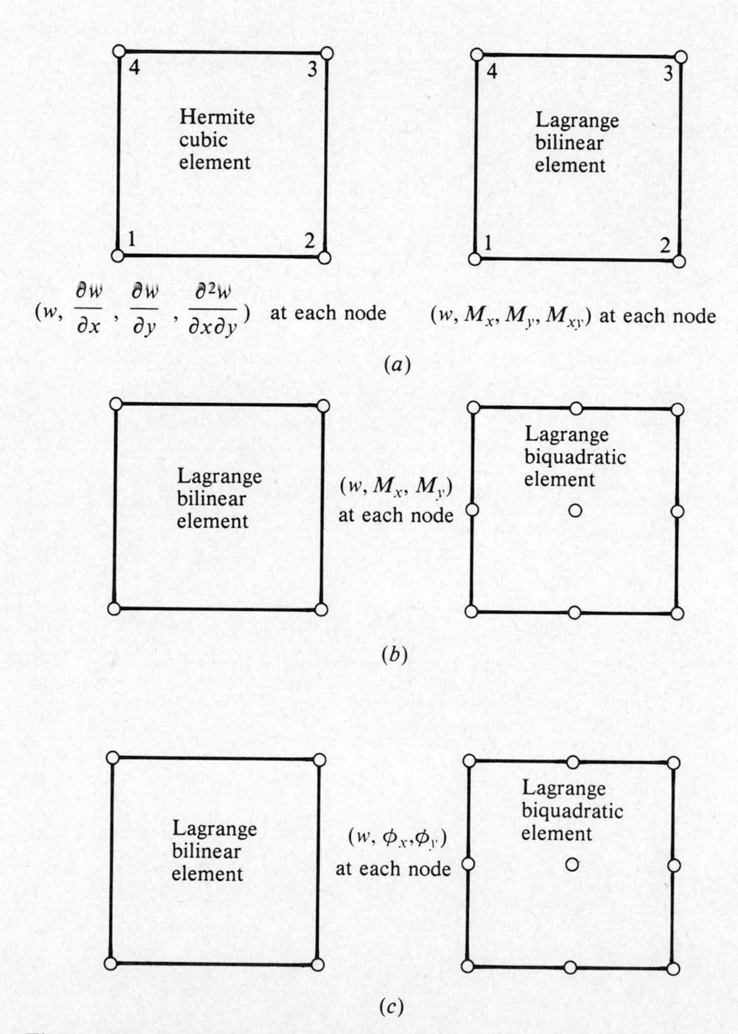

Figure 7-43 Finite elements for (*a*) the displacement model and mixed model I, (*b*) mixed model II and (*c*) shear deformation models of plates.

Mixed models The continuity requirements on the interpolation functions can be relaxed by reformulating the fourth-order equation (7-190) as a set of second-order equations in terms of the transverse deflection and bending moments. Such a formulation is called a *mixed formulation.* We discuss two such models here.

In view of the definitions in Eq. (7-194), Eq. (7-190) can be written as

$$-\left(\frac{\partial^2 M_x}{\partial x^2} + 2\,\frac{\partial^2 M_{xy}}{\partial x\,\partial y} + \frac{\partial^2 M_y}{\partial y^2}\right) = f \tag{7-197}$$

The first three equations of Eq. (7-194) and Eq. (7-197) define a system of four equations in the four unknowns, w, M_x, M_y, and M_{xy}. The finite-element model corresponding to these four equations is termed *mixed model I.*

To facilitate the variational formulation, Eqs. (7-194) are expressed in the alternative form,

$$\begin{aligned}
\frac{\partial^2 w}{\partial x^2} &= -\bar{D}_{22} M_x + \bar{D}_{12} M_y \\
\frac{\partial^2 w}{\partial y^2} &= \bar{D}_{12} M_x - \bar{D}_{11} M_y \\
2\,\frac{\partial^2 w}{\partial x\,\partial y} &= -(D_{66})^{-1} M_{xy}
\end{aligned} \tag{7-198}$$

where

$$\bar{D}_{ij} = D_{ij}/D_0 \qquad D_0 = D_{11}D_{22} - D_{12}^2 \tag{7-199}$$

The variational problem of Eqs. (7-197) and (7-198) over a typical element is to find $(w_e, M_x^e, M_y^e, M_{xy}^e) \in \mathbf{S}_1^h$ such that

$$\begin{aligned}
0 &= \int_{\Omega^e} \left(\frac{\partial \bar{w}_e}{\partial x}\frac{\partial M_x^e}{\partial x} + \frac{\partial \bar{w}_e}{\partial y}\frac{\partial M_{xy}^e}{\partial x} + \frac{\partial \bar{w}_e}{\partial x}\frac{\partial M_{xy}^e}{\partial y} + \frac{\partial \bar{w}_e}{\partial y}\frac{\partial M_y^e}{\partial y} - \bar{w}_e f_e\right) dx\,dy \\
&\quad - \oint_{\Gamma^e} Q_n^e \bar{w}_e\, ds \\
0 &= \int_{\Omega^e} \left(\frac{\partial \bar{M}_x^e}{\partial x}\frac{\partial w_e}{\partial x} - \bar{D}_{22}^e \bar{M}_x^e M_x^e + \bar{D}_{12}^e \bar{M}_x^e M_y^e\right) dx\,dy - \oint_{\Gamma^e} \bar{M}_x^e \frac{\partial w_e}{\partial x} n_x\, ds \\
0 &= \int_{\Omega^e} \left(\frac{\partial \bar{M}_y^e}{\partial y}\frac{\partial w_e}{\partial y} + \bar{D}_{12}^e \bar{M}_y^e M_x^e - \bar{D}_{11}^e \bar{M}_y^e M_y^e\right) dx\,dy - \oint_{\Gamma^e} \bar{M}_y^e \frac{\partial w_e}{\partial y} n_y\, ds \\
0 &= \int_{\Omega^e} \left(\frac{\partial \bar{M}_{xy}^e}{\partial x}\frac{\partial w_e}{\partial y} + \frac{\partial \bar{M}_{xy}^e}{\partial y}\frac{\partial w_e}{\partial x} - \frac{1}{D_{66}^e} \bar{M}_{xy}^e M_{xy}^e\right) dx\,dy \\
&\quad - \oint_{\Gamma^e} \bar{M}_{xy}^e \left(\frac{\partial w_e}{\partial x} n_y + \frac{\partial w_e}{\partial y} n_x\right) ds
\end{aligned} \tag{7-200}$$

hold for any arbitrary $(\bar{w}_e, \bar{M}_x^e, \bar{M}_y^e, \bar{M}_{xy}^e) \in \mathbf{S}_1^h \subset \mathbf{H}_1$, where

$$\mathbf{H}_1 = H^1(\Omega) \times H_x^1(\Omega) \times H_y^1(\Omega) \times H^1(\Omega)$$

$$H_x^1(\Omega) = \left\{u\colon\ u, \frac{\partial u}{\partial x} \in L_2(\Omega)\right\}$$

$$H_y^1(\Omega) = \left\{u\colon\ u, \frac{\partial u}{\partial y} \in L_2(\Omega)\right\}$$

Thus, only C°-approximations of w_e, M_x^e, M_y^e, and M_{xy}^e are required.

Let w_e, M_x^e, M_y^e, and M_{xy}^e be interpolated by expressions of the form

$$\begin{aligned} w_e &= \sum_{i=1}^{r} w_i^e \psi_i^{1e} \qquad & M_x^e &= \sum_{i=1}^{s} M_{xi}^e \psi_i^{2e} \\ M_y^e &= \sum_{i=1}^{p} M_{yi}^e \psi_i^{3e} \qquad & M_{xy}^e &= \sum_{i=1}^{q} M_{xyi}^e \psi_i^{4e} \end{aligned} \tag{7-201}$$

where $\psi_i^{\alpha e}$ ($\alpha = 1, 2, 3, 4$) are appropriate interpolation functions.

Substituting Eq. (7-201) into Eq. (7-200), we obtain

$$\begin{bmatrix} [K^{11}] & [K^{12}] & [K^{13}] & [K^{14}] \\ & [K^{22}] & [K^{23}] & [K^{24}] \\ \text{symm.} & & [K^{33}] & [K^{34}] \\ & & & [K^{44}] \end{bmatrix}_e \begin{Bmatrix} \{w\} \\ \{M_x\} \\ \{M_y\} \\ \{M_{xy}\} \end{Bmatrix}_e = \begin{Bmatrix} \{F^1\} \\ \{F^2\} \\ \{F^3\} \\ \{F^4\} \end{Bmatrix}_e \tag{7-202}$$

where

$$K_{ij}^{11} = 0 \qquad i, j = 1, 2, \ldots, r$$

$$K_{ij}^{12} = \int_{\Omega^e} \frac{\partial \psi_i^{1e}}{\partial x} \frac{\partial \psi_j^{2e}}{\partial x}\, dx\, dy \qquad i = 1, 2, \ldots, r; j = 1, 2, \ldots, s$$

$$K_{ij}^{13} = \int_{\Omega^e} \frac{\partial \psi_i^{1e}}{\partial y} \frac{\partial \psi_j^{3e}}{\partial y}\, dx\, dy \qquad i = 1, 2, \ldots, r; j = 1, 2, \ldots, p$$

$$K_{ij}^{14} = \int_{\Omega^e} \left(\frac{\partial \psi_i^{1e}}{\partial x} \frac{\partial \psi_j^{4e}}{\partial y} + \frac{\partial \psi_i^{1e}}{\partial y} \frac{\partial \psi_j^{4e}}{\partial x} \right) dx\, dy \qquad i = 1, 2, \ldots, r; j = 1, 2, \ldots, q$$

$$K_{ij}^{22} = \int_{\Omega^e} (-\bar{D}_{22}^e) \psi_i^{2e} \psi_j^{2e}\, dx\, dy \qquad i, j = 1, 2, \ldots, s$$

$$K_{ij}^{23} = \int_{\Omega^e} \bar{D}_{12}^e \psi_i^{2e} \psi_j^{3e}\, dx\, dy \qquad i = 1, 2, \ldots, s; j = 1, 2, \ldots, p$$

$$K_{ij}^{24} = 0 \qquad i = 1, 2, \ldots, s; j = 1, 2, \ldots, q$$

$$K_{ij}^{33} = \int_{\Omega^e} (-\bar{D}_{11}^e) \psi_i^{3e} \psi_j^{3e}\, dx\, dy \qquad i, j = 1, 2, \ldots, p \tag{7-203}$$

(*equation continued*)

[*equation* (7-203) *continued*]

$$K_{ij}^{34} = 0 \qquad i = 1, 2, \ldots, p; j = 1, 2, \ldots, q$$

$$K_{ij}^{44} = -\int_{\Omega^e} (D_{66}^e)^{-1} \psi_i^{4e} \psi_j^{4e} \, dx \, dy \qquad i, j = 1, 2, \ldots, q$$

$$F_i^1 = \int_{\Omega^e} f_e \psi_i^{1e} \, dx \, dy + \oint_{\Gamma^e} Q_n^e \psi_i^{1e} \, ds \qquad i = 1, 2, \ldots, r$$

$$F_i^2 = \oint_{\Gamma^e} \left(\frac{\partial w_e}{\partial x} n_x \right) \psi_i^{2e} \, ds \qquad i = 1, 2, \ldots, s$$

$$F_i^3 = \oint_{\Gamma^e} \left(\frac{\partial w_e}{\partial y} n_y \right) \psi_i^{3e} \, ds \qquad i = 1, 2, \ldots, p$$

$$F_i^4 = \oint_{\Gamma^e} \left(\frac{\partial w_e}{\partial x} n_y + \frac{\partial w_e}{\partial y} n_x \right) \psi_i^{4e} \, ds \qquad i = 1, 2, \ldots, q$$

An examination of the variational form in Eq. (7-200) shows that the following minimum continuity conditions are required of $\psi_i^{\alpha e}$ ($\alpha = 1, 2, 3, 4$):

$$\begin{aligned} \psi_i^{1e} &\in H^1(\Omega^e) = \text{linear in } x \text{ and linear in } y \\ \psi_i^{2e} &\in H_x^1(\Omega^e) = \text{linear in } x \text{ and constant in } y \\ \psi_i^{3e} &\in H_y^1(\Omega^e) = \text{linear in } y \text{ and constant in } x \\ \psi_i^{4e} &\in H^1(\Omega^e) = \text{linear in } x \text{ and linear in } y \end{aligned} \tag{7-204}$$

The interpolation functions $\psi_i^{\alpha e}$ that meet only the minimum requirements are

$$\begin{aligned} &\psi_i^{1e} = \text{the bilinear functions in Eq. (7-116)} \\ &\psi_1^{2e} = 1 - \frac{\xi}{a} \qquad \psi_2^{2e} = \frac{\xi}{a} \\ &\psi_1^{3e} = 1 - \frac{\eta}{b} \qquad \psi_2^{3e} = \frac{\eta}{b} \\ &\psi_i^{4e} = \text{the bilinear functions in Eq. (7-116)} \end{aligned} \tag{7-205a}$$

The coefficient matrices in Eq. (7-203) can be easily evaluated for this element. We have, for example,

$$[K^{12}] = \frac{b}{2a} \begin{bmatrix} 1 & -1 \\ -1 & 1 \\ -1 & 1 \\ 1 & -1 \end{bmatrix} \qquad [K^{13}] = \frac{a}{2b} \begin{bmatrix} 1 & -1 \\ 1 & -1 \\ -1 & 1 \\ -1 & 1 \end{bmatrix}$$

$$[K^{22}] = -\bar{D}_{66}\,\frac{a}{6}\begin{bmatrix} 2 & 1 \\ 1 & 2 \end{bmatrix} \qquad [K^{33}] = -\bar{D}_{11}\,\frac{b}{6}\begin{bmatrix} 2 & 1 \\ 1 & 2 \end{bmatrix}$$

etc. The element stiffness matrix is of the order 12 × 12.

Another choice of $\psi_i^{\alpha e}$ is provided by (see Fig. 7-43*a*):

$$\psi_i^{1e} = \psi_i^{2e} = \psi_i^{3e} = \psi_i^{4e} = \text{the bilinear functions in Eq. (7-116)} \qquad \text{(7-205}b\text{)}$$

Computation of the associated coefficient matrices is simple and straightforward. The resulting stiffness matrix is of the order 16 × 16.

A simplified mixed model can be derived by eliminating the twisting moment M_{xy} from equation (7-197) by using the last equation in Eq. (7-198). We obtain

$$\begin{aligned} -\left(\frac{\partial^2 M_x}{\partial x^2} - 4D_{66}\,\frac{\partial^4 w}{\partial x^2\,\partial y^2} + \frac{\partial^2 M_y}{\partial y^2}\right) &= f \\ \frac{\partial^2 w}{\partial x^2} &= -\bar{D}_{22}\,M_x + \bar{D}_{12}\,M_y \\ \frac{\partial^2 w}{\partial y^2} &= \bar{D}_{12}\,M_x - \bar{D}_{11} M_y \end{aligned} \qquad \text{(7-206)}$$

The finite-element model of Eqs. (7-206) is termed *mixed model II*.

The variational problem of Eqs. (7-206) over an element Ω^e is to find $(w_e, M_x^e, M_y^e) \in \mathbf{S}_2^h \subset \mathbf{H}_2$ such that

$$\begin{aligned} 0 &= \int_{\Omega^e}\left(\frac{\partial \bar{w}_e}{\partial x}\frac{\partial M_x^e}{\partial x} + 4D_{66}^e\,\frac{\partial^2 \bar{w}_e}{\partial x\,\partial y}\frac{\partial^2 w_e}{\partial x\,\partial y} + \frac{\partial \bar{w}_e}{\partial y}\frac{\partial M_y^e}{\partial y} - \bar{w}_e f_e\right) dx\,dy - \oint_{\Gamma^e} Q_n^e\,\bar{w}_e\,ds \\ 0 &= \int_{\Omega^e}\left(\frac{\partial \bar{M}_x^e}{\partial x}\frac{\partial w_e}{\partial x} - \bar{D}_{22}^e\,\bar{M}_x^e M_x^e + \bar{D}_{12}^e\,\bar{M}_x^e M_y^e\right) dx\,dy - \oint_{\Gamma^e} \bar{M}_x^e\,\frac{\partial w_e}{\partial x}\,n_x\,ds \\ 0 &= \int_{\Omega^e}\left(\frac{\partial \bar{M}_y^e}{\partial y}\frac{\partial w_e}{\partial y} + \bar{D}_{12}^e\,\bar{M}_y^e M_x^e - \bar{D}_{11}^e \bar{M}_y^e M_y^e\right) dx\,dy - \oint_{\Gamma^e} \bar{M}_y^e\,\frac{\partial w_e}{\partial y}\,n_y\,ds \end{aligned} \qquad \text{(7-207)}$$

hold for all $(\bar{w}_e, \bar{M}_x^e, \bar{M}_y^e) \in \mathbf{S}_2^h \subset \mathbf{H}_2$, where

$$\mathbf{H}_2 = H^1(\Omega) \times H_x^1(\Omega) \times H_y^1(\Omega)$$

The finite-element model of Eqs. (7-207) is obtained by substituting Eq. (7-201) into Eq. (7-207). We obtain

$$\begin{bmatrix} [K^{11}] & [K^{12}] & [K^{13}] \\ & [K^{22}] & [K^{23}] \\ \text{symm.} & & [K^{33}] \end{bmatrix}_e \begin{Bmatrix} \{w\} \\ \{M_x\} \\ \{M_y\} \end{Bmatrix}_e = \begin{Bmatrix} \{F^1\} \\ \{F^2\} \\ \{F^3\} \end{Bmatrix}_e \qquad \text{(7-208)}$$

where

$$
\begin{aligned}
K_{ij}^{11} &= 4D_{66}^{e} \int_{\Omega^e} \frac{\partial^2 \psi_i^{1e}}{\partial x\,\partial y} \frac{\partial^2 \psi_j^{1e}}{\partial x\,\partial y}\, dx\, dy \qquad i, j = 1, 2, \ldots, r \\
K_{ij}^{12} &= \int_{\Omega^e} \frac{\partial \psi_i^{1e}}{\partial x} \frac{\partial \psi_j^{2e}}{\partial x}\, dx\, dy \qquad i = 1, 2, \ldots, r;\ j = 1, 2, \ldots, s \\
K_{ij}^{13} &= \int_{\Omega^e} \frac{\partial \psi_i^{1e}}{\partial y} \frac{\partial \psi_j^{3e}}{\partial y}\, dx\, dy \qquad i = 1, 2, \ldots, r;\ j = 1, 2, \ldots, p \\
K_{ij}^{22} &= \int_{\Omega^e} (-\bar{D}_{22}^{e}) \psi_i^{2e} \psi_j^{2e}\, dx\, dy \qquad i, j = 1, 2, \ldots, s \\
K_{ij}^{23} &= \int_{\Omega^e} \bar{D}_{12}^{e} \psi_i^{2e} \psi_j^{3e}\, dx\, dy \qquad i = 1, 2, \ldots, s;\ j = 1, 2, \ldots, p \qquad (7\text{-}209) \\
K_{ij}^{33} &= \int_{\Omega^e} (-\bar{D}_{11}^{e}) \psi_i^{3e} \psi_j^{3e}\, dx\, dy \qquad i, j = 1, 2, \ldots, p \\
F_i^{1} &= \int_{\Omega^e} f \psi_i^{1e}\, dx\, dy + \oint_{\Gamma^e} Q_n^e \psi_i^{1e}\, ds \qquad i = 1, 2, \ldots, r \\
F_i^{2} &= \oint_{\Gamma^e} \left(\frac{\partial w_e}{\partial x} n_x \right) \psi_i^{2e}\, ds \qquad i = 1, 2, \ldots, s \\
F_i^{3} &= \oint_{\Gamma^e} \left(\frac{\partial w_e}{\partial y} n_y \right) \psi_i^{3e}\, ds \qquad i = 1, 2, \ldots, p
\end{aligned}
$$

The interpolation functions in Eq. (7-205) are also valid for mixed model II. The corresponding elements are shown in Fig. 7-43*b*. Of course, the quadratic elements, the Lagrange type or serendipity type, can also be used to model these elements.

Shear deformation theory Here we consider a first-order shear deformation theory in which the normality condition is relaxed (i.e., the normals to the mid-surface are not required to remain normal after deformation but still remain straight). This allows us to include a constant state of transverse shear stresses through the plate thickness, whereas the experimental evidence shows that the transverse shear stresses vary approximately parabolically through the thickness. To correct this discrepancy, certain coefficients are introduced in the calculation of the transverse shear stresses. These coefficients are called *shear correction factors*. For additional details the reader is referred to the book by the author (Reddy, 1984b), and the references therein.

Because of the relaxed normality condition, two additional dependent variables are introduced into the first-order theory. These variables denote the rotation of transverse normals about the x-axis and y-axis at the midplane. The governing equations of the theory are given by

$$A_{55} \frac{\partial}{\partial x}\left(\phi_x + \frac{\partial w}{\partial x}\right) + A_{44} \frac{\partial}{\partial y}\left(\phi_y + \frac{\partial w}{\partial y}\right) + f = 0$$

$$\frac{\partial}{\partial x}\left(D_{11} \frac{\partial \phi_x}{\partial x} + D_{12} \frac{\partial \phi_y}{\partial y}\right) + D_{66} \frac{\partial}{\partial y}\left(\frac{\partial \phi_x}{\partial y} + \frac{\partial \phi_y}{\partial x}\right) - A_{55}\left(\phi_x + \frac{\partial w}{\partial x}\right) = 0 \quad (7\text{-}210)$$

$$D_{66} \frac{\partial}{\partial x}\left(\frac{\partial \phi_x}{\partial y} + \frac{\partial \phi_y}{\partial x}\right) + \frac{\partial}{\partial y}\left(D_{12} \frac{\partial \phi_x}{\partial x} + D_{22} \frac{\partial \phi_y}{\partial y}\right) - A_{44}\left(\phi_y + \frac{\partial w}{\partial y}\right) = 0$$

where ϕ_x and ϕ_y denote the rotations of the transverse normal at point $(x, y, 0)$ about the y-axis and x-axis, respectively, and A_{44} and A_{55} are the shear rigidities

$$A_{44} = G_{23} tk \qquad \text{and} \qquad A_{55} = G_{13} tk \quad (7\text{-}211)$$

where k is the shear correction coefficient ($k = 5/6$).

The variational problem of Eq. (7-210) over an element is to find $(w, \phi_x, \phi_y) \in S^h$ such that

$$0 = \int_{\Omega^e} \left[A_{55}^e \frac{\partial \bar{w}_e}{\partial x}\left(\phi_x + \frac{\partial w_e}{\partial x}\right) + A_{44}^e \frac{\partial \bar{w}_e}{\partial y}\left(\phi_y^e + \frac{\partial w_e}{\partial y}\right) - \bar{w}_e f_e \right] dx\, dy - \oint_{\Gamma^e} Q_n^e \bar{w}_e\, ds$$

$$0 = \int_{\Omega^e} \left[\frac{\partial \bar{\phi}_x^e}{\partial x}\left(D_{11}^e \frac{\partial \phi_x^e}{\partial x} + D_{12}^e \frac{\partial \phi_y^e}{\partial y}\right) + D_{66}^e \frac{\partial \bar{\phi}_x^e}{\partial y}\left(\frac{\partial \phi_x^e}{\partial y} + \frac{\partial \phi_y^e}{\partial x}\right) + A_{55}^e \bar{\phi}_x^e\left(\phi_x^e + \frac{\partial w_e}{\partial x}\right)\right] dx\, dy - \oint_{\Gamma^e} M_n^e \bar{\phi}_x^e\, ds \quad (7\text{-}212)$$

$$0 = \int_{\Omega^e} \left[D_{66}^e \frac{\partial \bar{\phi}_y^e}{\partial x}\left(\frac{\partial \phi_x^e}{\partial y} + \frac{\partial \phi_y^e}{\partial x}\right) + \frac{\partial \bar{\phi}_y^e}{\partial y}\left(D_{12}^e \frac{\partial \phi_x^e}{\partial x} + D_{22}^e \frac{\partial \phi_y^e}{\partial y}\right) + A_{44}^e \bar{\phi}_y^e\left(\phi_y^e + \frac{\partial w_e}{\partial y}\right)\right] dx\, dy - \oint_{\Gamma^e} M_s^e \bar{\phi}_y^e\, ds$$

hold for all $(\bar{w}_e, \bar{\phi}_x^e, \bar{\phi}_y^e) \in \mathbf{S}^h \subset \mathbf{H}$, where

$$\mathbf{H} = H^1(\Omega) \times H^1(\Omega) \times H^1(\Omega)$$

and Q_n, M_n, and M_s are given by Eq. (7-193) with

$$M_x = D_{11} \frac{\partial \phi_x}{\partial x} + D_{12} \frac{\partial \phi_y}{\partial y} \qquad M_y = D_{12} \frac{\partial \phi_x}{\partial x} + D_{22} \frac{\partial \phi_y}{\partial y} \quad (7\text{-}213)$$

$$M_{xy} = D_{66}\left(\frac{\partial \phi_x}{\partial y} + \frac{\partial \phi_y}{\partial x}\right) \qquad Q_x = A_{55}\left(\phi_x + \frac{\partial w}{\partial x}\right) \qquad Q_y = A_{44}\left(\phi_y + \frac{\partial w}{\partial y}\right)$$

The finite-element model of Eqs. (7-212) can be derived by assuming interpolation of the form

$$w_e = \sum_{i=1}^{n} w_i^e \psi_i^e \qquad \phi_x^e = \sum_{i=1}^{n} S_x^{ie} \psi_i^e \qquad \phi_y^e = \sum_{i=1}^{n} S_y^{ie} \psi_i^e \tag{7-214}$$

Clearly, ψ_i^e can be the Lagrange type linear, quadratic, or cubic interpolation functions. Substituting (7-214) into Eq. (7-212), we obtain

$$\begin{bmatrix} [K^{11}] & [K^{12}] & [K^{13}] \\ & [K^{22}] & [K^{23}] \\ \text{symm.} & & [K^{33}] \end{bmatrix}_e \begin{Bmatrix} \{w\} \\ \{S_x\} \\ \{S_y\} \end{Bmatrix}_e = \begin{Bmatrix} \{F^1\} \\ \{F^2\} \\ \{F^3\} \end{Bmatrix}_e \tag{7-215}$$

where

$$K_{ij}^{11} = \int_{\Omega^e} \left(A_{55}^e \frac{\partial \psi_i^e}{\partial x} \frac{\partial \psi_j^e}{\partial x} + A_{44}^e \frac{\partial \psi_i^e}{\partial y} \frac{\partial \psi_j^e}{\partial y} \right) dx\, dy$$

$$K_{ij}^{12} = \int_{\Omega^e} A_{55}^e \frac{\partial \psi_i^e}{\partial x} \psi_j^e \, dx\, dy \qquad K_{ij}^{13} = \int_{\Omega^e} A_{44}^e \frac{\partial \psi_i^e}{\partial y} \psi_j^e \, dx\, dy$$

$$K_{ij}^{22} = \int_{\Omega^e} \left(D_{11}^e \frac{\partial \psi_i^e}{\partial x} \frac{\partial \psi_j^e}{\partial x} + D_{66}^e \frac{\partial \psi_i^e}{\partial y} \frac{\partial \psi_j^e}{\partial y} + A_{55}^e \psi_i^e \psi_j^e \right) dx\, dy$$

$$K_{ij}^{23} = \int_{\Omega^e} \left(D_{12}^e \frac{\partial \psi_i^e}{\partial x} \frac{\partial \psi_j^e}{\partial y} + D_{66}^e \frac{\partial \psi_i^e}{\partial y} \frac{\partial \psi_j^e}{\partial x} \right) dx\, dy \tag{7-216}$$

$$K_{ij}^{33} = \int_{\Omega^e} \left(D_{66}^e \frac{\partial \psi_i^e}{\partial x} \frac{\partial \psi_j^e}{\partial x} + D_{22}^e \frac{\partial \psi_i^e}{\partial y} \frac{\partial \psi_j^e}{\partial y} + A_{44}^e \psi_i^e \psi_j^e \right) dx\, dy$$

$$F_i^1 = \int_{\Omega^e} f^e \psi_i^e \, dx\, dy + \oint_{\Gamma^e} Q_n^e \psi_i^e \, ds$$

$$F_i^2 = \oint_{\Gamma^e} M_n^e \psi_i^e \, ds \qquad F_i^3 = \oint_{\Gamma^e} M_s^e \psi_i^e \, ds$$

The element stiffness matrix in Eq. (7-216) is of the order $3n \times 3n$, where n is the number of nodes per element. When the four-node rectangular element is used, the element stiffness matrix is of the order 12×12, and it is 27×27 for the nine-node element (see Fig. 7-43*c*).

It should be noted that the inclusion of the transverse shear strains (i.e., terms involving A_{44} and A_{55}) in the equations presents computational difficulties when the side-to-thickness ratio of the plate is large (i.e., when the plate becomes thin). For thin plates, the transverse shear strains, $2\varepsilon_{13} = \phi_x + \partial w/\partial x$ and $2\varepsilon_{23} = \phi_y + \partial w/\partial y$, are negligible, and consequently the element stiffness matrix becomes stiff (i.e., so-called locking occurs) and yields erroneous results for the generalized displacements (w_i, S_x^i, S_y^i). This phenomenon can be interpreted as one caused by the inclusion of the following constraints into the variational form by the penalty method [see Reddy (1979*b*, 1980)]:

$$\phi_x + \frac{\partial w}{\partial x} = 0 \qquad \phi_y + \frac{\partial w}{\partial y} = 0 \tag{7-217}$$

The penalty term in the total potential energy of the shear deformable plate is given by

$$\frac{1}{2}\int_{\Omega_h}\left[A_{44}\left(\phi_y + \frac{\partial w}{\partial y}\right)^2 + A_{55}\left(\phi_x + \frac{\partial w}{\partial x}\right)^2\right] dx\, dy$$

This represents the energy due to the transverse shear stresses. The locking observed in the finite-element model of the first-order theory is the result of the fact that the discrete form of the penalty term is not necessarily positive-definite for a fixed mesh size and thickness. Of course, when the plate is thick the relations (7-217) are not required to be satisfied and the locking does not occur (at least it is not severe enough to give completely wrong results). However, for thin plates, the constraints (7-217) are valid, and therefore we face the same problem as the one we did in the penalty finite-element model of the Stokes problem. Therefore, we use the same remedy as before: use reduced integration to evaluate stiffness coefficients involving the penalty terms. For example, when a four-node rectangular element is used, the one-point Gauss rule should be used to evaluate the shear energy terms (i.e., terms involving A_{44} and A_{55}) and the two-point Gauss rule should be used for all other terms. When an eight-node or nine-node rectangular element is used, the two-point and three-point Gauss rules should be used to evaluate the shear and bending terms, respectively.

This completes the development of the finite-element models of plates in bending. Next we consider several numerical examples. The first example deals with the effect of the reduced integration on the deflections and stresses of a simply supported plate as computed by the shear deformable plate element.

Example 7-24 Consider a simply supported, isotropic, square plate subjected to a uniformly distributed transverse load f_0. We shall solve the problem by the shear deformable element. Due to the biaxial symmetry, we need to model only a quadrant of the plate. The essential boundary conditions at simply supported edges ($x = a/2$ and $y = a/2$) are given by (see Fig. 7-44)

$$w = 0 \qquad \phi_y = 0 \text{ at } x = a/2$$

$$w = 0 \qquad \phi_x = 0 \text{ at } y = a/2$$

The essential boundary conditions along the symmetry lines ($x = 0$ and $y = 0$) are given by

$$\phi_x(0, y) = \phi_y(x, 0) = 0$$

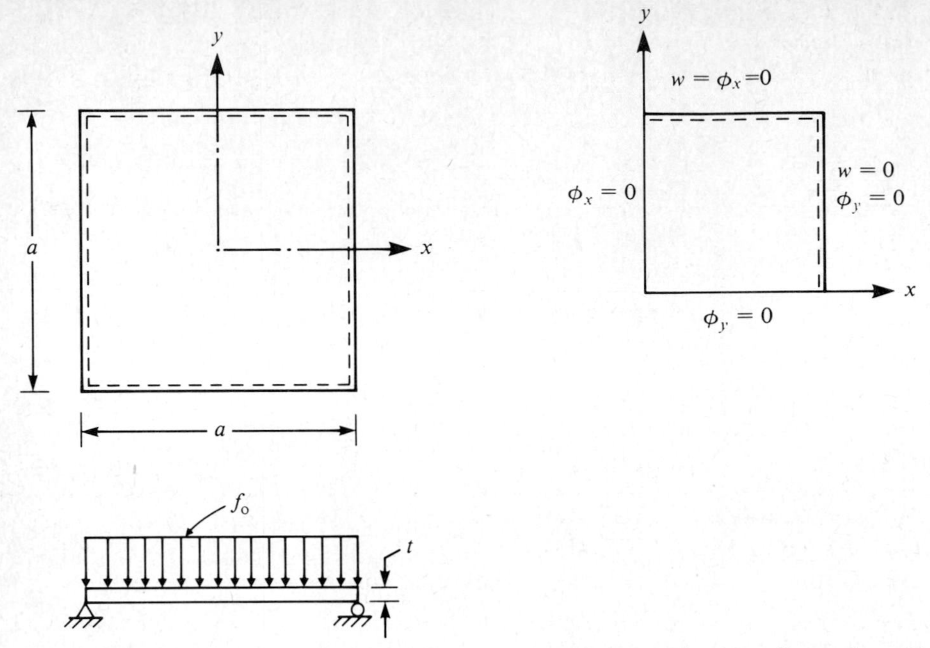

Figure 7-44 Domain and boundary conditions for a simply-supported square plate under uniform load.

The natural boundary conditions (which enter the finite-element equations through $\{F\}$) in the present case are given by

$$Q_n = 0 \text{ along } x = 0 \text{ and } y = 0$$

$$M_x = 0 \text{ along } y \text{ at } x = a/2$$

$$M_y = 0 \text{ along } x \text{ at } y = a/2$$

For a linear (four-node) element, the contribution of a uniformly distributed load to the nodes of the element is given by

$$\int_0^{a_e} \int_0^{b_e} f_0^e \psi_i^e \, dx \, dy = \frac{a_e b_e}{4} f_0$$

where a_e and b_e are the sides of the element Ω^e. This contribution goes only into the force degrees of freedom corresponding to the transverse deflection. Except for the above contribution, all other specified forces are zero.

The effect of the reduced integration, thickness, and mesh on the center deflection and stress is investigated and the results are presented in Table 7-20. Note that, in general, mixed (M) integration (i.e., full integration on bending energy terms and reduced integration on shear energy terms) gives

Table 7-20 The effect of reduced integration, thickness, and mesh refinement on the center deflection and stress* of a simply supported, isotropic ($\nu = 0.25$), square plate under uniform transverse load, f_0 (shear deformation plate theory)

		1 × 1 linear		2 × 2 linear		4 × 4 linear		2 × 2 quadratic		Exact	
a/h	integration	$\bar{w}_x$	$\bar{\sigma}_x$	$\bar{w}$	$\bar{\sigma}_x$	$\bar{w}$	$\bar{\sigma}_x$	$\bar{w}$	$\bar{\sigma}_x$	$\bar{w}$	$\bar{\sigma}_x$
10	F	0.964	0.0182	2.474	0.1185	3.883	0.216	4.770	0.2899	4.791	0.2762
	M	3.950	0.0953	4.712	0.2350	4.773	0.2661	4.799	0.2715		
20	F	0.270	0.0053	0.957	0.0476	2.363	0.1375	4.57	0.2683	4.625	0.2762
	M	3.669	0.0954	4.524	0.2350	4.603	0.2660	4.633	0.2715		
40	F	0.0695	0.0014	0.279	0.0140	0.9443	0.0558	4.505	0.2699	4.584	0.2762
	M	3.599	0.0953	4.375	0.2349	4.560	0.2661	4.592	0.2714		
50	F	0.0045	0.0001	0.182	0.0092	0.6515	0.0386	4.496	0.2667	4.579	0.2762
	M	3.590	0.0953	4.472	0.2350	4.555	0.2660	4.587	0.2714		
100	F	0.011	0.0002	0.047	0.0024	0.182	0.0108	4.482	0.2664	4.572	0.2762
	M	3.579	0.0953	4.465	0.235	4.548	0.2661	4.580	0.2715		

* $\bar{w} = wEt^3(10^2)/f_0 a^4$, $\bar{\sigma}_x = \sigma_x(A, A, \pm t/2)(t^2/f_0 a^2)$, $A = a/4$ (1 × 1 linear), $a/8$ (2 × 2 linear), $a/16$ (4 × 4 linear), $0.05283a$ (2 × 2 quadratic).

more accurate results than the full (F) integration (i.e., full integration on both bending and shear energy terms). The effect of reduced integration on the accuracy diminishes (i.e., both full integration and reduced integrations give acceptable results) as the mesh is refined or an equivalent mesh of higher-order elements is used. Also, the effect of shear deformation (i.e., including the transverse shear strains in the theory) on the deflection is to increase the deflection.

The next example deals with a comparison of various elements, the classical displacement model (CPT), mixed models and shear deformation models (SDPT), when applied to a simply supported and clamped square plates under a distributed transverse load. In the subsequent computations, the mixed integration rule is used in the evaluation of stiffness coefficients of the SDPT model.

Example 7-25 Consider an isotropic ($\nu = 0.3$) square plate under uniform load of intensity f_0. We shall consider both simply supported and clamped boundary conditions (see Fig. 7-45). Note that in the case of the displacement model based on CPT, we must also specify the boundary conditions on the cross-derivative $\partial^2 w/\partial x\, \partial y$. Once again, we exploit the biaxial symmetry and model only a quadrant of the plate.

Tables 7-21 and 7-22 contain the center deflection, $\bar{w} = w(0, 0)D \times 10^2/f_0 a^4$ $[D = Et^3/12(1 - \nu^2)]$ and center normal stress, $\bar{\sigma}_x = \sigma_x(A, A) \times 10/f_0 a$ as obtained by using uniform meshes of various elements. In both mixed

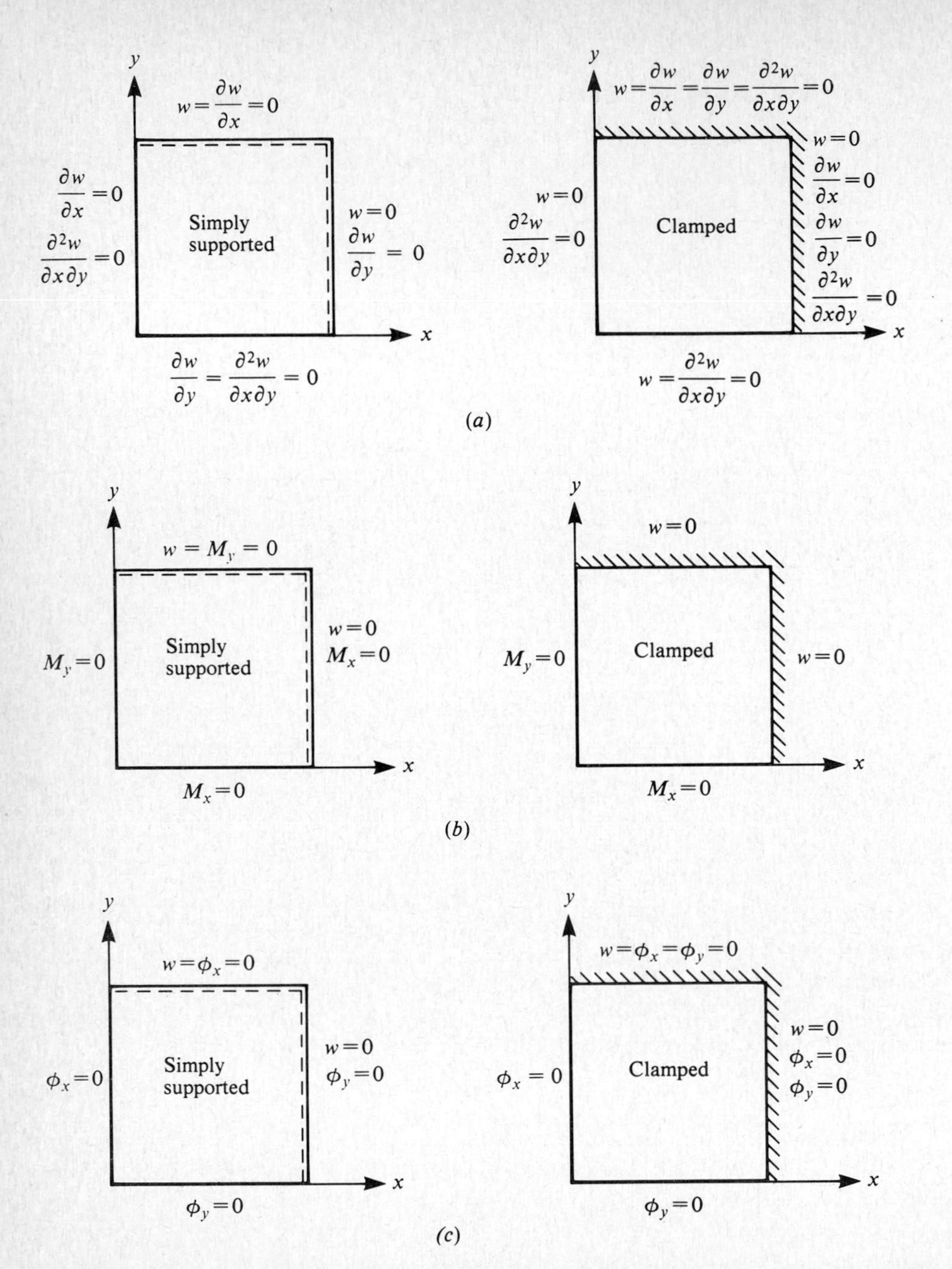

Figure 7-45 Boundary conditions for simply supported and clamped plates modeled by (*a*) displacement models based on CPT, (*b*) mixed models based on CPT, and (*c*) displacement models based on a shear deformation theory.

Table 7-21 Comparison of center deflection and normal stress of a simply supported square plate under uniformly distributed load as obtained by various finite element models

Mesh†	Displacement model*	Mixed model* I	Mixed model* II		Displacement model (SDPT; $a/t = 10$)	
		Linear	Linear	Quadratic	Linear	Quadratic
			Center deflection, $\bar{w}$			
1 × 1	0.4123	0.4613	0.3906	0.3867	0.3546	—
2 × 2	0.4065	0.4237	0.4082	0.4053	0.4207	0.4408
4 × 4	0.4062	0.4106	0.4069	0.4062	0.4257	0.4280
Exact	0.4062	0.4062	0.4062	0.4062	0.4259	0.4259
			Center stress, $\bar{\sigma}_x$			
1 × 1	3.088	4.3176	3.6564	2.2878	0.995	—
2 × 2	2.893	3.1476	3.0294	2.8908	2.445	2.735
4 × 4	2.876	2.9346	2.9094	2.8728	2.767	2.823
Exact			2.874			

* Based on the classical plate theory (CPT).

† Linear rectangular elements; equivalent mesh of nine-node elements is $N/2 \times N/2$ where $N \times N$ is the mesh of linear elements.

models, the point $(x, y) = (A, A)$ is defined by $A = 0$, whereas in the other models, $A \neq 0$ as defined by the following table:

Element type		Location, A			
		1 × 1	2 × 2(1 × 1)	4 × 4(2 × 2)	8 × 8(4 × 4)
CPT Model		$0.05635a$	$0.02817a$	$0.01409a$	$0.03125a$
SDPT Model	Linear	$0.25a$	$0.125a$	$0.0625a$	$0.03125a$
	Quad.	—	$0.1057a$	$0.0528a$	$0.02642a$

Table 7-22 Comparison of center deflection and normal stress of a clamped square plate under uniformly distributed load as obtained by various finite element meshes

Mesh	Displacement model	Mixed model I	Mixed model II		Displacement model (SDPT; $a/t = 10$)	
		Linear	Linear	Quadratic	Linear	Quadratic
			Center deflection, $\bar{w}$			
1 × 1	0.1943	0.1664	0.1563	0.1466	0.0357	—
2 × 2	0.1265	0.1529	0.1480	0.1260	0.1459	0.1757
4 × 4	0.1266	0.1339	0.1325	0.1264	0.1495	0.1586
Exact	0.1266	0.1266	0.1266	0.1266		
			Center stress, $\bar{\sigma}$			
1 × 1	2.443	3.116	2.925	1.234	0.000	—
2 × 2	1.415	1.900	1.739	1.349	1.142	1.321
4 × 4	1.381	1.487	1.466	1.372	1.333	1.345

The distance is measured from the center of the plate. The numbers in parenthesis denote the mesh of quadratic elements. Clearly, all finite-element models exhibit good convergence characteristics. The Hermite cubic element has a faster rate of convergence than the mixed elements. This is expected because of the higher degree of approximation used in the displacement model based on CPT. The difference between the CPT and SDPT is attributed to the inclusion of transverse shear deformation, which models the stiffness of the plate more correctly.

The last example of this section deals with simply supported orthotropic plates under distributed transverse load.

Example 7-26 Here we consider orthotropic plates with two different material properties ($\nu_{21} = \nu_{12} E_2/E_1$):

Glass-epoxy material:

$$E_1 = 7.8 \times 10^6 \text{ psi} \qquad E_2 = 2.6 \times 10^6 \text{ psi} \qquad \nu_{12} = 0.25$$
$$G_{12} = G_{23} = G_{13} = 1.3 \times 10^6 \text{ psi}$$

Graphite-epoxy:

$$E_1 = 31.8 \times 10^6 \text{ psi} \qquad E_2 = 1.02 \times 10^6 \text{ psi} \qquad \nu_{12} = 0.31$$
$$G_{12} = G_{23} = G_{13} = 0.96 \text{ psi}$$

Table 7-23 Comparison of center deflection ($\bar{w} = wH \times 10^3/f_0 a^4$) and normal stress ($\bar{\sigma}_x = \sigma_x \times 10/f_0 a^2$) of a glass-epoxy, simply-supported square plate under uniform transverse load

Mesh	Displacement model	Mixed model II		Displacement model (SDPT; $a/t = 10$)	
		Linear	Quadratic	Linear	Quadratic
		Center deflection, $\bar{w}$			
2 × 2	3.1421	3.1568	3.3642	3.2888	3.4547
4 × 4	3.1394	3.1455	3.1396	3.3330	3.3352
8 × 8	3.1394	3.1414	3.1400	3.3440	3.3469
Exact*		3.1400		3.3470	
		Center stress, $\bar{\sigma}_x$			
2 × 2	4.618	4.846	4.611	3.744	4.231
4 × 4	4.585	4.641	4.575	4.338	4.442
8 × 8	4.579	4.592	4.576	4.482	4.506
Exact*		4.576		4.529	

* From Reddy (1984b); $H = D_{12} + 2D_{66}$.

Table 7-24 Comparison of center deflection ($\bar{w} = wH \times 10^3/f_0 a^4$) and normal stress ($\bar{\sigma}_x = \sigma_x \times 10/f_0 a^2$) of a graphite-epoxy, simply-supported square plate under uniform transverse load

Mesh	Displacement model	Mixed model II		Displacement model (SDPT; $a/t = 10$)	
		Linear	Quadratic	Linear	Quadratic
		Center deflection, $\bar{w}$			
2 × 2	0.9220	0.9371	0.9249	1.2545	1.2715
4 × 4	0.9224	0.9245	0.9223	1.2186	1.2147
8 × 8	0.9224	0.9229	0.9225	1.2152	1.2147
Exact*		0.9225		1.215	
		Center stress, $\bar{\sigma}_x$			
2 × 2	7.678	8.142	7.856	6.277	7.192
4 × 4	7.616	7.707	7.596	7.256	7.399
8 × 8	7.600	7.622	7.595	7.449	7.478
Exact*		7.595		7.512	

* From Reddy (1984b); $H = D_{12} + 2D_{66}$.

The nondimensionalized center deflection $\bar{w}$ and normal stress $\bar{\sigma}_x$ obtained using the Hermite cubic displacement model (CPT), mixed model II (CPT) and displacement model based on SDPT are compared in Tables 7-23 and 7-24 for glass-epoxy and graphite-epoxy material plates. Once again, we observe the numerical convergence of the three finite-element models. It is clear that the effect of shear deformation increases with the degree of orthotropy.

This completes the study of coupled partial differential equations arising in plane elasticity, viscous incompressible fluid flow and plate bending.

7-8 TIME-DEPENDENT PROBLEMS IN TWO DIMENSIONS

7-8-1 Introduction

This section is devoted to the finite-element analysis of time-dependent problems in two dimensions. Once again, like in Section 7-4, we employ the results of Section 6-5 to fully discretize the semidiscrete finite-element models of time-dependent problems. Since the temporal approximations are already discussed in detail in Section 6-5, here we illustrate the procedure via the parabolic equation governing heat transfer-like problems. Hyperbolic equations can be modeled as described in Section 6-5.

7-8-2 Semidiscrete Approximations

Consider the partial differential equation governing the transient heat transfer (and like problems) in a two-dimensional region Ω with total boundary Γ

$$c_1 \frac{\partial u}{\partial t} - \frac{\partial}{\partial x}\left(k_1 \frac{\partial u}{\partial x}\right) - \frac{\partial}{\partial y}\left(k_2 \frac{\partial u}{\partial y}\right) = f \qquad \text{in } \Omega \qquad 0 < t \leq t_0 \qquad (7\text{-}218)$$

with boundary conditions

$$k_1 \frac{\partial u}{\partial x} n_x + k_2 \frac{\partial u}{\partial y} n_y + \beta(u - u_\infty) - q = 0 \text{ on } \Gamma_1 \qquad t \geq 0$$

$$u = \hat{u} \text{ on } \Gamma_2 \qquad t \geq 0 \qquad (7\text{-}219)$$

and initial conditions

$$u = u_0 \text{ in } \Omega \qquad t = 0 \qquad (7\text{-}220)$$

Here t denotes time, and c_1, k_1, k_2, β, u_∞, $\hat{u}$, u_0, f, and q are known functions of position and/or time.

The semidiscrete variational formulation of Eqs. (7-218) to (7-220) over an element Ω^e is given by

$$0 = \int_{\Omega^e} \left(c_1^e v_e \frac{\partial u_e}{\partial t} + k_1^e \frac{\partial v_e}{\partial x}\frac{\partial u_e}{\partial x} + k_2^e \frac{\partial v_e}{\partial y}\frac{\partial u_e}{\partial y} - v_e f_e \right) dx\, dy + \oint_{\Gamma^e} [\beta v_e(u_e - u_\infty^e) - v_e q_e]\, ds \qquad (7\text{-}221)$$

Substituting the finite-element interpolation of the form

$$u_e(x, y, t) = \sum_{j=1}^{r} u_j^e(t)\psi_j^e(x, y) \qquad (7\text{-}222)$$

and $v_e = \psi_i^e$ into Eq. (7.221), we obtain

$$0 = \sum_{j=1}^{r} \left\{ \left(\int_{\Omega^e} c_1^e \psi_i^e \psi_j^e \, dx\, dy \right) \frac{du_j^e}{dt} + \left[\int_{\Omega^e} \left(k_1^e \frac{\partial \psi_i^e}{\partial x}\frac{\partial \psi_j^e}{\partial x} + k_2^e \frac{\partial \psi_i^e}{\partial y}\frac{\partial \psi_j^e}{\partial y} \right) dx\, dy + \oint_{\Gamma^e} \beta \psi_i^e \psi_j^e \, ds \right] u_j^e \right\} - \oint_{\Gamma^e} (\beta u_\infty^e + q_e)\psi_i^e \, ds - \int_{\Omega^e} \psi_i^e f_e \, dx\, dy$$

or, in matrix form,

$$[M^e]\{\dot{u}^e\} + [K^e]\{u^e\} = \{F^e\} \qquad (7\text{-}223)$$

where

$$M_{ij}^e = \int_{\Omega^e} c_1^e \psi_i^e \psi_j^e \, dx\, dy$$

$$K_{ij}^e = \int_{\Omega^e} \left(k_1^e \frac{\partial \psi_i^e}{\partial x}\frac{\partial \psi_j^e}{\partial x} + k_2^e \frac{\partial \psi_i^e}{\partial y}\frac{\partial \psi_j^e}{\partial y} \right) dx\, dy + \oint_{\Gamma^e} \beta \psi_i^e \psi_j^e \, ds \qquad (7\text{-}224)$$

$$F_i^e = \int_{\Omega^e} \psi_i^e f_e \, dx\, dy + \beta u_\infty^e \oint_{\Gamma^e} \psi_i^e \, ds + \oint_{\Gamma^e} q_e \psi_i^e \, ds$$

7-8-3 Time Approximations

The fully discretized element equations associated with Eq. (7-223) can be expressed in the form (see Sec. 6-5)

$$[\hat{K}]\{u\}^{n+1} = \{\hat{F}\} \tag{7-225}$$

where $[\hat{K}]$ and $\{\hat{F}\}$ are given by

$$\begin{aligned} [\hat{K}] &= [M] + a_4[K] \\ \{\hat{F}\} &= ([M] - a_3[K])\{u\}^n + a_4\{F\}^n + a_3\{F\}^{n+1} \end{aligned} \tag{7-226}$$

Here a_3 and a_4 are given by

$$a_3 = (1-\alpha)\,\Delta t \qquad a_4 = \alpha\,\Delta t \tag{7-227}$$

Δt being the time step.

Equation (7-225) can be solved for $\{u\}$ at time $t = t_{n+1} \equiv (n+1)\,\Delta t$ by using known $\{u\}$ at $t = t_n$. At time $t = 0$, the initial conditions of the problem are used to initiate the time-marching scheme.

Example 7-27 We wish to solve the time-dependent heat conduction equation [i.e., time-dependent version of Eq. (7-125) in Example 7-14],

$$\frac{\partial T}{\partial t} - \nabla^2 T = 1 \text{ in } \Omega = \{(x, y)\colon 0 < (x, y) < 1\} \tag{7-228}$$

subject to the boundary conditions

$$\begin{aligned} T &= 0 \text{ on } \Gamma_1 = \{\text{lines } x = 1 \text{ and } y = 1\} \qquad t \geq 0 \\ \frac{\partial T}{\partial n} &= 0 \text{ on } \Gamma_2 = \{\text{lines } x = 0 \text{ and } y = 0\} \qquad t \geq 0 \end{aligned} \tag{7-229}$$

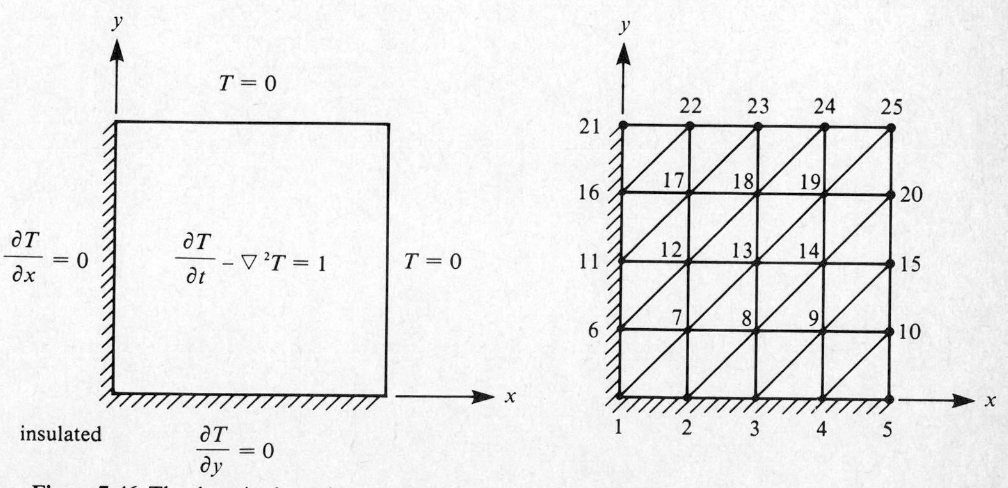

Figure 7-46 The domain, boundary conditions and the finite-element mesh for the transient analysis of a heat conduction problem.

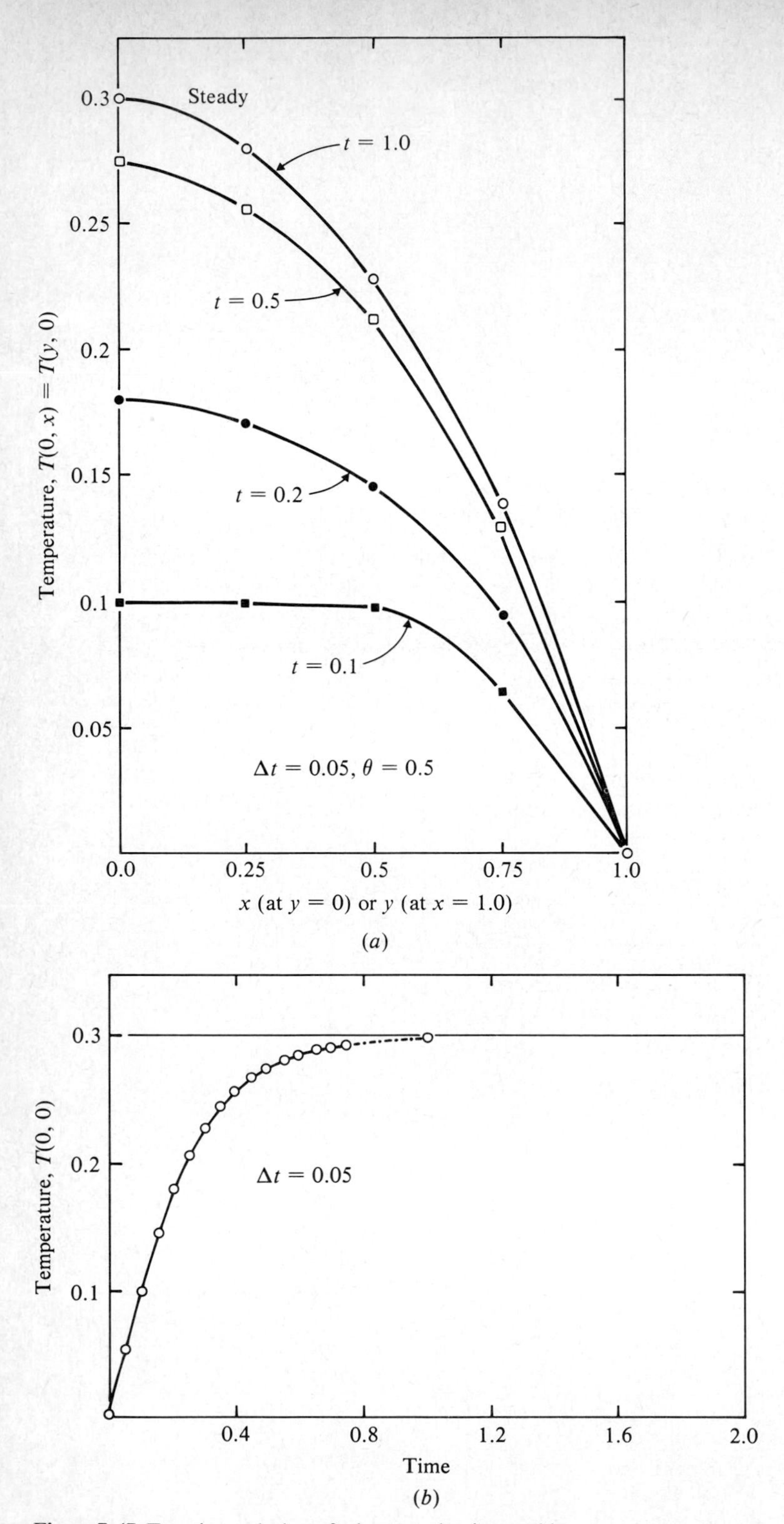

Figure 7-47 Transient solution of a heat conduction problem.

and the initial condition

$$T = 0 \text{ in } \Omega \tag{7-230}$$

We choose a 4 × 4 mesh of linear triangular and rectangular elements (see Fig. 7-46) to model the domain and use the Crank–Nicholson method (that is, $\theta = 0.5$) for the time approximation. Since the Crank–Nicholson method is unconditionally stable, one can choose any value of Δt (however, for large values of Δt the solution may not be accurate).

The element equations are given by Eq. (7-223), wherein $c_1 = 1$, $k_1 = k_2 = 1$, $\beta = 0$, and $f = 1$. The boundary conditions of the problem are given by

$$U_5 = U_{10} = U_{15} = U_{20} = U_{21} = U_{22} = U_{23} = U_{24} = U_{25} = 0$$

Beginning with the initial conditions, $U_i = 0$ ($i = 1, 2, \ldots, 25$), we solve the assembled set of equations. The temperature $T(x, 0, t)$ versus x for various values of time are shown in Fig. 7-47*a*. The steady state is reached at time $t = 1.0$. The temperature $T(0, 0, t)$ versus time is shown in Fig. 7.47*b*, which indicates the evolution of the temperature from zero to the steady state. The effect of the time step on the solution can be seen from the results presented in Table 7-25.

Table 7-25 Comparison of the transient solutions of Eq. (7-228) obtained by a mesh of triangular and rectangular element meshes

Time	Temperature along the $y = 0$ line, $T(x, 0, t) \times 10$				
t	Element*	$x = 0.0$	$x = 0.25$	$x = 0.5$	$x = 0.75$
0.1	$T1$	0.9758	0.9610	0.9063	0.7104
	$R1$	0.9832	0.9609	0.9172	0.7105
	$T2$	0.9928	0.9798	0.9168	0.6415
	$R2$	0.9945	0.9853	0.9264	0.6360
0.2	$T1$	1.8003	1.7238	1.4891	0.9321
	$R1$	1.8101	1.7507	1.5159	0.9415
	$T2$	1.7979	1.7060	1.4644	0.9462
	$R2$	1.8115	1.7329	1.4997	0.9612
0.3	$T1$	2.3130	2.1671	1.7961	1.1466
	$R1$	2.3312	2.2043	1.8415	1.1763
	$T2$	2.2829	2.1448	1.7943	1.1249
	$R2$	2.3011	2.1829	1.8434	1.1522
1.0	$T1$	2.9960	2.7871	2.2804	1.3843
	$R1$	3.0122	2.8408	2.3473	1.4229
	$T2$	2.9925	2.7862	2.2776	1.3849
	$R2$	3.0091	2.8398	2.3416	1.4248

* $T1$ = triangular element mesh with $\Delta t = 0.1$.
$T2$ = triangular element mesh with $\Delta t = 0.05$.
$R1$ = rectangular element mesh with $\Delta t = 0.1$.
$R2$ = rectangular element mesh with $\Delta t = 0.05$.

PROBLEMS 7-3

The solution of some of the problems might require the use of a computer.

7-3-1 Consider the Poisson equation $-\text{div}\,(k \text{ grad } u) = f$ in Ω. Construct the mixed finite element model of the equation by writing it as the pair of equations,

$$\left.\begin{aligned} -\frac{\boldsymbol{\sigma}}{k} + \text{grad } u &= 0 \\ -\text{div } \boldsymbol{\sigma} &= f \end{aligned}\right\} \text{ in } \Omega$$

7-3-2 to 7-3-4 For the plane elasticity problems shown in Figs. P2–P4, give the boundary degrees of freedom and compute the contribution of the specified forces to the nodes.

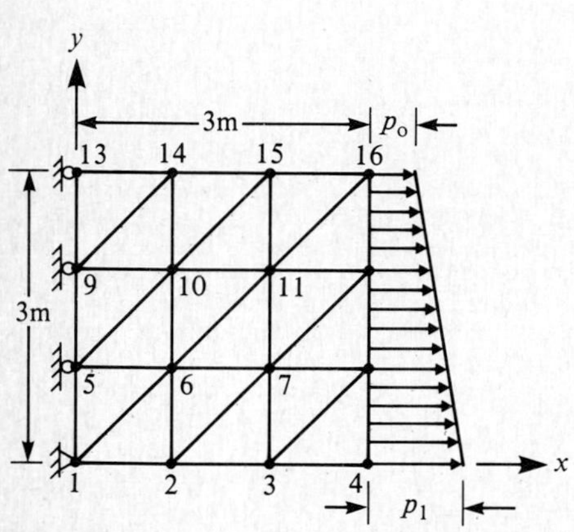

Figure P2

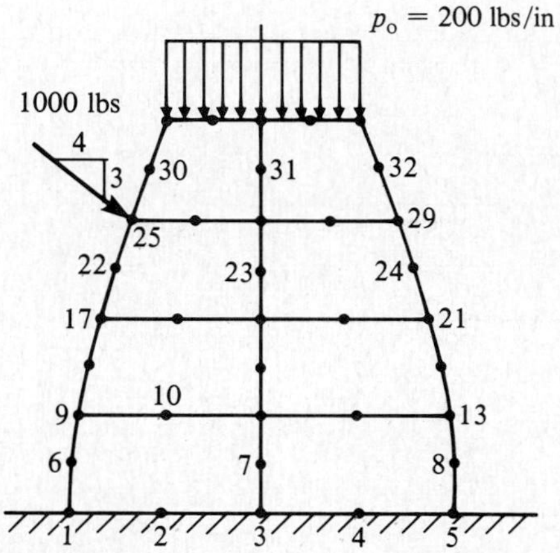

p_0 = 200 lbs/in

1000 lbs

4

3

30 31 32 29 25 22 23 24 17 21 10 9 13 6 7 8 1 2 3 4 5

Figure P3

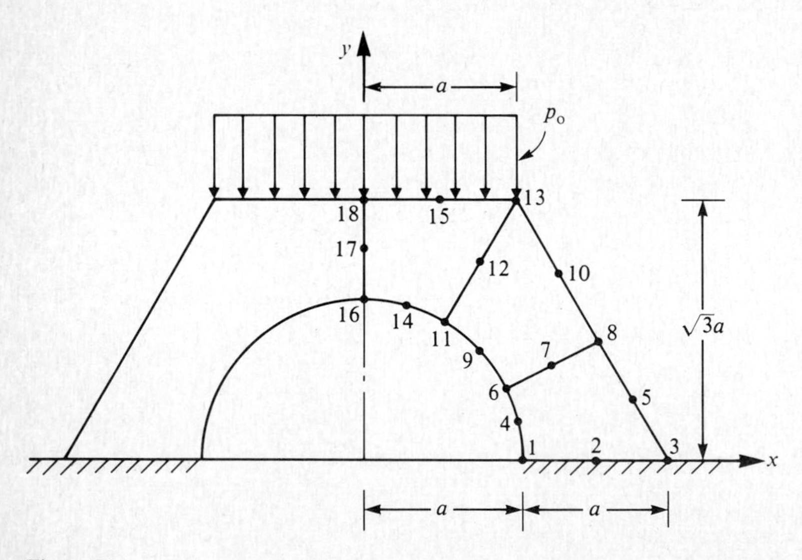

Figure P4

7-3-5 to 7-3-7 For the fluid problems shown in Figs. P5–P7, give the boundary degrees of freedom on the velocities and compute the contribution of the specified forces to the nodes.

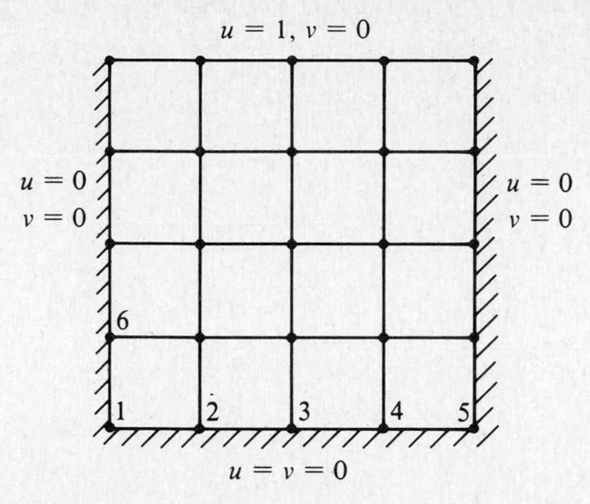

Figure P5

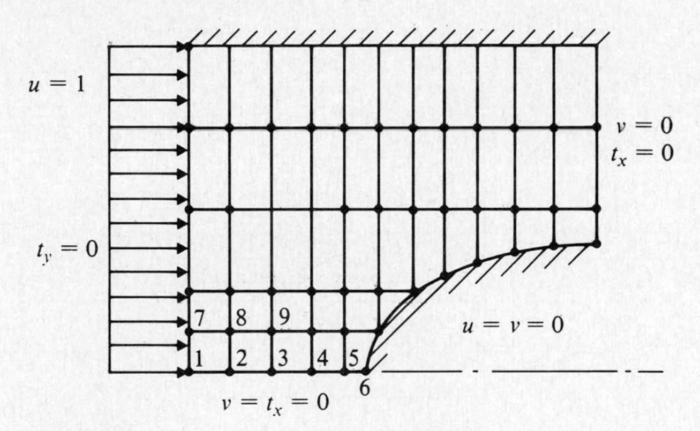

Figure P6

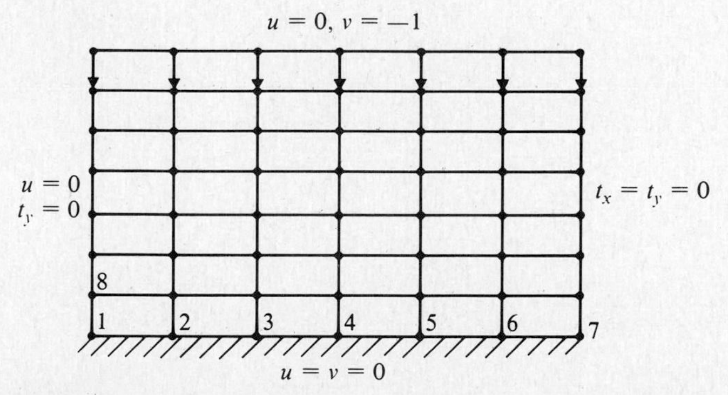

Figure P7

7-3-8 For mixed model II, compute the coefficient matrices $[K^{\alpha\beta}]$ in Eq. (7-209) using the linear rectangular element of sides a and b.

7-3-9 to 7-3-14 For the plate bending problems shown in Figs. P9–P14, give the specified displacement degrees of freedom and specified forces for (*a*) displacement model (CPT), (*b*) mixed model II and (*c*) displacement model (SDPT).

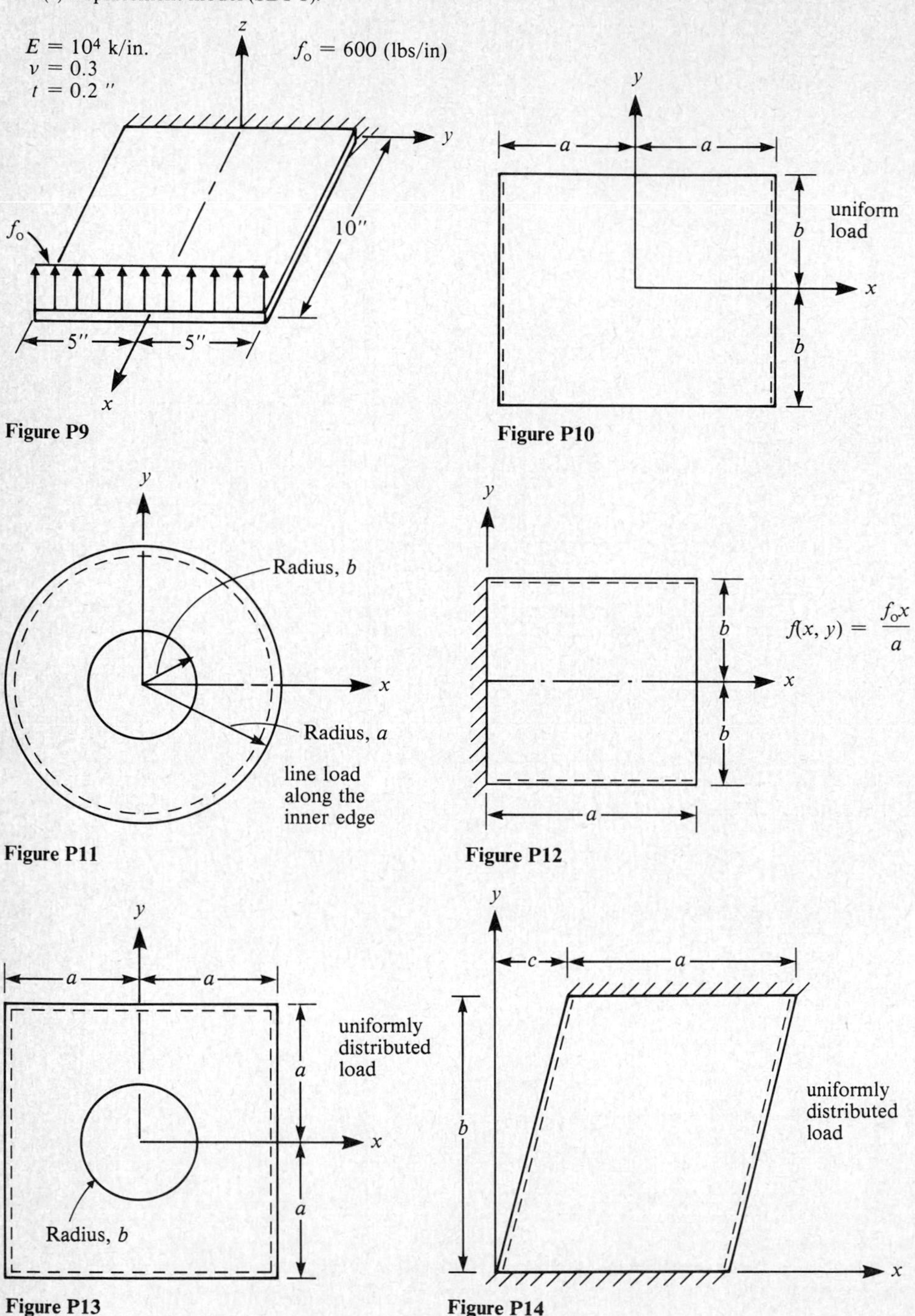

Figure P9

Figure P10

Figure P11

Figure P12

Figure P13

Figure P14

7-3-15 Compute the element stiffness matrix associated with Eq. (7-156*a*) for the linear triangular element shown in Fig. P18 of Prob. 7-2-18.

7-3-16 Repeat Prob. 7-3-15 for the linear rectangular element shown in Fig. P19 of Prob. 7-2-19.

7-3-17 Evaluate the coefficient matrices in Eq. (7-170) when ψ_i^e are the interpolation functions of the linear rectangular element (see Fig. P19 of Prob. 7-2-19) and ϕ_1^e = constant.

7-3-18 Evaluate the coefficient matrices in Eq. (7-203) for the choice of interpolation functions in Eq. (7-205*a*).

7-3-19 Consider the problem

$$\frac{\partial u}{\partial t} - \frac{\partial}{\partial x}\left[\left(4 + \frac{y^2}{8}\right)\frac{\partial u}{\partial x}\right] - \frac{\partial}{\partial y}\left[\left(4 + \frac{x^2}{25}\right)\frac{\partial u}{\partial y}\right] = 1 - \frac{x^2}{16} - \frac{y^2}{4}$$

in an ellipse with half axes $a = 4$ and $b = 2$ along the coordinate axes x and y. The boundary and initial conditions are $u = 0$ on Γ, $u(x, y, 0) = 0$ in Ω. Develop the finite-element model of the problem with (*a*) Crank–Nicholson method and (*b*) backward difference method for the first-order time derivatives.

7-3-20 Evaluate the coefficient matrix in Prob. 7-3-1 for a linear rectangular element.

REFERENCES

Babuska, I.: The Finite Element Method with Lagrange Multipliers, *Numer. Math.*, vol. 20, pp. 179–192, 1973a.

———: The Finite Element Method with Penalty, *Math. Comp.*, vol. 27, pp. 221–228, 1973b.

——— and A. K. Aziz: Survey Lectures on the Mathematical Foundations of the Finite Element Method, in A. K. Aziz (ed.), "The Mathematical Foundations of the Finite Element Method with Applications to Partial Differential Equations," Academic Press Inc., New York, 1972.

——— and M. Zlamal: Nonconforming Elements in the Finite Element Method with Penalty, *SIAM J. Numer. Anal.*, vol. 10, pp. 863–875, 1973.

Bathe, K. J., and E. L. Wilson: Stability and Accuracy Analysis of Direct Integration Methods, *Int. J. Earthquake Engng & Struct. Dynamics*, vol. 1, pp. 283–291, 1973.

Bercovier, M., and O. Pironneau: Estimations d'erreur pour la résolution du problème de Stokes en éléments finis conformes de Lagrange, *C. R. Acad. Sci. Paris, Ser. A*, vol. 285, pp. 1085–1087, 1977.

Ciarlet, P. G.: "The Finite Element Method for Elliptic Problems," North-Holland, Amsterdam, 1978.

Courant, R.: Variational Methods for the Solution of Problems of Equilibrium and Vibrations, *Bull. Amer. Math. Soc.*, vol. 49, pp. 1–23, 1943.

Hestenes, M. R.: "Calculus of Variations and Optimal Control Theory," John Wiley & Sons, Inc., New York, 1966.

———: "Optimization Theory: the Finite Dimensional Case," John Wiley & Sons, Inc., New York, 1975.

Hlavacek, I., and J. Necas: On Inequalities of Korn's Type, *Archive for Ratl. Mech. Anal.*, vol. 36, pp. 305–334, 1970.

Kantorovich, L. V., and V. I. Krylov: "Approximate Methods of Higher Analysis" (translated by C. D. Benster), Interscience (Noordhoff, The Netherlands), New York, 1964.

Kolmogorov, A. N., and S. V. Fomin: "Elements of the Theory of Functions and Functional Analysis," vols. I and II, Graylock, Albany, N.Y., 1957.

Kondratev, V. A.: Boundary Problems for Elliptic Equations with Conical or Angular Points, *Trans. Moscow Math. Soc.*, vol. 16, pp. 227–313, 1967.

Ladyzhenskaya, O. A.: "The Mathematical Theory of Viscous Incompressible Flow," 2d ed. (translated from the Russian by R. A. Silverman), Gordon and Breach, New York, 1969.

Lapidus, L., and G. F. Pinder: "Numerical Solution of Partial Differential Equations in Science and Engineering," John Wiley & Sons, Inc., New York, 1982.

Lions, J. L., and E. Magenes: "Non-Homogeneous Boundary Value Problems and Applications," vol. I (translated from the French by P. Kenneth), Springer-Verlag, Berlin, 1972.

Luenberger, D. G.: "Optimization by Vector Space Methods," John Wiley & Sons, Inc., New York, 1969.

Mikhlin, S. G.: "Variational Methods in Mathematical Physics" (translated from the Russian by T. Boddington), Pergamon Press, Ltd., Oxford, 1964.

———: "The Numerical Performance of Variational Methods" (translated from the Russian by R. S. Anderssen), Wolters-Noordhoff, The Netherlands, 1971.

Naylor, A. W., and G. R. Sell: "Linear Operator Theory in Engineering and Science," Holt, Rinehart and Winston, New York, 1971.

Necas, J.: "Les méthodes directes en théorie des équations elliptiques," Masson, Paris, 1967.

Oden, J. T.: RIP methods for Stokesian flows, in R. H. Gallagher, O. C. Zienkiewicz, J. T. Oden, and D. Norrie (eds.), "Finite Element Method in Flow Problems," vol. IV, John Wiley & Sons, Inc., London, 1982.

——— and G. F. Carey: "Finite Elements, Mathematical Aspects," vol. IV, Prentice-Hall, Inc., Englewood Cliffs, N.J., 1983.

——— and J. N. Reddy: "Variational Methods in Theoretical Mechanics," 2d ed., Springer-Verlag, Berlin, 1983.

Polyak, B. T.: The Convergence Rate of the Penalty Function Method, *Zh. vychisl. Mat. fiz.*, vol. 11, pp. 3–11, 1971 (English translation: *USSR Comput. Math. Mathematical Phy.*, vol. 11, pp. 1–12, 1971).

Raviart, P. A., and J. M. Thomas: Dual Finite Element Models for Second Order Elliptic Problems, Chapter 9 in R. Glowinski, E. Y. Rodin and O. C. Zienkiewicz (eds.) "Energy Methods in Finite Element Analysis," John Wiley & Sons, Inc., New York, pp. 175–191, 1979.

Reddy, J. N.: On the Accuracy and Existence of Solutions to Primitive Variable Models of Viscous Incompressible Fluids, *Int. J. Engng Sci.*, vol. 16, pp. 921–929, 1978.

———: On the Finite Element Method with Penalty for Incompressible Fluid Flow Problems, in J. R. Whiteman (ed.), "The Mathematics of Finite Elements and Applications III," Academic Press Inc., New York, 1979a.

———: Simple Finite Elements with Relaxed Continuity for Nonlinear Analysis of Plates, in A. P. Kabaila and V. A. Pulmano (eds.), "Finite Element Methods in Engineering," The University of New South Wales, Sydney, pp. 265–281, 1979b.

———: A Penalty Plate Bending Element for the Analysis of Laminated Anisotropic Composite Plates, *Int. J. Numer. Methods Eng.*, vol. 15, pp. 1187–1206, 1980.

———: "An Introduction to the Finite Element Method," McGraw-Hill Book Company, New York, 1984a.

———: "Energy and Variational Methods in Applied Mechanics" (with an Introduction to the Finite Element Method), John Wiley & Sons, Inc., New York, 1984b.

——— and M. L. Rasmussen: "Advanced Engineering Analysis," John Wiley & Sons, Inc., New York, 1982.

Rektorys, K.: "Variational Methods in Mathematics, Science and Engineering," D. Reidel Publishing Co., Boston, Mass., 1980.

———: "The Method of Discretization in Time," D. Reidel Publishing Co., Boston, Mass., 1982.

Ritz, W.: Ueber eine neue Methode zur Losung gewisser Variations probleme der mathematischen Physik, *J. Reine Angew. Math.*, vol. 135, pp. 1–61, 1908.

Temam, R.: "Theory and Numerical Analysis of the Navier–Stokes Equations," North-Holland, Amsterdam, 1977.

BIBLIOGRAPHY

FUNCTIONAL ANALYSIS

Adams, R. A.: "Sobolev Spaces," Academic Press Inc., New York, 1975.

Aubin, J. P.: "Applied Functional Analysis," John Wiley & Sons, Inc., New York, 1979.

Bachman, G., and L. Narici: "Functional Analysis," Academic Press Inc., New York, 1966.

Edwards, R. E.: "Functional Analysis," Holt, Rinehart and Winston, New York, 1965.

Griffel, D. H.: "Applied Functional Analysis," John Wiley & Sons, Inc., New York, 1981.

Kantorovich, L. V., and G. P. Akilov: "Functional Analysis in Normed Spaces," Pergamon Press, Ltd., New York, 1964.

Kolmogorov, A. N., and S. V. Fomin: "Elements of the Theory of Functions and Functional Analysis," vols. I and II, Graylock, Albany, N.Y., 1957.

Luisternik, L. A., and V. J. Sobolev: "Elements of Functional Analysis" (third English translation from the second Russian edition), John Wiley & Sons, Inc., New York, 1974.

Naylor, A. W., and G. R. Sell: "Linear Operator Theory in Engineering and Science," Holt, Rinehart and Winston, New York, 1971.

Nowinski, J. L.: "Applications of Functional Analysis in Engineering," Plenum Press, New York, 1981.

Oden, J. T.: "Applied Functional Analysis," Prentice-Hall, Inc., Englewood Cliffs, N.J., 1979.

Showalter, R. E.: "Hilbert Space Methods in Partial Differential Equations," Pitman Publishing Ltd., London, 1977.

Smirnov, V. I.: "A Course of Higher Mathematics," vol. V: Integration and Functional Analysis (translated from the Russian by D. E. Brown), Pergamon Press, Ltd., New York, 1964.

Sobolev, S. L.: "Applications of Functional Analysis in Mathematical Physics," American Math. Soc. Transl. Math. Monographs, vol. 7, Providence, R.I., 1963.

Vilenkin, N. Ya.: "Functional Analysis" (translated from the Russian by R. E. Flaherty), Wolters-Noordhoff, The Netherlands, 1972.

Vulikh, B. Z.: "Introduction to Functional Analysis for Scientists and Technologists," Pergamon Press, Ltd., Oxford, 1963.

VARIATIONAL FORMULATIONS AND METHODS

Agmon, S.: "Lectures on Elliptic Boundary Value Problems," Van Nostrand, Princeton, N.J., 1965.
Aubin, J.-P.: "Approximation of Elliptic Boundary-Value Problems," John Wiley & Sons, Inc., New York, 1972.
Crandall, S. H.: "Engineering Analysis," McGraw-Hill Book Company, New York, 1956.
Ekeland, I., and R. Temam: "Convex Analysis and Variational Problems," North-Holland, Amsterdam, 1976.
Finlayson, B. A.: "The Method of Weighted Residuals and Variational Principles," Academic Press Inc., New York, 1972.
Forray, M. J.: "Variational Calculus in Science and Engineering," McGraw-Hill Book Company, New York, 1969.
Fortin, M., and R. Glowinski: "Augmented Lagrangian Methods: Applications to the Numerical Solution of Boundary-Value Problems," North-Holland, Amsterdam, 1983.
Friedman, A.: "Variational Principles and Free-Boundary Problems," John Wiley & Sons, Inc., New York, 1982.
Gelfand, I. M., and S. V. Fomin: "Calculus of Variations" (translated and edited by R. A. Silverman), Prentice-Hall, Inc., Englewood Cliffs, N.J., 1963.
Glowinski, R.: "Numerical Methods for Nonlinear Variational Problems," Springer-Verlag, New York, 1984.
———, J. L. Lions, and R. Tremolieres: "Numerical Analysis of Variational Inequalities," North-Holland, Amsterdam, 1981.
Hildebrand, F. B.: "Methods of Applied Mathematics," 2d ed., Prentice-Hall, Inc., Englewood Cliffs, N.J., 1965.
Kantorovich, L. V., and V. I. Krylov: "Approximate Methods of Higher Analysis" (translated by C. D. Benster), Interscience (Noordhoff, The Netherlands), New York, 1964.
Kinderlehrer, D., and G. Stampacchia: "An Introduction to Variational Inequalities and Their Applications," Academic Press Inc., New York, 1980.
Ladyzhenskaya, O. A.: "The Mathematical Theory of Viscous Incompressible Flow," 2d ed. (translated from the Russian by R. A. Silverman), Gordon and Breach, New York, 1969.
Lanczos, C.: "The Variational Principles of Mechanics," 4th ed., University of Toronto Press, 1970.
Lascaux, P.: "Numerical Methods for Time Dependent Equations. Application to Fluid Flow Problems," Lecture Notes, vol. 52, Tata Institute of Fundamental Research, Bombay, 1976.
Lawson, C. L., and R. J. Hanson: "Solving Least Squares Problems," Prentice-Hall, Inc., Englewood Cliffs, N.J., 1974.
Leipholz, H.: "Direct Variational Methods and Eigenvalue Problems in Engineering," Noordhoff, The Netherlands, 1977.
Lions, J. L.: "Quelques méthodes de résolution des problèmes aux limites non linéaires," Dunod, Gauthier-Villars, 1969.
——— and E. Magenes: "Non-Homogeneous Boundary Value Problems and Applications," vol. I (translated from the French by P. Kenneth), Springer-Verlag, New York, 1972.
Lippmann, H.: "Extremum and Variational Principles in Mechanics," Springer-Verlag, New York, 1972.
Mikhlin, S. G.: "Variational Methods in Mathematical Physics" (translated from the Russian by T. Boddington), Pergamon Press, Ltd., Oxford, 1964.
———: "Mathematical Physics, An Advanced Course," North-Holland, Amsterdam, 1970.
———: "The Numerical Performance of Variational Methods" (translated from the Russian by R. S. Anderssen), Wolters-Noordhoff, The Netherlands, 1971.
———: "Approximation on a Rectangular Grid" (translated and edited by R. S. Anderssen and T. O. Shaposhnikova), Sijthoff & Noordhoff, The Netherlands, 1979.
Oden, J. T., and J. N. Reddy: "Variational Methods in Theoretical Mechanics," 2d ed., Springer-Verlag, Berlin, 1983.
Parkus, H.: "Variational Principles in Thermo- and Magneto-Elasticity," Lecture Notes, no. 58, Springer-Verlag, New York, 1972.

Petrov, I. P.: "Variational Methods in Optimal Control Theory" (translated from the 1965 Russian edition), Academic Press Inc., New York, 1968.

Prenter, P. M.: "Splines and Variational Methods," John Wiley & Sons, Inc., New York, 1975.

Reddy, J. N.: "Energy and Variational Methods in Applied Mechanics (with an introduction to the finite element method)," John Wiley & Sons, Inc., 1984.

——— and M. L. Rasmussen: "Advanced Engineering Analysis," John Wiley & Sons, Inc., New York, 1982.

Rektorys, K.: "Variational Methods in Mathematics, Science and Engineering," 2d ed., D. Reidel Publishing Co., Boston, Mass., 1980.

———: "The Method of Discretization in Time," D. Reidel Publishing Co., Boston, Mass., 1982.

Schechter, R. S.: "The Variational Methods in Engineering," McGraw-Hill Book Company, New York, 1967.

Stacy, W. M.: "Variational Methods in Nuclear Reactor Physics," Academic Press Inc., New York, 1974.

Strieder, W., and R. Aris: "Variational Methods Applied to Problems of Diffusion and Reaction," Springer-Verlag, Berlin, 1973.

Temam, R.: "Theory and Numerical Analysis of the Navier–Stokes Equations," North-Holland, Amsterdam, 1977.

Treves, F.: "Basic Linear Partial Differential Equations," Academic Press Inc., New York, 1975.

Vainberg, M. M.: "Variational Methods for the Study of Nonlinear Operators" (translated and supplemented to the 1956 Russian edition by A. Feinstein), Holden-Day, San Francisco, Calif., 1964.

———: "Variational Methods and Method of Monotone Operators in the Theory of Nonlinear Equations" (translated from the Russian by A. Libin), John Wiley & Sons, Inc., New York, 1973.

Washizu, K.: "Variational Methods in Elasticity and Plasticity," 3d ed., Pergamon, New York, 1982.

Weinberger, H. F.: "Variational Methods for Eigenvalue Problems," University of Minnesota, Minneapolis, Minn., 1962.

Weinstock, R.: "Calculus of Variations, with applications to Physics and Engineering," McGraw-Hill Book Company, New York, 1952.

Yourgrau, W., and S. Mandelstam: "Variational Principles in Dynamics and Quantum Theory," 3d. ed., W. B. Sanders, Philadelphia, Pa., 1968.

OPTIMIZATION (Lagrange Multiplier and Penalty Function Methods)

Bertsekas, D. P.: "Constrained Optimization and Lagrange Multiplier Methods," Academic Press Inc., New York, 1982.

Denn, M. M.: "Optimization by Variational Methods," Robert E. Krieger Publishing Co., Huntington, 1978 (reprint of the edition published by McGraw-Hill Book Company, New York, 1969).

Fiacco, A. V., and G. P. McCormick: "Nonlinear Programming: Sequential Unconstrained Minimization Techniques," John Wiley & Sons, Inc., New York, 1968.

Gill, P. E., W. Murray, and M. H. Wright: "Practical Optimization," Academic Press Inc., London, 1981.

Hestenes, M. R.: "Calculus of Variations and Optimal Control Theory," John Wiley & Sons, Inc., New York, 1966.

———: "Optimization Theory: the Finite Dimensional Case," John Wiley & Sons, Inc., New York, 1975.

Leitmann, G.: "Optimization Techniques: With Applications to Aerospace Systems," Academic Press Inc., New York, 1962.

Luenberger, D. G.: "Optimization by Vector Space Methods," John Wiley & Sons, Inc., New York, 1969.

Polak, E.: "Computational Methods in Optimization; A Unified Approach," Academic Press Inc., New York, 1971.

Smith, D. R.: "Variational Methods in Optimization," Prentice-Hall, Inc., Englewood Cliffs, N.J., 1974.

Wilde, D. J., and C. S. Beightler: "Foundations of Optimization," Prentice-Hall, Inc., Englewood Cliffs, N.J., 1967.

NUMERICAL METHODS

Atluri, S. N., R. H. Gallagher, and O. C. Zienkiewicz (eds.): "Hybrid and Mixed Finite Element Methods," John Wiley & Sons, Inc., Chichester, 1983.

Banerjee, P. K., and R. Butterfield: "Boundary Element Methods in Engineering Science," McGraw-Hill Book Company, London, 1981.

——— and R. P. Shaw (eds.), "Developments in Boundary Element Methods—1," Applied Science, London, 1979.

——— and ——— (eds.), "Developments in Boundary Element Methods—2," Applied Science, London, 1982.

Becker, E. B., G. F. Carey, and J. T. Oden: "Finite Elements, An Introduction," vol. I, Prentice-Hall, Inc., Englewood Cliffs, N.J., 1981.

Belytschko, T., and T. J. R. Hughes (eds.), "Computational Methods for Transient Analysis," vol. 1 in Computational Methods in Mechanics, North-Holland, Amsterdam, 1983.

Brebbia, C. A.: "The Boundary Element Method for Engineers," Pentech, London, 1980.

Ciarlet, P. G.: "The Finite Element Method for Elliptic Problems," North-Holland, Amsterdam, 1978.

Cruse, T. A., and F. J. Rizzo (eds.), "Boundary-Integral Equation Method: Computational Applications in Applied Mechanics," AMD-Vol. 11, 1975 Applied Mechanics Conference (ASME), RPI, Troy, 1975.

Fletcher, C. A. J.: "Computational Galerkin Methods," Springer-Verlag, New York, 1984.

Girault, V., and P. A. Raviart: "Finite Element Approximation of the Navier–Stokes Equations," Springer-Verlag, Berlin, 1979.

Glowinski, R., E. Y. Rodin, and O. C. Zienkiewicz (eds.): "Energy Methods in Finite Element Analysis," John Wiley & Sons, Inc., Chichester, 1979.

Lapidus, L., and G. F. Pinder: "Numerical Solution of Partial Differential Equations in Science and Engineering," John Wiley & Sons, Inc., New York, 1982.

Lewis, R. W., P. Bettess, and E. Hinton (eds.), "Numerical Methods in Coupled Systems," John Wiley & Sons, Inc., Chichester, 1984.

Oden, J. T., and G. F. Carey: "Finite Elements, Mathematical Aspects," vol. IV, Prentice-Hall, Inc., Englewood Cliffs, N.J., 1983.

——— and J. N. Reddy: "An Introduction to the Mathematical theory of Finite Elements," John Wiley & Sons, Inc., New York, 1976.

Reddy, J. N. (ed.): "Penalty-Finite Element Methods in Mechanics," AMD-Vol. 51, The American Society of Mechanical Engineers, New York, 1982.

———: "An Introduction to the Finite Element Method," McGraw-Hill Book Company, New York, 1984.

Strang, G., and G. Fix: "An Analysis of the Finite Element Method," Prentice-Hall, Inc., Englewood Cliffs, N.J., 1973.

Thomasset, F.: "Implementation of Finte Element Methods for Navier–Stokes Equations," Springer-Verlag, New York, 1981.

Zienkiewicz, O. C.: "The Finite Element Method," 3rd., McGraw-Hill, London, 1977.

JOURNAL ARTICLES

Agmon, S., A. Douglis, and L. Nirenber: Estimates near the boundary for solution of elliptic partial differential equations satisfying general boundary conditions, *Commun. Pure Appl. Math.*, vol. 12, pp. 623–727, 1959.

Alt, H. W.: A free boundary problem associated with the flow of ground water, *Arch. Rat. Mech. Anal.*, vol. 64, pp. 111–126, 1977.

Aziz, A. K., and I. Babuska: Survey lectures on the mathematical foundations of the finite element method, in A. K. Aziz (ed.), "The Mathematical Foundations of the Finite Element Method with Applications to Partial Differential Equations," Academic Press Inc., New York, pp. 3-359, 1972.

Babuska, I.: The rate of convergence for the finite element method, *SIAM J. Numer. Anal.*, vol. 8, pp. 304–315, 1971.

———: Error bounds for finite element method, *Numer. Math.*, vol. 16, pp. 322–333, 1971.

———: Finite element schemes for domains with corners, *Numer. Math.*, vol. 20, pp. 1–21, 1972.

———: The finite element method with Lagrange multipliers, *Numer. Math.*, vol. 20, pp. 179–192, 1973.

———: The finite element method with penalty, *Math. Comput.*, vol. 27, pp. 221–228, 1973.

——— and M. Zlamal: Nonconforming elements in the finite element method with penalty, *SIAM J. Numer. Anal.*, vol. 10, pp. 863–875, 1973.

Baiocchi, C., and G. A. Pozzi: Error estimates and free-boundary convergence for a finite-difference discretization of a parabolic variational inequality, *Rev. Franc. Autom. Info. Rech. Op., Anal. Num.*, vol. 11, pp. 315–340, 1977.

———, V. Comincioli, E. Magenes, and G. A. Pozzi: Free boundary problems in the theory of fluid flow through porous media. Existence and uniqueness theorems, *Annali di Mat. Pura ed Appli.*, vol. XCVII, pp. 1–82, 1973.

Bathe, K. J., and E. L. Wilson: Stability and accuracy analysis of direct integration methods, *Int. J. Solids Struct.*, vol. 7, pp. 301–319, 1973.

Belytschko, T., H. J. Yen, and R. Mullen: Mixed methods for time integration, *J. Comp. Meth. Appl. Mech. Engng*, vol. 17/18, pp. 259–275, 1979.

Bercovier, M.: Perturbation of a mixed variational problem; application of mixed finite element methods, R.A.I.R.O., *Numer. Anal.*, vol. 12, pp. 211–236, 1978.

——— and M. Engelman: A finite element method for the solution of viscous incompressible flows, *J. Comput. Phys.*, vol. 30, pp. 181–201, 1979.

———, Y. Hasbani, Y. Gilon, and K. J. Bathe: On a finite element procedure for non-linear incompressible elasticity, Chapter 26 in S. N. Atluri, R. H. Gallagher, and O. C. Zienkiewicz (eds.), "Hybrid and Mixed Finite Element Methods," John Wiley & Sons, Inc., Chichester, 1983, pp. 497–517.

——— and O. Pironneau: Error estimates for finite element method solution of the Stokes problem in the primitive variables, *Numer. Mathematik*, vol. 33, pp. 211–224, 1979.

Berezanskii, J. M.: Expansions in eigenfunctions of self-adjoint operators, *Trans. Math. Monographs*, vol. 17, Amer. Math. Soc., Providence, R.I., 1968.

Bramble, J. H.: The Lagrange multiplier method for Dirichlet's problem, *Mathematics of Computation*, vol. 37, pp. 1–11, 1981.

——— and S. R. Hilbert: Bounds for a class of linear functionals with applications to Hermite interpolation, *Numer. Math.*, vol. 16, pp. 362–369, 1971.

——— and A. M. Schatz: Rayleigh–Ritz–Galerkin methods for Dirichlet's problem using subspaces without boundary conditions, *Comm. Pure and Appl. Math.*, vol. 23, pp. 115–175, 1970.

——— and ———: On the numerical solution of elliptic boundary value problems by least squares approximation of the data, Numerical Solution of Partial Differential Equations, II (SYNSPADE 1970), Academic Press Inc., New York, pp. 107–131, 1971.

——— and M. Zlamal: Triangular elements in the finite element method, *Math. of Comp.*, vol. 24, pp. 809–821, 1970.

Brezzi, F.: On the existence, uniqueness and approximation of saddle point problems arising from Lagrange multipliers, R.A.I.R.O., *Numer. Anal.*, vol. 8, pp. 129–151, 1974.

———, C. Johnson, and B. Mercier: Analysis of a mixed finite element method for elasto-plastic plates, *Math. Comput.*, vol. 31, pp. 809–817, 1977.

———, W. W. Hager, and P. A. Raviart: Error estimates for the finite element solution of variational inequalities, Part I: Primal theory, *Numer. Math.*, vol. 28, pp. 431–443, 1977; Part II: Mixed methods, *Numer. Math.*, vol. 31, pp. 1–16, 1978.

Bristeau, M. O., R. Glowinski, J. Periaux, P. Perrier, and O. Pironneau: Application of optimal control and finite element methods to calculation of transonic flows and incompressible flows, in

B. Hunt (ed.), "Numerical Methods in Applied Fluid Dynamics," Academic Press Inc., London, pp. 203–312, 1980.

Chorin, A. J.: A numerical method for solving incompressible viscous flow problems, *J. Compt. Phys.*, vol. 2, pp. 12–26, 1967.

———: On the convergence and approximation of discrete approximation to the Navier–Stokes equations, *Math. Compt.*, vol. 23, pp. 341–353, 1968.

Christie, I., and A. R. Mitchell: Upwinding of high order Galerkin methods in conduction-convection problems, *Int. J. Numer. Methods Eng.*, vol. 12, pp. 1764–1771, 1978.

Ciarlet, P. G., and P. Destuynder: Approximation of three-dimensional models by two-dimensional models in plate theory, in R. Glowinski, E. Y. Rodin, and O. C. Zienkiewicz (eds.), "Energy Methods in Finite Element Analysis," John Wiley & Sons, Inc., Chichester, pp. 33–45, 1979.

——— and P. A. Raviart: General Lagrange and Hermite interpolation in $\mathbb{R}^n$ with applications to finite element methods, *Archive Rat. Mech. Anal.*, vol. 46, pp. 177–179, 1972.

——— and ———: Interpolation theory over curved elements, with applications to finite element methods, *Computer Meth. in Appl. Mech. and Engng*, vol. 1, pp. 217–249, 1972.

——— and ———: A mixed finite element method for the biharmonic equation, in C. de Boor (ed.), "Mathematical Aspects of Finite Elements in Partial Differential Equations," Academic Press Inc., New York, pp. 125–145, 1974.

Courant, R.: Variational methods for the solution of problems of equilibrium and vibrations, *Bull. Amer. Math. Soc.*, vol. 49, pp. 1–23, 1943.

———, K. Friedrichs, and H. Lewy: On the partial difference equations of mathematical physics, *IBM Journal*, vol. 11, pp. 215–234, 1967.

Crouzeix, M., and P. A. Raviart: Conforming and nonconforming finite element methods for solving the stationary Stokes equations, I. *Rev. Fr. Autom. Inf. Rech. Oper.*, R-3, pp. 33–76, 1973.

Douglas, J., and T. Dupont: Galerkin methods for parabolic equations, *SIAM J. Numer. Anal.*, pp. 575–626, 1970.

——— and ———: Interior penalty procedures for elliptic and parabolic Galerkin methods, in R. Glowinski and J. L. Lions (eds.), "Computing Methods in Applied Sciences and Engineering," Springer-Verlag, Berlin, pp. 207–216, 1976.

Ehrlich, L. W., Solving the biharmonic equation as coupled finite difference equations, *SIAM J. Num. Anal.*, vol. 8, pp. 278–287, 1971.

Engquist, B., and H. O. Kreiss: Difference and finite element methods for hyperbolic differential equations, *J. Comp. Meth. Appl. Mech. Eng.*, vol. 17/18, pp. 581–596, 1979.

Falk, R. S.: Approximate solutions of some variational inequalities with order of convergence estimates, Ph.D. Thesis, Cornell University, Ithaca, 1971.

———: Error estimates for the approximation of a class of variational inequalities, *Math. Comput.*, vol. 28, pp. 963–971, 1974.

——— and B. Mercier: Error estimates for elasto-plastic problems, *Rev. Fr. Autom. Inf. Rech. Oper.*, vol. 11, pp. 135–144, 1977.

Fiacco, A. V., and G. P. McCormick: The sequential unconstrained minimization technique for nonlinear programming, a primal-dual method, *Management Sci.*, vol. 10, pp. 360–366, 1964.

Fix, G., and G. Strang: A Fourier analysis of the finite element method in Ritz-Galerkin theory, *Studies in Appl. Math.*, vol. 48, pp. 265–273, 1969.

Fortin, M.: Minimization of some non-differentiable functionals by the augmented Lagrangian method of Hestenes and Powell, *Appl. Math. Optim.*, vol. 2, pp. 236–250, 1976.

——— and F. Thomasset: Mixed finite element methods for incompressible flow problems, *J. Compt. Phy.*, vol. 31, pp. 173–215, 1979.

Fried, I.: Condition of finite element matrices generated from non-uniform meshes, *AIAA J.*, vol. 10, pp. 219–221, 1971.

———: Accuracy of finite element eigenproblems, *J. Sound and Vib.*, vol. 18, pp. 289–295, 1971.

———: Discretization and Computational errors in higher-order finite elements, *AIAA J.*, vol. 9, pp. 2071–2073, 1971.

Friedrichs, K., and H. B. Keller: A finite difference scheme for generalized Neumann problems, in J.

Bramble (ed.), "Numerical Solution of Partial Differential Equations," Academic Press Inc., New York, 1966.

Girault, V., and P. A. Raviart: An analysis of upwind schemes for the Navier–Stokes equations, *SIAM J. Numer. Anal.*, vol. 9, pp. 312–333, 1982.

Glowinski, R., and A. Morracco: Numerical solution of two-dimensional magneto-static problems by augmented Lagrangian methods, *Comput. Meth. Appl. Mech. Eng.*, vol. 12, pp. 33–46, 1977.

——— and O. Pironneau: Numerical methods for the first biharmonic equation and for the two-dimensional Stokes problem, *SIAM Rev.*, vol. 21, pp. 167–212, 1979.

——— and ———: On a mixed finite element approximation of the Stokes problem (I). Convergence of the approximate solution, *Numer. Math.*, vol. 33, pp. 397–424, 1979.

———, B. Mantel, J. Periaux, and O. Pironneau: A finite element approximation of Navier-Stokes equations for incompressible viscous fluids. Functional least squares methods of solution, in K. Morgan, C. Taylor, and C. A. Brebbia (eds.), "Computer Methods in Fluids," Pentech, London, pp. 84–133, 1980.

——— and P. Le Tellac: Numerical solution of problems in incompressible finite elasticity by augmented Lagrangian methods (I) Two-dimensional and axisymmetric problems, *SIAM J. Appl. Math.*, vol. 42, pp. 400–429, 1982.

Goudreau, G. L., and R. L. Taylor: Evaluation of numerical integration methods in elastodynamics, *J. Comp. Meth. Appl. Mech. Engng*, vol. 2, pp. 69–97, 1973.

Gurtin, M. E.: Variational principles for initial-value problems, *Q. Appl. Math.*, vol. 22, pp. 252–256, 1964.

Hlavacek, I., and J. Necas: On inequalities of Korn's type, *Archive for Ratl. Mech. Anal.*, vol. 36, pp. 305–334, 1970.

Houbolt, J. C.: A recurrence matrix solution for the dynamic response of elastic aircraft, *J. Aeronautical Science*, vol. 17, pp. 540–550, 1950.

Hughes, T. J. R., W. K. Liu, and A. Brooks: Finite element analysis of incompressible viscous flows by the penalty function formulation, *J. Comput. Phy.*, vol. 30, pp. 1–40, 1979.

——— and D. S. Malkus: A general penalty/mixed equivalence theorem for anisotropic, incompressible finite elements, Chapter 25 in S. N. Atluri, R. H. Gallagher, and O. C. Zienkiewicz (eds.), "Hybrid and Mixed Finite Elements," John Wiley & Sons, Inc., Chichester, 1983, pp. 487–496.

Johnson, C.: A convergence estimate for an approximation of a parabolic variational inequality, *SIAM J. Numer. Anal.*, vol. 34, pp. 599–606, 1976.

———: A mixed finite element method for the Navier–Stokes equations, *Rev. Fr. Autom. Inf. Rech. Oper. Ser. Anal. Numer.*, vol. 12, pp. 335–348, 1978.

Kalker, J. J.: Variational principles of contact elastostatics, *J. Inst. Maths. Applics.*, vol. 20, pp. 199–219, 1977.

———: A minimum principle for the law of dry friction with application to elastic cylinders in rolling contact, *J. Appl. Mech.*, vol. 38, pp. 875–887, 1971.

Kikuchi, N.: Beam bending problems on a Pasternak foundation using reciprocal variational inequalities, *Q. Appl. Math.*, pp. 91–108, 1980.

——— and Y. J. Song: Penalty/finite-element approximations of a class of unilateral problems in linear elasticity, *Q. Appl. Math.*, vol. 39, pp. 1–22, 1981.

Knops, R. J., and L. E. Payne: Uniqueness theorems in linear elasticity, *Springer Tracts in Natural Philosophy*, vol. 19, Springer-Verlag, New York, 1971.

Kondrat'ev, V. A.: Boundary problems for elliptic equations with conical or angular points, *Trans. Moscow Math. Soc.*, vol. 16, pp. 227–313, 1967.

Lax, P. D., and R. D. Richtmyer: Survey of the stability of finite difference equations, *Comm. Pure Appl. Math.*, vol. 9, pp. 267–293, 1956.

Le Tellac, P.: A mixed finite element approximation of the Navier–Stokes equations, *Numer. Math.*, vol. 35, pp. 381–404, 1980.

Lewy, H., and G. Stampacchia: On the regularity of the solution of a variational inequality, *Comm. Pure Appl. Math.*, vol. 22, pp. 153–188, 1969.

——— and ———: On the existence and smoothness of solutions of some non-coercive variational inequalities, *Arch. Rat. Mech. Anal.*, vol. 41, pp. 241–253, 1971.

Lions, J. L., and G. Stampacchia: Variational Inequalities, *Pure Appl. Math.*, vol. 20, 493–519, 1967.

Mikhlin, S. G.: The stability of the Ritz method, *Soviet Math. Dokl.*, vol. 1, pp. 1230–1233, 1966.

Mitchell, A. R., G. Phillips, R. Wachpress: Forbidden shape in the finite element method, *J. Inst. Maths. Applic.*, vol. 8, 1971.

Mossolov, P. P., and V. P. Miasnikov: Variational methods in the theory of fluidity of a viscous-plastic medium, *PMM*, vol. 29, pp. 468–492, 1965.

Necas, J.: Les méthodes directes et théorie des equations elliptiques, Prague, Academia, 1967.

Newmark, N. M.: A method for computation for structural dynamics, *ASCE J. Engineering Mechanics Div.*, vol. 85, pp. 67–94, 1959.

Nickell, R. E.: On the stability of approximation operators in problems of structural dynamics, *Int. J. Solids and Structures*, vol. 7, pp. 301–319, 1971.

Oden, J. T.: Mixed finite element approximations via interior and exterior penalties for contact problems in elasticity, Chapter 24 in S. N. Atluri, R. H. Gallagher, and O. C. Zienkiewicz (eds.), "Hybrid and Mixed Finite Element Methods," John Wiley & Sons, Inc., Chichester, 1983, pp. 467–486.

——— and N. Kikuchi: Theory of Variational Inequalities with application to problems of flow through porous media., *Int. J. Eng. Sci.*, vol. 18, pp. 1173–1284, 1980.

——— and J. N. Reddy: On dual-complementary variational principles in mathematical physics, *Int. J. Eng. Sci.*, vol. 12, pp. 1–29, 1974.

Ohtake, K., J. T. Oden, and N. Kikuchi: Analysis of certain unilateral problems in von Karman plate theory by a penalty method, Parts 1 and 2, *J. Comp. Meth. Appl. Mech. Eng.*, vol. 24, pp. 187–213 and 317–337, 1980.

Polyak, B. T.: Existence theorems and convergence of minimizing sequences in extremum problems with restrictions, *Soviet Math. Dokl.*, vol. 7, pp. 72–75, 1966.

———: The convergence rate of the penalty function method, *Zh. vychisl. Mat. fiz.*, vol. 11, pp. 3–11, 1971 (English translation: *USSR Comput. Math. Mathematical Phy.*, vol. 11, pp. 1–12, 1971).

Raviart, P. A., and J. M. Thomas: Dual finite element models for second order elliptic problems, Chapter 9 in R. Glowinski, E. Y. Rodin, and O. C. Zienkiewicz (eds.), "Energy Methods in Finite Element Analysis," John Wiley & Sons, Inc., 1979, pp. 175–191.

Reddy, J. N.: A note on mixed variational principles for initial-value problems, *Q. J. Mech. and Appl. Math.*, vol. 28, pp. 123–132, 1975.

———: On the accuracy and existence of solutions to primitive variable models of viscous incompressible fluids, *Int. J. Eng. Sci.*, vol. 16, pp. 921–929, 1978.

———: On the finite element method with penalty for incompressible fluid flow problems, in J. R. Whiteman (ed.), "The Mathematics of Finite Elements and Applications, III," Academic Press Inc., London, pp. 227–235, 1979.

———: On penalty function methods in the finite-element analysis of flow problems, *Int. J. Numer. Meth. Fluids*, vol. 2, pp. 151–171, 1982.

Reid, J. K.: On the construction and convergence of a finite-element solution of Laplace's equation, *J. Inst. Math. Appl.*, vol. 9, pp. 1–13, 1972.

Samuelsson, A., and M. Froier: Finite elements in plasticity, A variational inequality approach, in J. R. Whiteman (ed.), "The Mathematics of Finite Elements and Applications, III," Academic Press Inc., London, pp. 105–115, 1979.

Schultz, M. H.: Rayleigh–Ritz methods for multidimensional problems, *SIAM J. Numer. Anal.*, vol. 6, pp. 523–538, 1969.

———: L_2 error bounds for the Rayleigh–Ritz–Galerkin method, *SIAM J. Numer. Anal.*, vol. 8, pp. 737–748, 1971.

Slobodeckii, M. I.: Generalized Sobolev spaces and their applications to boundary problems for partial differential equations, *Amer. Math. Soc. Transl.*, vol. 21, pp. 207–275, 1966.

Smith, J.: The coupled equation approach to the numerical solution of the biharmonic equation by finite differences, I, *SIAM J. Num. Anal.*, vol. 5, pp. 323–339, 1968.

———: The coupled equation approach to the numerical solution of the biharmonic equation by finite differences, II, *SIAM J. Numer. Anal.*, vol. 7, pp. 104–111, 1970.

Strang, G., and A. E. Berger: The change in solution due to change in domain, *Proc. AMS Symp. on Partial Diff. Eqns.*, Berkeley, Calif., 1971.

Swartz, B., and B. Wendroff: Generalized finite difference schemes, *Math. of Comp.*, vol. 23, pp. 37–50, 1969.

Taylor, R. L., and O. C. Zienkiewicz: Complementary energy with penalty functions in finite element analysis, Chapter 8 in R. Glowinski, E. Y. Rodin, and O. C. Zienkiewicz (eds.), "Energy Methods in Finite Element Analysis," John Wiley & Sons, Inc., 1979, pp. 154–174.

Thomee, V.: Elliptic difference operators and Dirichlet's problem, *Diff. Eqns.*, vol. 3, pp. 301–324, 1964.

Ting, T. W.: Elastic-Plastic torsion problem III, *Archive Rat. Mech. Anal.*, vol. 34, pp. 228–243, 1969.

———: Elastic-plastic torsion of simply-connected cylindrical bars, *Indiana Univ. Math. J.*, vol. 20, pp. 1047–1076, 1971.

Tong, P., and T. H. H. Pian: The convergence of finite element method in solving linear elastic problems, *Int. J. Solids and Struct.*, vol. 3, pp. 865–879, 1969.

Trefftz, E.: Ein gegenstück zum Ritzsche verfahren, Second Congress of Applied Mathematics, Zurich, 1926.

Wheeler, M. F.: An elliptic collocation finite-element method with interior penalties, *SIAM J. Numer. Anal.*, vol. 15, pp. 152–161, 1978.

Zangwill, W. I.: Nonlinear programming via penalty functions, *Management Science*, vol. 13, pp. 344–358, 1967.

Zienkiewicz, O. C.: Constrained variational principles and penalty function methods in finite element analysis, in G. A. Watson (ed.) Lecture Notes in Mathematics: Conf. on Numer. Sol. of Differential Eqns., Springer-Verlag, Berlin, pp. 207–214, 1971.

Zlamal, M.: On the finite element method, *Numer. Math.*, vol. 12, pp. 394–409, 1968.

———: A finite element procedure of the second order accuracy, *Numer. Math.*, vol. 14, pp. 394–402, 1970.

ANSWERS TO SELECTED PROBLEMS

Chapter One

1-4 (i) $\delta_{ij}\delta_{ik} = \delta_{1j}\delta_{1k} + \delta_{2j}\delta_{2k} + \delta_{3j}\delta_{3k}$

Clearly, the sum is zero if $j \neq k$, and it is equal to 1 if $j = k$ for any $k = 1, 2, 3$. This is the same as δ_{jk}.

(ii) $\varepsilon_{ijk}\varepsilon_{ijk} = \delta_{jj}\delta_{kk} - \delta_{jk}\delta_{jk}$ (by $\varepsilon - \delta$ identity with i as the repeated index)

$= \delta_{jj}\delta_{kk} - \delta_{kk}$ (by part (i))

$= 3 \times 3 - 3 = 6$

(iii) $2A_{ij}\varepsilon_{ijk} = A_{ij}(\varepsilon_{ijk} + \varepsilon_{ijk})$

$= A_{ij}(\varepsilon_{ijk} - \varepsilon_{jik})$ (interchanged i and j)

$= A_{ij}\varepsilon_{ijk} - A_{ij}\varepsilon_{jik}$

$= (A_{ij} - A_{ji})\varepsilon_{ijk}$ (renamed i to be j, and j to be i in the second term)

$= 0$, if $A_{ij} = A_{ji}$ (that is, A_{ij} is symmetric)

Alternatively,

$$0 = A_{ij}\varepsilon_{ijk} \qquad \text{for any } k$$

In particular, we have,

$$\begin{aligned} 0 &= A_{23} - A_{32} && \text{for } k = 1 \\ &= A_{13} - A_{31} && \text{for } k = 2 \\ &= A_{12} - A_{21} && \text{for } k = 3 \end{aligned}$$

or $A_{ij} = A_{ji}$.

1-5 (i) $\text{curl}\,(\phi\mathbf{A}) = \hat{\mathbf{e}}_i \dfrac{\partial}{\partial x_i} \times (\phi\hat{\mathbf{e}}_j A_j)$

$$= (\hat{\mathbf{e}}_i \times \hat{\mathbf{e}}_j) \frac{\partial}{\partial x_i} (\phi A_j)$$

$$= \varepsilon_{ijk}\,\hat{\mathbf{e}}_k\left(\frac{\partial\phi}{\partial x_i} A_j + \phi\,\frac{\partial A_j}{\partial x_i}\right)$$

$$= \varepsilon_{ijk}\,\hat{\mathbf{e}}_k \frac{\partial\phi}{\partial x_i} A_j + \varepsilon_{ijk}\,\hat{\mathbf{e}}_k\,\phi\,\frac{\partial A_j}{\partial x_i}$$

$$= -A_j\varepsilon_{jik}\frac{\partial\phi}{\partial x_i}\hat{\mathbf{e}}_k + \left(\hat{\mathbf{e}}_i\frac{\partial}{\partial x_i}\right) \times (A_j\hat{\mathbf{e}}_j)\phi$$

$$= (-A_j\hat{\mathbf{e}}_j) \times \left(\hat{\mathbf{e}}_i\frac{\partial}{\partial x_i}\right)\phi + (\boldsymbol{\nabla} \times \mathbf{A})\phi$$

$$= -\mathbf{A} \times \boldsymbol{\nabla}\phi + (\boldsymbol{\nabla} \times \mathbf{A})\phi$$

$$= (\boldsymbol{\nabla}\phi) \times \mathbf{A} + \phi(\boldsymbol{\nabla} \times \mathbf{A})$$

1-6 (i) $\text{div}\,(r^n\mathbf{r}) = \left(\hat{\mathbf{e}}_i\dfrac{\partial}{\partial x_i}\right)\cdot(r^n x_j\hat{\mathbf{e}}_j)$

$$= \hat{\mathbf{e}}_i\cdot\hat{\mathbf{e}}_j\frac{\partial}{\partial x_i}(r^n x_j)$$

$$= \delta_{ij}\left(nr^{n-1}\frac{\partial r}{\partial x_i}x_j + r^n\delta_{ij}\right)$$

$$= nr^{n-1}\frac{x_i}{r}x_i + 3r^n$$

$$= (n+3)r^n$$

(iii) $\text{grad}\,(r^n) = \left(\hat{\mathbf{e}}_i\dfrac{\partial}{\partial x_i}\right)(r^n)$

$$= \hat{\mathbf{e}}_i\,nr^{n-1}\frac{\partial r}{\partial x_i}$$

$$= \hat{\mathbf{e}}_i\,nr^{n-1}\frac{x_i}{r}$$

$$= nr^{n-2}\mathbf{r}$$

Chapter Two

2-1 (*a*) $\dfrac{D\mathbf{v}}{Dt} = \dfrac{\partial\mathbf{v}}{\partial t} + \mathbf{v}\cdot\text{grad}\,\mathbf{v}$

$$= (2x_1\hat{\mathbf{e}}_1 + 3x_2\hat{\mathbf{e}}_2) + (2x_1t\hat{\mathbf{e}}_1 + 3x_2t\hat{\mathbf{e}}_2)\cdot(2t\hat{\mathbf{e}}_1\hat{\mathbf{e}}_1 + 3t\hat{\mathbf{e}}_2\hat{\mathbf{e}}_2)$$

$$= (2x_1 + 4x_1t^2)\hat{\mathbf{e}}_1 + (3x_2 + 9x_2t^2)\hat{\mathbf{e}}_2$$

(*b*) $\dfrac{D\mathbf{v}}{Dt} = C(C-1)x_1\hat{\mathbf{e}}_1/t^2 + (1+2C)Dx_1^2\hat{\mathbf{e}}_2$

2-2 $x_1 = X_1 + \frac{e_0}{b} X_2 \qquad x_2 = X_2 + \frac{e_0}{a} X_1 \qquad x_3 = X_3$

$$u_1 = \frac{e_0}{b} X_2 \qquad u_2 = \frac{e_0}{a} X_1 \qquad u_3 = 0$$

$$E_{11} = \frac{1}{2}\left(\frac{e_0}{a}\right)^2 \qquad E_{12} = \frac{e_0}{2}\left(\frac{a+b}{ab}\right) \qquad E_{22} = \frac{1}{2}\left(\frac{e_0}{b}\right)^2$$

$$E_{13} = E_{23} = E_{33} = 0$$

2-4 (a) $\sigma_{ij} v_{i,j} = \frac{1}{2}\sigma_{ij}[(v_{i,j} + v_{j,i}) + (v_{i,j} - v_{j,i})]$

$$= \sigma_{ij} D_{ij} + \sigma_{ij}\omega_{ij}$$

$$= \sigma_{ij} D_{ij}$$

because $\sigma_{ij}\omega_{ij} = 0 \quad (\omega_{ij} = -\omega_{ji}$ and $\sigma_{ij} = \sigma_{ji})$

(b) $\frac{1}{2}\varepsilon_{ijk}\varepsilon_{mnk}\frac{\partial v_m}{\partial x_n} = \frac{1}{2}\varepsilon_{ijk}\varepsilon_{mnk}\omega_{mn}$ because $\varepsilon_{mnk} D_{mn} = 0$ by Prob. 1-4 (iii)

$$= \tfrac{1}{2}(\delta_{im}\delta_{jn} - \delta_{in}\delta_{jm})\omega_{mn}$$

$$= \tfrac{1}{2}(\omega_{ij} - \omega_{ji}) = \omega_{ij}$$

(c) $\mathbf{\Omega} = \frac{1}{2}$ curl $\mathbf{v}$

$$= \tfrac{1}{2}\varepsilon_{ijk}\frac{\partial v_j}{\partial x_i}\hat{\mathbf{e}}_k$$

$$= \tfrac{1}{2}\varepsilon_{ijk}\hat{\mathbf{e}}_k(D_{ji} + \omega_{ji})$$

$= \frac{1}{2}\omega_{ji}\varepsilon_{ijk}\hat{\mathbf{e}}_k$ because $\varepsilon_{ijk} D_{ji} = 0$ by Prob. 1-4 (iii)

or $\Omega_k = -\frac{1}{2}\omega_{ij}\varepsilon_{ijk}$

2-5 $e_{11} = e_{22} = e_{33} = e_{12} = 0$

$$2e_{13} = \alpha\left(-x_2 + \frac{\partial\phi}{\partial x_1}\right) \qquad 2e_{23} = \alpha\left(x_1 + \frac{\partial\phi}{\partial x_2}\right)$$

2-6 $e_{11} = \frac{\partial u_0}{\partial x_1} + x_3\frac{\partial\psi}{\partial x_1} \qquad e_{22} = \frac{\partial v_0}{\partial x_2} + x_3\frac{\partial\phi}{\partial x_2} \qquad e_{33} = 0$

$$2e_{12} = \frac{\partial u_0}{\partial x_2} + \frac{\partial v_0}{\partial x_1} + x_3\left(\frac{\partial\psi}{\partial x_2} + \frac{\partial\phi}{\partial x_1}\right)$$

$$2e_{13} = \psi + \frac{\partial w_0}{\partial x_1} \qquad 2e_{23} = \phi + \frac{\partial w_0}{\partial x_2}$$

2-7 $0 = u_{i,jk}\varepsilon_{jkl}$

$$= (e_{ij,k} + \omega_{ij,k})\varepsilon_{jkl}$$

Differentiate with respect to x_m and multiply with ε_{imn} to obtain $0 = \varepsilon_{imn}\varepsilon_{jkl}e_{ij,km} + \omega_{ij,km}\varepsilon_{jkl}\varepsilon_{imn}$. Then show that the second term is zero to obtain the required equality.

2-9 (a) $0 = \frac{\partial^2 e_{11}}{\partial x_2^2} + \frac{\partial^2 e_{22}}{\partial x_1^2} - 2\frac{\partial^2 e_{12}}{\partial x_1\,\partial x_2}$

$= 2\alpha + 0 - 2\alpha x_3$; not satisfied

(*b*) One of the six equations is not satisfied

$$0 = \frac{\partial^2 e_{22}}{\partial x_1\, \partial x_3} + \frac{\partial^2 e_{13}}{\partial x_2^2} - \frac{\partial^2 e_{23}}{\partial x_2\, \partial x_1} - \frac{\partial^2 e_{21}}{\partial x_2\, \partial x_3}$$

$$= 0 + 0 - 0 - 2\alpha x_1$$

2-10 $\frac{1}{r}\frac{\partial}{\partial r}\left(r\frac{\partial E_\theta}{\partial r} + E_\theta - E_r\right) = 0$

$$-\frac{\partial^2 E_\theta}{\partial r\,\partial z} + \frac{1}{r}\frac{\partial E_r}{\partial z} - \frac{1}{r}\frac{\partial E_\theta}{\partial z} = 0$$

$$\frac{\partial}{\partial z}\left(\frac{\partial E_\theta}{\partial r} + \frac{E_\theta - E_r}{r}\right) = 0$$

$$\frac{\partial^2 E_r}{\partial z^2} = \frac{\partial^2 E_\theta}{\partial z^2} = 0$$

2-11 See the book by Reddy and Rasmussen (1982) or by Reddy (1984b).

2-12 (*a*) $T_i = n_j \sigma_{ji} \qquad \hat{\mathbf{n}} = 2\hat{\mathbf{e}}_1 + 3\hat{\mathbf{e}}_2 + \sqrt{3}\,\hat{\mathbf{e}}_3$

$$\mathbf{T} = -4\hat{\mathbf{e}}_1 - 4\hat{\mathbf{e}}_2 + 3\sqrt{3}\,\hat{\mathbf{e}}_3$$

(*b*) $\mathbf{T} = 5\hat{\mathbf{e}}_1 + \hat{\mathbf{e}}_3$

(*c*) $\mathbf{T} = (7 + \sqrt{2})\hat{\mathbf{e}}_1 - (8 + \sqrt{2})\hat{\mathbf{e}}_2 + 4(1 + \sqrt{2})\hat{\mathbf{e}}_3$

2-13 $(\sigma_{ij} - \lambda\delta_{ij})n_j = 0$

Since σ_{ij} are real, any complex roots should occur in pairs of the form $\lambda_1 = a + ib$ and $\lambda_2 = a - ib$. Then $(\sigma_{ij} - \lambda_1\delta_{ij})n_j^{(1)} = 0$, $(\sigma_{ij} - \lambda_2\delta_{ij})n_j^{(2)} = 0$. Multiply the first equation with $n_i^{(2)}$ and the second by $n_i^{(1)}$ and subtract the resulting equations one from the other: $0 = (\lambda_1 - \lambda_2)n_i^{(1)}n_i^{(2)} = 2ibn_i^{(1)}n_i^{(2)}$. The expression $n_i^{(1)}n_i^{(2)}$ is always positive because $\mathbf{n}^{(1)}$ and $\mathbf{n}^{(2)}$ are complex conjugate vectors associated with λ_1 and λ_2. Then it follows that $b = 0$, giving that the eigenvalues are real.

2-14 If λ_1 and λ_2 are real and distinct, from Prob. 2-13 it follows that (note that $n_i^{(1)}$ and $n_i^{(2)}$ are real eigenvectors) $(\lambda_1 - \lambda_2)n_i^{(1)}n_i^{(2)} = 0$ or $n_i^{(1)}n_i^{(2)} = 0$. Thus $\hat{\mathbf{n}}^{(1)}$ is orthogonal to $\hat{\mathbf{n}}^{(2)}$.

2-15 $\lambda_1 = 2 + \sqrt{20}$, $\lambda_2 = 3$, $\lambda_3 = 2 - \sqrt{20}$, $\mathbf{n}^{(1)} = \pm(-0.8506, 0.526, 0.0)$.

2-17 $\lambda_1 = 3$, $\lambda_2 = 2$, $\lambda_3 = -1$, $\hat{\mathbf{n}}^{(1)} = \pm(1/\sqrt{2})(1, 0, 1)$, $\hat{\mathbf{n}}^{(2)} = \pm(1/\sqrt{3})(-1, 1, 1)$, $\hat{\mathbf{n}}^{(3)} = \pm(1/\sqrt{6})(-1, -2, 1)$, clearly $\hat{\mathbf{n}}^{(i)} \cdot \hat{\mathbf{n}}^{(j)} = \delta_{ij}$.

2-18 and **2-19** See Reddy (1984b), pp. 22–24.

2-20 Use the stress transformation equation, $[\sigma] = [A]^T[\bar{\sigma}][A]$, where σ_{ij} are referred to $(\hat{\mathbf{e}}_1, \hat{\mathbf{e}}_2, \hat{\mathbf{e}}_3)$ system (old), $\bar{\sigma}_{ij}$ are referred to the $(\hat{\mathbf{n}}^{(1)}, \hat{\mathbf{n}}^{(2)}, \hat{\mathbf{n}}^{(3)})$ system (new) and A_{ij} are the direction cosines defined by $A_{ij} = \hat{\mathbf{n}}^{(i)} \cdot \hat{\mathbf{e}}_j$. Note that $\hat{\mathbf{n}}^{(3)} = \hat{\mathbf{n}}^{(1)} \times \hat{\mathbf{n}}^{(2)}$.

2-21 (*b*) $[\sigma] = \begin{bmatrix} 2 & -1 & 1 \\ -1 & 0 & 1 \\ 1 & 1 & 2 \end{bmatrix}$

2-24 See Prob. 2-4.

2-26 div (curl $\boldsymbol{\psi}$) = **0**

2-28 $\operatorname{div} \boldsymbol{\sigma} + \mathbf{f} = \rho \dfrac{D\mathbf{v}}{Dt}$

$$-\nabla P + \mathbf{f} = \rho \frac{D\mathbf{v}}{Dt}$$

2-30 $\rho \mathbf{v} \cdot \dfrac{D\mathbf{v}}{Dt} = \dfrac{\rho}{2} \dfrac{D}{Dt} (\mathbf{v} \cdot \mathbf{v})$

2-33 For the isotropic case the strains are related to the stresses by

$$e_{11} = \frac{1}{E} (\sigma_{11} - \nu \sigma_{22})$$

$$e_{22} = \frac{1}{E} (\sigma_{22} - \nu \sigma_{11})$$

$$2e_{12} = \frac{\sigma_{12}}{G}$$

Substituting these strains into Eq. (2-27) and using the 2-D equilibrium equations, we obtain $(1 - \nu)\nabla^2(\sigma_{11} + \sigma_{22}) = 0$. Use the result of Prob. 2-27 to show that $\nabla^4 \phi = 0$.

2-34 Satisfies all boundary conditions, except on edge $x = L$.

2-35 $u_1 = \dfrac{P}{2EI} (L^2 - x_1^2) x_2 + \dfrac{(2 + \nu) P x_2^3}{6EI}$

$$u_2 = \frac{Ph^2(1 + \nu)}{EI} (L - x_1) + \frac{\nu P x_1 x_2^2}{2EI} + \frac{P x_1^3}{6EI} - \frac{PL^2 x_1}{2EI} + \frac{PL^3}{3EI}$$

2-37 Under zero body forces, the equation is given by $\nabla^4 \phi + \alpha E \nabla^2 \theta = 0$, where α is the coefficient of linear thermal expansion.

2-39 Assume $v_1 = v_1(x_3)$ and $v_3 = 0$, and simplify Eq. (2-100) to obtain the required equations.

2-40 $\mu v_1 = \dfrac{x_3^2}{2} c_1 + \dfrac{x_3}{h} \left(\mu V_0 - \dfrac{h^2}{2} c_1 \right)$

where h is the distance between the plates and $c_1 = \partial P / \partial x_1$.

2-42 $u(x_1, x_2) = \dfrac{4}{\pi} \displaystyle\sum_{n=0}^{\infty} \dfrac{\sin [(2n + 1)\pi x_1] \sinh [(2n + 1)\pi x_2]}{(2n + 1) \sinh [(2n + 1)\pi]}$

2-43 $u(x_1, x_2) = \dfrac{\sin \pi x_1 \sin \pi x_2}{\sinh \pi}$

2-44 $u(x_1, x_2) = \dfrac{1}{\alpha} \left\{ \tfrac{1}{2}(x_1 - x_2^2) - \dfrac{4}{\pi^3} \displaystyle\sum_{n=0}^{\infty} \dfrac{\sin [(2n + 1)\pi x_1]}{(2n + 1)^3} \cdot F(x_2) \right\}$

$$F(x_2) = \frac{\sinh [(2n + 1)\pi(1 - x_2)] + \sinh [(2n + 1)\pi x_2]}{\sinh (2n + 1)\pi}$$

2-45 Can be obtained from the solution of Prob. 2-44 by considering a quadrant.

2-46 $u(x_1, t) = 2 \displaystyle\sum_{n=1}^{\infty} \dfrac{\lambda_n + h \sin \lambda_n}{\lambda_n^2 + h \sin^2 \lambda_n} \sin \lambda_n x_1 \, e^{-\lambda_n{}^2 t} \cdot \dfrac{4}{\pi} \displaystyle\sum_{m=0}^{\infty} \dfrac{\sin [(2m + 1)\pi x_1]}{2m + 1} e^{-(2m+1)^2 \pi^2 t}$

where the λ_n are the roots of the equation $\lambda_n \cot \lambda_n + h = 0$.

Chapter Three

Problems 3-1

3-1-1 $\mathbf{a} + \boldsymbol{\theta} = (a_1\theta_1, a_2\theta_2) = (a_1, a_2) \to \theta_1 = \theta_2 = 1$

Thus the identity element is given by $\boldsymbol{\theta} = (1, 1)$.

$$\mathbf{I}a = (a_1 I_1, a_2 I_2) = (1, 1) \to I_1 = \frac{1}{a_1} \qquad I_2 = \frac{1}{a_2}$$

Thus the inverse element associated with an element $\mathbf{a}$ is given by $\mathbf{I}a = (1/a_1, 1/a_2)$. Note that $(\alpha + \beta)\mathbf{a} = ((\alpha + \beta)a_1, (\alpha + \beta)a_2)$.

$$\begin{aligned}\alpha\mathbf{a} + \beta\mathbf{a} &= (\alpha a_1, \alpha a_2) + (\beta a_1, \beta a_2)\\ &= (\alpha\beta a_1^2, \alpha\beta a_2^2) \neq (\alpha + \beta)\mathbf{a}\end{aligned}$$

Hence V is not a linear vector space.

3-1-2 All except 2*a* and 2*d* are violated.

3-1-3 (*a*) yes; (*b*) no, because the sum of $p_1 = x^n$ and $p_2 = x^{n-1} - x^n$ is not in p_2.

3-1-4 The general solution is of the form

$$u(x, y) = \sum_{n,m=0}^{\infty} \cos n\pi x \cos m\pi y \qquad \lambda_{mn} = (n^2 + m^2)\pi^2$$

3-1-5 Let S_1, S_2 be subspaces of V, and let $S = S_1 \cap S_2$. Let $x, y \in S$. Then $x, y \in S_1$ and $x, y \in S_2$. Also $(\alpha x + \beta y) \in S_1$ and $(\alpha x + \beta y) \in S_2$. Hence $(\alpha x + \beta y) \in S$, and therefore S is a subspace of V.

3-1-6 (*a*) Not a subspace.
(*b*) A subspace.
(*c*) Not a subspace.
(*d*) A subspace.

3-1-7 (*a*) A subspace.
(*b*) Not a subspace.

3-1-9 (*a*)–(*b*): If $W = U \oplus V$, by definition we have $W = U + V$ and $U \cap V = \{0\}$. Then every element w in W is of the form $w = u + v$, $u \in U$ and $v \in V$. To show that this representation is unique, assume that $w = u_1 + v_1 = u_2 + v_2$, $u_1, u_2 \in U$ and $v_1, v_2 \in V$. Then

$$u_2 - u_2 = v_1 - v_2 \tag{a}$$

Since $u_1 - u_2 \in U$ and $v_1 - v_2 \in V$ and the fact that Eq. (*a*) holds implies that $u_1 - u_2 = v_1 - v_2 = 0$ because the only element common to both U and V is the zero element. This leads to the assertion in part *b*.

3-1-10 See Prob. 3-1-9.

3-1-11 (*a*) A direct sum.
(*b*) $U \cap V = \{\mathbf{x} = (0, x_2, 0)\}$; $U \cup V = \mathbb{R}^3$
(*c*) $U \cap V = \{\mathbf{x} = (x_1 - 2x_1, x_1)\}$; $U \cup V = \mathbb{R}^2$
(*d*) A direct sum.

3-1-12 (*a*) Linearly dependent.
(*b*) Linearly independent.
(*c*) Linearly independent.
(*d*) Linearly dependent.

3-1-14 (*a*) Does not span.
(*b*) Spans.
(*c*) Spans.
(*d*) Spans.

3-1-15 $S = \{\mathbf{x} \in \mathbb{R}^4 : x_1 = x_3\}$

3-1-16 $S = \{\mathbf{x} \in \mathbb{R}^4 : x_3 = 2x_2 + 3x_1, x_4 = x_1 + x_2 + x_3\}$

3-1-18 The basis is $\{e^{-kx}, e^{kx}, e^{ikx}, e^{-ikx}\}$.

3-1-19 A basis for $S_1 \cap S_2$ is $(-1, 2, 1, 2)$, and a basis for $S_1 + S_2$ is given by $\{(-1, 2, 1, 2), (1, 2, 3, 6), (1, -1, 1, 1)\}$.

3-1-21 $\{(0, 0, 1, 1), (5, 1, -8, 0)\}$

3-1-23 $\{(1, 1, -1, 0), (0, 0, 0, 1)\}$

Problems 3-2

3-2-1 Not linear.

3-2-2 Onto; not one-to-one.

3-2-3 $T = k + (d^2/dx^2)[b(d^2/dx^2)]$; T is linear.

The domain is a linear space that contains functions that are continuous with their derivatives up to and including order four.

3-2-5 T is a linear matrix operator

$$T = \left[\begin{array}{c|c|c} -2\mu \dfrac{\partial^2}{\partial x^2} - \mu \dfrac{\partial^2}{\partial y^2} & -\mu \dfrac{\partial^2}{\partial x \, \partial y} & \dfrac{\partial}{\partial x} \\ -\mu \dfrac{\partial^2}{\partial x \, \partial y} & -2\mu \dfrac{\partial^2}{\partial y^2} - \mu \dfrac{\partial^2}{\partial x^2} & \dfrac{\partial}{\partial y} \\ \dfrac{\partial}{\partial x} & \dfrac{\partial}{\partial y} & 0 \end{array}\right]$$

The domain is a linear (product) space of elements of the type (u, v, P), where u and v are twice-differentiable and P is once-differentiable with respect to x and y.

3-2-7 Not a linear operator.

3-2-10 $(T_1 + T_2)\mathbf{x} = (3x_1 - x_2 - x_3, x_1 + x_2 + x_3, x_2 - x_3)$

$$T_1 T_2 \mathbf{x} = (2x_1 - 2x_2 + x_3, x_1 + x_2 - x_3, 0)$$

$$T_2 T_1 \mathbf{x} = (2x_1 - x_2 - 3x_3, x_1 - x_3, x_2 + x_3)$$

3-2-11 (*a*) $DTp(t) = D[tp(t)] = p(t) + t(dp(t)/dt) = [1 + t(d/dt)]p(t)$ or $DT = I + TD$.

3-2-13 $\mathcal{N} = \{0\}$

3-2-14 $\mathcal{N} = \{\mathbf{x} = (x_1, -x_1, x_1)\}$; $\mathcal{N} = \{0\}$

3-2-15 (*a*) $[T_1] = \begin{bmatrix} 1 & 1 & 1 & 1 \\ 0 & 1 & 2 & 3 \\ 0 & 0 & 1 & 3 \\ 0 & 0 & 0 & 1 \end{bmatrix}$

3-2-16 $B(v, u) = \int_0^L a \frac{dv}{dx}\frac{du}{dx}\,dx, \qquad l(v) = \int_0^L vf\,dx + v(0)P_1 + v(L)P_2$

3-2-17 $l(v) = \int_\Omega vf\,dx\,dy + \oint_\Gamma vq\,ds$

3-2-18 $l(v) = \int_0^a \int_0^b vq\,dx\,dy$

3-2-19 Quadratic form.

3-2-20 Quadratic form.

3-2-21 Neither completely quadratic nor linear.

3-2-22 Bilinear; yes.

3-2-23 $[B] = \frac{2\alpha}{L^3}\begin{bmatrix} 6 & -3L & -6 & -3L \\ & 2L^2 & 3L & L^2 \\ \text{symm.} & & 6 & 3L \\ & & & 2L^2 \end{bmatrix} \qquad (\alpha = \text{constant})$

3-2-26 (*a*) $I - P$ is linear and $(I - P)^2 = I - 2P + P^2 = I - P$, (*b*) $\mathscr{R}(P)$, (*c*) $P(I - P) = P - P^2 = 0$.

Problems 3-3

3-3-1 No.

3-3-4 $\|u\|_0 = \sqrt{\frac{5}{6} - 2/\pi}$

3-3-5 $\|u\|_0 = \sqrt{\frac{3}{5}}$

3-3-7 $|u|_1^2 = 10 - \pi - 2\sqrt{2}$

3-3-8 $\|u\|_0 = \frac{1}{30}$

3-3-10 $|u|_1^2 = 1 + \pi^2/2$

3-3-12 $|u|_{1,p} = 0$ implies u = constant in Ω. Since $u \in C_0^1(\Omega)$ it follows that this constant be zero.

3-3-14 $(Tu)(x) = \int_0^1 K(x, y)u(y)\,dy$

Let $|K(x, y)| \le M$. Then

$$|(Tu_1)(x) - (Tu_2)(x)| = \left|\int_0^1 K(x, y)[u_1(y) - u_2(y)]\,dy\right|$$

$$\le M \sup |u_1(y) - u_2(y)|$$

$$\|Tu_1 - Tu_2\| = \sup |Tu_1(x) - Tu_2(x)| \le M\|u_1 - u_2\|$$

3-3-16 $|Tu_1 - Tu_2| = \left|\frac{1}{x}\int_0^x (u_1 - u_2)\,dy\right|$

$$\le \int_0^x |u_1 - u_2|\,dy$$

$$\|Tu_1 - Tu_2\| \le \left[\int_0^x |u_1 - u_2|^p\,dy\right]^{1/p} \le \|u_1 - u_2\|_p$$

3-3-18 $$|l(u)| = \left|\int_a^b fu\,dx\right| \le \left[\int_a^b |f|^2\,dx\right]^{1/2}\left[\int_a^b |u|^2\,dx\right]^{1/2}$$
$$\le M\|u\|_0 \qquad M = \|f\|_0$$

$$|B(u, v)| \le M\left\{\int_a^b \left[|u|^2 + \left|\frac{du}{dx}\right|^2\right] dx\right\}^{1/2} \cdot \left\{\int_a^b \left[|v|^2 + \left|\frac{dv}{dx}\right|^2\right] dx\right\}^{1/2}$$
$$= M\|u\|_1\|v\|_1 \qquad M = \max_{a \le x \le b} \{r(x), s(x)\}$$

Problems 3-4

3-4-1 (*a*) $$\|u + v\|^2 + \|u - v\|^2 = \|u\|^2 + \|v\|^2 + 2(u, v) + \|u\|^2 + \|v\|^2 - 2(u, v)$$
$$= 2(\|u\|^2 + \|v\|^2)$$

(*b*) $$(u, v) = \tfrac{1}{2}[(u + v, u + v) - (u, u) - (v, v)]$$
$$(u, v) = -\tfrac{1}{2}[(u - v, u - v) - (u, u) - (v, v)]$$

Adding the two equations and dividing throughout by 2, we obtain the result [subtracting the second from the first gives part (*a*)].

(*c*) $$(u - v, u - v) = \|u\| + \|v\| - 2(u, v)$$
$$\ge \|u\| + \|v\| - 2\|u\|\,\|v\|$$
$$= (\|u\| - \|v\|)^2$$

where the Schwarz inequality is used in the second line.

3-4-2 We must check for the symmetry, $(u, v) = \overline{(v, u)}$, and positive-definiteness. Let $u = u_1 + iu_2$, $v = v_1 + iv_2$.

$$(u, v) = \int_0^1 u(x)v(x)\,dx = u_1v_1 + u_2v_2 + i(u_2v_1 - u_1v_2)$$

$$(v, u) = \int_0^1 v(x)u(x)\,dx = v_1u_1 + v_2u_2 + i(v_2u_1 - v_1u_2)$$
$$= v_1u_1 + v_2u_2 - i(u_2v_1 - u_1v_2)$$

Clearly $(u, v) = \overline{(v, u)}$; hence symmetric.

Next, let $u = v$; we have $(u, u) = u_1^2 + u_2^2 \ge 0$ and equal to zero if and only if $u_1 = u_2 = 0$. $f(x) = (-a + ib)\sin \pi x$ only if $a = b$.

3-4-4 $(u, v)_0 = 4/\pi^3$

3-4-5 $(u, v)_0 = 0$

3-4-6 $(u, v)_1 = 0$

3-4-7 $(u, v)_1 = 3/\pi$

3-4-9 (*a*) $S = \{\mathbf{x} = (x_1, x_2, -x_2)\}$
(*c*) $S = \{\mathbf{x} = (-2x_2 - 3x_3, x_2, x_3)\}$

3-4-10 $p(x) = 85 - 152x + 225x^2$

3-4-12 $\{1/\sqrt{2}, \sqrt{\tfrac{3}{2}}\,x, \sqrt{\tfrac{5}{8}}(3x^2 - 1)\}$

3-4-14 The subspace of all even functions, $f(-x) = f(x)$.

3-4-15 $T = \sum_{i=1}^{m} P_i$; $\quad T$ is linear.

$$T^2 = \left(\sum_{i=1}^{m} P_i\right)\left(\sum_{j=1}^{m} P_j\right) = \sum_{i,j=1}^{m} P_i P_j = \sum_{i=1}^{m} P_i^2 = \sum_{i=1}^{m} P_i = T$$

3-4-16 (*a*) $v(x) = x^2 - \frac{1}{3}$
(*b*) $v(x) = \frac{1}{2} - x$
(*c*) $v(x) = x(1 - x) - \frac{1}{6}$

3-4-18 (*b*) $x^2 = (x^2 - \frac{1}{3}) + \frac{1}{3}$

3-4-19 $p(x) = -105 + 39e + (588 - 216e)x + (-570 + 210e)x^2$

3-4-21 See Prob. 3-4-1(*b*).

3-4-22 T is bounded below (see Example 4-4).

Chapter Four

Problems 4-1

4-1-1 (*a*) Let $A + B = C$. Then

$$\begin{aligned}(u, C^*v) = (Cu, v) &= ((A + B)u, v) = (Au, v) + (Bu, v)\\ &= (u, A^*v) + (u, B^*v)\\ &= (u, (A^* + B^*)v)\end{aligned}$$

This implies that $C^* \equiv (A + B)^* = A^* + B^*$.

4-1-3 $A = d/dt$; $A^* = A$ with respect to the scalar product given.

4-1-5 By definition $\mathscr{R}(A)$ and $\mathscr{N}(A)$ are subsets of H. For $\alpha u + \beta v \in H$, we have $A(\alpha u + \beta v) = \alpha A(u) + \beta A(v) \in \mathscr{R}(A)$. Also for $u = 0 \in H$, $Au = 0 \in \mathscr{R}(A)$. Hence $\mathscr{R}(A)$ is a linear subspace. Similarly, show that $\mathscr{N}(A)$ is a linear subspace of H.

By definition, we have

$$\begin{aligned}[\mathscr{R}(A)]^{\perp} &= \{v: v \in H, (Au, v) = 0 \text{ for all } u \in H\}\\ &= \{v: v \in H, (u, A^*v) = 0 \text{ for all } u \in H\}\\ &= \mathscr{N}(A^*)\end{aligned}$$

4-1-6
$$\begin{aligned}(Au, v) &= \int_0^L \left[\frac{d^2}{dx^2}\left(b\,\frac{d^2u}{dx^2}\right) + ku\right]v\,dx\\ &= \int_0^L u\left[\frac{d^2}{dx^2}\left(b\,\frac{d^2v}{dx^2}\right) + kv\right]dx + \left[\frac{d}{dx}\left(b\,\frac{d^2u}{dx^2}\right)v - b\,\frac{d^2u}{dx^2}\,\frac{dv}{dx}\right.\\ &\qquad \left. + b\,\frac{du}{dx}\,\frac{d^2v}{dx^2} - \frac{d}{dx}\left(b\,\frac{d^2v}{dx^2}\right)u\right]_{x=0}^{x=L}\\ &= \int_0^L u\left[\frac{d^2}{dx^2}\left(b\,\frac{d^2v}{dx^2}\right) + kv\right]dx\\ &= (u, A^*v) \qquad \text{for all } u \in \mathscr{D}_A\end{aligned}$$

$$\mathscr{D}_A = \left\{u \colon u \in C^4[0, L],\ u = \frac{du}{dx} = 0 \text{ at } x = 0, L\right\}$$

$A = A^*$ and $\mathscr{D}_A = \mathscr{D}_{A^*}$

4-1-7 Let P be a projection on H (that is, P is linear and $P^2 = P$). Let P be orthogonal (that is, $\mathscr{R}(P) \perp \mathscr{N}(P)$). Then

$$0 = (Pu, v) \text{ for } u \in H,\ v \in \mathscr{N}(P) = \mathscr{R}(P)^{\perp} \text{ (by Theorem 3-9)}$$

$$= (u, P^*v)$$

This gives $P^*v = 0$ or $v \in \mathscr{N}(P^*)$. By Prob. 4-1-5, $\mathscr{N}(P^*) = [\mathscr{R}(P)]^{\perp}$. Thus $\mathscr{N}(P) = \mathscr{N}(P^*)$ which implies $P = P^*$. The converse follows by reversing the arguments.

4-1-9 Since $T = \lambda_1 P_1 + \lambda_2 P_2$ is self adjoint, if follows that $T^*T = TT^*$ (that is, T is normal).

4-1-11 Since the operator $-d^2/dx^2$ is positive on $\mathscr{D}_A$, A is positive on $\mathscr{D}_A$.

4-1-12 We have

$$(Au, v) = \int_0^L \frac{d^2}{dx^2}\left(EI\,\frac{d^2u}{dx^2}\right) dx$$

$$= \int_0^L EI\,\frac{d^2v}{dx^2}\frac{d^2u}{dx^2}\,dx + \left[\frac{d}{dx}\left(EI\,\frac{d^2u}{dx^2}\right)v - EI\,\frac{d^2u}{dx^2}\frac{dv}{dx}\right]_0^L$$

The boundary term is zero for all $u, v \in \mathscr{D}_A$. Hence $(Au, u) = \int_0^L EI[(d^2u/dx^2)]^2\,dx \geq 0$, $(Au, u) = 0$ implies $d^2u/dx^2 = 0$ or $u = c_1x + c_2$. Since $u = du/dx = 0$ at $x = 0, L$, we have $c_1 = c_2 = 0$. Hence A is positive.

4-1-14 Follows from Prob. 4-1-6.

Problems 4-2

4-2-3 $I(u) = \displaystyle\int_0^L \left[\frac{EI}{2}\left(\frac{d^2u}{dx^2}\right)^2 - uf\right] dx$

4-2-4 $I(u) = \displaystyle\int_0^1 \int_0^1 \left\{\frac{1}{2}\left[\left(\frac{\partial u}{\partial x}\right)^2 + \left(\frac{\partial u}{\partial y}\right)^2\right] - uf\right\} dx\,dy$

4-2-5 $-\nabla^2 u = f$ in $0 \leq x, y \leq 1$

Natural:

$$-\frac{\partial u}{\partial x} + \hat{t}_1 = 0 \quad \text{on} \quad x = 0 \qquad \frac{\partial u}{\partial x} = 0 \quad \text{on} \quad x = 1$$

$$-\frac{\partial u}{\partial y} + \hat{t}_2 = 0 \quad \text{on} \quad y = 0 \qquad \frac{\partial u}{\partial y} = 0 \quad \text{on} \quad y = 1$$

Essential:

$$u = 0 \quad \text{on} \quad x = 1 \quad \text{and} \quad y = 1 \quad \left(\text{if} \quad \frac{\partial u}{\partial x}(1, y) \neq 0, \quad \frac{\partial u}{\partial y}(x, 1) \neq 0\right)$$

4-2-6 Eqs. (4-95), with $\mu = -\alpha$ and $P = -w$; and

$$\left.\begin{aligned} -\left(2\mu \frac{\partial u}{\partial x} - P\right)n_x + \mu\left(\frac{\partial u}{\partial y} + \frac{\partial v}{\partial x}\right)n_y + \hat{t}_1 = 0 \\ -\mu\left(\frac{\partial u}{\partial y} + \frac{\partial v}{\partial x}\right)n_x + \left(2\mu \frac{\partial v}{\partial y} - P\right)n_y + \hat{t}_2 = 0 \end{aligned}\right\} \text{ on } \Gamma_2 \text{ (natural)}$$

$u = v = 0$ on Γ_1 (essential)

4-2-7 $$-\frac{1}{4}\left(\frac{du}{dx}\right)^2 \frac{d^2u}{dx^2} - 2\frac{d^2u}{dx^2} + 6u^5 - 6 = 0 \quad \text{in } (0, 1)$$

$u(0) = u(1) = 0$ (essential)

or

$$\frac{1}{12}\left(\frac{du}{dx}\right)^3 + 2\frac{du}{dx} = 0 \quad \text{at} \quad x = 0, 1 \quad \text{(natural)}$$

4-2-8 $$\frac{\partial}{\partial x}\left(EA\frac{\partial u}{\partial x}\right) + \frac{\partial}{\partial x}\left(EA_1\frac{\partial \psi}{\partial x}\right) - \rho A\frac{\partial^2 u}{\partial t^2} - \rho A_1 \frac{\partial^2 \psi}{\partial t^2} = 0$$

$$\frac{\partial}{\partial x}\left[GA\left(\frac{\partial w}{\partial x} + \psi\right)\right] - \rho A\frac{\partial^2 w}{\partial t^2} = 0$$

$$\frac{\partial}{\partial x}\left(EA_2\frac{\partial \psi}{\partial x}\right) + \frac{\partial}{\partial x}\left(EA_1\frac{\partial u}{\partial x}\right) - AG\left(\psi + \frac{\partial w}{\partial x}\right) - \rho A_2\frac{\partial^2 \psi}{\partial t^2} - \rho A_1\frac{\partial^2 u}{\partial t^2} = 0$$

where A is the area of cross-section, and $A_1 = \int_A z\, dA$, $A_2 = \int_A z^2\, dA$.

Natural:	or	*Essential:*
$\frac{\partial u}{\partial x} = 0$		$u = 0$ at $x = 0, L$
$\frac{\partial \psi}{\partial x} = 0$		$\psi = 0$ at $x = 0, L$
$GA\left(\frac{\partial w}{\partial x} + \psi\right)\Big\vert_{x=L} = P$, $\quad GA\left(\frac{\partial w}{\partial x} + \psi\right)\Big\vert_{x=0} = 0$,		$w(0) = 0$

4-2-9 Same as those in Eq. (4-95); for the natural B.C., see Prob. 4-2-6.

4-2-10 $$\frac{d^2}{dr^2}\left(rD_{11}\frac{d^2w}{dr^2}\right) - D_{22}\frac{d}{dr}\left(\frac{1}{r}\frac{dw}{dr}\right) - f = 0 \quad (D_{ij} = \text{constants})$$

4-2-11 $$-\frac{d}{dr}\left(rA_{11}\frac{du}{dr}\right) + A_{22}\frac{u}{r} = 0$$

$$-\frac{d}{dr}\left[rA_{55}\left(\psi + \frac{dw}{dr}\right)\right] - fr = 0$$

$$rA_{55}\left(\psi + \frac{dw}{dr}\right) - \frac{d}{dr}(rM_r) + \frac{M_\theta}{r} = 0 \quad \text{in} \quad (0, a)$$

$$r\left(\frac{d\psi}{dr} + \bar{D}_{22}M_r - \bar{D}_{12}M_\theta\right) = 0$$

$$r\left(\frac{\psi}{r} - \bar{D}_{12}M_r + \bar{D}_{11}M_\theta\right) = 0$$

4-2-12 $B(v, w) = \int_0^L \left(EI \frac{d^2v}{dx^2} \frac{d^2w}{dx^2} + a \frac{dv}{dx} \frac{dw}{dx} \right) dx$

$l(v) = \int_0^L vf \, dx$

$H_A = H_0^2[0, L]$

4-2-13 $B(v, u) = \int_\Omega k \text{ grad } v \cdot \text{grad } u \, d\mathbf{x} + \oint_\Gamma hvu \, ds$

$l(v) = -\int_\Omega fv \, d\mathbf{x} + \int_{\Gamma_2} v(hu_\infty - q_n) \, ds$

$H_A = \{u \in H^1(\Omega): u = 0 \text{ on } \Gamma_1\}$

4-2-14 $B(\mathbf{v}, \mathbf{u}) = -\int_\Omega [(\lambda + \mu) \text{ div } \mathbf{v} \text{ div } \mathbf{u} + \mu \text{ grad } \mathbf{v}\text{: grad } \mathbf{u}] \, d\mathbf{x}$

$l(\mathbf{v}) = \int_\Omega \mathbf{v} \cdot \mathbf{f} \, d\mathbf{x} - \int_{\Gamma_2} \mathbf{v} \cdot \hat{\mathbf{t}} \, ds$

$H_A = H_*^1(\Omega) \times H_*^1(\Omega)$

$H_*^1(\Omega) = \{u\text{: } u \in H^1(\Omega), \quad u = 0 \text{ on } \Gamma_1\}$

4-2-16 $B(\mathbf{v}, \mathbf{u})$ is given by Eq. (4-158) with $K = \gamma$, and $l(\mathbf{v})$ is given by Eq. (4-98).

Problems 4-3

4-3-4 $x = -1, \quad \lambda = 2$

4-3-5 $L(\mathbf{x}) = 2(x_1 + x_2) + \lambda(x_1 x_2 - A)$

$\delta L = 0$ gives $2 + \lambda x_2 = 0$ and $2 + \lambda x_1 = 0$ $(x_1 x_2 - A = 0)$.

This solution of the three equations is given by $x_1 = x_2 = -2/\lambda$ and $\lambda = 2/\sqrt{A}$ or $-2/\sqrt{A}$. Since $x_1, x_2 > 0$, we have $x_1 = x_2 = \sqrt{A}, \lambda = -(2/\sqrt{A})$.

4-3-6 $x_1 = 0, x_2 = 0, \lambda = -1$ or $x_1 = -\frac{4}{9}, \quad x_2 = -\frac{2}{3}, \quad \lambda = -\frac{1}{3}$

4-3-8 $x_1 = 11, \quad x_2 = 15, \quad x_3 = 9, \quad \lambda = -21$

4-3-9 Use $u_0 = -\frac{1}{2}\left(\frac{1}{\lambda_0} + x\right)$ in $\int_0^1 (u_0^2 + u_0 x) \, dx = \frac{47}{12}$ to compute λ_0;

show that $J(u_0) \geq J(u_0 + v)$ for an *admissible* v.

4-3-10 $u_0 = -(x/2), \quad \lambda_0 = 1$

4-3-11 $1 + \lambda(2u + x) = 0, \quad x + 2u\lambda = 0$

4-3-12 $(x_1 - a)^2 + (y - b)^2 = c^2$

4-3-14 See Reddy (1984b).

4-3-15 See Reddy (1984b).

4-3-16 $x_K = -\frac{K}{1 + K}; \quad \lim_{K \to \infty} x_K = x_0 = -1$

4-3-17 $x_1(K) = -[1 + (1 + 2K)x_2(K)]/2K, \quad x_2^2(K) + bx_2(K) + c = 0,$

$$b = (6K + 1)/[4K(4K + 1)], \quad c = 1/[4K(4K + 1)]$$

In the limit we obtain $x_1 = x_2 = 0$.

4-3-20 $-2\mu \dfrac{\partial^2 u_1}{\partial x_1^2} - \mu \dfrac{\partial^2 u_1}{\partial x_2^2} - K \dfrac{\partial^2 u_1}{\partial x_1^2} = f_1$

$$-2\mu \frac{\partial^2 u_1}{\partial x_1\, \partial x_2} - K \frac{\partial^2 u_1}{\partial x_1\, \partial x_2} = f_2$$

For $f_1 = f_2 = 0$, the equations do not have a solution.

Chapter Five

Problems 5-1

5-1-1 $x_1 = \frac{5}{9}, \quad x_2 = -\frac{23}{9}, \quad x_3 = \frac{2}{3}$

5-1-2 No; $\beta_1 = 2\beta_2 = -\frac{2}{3}\beta_3$; solution does not exist.

5-1-3 $x_1 = -\frac{1}{4}, \quad x_2 = -\frac{5}{4}, \quad x_3 = \frac{3}{4}$

5-1-4 $x_1 = 3, \quad x_2 = -14, \quad x_3 = 15$

5-1-5 No; $\beta_1 = -3\beta_2, \quad \beta_3 = 5\beta_2$; many solutions exist.

5-1-6 $x_1 = \frac{4}{3}, \quad x_2 = 0, \quad x_3 = \frac{1}{3}$

5-1-8 Initial guess $\mathbf{x}^{(3)} = (0, 0, 0)$ gives $x_1^{(4)} = 0.9770$, $x_2^{(4)} = 2.009$, $x_3^{(4)} = 3.034$.

5-1-9 $\mathbf{x}^{(0)} = (0, 0, 0)$ gives $x_1^{(10)} = 2.678$, $x_2^{(10)} = 4.928$, $x_3^{(10)} = 1.923$ and $x_1^{(13)} = 2.692$, $x_2^{(13)} = 4.937$, $x_3^{(13)} = 1.928$.

5-1-10 $\mathbf{x}^{(0)} = (0, 0, 0)$ gives $x_1^{(9)} = -5.001$, $x_2^{(9)} = 2.000$, $x_3^{(9)} = -2.999$ and the same result is obtained using $\mathbf{x}^{(0)} = (1, 1, 1)$.

5-1-11 $\mathbf{x}(0) = (1, 1, 1)$ gives $x_1^{(9)} = -2.000$, $x_2^{(9)} = -0.000927$, $x_3^{(9)} = -0.9998$.

5-1-12 (1, 1) in the first quadrant, (−1, −1) in the third quadrant; and does not converge in the second and fourth quadrants.

Problems 5-2

5-2-1 $w = a + bx$, $w(0) = 1$ gives $a = 1$ and $(dw/dx)(1) = 2$ gives $b = 2$

$$-\frac{d}{dx}\left[(1 + 2x^2)\frac{du}{dx}\right] + u = -1 + 6x + x^2$$

$$u(0) = 0, \quad (du/dx)(1) = 0$$

5-2-2 $I(u) = \displaystyle\int_a^b \left[c_1 \frac{du}{dx} + c_0 u - f\right]^2 dx$

$$B(v, u) = \int_a^b \left(c_1 \frac{dv}{dx} + c_0 v\right)\left(c_1 \frac{du}{dx} + c_0 u\right) dx$$

$$l(v) = \int_a^b \left(c_1 \frac{dv}{dx} + c_0 v\right) f\, dx$$

Solution exists in $H_*^1[a, b] = \{u : u \in H^1[a, b], u \text{ satisfies the specified boundary condition}\}$.

5-2-4 $B(v, u) = \int_a^b \left[p(x) \frac{dv}{dx} \frac{du}{dx} + q(x)vu \right] dx$

$$l(v) = \int_a^b vf \, dx$$

$$M = \max_{a \le x \le b} \{p(x), q(x)\}, \alpha = \min_{a \le x \le b} \{p(x), q(x)\}$$

Solution exists in $H_0^1(\Omega)$.

5-2-5 No.

5-2-6 $B(v, u) = \int_\Omega \left[\sum_{i,j=1}^n a_{ij} \frac{\partial v}{\partial x_i} \frac{\partial u}{\partial x_j} + a_0 vu \right] d\mathbf{x}$

$$l(v) = \int_\Omega fv \, d\mathbf{x} + \oint_\Gamma vq \, ds$$

$B(v, u)$ is positive-definite because $a_0 > 0$.

5-2-9 (i) Elliptic; (ii) Hyperbolic.

Problems 5-3

5-3-1 $B(\lambda, \mu; w, M) = \int_0^L \left(\frac{d\lambda}{dx} \frac{dw}{dx} + \frac{d\mu}{dx} \frac{dM}{dx} + \frac{\mu M}{EI} \right) dx$

$$l(\lambda, \mu) = -\int_0^L \lambda f \, dx$$

5-3-2 $B(u, \phi, \mu; w, \psi, \lambda) = \int_0^L \left[EI \frac{d\phi}{dx} \frac{d\psi}{dx} + \lambda\left(\phi + \frac{du}{dx} \right) + \mu\left(\psi + \frac{dw}{dx} \right) \right] dx$

$$l(u, \phi, \mu) = \int_0^L uf \, dx$$

5-3-3 $B_K(u, \phi; w, \psi) = \int_0^L \left[EI \frac{d\phi}{dx} \frac{d\psi}{dx} + K\left(\phi + \frac{du}{dx} \right)\left(\psi + \frac{dw}{dx} \right) \right] dx$

$$l_K(u, \phi) = \int_0^L uf \, dx$$

5-3-9 $B_K(v, u) = \int_\Omega [\text{grad } v \cdot \text{grad } u + K(\nabla^2 v \nabla^2 u)] \, d\mathbf{x}$

$$l_K(v) = -\int_\Omega (vf + K\nabla^2 vf) \, d\mathbf{x}$$

5-3-13 The approximate problem is equivalent to $\hat{A}w_j = \hat{f}_j$ for each j, where $\hat{A} = A + (1/\Delta t)$, $\hat{f}_j = f + (w_{j-1}/\Delta t)$. Since A is positive-definite, $\hat{A}$ is positive definite for each j.

5-3-15 See the comments of Prob. 5-3-13.

5-3-18 (*a*) $\lambda_1 = 3, \quad \lambda_2 = 2, \quad \lambda_3 = -1$

$$\hat{\mathbf{n}}^{(1)} = \pm \frac{1}{\sqrt{2}} (1, 0, 1), \quad \hat{\mathbf{n}}^{(2)} = \pm \frac{1}{\sqrt{3}} (-1, 1, 1), \quad \hat{\mathbf{n}}^{(3)} = \pm \frac{1}{\sqrt{6}} (-1, -2, 1)$$

Chapter Six

Problems 6-1

6-1-1 (a) $\mathscr{D}_A = H^2(0, 1) \cap H_0^1(0, 1)$, $H_A = H_0^1(0, 1)$; $\phi_i = x^i(1 - x)$

b_{ij} is defined by Eq. (6-43a) and

$$l(\phi_i) = \frac{3}{i+2} - \frac{2}{i+3} - \frac{1}{i+1} \quad [l(\phi_1) = 0, \quad l(\phi_2) = \tfrac{1}{60}]$$

$$u_2(x) = -\frac{x}{6} + \frac{x^2}{2} - \frac{x^3}{3} \quad (c_1 = -\tfrac{1}{6}, c_2 = \tfrac{1}{3})$$

which coincides with the exact solution. The error is zero.

6-1-2 $u_N = \sum_{i=1}^{N} c_i \phi_i$, $\phi_i = x^i \in H_A(\phi_i \notin \mathscr{D}_A)$

$$b_{ij} = \frac{ij}{i+j-1}, \quad l(\phi_i) = (-1)^i \left[\log 2 + \sum_{j=1}^{i} \frac{1}{j}(-1)^j \right]$$

$$u_1 = 0.3469x$$

$$u_2 = 0.648x - 0.3411x^2$$

$$u_3 = 0.6869x - 0.4579x^2 + 0.07784x^3$$

$$u_4 = 0.6922x - 0.49x^2 + 0.1314x^3 - 0.02678x^4$$

$$u_5 = 0.6931x - 0.4989x^2 + 0.1579x^3 - 0.0577x^4 + 0.01236x^5$$

$$\vdots$$

$$u_8 = 0.6967x - 0.5606x^2 + 0.4825x^3 - 0.7141x^4 - 0.3038x^5 + 0.7256x^6 - 0.9801x^7 + 0.3531x^8$$

6-1-3 Use $\phi_i = x^{i-1}(1 - x^2)$

$$u_2(x) = \tfrac{15}{56}(1 - x^2) - \tfrac{5}{56}(x^2 - x^3)$$

The exact solution is given by $u(x) = -(1 + x) + 2 \log (2 + x)/\log 3$.

6-1-4 $m_{11} = \frac{8}{9}$, $m_{12} = 0$, $m_{22} = \frac{16}{105}$; $\lambda_1 = 5.424$, $\lambda_2 = 20.58$

6-1-5
$$b_{ij} = \int_a^b \left(p \frac{d\phi_i}{dx} \frac{d\phi_j}{dx} + q\phi_i \phi_j \right) dx$$

$$l(\phi_i) = \int_a^b \left(f\phi_i - p \frac{d\phi_0}{dx} \frac{d\phi_i}{dx} - q\phi_0 \phi_i \right) dx$$

where $u_N = \sum_{j=1}^{N} c_j \phi_j + \phi_0$.

(a) $\phi_i = \sin i\pi x$, $\phi_0 = 0$

$$b_{ii} = \tfrac{1}{2}[(i\pi)^2 - 1], \quad l(\phi_i) = -\frac{(-1)^i}{i\pi}$$

$$c_i = \frac{2(-1)^{i+1}}{i\pi[(i\pi)^2 - 1]}$$

(*b*) $\phi_i = x^i(1 - x), \quad \phi_0 = 0$

$$b_{ij} = \frac{ij}{i + j - 1} - \frac{2ij + i + j}{i + j} + \frac{ij + i + j}{i + j + 1} - \frac{1}{i + j + 3} + \frac{2}{i + j + 2}$$

$$l(\phi_i) = \frac{1}{(i + 2)(i + 3)}$$

$N = 2$: $c_1 = \frac{71}{369}$ and $c_2 = \frac{7}{41}$

6-1-6 Let $u = c_1(x - 1)(2 - x)$ and obtain $c_1 = 0.811$.

6-1-7 $\phi_1(r) = (r - 20)(r - 50), \quad \phi_0 = \frac{1}{3}(500 - 10r)$

$$b_{11} = \int_{20}^{50} r\left(\frac{d\phi_1}{dr}\right)^2 dr, \quad l(\phi_1) = -\int_{20}^{50} r \frac{d\phi_0}{dr}\frac{d\phi_1}{dr}\, dr$$

$c_1 = -7/300$

6-1-8 $N = 1$: Let $u = c_1 \cos(\pi r/2a)$ and $(\phi_1, \phi_1) = \int_0^a \phi_1^2 r\, dr$, and obtain

$$\lambda_1 = \frac{\pi^2}{4a^2}\left(\frac{\pi^2 + 4}{\pi^2 - 4}\right) \approx \frac{5.832}{a^2}$$

$N = 2$: Let $u = c_1 \cos(\pi r/2a) + c_2 \cos(3\pi r/2a)$ and obtain the equations,

$$\begin{bmatrix} 1.7337 - 0.29736\lambda a^2 & 0.20264\lambda a^2 - 1.5 \\ 0.20264\lambda a^2 - 1.5 & 11.603 - 0.47748\lambda a^2 \end{bmatrix}\begin{Bmatrix} c_1 \\ c_2 \end{Bmatrix} = \begin{Bmatrix} 0 \\ 0 \end{Bmatrix}$$

The characteristic polynomial is $0.10092\lambda^2 a^4 - 3.6701\lambda a^2 + 17.866 = 0$. The smallest root is $\lambda_1 = 5.792/a^2$. The exact value is $\lambda = 5.779/a^2$.

6-1-10 The equation can be written as

$$-\frac{d}{dx}\left[(1 + x)\frac{du}{dx}\right] = x$$

For $\phi_i = x^i(1 - x)$, we have

$$b_{ij} = \frac{ij}{(i + j - 1)} - \frac{(i + j + ij)}{(i + j)} + \frac{(1 - ij)}{(i + j + 1)} + \frac{(i + 1)(j + 1)}{(i + j + 2)}$$

$$l(\phi_i) = \frac{1}{(i + 2)(i + 3)}$$

For $N = 2$ we obtain $c_1 = \frac{19}{131}$ and $c_2 = \frac{5}{131}$.

6-1-11 Let $\phi_i = x^i$. The bilinear forms are given by

$$b_{ij} = \int_0^1 \frac{d\phi_i}{dx}\frac{d\phi_j}{dx}\, dx + 2\phi_i(1)\phi_j(1) = \frac{ij}{i + j - 1} + 2$$

$$l(\phi_i) = \int_0^1 x^2\phi_i\, dx + \phi_i(1) = \frac{1}{i + 3} + 1$$

$N = 2$: $c_1 = \frac{17}{30}$, $c_2 = -\frac{3}{20}$; $N = 3$: $c_1 = 0.4833$, $c_2 = 0.1$, $c_3 = -0.1667$; $N = 4$: $c_1 = 0.5$, $c_2 = 0.0001$, $c_3 = -0.0002$, $c_4 = -0.0833$.

The exact solution is $u(x) = \frac{1}{12}(6x - x^4)$.

6-1-12 $\lambda_n = (2n - 1)^2(\pi^2/4) + 2\alpha$

6-1-14 $$b_{ij} = \int_0^L EI \frac{d^2\phi_i}{dx^2}\frac{d^2\phi_j}{dx^2}\,dx,$$

$$m_{ij} = \int_0^L \phi_i \phi_j \, dx,$$

$$w_2(x) = c_1\phi_1 + c_2\phi_2, \quad \phi_i = x^{i+1}\left(1 - \frac{x}{L}\right)$$

This choice gives

$$4EIL\begin{bmatrix} 1 & L \\ L & \frac{6}{5}L^2 \end{bmatrix} - \lambda L^5 \begin{bmatrix} \frac{1}{105} & \frac{L}{168} \\ \frac{L}{168} & \frac{L^2}{252} \end{bmatrix} \begin{Bmatrix} c_1 \\ c_2 \end{Bmatrix} = \begin{Bmatrix} 0 \\ 0 \end{Bmatrix}$$

The characteristic polynomial is given by $52.5\bar{\lambda}^2 - 77616\bar{\lambda} + 4445280 = 0$, $\bar{\lambda} = \lambda L^4/4EI$. The roots are given by $\bar{\lambda}_1 = 59.682$ and $\bar{\lambda}_2 = 1418.718$.

6-1-15 $$\sum_{m,n=0}^{\infty} B(\phi_{ij}, \phi_{mn})c_{mn} = l(\phi_{ij})$$

$$B(\phi_{ij}, \phi_{mn}) = \int_0^1 \int_0^1 \left(\frac{\partial \phi_{ij}}{\partial x}\frac{\partial \phi_{mn}}{\partial x} + \frac{\partial \phi_{ij}}{\partial y}\frac{\partial \phi_{mn}}{\partial y}\right) dx\,dy$$

$$= \frac{\pi^2}{4}(in + jm)\delta_{in}\delta_{jm}$$

$$l(\phi_{ij}) = \int_0^1 \int_0^1 \phi_{ij}\,dx\,dy$$

$$= \begin{cases} \dfrac{1}{\pi}\dfrac{2i}{i^2 - j^2} & \text{if } i - j \text{ is odd} \\ 0 & \text{if } i - j \text{ is even} \end{cases}$$

where $\phi_{ij} = \sin i\pi x \cos j\pi y$. Hence

$$c_{ij} = \begin{cases} \dfrac{8i}{\pi^3}\dfrac{1}{(i - j)(i^2 + j^2)} & \text{if } i - j \text{ is odd} \\ 0 & \text{if } i - j \text{ is even} \end{cases}$$

6-1-16 $u_2(x, y) = c_1\phi_1(x, y) + c_2\phi_2(x, y) + \phi_0(x, y)$

Use $\phi_0 = x^2 + y^2$, $\phi_1 = xy$, $\phi_2 = xy(x + y)$,

$$b_{ij} = \int_0^1 \int_0^1 \left(\frac{\partial \phi_i}{\partial x}\frac{\partial \phi_j}{\partial x} + \frac{\partial \phi_i}{\partial y}\frac{\partial \phi_j}{\partial y}\right) dx\,dy$$

$$l(\phi_i) = \int_0^1 \int_0^1 \left\{[2(x + y) - 4]\phi_i - \left(\frac{\partial \phi_i}{\partial x}\frac{\partial \phi_0}{\partial x} + \frac{\partial \phi_i}{\partial y}\frac{\partial \phi_0}{\partial y}\right)\right\} dx\,dy$$

$$+ \int_0^1 \phi_i(2 - 2y - y^2)\,dy + \int_0^1 \phi_i(2 - 2x - x^2)\,dx$$

We obtain

$$\begin{bmatrix} \frac{2}{3} & \frac{7}{6} \\ \frac{7}{6} & \frac{103}{45} \end{bmatrix} \begin{Bmatrix} c_1 \\ c_2 \end{Bmatrix} = \begin{Bmatrix} -\frac{1}{3} - 1 + \frac{1}{6} \\ -\frac{7}{18} - 2 + \frac{1}{10} \end{Bmatrix}$$

whose solution is given by $c_1 = 0, c_2 = -1.0$.

6-1-18 $\lambda_{ij} = \pi^2(i^2 + j^2)[\lambda_{10} = 9.8696, \lambda_{11} = 19.7392]$

6-1-20 We obtain the equations

$$\begin{bmatrix} 1152 & \frac{8064}{3773}a^2 & \frac{8064}{3773}a^2 \\ \frac{8064}{3773}a^2 & \frac{32128}{7007}a^2 & \frac{128}{539}a^2 \\ \frac{8064}{3773}a^2 & \frac{128}{539}a^2 & \frac{32128}{7007}a^2 \end{bmatrix} \begin{Bmatrix} c_1 \\ c_2 \\ c_3 \end{Bmatrix} = \begin{Bmatrix} \frac{f_0}{2a^4} \\ \frac{f_0}{14a^4} \\ \frac{f_0}{14a^4} \end{Bmatrix}$$

whose solution is given by $c_1 = 0.0202 f_0/a^4, \quad c_2 = c_3 = 0.00587 f_0/a^6$.

Problems 6-2

6-2-1 For $a = 2$ and $b = 1$, we have

$$A\phi_1 = 2304r^4 - 5760r^2 + 2112$$

$$A\phi_2 = 6400r^6 - 23040r^4 + 19008r^2 - 2560$$

Using the inner product

$$(\phi_1, \phi_2)_0 = \int_1^2 \int_0^{2\pi} \phi_1 \phi_2 r \, dr \, d\theta.$$

we obtain

$$(\phi_1, A\phi_1)_0 = 72{,}588.95, \quad (\phi_1, A\phi_2)_0 = (\phi_2, A\phi_1)_0 = 228{,}434.71$$

$$(\phi_2, A\phi_2)_0 = 805{,}459.24, \quad (\phi_1, 1)_0 = 25.45, \quad (\phi_2, 1)_0 = 63.62$$

The solution becomes

$$u_2(r) = 10^{-4}(9.49 - 1.9r^2)(r^2 - 1)^2(4 - r^2)^2$$

6-2-2 $\phi_1 = \cos \frac{\pi r}{2a}, \quad A\phi_1 = \frac{\pi}{2a}\frac{1}{r}\sin\frac{\pi r}{2a} + \left(\frac{\pi}{2a}\right)^2 \cos\frac{\pi r}{2a}$

$$(A\phi_1, \phi_1)_0 = \int_0^a \int_0^{2\pi} A\phi_1 \cos\frac{\pi r}{2a}\, dr = 2\pi\left(\frac{\pi^2}{16} + \frac{1}{4}\right)$$

$$(\phi_1, \phi_1)_0 = 2\pi\left(\frac{a^2}{4} - \frac{a^2}{\pi^2}\right)$$

Hence $\lambda_1 = (\pi^2/4a^2)[(\pi^2 + 4)/(\pi^2 - 4)]$ (same as the Ritz solution).

6-2-3 Same as the Ritz solution.

6-2-4 $\phi_i = x^i(1 - x)$

$$b_{ij} = \left[\frac{ij(i-1)(j-1)}{i+j-3}\right] - \frac{ij(j-1)}{i+j-2} - \frac{ij(i-1)}{i+j-2} - \frac{(i+1)^2(j-1)j - ij + (j+1)^2(i-1)i}{i+j-1}$$

$$+ \frac{i(j+1)^2 + j(i+1)^2}{i+j} + \frac{(i+1)^2(j+1)^2}{i+j+1}$$

The bracketed term is not to be used for $i = 1$ or $j = 1$.

$$l(\phi_i) = \frac{i}{i+1} - (i-1) + \frac{(i+1)^2}{i+2}$$

For $N = 2$, the parameters are $c_1 = 0.1431$, $c_2 = 0.03672$.

6-2-5 Using collocation points $x^{(1)} = \frac{1}{3}$ and $x^{(2)} = \frac{2}{3}$ in the residual $R = c_1(1 + 4x) + c_2(-2 + 2x + 9x^2) - x$, we obtain

$$\frac{1}{3}\begin{bmatrix} 7 & -1 \\ 11 & 10 \end{bmatrix}\begin{Bmatrix} c_1 \\ c_2 \end{Bmatrix} = \frac{1}{3}\begin{Bmatrix} 1 \\ 2 \end{Bmatrix} \qquad c_1 = \tfrac{4}{27} \qquad \text{and} \qquad c_2 = \tfrac{1}{27}$$

6-2-6 For weight functions $\psi_1 = 1$ and $\psi_2 = x$, the equations are

$$\begin{bmatrix} 3 & 2 \\ \frac{11}{6} & \frac{23}{12} \end{bmatrix}\begin{Bmatrix} c_1 \\ c_2 \end{Bmatrix} = \begin{Bmatrix} \frac{1}{2} \\ \frac{1}{3} \end{Bmatrix}$$

and the parameters are $c_1 = \frac{7}{50}$ and $c_2 = \frac{1}{25}$.

6-2-7 We must choose ϕ_i from $\mathscr{D}_A$. We select $\phi_1 = 4x - 3x^2$, $\phi_2 = 5x - 3x^3$, $\phi_0 = x/3$. This gives

$$\begin{bmatrix} 6 & \frac{21}{2} \\ \frac{21}{2} & \frac{96}{5} \end{bmatrix}\begin{Bmatrix} c_1 \\ c_2 \end{Bmatrix} = \begin{Bmatrix} \frac{2}{5} \\ \frac{3}{4} \end{Bmatrix} \qquad c_1 = -\tfrac{13}{330} \qquad \text{and} \qquad c_2 = \tfrac{2}{33}$$

The solution becomes $u_2(x) = 0.5091x + 0.1182x^2 - 0.1818x^3$.

6-2-8 $c_1 = -\frac{1}{36}$, $\quad c_2 = \frac{1}{18}$

6-2-9 Collocation at $x = \frac{1}{3}$ and $\frac{2}{3}$ gives $c_1 = -\frac{1}{27}$ and $c_2 = \frac{1}{18}$.

6-2-10 Using the subdomains $(0, \frac{1}{2})$ and $(\frac{1}{2}, 1)$, we obtain $c_1 = -\frac{1}{36}$ and $c_2 = \frac{1}{18}$.

6-2-11 (*a*) Collocation at $x = \frac{1}{3}$ and $\frac{2}{3}$ gives $c_1 = \frac{81}{416}$ and $c_2 = \frac{9}{52}$.
(*b*) Subdomains $(0, \frac{1}{2})$ and $(0, 1)$ give $c_1 = \frac{97}{517}$ and $c_2 = \frac{88}{517}$.

6-2-13 (*a*) $u(x) = c_1(3x - 2x^2) + c_2(x + 2x^2 - 2x^3)$ gives $\lambda_1 = 4.121$ (and $\lambda_2 = 25.48$). The exact solution is given by the roots of the transcendental equation $\lambda + \tan \lambda = 0$ ($\lambda_1 = 4.116$).

6-2-15 For $u = c_1(1 - x^2)$ we obtain $c_1 = \frac{3}{16}$.

6-2-17 For $u = c_1\left(1 - \dfrac{x^2}{a^2} - \dfrac{y^2}{b^2}\right)$ we obtain $c_1 = \dfrac{a^2 + b^2}{2(a^2 + b^2)}$.

6-2-18 $c_1 = \dfrac{1295}{1416a^2}$ and $c_2 = \dfrac{525}{4432a^4}$

The moment value is $M = 0.1404(2a)^4 G\theta$.

6-2-20 The Bubnov–Galerkin solution coincides with the exact solution.

6-2-24 See Reddy and Rasmussen (1982), pp. 452 and 475. Use $\phi_0 = 1 - x$ and $\phi_i = \sin (2i - 1)\pi x$.

Problems 6-3

6-3-1 $$u(x, y) = \frac{32a^2}{\pi^3}\left\{\left[1 - \frac{\cosh (\pi y/2a)}{\cosh (\pi/2a)}\right]\cos\frac{\pi x}{2a} + \frac{1}{27}\left[1 - \frac{\cosh (3\pi y/2a)}{\cosh (3\pi/2a)}\right]\cos\frac{3\pi x}{2a}\right\}$$

6-3-2 See Reddy (1984b), Example 4-8 on pp. 343–346.

6-3-3 Use $u(x, y) = c_1(x) \cos(\pi y/2b)$ and obtain

$$c_1'' + \left[\lambda - \left(\frac{\pi}{2b}\right)^2\right]c_1 = 0$$

$$c_1(a) = c_1(-a) = 0$$

The solution of these equations requires λ be given by $\sqrt{\lambda - (\pi/2b)^2}\, a = (2n - 1)(\pi/2)$, $n = 1, 2, \ldots$. For $n = 1$, we find $\lambda_1 = (\pi/2a)^2 + (\pi/2b)^2$, which coincides with the exact solution.

6-3-5 $u(x, y) = x(1 - x)e^{-3y}$

6-3-8 $c_1'' + \left(\lambda - \dfrac{153}{62b^2}\right)c_1 = 0$

Problems 6-4

6-4-1 $u_3(x, t) = 4(1 - e^{-t}) \sin x + \dfrac{4}{27 \times 9}(1 - e^{-9t}) \sin 3x$

$$\left[c_1(t) = 4(1 - e^{-t}), \quad c_2 = 0, \quad c_3 = \frac{4}{27 \times 9}(1 - e^{-9t})\right]$$

6-4-2 The differential equations in time are given by

$$\begin{bmatrix} 1 & 0 & 0 \\ 0 & 1 & 0 \\ 0 & 0 & 1 \end{bmatrix} \begin{Bmatrix} \dot{c}_1 \\ \dot{c}_2 \\ \dot{c}_3 \end{Bmatrix} + \begin{bmatrix} 1 & 0 & 0 \\ 0 & 4 & 0 \\ 0 & 0 & 9 \end{bmatrix} \begin{Bmatrix} c_1 \\ c_2 \\ c_3 \end{Bmatrix} = \begin{Bmatrix} 4 \\ 0 \\ \frac{4}{27} \end{Bmatrix}$$

Use the Crank–Nicholson method to solve them.

6-4-3 The variational formulation is given by

$$0 = \int_0^1 \left(v\frac{\partial u}{\partial t} + \frac{\partial v}{\partial x}\frac{\partial u}{\partial x}\right) dx + v(0)u(0, t)$$

The Ritz approximation of the form, $u(x, t) = c_1(t) + c_2(t)x$ gives the equations

$$\begin{bmatrix} 1 & \frac{1}{2} \\ \frac{1}{2} & \frac{1}{3} \end{bmatrix} \begin{Bmatrix} \dot{c}_1 \\ \dot{c}_2 \end{Bmatrix} + \begin{bmatrix} 1 & 0 \\ 0 & 1 \end{bmatrix} \begin{Bmatrix} c_1 \\ c_2 \end{Bmatrix} = \begin{Bmatrix} 0 \\ 0 \end{Bmatrix}$$

The Galerkin approximation of the initial condition $u(x, 0) = 1$ gives $c_1(0) = 1$ and $c_2(0) = 0$. Use the Laplace transform to obtain the solution to the above ordinary differential equations.

6-4-4 (*a*) $u(x, t) = x(1 - x) \cos \sqrt{10}\, t$
(*b*) $u(x, t) = x(1 - x) \cos \sqrt{8}\, t$

6-4-5 Use $u(x, t) = c_1(t)x(1 - x) + c_2(t)x^2(1 - x)$

6-4-11 For $u(x, t) = c_1(t) \sin 2x$, we obtain the exact solution.

6-4-12 For $u(x, t) = c_1(t) \sin x$, we obtain the exact solution.

6-4-13 For $u(x, t) = c_1(t) \sin x$, we obtain the exact solution, $u(x, t) = (1 - \cos t) \sin x$.

6-4-15 $-z_j'' + \dfrac{1}{\Delta t}(z_j - z_{j-1}) = \sin nx$

$$z_j(0) = 0, \quad z_j(\pi) = 0 \quad \text{with} \quad z_0(x) = 0$$

The exact solution of this equation is

$$z_j(x) = \frac{1}{n^2}\left[1 - \frac{1}{(1 + n^2\,\Delta t)^j}\right]\sin nx$$

6-4-16 $-z_j'' + \dfrac{z_j - z_{j-1}}{\Delta t} = 0$

$$z_j(0) = 0, \quad z_j(\pi) = 0 \quad \text{with} \quad z_0(x) = \sin x$$

For $j = 1$, the equation can be solved exactly: $z_1 = 1/(1 + \Delta t)\sin x$. Then for $j = 2$, we have $z_2 = 1/(1 + \Delta t)\sin x$. Proceeding in this manner, we obtain $z_j = 1/(1 + \Delta t)^j \sin x$.

6-4-17 The discretization in time leads to the equation, $Az_j + (z_j - z_{j-1})/\Delta t = f$ in Ω, $z_j = 0$ on Γ, where A is the spatial operator defined by the governing equation and $f = 1 - x^2/16 - y^2/4$. The above equation can be solved using the Ritz method. See Rektorys (1982), pp. 107–117.

6-4-18 See Rektorys (1982), pp. 161–169.

Chapter Seven

Problems 7-1

7-1-1 $0 = \displaystyle\int_{r_e}^{r_{e+1}} -\psi_i^e\left[\frac{d}{dr}\left(rk_e \sum_{j=1}^{n} u_j^e \frac{d\psi_j^e}{dr}\right) - f_e\right] r\,dr$

$$= \sum_{j=1}^{n} K_{ij}^e u_j^e - F_i^e$$

where

$$K_{ij}^e = -\int_{r_e}^{r_{e+1}} r\psi_i^e \frac{d}{dr}\left(rk_e \frac{d\psi_j^e}{dr}\right) dr, \qquad F_i^e = \int_{r_e}^{r_{e+1}} rf_e \psi_i^e \, dr$$

Hermite cubic polynomials must be used for ψ_j^e.

7-1-2 (*a*) Solve the eigenvalue problem,

$$\frac{4}{L}\begin{bmatrix} 2 & -1 & 0 \\ -1 & 2 & -1 \\ 0 & -1 & 2 \end{bmatrix}\begin{Bmatrix} U_2 \\ U_3 \\ U_4 \end{Bmatrix} = \frac{\lambda L}{24}\begin{bmatrix} 4 & 1 & 0 \\ 1 & 4 & 1 \\ 0 & 1 & 4 \end{bmatrix}\begin{Bmatrix} U_2 \\ U_3 \\ U_4 \end{Bmatrix}$$

to obtain

$$\lambda_1 = \frac{10.387}{L^2}, \qquad \lambda_2 = \frac{48}{L^2}, \qquad \lambda_3 = \frac{126.72}{L^2}$$

(*b*) Solve the equations

$$\frac{2}{3L}\begin{bmatrix} 16 & -8 & 0 \\ -8 & 14 & -8 \\ 0 & -8 & 16 \end{bmatrix}\begin{Bmatrix} U_2 \\ U_3 \\ U_4 \end{Bmatrix} = \frac{\lambda L}{60}\begin{bmatrix} 16 & 2 & 0 \\ 2 & 8 & 2 \\ 0 & 2 & 16 \end{bmatrix}\begin{Bmatrix} U_2 \\ U_3 \\ U_4 \end{Bmatrix}$$

to obtain

$$\lambda_1 = \frac{9.944}{L^2}, \qquad \lambda_2 = \frac{40}{L^2}, \qquad \lambda_3 = \frac{128.72}{L^2}$$

The exact solution is given by (see Example 6-3), $\lambda_n = (n\pi/L)^2$.

7-1-3

$$2\begin{bmatrix} 1 & -1 & 0 \\ -1 & 2 & -1 \\ 0 & -1 & 1 \end{bmatrix}\begin{Bmatrix} U_1 \\ U_2 \\ U_3 \end{Bmatrix} = \frac{\lambda}{12}\begin{bmatrix} 2 & 1 & 0 \\ 1 & 4 & 1 \\ 0 & 1 & 2 \end{bmatrix}\begin{Bmatrix} U_1 \\ U_2 \\ U_3 \end{Bmatrix} + \begin{Bmatrix} P_1^1 \\ P_2^1 + P_1^2 \\ P_2^2 \end{Bmatrix}$$

use $U_1 = 0$, $P_2^1 + P_1^2 = 0$, and $P_2^2 \equiv (du/dx)|_{x=1} = -U_3$ and solve the eigenvalue problem to obtain $\lambda_1 = 7.603$ and $\lambda_2 = 54.12$.

7-1-4 The element matrices and column vectors are given by

$$[K^1] = \begin{bmatrix} 3.0082 & -2.9928 \\ -2.9928 & 3.0206 \end{bmatrix} \qquad \{f^1\} = \begin{Bmatrix} 0.0551 \\ 0.1041 \end{Bmatrix}$$

$$[K^2] = \begin{bmatrix} 3.0267 & -2.9866 \\ -2.9866 & 3.0267 \end{bmatrix} \qquad \{f^2\} = \begin{Bmatrix} 0.1592 \\ 0.1592 \end{Bmatrix}$$

$$[K^3] = \begin{bmatrix} 3.0206 & -2.9928 \\ -2.9928 & 3.0206 \end{bmatrix} \qquad \{f^3\} = \begin{Bmatrix} 0.1041 \\ 0.0551 \end{Bmatrix}$$

The solution is given by $U_2 = U_3 = 0.0860$.

7-1-5 Assume (the nodes are equally spaced)

$$\psi_1^e(\xi) = c_1\left(1 - \frac{\xi}{h_e}\right)\left(1 - \frac{3\xi}{2h_e}\right)\left(1 - \frac{3\xi}{h_e}\right)$$

(so that it vanishes at nodes 2, 3 and 4) and evaluate c_1 by requiring $\psi_1^e(0) = 1$, which gives $c_1 = 1$. Here h_e denotes the length of the element and ξ is the local coordinate with the origin at node 1. Similarly,

$$\psi_2^e(\xi) = \frac{9\xi}{h_e}\left(1 - \frac{\xi}{h_e}\right)\left(1 - \frac{3\xi}{2h_e}\right)$$

$$\psi_3^e(\xi) = -\frac{9}{2}\frac{\xi}{h_e}\left(1 - \frac{\xi}{h_e}\right)\left(1 - \frac{3\xi}{h_e}\right)$$

$$\psi_4^e(\xi) = \frac{\xi}{h_e}\left(1 - \frac{3}{2}\frac{\xi}{h_e}\right)\left(1 - \frac{3\xi}{h_e}\right)$$

7-1-6 (*a*)

$$\left(\frac{1}{2}\begin{bmatrix} 0 & 1 & 0 \\ -1 & 0 & 1 \\ 0 & -1 & 1 \end{bmatrix} + \frac{1}{9}\begin{bmatrix} 4 & 1 & 0 \\ 1 & 4 & 1 \\ 0 & 1 & 2 \end{bmatrix}\right)\begin{Bmatrix} U_2 \\ U_3 \\ U_4 \end{Bmatrix} = \frac{1}{6}\begin{Bmatrix} 2 \\ 2 \\ 1 \end{Bmatrix} + \begin{Bmatrix} \frac{1}{2} - \frac{1}{9} \\ 0 \\ 0 \end{Bmatrix}$$

7-1-7 The element matrices are

$$[K^1] = \begin{bmatrix} 1.5288 & -1.4794 \\ -1.4794 & 1.5535 \end{bmatrix} \qquad [K^2] = \begin{bmatrix} 1.5288 & -1.5041 \\ -1.5041 & 1.4547 \end{bmatrix}$$

$$[K^3] = \begin{bmatrix} 1.3313 & -1.6276 \\ -1.6276 & 1.1584 \end{bmatrix} \qquad \{f^e\} = \frac{1}{3}\begin{Bmatrix} 1 \\ 1 \end{Bmatrix}$$

The solution is given by $U_2 = 0.4522$ and $U_3 = 0.4834$.

7-1-9
7-1-10 } See Reddy (1984a), p. 130 and pp. 450–451.
7-1-12

7-1-14 Use the inner product $(u, v) = \int_{r_e}^{r_{e+1}} uvr\, dr$ to formulate the equation variationally. The element coefficient matrix and column vector are

$$K_{ij}^e = \int_{r_e}^{r_{e+1}} rT(\psi_i^e)T(\psi_j^e)\, dr \qquad F_i^e = \int_{r_e}^{r_{e+1}} rf(r)\, dr + P_i^e$$

where $T(w) = (d^2w/dr^2 + (1/r)\, dw/dr)$.

7-1-15

	Two Elements		Four Elements		
t	$\Delta t = 0.1$	$\Delta t = 0.05$	$\Delta t = 0.1$	$\Delta t = 0.05$	Exact
0.1	0.8854	0.8855	0.9000	0.9001	0.9048
0.2	0.7839	0.7841	0.8100	0.8102	0.8187
0.3	0.6941	0.6943	0.7291	0.7292	0.7408
0.4	0.6145	0.6148	0.6562	0.6564	0.6703
0.5	0.5441	0.5444	0.5906	0.5908	0.6065
0.6	0.4817	0.4820	0.5315	0.5318	0.5488
0.7	0.4265	0.4268	0.4784	0.4786	0.4966
0.8	0.3776	0.3780	0.4306	0.4308	0.4493
0.9	0.3343	0.3347	0.3875	0.3878	0.4066
1.0	0.2960	0.2964	0.3488	0.3490	0.3679

7-1-16

	Two Elements		Four Elements		
t	$\Delta t = 0.1$	$\Delta t = 0.05$	$\Delta t = 0.1$	$\Delta t = 0.05$	Exact*
0.1	0.8854	0.8855	1.0413	1.0318	0.9991
0.2	0.7839	0.7841	0.9680	0.9635	0.9740
0.3	0.6941	0.6943	0.8779	0.8764	0.9149
0.4	0.6145	0.6148	0.7916	0.7913	0.8419
0.5	0.5441	0.5444	0.7128	0.7129	0.7675
0.6	0.4817	0.4820	0.6416	0.6418	0.6969
0.7	0.4265	0.4268	0.5775	0.5777	0.6315
0.8	0.3771	0.3780	0.5197	0.5200	0.5718
0.9	0.3343	0.3347	0.4678	0.4681	0.5175
1.0	0.2960	0.2964	0.4210	0.4213	0.4683

* $n = 1, 3, \ldots, 9$

7-1-18 Linear: $\psi_1 = \frac{1}{2}(1 - \xi)$, $\psi_2 = \frac{1}{2}(1 + \xi)$

Quadratic: $\psi_1 = -\frac{1}{2}\xi(1 - \xi)$, $\psi_2 = (1 - \xi^2)$, $\psi_3 = \frac{1}{2}\xi(1 + \xi)$

Cubic: $\psi_1 = -\frac{1}{16}(1 - \xi)(1 - 9\xi^3)$

$$\psi_2 = \frac{9}{16}(1 - \xi^2)(1 - 3\xi)$$

$$\psi_3 = \frac{9}{16}(1 - \xi^2)(1 + 3\xi)$$

$$\psi_4 = -\frac{1}{16}(1 - 9\xi^2)(1 + \xi)$$

7-1-19 $\psi_1 = \frac{1}{4}(2 - 3\xi + \xi^3)$ $\qquad \psi_2 = -\dfrac{h}{8}(1 - \xi)(1 - \xi^2)$

$\psi_3 = \frac{1}{4}(2 + 3\xi - \xi^3)$ $\qquad \psi_4 = \dfrac{h}{8}(1 + \xi)(1 - \xi^2)$

7-1-20 Let $\Delta t_e = t_{e+1} - t_e$. The substituting $v = \psi_i^e$ and $u = \sum_{j=1}^{2} u_j^e \psi_j^e$ in $0 = \int_{t_e}^{t_{e+1}} [a(du/dt) + bu - f]\, dt$ gives the finite-element equations

$$\left(\frac{a}{2}\begin{bmatrix} -1 & 1 \\ -1 & 1 \end{bmatrix} + \frac{b\,\Delta t_e}{6}\begin{bmatrix} 2 & 1 \\ 1 & 2 \end{bmatrix}\right)\begin{Bmatrix} u_1^e \\ u_2^e \end{Bmatrix} = \frac{f\,\Delta t_e}{2}\begin{Bmatrix} 1 \\ 1 \end{Bmatrix}$$

for constant a, b, and f. Using the notation $u_1^e = u^n$ and $u_2^e = u^{n+1}$ (and u^n is known), we can write the second equation as

$$\left(\frac{a}{2} + \frac{b}{3}\Delta t_n\right)u^{n+1} = \left(\frac{a}{2} - \frac{b}{6}\Delta t_n\right)u^n + \frac{f}{2}\Delta t_n$$

or

$$(a + \tfrac{2}{3}b\,\Delta t_n)u^{n+1} = \left(a - \frac{b}{3}\Delta t_n\right)u^n + f\,\Delta t_n \tag{i}$$

If the θ-family of approximation is applied to the equation $a(du/dt) + bu = f$, we obtain

$$(a + \theta b\,\Delta t_{n+1})u^{n+1} = [a - (1-\theta)b\,\Delta t_n]u^n + f\,\Delta t_n \tag{ii}$$

Comparing (i) with (ii), we note that for $\theta = \frac{2}{3}$ (ii) gives (i).

7-1-22 (*b*) For four elements, the solution is given by $u(0.25) = 0.3128$, $u(0.5) = 0.61075$, $u(1.0) = 1.1439$. The corresponding values of the exact solution are 0.31304, 0.61116, and 1.14422, respectively.

7-1-23 and **7-1-24** See Problems 7-1-15 and 7-1-16.

Problems 7-2

7-2-1 $$K_{ij}^e = \int_{\Omega^e}\left(\frac{\partial\psi_i^e}{\partial x}\frac{\partial\psi_j^e}{\partial x} + \frac{\partial\psi_i^e}{\partial y}\frac{\partial\psi_j^e}{\partial y} + c_0\psi_i^e\psi_j^e\right)dx\,dy + \oint_{\Gamma^e}\beta_e\psi_i^e\psi_j^e\,ds$$

$$F_i^e = \int_{\Omega^e}\psi_i^e f_e\,dx\,dy + \oint_{\Gamma^e} g\psi_i^e\,ds$$

7-2-2 $$K_{ij}^e = \int_{\Omega^e}(-\nabla^2\psi_i^e + c_0\psi_i^e)(-\nabla^2\psi_j^e + c_0\psi_j^e)\,dx\,dy + \oint_{\Gamma^e}\left(\frac{\partial\psi_i^e}{\partial n} + \beta_e\psi_i^e\right)\left(\frac{\partial\psi_j^e}{\partial n} + \beta_e\psi_j^e\right)ds$$

$$F_i^e = \int_{\Omega^e}(-\nabla^2\psi_i^e + c_0\psi_i^e)f\,dx\,dy + \oint_{\Gamma^e}\left(\frac{\partial\psi_i^e}{\partial n} + \beta_e\psi_i^e\right)g\,ds$$

7-2-3 $$M_{ij} = \frac{1}{4A_e^2}\int_{\Omega^e}(\alpha_i^e + \beta_i^e x + \gamma_i^e y)(\alpha_j^e + \beta_j^e x + \gamma_j^e y)\,dx\,dy$$

$$= \frac{1}{4A_e^2}[I_{00}\alpha_i^e\alpha_j^e + (\alpha_i^e\beta_j^e + \beta_i^e\alpha_j^e)I_{10}$$

$$+ (\alpha_i^e\gamma_j^e + \gamma_i^e\alpha_j^e)I_{01} + (\beta_i^e\gamma_j^e + \gamma_i^e\beta_j^e)I_{11} + \beta_i^e\beta_j^e I_{20} + \gamma_i^e\gamma_j^e I_{02}]$$

where $I_{mn} = \int_{\Omega^e} x^m y^n\,dx\,dy$. See Eq. (4-37) on page 205 of Reddy (1984a).

7-2-4 $$[M^e] = \frac{ab}{36}\begin{bmatrix} 4 & 2 & 1 & 2 \\ 2 & 4 & 2 & 1 \\ 1 & 2 & 4 & 2 \\ 2 & 1 & 2 & 4 \end{bmatrix}$$

7-2-5 $f_i^e = \begin{cases} \dfrac{A_e}{3} \text{ for a linear triangle} \\ \dfrac{A_e}{4} \text{ for a linear rectangle} \end{cases}$

where A_e denotes the area of the element.

7-2-6 $K_{ij}^e = \displaystyle\int_{\Omega^e} r\left(\frac{\partial \psi_i^e}{\partial r}\frac{\partial \psi_j^e}{\partial r} + \frac{\partial \psi_i^e}{\partial z}\frac{\partial \psi_j^e}{\partial z}\right) dr\, dz$

$$f_i^e = \int_{\Omega^e} rf\psi_i^e \, dr\, dz$$

7-2-7 See pp. 248–252 of Reddy (1984a) for the procedure and Table 7-9 for the functions.

7-2-8 *Triangular element:* $2A = 8.25$

$$\psi_1 = \frac{1}{8.25}(12.25 - 2.5x - 1.5y); \qquad Q_1 = Q_0/3$$

$$\psi_2 = \frac{1}{8.25}(-1.5 + 3x - 1.5y); \qquad Q_2 = 0$$

$$\psi_3 = \frac{1}{8.25}(-2.5 - 0.5x + 3y); \qquad Q_3 = 2Q_0/3$$

Rectangular element: $a = 3.5, \quad b = 2.5, \quad \xi = x - 1, \quad \eta = y - 1.$

$$\psi_1 = \left(1 - \frac{\xi}{3.5}\right)\left(1 - \frac{\eta}{2.5}\right); \qquad Q_1 = Q_0/7$$

$$\psi_2 = \frac{\xi}{3.5}\left(1 - \frac{\eta}{2.5}\right); \qquad Q_2 = 2Q_0/35$$

$$\psi_3 = \frac{\xi}{3.5}\frac{\eta}{2.5}; \qquad Q_3 = 8Q_0/35$$

$$\psi_4 = \left(1 - \frac{\xi}{3.5}\right)\frac{\eta}{2.5}; \qquad Q_4 = 4Q_0/7$$

7-2-9 Evaluate:

$$u(x, y) = \sum_{j=1}^{n} u_j \psi_j^e(x, y) \qquad \frac{\partial u}{\partial x}(x, y) = \sum_{j=1}^{n} u_j \frac{\partial \psi_j}{\partial x}(x, y)$$

$$\frac{\partial u}{\partial y}(x, y) = \sum_{j=1}^{n} u_j \frac{\partial \psi_j}{\partial y}(x, y)$$

at $x = y = 3.75$ for the triangular and rectangular elements.

7-2-13 $x = \frac{5}{2} + 2\xi - \frac{1}{2}\xi\eta,\ y = 2 + \frac{5}{2}\eta + \frac{1}{2}\eta\xi$

$$J = 5 - \tfrac{5}{4}\eta + \xi > 0 \qquad \text{for all } \xi, \text{ in } [-1, 1]$$

7-2-14 $\dfrac{\partial \psi_i}{\partial x} = J_{11}^* \dfrac{\partial \psi_i}{\partial \xi} + J_{12}^* \dfrac{\partial \psi_i}{\partial \eta},$

where J^*_{ij} are the elements of the inverse of the Jacobian:

$$[J^*] = \frac{1}{J}\begin{bmatrix} \frac{1}{2}(5+\xi) & -\frac{1}{2}\eta \\ \frac{1}{2}\xi & \frac{1}{2}(4-\eta) \end{bmatrix}, \qquad J = 5 - \tfrac{5}{4}\eta + \xi$$

$$dx\,dy = J\,d\xi\,d\eta$$

7-2-15 $x = \frac{1}{4}[-10 + 9\xi + 4\eta - \xi\eta + 7\xi^2\eta - \xi^2 - 6\eta^2 - 3\eta^2\xi + 7\eta^3\xi]$

7-2-16 Set $\eta = -1$ in $x = \sum_{i=1}^{8} x_i^e \psi_i^e(\xi, \eta)$ and obtain the required relations. Note also that

$$1 + \xi = 2\sqrt{\frac{x}{a}}, \qquad 1 - \xi = 2\left(1 - \sqrt{\frac{x}{a}}\right), \qquad 1 - \xi^2 = 4\left(\sqrt{\frac{x}{a}} - \frac{x}{a}\right)$$

7-2-18 $\psi_1^e = \left(1 - \frac{x}{a} - \frac{y}{b}\right) \qquad \psi_2^e = \frac{x}{a} \qquad \psi_3^e = \frac{y}{b}$

$$K_{11}^e = \int_{\Omega^e}\left(\frac{1}{a^2} + \frac{1}{b^2}\right)dx\,dy = \frac{ab}{2}\left(\frac{1}{a^2} + \frac{1}{b^2}\right) = \tfrac{1}{2}(\alpha + \beta)$$

7-2-19 $[K^e] = [S^{xe}] + [S^{ye}]$, where $[S^{xe}]$ and $[S^{ye}]$ are given in Eq. (7-118).

7-2-20 $$[K^e] = \frac{1}{2}\begin{bmatrix} \alpha+\beta & \alpha & 0 & -\beta \\ \alpha & \alpha+\beta & -\beta & 0 \\ 0 & -\beta & \alpha+\beta & -\alpha \\ -\beta & 0 & -\alpha & \alpha+\beta \end{bmatrix}$$

7-2-21 In view of the zero specified boundary conditions $U_1 = U_2 = U_3 = U_6 = U_9 = 0$, we need to assemble coefficients corresponding to the nodes 4, 5, 7, and 8. We have

$$\begin{bmatrix} K_{44} & K_{45} \\ K_{54} & K_{55} \end{bmatrix}\begin{Bmatrix} U_4 \\ U_5 \end{Bmatrix} = -\begin{Bmatrix} K_{47}U_7 + K_{48}U_8 \\ K_{57}U_7 + K_{58}U_8 \end{Bmatrix}$$

where

$$K_{44} = \alpha + \beta, \qquad K_{45} = K_{54} = -\alpha, \qquad K_{55} = 2(\alpha + \beta)$$

$$K_{47} = -\frac{\beta}{2}, \qquad K_{48} = 0, \qquad K_{57} = 0, \qquad K_{58} = -\beta$$

and $\alpha = (2b/a)$, and $\beta = (a/2b)$. Solve the equations

$$\begin{bmatrix} \alpha+\beta & -\alpha \\ -\alpha & 2(\alpha+\beta) \end{bmatrix}\begin{Bmatrix} U_4 \\ U_5 \end{Bmatrix} = \frac{\beta}{2}\begin{Bmatrix} 1 \\ 2 \end{Bmatrix}$$

7-2-22 Solve the equations

$$2\begin{bmatrix} 2(\alpha+\beta) & -2\alpha+\beta \\ -2\alpha+\beta & 4(\alpha+\beta) \end{bmatrix}\begin{Bmatrix} U_4 \\ U_5 \end{Bmatrix} = \begin{Bmatrix} -3\beta \\ \alpha - 5\beta \end{Bmatrix}$$

7-2-23 $$\frac{1}{6}\begin{bmatrix} 4 & -1 & -2 & -1 \\ -1 & 4 & -1 & -2 \\ -2 & -1 & 4 & -1 \\ -1 & -2 & -1 & 4 \end{bmatrix}\begin{Bmatrix} U_1 \\ U_2 \\ U_3 \\ U_4 \end{Bmatrix} = \begin{Bmatrix} P_1^1 \\ P_2^1 \\ P_3^1 \\ P_4^1 \end{Bmatrix} \qquad \text{(i)}$$

The specified boundary conditions are,

$$U_1 = U_2 = 1, \tag{ii}$$

$$P_3^1 \equiv \int_0^1 \left[\frac{\partial u}{\partial y}\right]_{y=1} dx = \int_0^1 [\psi_3(2-u)]\bigg|_{y=1} dx$$

$$\equiv 1 - (\tfrac{1}{3}U_3 + \tfrac{1}{6}U_4) \tag{iii}$$

$$P_4^1 = 1 - (\tfrac{1}{6}U_3 + \tfrac{1}{3}U_4) \tag{iv}$$

Substituting Eqs. (ii)–(iv) in Eq. (i), and solving for U_3 and U_4, we obtain $U_3 = U_4 = 1.5$. The exact solution is $u(x, y) = 1 + 0.5y$.

7-2-25 The specified boundary conditions on U's are: $U_1 = 0$, $U_2 = 0.25$, $U_3 = 1$, $U_4 = 0.25$, $U_6 = 0.5$, $U_7 = 1$, $U_8 = 0.5$, $U_9 = 0.0$. We need to solve only the fifth equation of the assembled equations for U_5: $K_{55}U_5 = -\sum_{i=1}^{9} K_{5i}U_i + F_5$, $\frac{16}{6}U_5 = \frac{1}{6}(0.5 + 2 + 0.5 + 1 + 2 + 1) + \frac{1}{2}$ or $U_5 = 0.625$.

7-2-26 $U_1 = 0, \quad U_2 = 0.25, \quad U_4 = 1, \quad U_5 = 0.5, \quad U_6 = 0.0$

Solving for U_3, we obtain $U_3 = 0.5$.

7-2-27 $U_1 = 0, \quad U_2 = 1, \quad U_4 = 1, \quad U_5 = 0.5, \quad U_6 = 0$

Solving for U_3, we obtain $U_3 = 0.675$.

Problems 7-3

7-3-1

$$-\sum_{j=1}^{m} K_{ij}^{11}\sigma_1^j + \sum_{j=1}^{n} K_{ij}^{13}u_j = 0, \qquad i = 1, 2, \ldots, m$$

$$-\sum_{j=1}^{m} K_{ij}^{22}\sigma_2^j + \sum_{j=1}^{n} K_{ij}^{23}u_j = 0, \qquad i = 1, 2, \ldots, m$$

$$\sum_{j=1}^{m} (K_{ji}^{13}\sigma_1^j + K_{ji}^{23}\sigma_2^j) = F_i, \qquad i = 1, 2, \ldots, n$$

where

$$K_{ij}^{11} = \int_{\Omega^e} \frac{1}{k}\psi_i^1\psi_j^1\, d\mathbf{x}, \qquad K_{ij}^{22} = \int_{\Omega^e} \frac{1}{k}\psi_i^1\psi_j^1\, d\mathbf{x}$$

$$K_{ij}^{13} = \int_{\Omega^e} \psi_i^1 \frac{\partial \psi_j^2}{\partial x_1}\, d\mathbf{x}, \qquad K_{ij}^{23} = \int_{\Omega^e} \psi_i^1 \frac{\partial \psi_j^2}{\partial x_2}\, d\mathbf{x}$$

$$F_i = \int_{\Omega^e} \psi_i^2 f\, d\mathbf{x} + \oint_{\Gamma^e} t\psi_i^2\, ds$$

$$t \equiv (\sigma_1 n_1 + \sigma_2 n_2), \qquad \sigma_\alpha = \sum_{j=1}^{m} \sigma_\alpha^j \psi_j^1, \qquad u = \sum_{j=1}^{n} u_j \psi_j^2$$

7-3-2 $u_1 = v_1 = 1, \qquad u_5 = u_9 = u_{13} = 0$

$$F_{4x} = \frac{8p_1 + p_0}{18} \qquad F_{8x} = \frac{2p_1 + p_0}{3} \qquad F_{12x} = \frac{2p_0 + p_1}{3} \qquad F_{16x} = \frac{8p_0 + p_1}{18}$$

7-3-4 $u_1 = v_1 = u_2 = v_2 = u_3 = v_3 = u_{16} = u_{17} = u_{18} = 0$

$$F_{18y} = -\frac{p_0 a}{6} \qquad F_{15y} = -\frac{4p_0 a}{6} \qquad F_{13y} = -\frac{p_0 a}{6}$$

7-3-5
7-3-6 The boundary conditions on the velocities are obvious from the figures.
7-3-7

7-3-8 $$[K^{11}] = \frac{1}{ab}\begin{bmatrix} 1 & -1 & 1 & -1 \\ -1 & 1 & -1 & 1 \\ 1 & -1 & 1 & -1 \\ -1 & 1 & -1 & 1 \end{bmatrix}$$

$$[K^{12}] = [S^{xe}], \qquad [K^{13}] = [S^{ye}], \qquad [K^{22}] = (-\bar{D}^e_{22})[S^e]$$

$$[K^{23}] = \bar{D}^e_{12}[S^e], \qquad [K^{33}] = (-\bar{D}^e_{11})[S^e]$$

where $[S^{xe}]$, $[S^{ye}]$, and $[S^e]$ are defined in Eq. (7-118).

7-3-9 (*a*) $w = 0$, $\partial w/\partial x = 0$, $\partial w/\partial y = 0$ along $x = 0$; $\partial w/\partial y = 0$ along the symmetry line, $y = 0$.
(*b*) $w = 0$ along $x = 0$; $M_x = 0$ along $y = \pm 5$ and $M_y = 0$ at $x = 10$.
(*c*) $w = 0$, $\psi_x = 0$, $\psi_y = 0$ along $x = 0$; $\psi_y = 0$ along the symmetry line, $y = 0$.

7-3-10 (*c*) $w = 0$, $\psi_y = 0$ along $x = a$; $\psi_y = 0$ along $y = 0$ and $\psi_x = 0$ along $x = 0$.

7-3-12 (*a*) $w = 0$, $\partial w/\partial x = 0$, $\partial w/\partial y = 0$ along $x = 0$; $w = 0$, $\partial w/\partial x = 0$ along $y = b$; $\partial w/\partial y = 0$ along $y = 0$.
(*b*) $w = 0$ along $x = 0$; $w = 0$, $M_y = 0$ along $y = b$; $M_x = 0$ along $x = a$.

7-3-14 (*a*) $w = 0$, $\partial w/\partial x = 0$, $\partial w/\partial y = 0$ along $y = 0$ and $y = b$; $w = 0$, $M_n = 0$, $\partial w/\partial s = 0$ along the slant edges.

7-3-15 $$[K^{11}] = \frac{t}{2}\begin{bmatrix} c^e_{11}\alpha + c^e_{66}\beta & -c^e_{11}\alpha & -c^e_{66}\beta \\ -c^e_{11}\alpha & c^e_{11}\alpha & 0 \\ -c^e_{66}\beta & 0 & c^e_{66}\beta \end{bmatrix}$$

$$[K^{12}] = [K^{21}]^T = \frac{t}{2}\begin{bmatrix} c^e_{11} + c^e_{66} & -c^e_{66} & -c^e_{12} \\ -c^e_{66} & 0 & c^e_{11} \\ -c^e_{12} & c^e_{11} & 0 \end{bmatrix}$$

$$[K^{22}] = \frac{t}{2}\begin{bmatrix} c^e_{66}\alpha + c^e_{22}\beta & -c^e_{66}\alpha & -c^e_{22}\beta \\ -c^e_{66}\alpha & c^e_{66}\alpha & 0 \\ -c^e_{22}\beta & 0 & c^e_{22}\beta \end{bmatrix}$$

where $\alpha = b/a$ and $\beta = a/b$.

7-3-16 $$[K^{11}] = t(c^e_{11}[S^{xe}] + c^e_{66}[S^{ye}])$$

$$[K^{12}] = t(c^e_{12}[S^{xye}] + c^e_{66}[S^{xye}]^T)$$

$$[K^{22}] = t(c^e_{66}[S^{xe}] + c^e_{22}[S^{ye}])$$

where $[S^{xe}]$, $[S^{ye}]$ and $[S^{xye}]$ are defined in Eq. (7-118).

INDEX

C

D

F

G

P

T